USEFUL RELATIONS

$$\tilde{f}(\omega) = \frac{1}{\sqrt{2\pi}} \int_{-\infty}^{\infty} f(t) e^{i\omega t} dt$$

$$f(t) = \frac{1}{\sqrt{2\pi}} \int_{-\infty}^{\infty} \tilde{f}(\omega) e^{-i\omega t} d\omega$$

$$\int_{a}^{b} f(x)\delta(x - x_0) dx = f(x_0)$$

$$\int_{-\infty}^{\infty} e^{i\omega t} dt = 2\pi\delta(\omega)$$

$$\delta(x^2 - a^2) = \frac{1}{2|a|}[\delta(x + a) + \delta(x - a)]$$

$$\delta(g(x)) = \frac{\delta(x - x_0)}{|dg/dx|}$$

$$\int_{-\infty}^{\infty} \delta(x - a)\delta(x - b) dx = \delta(a - b)$$

$$\delta(x) = -x\delta'(x)$$

$$\int \frac{dx}{\sqrt{a^2 + x^2}} = \ln(x + \sqrt{a^2 + x^2})$$

$$\int \frac{x \, dx}{\sqrt{a^2 + x^2}} = \sqrt{a^2 + x^2}$$

$$\int \frac{dx}{(a^2 + x^2)^{3/2}} = \frac{x}{a^2\sqrt{a^2 + x^2}}$$

$$\int \frac{x \, dx}{(a^2 + x^2)^{3/2}} = \frac{-1}{\sqrt{a^2 + x^2}}$$

$$\int \frac{x^2 \, dx}{(a^2 + x^2)^{3/2}} = \frac{-x}{\sqrt{a^2 + x^2}} + \ln(x + \sqrt{a^2 + x^2})$$

$$\int_{-\infty}^{\infty} \frac{e^{i\omega t} \, dt}{(\tau^2 + t^2)^{3/2}} = 2\frac{\omega}{\tau} K_1(\omega\tau)$$

$$\int_{-\infty}^{\infty} \frac{e^{i\omega t} t \, dt}{(\tau^2 + t^2)^{3/2}} = i2\omega K_0(\omega\tau)$$

RELATIVISTIC KINEMATICS

$$ds^2 = c^2 dt^2 - dx^2 - dy^2 - dz^2$$

$$ds^2 = r^\alpha r_\alpha = c^2 d\tau^2$$

$$g^{\alpha\beta} = \begin{bmatrix} 1 & 0 & 0 & 0 \\ 0 & -1 & 0 & 0 \\ 0 & 0 & -1 & 0 \\ 0 & 0 & 0 & -1 \end{bmatrix} = g_{\alpha\beta}$$

$$L^\alpha{}_\beta = \begin{bmatrix} \gamma & -\beta\gamma & 0 & 0 \\ -\beta\gamma & \gamma & 0 & 0 \\ 0 & 0 & 1 & 0 \\ 0 & 0 & 0 & 1 \end{bmatrix}$$

$$ct' = \gamma(ct - \beta x)$$

$$x' = \gamma(x - \beta ct)$$

$$y' = y$$

$$z' = z$$

$$U^\alpha = \gamma(c, \mathbf{v})$$

$$U^\alpha U_\alpha = c^2$$

$$v_x = \frac{v + v_x'}{1 + \dfrac{vv_x'}{c^2}}$$

$$v_y = \frac{v_y'}{\gamma\left(1 + \dfrac{vv_x'}{c^2}\right)}$$

$$k^\alpha = \left(\frac{\omega}{c}, \mathbf{k}\right)$$

$$\omega = \frac{\omega'}{\gamma(1 - \beta\cos\theta)}$$

MODERN PROBLEMS IN CLASSICAL ELECTRODYNAMICS

MODERN PROBLEMS IN

CLASSICAL ELECTRODYNAMICS

Charles A. Brau, 1938–

Vanderbilt University, Nashville, TN

New York ■ Oxford
OXFORD UNIVERSITY PRESS
2004

Oxford University Press

Oxford   New York
Auckland   Bangkok   Buenos Aires   Cape Town   Chennai
Dar es Salaam   Delhi   Hong Kong   Istanbul   Karachi   Kolkata
Kuala Lumpur   Madrid   Melbourne   Mexico City   Mumbai
Nairobi   São Paulo   Shanghai   Taipei   Tokyo   Toronto

Published by Oxford University Press, Inc.
198 Madison Avenue, New York, New York, 10016
http://www.oup-usa.org

Library of Congress Cataloging-in-Publication Data

Brau, Charles A., 1938–
    Modern problems in classical electrodynamics / C.A. Brau.
        p. cm.
    Includes bibliographical references and index.
    ISBN 0–19–514665–4
    1. Electrodynamics.   I. Title.

QC631 .B73 2003
537. 6—dc21   2002070369

Printing number: 9 8 7 6 5 4 3 2 1

Printed in the United States of America
on acid-free paper

Contents

Preface

For the past eight years I have taught a course on classical electrodynamics, mostly to first-year and second-year graduate students but also to a few advanced undergraduates. This book represents my thoughts on that subject and is addressed to them. It also represents a collection of things I have learned and found useful in my research and serves me as a reference book for that part of my life. I hope that it can serve others as well.

Given the large number of texts in this field, the question "why another book?" immediately presents itself. The answer to this question, insofar is there is an answer beyond the simple human imperative to write a book, has several parts. In the first place, the longer I taught the course the more I found myself drawing on multiple texts instead of just one, and during this period one of the texts I used (*The Classical Theory of Fields* by Landau and Lifshitz) became unavailable. This is actually what precipitated the writing of this book. In the second place, the classical theory of electricity and magnetism is not just the precursor to quantum mechanics, it is an evolving subject that plays an important role in modern physics, including, in particular, lasers and nonlinear optics. None of the major electrodynamics texts adequately represents this aspect of the subject. In the third place, I am principally an experimenter, whereas most of the available books have been written by theorists. This inevitably gives the present text a different flavor that some might find comfortable and useful.

From a pedagogical standpoint I believe in the importance of homework. The most I can realistically hope to do is be present and offer encouragement while the students teach themselves physics. I have therefore included exercises at the end of almost every section. In addition, I find that given the correct answer, a student will keep on working until he or she figures out how to get it. This is the way to learn physics. Therefore, most of the questions are of the form "show that (something) is true." Somewhat surprisingly, this has three other beneficial effects. The questions are less ambiguous, since the intended objective is clear; the questions frequently have useful results, so the answers in the text can be used for reference afterward; and since the student usually gets the correct answer, I don't have to reconstruct the algebra to discover where a mistake has been made. This makes the homework assignments easier to grade.

The arrangement of the book is somewhat unorthodox, although it follows the spirit of Landau and Lifshitz' *The Classical Theory of Fields*. I begin with special relativity and classical field theory in Chapters 1 and 2 before moving on to the details and applications of electrodynamics. I do this for several reasons. In the first place, electrodynamics is essentially a relativistic theory. Without the theory of relativity, electrodynamics becomes disjointed and unsymmetrical, and seems to me harder to comprehend and less appealing. In addition, the 4-vector tools we develop are a great help in our later work, and by introducing them early we not only have them available when we need them for classical electrodynamics, we become comfortable enough to use them in other subjects, such as quantum field theory. As a life preserver offered to those who begin classical electrodynamics with less preparation, the publisher persuaded me to include the Prologue, which reviews the subject from a conventional, historical perspective. But the book really begins with Chapter 1. This chapter, and the rest of the book, should be accessible to anyone who understands the material in one of the excellent undergraduate texts that are now available, such as Griffiths' *Introduction to Electrodynamics*.

Chapters 3, 4, and 5 develop the tools, as it were, of the trade. Chapter 3 covers all of electrostatics and magnetostatics, including Green functions. These are important subjects, but not ones that I find very exciting, so I have limited their discussion to a single chapter. Most students seem to share my feelings. Chapter 4 discusses electromagnetic waves, and Chapter 5 (Fourier techniques) shows how waves can be used to analyze a multitude of phenomena that we don't ordinarily think of as wavelike.

Chapters 6, 7, and 8 extend the machinery we have developed to macroscopic materials. Considerable time is spent on linear materials, since most materials are roughly linear and linear theory ties so much together in an elegant and general way. But the exciting areas of modern physics include nonlinear optics, so these are discussed in Chapter 8. Chapter 9 discusses diffraction, which in modern physics includes laser beams and optical resonators. Chapter 10 describes the radiation produced by charges in motion, both relativistic and nonrelativistic. It also forms a lead-in to Chapter 11.

Chapter 11, fundamental particles, was arguably the most interesting chapter to write, and I hope that you have the opportunity to get this far. The theory of fundamental particles faces squarely all the limitations and inconsistencies in the beautiful edifice we call classical physics. For this reason, I find it both disturbing and surprising. It is also an exercise of the mind that may serve as preparation for the construction of new theories, perhaps the "theory of everything," in Hawking's words.

In keeping with recent trends, and with my nature as an experimenter, I have elected to use SI (MKSA) units throughout. Although Gaussian units are arguably more "beautiful," most experimenters actually use mixed units, such as watts per square centimeter, because statvolts and ergs per second are just too far removed from the real world in which we live and work. In the search for consistency it seems better to go to meters than to statvolts. It has even been suggested that the book be published in two versions, one using Gaussian and the other SI units, and see which sells more copies. In the meantime, the conversion between Gaussian and SI units is described in the obligatory appendix.

I should also point out some idiosyncrasies of style that I use. In particular, I use the symbol q for charge, even for the charge on the electron, instead of the commonly used e. This is in small part to distinguish the charge from the base of natural logarithms, but mostly to avoid confusion about whether the unit e of charge on an electron is positive or negative. Besides, most of the formulas apply equally to other charges, so q seems more general. In a few places (as in the definition of the Bohr magneton) I do use the symbol e, and it is positive for an electron. I use script letters (as in $\mathcal{E}$) for energy or energylike quantities and for the dual of tensors, and reserve Times Roman (as in E) for fields (like the electric field). Most of the remaining notation is pretty standard. I use boldface (as in $\mathbf{E}$) to indicate vectors, and the caret (as in $\hat{\mathbf{k}}$) to indicate a unit vector (in this case $\mathbf{k}/k$). In 4-vectors I use $\alpha = 0 \ldots 3$ for the indices, with the timelike index first, and

$ds^2 = c^2\,dt^2 - dr^2$ for the metric. I usually use a tilde (as in $\tilde{f}$) to denote a Fourier transform, with the factor $1/\sqrt{2\pi}$ appearing symmetrically in the transform and its inversion. I'm not trying to break any new ground here. However, since I discuss nonlinear optics using Fourier transforms, I define the nth-order nonlinear susceptibility $\chi_{i;j...}^{(n)}$ with some extra factors of $\sqrt{2\pi}$ compared with the conventions used by other authors. But I do include the rules for converting to one of the more conventional definitions (at least two different conventions are common in the literature).

It will be clear upon reading the book that I owe a great deal to those who have gone before, not only in matters of theory but also of style. I owe my greatest debt to Landau and Lifshitz' *The Classical Theory of Fields,* not only for the overall approach and for much of the content of Chapters 1 and 2 on field theory, but also for the inspiration they provide by their graceful exposition of the subject. The material in Chapter 3, electrostatics and magnetostatics, comes from many places, as does that in Chapters 4 and 5. Chapters 6 and 7, which discuss macroscopic materials, received considerable inspiration from Landau, Lifshitz, and Pitaevskii's *Electrodynamics of Continuous Media.* Chapter 8, nonlinear optics, owes much to Boyd's book with the same title, although the approach is rather different. Chapter 9, again, borrows from many places, but especially from Born and Wolf's *Principles of Optics* and Goodman's *Introduction to Fourier Optics.* Chapter 10, radiation theory, owes much to Jackson's clear introduction to the subject. Chapter 11, fundamental particles, was inspired by Kim and Sessler's paper on the subject, but again the material comes from many places.

In addition, I owe a great deal to the people who have directly helped, inspired, corrected, and supported me in so many ways. In this connection I gratefully acknowledge many discussions about the physics and about the manuscript with Evgeni Bessonov, Leonard Feldman, Eric Goff, Richard Haglund, Marcin Jankiewicz, Thomas Kephart, Kwang-Je Kim, Vladimir Litvinenko, Patrick O'Shea, and Thomas Weiler. I want to acknowledge my debt to all the students who suffered through the early versions of the manuscript and helped me to find many of the mistakes. In the end, the remaining mistakes are solely my responsibility. I also want to thank Peter Gordon, at Oxford University Press, for his advice and his enthusiasm, and Justin Collins for his expertise that produced the book as you see it. And finally, I want to express my sincere appreciation to my wife, Virginia Shepherd, an outstanding scientist and educator in her own field of biochemistry, who put up with me through the many long hours and months of preparation. I look forward to writing the next book with her.

Charles A. Brau
Vanderbilt University

Prologue

0.1 INTRODUCTION

In Chapters 1 and 2 we see how the incredible insights of Maxwell and Einstein reveal the unity and simplicity of all of classical physics. Everything in classical electrodynamics and classical mechanics, save for the values of certain constants such as the charge and mass of the fundamental particles, can be derived from Einstein's postulates and Hamilton's principle. But historically, classical physics was not discovered this way. It was discovered piecemeal and assembled bit by bit, and as each piece was seen to fit with another piece, the picture gradually became clearer until the whole was understood. For hundreds of years electricity and magnetism were thought to be unrelated phenomena, until the discoveries of Faraday and Oersted tied them together. Then Maxwell, by identifying and removing the last inconsistency in the theory of electrodynamics, completed the picture and triumphantly showed that light itself is a traveling electromagnetic field. But mechanics was developed separately, with no awareness that all the (then) known forces of nature, save gravity, were electromagnetic in origin, and classical mechanics contained within it an implicit inconsistency with electrodynamics. Finally Einstein, by eliminating this last inconsistency, changed our understanding of space and time and showed that, at least classically, the laws of nature are very strictly circumscribed.

But before we see classical physics revealed in all its beauty, we should look back briefly on how we got to this point. In the first place, this allows us to introduce a great many terms and concepts that are worth understanding in their physical and historical context. In the second place, this sets the stage for what follows by illustrating the lack of unity and consistency in the theory as it developed. And finally, history is interesting. The presentation here is rather brief, and is not intended to serve as an introduction to electrodynamics. Excellent and thorough introductions can be found in some of the references cited at the end of the chapter. Rather, the treatment here is intended to review the subject quickly with the purpose of providing an overview of the subject as a whole, like viewing the landscape from a distance at the expense of some of the details that would be revealed by a closer look.

In the following sections we take up, in turn, electrostatics, magnetostatics, and their unification in electrodynamics, but we leave the unification of mechanics and electrodynamics to Chapters 1 and 2.

0.2 ELECTROSTATICS

Thales of Miletus, founder of the school that produced Socrates and a contemporary of King Nebuchadnezzar of Babylon, is credited with the discovery around 600 B.C. that when rubbed with fur, amber attracts nearby light objects to itself. So far as was known then, the phenomenon was limited to amber and was believed to be a property of that substance. Since action at a distance was contrary to experience, the Greeks believed that forces of attraction are exercised through the particles that must fill all space, for they argued, using reasoning that seems now to be somewhat circular, that all space must be filled with some sort of matter, else space would have no form or definition. This belief persisted even to the time of Newton, who despite assessing correctly the $1/r^2$ dependence of the gravitational force between celestial bodies did not accept the notion of action at a distance. Even Maxwell held fast to the belief that the electromagnetic forces are transmitted mechanically by the ether that permeates all space. And when Einstein, in that final act of reductionism, eliminated even the ether, space and time, with nothing to support them, lost their definition and collapsed like a balloon robbed of its air, able to be stretched and distorted by motion through them—and even by the mere presence of matter. In this sense the theory of relativity seems almost like the catastrophe the ancient Greek philosophers would have predicted. But in fact relativity has an order and beauty of its own, and we have learned to be comfortable with action at a distance.

0.2.1 Charges

The *science* of electrostatics (and magnetism, for that matter) may be said to have begun with the research of Gilbert, who in 1600 discovered that the phenomenon of electrostatics is not limited to amber but is exhibited by many other substances including glass, sulfur, sealing wax, and various precious stones. He called the phenomenon electricity, "amberness," in honor of the Greek word ηλεκτρον (electron), meaning "amber."

The phenomenon of repulsion, in addition to attraction, was discovered in 1733 by du Fay, the king's superintendent of gardens, and this discovery led him to identify two types of electrical charge. Referring to the types of substances in which he could generate these charges, du Fay called them vitreous and resinous electricity. Watson, in 1746, proposed that electricity is "a very subtil and elastic fluid occupying all bodies in contact with the terraqueous globe," and that a body becomes charged not by the creation of electricity but by its transfer from one body to another. Franklin came to the same conclusion independently in 1747 and argued that vitreous electricity corresponded to an excess of electric fluid and resinous electricity to a deficiency of this fluid. These were the first statements of the conservation of charge. Franklin called an excess of charge (vitreous electricity) positive and a deficiency (resinous electricity) negative. Unfortunately he chose the convention that places a negative charge on electrons, so the current they carry goes in the direction opposite their motion. But he can hardly be blamed for that.

Nowadays, of course, we know (or believe) that all charges come in multiples (or simple submultiples like $\frac{1}{3}$) of the electron charge $e = 1.602 \times 10^{-19}$ C. We adopt here the convention that e is a positive number, and whenever we refer to the charge on any particle, including an electron, we use the symbol q, giving q the value $q = -e$ when the formula is applied to an electron. We also believe nowadays that the elementary charges are points, having no spatial extent. This makes the mathematics simple in some ways but gives rise to infinities that we haven't yet learned to deal with completely. In Franklin's time this was not yet understood, and charge was believed to be a continuous fluid with a variable density. This is still a useful approximation, and we have occasion to use it frequently.

0.2.2 Forces and Electric Fields

The notion of action at a distance, that is, the rejection of intermediate particles filling all space, was first defended by Aepinius around 1759, but it was left to Priestly in 1767 to deduce the law of force. The discoverer of oxygen, Priestly was also a good friend of Franklin, who described to him some puzzling experiments that Priestly reproduced. Priestly was able to explain the results by analogy with Newton's theory of universal gravity, and from this he inferred a $1/r^2$ law for the dependence of the electrostatic force on distance. In 1769 Robinson and in 1771 Cavendish did direct experiments confirming the $1/r^2$ force law, and in 1785 Coulomb, using his torsion balance, carried out the decisive experiments. He further argued that both positive and negative charges exist; that is, negative charge is not merely the absence of positive charge. This makes it possible to write the law for the force exerted by one point charge on another in the form

$$\mathbf{F} = \frac{qq'}{4\pi\varepsilon_0} \frac{\mathbf{r}-\mathbf{r}'}{|\mathbf{r}-\mathbf{r}'|^3} \tag{0.1}$$

where $\mathbf{F}$ is the force on the charge q at position $\mathbf{r}$ due to the charge q' at position $\mathbf{r}'$. As shown in Figure 0.1, the force $\mathbf{F}$ acts in the direction along the line through the centers of the charges, as symmetry says it must, for in the absence of any motion this is the only identifiable direction. The constant of proportionality ε_0 depends on the system of units used. In SI units the unit of force is the newton (N), the unit of charge is the coulomb (C), and the unit of distance is the meter (m). The value of ε_0, called the permittivity of free space, is 8.854×10^{-12} C^2/N-m^2. Equation (0.1) is known as Coulomb's law.

The force on a charge q at position $\mathbf{r}$ due to the presence of several charges q_i at positions $\mathbf{r}_i$ is just the linear vector sum of the forces due to the individual charges. Thus we may write

$$\mathbf{F} = \frac{q}{4\pi\varepsilon_0} \sum_i \frac{q_i(\mathbf{r}-\mathbf{r}_i)}{|\mathbf{r}-\mathbf{r}_i|^3} \tag{0.2}$$

or for a continuous distribution of charge

$$\mathbf{F} = \frac{q}{4\pi\varepsilon_0} \int \frac{\rho(\mathbf{r}')(\mathbf{r}-\mathbf{r}')\,d^3\mathbf{r}'}{|\mathbf{r}-\mathbf{r}'|^3} \tag{0.3}$$

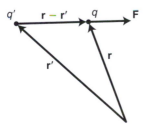

Figure 0.1 Coulomb's law.

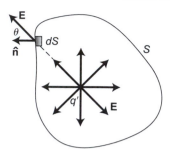

Figure 0.2 Electric field of a point charge.

where $\rho(\mathbf{r}')$ is the density of charge at point $\mathbf{r}'$. The statement that the total force is the linear combination of the individual forces is known as the principle of superposition.

Since the force on a particle is proportional to its charge q, it is useful to define the coefficient of proportionality

$$\mathbf{E} = \frac{\mathbf{F}}{q} = \frac{q'}{4\pi\varepsilon_0}\frac{\mathbf{r} - \mathbf{r}'}{|\mathbf{r} - \mathbf{r}'|^3} \tag{0.4}$$

or, for a distributed charge,

$$\mathbf{E}(\mathbf{r}) = \frac{1}{4\pi\varepsilon_0}\int\frac{\rho(\mathbf{r}')(\mathbf{r} - \mathbf{r}')\,d^3\mathbf{r}'}{|\mathbf{r} - \mathbf{r}'|^3} \tag{0.5}$$

and call it the electric field $\mathbf{E}$ at the point $\mathbf{r}$. The units of electric field are N/C.

The electric field surrounding a point charge q' is shown in Figure 0.2. If we place a surface S in the electric field, the number of electric field lines passing through the surface is, at least conceptually, the electric flux through the surface. Mathematically, the electric flux through the infinitesimal surface $d\mathbf{S} = \hat{\mathbf{n}}\,dS$, where $\hat{\mathbf{n}}$ is a unit vector normal to the area dS, is $\mathbf{E}\cdot d\mathbf{S}$. If we draw a surface S completely enclosing the charge q' and integrate the field outwardly normal to the surface over the entire surface we get

$$\oint_S \mathbf{E}\cdot d\mathbf{S} = \frac{q'}{4\pi\varepsilon_0}\oint_S\frac{(\mathbf{r} - \mathbf{r}')\cdot d\mathbf{S}}{|\mathbf{r} - \mathbf{r}'|} \tag{0.6}$$

where $\mathbf{r}$ is on the surface S. But

$$d\Omega = \frac{(\mathbf{r} - \mathbf{r}')}{|\mathbf{r} - \mathbf{r}'|^3}\cdot d\mathbf{S} = \frac{dS\cos\theta}{|\mathbf{r} - \mathbf{r}'|^2} \tag{0.7}$$

is just the infinitesimal solid angle subtended by the infinitesimal surface dS viewed from the point $\mathbf{r}'$, where $\cos\theta$ is the angle between the radius vector and the normal to the surface. Thus,

$$\oint_S \mathbf{E}\cdot d\mathbf{S} = \frac{q'}{4\pi\varepsilon_0}\oint_S d\Omega = \frac{q'}{\varepsilon_0} \tag{0.8}$$

independent of the shape of the surface S or the position $\mathbf{r}'$ of the charge inside the surface. This is called Gauss's law, and he developed it in 1839 as part of his program to study terrestrial magnetism. Gauss's law holds a fundamental position in the theory of electrodynamics, since it applies to time-dependent phenomena and is actually more general than Coulomb's law of electrostatics, from which it is derived here. Yet it represents only an incidental part of Gauss's contributions to potential theory and in any event is completely overshadowed by his enormous contributions to number theory. It is not even mentioned in most of his biographies.

From the principle of superposition we see that for a collection of charges, (0.8) becomes

$$\oint_S \mathbf{E} \cdot d\mathbf{S} = \frac{1}{\varepsilon_0} \sum_i q_i = \frac{1}{\varepsilon_0} \int_V \rho(\mathbf{r}) \, d^3\mathbf{r} \tag{0.9}$$

where V is the volume inside the surface S. But from the divergence theorem we know that for any vector field $\mathbf{E}$,

$$\oint_S \mathbf{E} \cdot d\mathbf{S} = \int_V \nabla \cdot \mathbf{E} \, d^3\mathbf{r} \tag{0.10}$$

Comparing (0.9) and (0.10), and recognizing that the volume V is arbitrary, we see that the integrands must be equal and we get

$$\nabla \cdot \mathbf{E} = \frac{\rho}{\varepsilon_0} \tag{0.11}$$

This is the differential form of Gauss's law.

0.2.3 Electric Potential

Since the distance between $\mathbf{r}$ and $\mathbf{r}'$ is $|\mathbf{r} - \mathbf{r}'| = \sqrt{(x - x')^2 + (y - y')^2 + (z - z')^2}$, we may evaluate the appropriate derivatives to show that

$$\nabla \frac{1}{|\mathbf{r} - \mathbf{r}'|} = -\frac{\mathbf{r} - \mathbf{r}'}{|\mathbf{r} - \mathbf{r}'|^3} \tag{0.12}$$

Comparing this with (0.4), we see that the electric field of a point charge q' at the point $\mathbf{r}'$ can be expressed by

$$\mathbf{E}(\mathbf{r}) = -\frac{q'}{4\pi \varepsilon_0} \nabla \frac{1}{|\mathbf{r} - \mathbf{r}'|} \tag{0.13}$$

By superposition we see that for a distribution of charges the electric field may be expressed as the gradient

$$\mathbf{E}(\mathbf{r}) = -\nabla \Phi(\mathbf{r}) \tag{0.14}$$

of some function $\Phi(\mathbf{r})$, where

$$\Phi(\mathbf{r}) = \frac{1}{4\pi \varepsilon_0} \int \frac{\rho(\mathbf{r}')}{|\mathbf{r} - \mathbf{r}'|} \, d^3\mathbf{r}' \tag{0.15}$$

The function $\Phi(\mathbf{r})$ is called the scalar potential. The first to realize that the field could be expressed as the gradient (0.14) seems to have been Laplace (in connection with gravitation), but the name "potential" was first coined by Green, in 1828, in his development of the mathematical theory of electricity and magnetism. It follows from (0.14) that

$$\nabla \times \mathbf{E} = 0 \tag{0.16}$$

since the curl of a gradient vanishes identically.

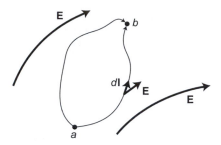

Figure 0.3 Work done by the electric field along the path from *a* to *b*.

When a point charge q moves from position $\mathbf{r}_a$ to position $\mathbf{r}_b$, as shown in Figure 0.3, the work done on the charge by the field is

$$W = q \int_a^b \mathbf{E} \cdot d\mathbf{r} = -q \int_a^b \nabla\Phi(\mathbf{r}) \cdot d\mathbf{r} = -q \int_a^b d\Phi = q[\Phi(\mathbf{r}_a) - \Phi(\mathbf{r}_b)] \quad (0.17)$$

Since the difference $\Phi(\mathbf{r}_a) - \Phi(\mathbf{r}_b)$ depends only on the end points and not on the path between them, the work done to move a charged particle along two different paths between the same points $\mathbf{r}_a$ and $\mathbf{r}_b$, as indicated in Figure 0.3, is the same. Since the work is independent of the path, it can be regarded as the change in potential energy between points $\mathbf{r}_a$ and $\mathbf{r}_b$. Thus, the scalar potential represents the potential energy per unit charge. Like the potential energy, the scalar potential is defined only to within a constant. That is, the electric field is unchanged by the transformation

$$\Phi(\mathbf{r}) \to \Phi'(\mathbf{r}) = \Phi(\mathbf{r}) + \Lambda \quad (0.18)$$

where Λ is a constant. The potential difference $\Phi(\mathbf{r}_b) - \Phi(\mathbf{r}_a)$ is likewise unaffected by the change.

In terms of the scalar potential, the differential form of Gauss's law (0.11) becomes

$$\nabla \cdot \nabla\Phi = \nabla^2\Phi = -\frac{\rho}{\varepsilon_0} \quad (0.19)$$

This is called Poisson's equation, after the French mathematician who introduced it in a memoir in 1813. Many powerful analytical and numerical methods have been developed for solving Poisson's equation, and we discuss some of these in Chapter 3.

0.2.4 Energy in the Electric Field

When a charge q moves from point a to point b, the work done on the particle by the field of the other charges is given by (0.17). But as the charge moves, the total electric field, including that of the charge q itself, changes, and it is possible to compute the work done on the charge from the change of some integral over the total electric field. This integral, then, represents the energy in the electric field, and the change in the potential energy of the particle is just the change in the energy of the field.

To see this we compute the total work required to assemble a set of charges and establish the electric field, as shown in Figure 0.4. We begin with a single charge q_1 at point $\mathbf{r}_1$, and define the potential $\Phi_1(\mathbf{r})$ due to this charge using the boundary condition that the potential vanishes at infinity, $\Phi_1(\mathbf{r}) \xrightarrow[r\to\infty]{} 0$. When we bring up a second charge q_2 from infinity to the point $\mathbf{r}_2$, the work done is $q_2\Phi_1(\mathbf{r}_2)$. But the work $q_1\Phi_2(\mathbf{r}_1)$ required to

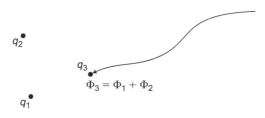

Figure 0.4 Work done to assemble the *i*th charge.

assemble these charges in the reverse order must be the same, so

$$W = \tfrac{1}{2}[q_1\Phi_2(\mathbf{r}_1) + q_2\Phi_1(\mathbf{r}_2)] \tag{0.20}$$

If we bring up a third charge q_3 to point $\mathbf{r}_3$, the additional work is

$$\Delta W = q_3[\Phi_1(\mathbf{r}_3) + \Phi_2(\mathbf{r}_3)] = q_1\Phi_3(\mathbf{r}_1) + q_2\Phi_3(\mathbf{r}_2) \tag{0.21}$$

since $q_3\Phi_1(\mathbf{r}_3) = q_1\Phi_3(\mathbf{r}_1)$ and $q_3\Phi_2(\mathbf{r}_3) = q_2\Phi_3(\mathbf{r}_2)$. Equivalently, the total work done is

$$W = \tfrac{1}{2}[q_1\Phi_{2,3}(\mathbf{r}_1) + q_2\Phi_{1,3}(\mathbf{r}_2) + q_3\Phi_{1,2}(\mathbf{r}_3)] \tag{0.22}$$

where $\Phi_{1,2}(\mathbf{r}_3) = \Phi_1(\mathbf{r}_3) + \Phi_2(\mathbf{r}_3)$ is the potential due to all the other charges besides q_3, and so on. But if the charge that we bring up is infinitesimal, then the potential due to all the other charges is just the potential at that point due to all the charges, so the total work done is evidently

$$W = \frac{1}{2}\int \rho(\mathbf{r})\Phi(\mathbf{r})\, d^3\mathbf{r} \tag{0.23}$$

If we now substitute (0.19) for $\rho(\mathbf{r})$ and integrate once by parts, keeping in mind that the potential vanishes at infinity, we get

$$W = -\frac{1}{2}\varepsilon_0 \int \Phi\nabla^2\Phi\, d^3\mathbf{r} = \frac{1}{2}\varepsilon_0 \int \nabla\Phi \cdot \nabla\Phi\, d^3\mathbf{r} = \frac{1}{2}\varepsilon_0 \int E^2\, d^3\mathbf{r} \tag{0.24}$$

That is, the work done to establish the field can be represented by an integral over the entire field that has been created, as was to be shown. We therefore identify the quantity

$$\mathcal{U} = \tfrac{1}{2}\varepsilon_0 E^2 \tag{0.25}$$

as the energy density in the electric field.

In some sense, we originally introduced the concept of the electric field $\mathbf{E}(\mathbf{r})$ as a bookkeeping technique, a measure of the force per unit charge for any charge placed at the point $\mathbf{r}$. Now we see that the energy of a set of charges can be computed from the energy in the electric field, suggesting that the fields have some intrinsic reality of their own. In fact, this is actually the case, for we see later that the fields can exist as waves (light) without the existence of any charges.

Clearly, the energy in the field is a positive definite quantity. However, we know that two charges of opposite sign attract one another, so the work done to move the second charge from infinity is negative. Where did we go wrong? The answer lies in the fact that when we consider a charge coming in from infinity there is a field surrounding each charge by itself, so the energy in the field has a finite value (actually an infinite value when the charges are point charges) even when the second charge is at infinity. The attraction of un-like charges comes from the fact that when the charges are close to one another their fields

cancel in some regions, and this reduces the total energy in the fields below the value it had when the charges were separated by a great distance. Since energy is equivalent to mass, the energy in the electric field of a charged particle must compose at least part of the observed mass of that particle. However, this is not the end of the matter, for as we see in Chapter 11, this energy by itself does not have the proper relativistic transformation properties. This is the so-called 4/3 problem. For point charges the problem is worse because the energy in the field is infinite, and the problem does not go away if we simply "renormalize" to a finite total mass by invoking an infinite negative intrinsic (nonelectric) contribution to the mass. This is discussed in Chapter 11.

0.3 MAGNETOSTATICS

We are told that in 2637 B.C., during the 61st year of the reign of Hoang-ti, the emperor's troops were pursuing the rebellious prince Tchi-yeou when they lost their way in the fog on the plains of Tchou-lou. Upon seeing this, the emperor had mounted on a chariot the figure of a woman whose arm always pointed to the south. Thus able to find their way, the emperor's troops captured the rebellious prince, who was put to death. This is the earliest historical allusion to the use of the compass. By the third century A.D. Chinese ships regularly used the compass to navigate the Indian Ocean. Early in the fourth century A.D. the Chinese physicist Kuopho, in his "Discourse on the Lodestone," speculated on the connection between magnetism and electricity, writing that "the magnet attracts iron as amber attracts mustard seeds. There is a breath of wind that promptly and mysteriously penetrates both bodies, uniting them imperceptibly with the rapidity of an arrow. It is incomprehensible." It would be another 1500 years before Oersted would make the connection.

The knowledge of magnetism appeared much later in the West. It was known to the ancient Greeks by 800 B.C. when, according to legend, the shepherd Magnes made the discovery because the nails of his shoes stuck fast to a "lodestone," a naturally magnetized sample of what is now called magnetite (Fe_2O_3), while he pastured his flock. The earliest recorded reference to use of the compass in Europe was made by Nekham, a monk of St. Albans, in 1186. One hundred years later, in 1286, Maricourt used a compass needle to map the magnetic field on the surface of a spherical lodestone. To his surprise he discovered that the needle traced out lines that converged like lines of latitude on the globe, we would now say, to two poles located opposite one another on the surface of the sphere. He noted that the pole of one lodestone would attract one pole and repel the other pole of another lodestone. These observations were extended to a scientific level in 1600 by Gilbert, who realized, among other things, that the directionality of the compass is owed to the fact that the earth itself is a magnet (the south magnetic pole of the earth is, of course, at the north pole, so it attracts the north magnetic pole of the compass needle).

Although it had been suspected for some time due to the interaction of lightening with compass needles, the connection between magnetism and electricity was first established experimentally by Oersted in 1820 when he discovered that an electric current flowing through a wire causes a compass needle to be deflected. A few months later (within a week of hearing the news of Oersted's experiments) Ampere showed that the magnetic field of one current-carrying wire would deflect a nearby current-carrying wire, and over the next few years he carried out elaborate experiments and carefully reasoned mathematical analyses from which he deduced the quantitative laws of the interaction. He called the new physics "electrodynamics."

0.3.1 Currents

Whereas electrostatics is associated with charges in fixed positions and is therefore intrinsically time independent, magnetism is associated with electrical currents, which are charges in motion. Magnetostatics deals with fixed currents, and if this seems like a contradiction it really reflects the fact that the circumstances (usually wires and batteries) that produce "fixed movements" of charge are really somewhat contrived and not terribly general in nature.

The discovery of conduction of electricity is attributed to Gray, who in 1729 observed that the electrostatic charge on one body could be transferred to another through a metallic wire or even a length of packing string. Watson and Franklin built on this with the principle of conservation of charge, which is the notion that charge is permanent and found everywhere, merely transferring from one body to another. But the real breakthrough in electrical current was the development of the so-called voltaic pile. The story begins with the accidental discovery by Galvani in 1780 that freshly killed frogs on which he was conducting experiments would convulse when they were hung by brass hooks against an iron lattice he used to hold a hanging garden. He noted the similarity between these convulsions and those observed in earlier experiments when an electric machine in his laboratory was accidentally discharged near a frog into which he had inserted a scalpel, and he advanced the hypothesis that the effect was electrical in nature. Although he attributed the effect to "animal electricity," others believed that it was chemical in origin and began experiments using various metals immersed in solutions. But the effects were at best feeble and of little interest until in 1800 Volta found that if a stack were constructed consisting of a copper disk, a moistened pasteboard disk, and a zinc disk, repeated as many times as desired, the effect could be magnified until a palpable shock, similar to an electrostatic shock, was produced. Unfortunately, we do not have time to pursue this fascinating story further to understand the workings of the voltaic pile or other aspects of electrochemistry. It is enough to recognize that the voltaic pile opened the way for serious research in magnetism, in which we have further interest.

Historically, electrical current was viewed as the flow of a fluid charge through wires, like water through a pipe, and just as a steady flow of water is possible, so is a steady current. Microscopically, of course, the current is carried by point charges, and surrounding each charge is a magnetic field that is fluctuating rapidly. Nevertheless, on a macroscopic scale the microscopic variations smooth out and we get a static magnetic field. In the following we view the current as a continuous fluid whose local velocity (or more specifically current density $\mathbf{J}$) is independent of time. Since the charge density is everywhere steady, the current flowing into any volume V bounded by the surface S must vanish. Using the divergence theorem, we get

$$\oint_S \mathbf{J} \cdot d\mathbf{S} = 0 = \int_V \nabla \cdot \mathbf{J} \, dV \tag{0.26}$$

But the volume is arbitrary, so the integrand must vanish identically, and we see that

$$\nabla \cdot \mathbf{J} = 0 \tag{0.27}$$

in magnetostatics. Since there are no "sources" of current in magnetostatics, the stream tubes that describe the flow of steady currents necessarily form closed loops, as shown in Figure 0.5.

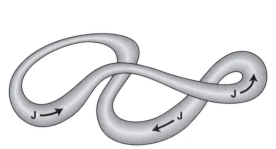

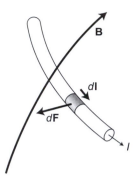

Figure 0.5 Stream tube in the flow of a steady current.

Figure 0.6 Force on a current-carrying wire.

0.3.2 Forces and Fields

In discussing electrostatics, we began with the forces between charged particles and then introduced the electric field almost as an afterthought. In magnetostatics, on the other hand, we begin with the magnetic field itself. This is partly historical, for the fields of lodestones and even the earth were known before the forces were understood. But in fact, the computation of the fields from the currents is an important, virtually necessary intermediate step before computing the forces, because the vector relationship between the currents and the forces is much more complicated than in the electrostatic case. We therefore begin the discussion by considering the force on an electric current due to a magnetic field and then take up the matter of computing the magnetic field from the electric current.

As illustrated in Figure 0.6, the force on the element $d\mathbf{l}$ of a wire carrying the current I in the magnetic field $\mathbf{B}$ is

$$d\mathbf{F} = I \, d\mathbf{l} \times \mathbf{B} \tag{0.28}$$

or in terms of the current density $\mathbf{J}$, the force on the volume element dV is

$$d\mathbf{F} = \mathbf{J} \times \mathbf{B} \, dV \tag{0.29}$$

This may be regarded as the definition of the magnetic field. In SI units the current has the dimensions coulombs per second, called amperes (A), and the force newtons (N), so the magnetic field has the dimensions N/A-m, which we call teslas (T).

If the infinitesimal current $\mathbf{J} \, dV$ arises from the motion of a small charged particle, we see that the force on the particle is

$$\mathbf{F} = q\mathbf{v} \times \mathbf{B} \tag{0.30}$$

where q is the total charge and $\mathbf{v}$ the velocity of the particle. Since the force is always perpendicular to the velocity, the work done on the particle by the force is always zero. Magneto*static* fields never do any work on a charged particle. When the magnetic field is time dependent, however, it can do work on the particle, as shown later, to accelerate or decelerate it and change its energy.

The law by which a steady current produces a magnetic field was discovered by the French physicists Biot and Savart in 1820, shortly after the announcement by Oersted.

Depending on whether you are discussing the current I in a wire or the current density $\mathbf{J}$ in a volume, the law that bears their names is expressed

$$\mathbf{B}(\mathbf{r}) = \frac{\mu_0 I}{4\pi} \oint \frac{d\mathbf{l} \times (\mathbf{r} - \mathbf{r}')}{|\mathbf{r} - \mathbf{r}'|^3} = \frac{\mu_0}{4\pi} \int \frac{\mathbf{J}(\mathbf{r}') \times (\mathbf{r} - \mathbf{r}')}{|\mathbf{r} - \mathbf{r}'|^3} d^3\mathbf{r}' \tag{0.31}$$

where the integral is around the entire circuit (which must be closed to permit a steady current to flow) or over all space. The coefficient of proportionality is $\mu_0 = 4\pi \times 10^{-7}$ newtons/ampere^2, exactly.

Although the Biot–Savart law is sometimes useful, it is more often useful to have differential equations for the magnetic field. If we compute the divergence of the Biot–Savart law using the vector identity

$$\nabla \cdot (\mathbf{a} \times \mathbf{b}) = \mathbf{b} \cdot (\nabla \times \mathbf{a}) - \mathbf{a} \cdot (\nabla \times \mathbf{b}) \tag{0.32}$$

we get

$$\nabla \cdot \mathbf{B}(\mathbf{r}) = \frac{\mu_0}{4\pi} \int \left[\frac{\mathbf{r} - \mathbf{r}'}{|\mathbf{r} - \mathbf{r}'|^3} \cdot (\nabla \times \mathbf{J}) - \mathbf{J} \cdot \left(\nabla \times \frac{\mathbf{r} - \mathbf{r}'}{|\mathbf{r} - \mathbf{r}'|^3} \right) \right] d^3\mathbf{r}' \tag{0.33}$$

But the first term in the integrand vanishes because $\mathbf{J}$ is a function of $\mathbf{r}'$ and its derivatives with respect to $\mathbf{r}$ vanish. In the second term we use (0.12) to write

$$\nabla \times \frac{\mathbf{r} - \mathbf{r}'}{|\mathbf{r} - \mathbf{r}'|^3} = \nabla \times \nabla \frac{1}{|\mathbf{r} - \mathbf{r}'|} = 0 \tag{0.34}$$

since the curl of a gradient vanishes identically. This leaves us with

$$\nabla \cdot \mathbf{B} = 0 \tag{0.35}$$

This is called Gauss's law for magnetism. Like Gauss's law for the electric field, Gauss's law for magnetism is more general than the Biot–Savart law from which it is derived here. Gauss's law holds for time-dependent magnetic fields, whereas the Biot–Savart law is valid only in magnetostatics. Since there are no sources of magnetic field, the lines of the magnetic field, like the streamlines of current flow in magnetostatics, must form closed loops as indicated in Figure 0.5. Note, however, that while $\nabla \cdot \mathbf{J} = 0$ only in magnetostatics, so the streamlines form closed loops only in this case, (0.35) is always true, so the magnetic field lines form closed loops even in the time-dependent case.

In the same fashion, we can compute the curl of the Biot–Savart law using the identity

$$\nabla \times (\mathbf{a} \times \mathbf{b}) = \mathbf{a}(\nabla \cdot \mathbf{b}) - \mathbf{b}(\nabla \cdot \mathbf{a}) + (\mathbf{b} \cdot \nabla)\mathbf{a} - (\mathbf{a} \cdot \nabla)\mathbf{b} \tag{0.36}$$

Since the derivatives of $\mathbf{J}$ vanish, as before, we are left with

$$\nabla \times \mathbf{B}(\mathbf{r}) = \frac{\mu_0}{4\pi} \int \left[\mathbf{J} \nabla \cdot \frac{\mathbf{r} - \mathbf{r}'}{|\mathbf{r} - \mathbf{r}'|^3} - (\mathbf{J} \cdot \nabla) \frac{\mathbf{r} - \mathbf{r}'}{|\mathbf{r} - \mathbf{r}'|^3} \right] d^3\mathbf{r}' \tag{0.37}$$

To evaluate the second term we note that

$$\nabla \frac{\mathbf{r} - \mathbf{r}'}{|\mathbf{r} - \mathbf{r}'|^3} = -\nabla' \frac{\mathbf{r} - \mathbf{r}'}{|\mathbf{r} - \mathbf{r}'|^3} \tag{0.38}$$

If we substitute this into (0.37) and integrate by parts we get

$$\int (\mathbf{J} \cdot \nabla) \frac{\mathbf{r} - \mathbf{r}'}{|\mathbf{r} - \mathbf{r}'|^3} d^3\mathbf{r}' = -\sum_{i=1}^{3} \int J_i \frac{\partial}{\partial r_i'} \frac{\mathbf{r} - \mathbf{r}'}{|\mathbf{r} - \mathbf{r}'|^3} d^3\mathbf{r}'$$

$$= -\sum_{i=1}^{3} \int J_i \frac{\mathbf{r} - \mathbf{r}'}{|\mathbf{r} - \mathbf{r}'|^3} d^2r_{j \neq i}' \Big|_{r_i'=-\infty}^{\infty} + \int \frac{\mathbf{r} - \mathbf{r}'}{|\mathbf{r} - \mathbf{r}'|^3} \nabla' \cdot \mathbf{J} d^3\mathbf{r}' \tag{0.39}$$

The first term vanishes because the current density vanishes at $r_i' = \pm\infty$ and the second term vanishes because $\nabla' \cdot \mathbf{J}(\mathbf{r}') = 0$ for steady currents, to which the Biot–Savart law is restricted in any event. To evaluate the first term in (0.37), we compute the derivative and find that

$$\sum_{j=1}^{3} \frac{\partial}{\partial r_j'} \frac{r_j' - r_j}{|\mathbf{r}' - \mathbf{r}|^3} = \sum_{j=1}^{3} \frac{1}{|\mathbf{r}' - \mathbf{r}|^3} - 3\sum_{j=1}^{3} \frac{(r_j' - r_j)^2}{|\mathbf{r}' - \mathbf{r}|^4} = 0 \tag{0.40}$$

except at $|\mathbf{r}' - \mathbf{r}| = 0$, where it is indeterminate. There is therefore no contribution to the integral except at the point $\mathbf{r}' = \mathbf{r}$, and we may shrink the volume of integration to a small volume V, bounded by the surface S, enclosing this point, as shown in Figure 0.7. Provided that $\mathbf{J}(\mathbf{r}')$ is continuous in this neighborhood, we may replace it by $\mathbf{J}(\mathbf{r})$ and move it outside the integral. To evaluate the remaining integral we use the divergence theorem to get

$$\int_V \nabla' \cdot \frac{\mathbf{r}' - \mathbf{r}}{|\mathbf{r}' - \mathbf{r}|^3} d^3\mathbf{r}' = \oint_S \frac{(\mathbf{r}' - \mathbf{r}) \cdot d\mathbf{S}'}{|\mathbf{r}' - \mathbf{r}|^3} = \oint_S d\Omega' = 4\pi \tag{0.41}$$

since $d\Omega' = (\mathbf{r}' - \mathbf{r}) \cdot d\mathbf{S}'/|\mathbf{r}' - \mathbf{r}|^3$ is the element of solid angle subtended by the surface element $d\mathbf{S}'$ viewed from the point $\mathbf{r}$, as discussed earlier. What we have showed, then, is that the function $\nabla' \cdot (\mathbf{r}' - \mathbf{r})/|\mathbf{r}' - \mathbf{r}|^3$ vanishes everywhere except the point $\mathbf{r}' = \mathbf{r}$ and has a finite value when integrated over this point. It therefore qualifies as the three-dimensional Dirac δ-function,

$$\nabla \cdot \frac{\mathbf{r} - \mathbf{r}'}{|\mathbf{r} - \mathbf{r}'|^3} = 4\pi\delta(\mathbf{r} - \mathbf{r}') \tag{0.42}$$

Using these results in (0.37) we find that

$$\nabla \times \mathbf{B} = \mu_0 \mathbf{J} \tag{0.43}$$

This is called Ampere's law. It was reported by him to the French Academy in 1827.

Upon applying Stokes' theorem to the surface S bounded by the curve C, as shown in Figure 0.8, we get

$$\int_S (\nabla \times \mathbf{B}) \cdot d\mathbf{S} = \oint_C \mathbf{B} \cdot d\mathbf{l} \tag{0.44}$$

and if we substitute (0.43) in the left-hand side we obtain the integral form of Ampere's law,

$$\oint_C \mathbf{B} \cdot d\mathbf{l} = \mu_0 I \tag{0.45}$$

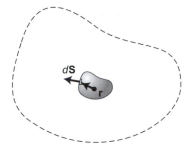

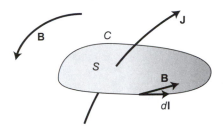

Figure 0.7 Integral around the point $\mathbf{r}' = \mathbf{r}$.

Figure 0.8 Integral form of Ampere's law.

where

$$I = \int_S \mathbf{J} \cdot d\mathbf{S} \tag{0.46}$$

is the total current flowing through the surface S. Note carefully the sign convention used in Figure 0.8 for positive current flowing through the surface S.

Comparing the laws of magnetostatics with those of electrostatics we see that the Biot–Savart Law (0.31) plays the same role in magnetostatics that Coulomb's law (0.5) plays in electrostatics. In the same way, we see that Gauss's law for magnetism (0.35) and Ampere's law (0.43) play the same roles as (0.16) and Gauss's law (0.11). There are, however, two important distinctions that make magnetostatic fields unlike electrostatic fields. The first is that magnetic field lines have no sources, so they always form closed loops. The second, of course, is the manner in which the fields cause forces on charged bodies.

0.3.3 Vector Potential

In electrostatics we take advantage of the fact that $\nabla \times \mathbf{E} = 0$ to introduce the scalar potential $\Phi(\mathbf{r})$. This has a number of advantages that include the fact that it is directly related to the energy and, being a scalar, is sometimes easier to compute than the electric field $\mathbf{E}$. We can do something similar in magnetostatics and use the fact that $\nabla \cdot \mathbf{B} = 0$ to introduce the vector potential $\mathbf{A}(\mathbf{r})$. As pointed out by Kirchhoff in 1857, this function is sometimes easier to compute than the magnetic field $\mathbf{B}$, although it is still a vector. It is also, as we see later, directly related to the momentum. However, the real importance of the vector potential is the fact that it can be combined with the scalar potential of the electric field to form a 4-vector potential. This final unification of the electromagnetic field becomes apparent only with the introduction of the special theory of relativity, so we save it for Chapter 1.

As we have seen, the electrostatic field $\mathbf{E}(\mathbf{r})$ satisfies the equations

$$\nabla \cdot \mathbf{E} = \frac{\rho}{\varepsilon_0} \tag{0.47}$$

and

$$\nabla \times \mathbf{E} = 0 \tag{0.48}$$

By introducing the scalar potential $\Phi(\mathbf{r})$, from which we find the electric field $\mathbf{E}(\mathbf{r})$ by the definition $\mathbf{E} = -\nabla\Phi$, we assure that the electric field satisfies (0.48), since $\nabla \times \nabla\Phi = 0$

identically. We then find the scalar potential from (0.47), which becomes

$$\nabla^2 \Phi = -\frac{\rho}{\varepsilon_0} \tag{0.49}$$

The magnetostatic field satisfies the corresponding equations

$$\nabla \cdot \mathbf{B} = 0 \tag{0.50}$$

and

$$\nabla \times \mathbf{B} = \mu_0 \mathbf{J} \tag{0.51}$$

If we write

$$\mathbf{B}(\mathbf{r}) = \nabla \times \mathbf{A}(\mathbf{r}) \tag{0.52}$$

for some vector potential $\mathbf{A}(\mathbf{r})$, then we automatically satisfy (0.50), since

$$\nabla \cdot (\nabla \times \mathbf{A}) = 0 \tag{0.53}$$

by a vector identity. The second equation, (0.51), then becomes

$$\nabla(\nabla \cdot \mathbf{A}) - \nabla^2 \mathbf{A} = \mu_0 \mathbf{J} \tag{0.54}$$

If this doesn't immediately strike you as an improvement over (0.51), consider the following. The definition (0.52) of the vector potential $\mathbf{A}(\mathbf{r})$ is arbitrary to the extent that we can add to it any vector function whose curl vanishes and the result gives the same magnetic field. But the curl of a gradient, $\nabla \times \nabla \Lambda(\mathbf{r}) = 0$, vanishes by a vector identity, so this is just the function we need. That is, the magnetic field is unaffected by the transformation

$$\mathbf{A}(\mathbf{r}) \rightarrow \mathbf{A}'(\mathbf{r}) = \mathbf{A}(\mathbf{r}) + \nabla \Lambda(\mathbf{r}) \tag{0.55}$$

The transformations (0.18) and (0.55) are simple examples of what are called gauge transformations of the potentials $\Phi(\mathbf{r})$ and $\mathbf{A}(\mathbf{r})$, and we can use them to simplify our lives. The function $\Lambda(\mathbf{r})$ is called a gauge function.

As an example of a gauge transformation we consider the vector potential

$$\mathbf{A} = \tfrac{1}{2} B_0 (x \hat{\mathbf{y}} - y \hat{\mathbf{x}}) \tag{0.56}$$

which describes the uniform magnetic field

$$\mathbf{B} = B_0 \hat{\mathbf{z}} \tag{0.57}$$

in the $\hat{\mathbf{z}}$ direction. The vector potential (0.56) consists of circles centered on the origin as shown in Figure 0.9. If we transform this with the gauge function

$$\Lambda = \tfrac{1}{2} B_0 y_0 x \tag{0.58}$$

for some constant y_0, we get the vector potential

$$\mathbf{A} = \tfrac{1}{2} B_0 [x \hat{\mathbf{y}} - (y - y_0) \hat{\mathbf{x}}] \tag{0.59}$$

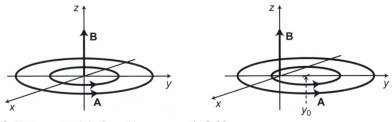

Figure 0.9 Vector potential of a uniform magnetic field.

This consists of circles centered on the point $(0, y_0)$, but it still corresponds to the same uniform magnetic field.

We can use the flexibility introduced by the gauge transformation to simplify the potential equation (0.54), for if we define our gauge function $\Lambda(\mathbf{r})$ by

$$\nabla^2 \Lambda = -\nabla \cdot \mathbf{A} \tag{0.60}$$

the divergence of the new vector potential vanishes:

$$\nabla \cdot \mathbf{A}' = \nabla \cdot \mathbf{A} + \nabla^2 \Lambda = 0 \tag{0.61}$$

But $\mathbf{A}'$ satisfies the same equation (0.54) that $\mathbf{A}$ satisfies, except that the first term vanishes. Then, since the vector potential $\mathbf{A}'(\mathbf{r})$ gives the same magnetic field as $\mathbf{A}(\mathbf{r})$, we can drop the prime and write

$$\nabla^2 \mathbf{A} = -\mu_0 \mathbf{J} \tag{0.62}$$

with the subsidiary condition

$$\nabla \cdot \mathbf{A} = 0 \tag{0.63}$$

which we get from (0.61). This is called the gauge condition, for it specifies the gauge function $\Lambda(\mathbf{r})$, or at least restricts it to within the addition of a function that satisfies the Laplace equation $\nabla^2 \Lambda = 0$.

Actually, as shown later, there are other useful choices for the gauge condition. The gauge condition (0.63) defines the so-called Coulomb gauge. The reason this gauge is useful is that the fundamental equation (0.62) for the vector potential is just Poisson's equation. Each vector component of the potential $\mathbf{A}$ depends only on the corresponding vector component of the current density $\mathbf{J}$, so (0.62) separates into three completely independent equations. That is, the equations are completely independent unless the boundary conditions couple them. In the case when the boundaries are at infinity and the current sources are specified, the solution to (0.62) and (0.63) is evidently

$$\mathbf{A}(\mathbf{r}) = \frac{\mu_0}{4\pi} \int \frac{\mathbf{J}(\mathbf{r}') \, d^3\mathbf{r}'}{|\mathbf{r} - \mathbf{r}'|} \tag{0.64}$$

Note that the vector potential more closely reflects the symmetry and orientation of the current distribution than does the magnetic field, and that the magnetic induction at each point is orthogonal to the vector potential. Thus, the vector potential of a line current is linear and parallel to the current, while the magnetic field lines are circles about the line current. On the other hand, the vector potential of a circular current loop consists of circles, as illustrated in Figure 0.10, while the magnetic field lines form loops in planes normal to the plane of the current loop.

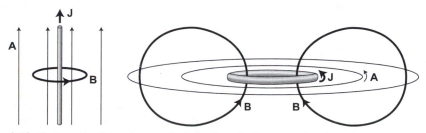

Figure 0.10 Vector potential and magnetic induction of a line current and a current loop.

0.4 ELECTRODYNAMICS

The next step in the development of electrodynamics came in 1831 when a self-educated apprentice bookbinder in England by the name of Faraday, and independently in the same year a physicist in America by the name of Henry, discovered that a changing electrical current in one circuit induces a current in a nearby circuit. In a rapid series of experiments guided by his powerful intuition, Faraday demonstrated that the equivalent effect could be produced by moving a magnet near a circuit or a circuit near a magnet, thereby demonstrating the obverse of Ampere's law, namely, that a moving magnetic field causes an electric current. The quantitative description of this phenomenon is now called Faraday's law of induction, while the unit of inductance is the henry (H). This essentially completed the unification of electricity and magnetism, although it would require the genius of Maxwell to finish the job.

0.4.1 Conservation of Charge

In 1847, Watson and Franklin stated that charge is neither created nor destroyed, but merely transferred from one place to another. Since that time we have observed processes like the annihilation of a positively charged positron by a negatively charged electron to form two neutral photons, but the overall charge before and after the event is zero, so the total charge is still conserved. So far as we know now, in all processes, the total charge is conserved. But the total charge within a volume V bounded by a surface S at time t is

$$Q(t) = \int_V \rho(\mathbf{r}', t)\, d\mathbf{r}' \tag{0.65}$$

while the rate at which charge flows out through the surface S is

$$I(t) = \oint_S \mathbf{J}(\mathbf{r}', t) \cdot d\mathbf{S}' = \int_V \nabla \cdot \mathbf{J}(\mathbf{r}', t)\, d^3\mathbf{r}' \tag{0.66}$$

by the divergence theorem. The conservation of charge requires that

$$\frac{dQ}{dt} + I(t) = \int_V \frac{\partial \rho(\mathbf{r}', t)}{\partial t}\, d^3\mathbf{r}' + \int_V \nabla \cdot \mathbf{J}(\mathbf{r}', t)\, d^3\mathbf{r}' = 0 \tag{0.67}$$

Since this is true for an arbitrary volume V, it must be true of the integrands everywhere, so we get

$$\frac{\partial \rho(\mathbf{r}, t)}{\partial t} + \nabla \cdot \mathbf{J}(\mathbf{r}, t) = 0 \tag{0.68}$$

This is the differential form of the statement of conservation of charge, also called the continuity relation.

0.4.2 Faraday's Law

Faraday's powerful experimental intuition and ignorance of mathematics stand in sharp contrast to Ampere's elegant theories and the elaborate experiments he designed to confirm them. Thus, in a relatively short time, unhindered by theoretical concerns and predispositions, Faraday was able to explain the experimental phenomenon of induction by means of the principle that the current is induced when the conductor cuts across the lines

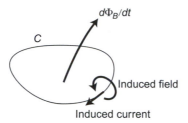

Figure 0.11 Lenz's law.

of force of the magnetic field, whether the magnetic field moves (or changes) or the conductor moves. But it was up to the German mathematical physicists Neumann, Weber, and finally Kirchhoff to construct the mathematical description of the phenomenon. Neumann actually began with Lenz's 1834 statement of the inductive effect, which is that the current induced in a circuit is such as to oppose the change in the magnetic flux through the circuit, as illustrated in Figure 0.11. In a quantitative form this law would now be stated

$$\mathcal{E} = -\frac{d\Phi_B}{dt} \tag{0.69}$$

where the so-called emf (or electromotive force) induced in the circuit is

$$\mathcal{E} = \oint_C \mathbf{E} \cdot d\mathbf{l} \tag{0.70}$$

and the magnetic flux [not to be confused with the scalar potential $\Phi(\mathbf{r}, t)$] is

$$\Phi_B(t) = \int_S \mathbf{B}(\mathbf{r}, t) \cdot d\mathbf{S} \tag{0.71}$$

for any surface S bounded by the circuit C. But if we apply Stokes' law to (0.70) and substitute into (0.69), we get

$$\int_S \nabla \times \mathbf{E}(\mathbf{r}, t) \cdot d\mathbf{S} = -\int_S \frac{\partial \mathbf{B}(\mathbf{r}, t)}{\partial t} \cdot d\mathbf{S} \tag{0.72}$$

Since this must be true for any bounded surface S, the integrands themselves must be equal and we find that

$$\nabla \times \mathbf{E} + \frac{\partial \mathbf{B}}{\partial t} = 0 \tag{0.73}$$

Actually, the law of electromagnetic induction was first stated in roughly this form by Kirchhoff in 1857, and Faraday's ignorance of mathematics would probably have prevented him from understanding it. Nevertheless, (0.73) [or the integral form (0.69)] is called Faraday's law of induction.

0.4.3 Energy in the Magnetic Field

At this point we are finally ready to discuss the energy in the magnetic field, for while we could discuss the energy in the electric field without explicit reference to its time dependence, we cannot do this with magnetic fields. The issue goes back fundamentally to the fact that the static magnetic field can do no work on a charged particle, since the magnetic force on a particle is orthogonal to its velocity. Thus, the work must be found from the

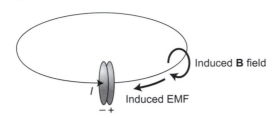

Figure 0.12 Circuit for computing the magnetic energy.

Induced **B** field

Induced EMF

induced electric field, which depends explicitly on the rate of change of the magnetic field. The argument goes as follows.

We consider the circuit shown in Figure 0.12, in which a current I is driven through the circuit by some voltage source. This voltage source is shown in Figure 0.12 as a charged capacitor, but it could be a battery or some other source. For example, if we were to consider the current due to a spinning ring of charge, the argument would involve the torque required to make the ring spin but the result would be the same. As the current in the ring increases, the magnetic field of the current also increases, as does the total magnetic flux through the ring. According to Lenz's law, this induces in the ring a voltage that opposes the increase in the current. The rate at which the voltage source must do work to increase the current is

$$\frac{dW}{dt} = -\mathcal{E}I = -\oint_C \mathbf{E} \cdot I \, d\mathbf{l} = -\int_V \mathbf{E} \cdot \mathbf{J} \, d^3\mathbf{r} \tag{0.74}$$

where we have converted the line integral of the current around the circuit to a volume integral by identifying the line current element $I \, d\mathbf{l}$ with the integral of the volume current element $\mathbf{J} \, d^3\mathbf{r}$ over the area of the wire. The volume V includes the entire circuit and as much of the surrounding volume as we want, since $\mathbf{J} = 0$ outside the wire. If we use Ampere's law (0.43) and the vector identity

$$\nabla \cdot (\mathbf{E} \times \mathbf{B}) = \mathbf{B} \cdot (\nabla \times \mathbf{E}) - \mathbf{E} \cdot (\nabla \times \mathbf{B}) \tag{0.75}$$

then (0.74) becomes

$$\frac{dW}{dt} = -\frac{1}{\mu_0} \int_V \mathbf{E} \cdot (\nabla \times \mathbf{B}) \, d^3\mathbf{r} = -\frac{1}{\mu_0} \int_V [\mathbf{B} \cdot (\nabla \times \mathbf{E}) - \nabla \cdot (\mathbf{E} \times \mathbf{B})] \, d^3\mathbf{r} \tag{0.76}$$

If we use Faraday's law in the first term and convert the second term to a surface integral by means of the divergence theorem, we get

$$\frac{dW}{dt} = \frac{1}{\mu_0} \int_V \mathbf{B} \cdot \frac{\partial \mathbf{B}}{\partial t} \, d^3\mathbf{r} + \int_S (\mathbf{E} \times \mathbf{B}) \cdot d\mathbf{S} \tag{0.77}$$

where S is the surface that bounds the volume V. But we have the freedom, as mentioned earlier, to extend the volume V as far as we like, and if we extend it far enough, then the fields vanish and the surface integral can be ignored. The total rate at which work is done is then

$$\frac{dW}{dt} = \frac{1}{2\mu_0} \frac{d}{dt} \int_V B^2 \, d^3\mathbf{r} \tag{0.78}$$

so the total work accomplished in establishing the field (referred to the condition when the field vanishes everywhere) is

$$W = \frac{1}{2\mu_0} \int_V B^2 \, d^3\mathbf{r} \tag{0.79}$$

That is, the work done to create the fields can be represented by an integral over the fields themselves. We therefore identify

$$\mathcal{U} = \frac{1}{2\mu_0} B^2 \tag{0.80}$$

as the energy density in the magnetic field.

0.5 THE MAXWELL EQUATIONS AND ELECTROMAGNETIC WAVES

0.5.1 The Maxwell–Ampere Law

There is a logical inconsistency in the arguments we used to compute the magnetic energy from the capacitive circuit in Figure 0.12, and it was first noticed by Maxwell. It didn't give us the wrong answer because we were clever enough to avoid it, but it goes like this. To eliminate the current density $\mathbf{J}$ we used Ampere's law (0.43), which in its integral form reads

$$\oint_C \mathbf{B} \cdot d\mathbf{l} = \mu_0 I \tag{0.81}$$

where

$$I = \int_S \mathbf{J} \cdot d\mathbf{S} \tag{0.82}$$

is the total current flowing through the surface S. But this can't be right, for as shown in Figure 0.13, if we draw the surface through the wire (surface S_1), then the total current is finite, but if we draw the surface through middle of the capacitor (surface S_2), the current vanishes, as must, therefore, the magnetic field.

Alternatively, we can find the flaw in Ampere's law directly, for if we take the divergence of (0.43), we get

$$\nabla \cdot \mathbf{J} = \frac{1}{\mu_0} \nabla \cdot (\nabla \times \mathbf{B}) = 0 \tag{0.83}$$

Figure 0.13 Inconsistency in Ampere's law.

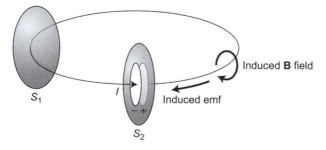

by a vector identity. But $\nabla \cdot \mathbf{J} = 0$ only in magnetostatics, so Ampere's law must also be restricted to this limit. However, this also suggests how to fix Ampere's law, for we know from Gauss's law that the correct answer is

$$\nabla \cdot \mathbf{J} = -\frac{\partial \rho}{\partial t} = -\varepsilon_0 \nabla \cdot \frac{\partial \mathbf{E}}{\partial t} \tag{0.84}$$

If we add this term to Ampere's law we get

$$\nabla \times \mathbf{B} = \mu_0 \mathbf{J} + \mu_0 \varepsilon_0 \frac{\partial \mathbf{E}}{\partial t} \tag{0.85}$$

which is called the Maxwell–Ampere law. Since the final term vanishes in the magnetostatic limit, we have done no violence to Ampere, and in the general case we get the correct expression for the continuity relation.

This last inconsistency was removed from the equations of electrodynamics by Maxwell in 1862, although he was not led by the arguments used here. Instead, he was seeking a mechanical model for the laws of electrodynamics, a connection with the ether that he pursued to the end of his life. Nonetheless, he got it right, and in fact the analogy with an elastic medium immediately suggested to him the possibility of propagating waves. As shown in the next section, the (now correct) equations of electrodynamics lead directly to the formation of waves that propagate at the velocity $c = 1/\sqrt{\mu_0 \varepsilon_0}$, which happens to be precisely the speed of light.

0.5.2 Electromagnetic Waves

In summary, then, the equations of electrodynamics, now called the Maxwell equations, are

$$\nabla \cdot \mathbf{E} = \frac{\rho}{\varepsilon_0} \qquad \text{(Gauss's law)} \tag{0.86}$$

$$\nabla \cdot \mathbf{B} = 0 \qquad \text{(Gauss's law for magnetism)} \tag{0.87}$$

$$\nabla \times \mathbf{E} + \frac{\partial \mathbf{B}}{\partial t} = 0 \qquad \text{(Faraday's law)} \tag{0.88}$$

and

$$\nabla \times \mathbf{B} = \mu_0 \mathbf{J} + \mu_0 \varepsilon_0 \frac{\partial \mathbf{E}}{\partial t} \qquad \text{(Maxwell–Ampere law)} \tag{0.89}$$

In empty space the charge density ρ and the current density $\mathbf{J}$ both vanish, and these equations simplify to

$$\nabla \cdot \mathbf{E} = 0 \tag{0.90}$$

$$\nabla \cdot \mathbf{B} = 0 \tag{0.91}$$

$$\nabla \times \mathbf{E} + \frac{\partial \mathbf{B}}{\partial t} = 0 \tag{0.92}$$

and

$$\nabla \times \mathbf{B} = \mu_0 \varepsilon_0 \frac{\partial \mathbf{E}}{\partial t} \tag{0.93}$$

To find wavelike solutions, we assume that the fields can be represented by the plane waves

$$\mathbf{E} = \mathbf{E}_0 e^{i(\mathbf{k}\cdot\mathbf{r}-\omega t)} \tag{0.94}$$

and

$$\mathbf{B} = \mathbf{B}_0 e^{i(\mathbf{k}\cdot\mathbf{r}-\omega t)} \tag{0.95}$$

where $\mathbf{k}$ is the wave vector, ω is the frequency, and the amplitudes $\mathbf{E}_0$ and $\mathbf{B}_0$ are constants. From Gauss's law for the electric and magnetic fields we find that

$$\nabla \cdot \mathbf{E} = i\mathbf{k} \cdot \mathbf{E}_0 e^{i(\mathbf{k}\cdot\mathbf{r}-\omega t)} = 0 \tag{0.96}$$

and

$$\nabla \cdot \mathbf{B} = i\mathbf{k} \cdot \mathbf{B}_0 e^{i(\mathbf{k}\cdot\mathbf{r}-\omega t)} = 0 \tag{0.97}$$

which show that the waves are purely transverse. From Faraday's law and the Maxwell–Ampere law we find that

$$i\mathbf{k} \times \mathbf{E}_0 e^{i(\mathbf{k}\cdot\mathbf{r}-\omega t)} - i\omega\mathbf{B}_0 e^{i(\mathbf{k}\cdot\mathbf{r}-\omega t)} = 0 \tag{0.98}$$

and

$$i\mathbf{k} \times \mathbf{B}_0 e^{i(\mathbf{k}\cdot\mathbf{r}-\omega t)} + i\mu_0\varepsilon_0\omega\mathbf{E}_0 e^{i(\mathbf{k}\cdot\mathbf{r}-\omega t)} = 0 \tag{0.99}$$

from which we see that the electric and magnetic fields are orthogonal to each other and to the wave vector, as shown in Figure 0.14. If we take the curl of (0.98) and combine it with (0.99) using the vector identity

$$\mathbf{k} \times (\mathbf{k} \times \mathbf{E}_0) = (\mathbf{k} \cdot \mathbf{E}_0)\mathbf{k} - (\mathbf{k} \cdot \mathbf{k})\mathbf{E}_0 = -k^2\mathbf{E}_0 \tag{0.100}$$

we obtain the dispersion relation

$$k^2 - \mu_0\varepsilon_0\omega^2 = 0 \tag{0.101}$$

The phase velocity of these waves is therefore

$$v_\phi = \frac{\omega}{k} = \frac{1}{\sqrt{\mu_0\varepsilon_0}} = c \tag{0.102}$$

where c is the velocity of light. The identification of light as electromagnetic waves was an enormous triumph for Maxwell's theory of electrodynamics, and it vindicated the work of Young, Fresnel, and others near the beginning of the 19th century when they defended the idea that light consists of transverse waves without knowing what the nature of these transverse waves might be. We see in Chapter 10 that Newton's objection that the behavior of light as rays denies their wave properties can be addressed by the eikonal approximation in the geometric-optics limit of small wavelengths. However, Maxwell's discovery also thrust

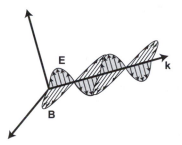

Figure 0.14 Electric and magnetic fields in a plane wave.

the aether theory back into the foreground, and the inconsistencies associated with the aether theory could not be resolved so straightforwardly.

0.5.3 Potentials and Gauges

In the electrostatic case we are able to define the scalar potential $\Phi(\mathbf{r})$ because the curl $\nabla \times \mathbf{E}(\mathbf{r})$ of the electric field vanishes. In the electrodynamic case the curl no longer vanishes, by virtue of Faraday's law, and things become more complicated. But for the magnetic field Gauss's law $\nabla \cdot \mathbf{B} = 0$ remains valid, so there is no immediate problem with the vector potential and we can still write

$$\mathbf{B}(\mathbf{r}, t) = \nabla \times \mathbf{A}(\mathbf{r}, t) \tag{0.103}$$

for some vector function $\mathbf{A}(\mathbf{r}, t)$. If we substitute this into Faraday's law, we see that

$$\nabla \times \left(\mathbf{E} + \frac{\partial \mathbf{A}}{\partial t} \right) = 0 \tag{0.104}$$

Since the curl of the function $\mathbf{E} + \partial \mathbf{A}/\partial t$ vanishes, we can express it as the gradient of some scalar potential $\Phi(\mathbf{r}, t)$ and write

$$\mathbf{E} + \frac{\partial \mathbf{A}}{\partial t} = -\nabla \Phi \tag{0.105}$$

or

$$\mathbf{E} = -\nabla \Phi - \frac{\partial \mathbf{A}}{\partial t} \tag{0.106}$$

That is, we can still represent the electric and magnetic fields in terms of the scalar and vector potentials $\Phi(\mathbf{r}, t)$ and $\mathbf{A}(\mathbf{r}, t)$, but of course the potentials satisfy different equations than they did in the electrostatic and magnetostatic cases.

To find the equations of motion of the scalar and vector potentials, we substitute (0.103) and (0.106) back into the Maxwell equations (0.86) to (0.89). But the potentials already satisfy the homogeneous equations (0.87) and (0.88) by design, so we have only to worry about the inhomogeneous equations (0.86) and (0.89). From these we obtain

$$\nabla^2 \Phi = -\frac{\rho}{\varepsilon_0} - \frac{\partial}{\partial t}(\nabla \cdot \mathbf{A}) \tag{0.107}$$

and

$$\nabla^2 \mathbf{A} - \mu_0 \varepsilon_0 \frac{\partial^2 \mathbf{A}}{\partial t^2} = -\mu_0 \mathbf{J} + \mu_0 \varepsilon_0 \nabla \frac{\partial \Phi}{\partial t} + \nabla(\nabla \cdot \mathbf{A}) \tag{0.108}$$

with the help of a vector identity. This reduces Maxwell's four equations to two, but you might say these equations look worse than the ones we started with. This is true, but we can improve the situation by taking advantage of the arbitrariness inherent in the definition of the potentials.

In the static case, the fields $\mathbf{E}(\mathbf{r})$ and $\mathbf{B}(\mathbf{r})$ are unaffected if we add a constant Λ to the scalar potential $\Phi(\mathbf{r})$ or the gradient of a function $\Lambda(\mathbf{r})$ to the vector potential $\mathbf{A}(\mathbf{r})$. In the general case, the fields are unaffected by the transformations

$$\Phi(\mathbf{r}, t) \rightarrow \Phi'(\mathbf{r}, t) = \Phi(\mathbf{r}, t) + \phi(\mathbf{r}, t) \tag{0.109}$$

and

$$\mathbf{A}(\mathbf{r}, t) \rightarrow \mathbf{A}'(\mathbf{r}, t) = \mathbf{A}(\mathbf{r}, t) + \mathbf{a}(\mathbf{r}, t) \tag{0.110}$$

provided that

$$\nabla\phi + \frac{\partial \mathbf{a}}{\partial t} = 0 \tag{0.111}$$

and

$$\nabla \times \mathbf{a} = 0 \tag{0.112}$$

The second requirement is satisfied if $\mathbf{a}(\mathbf{r}, t) = \nabla\Lambda(\mathbf{r}, t)$ for some function $\Lambda(\mathbf{r}, t)$, and the first is satisfied if $\phi(\mathbf{r}, t) = -\partial\Lambda/\partial t$, so the fields are unaffected by a transformation of the form

$$\Phi(\mathbf{r}, t) \rightarrow \Phi'(\mathbf{r}, t) = \Phi(\mathbf{r}, t) - \frac{\partial \Lambda(\mathbf{r}, t)}{\partial t} \tag{0.113}$$

$$\mathbf{A}(\mathbf{r}, t) \rightarrow \mathbf{A}'(\mathbf{r}, t) = \mathbf{A}(\mathbf{r}, t) + \nabla\Lambda(\mathbf{r}, t) \tag{0.114}$$

for any function $\Lambda(\mathbf{r}, t)$. Transformations (0.113) and (0.114) comprise what is called a gauge transformation, and $\Lambda(\mathbf{r}, t)$ is called a gauge function.

We can use the flexibility afforded by the gauge invariance of the electric and magnetic fields to simplify (0.107) and (0.108) in many ways. Two of these ways are of sufficiently general usefulness to deserve mention. The first is called the Coulomb gauge, and the second is the Lorentz gauge. If we start with the vector potential $\mathbf{A}(\mathbf{r}, t)$, we can transform it using the function $\Lambda(\mathbf{r}, t)$ that satisfies the Poisson equation

$$\nabla^2\Lambda = -\nabla \cdot \mathbf{A} \tag{0.115}$$

and vanishes at infinity. With this transformation we see that

$$\nabla \cdot \mathbf{A}' = \nabla \cdot \mathbf{A} + \nabla^2\Lambda = 0 \tag{0.116}$$

That is, we can always transform to a gauge in which $\nabla \cdot \mathbf{A}' = 0$. Equations (0.107) and (0.108) then become

$$\nabla^2\Phi = -\frac{\rho}{\varepsilon_0} \tag{0.117}$$

and

$$\nabla^2\mathbf{A} - \mu_0\varepsilon_0\frac{\partial^2\mathbf{A}}{\partial t^2} = -\mu_0\mathbf{J} + \mu_0\varepsilon_0\nabla\frac{\partial\Phi}{\partial t} \tag{0.118}$$

where we have left off the primes with the understanding that the potentials satisfy the additional condition

$$\nabla \cdot \mathbf{A} = 0 \tag{0.119}$$

which is always possible. This is called the Coulomb gauge condition, for reasons that are obvious from the form of (0.117). In the Coulomb gauge, the potential is just the electrostatic potential of the instantaneous charge distribution $\rho(\mathbf{r}, t)$. The usefulness of the Coulomb gauge lies in the fact that far from the sources the scalar potential Φ vanishes, so free radiation fields are described simply in terms of the vector potential $\mathbf{A}$. Because of its usefulness in problems of this type, the Coulomb gauge is sometimes called the radiation gauge.

The other important gauge, the Lorentz gauge, is obtained by choosing the gauge function that satisfies the wave equation

$$\nabla^2\Lambda - \mu_0\varepsilon_0\frac{\partial^2\Lambda}{\partial t^2} = -\nabla \cdot \mathbf{A} - \mu_0\varepsilon_0\frac{\partial\Phi}{\partial t} \tag{0.120}$$

and vanishes at infinity. In this case (0.107) and (0.108) become

$$\nabla^2 \Phi - \mu_0 \varepsilon_0 \frac{\partial^2 \Phi}{\partial t^2} = -\frac{\rho}{\varepsilon_0} \tag{0.121}$$

and

$$\nabla^2 \mathbf{A} - \mu_0 \varepsilon_0 \frac{\partial^2 \mathbf{A}}{\partial t^2} = -\mu_0 \mathbf{J} \tag{0.122}$$

The gauge condition for the Lorentz gauge is then

$$\nabla \cdot \mathbf{A} + \mu_0 \varepsilon_0 \frac{\partial \Phi}{\partial t} = 0 \tag{0.123}$$

The beauty of the Lorentz gauge lies in the fact that Φ and $\mathbf{A}$ are treated symmetrically, each satisfying an inhomogeneous wave equation. The scalar potential and the individual components of the vector potential depend on their own source terms and are independent (aside from the implicit coupling of the source terms imposed by the continuity relation) unless they are coupled through the boundary conditions, as frequently happens. But the full beauty and symmetry of the Lorentz gauge become apparent only in the context of the special theory of relativity.

0.6 CONSERVATION LAWS

0.6.1 Poynting's Theorem

In nonrelativistic theory, Newton's laws of motion conserve the energy and momentum of a closed set of interacting particles because the forces through which the particles interact are assumed to propagate with infinite velocity. This is not true in reality, and the energy and momentum of the particles by themselves are not conserved. Instead, the particles transfer energy and momentum to the fields and the fields transfer it back to the particles. Only the total energy and momentum of the fields and the particles together is conserved.

To describe the energy of the electromagnetic field, we begin with the Maxwell equations, (0.86)–(0.89). If we take the dot product of $\mathbf{E}$ with (0.89) and the dot product of $\mathbf{B}$ with (0.88) and combine the results, we get

$$\mathbf{E} \cdot (\nabla \times \mathbf{B}) - \mathbf{B} \cdot (\nabla \times \mathbf{E}) = \mu_0 \mathbf{J} \cdot \mathbf{E} + \frac{\partial}{\partial t} \left(\frac{B^2}{2} + \frac{E^2}{2c^2} \right) \tag{0.124}$$

To simplify the left-hand side we use the vector identity

$$\nabla \cdot (\mathbf{a} \times \mathbf{b}) = \mathbf{b} \cdot (\nabla \times \mathbf{a}) - \mathbf{a} \cdot (\nabla \times \mathbf{b}) \tag{0.125}$$

and get

$$\frac{\partial}{\partial t} \left(\frac{\varepsilon_0}{2} E^2 + \frac{1}{2\mu_0} B^2 \right) + \nabla \cdot \left(\frac{1}{\mu_0} \mathbf{E} \times \mathbf{B} \right) + \mathbf{J} \cdot \mathbf{E} = 0 \tag{0.126}$$

The fact that the sum of these three terms vanishes makes (0.126) look like some sort of conservation law, and in fact it expresses the conservation of energy. We define

$$\mathcal{U} = \frac{\varepsilon_0}{2} E^2 + \frac{1}{2\mu_0} B^2 \tag{0.127}$$

to be the energy density of the electromagnetic field and

$$\mathbf{S} = \frac{1}{\mu_0}\mathbf{E} \times \mathbf{B} \tag{0.128}$$

to be the so-called Poynting vector. In terms of these quantities, the energy conservation law becomes

$$\frac{\partial \mathcal{U}}{\partial t} + \nabla \cdot \mathbf{S} + \mathbf{E} \cdot \mathbf{J} = 0 \tag{0.129}$$

This is called Poynting's theorem, since it was discovered by him in 1884.

The Poynting vector $\mathbf{S}$ represents a flow of energy. To see this more clearly, we integrate (0.129) over the volume V bounded by the surface S, as shown in Figure 0.15. Using the divergence theorem to evaluate $\int \nabla \cdot \mathbf{S}\, dV$, we get

$$\frac{d}{dt}\int_V \mathcal{U}\, dV + \oint_S \mathbf{S} \cdot \hat{\mathbf{n}}\, dS + \int_V \mathbf{J} \cdot \mathbf{E}\, dV = 0 \tag{0.130}$$

We interpret this in the following way. The first term represents the rate of change of the total energy of the electromagnetic field in volume V. The second term represents the rate at which electromagnetic field energy flows out through surface S. The last term is the rate at which the field is doing work on the charges. Therefore, (0.130) states that the rate of change of the total energy of the electromagnetic field plus the rate at which work is done by the field on the charges in volume V plus the flow of energy out through surface S vanishes.

Poynting immediately applied his theorem to a resistive wire oriented parallel to an electric field E so that it carries a current $I = E/R$, where R is the resistance per unit length. The magnetic field of the current in the wire is azimuthally directed around the wire, so the Poynting vector $\mathbf{S} = (1/\mu_0)\mathbf{E} \times \mathbf{B}$ is directed radially inward into the wire, as shown in Figure 0.16. Poynting showed that the power flow into the surface of the wire represented by the Poynting vector is just equal to the rate $I^2 R$ at which the electrons in a unit length of the wire increase their energy by dissipation. Like Faraday, Poynting interpreted the physics of this situation mechanically in terms of the energy conveyed by motions of the lines of force as the electric and magnetic fields were established. We would view things more abstractly now, but he got the answer first.

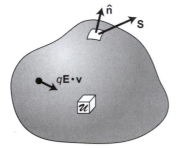

Figure 0.15 Conservation of energy in the volume V.

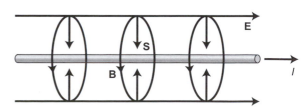

Figure 0.16 Electromagnetic fields and energy flow around a wire carrying an electric current.

0.6.2 Conservation of Momentum

The conservation of momentum can be treated in a similar manner, although the details are more complicated because it is a vector relation. If we take the cross product of $\mathbf{B}$ with (0.89) and $\mathbf{E}$ with (0.88) and combine the results, we get

$$\mathbf{B} \times (\nabla \times \mathbf{B}) + \frac{1}{c^2}\mathbf{E} \times (\nabla \times \mathbf{E}) + \frac{1}{c^2}\frac{\partial}{\partial t}\mathbf{E} \times \mathbf{B} + \mu_0 \mathbf{J} \times \mathbf{B} = 0 \qquad (0.131)$$

If we now subtract $\mathbf{B}(\nabla \cdot \mathbf{B}) = 0$ and $(\mathbf{E}/c^2)[\nabla \cdot \mathbf{E} - (\rho/\varepsilon_0)] = 0$ from the left-hand side we get

$$\frac{\partial}{\partial t}(\varepsilon_0 \mathbf{E} \times \mathbf{B}) + (\rho\mathbf{E} + \mathbf{J} \times \mathbf{B})$$

$$+ \left\{\frac{1}{\mu_0}[\mathbf{B} \times (\nabla \times \mathbf{B}) - \mathbf{B}(\nabla \cdot \mathbf{B})] + \varepsilon_0[\mathbf{E} \times (\nabla \times \mathbf{E}) - \mathbf{E}(\nabla \cdot \mathbf{E})]\right\} = 0 \qquad (0.132)$$

When we compare the structure of this expression with that of Poynting's theorem (0.129), we are led to identify the first term as the rate of change of the momentum density

$$\mathbf{g} = \varepsilon_0 \mathbf{E} \times \mathbf{B} \qquad (0.133)$$

The second term is evidently the force on the charges, so the third term in (0.132) must be the divergence of the flow of momentum. But the flux of momentum is a tensor quantity because its components are the flow in the ith direction of momentum in the jth direction. We therefore use tensor notation to write

$$[\mathbf{B} \times (\nabla \times \mathbf{B}) - \mathbf{B}(\nabla \cdot \mathbf{B})]_i = \sum_{j,k,l,m=1}^{3} \varepsilon_{ijk}\varepsilon_{klm} B_j \frac{\partial B_m}{\partial x_l} - \sum_{j=1}^{3} B_i \frac{\partial B_j}{\partial x_j} \qquad (0.134)$$

where ε_{ijk} is the Levi-Civita symbol. But $\varepsilon_{ijk}\varepsilon_{klm}$ vanishes unless $i, j = l, m$ or $i, j = m, l$, and we see that

$$\sum_{k=1}^{3} \varepsilon_{ijk}\varepsilon_{klm} = \sum_{k=1}^{3} \varepsilon_{ijk}\varepsilon_{lmk} = \delta_{il}\delta_{jm} - \delta_{im}\delta_{jl} \qquad (0.135)$$

where δ_{ij} is the Kronecker delta. If we substitute this back in (0.134) and do some algebra, we find that

$$[\mathbf{B} \times (\nabla \times \mathbf{B}) - \mathbf{B}(\nabla \cdot \mathbf{B})]_i = -\sum_{j=1}^{3} \frac{\partial}{\partial x_j}\left(B_i B_j - \frac{1}{2}B^2\delta_{ij}\right) \qquad (0.136)$$

with a corresponding expression for $[\mathbf{E} \times (\nabla \times \mathbf{E}) - \mathbf{E}(\nabla \cdot \mathbf{E})]$. We may therefore write the momentum conservation law in the form

$$\frac{\partial g_i}{\partial t} + F_i^{(L)} + \sum_{j=1}^{3} \frac{\partial T_{ij}^{(M)}}{\partial x_j} = 0 \qquad (0.137)$$

where

$$\mathbf{F}^{(L)} = \rho\mathbf{E} + \mathbf{J} \times \mathbf{B} \qquad (0.138)$$

is the Lorentz force per unit volume and

$$T_{ij}^{(M)} = -\varepsilon_0 E_i E_j - \frac{1}{\mu_0}B_i B_j + \mathcal{U}\delta_{ij} = T_{ji}^{(M)} \qquad (0.139)$$

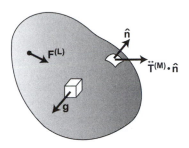

Figure 0.17 Conservation of momentum in the volume V.

is called the Maxwell stress tensor. Note that some authors define the Maxwell stress tensor with the opposite sign. In this case $T_{ij}^{(M)}$ represents the negative of the momentum flux but has the correct sign to represent the stress inward on a region of space.

To understand what is meant by the stress, we can integrate (0.137) over some finite volume V bounded by the surface S, as shown in Figure 0.17, and use the divergence theorem to get

$$\frac{dp_i}{dt} = -\sum_{j=1}^{3} \int_S T_{ij}^{(M)} \hat{n}_j \, dS \tag{0.140}$$

or in dyadic notation,

$$\frac{d\mathbf{p}}{dt} = -\int_S \overset{\leftrightarrow}{\mathbf{T}}^{(M)} \cdot \hat{\mathbf{n}} \, dS \tag{0.141}$$

where $\hat{\mathbf{n}}$ is a unit vector normal to the surface in the outward direction and

$$\frac{d\mathbf{p}}{dt} = \frac{d}{dt} \int_V \mathbf{g} \, dV + \int_V (\rho \mathbf{E} + \mathbf{J} \times \mathbf{B}) \, dV \tag{0.142}$$

is the total rate of change of momentum (including the particles and the fields) inside volume V. Expressed in terms of the vector fields themselves, this becomes

$$\frac{d\mathbf{p}}{dt} = -\oint_S \mathcal{U}\hat{\mathbf{n}} \, dS + \varepsilon_0 \oint_S \mathbf{E}(\mathbf{E} \cdot \hat{\mathbf{n}}) \, dS + \frac{1}{\mu_0} \oint_S \mathbf{B}(\mathbf{B} \cdot \hat{\mathbf{n}}) \, dS \tag{0.143}$$

The first term on the right looks like the surface integral of a pressure $p = \mathcal{U}$ pushing inward on volume V. The last two terms suggest that there is a tension $\varepsilon_0 E^2$ and B^2/μ_0 along the electric and magnetic lines of force, respectively, pulling on the surface of volume V. For simplicity we consider a static electric field. On surface S, we have, from the second term, the effect of the electrostatic pressure

$$\boldsymbol{\sigma} = -\mathcal{U}\hat{\mathbf{n}} = -\tfrac{1}{2}\varepsilon_0 E^2 \hat{\mathbf{n}} \tag{0.144}$$

which acts inward (due to the negative sign), normal to the surface. From the third term, we have the tension

$$\boldsymbol{\sigma} = \varepsilon_0 \mathbf{E}(\mathbf{E} \cdot \hat{\mathbf{n}}) \tag{0.145}$$

It acts in the direction of the electric field $\mathbf{E}$ with a magnitude proportional to the component of $\mathbf{E}$ normal to the surface. For an electric field parallel to the surface this stress vanishes, but for an electric field normal to the surface the stress is

$$\boldsymbol{\sigma} = \varepsilon_0 E^2 \hat{\mathbf{n}} \tag{0.146}$$

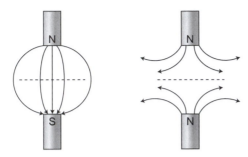

Figure 0.18 Interaction of permanent magnets.

which acts outward along the lines of force, independent of the sign of the electric field. Its magnitude is twice that of the electric pressure, so for a field normal to the surface the net result is a tension pulling outward across the surface. In the same way, we find that there is an effective pressure across the magnetic lines of force due to the magnetic energy density, but a tension along the lines of force. We see this very clearly in the behavior of permanent magnets. When the poles of two magnets are aligned north to south, as shown in Figure 0.18, the lines of force stretch from the north pole of one magnet to the south pole of the other. The tension along the lines of force causes the magnets to attract one another. When the poles are aligned north to north, the lines of force butt up against one another and the pressure transverse to the lines of force causes the magnets to repel one another. The concepts of tension along the magnetic lines of force and pressure across them were first recognized by Faraday and can be used to understand a variety of phenomena. For example, the current carried by an electron beam or through a plasma is encircled by magnetic lines of force. The tension in these lines of force compresses the electron beam or the plasma and creates the so-called pinch effect. On the other hand, the magnetic field inside a solenoid exerts an outward pressure on the windings that create the field. The stress on the windings can have serious consequences for the structural integrity of the magnet, especially in superconducting magnets, where the strain can cause the windings to leave the superconducting state. When this happens, the energy in the magnetic field is quickly converted to heat in the normal-conducting windings, and the magnet "quenches" with a sudden burst of boiling liquid helium!

BIBLIOGRAPHY

Excellent and thorough introductions to electrodynamics are to be found in

> D. J. Griffiths, *Introduction to Electrodynamics,* 3rd edition, Prentice Hall, Upper Saddle River, NJ (1999).
> P. Lorrain, D. R. Corson, and F. Lorrain, *Fundamentals of Electromagnetic Phenomena,* W. H. Freeman and Company, New York (2000).
> G. L. Pollack and D. R. Strump, *Electromagnetism,* Addison Wesley, San Francisco (2002).

For particularly entertaining discussions of the history of electrodynamics, the interested reader is referred to

> P. F. Mottelay, *Bibliographical History of Electricity and Magnetism,* Charles Griffin & Company, London (1922).
> E. T. Whittaker, *A History of the Theories of Aether and Electricity from the Age of Descartes to the Close of the Nineteenth Century,* Longmans, Green and Company, London (1910).

1   Relativistic Kinematics

The one truth of which the human mind can be certain—indeed, this is the meaning of consciousness itself—is the recognition of its own existence. That we may be secure in this truth is assured us by Descartes' famous axiom even if everything else, including Descartes, is a figment of our imagination. Nothing else can be proved. Our most fundamental *belief,* then, is that the universe exists around us, consisting of three spatial dimensions and time, and that while we can move about in the three spatial dimensions, time flows inexorably onward, everywhere the same. Newton himself said it: ". . . absolute, true, and mathematical time, of itself, and from its own nature, flows equably and without relation to anything external." But he was wrong. The theory of relativity shows us that time and space do not have the meaning we thought they had. In the words of Weyl, ". . . *we are to discard our belief in the objective meaning of simultaneity; it was the great achievement of Einstein in the field of the theory of knowledge that he banished this dogma from our minds,* and this is what leads us to rank his name with that of Copernicus" (italics his).

But the discovery of relativity by Einstein in 1905 was not a bolt from the blue. People had been concerned about the nature of space and time for at least hundreds of years before that, becoming more and more disturbed by the inconsistencies in our understanding of the physical world toward the end of the 19th century. Nevertheless, even after the true nature of space and time became clear, the theory of relativity so contradicted our most fundamental belief that it was rejected for years. Einstein himself never received the Nobel Prize for this work that was, in the words of Bertrand Russell, "probably the greatest synthetic achievement of the human intellect up to the present time." Some 16 years afterward he was reluctantly awarded the Nobel Prize for a lesser work because the greatest physicist of the century, known to more people than the president of the United States, could not be completely ignored. Yet, the truth Einstein taught us displayed once again nature's tendency to assume the most beautiful, symmetric form, in spite of our objections. And while the truth is a merger of space and time that prevents us from ordering events absolutely in time, it does not result in chaos but preserves those features that we cannot logically be denied. The principle of causality is never violated, and as each of us progresses through this four-dimensional space–time, our individual perception of time as moving always forward is not contradicted.

We begin our discussion of classical electricity and magnetism with the theory of relativity because the two are intimately connected. Not only can we not correctly describe electromagnetism outside the theory of relativity, but the theory of relativity was thrust upon us by problems of the propagation of light and, eventually, by the Maxwell equations themselves. Moreover, the concepts of relativity unite electricity and magnetism and give the subject a beauty and symmetry that it otherwise lacks. Finally, the theory of relativity provides us with a mathematical framework and a system of notation that simplify our lives and lend elegance to our work.

1.1 THE PRINCIPLES OF SPECIAL RELATIVITY

1.1.1 Historical Overview

The earliest theory of light that we discuss is that of Newton. Newton believed that light consists of particles, corpuscles, that obey the same laws of motion as other bodies. He stated as his first law of motion that unless acted upon by an outside force, a body at rest will remain at rest and a body in motion will remain in uniform motion. A frame of reference K in which this is true is called an inertial frame. Clearly, if another reference frame K' is also an inertial frame, it moves relative to K with at most a constant velocity, and vice versa. If time is absolute, then the coordinates $\mathbf{r}$ and $\mathbf{r}'$ and the times t and t' in the two frames are related by

$$\mathbf{r}' = \mathbf{r} - \mathbf{v}t \tag{1.1}$$

and

$$t' = t \tag{1.2}$$

These are known as a Galilean transformation. Newton's second law states that the acceleration of a body is proportional to the applied force $\mathbf{F}$:

$$\mathbf{F} = m\frac{d^2\mathbf{r}}{dt^2} \tag{1.3}$$

where the mass m is a constant. If we transform this equation to the moving frame K', using (1.1) and (1.2), we recover (1.3) with the same force $\mathbf{F}' = \mathbf{F}$. That is, Newton's equation of motion is invariant under Galilean transformations. Newton's third law states that for every action (force on one body) there is an equal and opposite reaction (force on the other body). This implicitly assumes that the interaction between distant objects is felt instantly. If we have a closed set of objects that interact with each other according to these laws and sum over all the objects i, then the forces cancel out by the third law and we find that

$$\frac{d}{dt}\sum_i m_i \frac{d\mathbf{r}_i}{dt} = \frac{d}{dt}\sum_i \mathbf{p}_i = 0 \tag{1.4}$$

where $\mathbf{p}_i = m_i\, d\mathbf{r}_i/dt$ is the momentum of particle i. This tells us that the total mechanical momentum of the particles is conserved if the influence of one particle is felt instantly by the others. There is then no intrinsic need for the concept of fields, except as a way of visualizing the forces. We can compute the momentum in the fields, but this is merely a bookkeeping device. The momentum of the particles is conserved without considering the fields. We shall see shortly, however, that action at a distance is in fact not instantaneous, so action and reaction do not cancel and the momentum of interacting particles is not conserved. What is left over is the momentum of the fields, and this momentum gives the fields a reality of their own.

Newton believed that these same laws, which obey Galilean relativity, describe the motion of light corpuscles. In fact, Descartes derived Snell's law of refraction on the basis of the corpuscular theory of light, and so in France it is known as Descartes' law. He explained that refraction is the acceleration of the corpuscles toward the denser medium (the one with a higher index of refraction) as they enter the surface of the medium. Thus, light travels faster inside a denser medium. Newton's theory is completely self-consistent, and its invariance under Galilean transformations shows that all inertial reference frames are equivalent. However, the velocity of light in free space, while finite, is not a universal constant except, perhaps, with respect to the source of the light.

Of course, this theory does not agree with reality. The most serious defect, and the earliest discovered, is that light actually consists of waves. Moreover, the speed of light is slower rather than faster in a medium with a higher index of refraction. At the beginning of the 19th century, Young discovered that light emerging from a pair of very narrow slits creates interference patterns on a screen some relatively large distance from the slits. From this he was actually able to determine the wavelength of the light. Around the same time, Fraunhofer and Fresnel also investigated the diffraction of light, both experimentally and mathematically, and firmly established the wave theory of light. Although we were to learn one hundred years later from Planck and Einstein of the particle aspects of light, the evidence at this time for the wave theory of light was incontrovertible. But it demolished the beautiful self-consistency of Newton's theory, for waves, it was believed, must propagate through a medium. This made it necessary to introduce the "luminiferous aether," a tenuous medium that permeates all space, imperceptible except that it carries light. Moreover, light travels more slowly in a denser medium, so the ether must interact with matter in some way. But the most bothersome aspect of this ether was the recognition that the ether must be at rest in some frame of reference, perhaps that of the fixed stars or the center of mass of the universe. If this is so, how does light propagate through a medium that is moving through the ether?

In a series of brilliant (especially when you consider the state of knowledge and the technology available at the time) experiments, Fizeau and Hoek set about to answer this question, without so much as a laser at their disposal! Fizeau used the apparatus shown in Figure 1.1. The beam splitter divides the light into two beams, one of which propagates

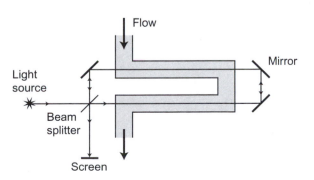

Figure 1.1 Fizeau's apparatus.

clockwise around the ring while the other propagates counterclockwise. The water flowing through the pipe circulates clockwise around the ring. If the apparatus is motionless in the ether and the water is not flowing through the pipe, light travels clockwise around the ring in the same time as light traveling counterclockwise and there is observed some fringe pattern where the beams interfere at the screen. When the water is flowing, however, light traveling clockwise moves with the water, while light traveling counterclockwise moves against it. If the ether is carried along with the water, one beam experiences a longer trip than the other, and the interference fringes shift. If we assume that the ether is dragged along at velocity αv, where v is the velocity of the water and α is some coefficient, then the velocity of propagation in the water is $c/n \pm \alpha v$ and the relative phase shift is

$$\Delta\phi = 2\pi \frac{c}{\lambda_0}\left(\frac{L}{\dfrac{c}{n}-\alpha v} - \frac{L}{\dfrac{c}{n}+\alpha v}\right) \approx 4\pi n^2 \frac{L}{\lambda_0}\frac{\alpha v}{c}\left[1 + \left(\frac{n\alpha v}{c}\right)^2 + \cdots\right] \quad (1.5)$$

where L is the length of the path through the water, n the index of refraction of the water (we ignore dispersion effects caused by the Doppler shift of the light traveling with respect to water moving in opposite directions), c the velocity of light in vacuum, and λ_0 the vacuum wavelength. The approximate expression is accurate only for $n\alpha v/c \ll 1$, but since Fizeau's experiment used velocities of the order of $n\alpha v/c = O(\lambda_0/nL) \ll 1$, this approximation is valid. Fizeau obtained good agreement with this theory provided he used the value $\alpha = 1 - n^{-2}$ for the drag coefficient, a value proposed by Fresnel. Unfortunately, this theory has the logical difficulty that the ether drag velocity, like $n(\lambda)$, is different for different wavelengths. There would have to be a separate ether for each wavelength!

Later, Hoek did similar experiments except that the water was stationary and the motion of the medium relative to the ether was provided by the velocity of the earth. The fringe shift could be detected by rotating the apparatus to change the direction of the motion of the ether relative to the experiment. No motion of the experiment through the ether could be detected. However, in this case the fringe shift appears only in second order in v/c, and Hoek's experiment was not sensitive enough to detect this. Finally, in 1889, Michelson and Morley repeated Hoek's experiment using a different geometry, as shown in Figure 1.2. In Michelson's version of the experiment, the incident light is split into two beams propagating along axes normal to one another. Depending on the orientation of the apparatus, one beam travels parallel and antiparallel to the motion of the earth through the ether, while the other travels across the motion through the ether. The effect on the ether drift is different for the two axes of the interferometer. Based on Fizeau's results, the ether drag in air ($n \approx 1$) should have been negligible and motion relative to the ether should have

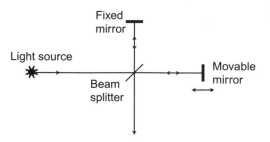

Figure 1.2 Michelson–Morley experiment.

been large. The sensitivity of the experiment was sufficient to detect motion much smaller than the velocity of the earth, but no motion was detected. Taken together, the Fizeau and the Michelson–Morley experiments are in contradiction with the ether theory.

In 1864, at about the same time as Hoek was doing his experiments, Maxwell showed theoretically that light is actually an electromagnetic wave. It was immediately apparent that the wave equation

$$\nabla^2 \mathbf{E} - \frac{1}{c^2} \frac{\partial^2 \mathbf{E}}{\partial t^2} = 0 \tag{1.6}$$

for the electric field $\mathbf{E}$ is not invariant under Galilean transformations. Thus, it can be valid only in a single reference frame, namely, the rest frame of the ether. Riemann and Gauss objected to the notion of the ether and proposed that all electric interactions propagate at the same speed as light. But Maxwell remained firmly convinced of the existence of the ether and wrote "There appears to be, in the minds of these eminent men, some prejudice, or a priori objection, against the hypothesis of a medium in which the phenomena of radiation of light and heat and the electric actions at a distance take place. . . . In fact, whenever energy is transmitted from one body to another in time, there must be a medium or substance in which the energy exists after it leaves the one body and before it reaches the other, for energy, as Torricelli remarked, 'is a quintessence of so subtle a nature that it cannot be contained in any vessel except the inmost substance of material things.' Hence all these theories lead to the conception of a medium in which the propagation takes place. . . ." Subsequently, Lorentz discovered a transformation that left the wave equation invariant, eliminating the attachment to one unique frame of reference, but it had the unphysical property of mixing time and space and leaving them stretched or compressed. However, FitzGerald showed that the contraction of space in the direction of motion relative to the ether could be used to explain the null result obtained by Michelson and Morley.

Finally, in 1905, Einstein provided a way out of the dilemma by discarding the ether altogether, but this "simplification" came at the cost of our preconceptions about the nature of space and time. In a sense, the theory of relativity did for electrodynamics what the first law of thermodynamics had done for heat a hundred years before when it eliminated the mysterious substance caloric as the origin of hotness and coldness. According to Einstein's theory of relativity, the speed of light is the same to all observers, so the Michelson–Morley experiment must give a null result. Moreover, a careful analysis of Fizeau's experiment using the special theory of relativity gives

$$\Delta\phi \approx 4\pi n^2 \frac{L}{\lambda_0} \frac{v}{c} \left(1 - \frac{1}{n^2}\right) \left[1 + \left(\frac{nv}{c}\right)^2 + \cdots\right] \tag{1.7}$$

To lowest order in v/c (the accuracy of Fizeau's experiment) this agrees with (1.5), but now the "drag coefficient" $\alpha = 1 - n^{-2}$ has a logical basis. That is, since it no longer represents the velocity of the ether, there is no difficulty associated with the fact that it is wavelength dependent.

1.1.2 Einstein's Postulates

Einstein's solution to the dilemma of the velocity of light was as beautiful as it was radical. He chose the most symmetric form for nature, stating that all inertial reference frames are equivalent. He embodied this concept in his two postulates of special relativity, which state

1. The laws of nature are identical in all inertial frames of reference. That is, if we transform the mathematical equations of physics from one inertial reference frame to another they remain in the same form.
2. The speed of light c is the same to all observers at rest in inertial frames of reference. A more general statement of this principle might be that the influence of one particle is not felt instantaneously by another. Instead, the influence propagates at some (maximum) velocity c.

These two postulates can be summed up in a single postulate that states that there is no experiment that can be done to distinguish the absolute velocity of any coordinate system.

Einstein's postulates appear beautifully simple and symmetric. In fact they are deceptively simple, since within them lie profound consequences and more than a few startling paradoxes. Fundamental to all the paradoxes is the fact that simultaneity is no longer an objective reality, as Weyl points out, but rather a subjective one that depends on the observer. A simple example, shown in Figure 1.3, illustrates this. As shown there, a radio transmitter has been placed on the moon equidistant between two colonies, called Alpha and Beta. At the conclusion of the annual green-cheese eating contest, the winner's name is sent out from the transmitter to both colonies. Since the speed of light is the same for all observers, the news reaches Alpha and Beta simultaneously. Everyone on the moon agrees with this. However, some vacationers returning to earth past the moon hear the same news. As viewed by them, the moon is moving past them at a high velocity, so Alpha passes them first and Beta a little while later. Clearly, from their point of view, the radio transmission takes longer to catch up to the colony of Alpha as it moves away from the transmitter, and the news reaches Beta first, not simultaneously. What went wrong in this story? Who is correct? In fact, both the colonists and the vacationers are correct, each in their own coordinate system. Simultaneity is simply not an objective reality. Nor is the ordering of events in time necessarily unique, for if we watch the same events from a spaceship departing the earth and passing the moon in the opposite direction from the vacationers, we correctly note that the Alpha colonists hear the news first. You may object that this isn't the way you want it to be, but you can live with the paradox because nothing completely illogical has happened. Causality has not been violated; that is, no one hears the news before it has been transmitted and each observer sees things happening in a time-ordered fashion in his own reference frame.

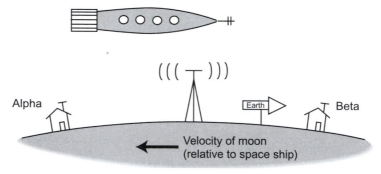

Figure 1.3 Simultaneity paradox.

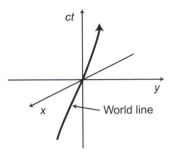

Figure 1.4 Minkowski diagram.

1.1.3 Intervals

To discuss the concepts of relativistic kinematics it is useful to introduce the world diagram, or Minkowski diagram. For two spatial dimensions and time this is shown in Figure 1.4. Each point $(t, \mathbf{r})$ in this four-dimensional space is called an event, and the path of a particle, $[t(s), \mathbf{r}(s)]$ for some parameter s, is called the world line of the particle.

We now consider the relationship between two events viewed in the reference frames K and K' when reference frame K' is moving at a constant velocity $\mathbf{v}$ relative to K. We note at the beginning that observers in both reference frames agree on the relative velocity $\mathbf{v}$. That is, they agree on the absolute magnitude $v = |\mathbf{v}|$, since by symmetry neither could observe a greater velocity than the other, but of course they differ on the sign of the vector $\mathbf{v}$. By symmetry, again, it is clear that the transformation from K to K' differs from the inverse transformation (from K' to K) only in the sign of $\mathbf{v}$.

Central to the discussion is what we call the interval between two events. In ordinary three-dimensional Euclidian geometry the distance between two infinitesimally separated points is

$$dl^2 = dx^2 + dy^2 + dz^2 > 0 \tag{1.8}$$

For two infinitesimally separated events in four-dimensional Minkowski space, however, we define the intervals

$$ds^2 = c^2 dt^2 - dx^2 - dy^2 - dz^2 \tag{1.9}$$

and

$$ds'^2 = c^2 dt'^2 - dx'^2 - dy'^2 - dz'^2 \tag{1.10}$$

in K and K', respectively. These expressions are called the metric equations for the two reference frames, and the quantity ds represents some sort of distance between the two events in four-dimensional space-time. Because of the minus signs in these expressions, the geometry of Minkowski space is not Euclidian but is called pseudo-Euclidian. Since the square of the interval can be positive or negative, some authors treat time as an imaginary coordinate, but we use a different approach here.

The rationale for defining the interval as we have done here clearly comes from the fact that the speed of light is the same in all inertial reference frames. If the two events separated by ds correspond to the passage of a signal traveling at the velocity of light, then the interval as we define it vanishes in all reference frames

$$ds^2 = ds'^2 = 0 \tag{1.11}$$

Put another way, $ds^2 = 0$ in one frame strictly implies that $ds'^2 = 0$ in any other, and vice versa. More generally we note that since uniform motion in one inertial reference frame implies uniform motion in another, the transformation between K and K' must be linear, so it must be true that

$$ds'^2 = a \, ds^2 \tag{1.12}$$

where a is some constant that depends, at most, on the relative motion of K and K'. But a cannot depend on the four-dimensional coordinates themselves, since space-time is presumed to be homogeneous, so it can depend at most on the velocity, $a = a(\mathbf{v})$. Since the velocity $\mathbf{v}$ introduces a special direction into the discussion, we might expect the interval to transform differently depending on the orientation of the interval relative to $\mathbf{v}$. But since all the spatial components of the interval enter (1.12) quadratically, the interval does not change if any their signs are reversed, which would change the orientation of the interval relative to $\mathbf{v}$. Equivalently, the transformation of the interval must be indifferent to the direction of $\mathbf{v}$, so that

$$a(\mathbf{v}) = a(v) \tag{1.13}$$

But as noted earlier, the inverse transformation is obtained from the forward transformation merely by changing the sign of $\mathbf{v}$, which does not affect $a(v)$. Therefore, we see that

$$ds^2 = a(v) \, ds'^2 = \frac{ds'^2}{a(v)} \tag{1.14}$$

from (1.12). It follows that

$$a^2(v) = 1 \tag{1.15}$$

or $a = \pm 1$. We can discard the negative sign because two successive transformations must give the same result as a single transformation with the same final velocity, so $a^2 = a$. Therefore, we conclude that $a = 1$ and the interval ds^2 is an invariant of the transformation between inertial coordinate systems. As we see shortly, this is all we need to define the Lorentz transformation between them.

It is convenient to classify intervals in the following way:

$$ds^2 > 0, \qquad \text{timelike interval} \tag{1.16}$$

$$ds^2 < 0, \qquad \text{spacelike interval} \tag{1.17}$$

$$ds^2 = 0, \qquad \text{lightlike (null) interval} \tag{1.18}$$

Since the interval is invariant, the characterization of an interval as timelike, spacelike, or lightlike is independent of the coordinate system in which the events are viewed.

For example, suppose that two events occur in the same place but at different times in the coordinate system K' so that $d\mathbf{r}' = 0$. Then $ds'^2 = c^2 dt'^2 > 0$, and the interval is timelike. In another reference frame K, the interval is still timelike, even though the events occur in different places at different times. Conversely, if the interval between two events is timelike in the reference frame K, then there exists a reference frame K' in which the events occur at the same place. Specifically, if

$$ds^2 = c^2 dt^2 - |d\mathbf{r}|^2 > 0 \tag{1.19}$$

in K, then in a coordinate system K' moving at velocity

$$\mathbf{v} = \frac{d\mathbf{r}}{dt} \tag{1.20}$$

relative to K, the events occur at the same place. In the moving reference frame K', the events are separated by time

$$dt'^2 = \frac{ds'^2}{c^2} = \frac{ds^2}{c^2} = dt^2 - \frac{dr^2}{c^2} \tag{1.21}$$

In the same way, if two events viewed in reference frame K' occur simultaneously at two different points, then $ds'^2 = -dr^2 < 0$, so the interval is spacelike. Viewed in reference frame K, the events appear at two different places and two different times but the interval is still spacelike. Conversely, if two events are separated by a spacelike interval in reference frame K, then in some other reference frame K' the events are simultaneous. In this reference frame, the separation of the events is

$$dr'^2 = -ds'^2 = -ds^2 = dr^2 - c^2 dt^2 \tag{1.22}$$

Based on classifications (1.16)–(1.18), we can divide Minkowski space into the regions shown in Figure 1.5. For an event located anywhere inside the light cone, the interval $s^2 = c^2 t^2 - r^2 > 0$ is timelike. Thus, in any other reference frame K', the event occurs in the same time order relative to the event at the origin as it does in frame K and may be regarded as in the absolute past or absolute future relative to the event at the origin. On the other hand, events located outside the light cone are related to the origin by spacelike intervals. Consequently, there is some reference frame K' in which the events are simultaneous. A second transformation to a frame K'' moving with respect to frame K' will separate the events in time. However, by symmetry we see that if a relative velocity in one direction places event A before B in K'', then a relative velocity in the opposite direction will place event B first. Thus, events separated by spacelike intervals cannot be absolutely time ordered. We say that events in the region outside the light cone are "elsewhere" relative to the event at the origin. For example, in the parable of the cheese-eating contest discussed earlier, the arrival of the news at Alpha and Beta represents two events separated by a spacelike interval. Thus, the news arrived simultaneously at Alpha and Beta as observed by the colonists on the moon but was observed by the space travelers to arrive first at Alpha or first at Beta depending on the relative velocity of their spaceship.

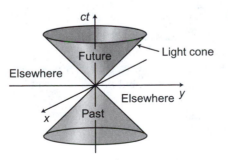

Figure 1.5 Minkowski space.

EXERCISE 1.1

Show that if anything (a force or a particle, called a tachyon) moves faster than the speed of light in some reference frame K, then in some other reference frame K' it is observed to arrive before it departs. Thus, if causality is true, then nothing can move faster than the speed of light. Do not use a Lorentz transformation to solve this problem. Simple arguments are sufficient.

1.1.4 Proper Time

For an object such as a clock at rest in an inertial reference frame K', the distance $|d\mathbf{r}'|^2$ between two events along its world line vanishes, so the invariant interval $ds'^2 = c^2 dt'^2$ is just the time between the events. Viewed from the laboratory frame K the interval is the same, so if the rest frame K' moves with the velocity $\mathbf{v}$ relative to the laboratory frame, then the interval in the laboratory frame is

$$ds^2 = c^2 dt^2 - dr^2 = (c^2 - v^2)\, dt^2 = c^2 dt'^2 \tag{1.23}$$

Therefore, compared with the time dt elapsed on a clock in the laboratory frame, the time elapsed on the moving clock is less, amounting to

$$dt' = \sqrt{1 - \beta^2}\, dt = \frac{dt}{\gamma} \le dt \tag{1.24}$$

where

$$\boldsymbol{\beta} = \frac{\mathbf{v}}{c} \tag{1.25}$$

and

$$\gamma = \frac{1}{\sqrt{1 - \beta^2}} \tag{1.26}$$

This shows that a clock at rest in the moving coordinate system indicates less elapsed time than clocks at rest in the laboratory reference frame to which it is compared. This phenomenon is called time dilation. Ample experimental evidence now exists to confirm this effect. It has, of course, nothing to do with the failure of moving clocks to perform correctly. It has to do with the nature of time itself, or, more precisely, the subjective nature of simultaneity. Many physical phenomena, such as the decay rate of subatomic particles, can be used as clocks. For example, when cosmic rays strike the upper atmosphere of the earth, they create a variety of particles including muons. Ordinarily, muons have a half-life of 2.2 μs, so at the velocity of light they would travel, on average, about 600 m. However, due to time dilation a muon traveling at $\beta = 0.999$, which corresponds to an energy of 2.4 GeV, lives for 50 μs and travels 15 km. This accounts for the fact that large numbers of muons are observed at the earth's surface. In the same way, the subatomic particles created by collisions in high-energy physics experiments would not be observable except that their brief lifetimes are extended by time dilation.

We call the time elapsed on a clock moving with an object the proper time $d\tau$ for the object. This is the time actually experienced by the object. For objects that are not in uniform motion, the motion within any brief interval may be regarded as uniform and the elapsed proper time between two points on the world line of the object is

$$\tau_2 - \tau_1 = \int_{t_1}^{t_2} \frac{dt}{\gamma(t)} \leq t_2 - t_1 \tag{1.27}$$

Note carefully that in this expression, τ_1 and τ_2 refer to the time elapsed in an accelerating coordinate system but t is the time in an inertial reference frame.

A few remarks are in order. In the first place, the proper time, defined by the invariant interval in the rest frame of the moving object, is an invariant. That is, all observers agree on the proper time elapsed along the world line of the object. Physically, this corresponds simply to the fact that the clock moving with the object has an indicator (hands, or even a digital readout) on it to indicate the time. All observers, regardless of their relative motion, agree on what the indicator shows. That is, all observers agree on the numbers showing on the digital readout. In the case of subatomic particles, the time is indicated by the number of particles that have—or have not—decayed. The number of particles is the same to all observers.

In the second place, when viewed from the rest frame of the moving object, clocks in the laboratory frame are moving with the velocity $v = |-\mathbf{v}|$, so they indicate an elapsed time that is less than that in the object's rest frame. Thus, to an observer in the laboratory frame, the clock in the object rest frame appears to be slow, while viewed from the object rest frame, a clock in the laboratory frame appears to be slow. The paradox is resolved by recognizing that when the time indicated on the clock in the rest frame is observed from the laboratory frame, the observations are made by two observers at different places in the laboratory frame. They observe the moving clock as it passes by and compare the time indicated on the moving clock with that on their individual clocks. This is shown in Figure 1.6. Conversely, when the elapsed time on a clock in the laboratory frame is measured by observers in the moving frame, the laboratory clock is compared with two separate clocks in the moving frame. Thus, the measurements are not identical, and for this reason they do not give the same results.

To avoid the problem of measuring elapsed time on a moving clock by comparing a single moving clock with two "stationary" clocks, we start with two clocks that are initially at rest in the laboratory frame. We then accelerate one clock to a high velocity, bring it to rest again, and then return it to its original position in the laboratory next to the other clock. When the two clocks are compared, it is found that the clock that has been accelerated and decelerated indicates a smaller elapsed time than the stationary clock, in accordance

Figure 1.6 Clock paradox.

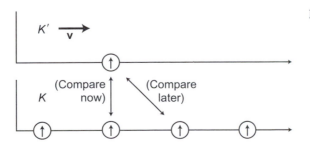

with (1.27). This is called the "twin paradox," since it is often stated in the form of an allegory in which two twins are compared. One twin stays on earth and grows old, while the other becomes an astronaut, flies off at high speed to a nearby star, and when she returns is younger than her twin sister. It is easy to understand that the earthbound twin observes the astronaut's clock as progressing too slowly, but why doesn't the astronaut observe her twin sister's clock as progressing too slowly? In this case, the paradox is resolved by recognizing that the astronaut twin must accelerate to leave the earth, decelerate and turn around after she arrives at the star, accelerate back toward earth, and then decelerate again upon reaching home. The other twin does not accelerate at all, and this is what breaks the symmetry. More to the point, (1.27) involves transformations from one inertial frame to a sequence of inertial frames, each of which describes the moving object for a brief period. Since the earthbound twin remains in an inertial frame of reference, this equation provides a valid description of her observations of her astronaut sister. On the other hand, the astronaut twin is not in an inertial frame and cannot use (1.27) to compute her sister's age. In actual fact, while the astronaut is traveling at constant velocity, she does see her sister aging more slowly than herself. However, when she accelerates to turn around at the outbound end of her trip, she observes her sister aging at an accelerated rate. Although technology has not reached the stage where the astronaut experiment can actually be tried, experimental confirmation of the twin paradox does exist. In careful experiments using atomic clocks, it has been observed that a clock that is flown around the world in an airplane arrives back at the laboratory "younger" than a clock that remains at home. However, the difference in this case is only hundreds of nanoseconds, and the effects of gravity (accounted for in general relativity) are of the same order of magnitude as the time dilation discussed here.

One final remark: If we draw the twins' world lines on a Minkowski diagram, we get paths like those shown in Figure 1.7. For the astronaut the elapsed time is given by (1.27). The elapsed time for the clock at rest is larger than this. In terms of the intervals,

$$\Delta s \text{(earth)} = \int_{t_1}^{t_2} c\, dt \geq \int_{t_1}^{t_2} \frac{c\, dt}{\gamma(t)} = \Delta s' \text{(astronaut)} \qquad (1.28)$$

That is, the interval along the straight line is larger. We see, therefore, that pseudo-Euclidian geometry is different from what we are used to in Euclidian geometry. A straight line in four-dimensional space-time (called a geodesic) is the *longest* interval between two events, rather than the shortest.

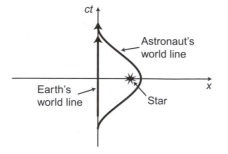

Figure 1.7 Twin paradox.

EXERCISE 1.2

Your kids are at State U, class of '31, and they are hungry. Domino's Pizza is made about 1 km from there, and they deliver in 1000 s (about 15 min). Guaranteed freshest. Mercurial Pizza is made on mercury, about 1.5×10^{11} m away, and is delivered in about 1500 s. Because of the speed at which they travel, time dilation slows down the rate at which the pizza gets cold. Guaranteed freshest! Whose pizza should the kids order to get it as fresh as possible? Price is no object for your kids in their quest for a good pizza.

EXERCISE 1.3

Here it is. You knew it was coming: Star Trek MCLXI—The Seventeenth Generation! Data and the Ferengi Emperor are locked in a deadly battle of Wheel of Fortune, starring Vanna White, who has been traveling around at nearly the speed of light for centuries trying to stay young. Vanna is in orbit 100 light seconds from earth with Data and the Ferengi nearby, while the host, Darth Vader, on loan from another studio, is on Earth. When Vanna gives the contestants the last question, the Ferengi attempts to cheat by speeding toward the earth at 0.99c while he is thinking, so that when he radios his answer to Earth it will arrive first. However, he forgets that when traveling at near the speed of light, his thinking (like all other processes on his ship) slows down due to time dilation. Data takes 10 s and the Ferengi 12 s to answer the question, each according to his own clock, so it takes, for example, 110 s for Data's answer to reach earth (10 s to think and 100 s to radio the answer back).

(a) Does cheating pay? That is, whose answer arrives first? Explain.
(b) What does MCLXI mean?

1.2 THE LORENTZ TRANSFORMATION

1.2.1 Rotation in 4-Space

To find the Lorentz transformation that relates the coordinates in two inertial reference frames, we look for the most general linear transformation that leaves the interval ds invariant. The transformation must be linear, because uniform motion in one frame must correspond to uniform motion in the other, as noted earlier. To reduce the algebra, we make two simplifications. In the first place, we assume that the axes of the two frames are parallel at all times and that the origins coincide at time $t = t' = 0$, as shown in Figure 1.8. In the second place, we assume that the reference frame K' moves at velocity v in the $\hat{\mathbf{x}}$ (and $\hat{\mathbf{x}}'$) direction relative to frame K. More general transformations can be obtained by rotations of K or K' and simple corrections to the origins of the times and distances.

On physical grounds we see that in the directions transverse to the relative motion, the coordinates y and z transform into themselves. Since the transformation is linear, we may write

$$y' = ay \qquad (1.29)$$

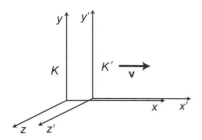

Figure 1.8 Reference frames.

and

$$z' = bz \tag{1.30}$$

for some constants a and b. But from the symmetry of the forward and backward transformations, we see that $a = b = 1$, so

$$y' = y \tag{1.31}$$

and

$$z' = z \tag{1.32}$$

In the longitudinal direction, the most general linear transformation is

$$ct' = mct + nx \tag{1.33}$$

$$x' = pct + qx \tag{1.34}$$

for some constants m, n, p, and q. To preserve the interval between the origin and point $(ct, \mathbf{r})$, we require that

$$c^2 t^2 - x^2 = c^2 t'^2 - x'^2 = (mct + nx)^2 - (pct + qx)^2$$
$$= (m^2 - p^2)c^2 t^2 + (n^2 - q^2)x^2 + 2(mn - pq)xct \tag{1.35}$$

From the first term on the right-hand side we see that

$$m^2 - p^2 = 1 \tag{1.36}$$

so we can write

$$m = \cosh \zeta, \qquad p = -\sinh \zeta \tag{1.37}$$

for some ζ, and from the second term we see that

$$n^2 - q^2 = -1 \tag{1.38}$$

so we can write

$$n = -\sinh \psi, \qquad q = \cosh \psi \tag{1.39}$$

for some ψ. From the third term we find that

$$mn - pq = -\cosh \zeta \sinh \psi + \sinh \zeta \cosh \psi = 0 \tag{1.40}$$

so that $\psi = \zeta$. The most general transformation that preserves the interval is therefore

$$ct' = ct \cosh \zeta - x \sinh \zeta \tag{1.41}$$

$$x' = x \cosh \zeta - ct \sinh \zeta \tag{1.42}$$

where ζ is called the "boost parameter," or the "rapidity." The transformation, (1.41) and (1.42), resembles a rotation of coordinates except that sin and cos are replaced by sinh and cosh. In fact, a rotation of coordinates is the most general linear homogeneous transformation that preserves the length $dx^2 + dy^2 = dx'^2 + dy'^2$ in Euclidian geometry. Thus, the Lorentz transformation has the form of a "pseudorotation" in pseudo-Euclidian space.

To determine the boost parameter ζ, we consider the motion of the origin of the K' frame, as viewed in the K frame. Since the origin ($x' = 0$) of the moving frame is at position $x = vt$ in the laboratory frame, we see from (1.42) that

$$x' = 0 = vt \cosh \zeta - ct \sinh \zeta \tag{1.43}$$

so

$$\tanh \zeta = \frac{v}{c} = \beta \tag{1.44}$$

Therefore, the coefficients in the transformation are

$$\cosh \zeta = \frac{1}{\sqrt{1 - \tanh^2 \zeta}} = \frac{1}{\sqrt{1 - \beta^2}} = \gamma \tag{1.45}$$

$$\sinh \zeta = \tanh \zeta \cosh \zeta = \beta \gamma \tag{1.46}$$

and the complete Lorentz transformation is

$$ct' = \gamma (ct - \beta x) \tag{1.47}$$

$$x' = \gamma (x - \beta ct) \tag{1.48}$$

$$y' = y \tag{1.49}$$

$$z' = z \tag{1.50}$$

For a boost to a coordinate system moving to the right, $\tanh \zeta = \beta > 0$. The relation between the new and old coordinate axes is shown in Figure 1.9, where we see that the new axes $x' = 0$ and $ct' = 0$ are tilted toward the light line in the upper right quadrant. For a boost to a frame moving to the left, the axes are tilted away from this same light line. From Figure 1.9, it is easy to see how events separated by a spacelike interval can be reordered in time. For example, point B is elsewhere with respect to the origin of the stationary system K and occurs later ($ct > 0$) in that system. In the K' system, point B lies below the axis $ct' = 0$ and therefore occurs earlier ($ct' < 0$).

The inverse transformation can be found by solving for x and ct, but it is easier simply to use the symmetry of the forward and inverse transformations. If we just change the

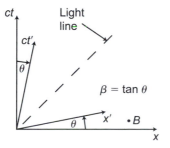

Figure 1.9 Lorentz transformation.

direction of motion to $-v$, we immediately get

$$ct = \gamma(ct' + \beta x') \tag{1.51}$$

$$x = \gamma(x' + \beta ct') \tag{1.52}$$

$$y = y' \tag{1.53}$$

$$z = z' \tag{1.54}$$

In the limit $c \to \infty$, we recover Galilean relativity. To see this, we write the Lorentz transformation explicitly in terms of the velocities and get

$$t' = \frac{t - \dfrac{v}{c^2}x}{\sqrt{1 - \dfrac{v^2}{c^2}}} \xrightarrow{c \to \infty} t \tag{1.55}$$

and

$$x' = \frac{x - vt}{\sqrt{1 - \dfrac{v^2}{c^2}}} \xrightarrow{c \to \infty} x - vt \tag{1.56}$$

Finally, we note that Galilean transformations commute but Lorentz transformations do not. That is, if we transform first into a frame K' moving at velocity $\mathbf{v}_1$ with respect to K and then into a frame K'' moving at velocity $\mathbf{v}_2$ with respect to K', we get a Galilean transformation directly into the frame moving at velocity $\mathbf{v}_3 = \mathbf{v}_1 + \mathbf{v}_2$. This is the group property. We get the same transformation if we reverse the order in which the individual transformations are performed:

$$G(\mathbf{v}_2)G(\mathbf{v}_1) = G(\mathbf{v}_1)G(\mathbf{v}_2) = G(\mathbf{v}_1 + \mathbf{v}_2) \tag{1.57}$$

The same is not true for Lorentz transformations. The group property holds, but the transformations do not, in general, commute:

$$L(\mathbf{v}_2)L(\mathbf{v}_1) = L(\mathbf{v}_3) \neq L(\mathbf{v}_1)L(\mathbf{v}_2) \tag{1.58}$$

and $\mathbf{v}_3 \neq \mathbf{v}_1 + \mathbf{v}_2$. This follows, in some sense, from the fact that the successive Lorentz transformations correspond to pseudorotations about different axes, which do not in general commute. The exception occurs when both velocities are in the same direction, which corresponds to two pseudorotations about the same axis. As we see in the next section, the final velocity is not exactly the sum of the two individual velocities, but it is symmetric in $\mathbf{v}_1$ and $\mathbf{v}_2$, and this is sufficient to make the transformations commute when the velocities are parallel.

EXERCISE 1.4

Political science/relativistic law question. A truce has been arranged between the Federation and the Romulans. According to the treaty, the first captain to fire in an engagement is executed! The evidence is as follows: Captain Picard of the Federation starship *Enterprise* detects emissions from a Romulan ship as it begins to uncloak and realizes that the Romulan is about to fire at him. The strength of the emissions shows that the Romulan was about 1000 light seconds away when they were radiated, so the Romulan must already have fired his photon torpedo. Picard therefore fires a photon torpedo (which travels at the speed of light) at the Romulan. One hundred seconds later, the *Enterprise* is hit by a photon torpedo fired by the Romulan, and some time after that, in the starship coordinate system, the Romulan is hit by the photon torpedo fired by the *Enterprise*. Both ships travel at constant velocity throughout the engagement. With both ships disabled, the engagement ends and a court of inquiry is convened. The two captains agree on the evidence presented here, but each captain argues, on the basis of his electronic logs, that the other fired first. Is it possible that both are telling the truth, at least as they see it in their own coordinate systems? If so, what was the minimum relative speed of the Romulan and the Enterprise, and in which direction (toward each other or away) were they traveling? If not, which captain should be executed? Draw a world diagram in the coordinate system of the *Enterprise* to explain this. This is a tricky question, and you need every bit of the evidence.

EXERCISE 1.5

An astronaut leaves the earth for a star $L = 10$ light years away. The spaceship accelerates quickly to half the speed of light ($\beta = 0.5$, $\gamma = 1/\sqrt{1 - \beta^2} = 1.155$) and coasts at uniform velocity to the star. When she reaches the star, the astronaut quickly reverses her velocity and coasts back to earth at half the speed of light.

(a) Show that when she reaches the star, immediately before she begins to turn around, the time elapsed on her clock is

$$ct_1' = \frac{L}{\beta\gamma} = 17 \text{ light years} \tag{1.59}$$

The earth is at position $x' = -\beta ct_1'$. Use the inverse Lorentz transformation to show that she would argue that at this point in her trip the time elapsed on the earthbound clock is

$$ct_{1-\varepsilon} = \frac{L}{\beta\gamma^2} = 15 \text{ light years} \tag{1.60}$$

Explain how she might check this prediction.

(b) Immediately after she has turned around and is about to leave the star, the elapsed time on her clock is (roughly) the same as it was immediately before she began to turn around, since she turns around quickly. In the new coordinate system K'', the time is $t'' = t_1'$, and the position of the earth is $x'' = -\beta ct_1'$. Use the inverse Lorentz transformation to show that she would now argue that the time elapsed on the earthbound clock is

$$ct_{1+\varepsilon} = \frac{L}{|\beta|}(1 + \beta^2) = 25 \text{ light years} \tag{1.61}$$

That is, in this brief (to her) interval the time on the earth clock has increased by the amount

$$\Delta ct = 2\,|\beta|L = 10 \text{ light years!} \tag{1.62}$$

Explain how she might check this prediction.

(c) Using the same Lorentz transformation, confirm that at time $t'' = 2t_1'$, when the astronaut returns to earth, the time on the earth clock (now at $x'' = 0$) is

$$t_2 = 2\frac{L}{\beta c} \tag{1.63}$$

(d) Explain what happened while she was accelerating around the star, as viewed by her and as viewed on earth. Using Einstein's equivalence principle, which states that gravity is equivalent to acceleration, explain what this means for clocks in a gravitational field. Show that the increment of time dt' on a clock at rest in a static gravitational potential $\Phi(\mathbf{r}) < 0$, relative to the increment of time dt on a clock at rest at $\Phi = 0$, is

$$dt' = \left(1 + \frac{\Phi}{c^2}\right) dt \tag{1.64}$$

The gravitational red shift is the reciprocal of this:

$$\frac{\omega}{\omega'} = \frac{dt'}{dt} = 1 + \frac{\Phi}{c^2} \tag{1.65}$$

where ω' is the frequency of an atom in the star ($\Phi < 0$), and $\omega < \omega'$ is the frequency observed on earth ($\Phi \approx 0$). As shown by the general theory of relativity, this result is valid only for $|\Phi/c^2| \ll 1$.

Hint: Consider an incremental acceleration that takes place over a short interval of time $\Delta t'$ in the moving frame, which corresponds to the time interval Δt on earth. Examine the earth time in the inertial reference frames in which the astronaut is at rest at the beginning and the end of the brief period of acceleration. The gravitational potential between two points in a uniform gravitational field is $\Phi = -gL$, where g is the acceleration due to gravity and L is the distance between the two points.

1.2.2 Time Dilation and Length Contraction

Time dilation is the phenomenon that makes a moving clock appear to go slower, and length contraction is the phenomenon that makes a moving object shrink in the direction of motion. To see how time dilation and length contraction arise, we examine the measurement process by which each is determined. To observe time dilation, we consider the progress of a moving clock, that is, a single clock at rest at the point x' in the frame K'. Differentiating the *inverse* transformation (1.51) with respect to t, keeping $x' = \text{constant}$, we get

$$\frac{dt'}{dt} = \frac{1}{\gamma} < 1 \tag{1.66}$$

That is, the moving clock always appears to go slower than the stationary clocks to which it is compared. Time dilation refers to the fact that the ticks of the moving clock appear farther apart as viewed from the stationary frame.

To observe length contraction, we measure the length of a moving rod by determining the positions of the two ends at a single time t in the stationary frame K. For an infinitesimal rod the corresponding length in the stationary frame is found by differentiating the *forward* transformation (1.48) with respect to x, holding $t = $ constant, to get

$$\frac{dx'}{dx} = \gamma > 1 \tag{1.67}$$

Length contraction refers to the fact that the tick marks on the moving length scale appear closer together than those on the stationary scale. That is, a rod of length dx' appears to have a length $dx = dx'/\gamma < dx'$ in the stationary frame.

Note that the factor $1/\gamma$ appears in time dilation where the factor γ appears in length contraction. This is because in the first case a coordinate (x') was held fixed in the moving frame, while in the second case a coordinate (ct) was held fixed in the stationary frame. If we actually look in detail at a moving coordinate system as it passes by, what we see is shown in Figure 1.10. Compared with the clocks in the stationary frame, the clocks in the moving frame appear to be too close together (length contraction) and are not synchronized.

We have already introduced the proper time $d\tau$, which in the present discussion corresponds to dt' ($x' = $ constant). We similarly define the proper length as the length observed in the reference frame in which the object is at rest, $d\lambda = dx'$ ($t' = $ constant). We saw previously that the proper time is invariant,

$$ds'^2 = c^2 dt'^2 - dx'^2 = c^2 d\tau^2 \quad \text{(holding } dx' = 0) \tag{1.68}$$

In the same way, we see that the proper length is an invariant,

$$ds'^2 = c^2 dt'^2 - dx'^2 = -d\lambda^2 \quad \text{(holding } dt' = 0) \tag{1.69}$$

Physically, the Lorentz invariance of proper time just means that all observers agree on what a clock at rest in the moving frame indicates. Likewise, the invariance of proper length just means that all observers agree on the numbers that appear on a ruler or dial gauge at rest in the moving frame.

1.2.3 Velocity Transformation

Clearly, relativistic velocities do not add in the same way that nonrelativistic velocities do, for if we are traveling at velocity $v = 0.75c$ through a railroad station and another train passes us at $0.75c$, the passengers waiting in the station do not see the faster train traveling at $1.5c$. To see how velocities add, we consider a reference frame K' that is moving at velocity $\mathbf{v}$ with respect to frame K, and a particle moving in frame K' at velocity $\mathbf{v}'$, as shown in Figure 1.11. The inverse transformation between K and K' is given by (1.51)–(1.54). For

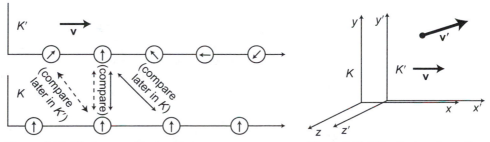

Figure 1.10 Moving coordinate system.

Figure 1.11 Velocity transformation.

an infinitesimal movement

$$d\mathbf{r}' = \mathbf{v}'dt' \qquad (1.70)$$

in the moving frame, the motion in the stationary frame is

$$dt = \gamma \, dt' \left(1 + \frac{vv'_x}{c^2}\right) \qquad (1.71)$$

$$dx = \gamma \, dt' \left(v + v'_x\right) \qquad (1.72)$$

$$dy = v'_y \, dt' \qquad (1.73)$$

$$dz = v'_z \, dt' \qquad (1.74)$$

Dividing these expressions we obtain the relations

$$v_x = \frac{dx}{dt} = \frac{v + v'_x}{1 + \frac{vv'_x}{c^2}} \qquad (1.75)$$

$$v_y = \frac{dy}{dt} = \frac{v'_y}{\gamma\left(1 + \frac{vv'_x}{c^2}\right)} \qquad (1.76)$$

$$v_z = \frac{dz}{dt} = \frac{v'_z}{\gamma\left(1 + \frac{vv'_x}{c^2}\right)} \qquad (1.77)$$

These are called the Einstein velocity-addition laws. Clearly, the addition of velocities transverse to one another is different from the addition of velocities that are parallel to one another.

As an example we consider a relativistic particle that in some decay process emits another particle with velocity v' at the angle θ' in the $x'y'$ plane. In the laboratory frame, the velocity of the secondary particle is

$$v_x = \frac{v + v' \cos\theta'}{1 + \frac{vv' \cos\theta'}{c^2}} \qquad (1.78)$$

$$v_y = \frac{v' \sin\theta'}{\gamma\left(1 + \frac{vv' \cos\theta'}{c^2}\right)} \qquad (1.79)$$

In the laboratory frame, the emission angle is

$$\tan\theta = \frac{v_y}{v_x} = \frac{v' \sin\theta'}{\gamma(v + v' \cos\theta')} \qquad (1.80)$$

For highly relativistic motion of the primary particle, $\gamma \gg 1$, we see that most of the emission appears in the forward direction in the laboratory frame with angles $\theta \le O(1/\gamma)$. When the emitted particles are photons, so that $v' = c$, all radiation emitted in the forward hemisphere ($\theta' \le \pi/2$) in the rest frame of the primary particle is emitted inside the cone

$$\tan\theta \le \frac{1}{\beta\gamma} \qquad (1.81)$$

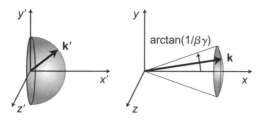

Figure 1.12 Relativistic shift of radiation to the forward direction.

in the laboratory frame, as illustrated in Figure 1.12. This effect is quite pronounced in the radiation from synchrotron radiation sources and high-energy physics experiments, where the particle energy corresponds to $\gamma \gg 1$ and often $\gamma > 10^3$.

EXERCISE 1.6

Use the relativistic velocity-addition law to discuss the Fizeau experiments on the propagation of light in moving liquids. Show that for liquid flowing at a speed v parallel and antiparallel to the path of the light over a total distance L, the phase shift ϕ is given by (1.7). It is assumed that the light travels with the speed $u' = c/n$ in the coordinate system of the moving liquid. Ignore the dependence of the index of refraction on the frequency.

EXERCISE 1.7

Assume that a rocket ship leaves the earth and travels to a star 10 light years away. The rocket is so constructed that it has an acceleration g in its own frame to make the astronaut feel right at home. The rocket accelerates in a straight-line path for the entire distance to the star to get there as quickly as possible.

  (a) Show that the elapsed time on the rocket (the proper time) for the trip is

$$\tau = \frac{c}{g} \operatorname{arccosh}\left(\frac{gL}{c^2} + 1\right) = 3 \text{ years} \tag{1.82}$$

      where L is the distance to the star in the stationary (earth and star) frame.

  (b) Show that the elapsed time on earth for the same trip is

$$t = \frac{c}{g} \sinh\left(\frac{g\tau}{c}\right) = 11 \text{ years} \tag{1.83}$$

1.3 4-VECTORS AND 4-TENSORS

1.3.1 Cartesian Tensors

The use of 4-vectors and 4-tensors for special relativity is such an enormous convenience that it is very nearly a necessity. For general relativity it is absolutely essential, and in that case it is necessary to use generalized tensors. For special relativity it is simpler to use

Cartesian tensors, since space is flat, and this is what we do. By introducing 4-tensors, we obtain a compact notation that facilitates our calculations in a natural and suggestive manner. Perhaps more important, physical laws that are expressed in 4-tensor form automatically satisfy Einstein's postulates of relativity. We say then that they are manifestly covariant. The principle of covariance can even be used to construct relativistic generalizations of nonrelativistic laws, and we shall have occasion to do this. Because of the enormous benefit to be gained from the use of 4-vectors, we devote this and the following two sections to a discussion of their properties. Aside from some examples introduced to illustrate the use of the notation, these sections are entirely mathematical and essentially devoid of physics content. The outstanding exception to this is the use of Einstein's second postulate of relativity to define the metric tensor. This one step insures that all physical laws expressed in 4-vector form are relativistically correct.

By definition, a Lorentz 4-vector is a set of four numbers that transform to another reference frame by a Lorentz transformation. Physically speaking, the four numbers must represent observable quantities. Moreover, they must have the property that when these same physical quantities are observed from another inertial reference frame, the values measured are related to those measured in the first reference frame in the same way that the coordinates and time are related between the two reference frames. Not all sets of four numbers form 4-vectors, and in fact not all sets of four physically observable quantities form 4-vectors. For example, you might think that the velocity $\mathbf{v}$ of a particle and the speed of light c would form a 4-vector $(c, \mathbf{v})$, since they are the derivatives with respect to the time t of the 4-vector position $(ct, \mathbf{r})$ in Minkowski space. But they do not. In another reference frame we would measure another velocity $\mathbf{v}'$ and the same speed of light c, which form the set $(c, \mathbf{v}')$, but the relationship between $(c, \mathbf{v})$ and $(c, \mathbf{v}')$ would not be the same as the relationship between the positions $(ct, \mathbf{r})$ and $(ct', \mathbf{r}')$ in the two reference frames. However, we see shortly that the set of observable quantities $\gamma(c, \mathbf{v})$ has just the transformation properties we seek, and this is the 4-vector velocity. Physically speaking, while $(c, \mathbf{v})$ is the time derivative of the position $(ct, \mathbf{r})$, it does not transform like a 4-vector because we've forgotten about time dilation. When the factor γ is included, this is accounted for. In the nonrelativistic limit, of course, the 4-vector velocity reduces to the ordinary velocity. Because it is sometimes difficult to determine whether some set of quantities is actually a 4-vector, we develop a rule, called the quotient rule, that we can use as a test.

We distinguish two types of 4-vectors and 4-tensors, called *contravariant* and *covariant*. In fact, as we shall see, any vector can be converted from its covariant form to its contravariant form, and vice versa. Contravariant vectors and tensors are identified by enumerating their elements with superscripts, while covariant vectors and tensors use subscripts. In a 4-dimensional Cartesian coordinate system K, the position in Minkowski space is given by the contravariant 4-vector

$$r^\alpha = (ct, x, y, z) = (ct, \mathbf{r}) \qquad \alpha = 0 \ldots 3 \tag{1.84}$$

The zeroth coordinate is the time, ct, and the remaining coordinates are the space coordinates $(x, y, z) = \mathbf{r}$. We use Greek letters to denote indices that run from 0 to 3 and Latin letters to denote indices that run from 1 to 3, corresponding to the spatial coordinates x, y, and z, respectively. The coordinates in another reference frame K' are related to those in K

by the Lorentz transformation

$$r'^\alpha = \sum_{\beta=0}^{3} L^\alpha{}_\beta r^\beta = L^\alpha{}_\beta r^\beta \tag{1.85}$$

In the last expression we have introduced Einstein's summation convention: any index that is repeated in a single term is assumed to be summed over the values $0 \ldots 3$. Note, however, that in this sum one index must always be a superscript and one a subscript. More about this shortly. Generalizing from (1.84), we define as a *contravariant vector* any set of four quantities A^α that transform like the coordinates of a position,

$$A'^\alpha = L^\alpha{}_\beta A^\beta \tag{1.86}$$

By extension, we define the *contravariant tensor* of rank n as any set of 4^n quantities that transform according to the rule

$$T'^{\alpha\beta\cdots} = L^\alpha{}_\kappa L^\beta{}_\lambda \ldots T^{\kappa\lambda\cdots} \tag{1.87}$$

where n is the number of indices in $T^{\alpha\beta\cdots}$. In particular, a tensor of rank $n = 0$ is called a scalar and is invariant under a Lorentz transformation. A tensor of rank $n = 1$ is called a vector.

In the same fashion, we define the *covariant vector* as any set of four quantities that transform according to the inverse rule

$$A'_\beta = L^{-1}{}^\alpha{}_\beta A_\alpha \tag{1.88}$$

and the *covariant tensor* of rank n as any set of 4^n quantities that transform according to the inverse rule

$$T'_{\alpha\beta\cdots} = L^{-1}{}^\kappa{}_\alpha L^{-1}{}^\lambda{}_\beta \ldots T_{\kappa\lambda\cdots} \tag{1.89}$$

where L^{-1} is the inverse Lorentz transformation. That is,

$$L^{-1}{}^\alpha{}_\gamma L^\gamma{}_\beta = \delta^\alpha{}_\beta = L^\alpha{}_\gamma L^{-1}{}^\gamma{}_\beta = \delta_\beta{}^\alpha \tag{1.90}$$

where $\delta_\alpha{}^\beta = \delta^\alpha{}_\beta$ is the four-dimensional extension of the Kronecker delta,

$$\delta_\alpha{}^\beta = \delta^\alpha{}_\beta = \begin{cases} 1 & \text{if } \alpha = \beta \\ 0 & \text{otherwise} \end{cases} \tag{1.91}$$

It may seem backward to define contravariant and covariant in this fashion, but this is the convention.

We define mixed tensors by their transformation properties. The mixed tensor $T^\alpha{}_\beta$ transforms like

$$T'^\alpha{}_\beta = L^\alpha{}_\kappa L^{-1}{}^\lambda{}_\beta T^\kappa{}_\lambda \tag{1.92}$$

and so on. Note that the order of the upper (contravariant) and lower (covariant) indices of the mixed tensor $T^\alpha{}_\beta$ is important unless the tensor is symmetric. Thus, in general

$$T^\alpha{}_\beta \neq T_\beta{}^\alpha \tag{1.93}$$

We define four tensor operations, as follows.

1. Sum of Two Tensors

If two tensors A and B have like valence (the same number and position of contravariant and covariant indices), the sum is defined by example as

$$C_\alpha{}^\beta = A_\alpha{}^\beta + B_\alpha{}^\beta \tag{1.94}$$

2. Outer Product of Two Tensors

If we have a tensor A of rank n and a tensor B of rank m, then the outer product is a tensor of rank $(n + m)$ defined by example as

$$C^{\alpha\beta}{}_{\gamma\delta} = A^{\alpha\beta} B_{\gamma\delta} \tag{1.95}$$

3. Contraction of a Tensor

If we sum over two indices of a tensor T of rank n of which one index is contravariant and one covariant, the result is a tensor S of rank $(n - 2)$ defined by example as

$$S^\alpha{}_\gamma = T^{\alpha\beta}{}_{\beta\gamma} \tag{1.96}$$

In particular, the trace of a second-rank tensor is a scalar:

$$A^\alpha{}_\alpha = A^0{}_0 + A^1{}_1 + A^2{}_2 + A^3{}_3 \tag{1.97}$$

That the result of contraction is a tensor is clear from the definition (1.92) of a tensor. Note, however, that the sum must always be over one contravariant and one covariant index. The sum over two contravariant or two covariant indices does not produce a tensor.

If we combine the outer product of two vectors with contraction we get the inner (or scalar) product of two vectors, defined by example as

$$A^\alpha B_\alpha = C^\alpha{}_\alpha = A^0 B_0 + A^1 B_1 + A^2 B_2 + A^3 B_3 = D \tag{1.98}$$

Clearly, the inner product of two vectors is a scalar (a Lorentz invariant), for

$$A'^\alpha B'_\alpha = A^\mu L^\alpha{}_\mu L^{-1}{}^\nu{}_\alpha B_\nu = A^\mu B_\mu \tag{1.99}$$

In the same way, we can form the scalar product of two tensors of higher (but equal) rank by contracting over all the indices.

4. Permutation of Indices

If we have a tensor of rank 2 or higher we may exchange (permute) two of the indices. In general the new tensor component will be different from the original. However, if the result is equal to the original tensor we say that the tensor is symmetric in the permuted indices. If the result of permutation has the opposite sign, the tensor is antisymmetric in that index. For example,

$$T^{\alpha\beta} = T^{\beta\alpha} \qquad \text{(symmetric tensor)} \tag{1.100}$$

$$T^{\alpha\beta\gamma} = -T^{\beta\alpha\gamma} \qquad \text{(antisymmetric in } \alpha \text{ and } \beta) \tag{1.101}$$

For mixed tensors the permutations are made without changing the nature (covariant or contravariant) of either index. A mixed tensor is therefore symmetric or antisymmetric if

$$T^\alpha{}_\beta = T_\beta{}^\alpha \qquad \text{(symmetric tensor)} \tag{1.102}$$

$$T^\alpha{}_{\beta\gamma} = -T_\beta{}^\alpha{}_\gamma \qquad \text{(antisymmetric in } \alpha \text{ and } \beta) \tag{1.103}$$

Some caution is required here, for if we examine the matrix representation of a symmetric mixed tensor we see that it is not symmetric in the usual (matrix) sense. By defining symmetry in this way we preserve the symmetry when we convert an index from covariant to contravariant, or vice versa, as described later.

Any second-order tensor can be written as the sum of a symmetric tensor and an antisymmetric tensor, since we can write, for example,

$$T^{\alpha\beta} = \tfrac{1}{2}(T^{\alpha\beta} + T^{\beta\alpha}) + \tfrac{1}{2}(T^{\alpha\beta} - T^{\beta\alpha}) \tag{1.104}$$

in which the first term is a symmetric tensor and the second term an antisymmetric tensor.

It is frequently useful to consider the behavior (parity) of a tensor under reflection or inversion of the space coordinates. The parity transformation is

$$P^{\alpha}{}_{\beta} = \begin{bmatrix} 1 & 0 & 0 & 0 \\ 0 & -1 & 0 & 0 \\ 0 & 0 & -1 & 0 \\ 0 & 0 & 0 & -1 \end{bmatrix} \tag{1.105}$$

If we apply this to a true vector $A^{\alpha} = (A^0, \mathbf{A})$ the result is

$$P^{\alpha}{}_{\beta} A^{\beta} = (A^0, -\mathbf{A}) \tag{1.106}$$

That is, the spacelike components change sign. If we apply the parity transformation to the true tensor $T^{\alpha\beta} = \begin{bmatrix} T^{00} & T^{0j} \\ T^{i0} & T^{ij} \end{bmatrix}$ we get

$$P^{\alpha}{}_{\gamma} P^{\beta}{}_{\delta} T^{\gamma\delta} = \begin{bmatrix} T^{00} & -T^{0j} \\ -T^{i0} & T^{ij} \end{bmatrix} \tag{1.107}$$

The zeroth row and zeroth column behave like the components of a true vector, but the space–space components T^{ij} are invariant under an inversion of the coordinates.

The usefulness of tensors, aside from the compact notation, lies in the main theorem of tensor calculus, which is at once trivial and profound. It states that if two tensors are equal in any coordinate system, then they are equal in any other coordinate system. Thus, if we express the laws of physics in the form of 4-tensors, then they are valid in any coordinate system. Laws expressed in this way are said to be manifestly covariant. They automatically satisfy Einstein's first postulate of special relativity, which states that the laws of physics are the same in all inertial frames of reference.

In trying to formulate the laws of physics in tensor form, we immediately face the difficulty of determining which physical quantities are in fact tensors. A useful guide to this identification is the so-called *quotient rule*. The name suggests that the quotient of two tensors is also a tensor. However, we cannot actually construct the quotient of two tensors, so the quotient rule states that *if the contraction of an entity $T^{\beta\cdots}_{\alpha\cdots}$ with an arbitrary tensor $B^{\nu\cdots}_{\mu\cdots}$ produces a tensor $A^{\lambda\cdots}_{\kappa\cdots}$, then $T^{\beta\cdots}_{\alpha\cdots}$ is a tensor*. To prove this we consider the example of a second-rank quantity $T^{\alpha\beta}$ for which

$$T^{\alpha\beta} B_{\beta} = A^{\alpha} \tag{1.108}$$

where B_{β} is an *arbitrary* 4-vector and A^{α} is known to be a 4-vector. However, the rule holds true for tensors of any rank and valence and for contraction over any number of indices. If we transform (1.108) to a new coordinate frame K' using (1.92) we get

$$T'^{\alpha\beta} L^{-1^\kappa}{}_\beta B_\kappa = L^\alpha{}_\beta A^\beta = L^\alpha{}_\beta T^{\beta\kappa} B_\kappa \tag{1.109}$$

But B_κ is arbitrary, so it must be true that

$$T'^{\alpha\beta} L^{-1^\kappa}{}_\beta = L^\alpha{}_\beta T^{\beta\kappa} \tag{1.110}$$

If we now multiply by $L^\mu{}_\kappa$ and use (1.90) we get

$$T'^{\alpha\mu} = L^\alpha{}_\beta L^\mu{}_\kappa T^{\beta\kappa} \tag{1.111}$$

which proves that $T^{\alpha\beta}$ is a tensor. Note carefully that the arbitrary tensor must be completely arbitrary, and not, for example, an arbitrary symmetric tensor.

To show how the quotient rule can be used to prove the tensor character of some entity, we consider the trivial example of the 4-vector velocity

$$U^\alpha = \frac{dr^\alpha}{d\tau} = \gamma(c, \mathbf{v}) \tag{1.112}$$

Rearranging this definition, we see that the infinitesimal displacement is

$$U^\alpha d\tau = dr^\alpha \tag{1.113}$$

Since the proper time $d\tau$ is an arbitrary scalar and the position dr^α is known to be a vector, the 4-vector velocity is a vector. Actually, division by a scalar is the one case when we can form the quotient of two tensors.

As a less trivial example we recall that the phase of a wave of frequency ω and wave vector $\mathbf{k}$ is given by the expression

$$\phi = \omega t - \mathbf{k} \cdot \mathbf{r} = k_\alpha r^\alpha \tag{1.114}$$

where we have introduced the covariant form of the wave 4-vector with the definition

$$k_\alpha = \left(\frac{\omega}{c}, -\mathbf{k}\right) \tag{1.115}$$

The minus sign appears on the right-hand side in this definition because this is the covariant form of the wave vector. We return to this point in the next section. But the phase ϕ is a scalar (the same in all coordinate systems), since, for a wave, all observers count the same number of phase fronts. Therefore, since r^α is an arbitrary vector, it follows from (1.114) that the wave vector k_α is a 4-vector, by the quotient rule.

EXERCISE 1.8

If $C^{\alpha\beta} A_\alpha B_\beta$ is a scalar for two arbitrary vectors A_α and B_β, use the quotient rule to prove that $C^{\alpha\beta}$ is a tensor.

EXERCISE 1.9

If $C^{\alpha\beta} A_\alpha A_\beta$ is a scalar for an arbitrary vector A_α, use the result of the previous exercise to prove that $C^{\alpha\beta} + C^{\beta\alpha}$ is a tensor.
Hint: consider the expression $C^{\alpha\beta}(A_\alpha + B_\alpha)(A_\beta + B_\beta)$.

1.3.2 Relativistic Metric and Lorentz Transformation

It is frequently necessary to convert a contravariant vector to covariant form or vice versa. To do this we introduce the metric tensor. Thus, for each contravariant vector A^α we define the associated covariant vector

$$A_\alpha = g_{\alpha\beta} A^\beta \tag{1.116}$$

where $g_{\alpha\beta}$ is called the metric tensor for reasons that will soon become apparent. Similarly, the contravariant vector associated with the covariant vector A_γ is

$$A^\beta = g^{\beta\gamma} A_\gamma \tag{1.117}$$

If we substitute (1.117) into (1.116) we find that

$$g_{\alpha\beta} g^{\beta\gamma} = \delta_\alpha{}^\gamma = \delta^\gamma{}_\alpha \tag{1.118}$$

That is, $g_{\alpha\beta}$ is its own inverse. More generally, we may use the metric tensor to raise and lower the indices of higher rank tensors as, for example,

$$T^{\cdot\alpha\beta}{}_\gamma = g^{\alpha\kappa} g^{\beta\lambda} T_{\kappa\lambda\gamma} \tag{1.119}$$

If we view (1.118) as an index-lowering operation on $g^{\beta\gamma}$, we see that the Kronecker delta $\delta^\alpha{}_\beta$ is the mixed tensor form of the metric tensor $g_{\alpha\beta}$. It is also called the unit tensor.

The magnitude, or norm, of a vector is defined by the inner product of the vector and its associate, so

$$|A|^2 = A^\alpha A_\alpha = g^{\alpha\beta} A_\alpha A_\beta = g_{\alpha\beta} A^\alpha A^\beta \tag{1.120}$$

For example, the magnitude of the invariant interval is given by

$$ds^2 = c^2 dt^2 - dx^2 - dy^2 - dz^2 \tag{1.121}$$

In 4-vector notation we may express this in the form

$$ds^2 = dr^\alpha dr_\alpha = g_{\alpha\beta} \, dr^\alpha dr^\beta \tag{1.122}$$

which is why $g_{\alpha\beta}$ is called the metric tensor. Therefore, in order to satisfy Einstein's second postulate of special relativity, as expressed mathematically by (1.11), the metric tensor must have the form

$$g_{\alpha\beta} = \begin{bmatrix} 1 & 0 & 0 & 0 \\ 0 & -1 & 0 & 0 \\ 0 & 0 & -1 & 0 \\ 0 & 0 & 0 & -1 \end{bmatrix} = g^{\alpha\beta} \tag{1.123}$$

The metric tensor is the same in all coordinate systems. In fact, this statement is enough to define the Lorentz transformation. Note that the metric tensor is not the same as the parity tensor (1.105), since that tensor is a mixed tensor, as defined there, whereas the metric tensor just defined is purely covariant or contravariant.

When we use the metric tensor to raise or lower the indices of a vector or tensor, the timelike components remain the same and the spacelike components change sign. Thus, the contravariant and covariant forms of the wave vector are

$$k^{\alpha} = (k^0, \mathbf{k}) = \left(\frac{\omega}{c}, \mathbf{k}\right) \Leftrightarrow k_{\alpha} = (k^0, -\mathbf{k}) = \left(\frac{\omega}{c}, -\mathbf{k}\right) \tag{1.124}$$

The magnitude of the wave vector is

$$k^{\alpha} k_{\alpha} = \frac{\omega^2}{c^2} - k^2 = k^2 \left(\frac{v_{\phi}^2}{c^2} - 1\right) \tag{1.125}$$

where the phase velocity of the wave is

$$v_{\phi} = \frac{\omega}{k} \tag{1.126}$$

If the phase velocity is the speed of light, the wave vector is a null vector, but the wave need not be a light wave to have a wave vector k^{α}.

When we use the metric tensor to raise or lower the indices of a higher rank tensor, the sign of each component remains the same or changes depending on how many of the raised or lowered indices are spacelike. Thus, in the example (1.119), we see that $T^{11}{}_1 = T_{111}$ because we have raised two spacelike indices and the sign changes cancel, but $T^{01}{}_1 = -T_{011}$ because the first index is timelike.

Since the metric tensor is the same in all inertial reference frames,

$$g'^{\alpha\beta} = L^{\alpha}{}_{\kappa} L^{\beta}{}_{\lambda} g^{\kappa\lambda} = g^{\alpha\beta} \tag{1.127}$$

Taking determinants of this equation we get

$$\det g^{\alpha\beta} = \det^2 L^{\alpha}{}_{\beta} \det g^{\alpha\beta} \tag{1.128}$$

or $\det^2 L^{\alpha}{}_{\beta} = 1$. Therefore, the determinant of the Lorentz transformation is

$$\det L^{\alpha}{}_{\beta} = \begin{cases} +1 & \text{(proper transformations)} \\ -1 & \text{(reflections)} \end{cases} \tag{1.129}$$

As shown previously, the Lorentz transformation is the most general transformation that leaves the interval invariant. In the language of 4-tensors, it is the most general transformation that leaves the metric tensor unchanged. It is not necessary to repeat the arguments here, and we merely quote the result. For a boost in the $\hat{\mathbf{x}}$ direction, the Lorentz transformation is, in matrix notation,

$$L^{\alpha}{}_{\beta} = \begin{bmatrix} \gamma & -\beta\gamma & 0 & 0 \\ -\beta\gamma & \gamma & 0 & 0 \\ 0 & 0 & 1 & 0 \\ 0 & 0 & 0 & 1 \end{bmatrix} \tag{1.130}$$

The inverse transformation is obtained by replacing β by $-\beta$. It is a simple matter to show that $L^{-1\alpha}{}_{\gamma} L^{\gamma}{}_{\beta} = \delta^{\alpha}{}_{\beta}$. That (1.130) is a tensor follows from (1.86) and the quotient rule. If $L^{\alpha}{}_{\beta}$ is the transformation from K to K' and $L'^{\alpha}{}_{\beta}$ is the transformation from K' to K'', then

$$L''^{\alpha}{}_{\beta} = L'^{\alpha}{}_{\kappa} L^{\kappa}{}_{\beta} \tag{1.131}$$

is the transformation from K to K''. This is the group property for the Lorentz transformation. The fact that the velocity-addition law (1.75) for parallel velocities is symmetric in $\mathbf{v}$ and $\mathbf{v}'$ reflects the fact that successive boosts in the same direction commute. Since the velocity addition law (1.76) for perpendicular velocities is not symmetric in $\mathbf{v}$ and $\mathbf{v}'$, successive boosts in different direction do not commute.

Finally, we introduce the Levi–Civita symbol, $\varepsilon^{\alpha\beta\gamma\delta}$, also known as the completely antisymmetric unit tensor of fourth rank. It has the values

$$\varepsilon^{\alpha\beta\gamma\delta} = \begin{cases} +1 & \text{if } (\alpha, \beta, \gamma, \delta) = (0, 1, 2, 3) \text{ or an even permutation} \\ -1 & \text{if } (\alpha, \beta, \gamma, \delta) = \text{an odd permutation of } (0, 1, 2, 3) \\ 0 & \text{otherwise.} \end{cases} \tag{1.132}$$

in all coordinate systems and represents the extension to Minkowski space of the symbol ε_{ijk} used to form the cross product of two vectors in three dimensions. We can raise or lower the indices in the usual way. We see immediately that

$$\varepsilon_{\alpha\beta\gamma\delta} = -\varepsilon^{\alpha\beta\gamma\delta} \tag{1.133}$$

since all the indices must be different (else $\varepsilon^{\alpha\beta\gamma\delta} = 0$), which makes three of the indices spacelike. The unit antisymmetric tensor is actually a pseudotensor. That is, under a reflection or inversion three of the coordinates change sign, so the components of $\varepsilon^{\alpha\beta\gamma\delta}$ should change sign. However, by definition $\varepsilon^{\alpha\beta\gamma\delta}$ has the same components in all coordinate systems. Since it behaves differently from a tensor under reflection and inversion, it is called a pseudotensor.

Using the Levi–Civita symbol, we define the *dual* tensor in the following way. If $A^{\alpha\beta}$ is an *antisymmetric* tensor of second rank, then the dual of this tensor is

$$\mathcal{A}^{\alpha\beta} = \tfrac{1}{2}\varepsilon^{\alpha\beta\gamma\delta} A_{\gamma\delta} \tag{1.134}$$

Since $\varepsilon^{\alpha\beta\gamma\delta}$ is a pseudotensor, the dual tensor $\mathcal{A}^{\alpha\beta}$ is a pseudotensor as well.

EXERCISE 1.11

Using the fact that $k^{\alpha} = (\omega/c, \mathbf{k})$ is a 4-vector, apply the Lorentz transformation to this quantity to derive the Doppler shift for a light wave, that is, the frequency observed in the laboratory for a source moving in the $\hat{\mathbf{x}}$ direction with velocity $v = \beta c$ for arbitrary $\mathbf{k}$. Show that the frequency ω observed in the laboratory is given by the formula

$$\frac{\omega}{\omega_0} = \frac{1}{\gamma(1 - \beta\cos\theta)} \tag{1.135}$$

where ω_0 is the frequency in the rest frame of the source and θ is the angle in the laboratory frame of reference.

EXERCISE 1.12

Consider the head-on collision of two photons of frequency ω_1 and ω_2 traveling in opposite directions along the x axis in the reference frame K. Use a Lorentz transformation to change to a new reference frame K' moving along the x axis in which the two photons have the same frequency ω. This is called the "center-of-mass" frame because the momenta of the colliding photons are equal and opposite one another. Using the Planck–Einstein quantization rule $\mathcal{E}' = \hbar\omega'$ for the energy of each photon in the center-of-mass frame, show that the total center-of-mass energy of the colliding photons is

$$\mathcal{E}'_{(COM)} = \hbar\omega'_1 + \hbar\omega'_2 = 2\hbar\sqrt{\omega_1\omega_2} \tag{1.136}$$

For a high-energy intergalactic photon of energy $\mathcal{E}_1 = \hbar\omega_1$ traveling through the cosmic background of photons for which $\mathcal{E}_2 = \hbar\omega_2 \approx 2.5 \times 10^{-4}$ eV, what photon energy is required to be above the threshold for electron–positron production? The electron rest energy is $m_e c^2 = 5 \times 10^5$ eV.

EXERCISE 1.13

Use the group property (1.131) to derive the Einstein velocity-addition law (1.75) for two parallel velocities.

EXERCISE 1.14

The 4-vector velocity and 4-vector acceleration are defined by the expressions

$$U^\alpha = \frac{dr^\alpha}{d\tau} \tag{1.137}$$

and

$$W^\alpha = \frac{d^2 r^\alpha}{d\tau^2} \tag{1.138}$$

The quantity

$$c^2 d\tau^2 = dr^\alpha dr_\alpha \tag{1.139}$$

is the invariant interval. Show that

$$U^\alpha U_\alpha = c^2 \tag{1.140}$$

and

$$U^\alpha W_\alpha = 0 \tag{1.141}$$

That is, the magnitude of the 4-vector velocity is always c and the 4-vector acceleration is orthogonal to the 4-vector velocity.

EXERCISE 1.15

If the 4-vectors

$$U^\alpha = \gamma(u)(c, \mathbf{u}) \tag{1.142}$$

and

$$V^\alpha = \gamma(v)(c, \mathbf{v}) \tag{1.143}$$

are the 4-vector velocities of two particles in some frame of reference, show that

$$U^\alpha V_\alpha = \gamma(w)c^2 \tag{1.144}$$

where w is the relative velocity of the two particles as viewed in the rest frame of one particle or the other. *Hint:* Evaluate the expression first in the rest frame of one particle, and then generalize.

EXERCISE 1.16

A particle moves at the relativistic velocity βc along the x axis and emits N photons with the distribution

$$\frac{dN}{d\Omega'} = f(\theta', \phi') \tag{1.145}$$

in the particle rest frame, where θ' is the angle between the photon and the direction of motion in the rest frame of the particle, and ϕ' is the azimuthal angle around the direction of motion. Use the Lorentz transformation of the wave vector k'^α of an emitted photon to show that for a photon emitted at the angle θ' relative to the direction of motion in the rest frame, the angle of emission in the laboratory frame satisfies the equation

$$\cos\theta' = \frac{\cos\theta - \beta}{1 - \beta\cos\theta} \tag{1.146}$$

Use this result to show that the distribution of photons in the laboratory frame is

$$\frac{dN}{d\Omega}(\theta, \phi) = \frac{f(\theta', \phi')}{\gamma^2(1 - \beta\cos\theta)^2} \tag{1.147}$$

EXERCISE 1.17

Show that the contraction of the Levi–Civita symbol gives

$$\varepsilon^{\alpha\beta\gamma\delta}\varepsilon_{\gamma\delta\mu\nu} = -2\big(\delta^\alpha{}_\mu\delta^\beta{}_\nu - \delta^\alpha{}_\nu\delta^\beta{}_\mu\big) \tag{1.148}$$

$$\varepsilon^{\alpha\beta\gamma\delta}\varepsilon_{\beta\gamma\delta\mu} = 6\delta^\alpha{}_\mu \tag{1.149}$$

$$\varepsilon^{\alpha\beta\gamma\delta}\varepsilon_{\alpha\beta\gamma\delta} = -24 \tag{1.150}$$

1.3.3 4-Vector Calculus

All the operations of 3-vector calculus can be extended to 4-vectors, with slight complications due to the covariant and contravariant forms of the vectors and operators. We define the derivative (the 4-vector gradient) of a scalar function $\phi(r^\alpha)$ with respect to the *contravariant* coordinates by

$$\frac{\partial \phi}{\partial r^\alpha} = \left(\frac{1}{c} \frac{\partial \phi}{\partial t}, \nabla \phi \right) = \partial_\alpha \phi \tag{1.151}$$

in a convenient notation. The differential $d\phi$ corresponding to the increment dr^α is then

$$d\phi = \partial_\alpha \phi \, dr^\alpha \tag{1.152}$$

Since $d\phi$ is a scalar, we see from the quotient rule that the derivative ∂_α with respect to a *contravariant* coordinate is a *covariant* vector operator, as suggested by the notation ∂_α. In the same way, we define the *contravariant* derivative by

$$\frac{\partial}{\partial r_\alpha} = \left(\frac{1}{c} \frac{\partial}{\partial t}, -\nabla \right) = \partial^\alpha \tag{1.153}$$

Note in (1.151) and (1.153) that the sign of the spacelike part of the covariant and contravariant operators is opposite that of covariant and contravariant vectors.

The 4-vector divergence of the 4-vector A^α is a scalar,

$$\partial_\alpha A^\alpha = \frac{1}{c} \frac{\partial A^0}{\partial t} + \nabla \cdot \mathbf{A} = \partial^\alpha A_\alpha \tag{1.154}$$

and the Laplacian of a scalar (or any component of a 4-tensor) is

$$\partial_\alpha \partial^\alpha \phi = \frac{\partial}{\partial r^\alpha} g^{\alpha\beta} \frac{\partial \phi}{\partial r^\beta} = \frac{1}{c^2} \frac{\partial^2 \phi}{\partial t^2} - \nabla^2 \phi = \partial^\alpha \partial_\alpha \phi \tag{1.155}$$

We note in passing that the volume in Minkowski space is an invariant under Lorentz transformations. To see this we recall that

$$c \, dt' \, d^3\mathbf{r}' = \mathcal{J} c \, dt \, d^3\mathbf{r} \tag{1.156}$$

where $\mathcal{J}$ is the Jacobian of the transformation. But the coefficients of the Lorentz transformation are just $L^\alpha{}_\beta = \partial r'^\alpha / \partial r^\beta$, so the Jacobian is

$$\mathcal{J} = \frac{\partial(r'^0 \ldots r'^3)}{\partial(r^0 \ldots r^3)} = \det \frac{\partial r'^\alpha}{\partial r^\beta} = \det L^\alpha{}_\beta = 1 \tag{1.157}$$

as shown previously. Thus, while a volume element in Minkowski space that is cubic in the moving frame K' appears distorted in the laboratory frame K, as shown in Figure 1.13 (drawn for $\gamma = 2$; see also Figure 1.9), the four-dimensional volume is conserved. Note that the three-dimensional volume is not conserved. In fact, due to length contraction and time dilation, (1.66) and (1.67), we see that

$$c \, dt' = \frac{c \, dt}{\gamma} \tag{1.158}$$

and

$$d^3\mathbf{r}' = dx' \, dy' dz' = \gamma \, dx \, dy \, dz = \gamma \, d^3\mathbf{r} \tag{1.159}$$

To illustrate the use of these concepts and 4-vector notation, we consider some examples. We begin by deriving the relativistically correct (and manifestly covariant) form of

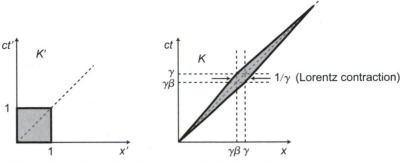

Figure 1.13 Transformation of a volume element in Minkowski space.

the continuity equation

$$\frac{\partial \rho}{\partial t} + \nabla \cdot (\rho \mathbf{v}) = 0 \qquad (1.160)$$

which expresses the conservation of charge. To do this, we consider a three-dimensional volume dV containing dN particles. Clearly, dN is a scalar, since the same number of particles is counted by all observers. We introduce some physics here and assert that the charge q on each particle is a scalar; that is, charge is a Lorentz invariant, the same to all observers. The total charge in the infinitesimal volume is then

$$\rho \, d^3\mathbf{r} = q \, dN = \text{scalar} \qquad (1.161)$$

where ρ is the average charge density in the volume. If dr^α is the displacement of each of the charges in $d^3\mathbf{r}$ in a short time dt, it follows that the quantity

$$\rho \, d^3\mathbf{r} \, dr^\alpha = \rho \, d^3\mathbf{r} \, dt \frac{dr^\alpha}{dt} \qquad (1.162)$$

is a 4-vector. Since the 4-volume $d^3\mathbf{r} \, c \, dt$ is a scalar, we see that the 4-vector current density

$$J^\alpha = \rho \frac{dr^\alpha}{dt} = \rho(c, \mathbf{v}) = (\rho c, \mathbf{J}) \qquad (1.163)$$

is also a 4-vector, where $\mathbf{J} = \rho \mathbf{v}$ is the ordinary 3-vector current density. Note that neither ρ nor $(c, \mathbf{v})$ by itself is a Lorentz tensor quantity, but their product is. We may therefore write the continuity equation (1.160) in the manifestly covariant form

$$\partial_\alpha J^\alpha = 0 \qquad (1.164)$$

which is valid in any inertial coordinate system.

We frequently are concerned with point charges, for which the charge density may be expressed in terms of the Dirac δ-function as

$$\rho(\mathbf{r}, t) = q \delta[\mathbf{r} - \mathbf{r}_0(t)] \qquad (1.165)$$

where q is the charge and $\mathbf{r}_0(t)$ the position of the charge, and we have used the shorthand notation

$$\delta(\mathbf{r}) = \delta(x)\delta(y)\delta(z) \qquad (1.166)$$

The Dirac δ-function is defined by its property that

$$\int_a^b f(x)\delta(x - x_0) \, dx = \begin{cases} f(x_0) & \text{if } a < x_0 < b \\ 0 & \text{otherwise} \end{cases} \qquad (1.167)$$

In this way it is really an operator that picks out the value of $f(x)$ at $x = x_0$ and not actually a function at all, although it can be envisioned as a function whose value is zero everywhere except at x_0, where it is infinite, with a unit area under the peak. A proper definition of the δ-function as a generalized function is discussed in Chapter 5, in connection with Fourier transforms.

Some useful properties of the δ-function can be established directly from the operational definition (1.167). In particular, we find that

$$\delta[g(x)] = \frac{\delta(x - x_0)}{|dg/dx|} \tag{1.168}$$

$$\delta(x^2 - a^2) = \frac{1}{2|a|}[\delta(x + a) + \delta(x - a)] \tag{1.169}$$

$$\int_{-\infty}^{\infty} \delta(x - a)\delta(x - b)\, dx = \delta(a - b) \tag{1.170}$$

$$\delta(x) = -x\delta'(x) \tag{1.171}$$

where $g(x)$ vanishes at x_0 and $\delta'(x)$ the derivative of the δ-function. Properties (1.168) and (1.169) follow directly from (1.167). Property (1.170) follows (at least formally) from (1.167) with $f(x) = \delta(x - a)$. Alternatively, it may be proved by multiplying (1.170) by $f(a)$ and integrating over a from a_1 to a_2. Property (1.171) is proved by multiplying by $f(x)$ and integrating by parts.

In terms of the Dirac δ-function, the 4-vector current density of a point charge is

$$J^\alpha = q(c, \mathbf{v}_0)\delta[\mathbf{r} - \mathbf{r}_0(t)] \tag{1.172}$$

where $\mathbf{v}_0 = d\mathbf{r}_0/dt$ is the velocity of the point charge. This expression doesn't look like a 4-vector, since $(c, \mathbf{v}_0)$ isn't a 4-vector, but it is. In fact, it has the same form in any coordinate system,

$$J'^\alpha = q(c, \mathbf{v}_0')\delta[\mathbf{r}' - \mathbf{r}_0'(t')] \tag{1.173}$$

This is intuitively evident, but we can see it explicitly if we consider a charge at rest in the reference frame K' moving at velocity $\mathbf{v}$ relative to the laboratory frame K. The current density in K' (and *only* in the rest frame K', where $\gamma = 1$) is then

$$J'^\alpha = q(c, 0)\delta(\mathbf{r}' - \mathbf{r}_0') = qU'^\alpha\delta(\mathbf{r}' - \mathbf{r}_0') \tag{1.174}$$

In the frame K the current density is then

$$J^\alpha = qU^\alpha\frac{\delta(\mathbf{r} - \mathbf{r}_0)}{|\partial x/\partial x'|} = q\gamma(c, \mathbf{v})\frac{\delta(\mathbf{r} - \mathbf{r}_0)}{\gamma} = q(c, \mathbf{v})\delta(\mathbf{r} - \mathbf{r}_0) \tag{1.175}$$

where γ now refers to the particle velocity, and we have used (1.168) to evaluate $\delta(\mathbf{r} - \mathbf{r}_0)$, and (1.52) (with t' fixed) to evaluate $\partial x/\partial x'$.

As another example of 4-vector calculus, we consider the Maxwell equations for the scalar potential Φ and vector potential $\mathbf{A}$, as discussed in the Prologue. In 3-vector notation, the Maxwell equations in the Lorentz gauge have the form

$$\frac{1}{c^2}\frac{\partial^2\Phi}{\partial t^2} - \nabla^2\Phi = \frac{\rho}{\varepsilon_0} = \mu_0 c^2\rho \tag{1.176}$$

for the scalar potential, where ε_0 is the vacuum permittivity and μ_0 the vacuum permeability, and

$$\frac{1}{c^2}\frac{\partial^2 \mathbf{A}}{\partial t^2} - \nabla^2 \mathbf{A} = \mu_0 \mathbf{J} \tag{1.177}$$

for the vector potential. The symmetry of these equations should not go unnoticed. In fact, if we divide (1.176) by c we see that the source terms on the right form the 4-vector current density $J^\alpha = (c\rho, \mathbf{J})$. This suggests that we should form the 4-vector potential

$$A^\alpha = \left(\frac{\Phi}{c}, \mathbf{A}\right) \tag{1.178}$$

The Maxwell equations (1.176) and (1.177) may then be expressed in the compact form

$$\partial_\alpha \partial^\alpha A^\beta = \mu_0 J^\beta \tag{1.179}$$

The 4-vector character of A^α is evident from the fact that the Laplacian operator is a scalar and the current density is a 4-vector.

Note carefully that while Φ is conventionally called the scalar potential, it is not a Lorentz scalar but the timelike component of a 4-vector. Likewise $\mathbf{A}$ is conventionally called the vector potential, but it is just the spacelike part of a 4-vector. In Chapter 2 we have occasion to introduce the Lorentz scalar potential ϕ, which should not be confused with Φ.

In addition, we have the Lorentz gauge condition,

$$\frac{1}{c^2}\frac{\partial \Phi}{\partial t} + \nabla \cdot \mathbf{A} = 0 \tag{1.180}$$

This may be expressed in covariant form as

$$\partial_\alpha A^\alpha = 0 \tag{1.181}$$

By writing the Maxwell equations in covariant form, we are able to express all of classical electricity and magnetism in an elegant, symmetric, and extremely compact form. More important, we see from the covariant form of the equations that they are relativistically correct. This is much easier and more satisfying than carrying out a detailed Lorentz transformation on the equations in 3-dimensional form, as Lorentz had to do!

1.4 ELECTROMAGNETIC FIELDS

1.4.1 The 4-Tensor Electromagnetic Field

In nonrelativistic theory the electric and magnetic fields are clearly related to one another through Faraday's law and the Maxwell–Ampere law. But the lack of symmetry in these relations conceals the real unity of the electric and magnetic fields. We see the beginning of this unification in the definition of the 4-vector potential, but a covariant description of the fields themselves reveals their true unity and symmetry.

If a particle with a charge q moves with velocity $\mathbf{v}_0$ through a uniform magnetic field $\mathbf{B}$, it experiences a force that depends on its velocity. If we view this particle in a reference frame K' in which the particle is at rest, the particle still experiences a force, but since the particle is now at rest, we attribute this force to an electric field $\mathbf{E}'$ in the K' frame. If the particle is allowed to move with velocity $\mathbf{v}'$ in the K' frame, it experiences an additional force that we ascribe to the magnetic field $\mathbf{B}'$ in the moving frame. In the nonrelativistic approximation, where velocities may be added linearly, the total force, called the Lorentz force, is

$$\mathbf{F} = q[\mathbf{E} + (\mathbf{v}_0 + \mathbf{v}') \times \mathbf{B}] = q(\mathbf{E}' + \mathbf{v}' \times \mathbf{B}') \tag{1.182}$$

Thus, the original magnetic field transforms into an electric field and a magnetic field:

$$\mathbf{E}' = \mathbf{E} + \mathbf{v}_0 \times \mathbf{B} \tag{1.183}$$

$$\mathbf{B}' = \mathbf{B} \tag{1.184}$$

in frame K'. For relativistic velocities these equations are no longer quite correct, but the mixing of the electric and magnetic fields still occurs. For the Lorentz transformation to mix the electric and magnetic fields in this way, they cannot separately be 4-vectors, although they are separately 3-vectors. Instead they must together be components of some 4-tensor entity. We can determine the 4-tensor of the electric and magnetic fields in the following way.

The electric and magnetic fields are derived from the scalar and vector potentials using the formulas

$$\mathbf{E} = -\frac{\partial \mathbf{A}}{\partial t} - \nabla \Phi \tag{1.185}$$

$$\mathbf{B} = \nabla \times \mathbf{A} \tag{1.186}$$

If we evaluate the x components, for example, we get

$$E_x = -\frac{\partial A_x}{\partial t} - \frac{\partial \Phi}{\partial x} = -c(\partial^0 A^1 - \partial^1 A^0) \tag{1.187}$$

$$B_x = \frac{\partial A_z}{\partial y} - \frac{\partial A_y}{\partial z} = -(\partial^2 A^3 - \partial^3 A^2) \tag{1.188}$$

and so on. Clearly, a pattern is emerging here. If we evaluate the remaining components we find that they form the antisymmetric, second-rank, contravariant tensor

$$F^{\alpha\beta} = \partial^\alpha A^\beta - \partial^\beta A^\alpha = \begin{bmatrix} 0 & -\dfrac{E_x}{c} & -\dfrac{E_y}{c} & -\dfrac{E_z}{c} \\[2mm] \dfrac{E_x}{c} & 0 & -B_z & B_y \\[2mm] \dfrac{E_y}{c} & B_z & 0 & -B_x \\[2mm] \dfrac{E_z}{c} & -B_y & B_x & 0 \end{bmatrix} \tag{1.189}$$

The field-strength tensor $F^{\alpha\beta}$ is a true tensor (not a pseudotensor) because the components with one spacelike index (F^{0i}) involve the electric field (a true vector) and change sign under inversion, while those with two spacelike indices (F^{ij}) involve the magnetic field (an axial vector) and are invariant under an inversion of the coordinates.

The covariant form of the electromagnetic field-strength tensor is found by lowering the superscripts. In this case only those components that have one timelike and one spacelike index change sign. These are the electric field components, and we get

$$F_{\alpha\beta} = g_{\alpha\gamma}g_{\beta\delta}F^{\gamma\delta} = \begin{bmatrix} 0 & \dfrac{E_x}{c} & \dfrac{E_y}{c} & \dfrac{E_z}{c} \\[2ex] -\dfrac{E_x}{c} & 0 & -B_z & B_y \\[2ex] -\dfrac{E_y}{c} & B_z & 0 & -B_x \\[2ex] -\dfrac{E_z}{c} & -B_y & B_x & 0 \end{bmatrix} \tag{1.190}$$

We can also construct the dual of the field-strength tensor

$$\mathcal{F}^{\alpha\beta} = \tfrac{1}{2}\varepsilon^{\alpha\beta\gamma\delta}F_{\gamma\delta} \tag{1.191}$$

Clearly, this tensor is antisymmetric, since it changes sign if α and β are interchanged. If we work out the components we find that

$$\mathcal{F}^{01} = \tfrac{1}{2}(F_{23} - F_{32}) = F_{23} = -B_x \tag{1.192}$$

and

$$\mathcal{F}^{23} = \tfrac{1}{2}(F_{01} - F_{10}) = F_{01} = E_x/c \tag{1.193}$$

for example. Comparing these results with (1.189), we see that they are related by the transformation

$$F^{\alpha\beta} \xrightleftharpoons[\mathbf{B} \leftrightarrow -\mathbf{E}/c]{\mathbf{E}/c \leftrightarrow \mathbf{B}} \mathcal{F}^{\alpha\beta} \tag{1.194}$$

Therefore, the dual of the field-strength tensor is

$$\mathcal{F}^{\alpha\beta} = \begin{bmatrix} 0 & -B_x & -B_y & -B_z \\[2ex] B_x & 0 & \dfrac{E_z}{c} & -\dfrac{E_y}{c} \\[2ex] B_y & -\dfrac{E_z}{c} & 0 & \dfrac{E_x}{c} \\[2ex] B_z & \dfrac{E_y}{c} & -\dfrac{E_x}{c} & 0 \end{bmatrix} = -\mathcal{F}^{\beta\alpha} \tag{1.195}$$

Actually, $\mathcal{F}^{\alpha\beta}$ is a pseudotensor, since $F^{\alpha\beta}$ is a true tensor but the Levi–Civita symbol is a pseudotensor.

If we form inner products of the field-strength tensors we obtain Lorentz scalars, or invariants. Thus,

$$F_{\alpha\beta}F^{\alpha\beta} = \text{invariant} = 2\left(B^2 - \frac{E^2}{c^2}\right) \tag{1.196}$$

Evidently, if the energy of the magnetic field is the same as that of the electric field in one reference frame ($B^2 = E^2/c^2$), so that $F^{\alpha\beta}F_{\alpha\beta} = 0$, then it is the same in all reference frames. This is true in a light wave, for example. In fact, if the magnetic energy exceeds the electric energy in any reference frame, then this is true in all reference frames, and vice versa. In the same way, we find that

$$F_{\alpha\beta}\mathcal{F}^{\alpha\beta} = \text{invariant} = -4\mathbf{E}\cdot\mathbf{B}/c \tag{1.197}$$

Therefore, if the electric and magnetic fields are perpendicular to one another in one reference frame so that $\mathbf{E} \cdot \mathbf{B} = 0$ (as in a light wave), then they are perpendicular in all frames, and if they form an acute (or obtuse) angle in any reference frame, then this is true in all reference frames. Strictly speaking, $F_{\alpha\beta}\mathcal{F}^{\alpha\beta}$ is a pseudoscalar, that is, it changes sign under inversions, since $\mathcal{F}^{\alpha\beta}$ is a pseudotensor. We can also see that this is true because $\mathbf{B}$ is an axial vector.

Using the electromagnetic field-strength tensor, we can write the Maxwell equations in covariant form. We begin with the inhomogeneous equations

$$\nabla \cdot \mathbf{E} = \frac{\rho}{\varepsilon_0} = \mu_0 c^2 \rho \tag{1.198}$$

$$\nabla \times \mathbf{B} - \frac{1}{c^2}\frac{\partial \mathbf{E}}{\partial t} = \mu_0 \mathbf{J} \tag{1.199}$$

We recognize on the right-hand side the components of the 4-vector current density J^α. In 4-tensor notation, Gauss's law (1.198) may be expressed

$$\partial_\alpha F^{\alpha 0} = \mu_0 J^0 \tag{1.200}$$

In the same way, the x component of the Maxwell–Ampere law becomes

$$\partial_\alpha F^{\alpha 1} = \mu_0 J^1 \tag{1.201}$$

and so on. Thus, Gauss's law and the Maxwell–Ampere law may together be expressed as the manifestly covariant, 4-vector equation

$$\partial_\alpha F^{\alpha\beta} = \mu_0 J^\beta \tag{1.202}$$

On the other hand, the homogeneous equations

$$\nabla \cdot \mathbf{B} = 0 \tag{1.203}$$

and

$$\nabla \times \mathbf{E} + \frac{\partial \mathbf{B}}{\partial t} = 0 \tag{1.204}$$

are obtained from (1.198) and (1.199) by the substitutions

$$\frac{\mathbf{E}}{c} \to \mathbf{B} \quad \text{and} \quad \mathbf{B} \to -\frac{\mathbf{E}}{c} \tag{1.205}$$

which also transform $F^{\alpha\beta} \to \mathcal{F}^{\alpha\beta}$. Thus, we see from (1.202) that the covariant form of the homogeneous Maxwell equations is

$$\partial_\alpha \mathcal{F}^{\alpha\beta} = 0 \tag{1.206}$$

This may be expressed in another form by using the definition of the dual tensor (1.191),

$$\partial_\alpha \tfrac{1}{2}\varepsilon^{\alpha\beta\gamma\delta} F_{\gamma\delta} = 0 = \tfrac{1}{2}\varepsilon^{\alpha\gamma\delta\beta}\partial_\alpha F_{\gamma\delta} \tag{1.207}$$

This expression corresponds to four equations, for $\beta = 0 \ldots 3$. But for each equation (each value of β), the only surviving terms are those for which α, β, γ, and δ are all different. In addition, when the indices γ and δ are interchanged, both $\varepsilon^{\alpha\beta\gamma\delta}$ and $F_{\gamma\delta}$ change sign. Therefore, for each value of β we get an equation of the form

$$\partial_\alpha F_{\gamma\delta} + \partial_\gamma F_{\delta\alpha} + \partial_\delta F_{\alpha\gamma} = 0 \qquad \alpha, \gamma, \text{ and } \delta \text{ all different} \tag{1.208}$$

EXERCISE 1.18

Use the invariants of the field-strength tensor to show that if the electric and magnetic field vectors are not perpendicular in reference frame K, then in some reference frame K' they are parallel and have the values

$$E' = \sqrt{K\mathbf{E} \cdot \mathbf{B}c} \tag{1.209}$$

and

$$B'c = \sqrt{\frac{\mathbf{E} \cdot \mathbf{B}c}{K}} \tag{1.210}$$

where

$$K = C + \sqrt{1 + C^2} \tag{1.211}$$

and

$$C = \frac{E^2 - B^2c^2}{4\mathbf{E} \cdot \mathbf{B}c} \tag{1.212}$$

EXERCISE 1.19

In the local rest frame of a conducting fluid, the charge density is ρ' and the 3-vector current density is given by Ohm's law, $\mathbf{J}' = \sigma \mathbf{E}'$, where σ is the conductivity and $\mathbf{E}'$ the electric field. At low (nonrelativistic) velocities, the charge density in the moving frame is nearly the same as that in the laboratory frame (the Lorentz contraction is of order $1/c^2$) and the electric field in the local rest frame of the fluid can be approximated by the Lorentz force law

$$\mathbf{E}' = \mathbf{E} + \mathbf{v} \times \mathbf{B} \tag{1.213}$$

where $\mathbf{E}$ and $\mathbf{B}$ are the electric and magnetic fields in the laboratory frame. The total 4-vector current density, including convection of the charge density, in the nonrelativistic approximation is therefore

$$J^\alpha = (\rho c, \mathbf{J}) = [\rho c, \rho \mathbf{v} + \sigma(\mathbf{E} + \mathbf{v} \times \mathbf{B})] \tag{1.214}$$

where $\mathbf{v}$ is the local velocity of the fluid. Show that the relativistic generalization of this is

$$J^\alpha = \frac{1}{c^2}(U_\beta J^\beta)U^\alpha + \sigma F^{\alpha\beta}U_\beta \tag{1.215}$$

by showing that (1.215) reduces to (1.214) in the nonrelativistic limit.

1.4.2 Transformation of Electromagnetic Fields

The Lorentz transformation of the electric and magnetic fields is more complex than that of the 4-vector potential due to the tensor character of the fields. To transform the field-strength tensor (1.189) for a boost in the $\hat{\mathbf{x}}$ direction, we use the Lorentz transformation

(1.130). This is a bit tedious, but for the (0, 1) component we get

$$F'^{01} = L^0{}_\alpha L^1{}_\beta F^{\alpha\beta} = L^0{}_\alpha \left(L^1{}_0 F^{\alpha 0} + L^1{}_1 F^{\alpha 1} + L^1{}_2 F^{\alpha 2} + L^1{}_3 F^{\alpha 3} \right)$$

$$= L^0{}_\alpha (-\beta\gamma F^{\alpha 0} + \gamma F^{\alpha 1}) = -\beta\gamma \left(L^0{}_0 F^{00} + L^0{}_1 F^{10} + L^0{}_2 F^{20} + L^0{}_3 F^{30} \right)$$

$$+ \gamma \left(L^0{}_0 F^{01} + L^0{}_1 F^{11} + L^0{}_2 F^{21} + L^0{}_3 F^{31} \right)$$

$$= \beta^2 \gamma^2 F^{10} + \gamma^2 F^{01} = \gamma^2 (1 - \beta^2) F^{01} = F^{01} \qquad (1.216)$$

That is, after all that effort there is no change in the longitudinal electric field. However, if we evaluate the rest of the components in the same way we get

$$E'_x = E_x \qquad (1.217)$$

$$E'_y = \gamma (E_y - v B_z) \qquad (1.218)$$

$$E'_z = \gamma (E_z + v B_y) \qquad (1.219)$$

$$B'_x = B_x \qquad (1.220)$$

$$B'_y = \gamma \left(B_y + \frac{v}{c^2} E_z \right) \qquad (1.221)$$

$$B'_z = \gamma \left(B_z - \frac{v}{c^2} E_y \right) \qquad (1.222)$$

The inverse transformation is obtained by changing $v \to -v$ in (1.217)–(1.222).

In the nonrelativistic limit $v/c \ll 1$ we recover the transformations (1.183) and (1.184), which we derived from the Lorentz force law. This does not mean that the Lorentz force law is valid only in the nonrelativistic limit. Rather, the factor γ and the higher-order terms in B'_y and B'_z appear in the transformed fields because the motions in the new coordinate system are altered by time dilation and length contraction, and the forces must change in the new coordinate system to reflect this.

In contrast with the case of 4-vectors, the longitudinal components of the electric and magnetic fields are unchanged by the boost while the transverse components are changed and mixed. That the longitudinal electric field should remain unchanged by the boost can be understood on physical grounds if we consider the field due to a parallel-plate capacitor whose axis (normal to the plates) is aligned parallel to the boost, as shown in Figure 1.14. Since transverse lengths are unaffected by the boost the charge density on the plates is unchanged, but the separation between the plates is reduced by length contraction. However, the field is independent of the separation of the plates so the electric field in the direction of the boost is unchanged. Similarly, if we consider the magnetic field of a solenoid aligned along the direction of the boost, as shown in Figure 1.15, we see that the winding density of the solenoid is increased by the Lorentz contraction, while the current in the solenoid is decreased by time dilation. The effects cancel and leave the longitudinal magnetic field unchanged.

The equations (1.218) and (1.219) for the transverse electric field in the new reference frame are similar to those we obtained by considering the nonrelativistic Lorentz force equation (1.182), except that they are modified by the factor γ. To understand the factor γ we consider the field of a parallel-plate capacitor aligned with its axis perpendicular to the boost, as shown in Figure 1.16. Due to the Lorentz contraction the charge density on

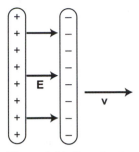

Figure 1.14 Capacitor in motion parallel to its axis.

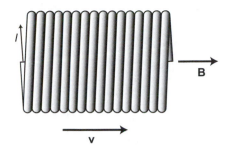

Figure 1.15 Solenoid in motion parallel to its axis.

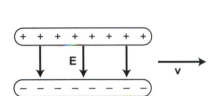

Figure 1.16 Capacitor in motion orthogonal to its axis.

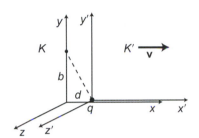

Figure 1.17 Charge in a moving frame.

the plates is increased by the factor γ, which increases the electric field accordingly. The magnetic field is slightly more complicated. Compared with the Lorentz force equation (1.184), we see that the equations (1.221) and (1.222) for the magnetic field in the reference frame K' have new terms that depend on the electric field in K. The new terms appear because the charges that give rise to the electric field in K are moving in K' and constitute a current. In the parallel-plate capacitor shown in Figure 1.16 there is no magnetic field in the rest frame of the capacitor, but in the laboratory frame the charged plates constitute a surface current density $\gamma\sigma\mathbf{v}$, where σ is the charge density of the plates in the rest frame. This is proportional to the transverse electric field and is responsible for the new terms in the magnetic fields (1.221) and (1.222). In the limit $c \to \infty$ the transformations (1.217)–(1.222) reduce to the nonrelativistic Lorentz force equations (1.183) and (1.186).

As an example of the transformation of electromagnetic fields we consider the field of a point charge in uniform motion. As shown in Figure 1.17, the charge is placed at the origin of the moving coordinate system K' and the observer is placed at the point $y = b$ in the stationary coordinate system K. The charge is therefore at the point $x = vt = d$ in the stationary frame. In the moving frame, the field at the observer is

$$E'_x = \frac{q}{4\pi\varepsilon_0}\frac{x'}{r'^3} = \frac{q}{4\pi\varepsilon_0}\frac{x'}{(x'^2 + b^2)^{3/2}} \tag{1.223}$$

$$E'_y = \frac{q}{4\pi\varepsilon_0}\frac{b}{r'^3} = \frac{q}{4\pi\varepsilon_0}\frac{b}{(x'^2 + b^2)^{3/2}} \tag{1.224}$$

$$E'_z = B'_x = B'_y = B'_z = 0 \tag{1.225}$$

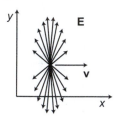

Figure 1.18 Electric field of a moving charge.

To transform the fields to the laboratory frame, we note first that the position of the observer is

$$x' = \gamma(x - \beta ct) = -\gamma vt = -\gamma d \tag{1.226}$$

since the observer is at $x = 0$. Using the inverse of the transformations (1.217)–(1.222) we find that

$$E_x = E'_x = -\frac{\gamma q}{4\pi\varepsilon_0} \frac{d}{(b^2 + \gamma^2 d^2)^{3/2}} \tag{1.227}$$

$$E_y = \gamma E'_y = \frac{\gamma q}{4\pi\varepsilon_0} \frac{b}{(b^2 + \gamma^2 d^2)^{3/2}} \tag{1.228}$$

$$B_z = \gamma \frac{v}{c^2} E'_y = \frac{\gamma q}{4\pi\varepsilon_0} \frac{v}{c^2} \frac{b}{(b^2 + \gamma^2 d^2)^{3/2}} \tag{1.229}$$

and

$$E_z = B_x = B_y = 0 \tag{1.230}$$

In these results we see several things. In the first place, the *effective* charge is increased by γ. This is not to be confused with the relativistic invariance of charge. We also see that the spatial extent of the field is compressed by γ in the longitudinal direction, since the field strength depends on the longitudinal position through the factor γd in the denominator. This is just the usual Lorentz contraction. However, the ratio of the x and y components of the electric field is

$$\frac{E_x}{E_y} = -\frac{d}{b} \tag{1.231}$$

Therefore, the electric field lines are directed radially away from the charge even when it is moving at a relativistic velocity. The form of the resulting field is suggested schematically in Figure 1.18. The magnetic field lines are circles about the direction of motion of the charge.

EXERCISE 1.20

The 4-vector potential surrounding a point charge at rest is

$$A^\alpha = \left(\frac{\Phi}{c}, \mathbf{A}\right) \tag{1.232}$$

where

$$\Phi = \frac{q}{4\pi \varepsilon_0 r} \tag{1.233}$$

and

$$\mathbf{A} = 0 \tag{1.234}$$

Find the 4-vector potential of a moving charge by placing the charge at the origin of a frame K' moving at velocity v in the $\hat{\mathbf{x}}$ direction with respect to frame K and applying a Lorentz transformation. Show that the components are given by the formulas

$$\Phi = \frac{\gamma q}{4\pi \varepsilon_0 [\gamma^2 (x - vt)^2 + y^2 + z^2]^{1/2}} \tag{1.235}$$

$$A_x = \frac{v}{c^2} \Phi \tag{1.236}$$

and

$$A_y = A_z = 0 \tag{1.237}$$

EXERCISE 1.21

In reference frame K', a parallel-plate capacitor is oriented normal to the y' axis. Inside the capacitor the electric field is E_y'. If the capacitor moves at velocity v_x in the $\hat{\mathbf{x}}$ direction (parallel to the plates of the capacitor) in the laboratory frame K, the charge on the plates becomes a current. Using simple arguments involving the integral forms of Gauss's law and Ampere's law, together with length contraction, show that the magnetic field in the laboratory frame is

$$B_z = \gamma \frac{v_x}{c^2} E_y' \tag{1.238}$$

Compare this with the inverse of the transformation law (1.222).

EXERCISE 1.22

The magnetic field surrounding a current I along the x axis is

$$B = \frac{\mu_0 I}{2\pi r} \tag{1.239}$$

where r is the distance from the axis. Transform this into the magnetic and electric fields in a coordinate system moving at velocity v in the $\hat{\mathbf{x}}$ direction. What is the current in the new coordinate system? What is the net charge density in the new coordinate system? Where does the net charge density come from?

EXERCISE 1.23

In its own rest frame, the potential of a point electric dipole $\mathbf{p}$ at the origin is

$$\Phi' = \frac{\mathbf{p} \cdot \mathbf{r}'}{r'^3} \tag{1.240}$$

$$\mathbf{A} = 0 \tag{1.241}$$

Show that if the dipole moves in the $\hat{\mathbf{x}}$ direction with velocity $\mathbf{v}_0 = \boldsymbol{\beta}_0 c$, then the 4-vector potential in the laboratory frame is

$$A^\alpha = \gamma \frac{\gamma p_x X + p_y Y + p_z Z}{c(\gamma^2 X^2 + Y^2 + Z^2)^{3/2}} (1, \boldsymbol{\beta}_0) \approx \frac{\mathbf{p} \cdot \mathbf{R}}{cR^3}(1, \boldsymbol{\beta}_0) \tag{1.242}$$

in the nonrelativistic limit, where

$$\mathbf{R} = (X, Y, Z) = \mathbf{r} - \mathbf{r}_0 \tag{1.243}$$

and $\mathbf{r}_0 = \mathbf{v}_0 t$ is the position of the dipole in the laboratory frame.

BIBLIOGRAPHY

The original papers describing the special theory of relativity are collected in

   A. Einstein, H. A. Lorentz, H. Minkowski, and H. Weyl, *The Principle of Relativity*, with notes by A. Sommerfeld, Dover Publications, New York (1952).

Good discussions of the covariant formulation of electrodynamics that parallel the discussion here are found in

   J. D. Jackson, *Classical Electrodynamics*, 3rd edition, John Wiley & Sons, New York (1999),
   L. D. Landau and E. M. Lifshitz, *The Classical Theory of Fields*, 4th edition, Pergamon Press, Oxford (1975),
   C. Moller, *The Theory of Relativity*, 2nd edition, Clarendon Press, Oxford (1991).

For an especially thorough introduction to 4-vectors, the reader is referred to

   W. Rindler, *Introduction to Special Relativity*, 2nd edition, Clarendon Press, Oxford (1991).

2 Relativistic Mechanics and Field Theory

Beauty, at least in theoretical physics, is perceived in the simplicity and compactness of the equations that describe the phenomena we observe about us. Dirac has emphasized this point and said, "It is more important to have beauty in one's equations than to have them fit experiment. . . . It seems that if one is working from the point of view of getting beauty in one's equations, and if one has really a sound insight, one is on a sure line of progress." In this sense the beauty of classical physics lies in the fact that it can all be derived from the postulates of relativity together with just one hypothesis, which we call Hamilton's principle. This includes all of classical mechanics and all of electricity and magnetism. In fact, if we postulate other interactions, such as the Yukawa potential, the mathematical form of these interactions is very restricted. The flexibility in the choice of natural laws is very limited.

In the future, as so-called grand unified theories are developed, it is expected that even this limited flexibility will be removed. One of the remarkable developments of modern physics has been the growing perception that the laws of physics are inevitable. Hawking may have gone beyond the realm of pure physics when he asked the question, "Did God have any choice?" in the way She wrote the laws of physics. However, it seems that if the universe consists of three spatial dimensions and time and we require causality, then there is little choice in the laws of physics.

We begin our discussion with the simplest problem, the motion of a free particle. From there we move to the motions of particles under the influence of forces (fields) and then on to the motions of the fields themselves.

2.1 RELATIVISTIC FREE PARTICLE

2.1.1 Hamilton's Principle and the Calculus of Variations

Hamilton's principle says that as a system moves from state a to state b, it does so along the trajectory that makes the action integral

$$S = \int_a^b \mathcal{L} \, dt \tag{2.1}$$

an extremum, generally a minimum, subject to the constraint that the end points a and b (including both the coordinates and the times) are fixed. The quantity $\mathcal{L}$ is called the Lagrangian for the system, and its form depends on the nature of the system under consideration. The task in classical mechanics and classical field theory therefore consists of two parts. First we must determine the Lagrangian $\mathcal{L}$ for the system, and second we must find the equations of motion that minimize the action S. As we shall see, the form of the Lagrangian for the particles and fields of classical electrodynamics follows from the postulates of relativity. Only the few parameters that appear in the equations must be determined from experiment. To find the equations of motion from the Lagrangian, we adopt the methods of the calculus of variations.

The calculus of variations traces its origin to 1696, when Johann Bernoulli challenged first his brother Jakob Bernoulli and later, in the pages of *Acta eruditorum,* "the acutest mathematicians of the world" to find the curve along which a frictionless bead must slide under the influence of gravity to go from point a to point b in the shortest possible time. This is called the brachistochrone problem. Ingenious solutions were offered by Leibniz, Newton, and L'Hôpital, but the most elegant solution was that of Euler, who in 1744 introduced the calculus of variations as we know it today.

We can illustrate the basic idea with a simple example. For this purpose we look for the surface of revolution having the smallest area subject to the constraint that the surface have a prescribed radius at its ends. Physically, this surface corresponds to the shape assumed by a soap film suspended from two circular loops positioned at $x = a$ and $x = b$, as shown in Figure 2.1. If the radius of the surface at point x is $r(x)$, the total area of the surface between points a and b is

$$A = \int_a^b 2\pi r \, ds = 2\pi \int_a^b \sqrt{1 + r'^2} \, r \, dx \tag{2.2}$$

where ds is the incremental distance along the curve $r(x)$, and $r'(x) = dr/dx$. The problem consists in finding the function $r(x)$ for which the integral is a minimum for fixed values of the end points $[a, r(a)]$ and $[b, r(b)]$. The method we use is as follows.

Since it depends on the entire shape of the film $r(x)$, the total surface area A is called a functional $A[r]$ of the function $r(x)$. A functional $F[y]$ is an entity whose value depends not on a simple variable but rather on all the values of the function $y(x)$ over some range $a \le x \le b$. Generally we are interested in functionals of the form

$$F[y] = \int_a^b f(y, y', \ldots, x) \, dx \tag{2.3}$$

in which derivatives

$$y^{(n)}(x) = \frac{d^n y}{dx^n} \tag{2.4}$$

up to the Nth order may appear as arguments. We seek the function $y(x)$ for which the functional $F[y]$ is an extremum subject to the boundary conditions

$$\left[y(a), \ldots, y^{(N-1)}(a) \right] = \text{constants} \tag{2.5}$$

$$\left[y(b), \ldots, y^{(N-1)}(b) \right] = \text{constants} \tag{2.6}$$

Figure 2.1 Minimum area surface of revolution.

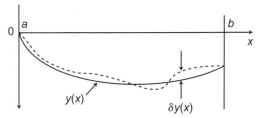

Figure 2.2 Variation of the argument of a functional.

as shown in Figure 2.2. To find the function $y(x)$ that makes $F[y]$ an extremum, we suppose that $y(x)$ is the function we seek and examine what happens if we replace $y(x)$ by the function $y(x) + \delta y(x)$, where $\delta y(x)$ is called the variation of $y(x)$. Because of the boundary conditions (2.5) and (2.6), we see that the variation $\delta y(x)$ is subject to the constraints

$$\delta y(a) = \cdots = \delta y^{(N-1)}(a) = 0 \qquad (2.7)$$
$$\delta y(b) = \cdots = \delta y^{(N-1)}(b) = 0 \qquad (2.8)$$

where the variation of the nth derivative is

$$\delta y^{(n)} = \delta \frac{d^n y}{dx^n} = \frac{d^n \delta y}{dx^n} \qquad (2.9)$$

Then to first order in the variation δy, the variation of the functional $F[y]$ is

$$\delta F[y] = \int_a^b [f(y + \delta y) - f(y)]\, dx = \int_a^b \left(\frac{\partial f}{\partial y}\, \delta y + \frac{\partial f}{\partial y'}\, \delta y' + \cdots \right) dx \qquad (2.10)$$

By analogy with ordinary calculus, we see that the condition for $F[y]$ to be an extremum is that to first order the variation must vanish; that is,

$$\delta F[y] = 0 \qquad (2.11)$$

To eliminate the derivatives $\delta y^{(n)}$ of the variation δy in (2.10), we integrate by parts as many times as necessary. When we do this we find that the variation $\delta F[y]$ is

$$\delta F = \frac{\partial f}{\partial y'}\, \delta y \bigg|_a^b + \cdots + \int_a^b \left[\frac{\partial f}{\partial y} - \frac{d}{dx}\left(\frac{\partial f}{\partial y'} \right) + \cdots \right] \delta y\, dx \qquad (2.12)$$

where it must be remembered that the total derivative with respect to x is

$$\frac{d}{dx}\left(\frac{\partial f}{\partial y'} \right) = \frac{\partial^2 f}{\partial y' \partial x} + \frac{\partial^2 f}{\partial y' \partial y}\frac{dy}{dx} + \frac{\partial^2 f}{\partial y'^2}\frac{dy'}{dx} + \cdots \qquad (2.13)$$

and so on. But the variations $\delta y(x)$ vanish at the end points, by (2.7) and (2.8), so we are left in (2.12) with

$$\delta F = \int_a^b \left[\frac{\partial f}{\partial y} - \frac{d}{dx}\left(\frac{\partial f}{\partial y'} \right) + \cdots \right] \delta y\, dx = 0 \qquad (2.14)$$

Since the variation $\delta y(x)$ is arbitrary over the rest of the interval $a < x < b$, the quantity in square brackets must vanish identically. This gives us the differential equation

$$\frac{d}{dx}\left(\frac{\partial f}{\partial y'}\right) + \cdots - \frac{\partial f}{\partial y} = 0 \tag{2.15}$$

which is called the Euler–Poisson equation. In mechanics this becomes the Euler–Lagrange equations of motion, and in electrodynamics it forms the Maxwell equations for the electric and magnetic fields.

To illustrate this general method with a specific example, we return to the original example of the soap film and seek the solution of the equation

$$\delta A[r] = 2\pi \delta \int_a^b \sqrt{1 + r'^2}\, r\, dx = 0 \tag{2.16}$$

subject to the constraints

$$\delta r(a) = \delta r(b) = 0 \tag{2.17}$$

Carrying out the computations as before and integrating by parts, we get

$$\delta A = 0 = 2\pi \int_a^b \left[\frac{rr'\delta r'}{\sqrt{1 + r'^2}} + \sqrt{1 + r'^2}\, \delta r\right] dx$$

$$= \frac{rr'\delta r}{\sqrt{1 + r'^2}}\bigg|_a^b - \int_a^b \left[\frac{d}{dx}\left(\frac{rr'}{\sqrt{1 + r'^2}}\right) - \sqrt{1 + r'^2}\right]\delta r\, dx \tag{2.18}$$

But the endpoint contributions must vanish because δr vanishes at the boundaries, so the integral must likewise vanish. It follows, then, that the quantity in square brackets must vanish because δr is arbitrary over the rest of the interval, which leaves us with the differential equation

$$\frac{d}{dx}\left(\frac{rr'}{\sqrt{1 + r'^2}}\right) - \sqrt{1 + r'^2} = 0 \tag{2.19}$$

The solution is

$$r = C_1 \cosh\left(\frac{x}{C_1} + C_2\right) \tag{2.20}$$

for some constants C_1 and C_2 that give the film the required radius at $x = a$ and $x = b$. This is the equation for a catenary, which is also the curve traced by a chain hanging in a gravitational field.

Returning to the problem with which we started at the beginning of this section, we see that in the notation of the calculus of variations, Hamilton's principle is stated

$$\delta S = \delta \int_a^b \mathcal{L}\, dt = 0 \tag{2.21}$$

where the variation of the trajectory is meant to be taken with the end points fixed. This is the fundamental law of classical physics.

EXERCISE 2.1

The brachistochrone problem, discussed earlier and illustrated in Figure 2.3, is still the archtype of all problems in the calculus of variations.

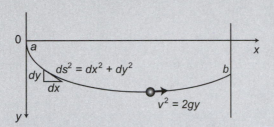

Figure 2.3 The brachistochrone problem.

(a) Show that the time for the bead to start from rest and move from point $[a, y(a)]$ to point $[b, y(b)]$ along the curve $y(x)$ (where $y \geq 0$ is the vertical distance the bead has descended from the starting point) is

$$T = \int_a^b \frac{ds}{v} = \frac{1}{\sqrt{2g}} \int_a^b \sqrt{\frac{1 + y'^2}{y}}\, dx \tag{2.22}$$

where v is the velocity of the bead, ds the increment of path length along the curve $y(x)$, and g the acceleration due to gravity.

(b) Show that the curve for which the total time is stationary with respect to arbitrary (small) variations $\delta y(x)$ satisfies the differential equation

$$\frac{d}{dx}\left[\frac{y'}{\sqrt{y(1 + y'^2)}}\right] + \sqrt{\frac{1 + y'^2}{4y^3}} = 0 \tag{2.23}$$

(c) As illustrated in Figure 2.4, a cycloid is the curve traced by point P on the rim of a circle as the circle rolls along the x axis. This curve can be represented in the parametric form

$$x = r(\theta - \sin\theta) \tag{2.24}$$

$$y = r(1 - \cos\theta) \tag{2.25}$$

where r is the radius of the circle and θ the angle through which the circle turns as it rolls. Show that this solution satisfies (2.23), which proves that the solution to the brachistochrone problem is a cycloid. The parameter r is chosen to make the curve pass through the point $[b, y(b)]$.

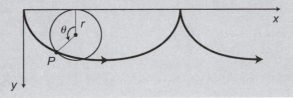

Figure 2.4 Geometric construction of a cycloid.

EXERCISE 2.2

In nonrelativistic mechanics the Lagrangian function for a particle of mass m at point $\mathbf{r}$ and time t in the potential $\Phi = \Phi(\mathbf{r}, t)$ is

$$\mathcal{L} = \tfrac{1}{2}mv^2 - \Phi \tag{2.26}$$

where $\mathbf{v} = d\mathbf{r}/dt$ is the velocity.

(a) Substitute this Lagrangian into (2.21) and compute the variation δS to show that the equation of motion of the particle is

$$m\frac{d\mathbf{v}}{dt} + \nabla\Phi = 0 \tag{2.27}$$

(b) Substitute the Lagrangian (2.26) into the Euler–Poisson equation (2.15) to derive the equation of motion (2.27) directly.

2.1.2 Lagrangian for a Free Particle

Up to this point we have not said anything about the physical system we are trying to describe, which may consist of matter, or fields, or both. In general, the constraints of relativity restrict material systems to either a continuous (fluid or elastic) medium or a collection of point particles that interact with each other by means of fields. We now recognize that matter is discontinuous, consisting of discrete particles (or strings, or . . .). However, rigid particles of nonvanishing extent cannot exist in relativistic physics, because for the particle to move as a rigid body, the information about the motion would have to be transmitted instantaneously to the entire structure, and this is impossible. Therefore, we turn our attention to the theory of structureless, point particles. Unfortunately, this introduces singularities into the theory, such as infinite energy in the field surrounding the point, and some other troubling issues. However, if we ignore these difficulties, considerable progress can be made. We return to these problems in Chapter 11. To satisfy the constraints of relativity, the particles cannot interact directly, but only by means of fields. Since information cannot be transmitted instantaneously from one particle to another, the influence of one particle must first be transmitted to the field around the particle, and then the field carries this influence through space to the other particle. Newton's third law (action equals reaction) and the conservation of mechanical momentum and energy, as discussed in Chapter 1, are no longer true. Momentum and energy lost by a particle are transferred to the field, which in turn has its own existence as it travels through space. Thus, in relativistic theory the fields are not merely a convenient way to describe the interaction between two particles, they are as real as the particles themselves.

As it stands, Hamilton's principle (2.21) is not expressed in a manifestly covariant form. Time is treated differently from the other coordinates. Nevertheless, the statement must be relativistically correct. That is, if $\delta S = 0$ for an observer in frame K, then it must be true for an observer in any other reference frame K'. Therefore, the action S must be a Lorentz scalar, as must the product $\mathcal{L}\,dt$. This puts enough constraints on the Lagrangian to determine its form with only minimal reference to experiment. We begin with the simplest system we can think of, a single free particle. In this case we may write (2.21) in the form

$$\delta S = \delta \int_a^b \mathcal{L} \, dt = \delta \int_a^b \gamma \mathcal{L} \, d\tau \qquad (2.28)$$

where τ is the proper time for the particle. Since τ and S are both scalars, we see that the product $\gamma \mathcal{L}$ must also be a scalar. But the Lagrangian for this simple system can depend only on the coordinates r^α. In fact, since space is homogeneous, the Lagrangian cannot depend explicitly on the coordinates themselves but only on the velocity. The only Lorentz scalar we can form from the 4-vector velocity $U^\alpha = \gamma(c, \mathbf{v})$ is $U^\alpha U_\alpha = c^2$, so we see that the Lagrangian must be simply

$$\mathcal{L} = K_1 \frac{c^2}{\gamma} \qquad (2.29)$$

where K_1 is a constant. We evaluate K_1 in Section 2.1.3.

Hamilton's principle for a free particle may now be stated in the form

$$\delta S = K_1 c^2 \delta \int_a^b \frac{dt}{\gamma} = K_1 c^2 \delta \int_a^b d\tau = 0 \qquad (2.30)$$

This means that for the action to be an extremum, the interval $\int_a^b d\tau$ must be an extremum. But the longest interval between two points in Minkowski space is a straight line, so this is the trajectory of a free particle. If we make $K_1 < 0$, the action is a minimum for this trajectory.

To show explicitly that the action is an extremum for a straight-line trajectory we appeal to the calculus of variations and proceed in the following way. Since $\mathbf{r}$ and t vary as the 4-vector trajectory is varied, except at the end points, and τ changes even at the end points, we introduce the variable $\xi_a < \xi < \xi_b$ to parametrize the trajectory. The end points ξ_a and ξ_b are arbitrary constants, such as 0 and 1, which are not varied as the trajectory is varied. The variation of the trajectory in Minkowski space (the world line) is then given by the functions $\delta r^\alpha(\xi)$, as indicated schematically in Figure 2.5, and the variation of the action is

$$\delta S = K_1 c^2 \delta \int_{\xi_a}^{\xi_b} \frac{d\tau}{d\xi} \, d\xi = K_1 c^2 \int_{\xi_a}^{\xi_b} \frac{d\delta\tau}{d\xi} \, d\xi \qquad (2.31)$$

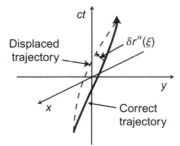

Figure 2.5 Variation of particle trajectory.

where the variation can be taken inside the integral because the end points ξ_a and ξ_b are fixed. From the definition of the interval

$$ds^2 = c^2 \, d\tau^2 = dr^\alpha \, dr_\alpha \tag{2.32}$$

we see that

$$\tfrac{1}{2}\delta(c^2 d\tau^2) = c^2 \, d\tau \, d\delta\tau = dr^\alpha d\delta r_\alpha \tag{2.33}$$

so

$$\frac{d\delta\tau}{d\xi} = \frac{1}{c^2}\frac{dr^\alpha}{d\tau}\frac{d\delta r^\alpha}{d\xi} = \frac{U^\alpha}{c^2}\frac{d\delta r_\alpha}{d\xi} \tag{2.34}$$

Substituting back into (2.31) and integrating once by parts to eliminate $d\delta r_\alpha/d\xi$, we arrive at the expression

$$\delta S = K_1 U^\alpha \delta r_\alpha \Big|_{\xi_a}^{\xi_b} - K_1 \int_{\xi_a}^{\xi_b} \frac{dU^\alpha}{d\xi} \delta r_\alpha \, d\xi = 0 \tag{2.35}$$

But $\delta r_\alpha = 0$ at the end points, so the first term vanishes. Moreover, between the end points δr_α is arbitrary, so the rest of the integrand must vanish identically. We therefore see that

$$\frac{dU^\alpha}{d\xi} = 0 \tag{2.36}$$

everywhere. That is, the trajectory of a free particle is a straight line, as was to be proved.

2.1.3 Energy and Momentum

We return to (2.35) for a moment and consider the case of a translation of the entire trajectory by the constant amount

$$\delta r_\alpha = \varepsilon_\alpha = \text{constant} \tag{2.37}$$

as illustrated in Figure 2.6. For a free particle the Lagrangian is independent of the coordinates, so the action is unchanged by the translation and $\delta S = 0$ for this variation of the trajectory. But the result of the translation remains a valid trajectory, so the integral in (2.35) still vanishes identically. However, the variation δr_α is no longer zero at the end points. Therefore, we see from (2.35) that

$$K_1 U^\alpha \varepsilon_\alpha \Big|_{\xi_a}^{\xi_b} = 0 = K_1 \varepsilon_\alpha [U^\alpha(\xi_b) - U^\alpha(\xi_a)] \tag{2.38}$$

Figure 2.6 Translation of a particle trajectory.

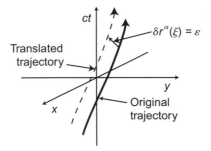

for arbitrary ε_α. Thus, the quantity U^α is conserved along the trajectory. We call those quantities that are conserved in a translationally (space and time) invariant system the momenta, so the 4-vector momentum must be U^α or, more correctly, something proportional to it. If we write

$$p^\alpha = mU^\alpha = \gamma m(c, \mathbf{v}) \tag{2.39}$$

then the spacelike components of the 4-vector momentum are

$$\mathbf{p} = \gamma m\mathbf{v} = \frac{m\mathbf{v}}{\sqrt{1 - \beta^2}} \xrightarrow[c \to \infty]{} m\mathbf{v} \tag{2.40}$$

Thus, if we identify the constant m with the mass of the particle, the spacelike components of the momentum 4-vector agree with the nonrelativistic momentum in the appropriate limit. The timelike component of the 4-vector momentum is

$$p^0 = \gamma mc = \frac{mc}{\sqrt{1 - \beta^2}} \xrightarrow[c \to \infty]{} \frac{1}{c}\left(mc^2 + \frac{1}{2}mv^2\right) \tag{2.41}$$

Thus, the energy forms the timelike component of the 4-vector momentum. For a particle at rest the energy is

$$\mathcal{E} = \gamma mc^2 \xrightarrow[v \to 0]{} mc^2 \tag{2.42}$$

which is arguably Einstein's most famous formula and possibly the most famous formula in all of physics. The complete momentum 4-vector is therefore

$$p^\alpha = \left(\frac{\mathcal{E}}{c}, \mathbf{p}\right) \tag{2.43}$$

Since the momentum is a 4-vector, its magnitude is an invariant. If we evaluate the magnitude of the momentum in the particle rest frame we get

$$p^\alpha p_\alpha = \frac{\mathcal{E}^2}{c^2} - \mathbf{p} \cdot \mathbf{p} = m^2c^2 \tag{2.44}$$

which may be rearranged to give

$$\mathcal{E}^2 = p^2c^2 + m^2c^4 \tag{2.45}$$

This is an extremely useful formula, for it gives the total energy (kinetic energy plus rest energy) of the particle in terms of its momentum. For a photon the rest mass vanishes, as far as we know, so the energy and momentum are related by

$$\mathcal{E}^2 = p^2c^2 \tag{2.46}$$

The 4-vector momentum is then

$$p^\alpha(\text{photon}) = \frac{\mathcal{E}(\text{photon})}{c}(1, \hat{\mathbf{n}}) \tag{2.47}$$

where $\hat{\mathbf{n}}$ is a unit vector in the direction of propagation of the photon.

To evaluate the constant K_1 in the Lagrangian, we consider a collection of noninteracting particles, for which the Lagrangian is

$$\mathcal{L} = \sum_i K_i \frac{c^2}{\gamma_i} \tag{2.48}$$

where the sum is over all the particles, $\gamma_i = 1/\sqrt{1 - \beta_i^2}$, and $\mathbf{v}_i$ is the velocity of the ith particle. In this case, if we follow through the arguments as before, we find that the conserved quantity is $\sum_i K_i U_i^\alpha$, where $U_i^\alpha = \gamma_i(c, \mathbf{v}_i) \xrightarrow[v/c \ll 1]{} (c, \mathbf{v}_i)$ is the 4-vector velocity of the ith particle. But we know that for a collection of particles in the nonrelativistic limit the conserved quantity is $\sum_i m_i \mathbf{v}_i$, where m_i is the mass of the ith particle, so it follows that $K_i = K m_i$, where K is some universal constant. We take this constant to be $K = -1$ to make the action a minimum along a straight-line trajectory, so the Lagrangian for a system of free particles is

$$\mathcal{L} = -\sum_i \frac{m_i c^2}{\gamma_i} \tag{2.49}$$

EXERCISE 2.3

A particle whose mass is m when it is in the ground state has an excited state that lies at an energy $\mathcal{E}^*$ above the ground state, where $\mathcal{E}^*$ is *not* small compared with mc^2. Thus, in the excited state the particle has the rest mass

$$m^* = m + \frac{\mathcal{E}^*}{c^2} \tag{2.50}$$

If a particle at rest in the ground state is struck by a gamma ray with 4-vector momentum

$$p^\alpha = \frac{\mathcal{E}_\gamma}{c} (1, \hat{\mathbf{n}}) \tag{2.51}$$

part of the energy goes into exciting the particle. The rest of the energy and all the momentum of the gamma ray go into the kinetic energy and momentum of the particle. Show that the gamma ray must have energy

$$\mathcal{E}_\gamma = \mathcal{E}^* \left(1 + \frac{\mathcal{E}^*}{2mc^2} \right) \tag{2.52}$$

to excite the particle.

EXERCISE 2.4

A particle with 4-momentum p^α approaches another particle moving with 4-velocity U_α. Show that the energy of the first particle moving relative to the second is

$$\mathcal{E} = U_\alpha p^\alpha \tag{2.53}$$

Hint: Start in the rest frame of one particle and generalize. Note that the energy $\mathcal{E}$ defined here is a Lorentz scalar, not the timelike component of a 4-vector.

EXERCISE 2.5

If we have two particles of mass m_1 and m_2, energy $\mathcal{E}_1$ and $\mathcal{E}_2$, and momentum $\mathbf{p}_1$ and $\mathbf{p}_2$, we may regard them as forming a composite particle for which the energy and momentum are

$$\mathcal{E} = \mathcal{E}_1 + \mathcal{E}_2 \tag{2.54}$$

and

$$\mathbf{p} = \mathbf{p}_1 + \mathbf{p}_2 \tag{2.55}$$

The magnitude of the 4-vector momentum of the composite particle is then

$$\mathcal{E}^2 = p^2 c^2 + m^2 c^4 \tag{2.56}$$

for some mass m. Show that m is an invariant, so that it may be regarded as the "mass" of the composite particle, and show that it has the value

$$m^2 = m_1^2 + m_2^2 + 2\left(\frac{\mathcal{E}_1 \mathcal{E}_2}{c^4} - \frac{\mathbf{p}_1 \cdot \mathbf{p}_2}{c^2}\right) \tag{2.57}$$

2.1.4 de Broglie Waves

As an example of the concept of a 4-vector momentum we consider the problem of de Broglie waves, the quantum-mechanical waves that describe the dynamics of material objects on a microscopic scale. Planck and Einstein had previously shown that light waves behave as particles with definite momentum and energy, and de Broglie argued by analogy that material particles might behave as waves. As we shall see, the analogy is rooted in the concept of the 4-vector momentum.

A nearly monochromatic wave packet is characterized by the wave vector

$$k^\alpha = \left(\frac{\omega}{c}, \mathbf{k}\right) \tag{2.58}$$

where for light waves the magnitude of the wave vector is

$$k^\alpha k_\alpha = \frac{\omega^2}{c^2} - k^2 = 0 \tag{2.59}$$

That is, the wave vector for light waves is a null vector. As shown in the Prologue, the electromagnetic field of the wave packet possesses an energy $\mathcal{E}$ and a momentum $\mathbf{p}$ that are related by the formula

$$\frac{\mathcal{E}}{c} = p \tag{2.60}$$

Together they form a 4-vector

$$p^\alpha = \left(\frac{\mathcal{E}}{c}, \mathbf{p}\right) \tag{2.61}$$

whose magnitude is

$$p^\alpha p_\alpha = \frac{\mathcal{E}^2}{c^2} - p^2 = 0 \tag{2.62}$$

That is, the 4-vector momentum of a light wave packet is also a null vector.

In 1900, to explain the spectrum of blackbody radiation, Planck proposed that the energy of light-wave packets is quantized into what we now call photons, each photon having the energy $\mathcal{E} = \hbar\omega$, where $\hbar$ is Planck's constant divided by 2π, and in 1905 Einstein used this idea to explain the photoelectric effect. From (2.59) and (2.62) it follows that the 3-vector momentum of a photon is $\mathbf{p} = \hbar\mathbf{k}$, so the complete 4-vector momentum is

$$p^\alpha = \hbar k^\alpha \tag{2.63}$$

In 1923 Compton siezed on the idea of a photon colliding with an electron and used the conservation of the 4-vector momentum to explain the wavelength shift (momentum shift) he observed in the scattering of x-rays.

In the same year, de Broglie proposed that if wave packets of light behave like particles, then material particles might also correspond to wave packets, and that the wave vector for material waves might be quantized by the same rule that applies to electromagnetic waves; that is,

$$p^\alpha = \hbar k^\alpha \tag{2.64}$$

However, for material particles the rest mass does not vanish, and we see from (2.44) that the magnitude of the momentum 4-vector for a de Broglie wave is

$$p^\alpha p_\alpha = \hbar^2 k^\alpha k_\alpha = \hbar^2 \left(\frac{\omega^2}{c^2} - k^2 \right) = m^2 c^2 \tag{2.65}$$

This dispersion relation is illustrated in Figure 2.7. The phase velocity of matter waves is

$$v_\phi = \frac{\omega}{k} = \frac{\mathcal{E}}{p} = c\sqrt{1 + \frac{m^2 c^2}{p^2}} > c \tag{2.66}$$

However, if we differentiate (2.65), we find from (2.61) that the group velocity is

$$v_g = \frac{d\omega}{dk} = \frac{c^2}{v_\phi} = c^2 \frac{k}{\omega} = c^2 \frac{p}{\mathcal{E}} = c^2 \frac{\gamma m v}{\gamma m c^2} = v \tag{2.67}$$

That is, the group velocity of the wave packet is just the particle velocity, as it should be. The fact that the phase velocity is greater than the speed of light is not a problem, since the phase of a matter wave is not observable. The group velocity, which is observable, is less than the speed of light and does not violate the postulates of relativity.

One of the first problems to which de Broglie applied his theory was that of an electron in a circular orbit of radius r around a nucleus. If we require that the circumference of

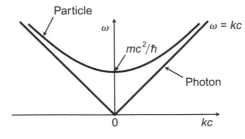

Figure 2.7 Dispersion relations for matter waves and light waves.

the orbit correspond to an integer number of wavelengths λ, then we get the quantization condition

$$2\pi r = n\lambda = n\frac{2\pi\hbar}{p} \tag{2.68}$$

or

$$l = rp = n\hbar \tag{2.69}$$

where $l = rp$ is the angular momentum. This is just the Bohr quantization condition, except that it is now justified by a more fundamental principle. The rest, as they say, is history.

EXERCISE 2.6

Consider the collision of a photon of wavelength λ and 4-vector momentum

$$p^\alpha = \hbar k^\alpha = \frac{h}{\lambda}(1, \hat{\mathbf{n}}) \tag{2.70}$$

where $h = 2\pi\hbar$ is Planck's constant, with an electron at rest, as shown in Figure 2.8. Show that if the photon is scattered into the angle θ, the wavelength λ' of the scattered photon is given by Compton's formula

$$\lambda' - \lambda = \lambda_C(1 - \cos\theta) \tag{2.71}$$

where $\lambda_C = h/mc = 2.426 \times 10^{-12}$ m is the Compton wavelength and m is the electron mass.

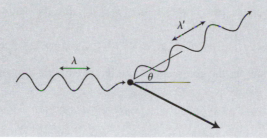

Figure 2.8 Scattering of a photon by an electron.

2.1.5 Rotational Invariance and Angular Momentum

Just as translational invariance is associated with linear momentum, rotational invariance is associated with angular momentum. The arguments are basically the same as those used in the previous sections. Before we begin, however, we need to establish some properties of rotations in 4-dimensional Minkowski space. For an infinitesimal rotation of the position of a particle, the increment in the coordinates is linear in the angle of rotation and in the coordinates. We may therefore write

$$\delta r^\alpha = \delta\omega^{\alpha\beta} r_\beta \tag{2.72}$$

where the coefficients $\delta\omega^{\alpha\beta}$ of the transformation are infinitesimals. The spatial components of $\delta\omega^\alpha{}_\beta$ correspond to rotations in space. The components involving the time correspond to

Lorentz boosts, which are pseudorotations in Minkowski space, as discussed in Chapter 1. In a rotation of the position of a particle, the length of the radius vector is preserved, so the increment in the coordinates must be orthogonal to the radius 4-vector. That is, $r_\alpha \delta r^\alpha = 0$. If we substitute (2.72) for δr^α, we get

$$\delta\omega^{\alpha\beta} r_\alpha r_\beta = 0 \tag{2.73}$$

Clearly, the contraction of a symmetric tensor with an antisymmetric tensor vanishes. Therefore, since $r_\alpha r_\beta = r_\beta r_\alpha$ is an arbitrary symmetric tensor, the 4-tensor $\delta\omega^{\alpha\beta}$ must be antisymmetric. We recognize, of course, that since $\delta\omega^{\alpha\beta}$ is the most general infinitesimal transformation that preserves lengths in Minkowski space, it must be related to the Lorentz transformation. In fact, the infinitesimal Lorentz transformation is

$$L^\alpha{}_\beta = \delta^\alpha{}_\beta - \delta\omega^\alpha{}_\beta \tag{2.74}$$

where $\delta^\alpha{}_\beta$ is the unit tensor. The negative sign appears because a rotation of the particle through $\delta\omega^\alpha{}_\beta$ corresponds to a rotation of the coordinate system in the opposite sense.

When the trajectory is varied as shown in Figure 2.9, the variation of the action is given by (2.35), as before. Substituting $\delta r_\alpha = \delta\omega_{\alpha\beta} r^\beta$, we now get

$$\delta S = -mU^\alpha r^\beta \delta\omega_{\alpha\beta}\Big|_{\xi_a}^{\xi_b} + m \int_{\xi_a}^{\xi_b} \frac{dU^\alpha}{d\xi}\, \delta r_\alpha \, d\xi = 0 \tag{2.75}$$

For a system that is rotationally invariant, the action is unaffected by this transformation, so $\delta S = 0$. But the trajectory remains valid, so the integral on the right still vanishes. We therefore find that

$$-mU^\alpha r^\beta \delta\omega_{\alpha\beta}\Big|_{\xi_a}^{\xi_b} = 0 \tag{2.76}$$

that is, $mU^\alpha r^\beta \delta\omega_{\alpha\beta}$ is conserved. If we separate $mU^\alpha r^\beta$ into its symmetric and antisymmetric parts, we get

$$mU^\alpha r^\beta = \tfrac{1}{2}(mU^\alpha r^\beta + mU^\beta r^\alpha) + \tfrac{1}{2}(mU^\alpha r^\beta - mU^\beta r^\alpha) \tag{2.77}$$

Since $\delta\omega_{\alpha\beta}$ is antisymmetric, the contraction with the symmetric part of $mU^\alpha r^\beta$ vanishes identically, and we see that

$$mU^\alpha r^\beta \delta\omega_{\alpha\beta} = \tfrac{1}{2} \delta\omega_{\alpha\beta}(mU^\alpha r^\beta - mU^\beta r^\alpha) \tag{2.78}$$

We call the conserved, antisymmetric, second-rank tensor quantity

$$M^{\alpha\beta} = m(r^\alpha U^\beta - r^\beta U^\alpha) = r^\alpha p^\beta - r^\beta p^\alpha \tag{2.79}$$

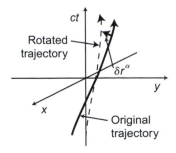

Figure 2.9 Rotation of a particle trajectory.

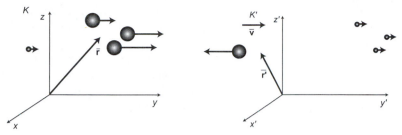

Figure 2.10 Motion of a collection of free particles.

the 4-vector angular momentum. This tensor has the components

$$
M^{\alpha\beta} =
\begin{bmatrix}
0 & \gamma mc(v_x t - x) & \gamma mc(v_y t - y) & \gamma mc(v_z t - z) \\
\gamma mc(x - v_x t) & 0 & l_z & -l_y \\
\gamma mc(y - v_y t) & -l_z & 0 & l_x \\
\gamma mc(z - v_z t) & l_y & -l_x & 0
\end{bmatrix}
\tag{2.80}
$$

where

$$
\mathbf{l} = \mathbf{r} \times \mathbf{p} \tag{2.81}
$$

is the relativistic generalization of the ordinary angular momentum. When the Lagrangian is invariant under a rotation in space, the angular momentum $\mathbf{l}$ is conserved.

For the collection of particles shown in Figure 2.10, the total 4-vector momentum is

$$
\bar{p}^\alpha = \sum_i p_i^\alpha = \sum_i \gamma_i m_i (c, \mathbf{v}_i) = \left(\frac{\bar{\mathcal{E}}}{c}, \bar{\mathbf{p}} \right) \tag{2.82}
$$

where

$$
\bar{\mathcal{E}} = \sum_i \gamma_i m_i c^2 \tag{2.83}
$$

and

$$
\bar{\mathbf{p}} = \sum_i \gamma_i m_i \mathbf{v}_i \tag{2.84}
$$

and the total 4-vector angular momentum is

$$
\bar{M}^{\alpha\beta} = \sum_i \left(r_i^\alpha p_i^\beta - r_i^\beta p_i^\alpha \right) \tag{2.85}
$$

where r_i^α and p_i^α are the position and momentum of the ith particle. Since in any given frame of reference $r_i^0 = ct$ is the same for all particles, the components in the top row of (2.85) are

$$
\bar{M}^{0\beta} = \left(0, \; \bar{\mathbf{p}} ct - \frac{\bar{\mathcal{E}}}{c} \bar{\mathbf{r}} \right) \tag{2.86}
$$

where we define the relativistic center of momentum $\bar{\mathbf{r}}$ of the system by

$$
\frac{\bar{\mathcal{E}}}{c^2} \bar{\mathbf{r}} = \sum_i \gamma_i m_i \mathbf{r}_i \tag{2.87}
$$

in analogy with the nonrelativistic definition of the center of mass. This is illustrated in Figure 2.10, where the size of the particles is scaled to indicate the weight $\gamma_i m_i$ given to each particle in (2.87). The components $\bar{M}^{0\beta}$ are conserved when the Lagrangian is invariant under a Lorentz boost. For example, a closed system of noninteracting particles is invariant under a Lorentz boost and the quantities $\bar{M}^{0\beta}$ are constants of the motion. Such a system is also translationally invariant, so the total linear momentum $\bar{p}^\alpha = (\bar{\mathcal{E}}/c, \bar{\mathbf{p}})$ is conserved and the center of momentum moves with uniform velocity $\bar{\mathbf{v}}$, where

$$\bar{\mathbf{r}} = \text{constant} + \frac{c^2}{\bar{\mathcal{E}}}\bar{\mathbf{p}}t = \text{constant} + \bar{\mathbf{v}}t \qquad (2.88)$$

In the nonrelativistic limit ($\gamma_i \to 1$) we see that

$$\bar{\mathbf{r}} = \frac{\sum_i \gamma_i m_i \mathbf{r}_i}{\sum_i \gamma_i m_i} \to \frac{\sum_i m_i \mathbf{r}_i}{\sum_i m_i} \qquad (2.89)$$

which corresponds to the usual definition of the center of mass. However, the components of $\bar{\mathbf{r}}$ are not the spacelike components of a 4-vector. Thus, if we perform a boost into another coordinate system, $\bar{\mathbf{r}}$ does not transform like the positions of the particles. This is related to the fact that the relativistic motion of the center of inertia (2.86) involves the time t. The center of inertia is defined by the particle positions at one specific time in some reference frame. In another reference frame, the time associated with each particle is different. That is, the idea of simultaneity does not carry over from one reference frame to another. In addition, if the particles are traveling at different velocities, the weight $\gamma_i m_i$ given each particle in (2.89) depends on the coordinate system chosen. This is indicated in Figure 2.10, where we see that in the center-of-momentum frame K', which moves at velocity $\bar{\mathbf{v}}$ relative to the laboratory frame, different particles receive the most weight.

In the center-of-momentum frame, the total momentum $\bar{\mathbf{p}}$ vanishes and the total energy of the system is $\bar{\mathcal{E}}'$. We call this the rest frame of the system. If we view the system of particles in its rest frame as a single composite particle with rest mass

$$\bar{m} = \frac{\bar{\mathcal{E}}'}{c^2} \qquad (2.90)$$

then the 4-vector momentum in any other frame is

$$\bar{p}^\alpha = \left(\frac{\bar{\mathcal{E}}}{c}, \bar{\mathbf{p}}\right) = \bar{\gamma}\bar{m}\,(c, \bar{\mathbf{v}}) = \bar{m}\bar{U}^\alpha \qquad (2.91)$$

where $\bar{\gamma} = 1/\sqrt{1 - \bar{\beta}^2}$, $\bar{\boldsymbol{\beta}} = \bar{\mathbf{v}}/c$, and $\bar{\mathbf{v}}$ is the velocity of the rest frame itself.

The 4-vector angular momentum defined by (2.85) is referred to the origin. Referred to another point R^μ in Minkowski space, the angular momentum is

$$\bar{M}^{\alpha\beta}(R^\mu) = \sum_i \left[(r_i^\alpha - R^\alpha)p_i^\beta - (r_i^\beta - R^\beta)p_i^\alpha\right] = \bar{M}^{\alpha\beta}(0) - (R^\alpha \bar{p}^\beta - R^\beta \bar{p}^\alpha) \qquad (2.92)$$

That is, the angular momentum about point R^μ is just the angular momentum about the origin less the angular momentum due to the total momentum acting about point R^μ. This is familiar from nonrelativistic mechanics. In the center-of-momentum frame the 4-vector total momentum is $\bar{p}^\alpha = (\bar{\mathcal{E}}/c,\ 0)$, so we see that in this reference frame

$$R^i \bar{p}^j - R^j \bar{p}^i = 0 \qquad (2.93)$$

identically for the space components ($i, j = 1 \ldots 3$). Therefore, in the center-of-momentum frame the total 3-vector angular momentum $\bar{\mathbf{l}}'$ is independent of the point of reference. We call the angular momentum in this frame the intrinsic angular momentum. If we view the collection of particles as a single composite particle, such as a nucleus, the total (intrinsic) angular momentum in the rest frame is called the spin $\mathbf{S}'$ of the particle. Referred to the origin, the total angular momentum tensor in the rest frame is

$$\bar{M}'^{\alpha\beta} = \begin{bmatrix} 0 & -\dfrac{\bar{\mathcal{E}}'}{c}\bar{x}' & -\dfrac{\bar{\mathcal{E}}'}{c}\bar{y}' & -\dfrac{\bar{\mathcal{E}}'}{c}\bar{z}' \\[2ex] \dfrac{\bar{\mathcal{E}}'}{c}\bar{x}' & 0 & S_z' & -S_y' \\[2ex] \dfrac{\bar{\mathcal{E}}'}{c}\bar{y}' & -S_z' & 0 & S_x' \\[2ex] \dfrac{\bar{\mathcal{E}}'}{c}\bar{z}' & S_y' & -S_x' & 0 \end{bmatrix} \qquad (2.94)$$

where $\bar{\mathbf{r}}$ is the position of the center of momentum in the rest frame.

In this context it is useful (as in the theory of electrons, for example) to introduce the Pauli–Lubanski vector

$$S^\alpha = -\frac{1}{2c}\varepsilon^{\alpha\beta\gamma\delta}\bar{U}_\beta \bar{M}_{\gamma\delta} \qquad (2.95)$$

which describes the intrinsic angular momentum, or spin, of a complex particle. In the rest frame (the center-of-momentum frame) the velocity is $\bar{U}'^\alpha = (c, 0)$, so if we carry out the multiplications, we find that the Pauli–Lubanski vector in the rest frame is just

$$S'^\alpha = -\frac{1}{2c}\varepsilon^{\alpha\beta\gamma\delta}\bar{U}'_\beta \bar{M}'_{\gamma\delta} = (0, \mathbf{S}') \qquad (2.96)$$

Clearly, S^α is a pseudovector because $\varepsilon^{\alpha\beta\gamma\delta}$ is a pseudotensor. This must be so, because the spin $\mathbf{S}$ is an axial vector. We see that the Pauli–Lubanski vector describes only the intrinsic angular momentum of a collection of particles or the spin of a composite particle. It discards information about the center-of-momentum motion. In other reference frames this is still true, although the timelike component of S^α vanishes only in the center-of-momentum frame (if it vanished in all frames the entire vector would vanish, by the zero-component lemma). The magnitude of the Pauli–Lubanski vector is

$$S^\alpha S_\alpha = S'^\alpha S_\alpha' = -\mathbf{S}' \cdot \mathbf{S}' \qquad (2.97)$$

in any coordinate system. This is just the magnitude of the intrinsic angular momentum, or spin, in the rest frame. Finally, we note that since it is true in the rest frame,

$$S^\alpha \bar{U}_\alpha = S'^\alpha \bar{U}'_\alpha = 0 \qquad (2.98)$$

in any reference frame. That is, the Pauli–Lubanski vector is orthogonal to the center-of-momentum velocity.

EXERCISE 2.7

Using (2.74) and the formulas in Chapter 1 for the Lorentz transformation corresponding to a boost $\delta\beta \ll 1$ in the $\hat{\mathbf{x}}$ direction, show that the infinitesimal pseudorotation is

$$\delta\omega^{\alpha\beta} = \begin{bmatrix} 0 & -\delta\beta & 0 & 0 \\ \delta\beta & 0 & 0 & 0 \\ 0 & 0 & 0 & 0 \\ 0 & 0 & 0 & 0 \end{bmatrix} \qquad (2.99)$$

in matrix form.

2.2 CHARGED PARTICLE IN A VECTOR POTENTIAL

2.2.1 Lagrangian Mechanics

In constructing Lagrangians to describe the interaction of a particle with a field we are guided by Dirac's advice to look for the simplest and most compact (most beautiful?) description of nature. The simplest potential we can conceive of is a Lorentz scalar function of the particle coordinates and time, and the theory of a 4-scalar potential is the subject of a series of exercises left to the reader. In the following sections we address the next simplest case, which is the Lagrangian for a 4-vector potential. This corresponds to the electromagnetic field.

We begin by hypothesizing that the particle interacts with a 4-vector potential

$$A^\alpha = \left(\frac{\Phi}{c}, \mathbf{A} \right) \qquad (2.100)$$

whose components we call the scalar potential Φ (not to be confused with a Lorentz scalar potential) and the vector potential $\mathbf{A}$. In general the potential A^α is the superposition of the externally applied field (which may consist of the fields of other particles) and the field of the particle itself. The field of the particle itself is, in turn, the superposition of what we call here the "self-field" of the particle and the radiation field. (In Chapter 10 we give these fields more precise definitions and call them the "near field" and the "far field.") Provided that the motions of the particle are not too violent, the near field looks much like the (Lorentz-transformed) Coulomb field of the particle, and the radiation field is small. We might, then, be inclined to ignore both these fields and consider only the effect of the externally applied field, and this is, in fact, precisely what we do. Well, almost. Actually, the Coulomb self-field of the particle contains energy, and when the particle is moving the field contains momentum. For present purposes we include the energy and momentum of the self-field in the energy and momentum of the particle by simply adjusting the mass of the particle and calling the result the "observed mass." Actually, the energy and momentum of the field of an electron do not transform covariantly (this is the so-called 4/3 problem), but we learn how to fix this in Chapter 11. Viewed another way, the self-field of an accelerating particle does not cancel out at the position of the particle, and the result is a retarding force on the particle when it accelerates. The effect is equivalent to an increase in the observed mass, as before. Finally, there is the problem of the energy and momentum of the radiation emitted by an accelerated particle, which must appear at the expense of the energy and momentum of the particle that radiates it. This is called the "radiation reaction." For present purposes we assume that this is small and ignore it. We therefore consider the motion of a point charge (with an appropriate mass) in an externally applied field. The problems introduced by these approximations are discussed in Chapter 11.

To find the Lagrangian for a particle interacting with the externally applied 4-vector potential (2.100), we start with the fact that the product $\gamma\mathcal{L}$, like the action S, must be a Lorentz scalar, as discussed previously. Since it cannot involve the coordinates explicitly, due to the homogeneity of space and time, the Lagrangian must be composed of the vectors A^α and U^α. The simplest scalars we can form from these are the products

$$U^\alpha U_\alpha = c^2 \tag{2.101}$$

$$A^\alpha U_\alpha = \gamma(\Phi - \mathbf{v} \cdot \mathbf{A}) \tag{2.102}$$

$$A^\alpha A_\alpha = \frac{\Phi^2}{c^2} - \mathbf{A} \cdot \mathbf{A} \tag{2.103}$$

We disregard the third product on two grounds. In the first place, the first two expressions are sufficient to provide an interaction between the particle and the field and we presume, a priori, that nature assumes the simplest form. In the second place, experiments indicate that the interaction is linear in the field. The simplest Lagrangian is therefore a linear combination of the first two products:

$$\gamma\mathcal{L} = K_1 U^\alpha U_\alpha + K_2 U^\alpha A_\alpha = K_1 c^2 + K_2 \gamma(\Phi - \mathbf{v} \cdot \mathbf{A}) \tag{2.104}$$

In the absence of the potential A^α, the particle becomes a free particle, so from the work of the previous sections we see that $K_1 = -m$, where m is the mass of the particle. We determine K_2 shortly.

As before, we introduce the independent variable ξ to parameterize the trajectory, so the action integral is

$$S = \int_a^b \gamma\mathcal{L}\,d\tau = -mc^2 \int_{\xi_a}^{\xi_b} \frac{d\tau}{d\xi}\,d\xi + K_2 \int_{\xi_a}^{\xi_b} A^\alpha \frac{dr_\alpha}{d\xi}\,d\xi \tag{2.105}$$

in which we have used $U_\alpha\,d\tau = dr_\alpha$. The variation of the action is therefore

$$\delta S = -m \int_{\xi_a}^{\xi_b} U^\alpha \frac{d\delta r_\alpha}{d\xi}\,d\xi + K_2 \int_{\xi_a}^{\xi_b} A^\alpha \frac{d\delta r_\alpha}{d\xi}\,d\xi + K_2 \int_{\xi_a}^{\xi_b} \delta A^\alpha \frac{dr_\alpha}{d\xi}\,d\xi \tag{2.106}$$

where we have substituted (2.34) in the first term. But $\delta A^\alpha = \partial^\beta A^\alpha \delta r_\beta$, so after integrating the first two terms by parts we get

$$\delta S = (-mU^\alpha + K_2 A^\alpha)\delta r_\alpha\Big|_a^b + m \int_{\xi_a}^{\xi_b} \delta r_\alpha \frac{dU^\alpha}{d\xi}\,d\xi$$

$$- K_2 \int_{\xi_a}^{\xi_b} \delta r_\alpha \frac{dA^\alpha}{d\xi}\,d\xi + K_2 \int_{\xi_a}^{\xi_b} \partial^\beta A^\alpha \delta r_\beta \frac{dr_\alpha}{d\xi}\,d\xi \tag{2.107}$$

Now that the variations of the trajectory have been completed, we are free to change the variable of integration from ξ to τ. Then, since $dA^\alpha = \partial^\beta A^\alpha dr_\beta$, we may write (2.107) in the form

$$\delta S = (-mU^\alpha + K_2 A^\alpha)\,\delta r_\alpha\big|_a^b + m\int_a^b \delta r_\alpha \frac{dU^\alpha}{d\tau}\,d\tau - K_2\int_a^b \delta r_\alpha (\partial^\beta A^\alpha - \partial^\alpha A^\beta)U_\beta\,d\tau$$
(2.108)

According to Hamilton's principle the first term vanishes because $\delta r_\alpha = 0$ at the end points, which leaves us with

$$\delta S = \int_a^b \delta r_\alpha\left[m\frac{dU^\alpha}{d\tau} - K_2(\partial^\beta A^\alpha - \partial^\alpha A^\beta)U_\beta\right]d\tau = 0$$
(2.109)

But the variation δr_α is arbitrary, so the rest of the integrand must vanish identically. We therefore obtain the equations of motion

$$m\frac{dU^\alpha}{d\tau} = -K_2(\partial^\alpha A^\beta - \partial^\beta A^\alpha)U_\beta = -K_2 F^{\alpha\beta}U_\beta$$
(2.110)

where $F^{\alpha\beta} = \partial^\alpha A^\beta - \partial^\beta A^\alpha$.

To describe a charged particle in an electromagnetic field, we identify $F^{\alpha\beta}$ with the antisymmetric field-strength tensor of Chapter 1. In the nonrelativistic limit, the spacelike components of this equation are then

$$m\frac{d\mathbf{v}}{dt} = -K_2(\mathbf{E} + \mathbf{v}\times\mathbf{B})$$
(2.111)

Comparing this with the known equation of motion of a nonrelativistic charged particle in an electromagnetic field, we make the identification

$$K_2 = -q$$
(2.112)

where q is the electric charge on the particle. The equation of motion of a relativistic charged particle is therefore

$$\frac{dp^\alpha}{d\tau} = q F^{\alpha\beta}U_\beta$$
(2.113)

where $p^\alpha = mU^\alpha$ is the 4-vector momentum.

The equations of motion may also be written in 3-vector notation. From the timelike component of (2.113) we get the energy equation

$$\frac{d\mathcal{E}}{dt} = q\mathbf{v}\cdot\mathbf{E} = mc^2\frac{d\gamma}{dt}$$
(2.114)

where $\mathcal{E} = \gamma mc^2$ is the total mechanical energy, including the rest energy. From the spacelike components of (2.113) we get the momentum equation

$$\frac{d\mathbf{p}}{dt} = q(\mathbf{E} + \mathbf{v}\times\mathbf{B}) = m\frac{d\gamma\mathbf{v}}{dt}$$
(2.115)

Note that (2.113) is written in terms of the proper time τ, whereas (2.114) and (2.115) are written in terms of the actual time t in some inertial reference frame.

If we define the force as the rate of change of the momentum, we get

$$\mathbf{F} = q(\mathbf{E} + \mathbf{v}\times\mathbf{B})$$
(2.116)

The force in this equation is just the ordinary Lorentz force, which we now see is relativistically correct. Note, however, that the Lorentz force $\mathbf{F}$ is not the spacelike part of a 4-vector. While the momentum $\mathbf{p}$ on the left-hand side of (2.115) is the spacelike part of a 4-vector, the time t is not a Lorentz scalar. This is also evident from (2.116) since while q is a good scalar, neither $\mathbf{E}$ nor $\mathbf{v} \times \mathbf{B}$ is the spacelike part of any 4-vector.

EXERCISE 2.8

As shown in Figure 2.11, calculate the relativistic corrections to a nearly circular orbit of the charge q in the Coulomb field

$$A^\alpha = \left(\frac{q_0}{4\pi \varepsilon_0 r}, 0 \right) \tag{2.117}$$

of the fixed charge q_0. Apply the result to an equivalent gravitational attraction and show that to lowest order in $\Delta \mathbf{r}$ the advance of the perihelion of Mercury is

$$\Delta \theta = 2\pi (\gamma_0 - 1) \approx \pi \frac{v_0^2}{c^2} \tag{2.118}$$

per orbit, where v_0 is the orbital velocity and $\gamma_0 = 1 / \sqrt{1 - v_0^2/c^2}$. The mean orbital velocity of Mercury is 47.9 km/s, and the period is 88 days. Compare your answer to the observed value of 43 arc seconds per century.

Figure 2.11 Relativistic perturbation of a nearly circular orbit.

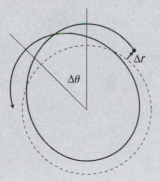

EXERCISE 2.9

We have used Hamilton's principle to derive the equations of motion for a particle moving in a 4-vector potential that we equate to the electromagnetic field. We may equally well postulate other fields and see if they correspond to any physical reality. For a particle moving in the *Lorentz scalar* potential ϕ, Hamilton's principle states that

$$\delta S = \delta \int_a^b \mathcal{L} \, dt = \delta \int_a^b \gamma \mathcal{L} \, d\tau = 0 \tag{2.119}$$

where the coordinates r^α at the end points of the trajectory are fixed in the variation. Since $d\tau$ is a scalar, a scalar action S can be formed by making $\gamma \mathcal{L}$ a scalar. We postulate that this scalar has the form

$$\gamma \mathcal{L} = K_1 U^\alpha U_\alpha + K_2 \phi = K_1 c^2 + K_2 \phi \qquad (2.120)$$

since this contains the only scalars that can be made up from the particle velocity and the field, up to lowest order in the field and the velocity. Use Hamilton's principle to show that a particle described by this Lagrangian satisfies the equations of motion

$$\frac{d}{d\tau}\left[\left(m + q_\phi \frac{\phi}{c^2}\right) U^\alpha\right] = q_\phi \partial^\alpha \phi \qquad (2.121)$$

where we define the "charge" interacting with this field by the formula

$$q_\phi = -K_2 \qquad (2.122)$$

Note that these equations of motion are not the same as the ones for an electric charge interacting with the scalar potential Φ because Φ is not a Lorentz scalar.

EXERCISE 2.10

Consider the electron gun in Figure 2.12. The electrons travel from the planar cathode to the planar anode a distance L away, and some of the beam is allowed to pass through the hole of radius $a \ll L$ into the field-free region on the right. In the vicinity of the hole, the field has radial components that deflect the electrons away from the axis.

(a) Use Gauss's law to develop an approximation for the radial components of the field near the axis in terms of the longitudinal field $E_z(z)$ and its derivatives on the axis.

(b) Assume that in the vicinity of the hole, the electrons follow nearly straight-line trajectories at constant velocity. Compute the change in the radial momentum as the electrons go through the hole to find the deflection of the electrons. Show that the electrons near the axis are defocused with a focal length

$$f = \frac{2\beta^2 \gamma}{\gamma - 1} L \approx 4L \qquad (2.123)$$

in the nonrelativistic limit, where βc is the final velocity and $\gamma m c^2$ the final energy of the electrons. Note that the focal length is independent of the electron charge, the electron mass, the radius of the hole, and (in the nonrelativistic limit) the electron final energy.

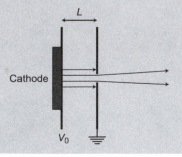

Figure 2.12 Electron gun.

2.2.2 Canonical Momentum

When an electromagnetic field is present, the concept of momentum can be generalized to a more powerful form. We return to (2.108) and consider a system that is invariant under the translation

$$\delta r_\alpha = \varepsilon_\alpha = \text{constant} \qquad (2.124)$$

Since the system is translationally invariant, $\delta S = 0$ for this variation. However, the new trajectory satisfies the equations of motion, so the integral still vanishes and we are left with

$$\delta S = (-mU^\alpha - qA^\alpha)\,\varepsilon_\alpha\big|_a^b = -(p^\alpha + qA^\alpha)\,\varepsilon_\alpha\big|_a^b = 0 \qquad (2.125)$$

We therefore find that the quantity

$$P^\alpha = p^\alpha + qA^\alpha \qquad (2.126)$$

is conserved along the trajectory in translationally invariant systems. We call this quantity the canonical momentum and denote it with a capital P^α.

The timelike component of the 4-vector canonical momentum is just the *total* energy

$$cP^0 = cp^0 + cqA^0 = \gamma mc^2 + q\Phi = \mathcal{E} + q\Phi \qquad (2.127)$$

where

$$\mathcal{E} = \gamma mc^2 = \sqrt{p^2 c^2 + m^2 c^4} \qquad (2.128)$$

is the mechanical, or *kinetic* energy (including the rest energy), and $q\Phi$ the *potential* energy. The *total* (kinetic plus potential) energy is conserved when the system is invariant under time translation, that is, when the potential has no explicit dependence on the time.

The spacelike components of the canonical momentum are

$$\mathbf{P} = \mathbf{p} + q\mathbf{\Lambda} \qquad (2.129)$$

where the first term is the kinetic momentum, as before, and the second is the vector potential. Thus, it is the canonical momentum, not the kinetic momentum, that is conserved in systems with translational symmetry.

As an example of a system with translational symmetry, we consider the motion of a charged particle in a plane electromagnetic wave, as shown in Figure 2.13. We see in Chapter 4 that in this case the vector potential is purely transverse and may be described by the expression

$$\mathbf{A} = \mathbf{A}_0 \exp(-ik^\alpha r_\alpha) \qquad (2.130)$$

Figure 2.13 Charged particle in a plane electromagnetic wave.

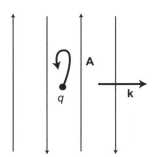

where $\mathbf{A}_0 \cdot \mathbf{k} = 0$. Since the system is translationally invariant in the directions transverse to the propagation of the wave, the transverse canonical momentum is conserved. For a particle whose initial canonical momentum is zero, the kinetic momentum transverse to the direction of propagation is therefore

$$\mathbf{p}_\perp = -q\mathbf{A} = -q\mathbf{A}_0 \exp(-ik^\alpha r_\alpha) \tag{2.131}$$

This is simpler than integrating the Lorentz-force equation (2.115), especially because the force depends on the position as well as the time.

EXERCISE 2.11

Show that the 4-vector canonical momentum of a particle in a Lorentz scalar potential ϕ is

$$P^\alpha = p^\alpha + \frac{q_\phi}{c^2}\phi U^\alpha = \left(m + \frac{q_\phi}{c^2}\phi\right)U^\alpha \tag{2.132}$$

where $p^\alpha = mU^\alpha$ is the 4-vector kinetic momentum, q_ϕ the charge, and U^α the 4-vector velocity of the particle.

EXERCISE 2.12

The concept of canonical momentum can be extended to include angular momentum by using the rotational symmetry arguments discussed earlier. For a system of particles, show that the 3-vector canonical angular momentum

$$\mathbf{L} = \sum_i \mathbf{r}_i \times \mathbf{P}_i = \sum_i \mathbf{r}_i \times \mathbf{p}_i + \sum_i \mathbf{r}_i \times \mathbf{A}_i = \mathbf{l} + \sum_i \mathbf{r}_i \times \mathbf{A}_i \tag{2.133}$$

is conserved in systems with rotational invariance, where $\mathbf{l} = \sum_i \mathbf{r}_i \times \mathbf{p}_i$ is the ordinary angular momentum.

2.2.3 Canonical Equations of Motion

In the Lagrangian formulation of mechanics, the variables in the Lagrangian are the coordinates r^α (or $\mathbf{r}$ and t) and the velocity $\mathbf{v} = d\mathbf{r}/dt$:

$$\mathcal{L} = -\frac{mc^2}{\gamma} - q(\Phi - \mathbf{v}\cdot\mathbf{A}) = \mathcal{L}(\mathbf{r}, \mathbf{v}, t) \tag{2.134}$$

The definition of the velocity $\mathbf{v} = d\mathbf{r}/dt$ is used explicitly in the variation of the trajectory to obtain the equations of motion (2.115) for the 3-vector momentum. The energy equation (2.114) is satisfied automatically if (2.115) is satisfied. But another approach, Hamiltonian mechanics, is possible. In the Hamiltonian formulation of mechanics, we change from the variables $\mathbf{r}$, $\mathbf{v}$, and t to the variables $\mathbf{r}$, $\mathbf{P}$, and t, and we give the coordinates and momenta equal standing. The space spanned by the coordinates $\mathbf{r}$ and the canonical momenta $\mathbf{P}$ is called the phase space of a system, and the motion of a system is described by a trajectory in phase space with t acting as a parameter along the trajectory. For example,

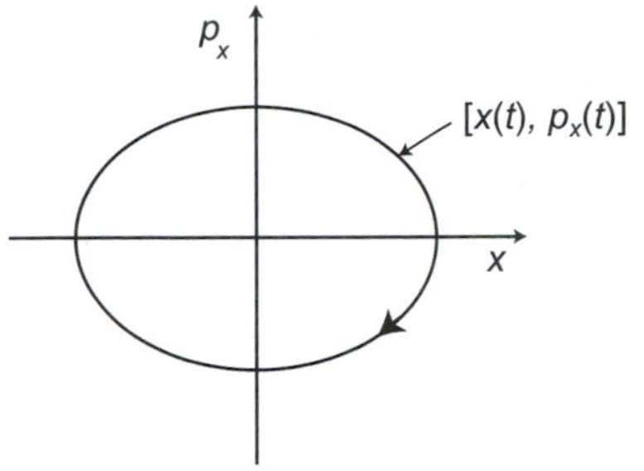

Figure 2.14 Phase plane of a harmonic oscillator.

the nonrelativistic motion of a one-dimensional harmonic oscillator is an ellipse in the x–p_x phase plane, as shown in Figure 2.14, with time acting as a parameter. To change variables from $(\mathbf{r}, \mathbf{v}, t)$ to $(\mathbf{r}, \mathbf{P}, t)$ we use what is called a Legendre transformation, and to make the new variables independent of one another, we carry out the variation of the trajectory separately for $\mathbf{r}$ and $\mathbf{P}$, ignoring the connection between position and momentum. The resulting equations of motion, called the canonical equations of motion, are more symmetric than the Lagrangian equations. However, there are now twice as many equations of motion, since each is a first-order differential equation instead of second order, and they are not manifestly covariant.

To begin, we must first define the canonical momentum $\mathbf{P}$. Since the canonical equations of motion are relativistically correct but not covariant, we can make the derivation of the canonical equations more general (that is, not specific to particles in electromagnetic fields) by introducing a more general, 3-vector definition of the canonical momentum using arguments similar to those we have used previously. This new definition is not covariant and defines only the spacelike components of the momentum, but it is relativistically correct and agrees with the previous definition for a particle in an electromagnetic field. For a Lagrangian $\mathcal{L}(\mathbf{r}, \mathbf{v}, t)$, the variation of the action is

$$\delta S = \int_a^b \left(\frac{\partial \mathcal{L}}{\partial \mathbf{r}} \cdot \delta \mathbf{r} + \frac{\partial \mathcal{L}}{\partial \mathbf{v}} \cdot \delta \mathbf{v} \right) dt \tag{2.135}$$

where we are now using t as the variable of integration and varying the 3-vector trajectory $\mathbf{r}(t)$. But

$$\delta \mathbf{v} = \frac{d \delta \mathbf{r}}{dt} \tag{2.136}$$

Substituting this into (2.135) and integrating once, by parts, in the usual fashion, we find that

$$\delta S = \frac{\partial \mathcal{L}}{\partial \mathbf{v}} \cdot \delta \mathbf{r} \Big|_a^b - \int_a^b \left(\frac{d}{dt} \frac{\partial \mathcal{L}}{\partial \mathbf{v}} - \frac{\partial \mathcal{L}}{\partial \mathbf{r}} \right) \cdot \delta \mathbf{r} \, dt = 0 \tag{2.137}$$

by Hamilton's principle. But the variation $\delta \mathbf{r}$ vanishes at the end points and is arbitrary in between. Therefore, the first term vanishes, and from the second term we obtain the equations of motion

$$\frac{d}{dt} \frac{\partial \mathcal{L}}{\partial \mathbf{v}} - \frac{\partial \mathcal{L}}{\partial \mathbf{r}} = 0 \tag{2.138}$$

These are known as the Euler–Lagrange equations, and they may be used to find the equations of motion for any Lagrangian $\mathcal{L}(\mathbf{r}, \mathbf{v}, t)$. The generalization to a system of many particles is straightforward.

Returning to (2.137), we consider a translation $\delta\mathbf{r} = \boldsymbol{\varepsilon}$ of the trajectory. For a system that is translationally invariant, we may use the arguments given previously to show that the quantity

$$\mathbf{P} = \frac{\partial\mathcal{L}}{\partial\mathbf{v}} \tag{2.139}$$

is conserved. We call this the canonical momentum. Although (2.139) defines only the 3-vector canonical momentum and is therefore not covariant, it is a simple matter to confirm that this definition agrees with (2.126) for the spacelike components of the 4-vector canonical momentum.

The change from the variables $\mathbf{r}$, $\mathbf{v}$, and t to the variables $\mathbf{r}$, $\mathbf{P}$, and t is accomplished by using a Legendre transformation. Legendre transformations are frequently used in thermodynamics to change from one set of variables to another. For example, a liquid or gas may be characterized by its pressure p, volume V, and temperature T. From the definition of the entropy S we see that the heat added to a system in a reversible process is $dQ = T\,dS$ and the work done by the system is $dW = p\,dV$. According to the first law of thermodynamics the change of the internal energy in a reversible process is

$$dU = dQ - dW = T\,dS - p\,dV \tag{2.140}$$

This shows that the internal energy does not change in a process that occurs at constant entropy and volume. Therefore, the internal energy U is a function of the variables S and V. For processes such as a change of state that take place at constant temperature and pressure, however, the Gibbs function is more useful. The Gibbs function is related to the internal energy by the Legendre transformation

$$G = U + pV - TS \tag{2.141}$$

Since

$$dG = dU + p\,dV + V\,dp - T\,dS - S\,dT = V\,dp - S\,dT \tag{2.142}$$

we see that the Gibbs function is a function of the variables p and T. For example, if we boil water at constant pressure, the temperature remains constant as we add heat, and the steam expands at constant pressure. In this case the internal energy of the system increases, but the Gibbs function, which is a function of the variables p and T, remains constant as the water changes to steam.

Returning to the equations of motion of a particle in a 4-vector potential, we see that to change the variables of the system from $\mathbf{r}$, $\mathbf{v}$, and t to $\mathbf{r}$, $\mathbf{P}$, and t, we use a Legendre transformation and introduce the Hamiltonian function

$$\mathcal{H} = \mathbf{P} \cdot \mathbf{v} - \mathcal{L} \tag{2.143}$$

Using the definition of the canonical momentum (2.139), we find that

$$d\mathcal{H} = \mathbf{P} \cdot d\mathbf{v} + \mathbf{v} \cdot d\mathbf{P} - d\mathcal{L} = \mathbf{v} \cdot d\mathbf{P} - \frac{\partial\mathcal{L}}{\partial\mathbf{r}} \cdot d\mathbf{r} - \frac{\partial\mathcal{L}}{\partial t}\,dt \tag{2.144}$$

Evidently, the Hamiltonian is the desired function of $\mathbf{r}$, $\mathbf{P}$, and t.

To derive the canonical equations of motion, we begin, as before, with Hamilton's principle,

$$\delta S = \delta \int_a^b \mathcal{L} \, dt = \delta \int_a^b (\mathbf{P} \cdot \mathbf{v} - \mathcal{H}) \, dt = 0 \tag{2.145}$$

but this time we use the time t as the parameter of the trajectory and vary only the space components $[\mathbf{r}(t), \mathbf{P}(t)]$ of the trajectory. Since the end points are fixed, we can take the variation inside the integral and get

$$\delta S = \int_a^b \delta \mathbf{P} \cdot \frac{d\mathbf{r}}{dt} \, dt + \int_a^b \mathbf{P} \cdot \frac{d\delta \mathbf{r}}{dt} \, dt - \int_a^b \delta \mathcal{H} \, dt = 0 \tag{2.146}$$

But the variation of the Hamiltonian is

$$\delta \mathcal{H} = \frac{\partial \mathcal{H}}{\partial \mathbf{r}} \cdot \delta \mathbf{r} + \frac{\partial \mathcal{H}}{\partial \mathbf{P}} \cdot \delta \mathbf{P} \tag{2.147}$$

If we substitute this into (2.146) and integrate the second term by parts, remembering that the variation $\delta \mathbf{r}$ vanishes at the end points, we find that the variation of the action is given by two terms,

$$\int_a^b \left(\frac{d\mathbf{r}}{dt} - \frac{\partial \mathcal{H}}{\partial \mathbf{P}} \right) \cdot \delta \mathbf{P} \, dt - \int_a^b \left(\frac{d\mathbf{P}}{dt} + \frac{\partial \mathcal{H}}{\partial \mathbf{r}} \right) \cdot \delta \mathbf{r} \, dt = 0 \tag{2.148}$$

Now, in the Hamiltonian formulation the coordinates $\mathbf{r}$ and $\mathbf{P}$ of phase space are given equal standing, so the variations $\delta \mathbf{P}$ and $\delta \mathbf{r}$ are individually arbitrary. Therefore, the quantities in parentheses must individually vanish, and we arrive at the canonical equations of motion

$$\frac{d\mathbf{r}}{dt} = \frac{\partial \mathcal{H}}{\partial \mathbf{P}} \tag{2.149}$$

$$\frac{d\mathbf{P}}{dt} = -\frac{\partial \mathcal{H}}{\partial \mathbf{r}} \tag{2.150}$$

We also see that the total time derivative of the Hamiltonian is

$$\frac{d\mathcal{H}}{dt} = \frac{\partial \mathcal{H}}{\partial \mathbf{r}} \cdot \frac{d\mathbf{r}}{dt} + \frac{\partial \mathcal{H}}{\partial \mathbf{P}} \cdot \frac{d\mathbf{P}}{dt} + \frac{\partial \mathcal{H}}{\partial t} = \frac{\partial \mathcal{H}}{\partial t} \tag{2.151}$$

Therefore, unless the Hamiltonian is explicitly time dependent, it is a constant of the motion. Since it is conserved in a time-invariant system, the Hamiltonian must be the total energy, or something proportional to it. For electromagnetic fields, we can see this explicitly. The Lagrangian is given by (2.134) and the canonical momentum by (2.129), so we compute

$$\mathcal{H} = \mathbf{P} \cdot \mathbf{v} - \mathcal{L} = (\gamma m \mathbf{v} + q\mathbf{A}) \cdot \mathbf{v} + \frac{mc^2}{\gamma} + q(\Phi - \mathbf{v} \cdot \mathbf{A}) = \gamma mc^2 + q\Phi \tag{2.152}$$

Comparing this with (2.127), we see that $\mathcal{H} = cP^0$, which is the total energy of the particle. Note that this makes the Hamiltonian the timelike component of a 4-vector, not a scalar. To express the Hamiltonian in terms of the coordinates and the canonical momentum, we use (2.127) and (2.128) to write

$$\mathcal{H} = cP^0 = \sqrt{(\mathbf{P} - q\mathbf{A})^2 c^2 + m^2 c^4} + q\Phi \qquad (2.153)$$

While they are not covariant, the Euler–Lagrange equations (2.138), the definition of the canonical momentum (2.139), and the canonical equations of motion (2.149) and (2.150) are relativistically correct and quite general. That is, they are valid not only for particles in an electromagnetic field but for other systems as well. Note carefully, however, that when using the canonical equations of motion, it is important that the Hamiltonian be expressed in terms of the coordinates and the canonical momenta, as in (2.153), not in terms of the velocities or the ordinary momenta.

EXERCISE 2.13

A particle with charge q and mass m moves in the potential $\Phi = -Ez$, which corresponds to the uniform electric field $\mathbf{E} = E\hat{\mathbf{z}}$, where E is a constant. If the particle has initial momentum $\mathbf{p}_0$ in an arbitrary direction, how long does it take for the particle to travel a distance L in the direction of the field? Use Hamiltonian mechanics and the conservation of energy.

EXERCISE 2.14

A uniform magnetic field $\mathbf{B} = B\hat{\mathbf{z}}$ parallel to the z axis is described by the 4-vector potential

$$A^\alpha = \left(0, \tfrac{1}{2}\mathbf{B} \times \mathbf{r}\right) \qquad (2.154)$$

which consists of circles as shown in Figure 2.15. A particle of mass m and charge q moves in the x–y plane.

  (a) Expressed in terms of the canonical coordinates and momenta, what is the Hamiltonian?
  (b) What are the canonical equations of motion for P_x, P_y, x, and y?
  (c) Show that the motion in the x–y plane is described by the equation

$$\frac{d\mathbf{p}}{dt} = \frac{q}{\gamma m}\mathbf{p} \times \mathbf{B} \qquad (2.155)$$

which describes a circle in the p_x–p_y plane.
  (d) What is the cyclotron frequency?

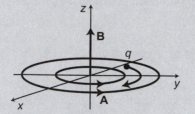

Figure 2.15 Charged particle in a uniform magnetic field.

EXERCISE 2.15

Consider a particle moving in a Lorentz scalar potential ϕ, as discussed in earlier exercises.

(a) Show that the Hamiltonian expressed in terms of $\mathbf{r}$ and $\mathbf{v}$ (through the function γ) has the form

$$\mathcal{H} = \gamma\left(m + \frac{q_\phi}{c^2}\phi\right)v^2 - \mathcal{L} = \gamma\left(mc^2 + q_\phi\phi\right) \tag{2.156}$$

and show that this is also the particle energy, that is, the timelike component of the 4-vector canonical momentum.

(b) Show that the Hamiltonian expressed in terms of its canonical variables r^α and P^α is

$$\mathcal{H} = \sqrt{\mathbf{P}\cdot\mathbf{P}c^2 + (mc^2 + q_\phi\phi)^2} \tag{2.157}$$

where the canonical momentum is

$$P^\alpha = p^\alpha + \frac{q_\phi}{c^2}\phi U^\alpha = \left(m + \frac{q_\phi}{c^2}\phi\right)U^\alpha \tag{2.158}$$

(c) From this show that the canonical equations of motion are given by the expressions

$$\frac{d\mathbf{r}}{dt} = \frac{\partial \mathcal{H}}{\partial \mathbf{P}} = \mathbf{v} \tag{2.159}$$

and

$$\frac{d\mathbf{P}}{dt} = -\frac{\partial \mathcal{H}}{\partial \mathbf{r}} = -\frac{q_\phi}{\gamma}\frac{\partial \phi}{\partial \mathbf{r}} \tag{2.160}$$

EXERCISE 2.16

Beginning with the Euler–Lagrange equations, show that the equations of motion of a particle in a 4-vector potential are given by (2.113).

2.3 THE MAXWELL EQUATIONS

2.3.1 Equations of Motion of a Vector Field

In the preceding discussion we have used Hamilton's principle to derive the relativistic equations of motion of a particle interacting with a 4-vector potential. Here we see how this same principle can be used to derive the equations of motion of the field itself. As before, we proceed by hypothesizing the existence of a Lagrangian and then determining its form from the principles of relativistic covariance.

For a system of particles, the total Lagrangian is the sum of the Lagrangians of the individual particles, $\mathcal{L} = \sum_i \mathcal{L}_i$. By extension, if the particles are replaced by a continuous distribution of matter with a mass density $\mu(\mathbf{r}, t)$ and a charge density $\rho(\mathbf{r}, t)$, the sum becomes an integral over all space of some Lagrangian density $\mathcal{D}(\mathbf{r}, t)$ that we must determine:

$$L(t) = \int \mathcal{D}(\mathbf{r}, t) \, d^3\mathbf{r} \tag{2.161}$$

In this expression the coordinate $\mathbf{r}$ plays a new role. It no longer describes the position of an identifiable particle. Instead, $\mathbf{r}$ has become an accounting device: it is the dummy variable used to integrate over all space. The "trajectory" of the system is replaced by the velocity field $\mathbf{v}(\mathbf{r}, t)$ of the material filling all space. When we vary the "trajectory" of the system, we do not vary $\mathbf{r}$ or, for that matter, t. Instead, we vary the velocity field itself throughout space. The same is true for fields of other types, such as the electromagnetic field. The "trajectory" of the field is the value of the field at each point in space, and the Lagrangian is the integral (2.161) of the Lagrangian density over all space at the time t.

In terms of the Lagrangian density $\mathcal{D}$, the action is then

$$S = \int_a^b L \, dt = \int_a^b \int \mathcal{D} \, d^3\mathbf{r} \, dt = \frac{1}{c} \int_a^b \mathcal{D} \, d^4 r^\gamma \tag{2.162}$$

where $d^4 r^\gamma = c \, dt \, dV$ is the differential volume in Minkowski space and the integral $\int_a^b d^4 r^\gamma$ extends over all of Minkowski space between the hypersurfaces $r^0 = ct_a$ and $r^0 = ct_b$. But the differential volume $d^4 r^\gamma$ is invariant under Lorentz transformations, as discussed in Chapter 1. Therefore, since the action S is a scalar, the Lagrangian density $\mathcal{D}$ must also be a scalar. This is enough to determine its form.

We consider the case when the field is a 4-vector potential A^α in the presence of a 4-vector source J^α. Physically, we identify A^α with the electromagnetic 4-vector potential and J^α with the 4-vector electric current density, but this matters only at the end, when we evaluate the constants that appear in the Lagrangian. The Lagrangian density for the field can depend on A^α and, in principle, all its derivatives. In fact, it is enough to include only the first derivatives $\partial^\alpha A^\beta$. To determine the functional form of $\mathcal{D}$ we use the fact that $\mathcal{D}$ must be a scalar function of A^α and $\partial^\alpha A^\beta$. If, for the moment, we keep only terms up to second order in A^α, we see from the discussion in Chapter 1 that the only possible scalars we can form from A^α and $\partial^\alpha A^\beta$ are

$$A^\alpha A_\alpha = \frac{\Phi^2}{c^2} - \mathbf{A} \cdot \mathbf{A} \tag{2.163}$$

$$F^{\alpha\beta} F_{\alpha\beta} = 2\left(B^2 - \frac{E^2}{c^2} \right) \tag{2.164}$$

and

$$\mathcal{F}^{\alpha\beta} F_{\alpha\beta} = -\frac{4}{c} \mathbf{E} \cdot \mathbf{B} \tag{2.165}$$

where $F^{\alpha\beta} = \partial^\alpha A^\beta - \partial^\beta A^\alpha$ is the field-strength tensor and $\mathcal{F}^{\alpha\beta}$ is its dual tensor. We disqualify the last expression, since it is actually a pseudoscalar and has the wrong symmetry under inversion. Thus, it cannot appear linearly in the same expression with true scalar quantities. The simplest scalar we can form from the field and the current density is the product

$$A_\alpha J^\alpha = \Phi\rho - \mathbf{A} \cdot \mathbf{J} \tag{2.166}$$

We therefore argue that the Lagrangian density must have the form

$$\mathcal{D} = K_1\left(\tfrac{1}{2}F^{\alpha\beta}F_{\alpha\beta} - \mu_\gamma^2 A^\alpha A_\alpha\right) + K_2 A_\alpha J^\alpha \tag{2.167}$$

for some constants K_1, K_2, and μ_γ. The action for a 4-vector field is then

$$S = \int\left[\frac{K_1}{c}\left(\frac{1}{2}F^{\alpha\beta}F_{\alpha\beta} - \mu_\gamma^2 A^\alpha A_\alpha\right) + \frac{K_2}{c}A_\alpha J^\alpha\right] d^4r^\gamma \tag{2.168}$$

To find the equations of motion for the vector field A^α, we invoke Hamilton's principle in the form

$$\delta S = \delta\frac{1}{c}\int_a^b \mathcal{D}\,d^4r^\gamma = \frac{1}{c}\int_a^b \delta\mathcal{D}\,d^4r^\gamma = 0 \tag{2.169}$$

We can take the variation inside the integral because the limits of integration for r^0 are fixed and the integral over $\mathbf{r}$ extends to infinity, where the Lagrangian density is presumed to vanish. In this variation, the current density J^α is fixed and the configuration $A_\alpha(\mathbf{r}, t)$ of the system is fixed at the end points t_a and t_b, as it is in the case of particles moving along a trajectory. Therefore, the variation δA_α of the field vanishes on the hypersurfaces $r^0 = ct_a$ and $r^0 = ct_b$. The variation of the Lagrangian density is

$$\delta\mathcal{D} = K_1\left(F^{\alpha\beta}\delta F_{\alpha\beta} - 2\mu_\gamma^2 A^\alpha \delta A_\alpha\right) + K_2 J^\alpha \delta A_\alpha$$
$$= -2K_1\left(F^{\alpha\beta}\partial_\beta\delta A_\alpha + \mu_\gamma^2 A^\alpha \delta A_\alpha\right) + K_2 J^\alpha \delta A_\alpha \tag{2.170}$$

since $F^{\alpha\beta}$ is antisymmetric. The variation of the action is therefore

$$\delta S = \int_a^b\left[-\frac{2K_1}{c}\left(F^{\alpha\beta}\partial_\beta\delta A_\alpha + \mu_\gamma^2 A^\alpha \delta A_\alpha\right) + \frac{K_2}{c}J^\alpha \delta A_\alpha\right] d^4r^\gamma = 0 \tag{2.171}$$

To convert the factor $\partial_\beta\delta A_\alpha$ to a more manageable form, we integrate the first term by parts, as we always do in the calculus of variations, or equivalently we use the 4-vector form of the divergence theorem to convert the volume integral to a surface integral. For this purpose we write

$$F^{\alpha\beta}\partial_\beta\delta A_\alpha = \partial_\beta(F^{\alpha\beta}\delta A_\alpha) - \delta A_\alpha\partial_b F^{\alpha\beta} \tag{2.172}$$

Using the divergence theorem for the first term, we get

$$\int F^{\alpha\beta}\partial_\beta\delta A_\alpha\,d^4r^\gamma = \oint_S F^{\alpha\beta}\delta A_\alpha\hat{n}_\beta\,d\sigma - \int \delta A_\alpha\partial_b F^{\alpha\beta}\,d^4r^\gamma \tag{2.173}$$

where the first integral on the right-hand side is over the four-dimensional hypersurface S enclosing all of Minkowski space between t_a and t_b, as indicated in Figure 2.16, $\hat{n}_\beta = (0, \hat{\mathbf{n}})$ is a unit vector normal to the hypersurface, and $d\sigma$ an infinitesimal area on that hypersurface. But $F^{\alpha\beta}$ vanishes at infinity, so the only contribution to the surface integral comes from the integrals over the hypersurfaces $r^0 = ct_a$ and $r^0 = ct_b$, which are integrals over all space, and we are left with

$$\int_a^b F^{\alpha\beta}\partial_\beta\delta A_\alpha\,d^4r^\gamma = \int F^{\alpha 0}\delta A_\alpha\,d^3\mathbf{r}\,\bigg|_{t_a}^{t_b} - \int_a^b \partial_\beta F^{\alpha\beta}\,\delta A_\alpha\,d^4r^\gamma \tag{2.174}$$

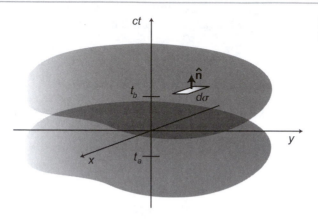

Figure 2.16 Integral over the hypersurfaces S.

Substituting back into (2.171) we get

$$\delta S = \int F^{\alpha 0}\,\delta A_\alpha\,d^3\mathbf{r}\Big|_{t_a}^{t_b} + \int \left[\frac{2K_1}{c}\left(\partial_\beta F^{\alpha\beta} - \mu_\gamma^2 A^\alpha\right) + \frac{K_2}{c}J^\alpha\right]\delta A_\alpha\,d^4 r^\gamma = 0 \quad (2.175)$$

But the variation δA_α vanishes at the end points t_a and t_b, so the first integral vanishes, and it is arbitrary everywhere else, so the expression in brackets must vanish identically. Then, since $F^{\alpha\beta} = -F^{\beta\alpha}$, we obtain the equations of motion

$$\partial_\beta F^{\beta\alpha} + \mu_\gamma^2 A^\alpha = \frac{K_2}{2K_1}J^\alpha \quad (2.176)$$

These are the simplest equations possible for a relativistic 4-vector potential field with a 4-vector source. Except for the term involving μ_γ^2 they are equivalent to the inhomogeneous Maxwell equations,

$$\partial_\beta F^{\beta\alpha} = \mu_0 J^\alpha \quad (2.177)$$

where μ_0 is the permeability of free space in SI units. Therefore, if they are used to describe the electromagnetic field, we see that the ratio of the constants in the Lagrangian must be

$$\frac{K_2}{2K_1} = \mu_0 = \frac{1}{\varepsilon_0 c^2} \quad (2.178)$$

where ε_0 is the permittivity of free space. We discuss the significance of μ_γ as the mass of a photon in Section 2.3.2, but for now we observe that if we take the 4-vector divergence of (2.176), we get

$$\partial_\alpha\partial_\beta F^{\beta\alpha} + \mu_\gamma^2\partial_\alpha A^\alpha = \mu_0\partial_\alpha J^\alpha \quad (2.179)$$

The first term vanishes because $\partial_\alpha\partial_\beta$ is symmetric in α and β, while $F^{\alpha\beta}$ is antisymmetric. Thus, if the photon mass μ_γ vanishes, we see that charge conservation is implicit in the Maxwell equations. If not, then only the combination $\mu_0 J^\alpha - \mu_\gamma^2 A^\alpha$ is conserved, and this depends on the gauge used to define the potential.

To complete the discussion, we note that the 4-vector curl of the field-strength tensor vanishes identically, for we see from the definition of $F^{\alpha\beta}$ that

$$\varepsilon_{\alpha\beta\gamma\delta}\partial^\beta F^{\gamma\delta} = \varepsilon_{\alpha\beta\gamma\delta}\partial^\beta\partial^\gamma A^\delta - \varepsilon_{\alpha\beta\gamma\delta}\partial^\beta\partial^\delta A^\gamma \quad (2.180)$$

But the order of differentiation is immaterial, so if, in the first term, we permute β and γ and then rename the dummy indices, we find that $\varepsilon_{\alpha\beta\gamma\delta}\partial^\beta\partial^\gamma A^\delta = -\varepsilon_{\alpha\beta\gamma\delta}\partial^\beta\partial^\gamma A^\delta$. Therefore, this term must vanish identically. The second term is handled in the same way, and we see that

$$\tfrac{1}{2}\varepsilon_{\alpha\beta\gamma\delta}\partial^\beta F^{\gamma\delta} = 0 = \partial^\beta \mathcal{F}_{\beta\alpha} \qquad (2.181)$$

identically. As shown in Chapter 1, (2.181) corresponds to the homogeneous Maxwell equations.

EXERCISE 2.17

In the derivation of the Maxwell equations we restrict ourselves to terms in the Lagrangian density that are at most quadratic in the fields. This ensures that the Maxwell equations are linear, so the fields satisfy the principle of superposition. However, we may construct a relativistically covariant theory by choosing for the Lagrangian density any function of the Lorentz scalars (2.163)–(2.165).

(a) For a Lagrangian density of the form

$$\mathcal{D} = -\frac{1}{4\mu_0} D(F^{\alpha\beta} F_{\alpha\beta}) - K A_\alpha J^\alpha \qquad (2.182)$$

where K is a constant and $D(x)$ is some function of x, use Hamilton's principle to show that the equations of motion of the field are

$$\partial_\alpha[D'(F^{\mu\nu} F_{\mu\nu})F^{\alpha\beta}] = \mu_0 K J^\beta \qquad (2.183)$$

where $D'(x) = dD/dx$. Thus, $D'/\mu_0 c^2 K$ plays the role of a field-strength-dependent permittivity of free space. To recover the Maxwell equations in the limit of weak fields, we require that

$$D'(x) \xrightarrow[x\to 0]{} K \qquad (2.184)$$

(b) For a point charge q at the origin, show that the field is

$$E = \frac{q}{4\pi\varepsilon_0 r^2} \frac{K}{D'(F^{\alpha\beta} F_{\alpha\beta})} \qquad (2.185)$$

(c) Born and Infeld have proposed the function

$$D(F^{\alpha\beta} F_{\alpha\beta}) = 4K\frac{E_0^2}{c^2}\left(\sqrt{1 + \frac{c^2}{2E_0^2} F^{\alpha\beta} F_{\alpha\beta}} - 1\right) \qquad (2.186)$$

where E_0 is a constant. Show that the electric field of a point charge q is

$$E = \frac{q}{4\pi\varepsilon_0 r^2} \frac{1}{\sqrt{1 + \left(\dfrac{q}{4\pi\varepsilon_0 r^2 E_0}\right)^2}} \qquad (2.187)$$

This has the finite limit

$$E \xrightarrow[r \to 0]{} E_0 \qquad (2.188)$$

at the origin, independent of the charge, and avoids the singularities normally associated with point charges.

EXERCISE 2.18

A relativistic Lagrangian density for a *Lorentz scalar* field ϕ that is no more than quadratic in the field and its derivatives and sources has the form

$$\mathcal{D} = K_1\phi + K_2\partial^\alpha\partial_\alpha\phi + K_3\phi^2 + K_4\partial^\alpha\phi\partial_\alpha\phi + K_5\rho_\phi\phi \qquad (2.189)$$

where ρ_ϕ is a scalar source density.

(a) Show that the equation of motion for the field is

$$\partial^\alpha\partial_\alpha\phi + \mu_\phi^2\phi = \chi_0\rho_\phi \qquad (2.190)$$

for some constants μ_ϕ and χ_0.

(b) Show that for a point charge q_ϕ at the point $\mathbf{r}_0(t)$, the wave equation is

$$\partial^\alpha\partial_\alpha\phi + \mu_\phi^2\phi = \chi_0\frac{q_\phi}{\gamma}\delta(\mathbf{r} - \mathbf{r}_0) \qquad (2.191)$$

where q_ϕ is a Lorentz scalar charge and γ is the relativistic factor for the moving charge.

2.3.2 Proca Mass Term

The term in (2.170) and (2.176) involving μ_γ^2 is called the Proca mass term. To see why this is so and to understand its possible significance, we consider the case of a free electromagnetic field; that is, $J^\alpha = 0$. The Maxwell equations (2.176) may then be expressed in the form

$$\partial_\beta\partial^\beta A^\alpha - \partial_\beta\partial^\alpha A^\beta + \mu_\gamma^2 A^\alpha = 0 \qquad (2.192)$$

If we look for a solution in the Lorentz gauge, we have the auxiliary condition

$$\partial_\beta A^\beta = 0 \qquad (2.193)$$

so the second term vanishes and we get the wave equation

$$\partial_\beta\partial^\beta A^\alpha + \mu_\gamma^2 A^\alpha = 0 \qquad (2.194)$$

or in 3-dimensional notation,

$$\frac{1}{c^2}\frac{\partial^2 A^\alpha}{\partial t^2} - \nabla^2 A^\alpha + \mu_\gamma^2 A^\alpha = 0 \qquad (2.195)$$

If we look for a wavelike solution of the form

$$A^\alpha = a^\alpha \exp[i(\omega t - \mathbf{k}\cdot\mathbf{r})] = a^\alpha \exp(ik^\beta r_\beta) \qquad (2.196)$$

for some constants a^α and k^α, we obtain the dispersion relation

$$\frac{\omega^2}{c^2} - k^2 = \mu_\gamma^2 \tag{2.197}$$

Comparing this with the dispersion relation (2.65) for de Broglie waves, we see that this is the dispersion relation for a wave packet with the mass

$$m_\gamma = \frac{\hbar \mu_\gamma}{c} \tag{2.198}$$

Thus if $\mu_\gamma \neq 0$, m_γ becomes the mass of a photon. We also see that to keep the group velocity from exceeding the speed of light, we require that $\mu_\gamma^2 \geq 0$. That is, we have no choice in the sign of the Proca term in the expression (2.167) for the Lagrangian density.

Many experiments have been done to determine the photon mass m_γ, or at least to place an upper bound on it. The most direct experiments are those that measure the dispersion of photon velocity from cosmic gamma-ray bursts. The most distant bursts (as determined from the cosmological red shift) occurred several billion years ago, at a distance corresponding to a significant fraction of the size of the universe. Despite the long travel times, the gamma rays are observed to arrive within milliseconds of one another. Dispersion in the speed of light, if it exists at all, must be exceedingly small. These observations place an upper bound on the photon mass on the order of 3×10^{-11} eV/c^2. For comparison, the mass of an electron is 5.11×10^5 eV/c^2, which is sixteen orders of magnitude larger.

Less direct observations of photon dispersion produce an even lower bound. We see from (2.197) that the lowest possible frequency of an electromagnetic wave is $\omega_0 = \mu_\gamma c$, so the mass of a photon is

$$m_\gamma = \frac{\hbar \omega_0}{c^2} \tag{2.199}$$

The lowest frequencies that have been observed so far are resonances in the (low-Q) cavity formed between the earth and the ionosphere. Resonant frequencies $\omega_0 < 50$ radians/s have been observed. Although the free-space dispersion relation (2.197) is not directly applicable to this geometry, when the differences are accounted for, the existence of low-frequency resonances puts an upper bound on the mass of a photon on the order of $m_\gamma < 4 \times 10^{-13}$ eV/c^2.

The Proca mass term also leads to deviations from Gauss's law and the electrostatic and magnetostatic fields of localized sources. The most sensitive laboratory tests of Gauss's law rely on the fact that the charge on a conductor resides entirely on the outside surface of the conductor, even if the conductor is hollow. Experiments to find deviations from this use a hollow conductor with a smaller conductor inside that is electrically connected to the outer conductor, as shown in Figure 2.17. When a charge is placed on the inner conductor, it should move entirely to the outer conductor. The test of Gauss's law, then, rests on measuring the charge remaining on the inner conductor when it is disconnected from the outer one. The first measurements of this type were conducted by Cavendish in 1772. He interpreted his results by assuming that the potential could be expressed in the form

$$F \propto \frac{1}{r^{2+\varepsilon}} \tag{2.200}$$

Figure 2.17 Schematic diagram of Cavendish's experiment.

and determined that to the accuracy of his experiments, $\varepsilon < 0.02$. However, the field of a stationary point charge that satisfies the Proca wave equation (2.195) actually has the form

$$\Phi \propto \frac{e^{-\mu_\gamma r}}{r} \qquad (2.201)$$

which shows that $\mu_\gamma \geq 0$ if the fields are to vanish at infinity. We can convert Cavendish's results to this form by making the rough equivalence $\nabla^2(e^{-\mu_\gamma r}/r) \approx \nabla^2(1/r^\varepsilon)$, which for $\varepsilon \ll 1$ and $\mu_\gamma R \ll 1$ gives

$$L = \frac{1}{\mu_\gamma} \approx O\left(\frac{R}{\sqrt{\varepsilon}}\right) \qquad (2.202)$$

where R is a characteristic size of the experiment and $L = \mu_\gamma^{-1}$ is the length scale over which deviations from Gauss's law occur. The radius of his experiment was $R \approx 0.3$ m, which corresponds to $L > O(2 \text{ m})$, or $m_\gamma < O(10^{-7} \text{ eV}/c^2)$. Modern versions of the Cavendish experiment place a lower bound on the scale length $L > 3 \times 10^7$ m, which corresponds to a photon mass $m_\gamma < 7 \times 10^{-15} \text{ eV}/c^2$.

Because of the large length scales over which deviations from Gauss's law occur, laboratory experiments are at a distinct disadvantage. Thus, it is not surprising that better tests of deviations from Gauss's law come from measurements of large-scale magnetic fields in space. The best tests within our solar system come from measurements of the Jovian magnetic field. These place a lower bound on the scale length $L > 5 \times 10^8$ m, which corresponds to a photon mass $m_\gamma < 4 \times 10^{-16} \text{ eV}/c^2$. But the best tests of all come from measurements of the intergalactic vector potential. These place a lower bound on the scale length $L > 10^9$ m, which corresponds to an upper bound on the photon mass $m_\gamma < 2 \times 10^{-16} \text{ eV}/c^2$. This is about five orders of magnitude smaller than the upper bound implied by the millisecond dispersion of gamma-ray bursts emitted billions of years ago and twenty-one orders of magnitude smaller than the mass of an electron.

The existence of a nonvanishing photon mass would have other effects as well. These include the failure of gauge invariance, and a contribution to the electromagnetic energy from the vector potential itself, rather than from merely its derivatives (that is, the electric and magnetic fields). These effects are discussed later in this chapter.

EXERCISE 2.19

As shown in a previous exercise, the equation of motion for a Lorentz scalar field ϕ in the presence of a scalar charge q_ϕ at point $\mathbf{r}_0$ is

$$\partial^\alpha \partial_\alpha \phi + \mu_\phi^2 \phi = \chi_0 \frac{q_\phi}{\gamma} \delta (\mathbf{r} - \mathbf{r}_0) \qquad (2.203)$$

for some constants μ_ϕ and χ_0, where γ is the Lorentz factor if the charge is moving.

(a) What are the phase velocity and group velocity of waves of this field? What do these tell us about the sign of the constant μ_ϕ^2 (that is, could we have chosen the constant to be $-\mu_\phi^2$ for a real μ_ϕ)?

(b) For a stationary charge at the origin, show that the potential has the form

$$\phi = -\frac{q_\phi}{4\pi \chi_0} \frac{\exp(-\mu_\phi r)}{r} \tag{2.204}$$

This is the form of the Yukawa potential postulated for meson fields and shows that $\mu_\phi \geq 0$ if the fields are to vanish at infinity.

EXERCISE 2.20

As an alternative to the general theory of relativity, it is tempting to try to describe gravity within the special theory of relativity. One possibility is to treat gravity as a Lorentz scalar field. Use the equations of motion developed in a preceding exercise to compute the advance of the perihelion of Mercury. Assume that the velocity of propagation of gravity waves is c, so that the mass $\mu_\phi = 0$.

(a) Show that the conservation of energy is expressed by the relation

$$\left(1 - \frac{GM_0}{c^2 r}\right)\gamma = \text{constant} \tag{2.205}$$

while the equation of motion for the spacelike coordinates is

$$\frac{d}{d\tau}\left[\left(1 - \frac{GM_0}{c^2 r}\right)\frac{d\mathbf{r}}{d\tau}\right] = GM_0 \nabla\left(\frac{1}{r}\right) \tag{2.206}$$

where G is the universal gravitational constant and M_0 is the mass of the sun. By taking the dot product and the cross product of $\mathbf{r}$ with (2.206), show that the orbit lies in a plane and obtain the equations of motion for the radial and azimuthal coordinates.

(b) To solve these equations, assume that the orbit is nearly circular, so that the radius is

$$r = R_0(1 + \varepsilon) \qquad \varepsilon \ll 1 \tag{2.207}$$

for some constant R_0. Show that in the slightly relativistic limit the advance of the perihelion is

$$\Delta\theta = -\frac{4\pi^3 R_0^2}{c^2 T_0^2} = -\pi\frac{v_0^2}{c^2} \tag{2.208}$$

where T_0 is the orbital period and v_0 the orbital velocity. Compare your answer to the observed value of 43 arc seconds per century. Can gravity be described by a Lorentz scalar field for which $\mu_\phi \neq 0$?

2.4 INVARIANCE AND CONSERVATION LAWS

As we observed in the mechanics of particles, invariance under translation and rotation implies conservation of momentum and angular momentum for fields. But for 4-vector fields there is another invariance, called gauge invariance, that implies another conservation law, the conservation of charge.

2.4.1 Gauge Transformations

Although particles in an electromagnetic field are hypothesized to interact with a 4-vector potential, they actually do so only through the derivatives of the potential, as shown by (2.110), or equivalently through the electric and magnetic field tensor, as in (2.113). Since the potentials themselves do not appear in the equations of motion, the equations of motion are invariant under transformations of the potentials that leave certain combinations of the derivatives unchanged. These are called gauge transformations.

Specifically, we consider the transformation

$$A_\alpha \rightarrow A'_\alpha = A_\alpha - \partial_\alpha \Lambda \tag{2.209}$$

where $\Lambda(r^\alpha)$ is called the gauge function. In 3-vector notation, the gauge transformation is written

$$\Phi \rightarrow \Phi' = \Phi - \frac{\partial \Lambda}{\partial t} \tag{2.210}$$

$$\mathbf{A} \rightarrow \mathbf{A}' = \mathbf{A} + \nabla \Lambda \tag{2.211}$$

Under the gauge transformation (2.209), the Lagrangian (2.104) for a particle in a 4-vector potential becomes

$$\gamma \mathcal{L} \rightarrow \gamma \mathcal{L}' = \gamma \mathcal{L} + q U^\alpha \partial_\alpha \Lambda = \gamma \mathcal{L} + \gamma q \frac{d\Lambda}{dt} \tag{2.212}$$

where

$$\frac{d\Lambda}{dt} = \partial_\alpha \Lambda \frac{dr^\alpha}{dt} = \partial_\alpha \Lambda \frac{U^\alpha}{\gamma} \tag{2.213}$$

is the total time derivative of the gauge function Λ at the position of the particle. But adding a total time derivative to the Lagrangian does not affect the variation of the action, since

$$\delta S' = \delta S + q \delta \int_a^b \frac{d\Lambda}{dt}\, dt = \delta S + q \delta \Lambda \big|_a^b \tag{2.214}$$

The last term vanishes because the trajectory is fixed at the end points, leaving the equations of motion unchanged. The gauge invariance of the equations of motion can also be seen by examining the electric and magnetic fields. Under the gauge transformation (2.209) the field-strength tensor becomes

$$F'^{\alpha\beta} = F^{\alpha\beta} - \partial^\alpha \partial^\beta \Lambda + \partial^\beta \partial^\alpha \Lambda = F^{\alpha\beta} \tag{2.215}$$

since the order of differentiation is immaterial. Thus, the electric and magnetic fields are unchanged, leaving the particle equations of motion (2.113) invariant under a gauge transformation.

The invariance of the equations of motion under a gauge transformation is also true in quantum mechanics. Of course, this might be expected, since the correspondence principle requires that any differences between the predictions of quantum mechanics and those of classical mechanics must vanish in the limit $\hbar \to 0$, where $\hbar$ is Planck's constant divided by 2π. However, the invariance is easy to prove directly. The Schroedinger equation for the wave function ψ of a particle in a vector potential is

$$i\hbar \frac{\partial \psi}{\partial t} = \frac{1}{2m}(-i\hbar\nabla - q\mathbf{A})^2 \psi + q\Phi\psi \tag{2.216}$$

where m is the mass and q the charge of the particle, and Φ and $\mathbf{A}$ are the scalar and vector components of the potential. If we make the gauge transformation (2.210)–(2.211) for some gauge function Λ, the Schroedinger equation becomes

$$i\hbar \frac{\partial \psi}{\partial t} = \frac{1}{2m}(-i\hbar\nabla - q\mathbf{A} - q\nabla\Lambda)^2 \psi + q\Phi\psi - q\frac{\partial \Lambda}{\partial t}\psi \tag{2.217}$$

If we now make the transformation

$$\psi \to \psi' = e^{iq\Lambda/\hbar}\psi \tag{2.218}$$

and substitute into (2.217), we recover (2.216). Since the observable quantities in quantum mechanics are independent of the complex phase of the wave function, this proves that the equations of motion are invariant under a gauge transformation. The arguments are similar for the Dirac equation.

If we ignore the Proca mass term in the Lagrangian, the equations of motion for the fields (the Maxwell equations) are also invariant under gauge transformations, provided that charge is conserved. To see this, we begin with the expression (2.168) for the action of a 4-vector field in the presence of sources:

$$S = -\frac{1}{\mu_0 c}\int \left(\frac{1}{4}F^{\alpha\beta}F_{\alpha\beta} + \mu_0 A_\alpha J^\alpha\right) d^4 r^\gamma \tag{2.219}$$

If we make the gauge transformation (2.209) and use (2.215), the action becomes

$$S \to S' = S + \frac{1}{c}\int J^\alpha \partial_\alpha \Lambda \, d^4 r^\gamma \tag{2.220}$$

If we write

$$J^\alpha \partial_\alpha \Lambda = \partial_\alpha(\Lambda J^\alpha) - \Lambda \partial_\alpha J^\alpha \tag{2.221}$$

substitute into (2.220), and apply the divergence theorem to the first term, we get

$$S' = S + \frac{1}{c}\oint_S \Lambda J^\alpha \hat{n}_\alpha \, d\sigma - \frac{1}{c}\int \partial_\alpha J^\alpha \Lambda \, d^4 r^\gamma \tag{2.222}$$

where the surface S encloses all of Minkowski space between the times t_a and t_b, $\hat{n}_\alpha$ is a unit vector normal to the surface, and $d\sigma$ is an infinitesimal area on the surface. But the current density J^α vanishes at infinity, so the only parts of the surface integral that survive are the integrals over the hypersurfaces $r^0 = ct_a$ and $r^0 = ct_b$, which are integrals over all space. This leaves us with

$$S' = S + \frac{1}{c}\int J^0 \Lambda \, d^3\mathbf{r}\Big|_{t_a}^{t_b} - \frac{1}{c}\int \partial_\alpha J^\alpha \Lambda \, d^4 r^\gamma \tag{2.223}$$

When we calculate the variation of this expression keeping the source term fixed we get

$$\delta S' = \delta S + \frac{1}{c} \int J^0 \delta \Lambda \, d^3\mathbf{r} \Big|_{t_a}^{t_b} - \int \partial_\alpha J^\alpha \delta \Lambda \, d^4 r^\gamma \tag{2.224}$$

The second term vanishes because the end points are fixed, which makes $\delta \Lambda = 0$ at $t = t_a$ and t_b. For the equations of motion to remain invariant for arbitrary variations $\delta \Lambda$, then, we see that the continuity equation

$$\partial_\alpha J^\alpha = 0 \tag{2.225}$$

must be satisfied. That is, charge conservation is implied by gauge invariance.

The same fundamental connection between gauge invariance and conservation of charge is implicit in the Prologue, where we discuss the Maxwell–Ampere law. The Maxwell equations are explicitly gauge invariant and charge conservation follows from them as shown there. When the Proca mass term is included in the Lagrangian for the fields, the Maxwell equations are no longer gauge invariant. If it is discovered that photons have mass, we will have to relinquish the principle of gauge invariance.

Since the Maxwell equations are invariant under a gauge transformation, we are free to choose the gauge function Λ for our convenience. When dealing with the equations for the 4-vector potential A^α, then, we may use this freedom to impose an additional condition on A^α, called the gauge condition. To see how this affects the equations for the potential, we begin with the Maxwell equation (2.177). In terms of the potential A^α, this may be expressed

$$\partial^\alpha \partial_\alpha A_\beta - \partial^\alpha \partial_\beta A_\alpha = \mu_0 J_\beta \tag{2.226}$$

In 3-vector notation, these equations are

$$\frac{1}{c^2} \frac{\partial^2 \Phi}{\partial t^2} - \nabla^2 \Phi - \frac{\partial}{\partial t} \left(\frac{1}{c^2} \frac{\partial \Phi}{\partial t} + \nabla \cdot \mathbf{A} \right) = \frac{\rho}{\varepsilon_0} \tag{2.227}$$

and

$$\frac{1}{c^2} \frac{\partial^2 \mathbf{A}}{\partial t^2} - \nabla^2 \mathbf{A} + \nabla \left(\frac{1}{c^2} \frac{\partial \Phi}{\partial t} + \nabla \cdot \mathbf{A} \right) = \mu_0 \mathbf{J} \tag{2.228}$$

Two particular gauge conditions suggest themselves naturally and are widely useful. The first is called the Lorentz gauge. In it we use the auxiliary condition

$$\partial^\alpha A_\alpha = 0 = \frac{1}{c^2} \frac{\partial \Phi}{\partial t} + \nabla \cdot \mathbf{A} \tag{2.229}$$

The Maxwell equation (2.226) then simplifies to the wave equation

$$\partial^\alpha \partial_\alpha A_\beta = \mu_0 J_\beta \tag{2.230}$$

or, in 3-vector notation,

$$\frac{1}{c^2} \frac{\partial^2 \Phi}{\partial t^2} - \nabla^2 \Phi = \frac{\rho}{\varepsilon_0} \tag{2.231}$$

and

$$\frac{1}{c^2} \frac{\partial^2 \mathbf{A}}{\partial t^2} - \nabla^2 \mathbf{A} = \mu_0 \mathbf{J} \tag{2.232}$$

The beauty of the Lorentz gauge lies in the fact that equations (2.229) (in tensor form) and (2.230) are manifestly covariant, and the vector and scalar potentials are treated equivalently. Each of the components of A_α satisfies the same equation and depends only on its corresponding source term.

Another useful gauge, called the Coulomb gauge, is not covariant. Nevertheless, it is very useful, especially in problems dealing with electromagnetic waves, for which reason it is sometimes called the radiation gauge. The Coulomb gauge is defined by the auxiliary condition

$$\nabla \cdot \mathbf{A} = 0 \tag{2.233}$$

That is, the 3-vector divergence is set equal to zero. From (2.227) and (2.228) we obtain the equations

$$\nabla^2 \Phi = -\frac{\rho}{\varepsilon_0} \tag{2.234}$$

and

$$\frac{1}{c^2} \frac{\partial^2 \mathbf{A}}{\partial t^2} - \nabla^2 \mathbf{A} = \mu_0 \mathbf{J} - \frac{1}{c^2} \nabla \frac{\partial \Phi}{\partial t} \tag{2.235}$$

The first equation is just Coulomb's law for the electrostatic potential. Its solution is

$$\Phi(\mathbf{r}, t) = \frac{1}{4\pi\varepsilon_0} \int \frac{\rho(\mathbf{r}', t)}{|\mathbf{r} - \mathbf{r}'|} d^3\mathbf{r}' \tag{2.236}$$

This expression has the curious behavior that the potential appears instantly at infinity, seemingly in defiance of Einstein's postulates of relativity. However, the equations of motion (2.113) depend only on the electric and magnetic fields, and these, as we see shortly, do not propagate instantly to infinity.

The second equation, (2.235), is just the wave equation for $\mathbf{A}$, but the source term on the right-hand side is not simply the current density. To evaluate the second term in the source density, we differentiate (2.236) with respect to time and use the continuity relation (2.225) to show that

$$-\frac{1}{c^2} \frac{\partial \Phi}{\partial t} = -\frac{1}{c^2} \frac{\partial}{\partial t} \int \frac{\rho(\mathbf{r}', t)}{|\mathbf{r} - \mathbf{r}'|} d^3\mathbf{r}' = \frac{1}{c^2} \int \frac{\nabla' \cdot \mathbf{J}}{|\mathbf{r} - \mathbf{r}'|} d^3\mathbf{r}' \tag{2.237}$$

Substituting this into (2.235), we find that the vector potential satisfies the wave equation

$$\frac{1}{c^2} \frac{\partial^2 \mathbf{A}}{\partial t^2} - \nabla^2 \mathbf{A} = \mu_0 \mathbf{J}_T \tag{2.238}$$

where the so-called "transverse" current density is

$$\mathbf{J}_T = \mathbf{J} - \frac{1}{4\pi} \frac{\partial}{\partial t} \nabla \int \frac{\rho(\mathbf{r}', t)}{|\mathbf{r} - \mathbf{r}'|} d^3\mathbf{r}' = \mathbf{J} + \frac{1}{4\pi} \nabla \int \frac{\nabla' \cdot \mathbf{J}}{|\mathbf{r} - \mathbf{r}'|} d^3\mathbf{r}' \tag{2.239}$$

As shown in Chapter 4, the field of a plane electromagnetic wave is purely transverse and the transverse part of the current density is solely responsible for interacting with the wave. This is the origin of the name transverse current, and the Coulomb gauge is sometimes referred to as the transverse gauge. Using the continuity relation and the fact that

$$\nabla^2 \left(\frac{1}{|\mathbf{r} - \mathbf{r}'|} \right) = -4\pi \delta (\mathbf{r} - \mathbf{r}')   \tag{2.240}$$

it is not difficult to show that

$$\nabla \cdot \mathbf{J}_T = 0   \tag{2.241}$$

as must be true, in view of (2.238), since $\nabla \cdot \mathbf{A} = 0$ by the gauge condition.

Note that while the vector potential $\mathbf{A}$ satisfies the wave equation and would seem to satisfy Einstein's postulates regarding the velocity of propagation of fields, this is not really true, since the transverse current extends to infinity. Therefore, the vector potential fields generated by the transverse current $\mathbf{J}_T$ also extend instantly to infinity. In fact, they are just sufficient to cancel the effects of the electrostatic potential (2.236). Thus, while the potentials in the Coulomb gauge do not satisfy the postulates of relativity, the electric and magnetic fields do.

We see, therefore, that the Coulomb gauge has the disadvantage that it has potentials that propagate instantly to infinity, uses a complicated source for the vector potential, and is not covariant. That is, the potentials satisfy neither the Coulomb gauge condition nor the equations of motion (2.234) and (2.238) when they are transformed to another reference frame. But far from the sources the scalar potential vanishes and the fields are described completely by the vector potential $\mathbf{A}$. The Coulomb gauge is particularly convenient for describing free fields, that is, when the sources vanish. In this case, we may set

$$\Phi = 0   \tag{2.242}$$

and have only the wave equation

$$\frac{1}{c^2} \frac{\partial^2 \mathbf{A}}{\partial t^2} - \nabla^2 \mathbf{A} = 0   \tag{2.243}$$

and the gauge condition (2.233) to deal with.

As a final remark it should be mentioned that the potentials that satisfy the Lorentz and Coulomb gauge conditions (or any other gauge condition) are not unique. Any function that satisfies the gauge condition itself may be added to the potential in what is called a restricted gauge transformation without affecting either the equations of motion of the potentials or, of course, the electric and magnetic fields.

EXERCISE 2.21

Consider the motion of a charged particle of mass m and charge q in a uniform magnetic field $\mathbf{B} = B_0 \hat{\mathbf{z}}$ parallel to the z axis. The field may be described by the vector potential

$$A^\alpha = B_0(0, x\hat{\mathbf{y}})   \tag{2.244}$$

where $\hat{\mathbf{y}}$ is a unit vector, as shown in Figure 2.18.

(a) Since the vector potential is uniform in the $\hat{\mathbf{y}}$ direction, the canonical momentum is conserved. Use this and the conservation of energy to show that for $\mathbf{P} \cdot \hat{\mathbf{y}} = 0$ the

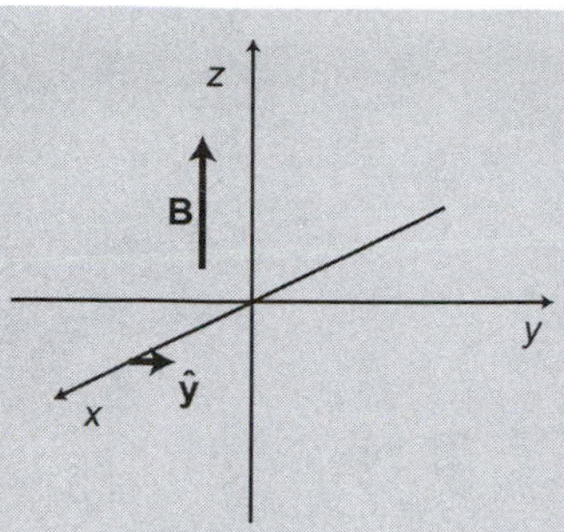

Figure 2.18 Coordinates of the magnetic field.

trajectories are circles around a point on the y axis, and prove that the cyclotron frequency is

$$\omega = \frac{qB_0}{\gamma m} \tag{2.245}$$

(b) In the more general case $\mathbf{P} \cdot \hat{\mathbf{y}} = \bar{P} \neq 0$, use the gauge transformation

$$A^\alpha \to A'^\alpha = A^\alpha - \bar{P}^\alpha = A^\alpha - \partial^\alpha \bar{P}^\beta r_\beta \tag{2.246}$$

to transform the problem into the one we solved in (a), and show that the trajectories are now circles about a point on the line

$$\bar{x} = -\frac{\bar{P}}{qB_0} \tag{2.247}$$

EXERCISE 2.22

The equations of motion of the 4-vector potential are invariant under the gauge transformation

$$A_\alpha \to A'_\alpha = A_\alpha - \partial_\alpha \Lambda \tag{2.248}$$

where $\Lambda\,(r^\alpha)$ is called the gauge function.

(a) Find the differential equation for the gauge function Λ for which the transformed vector potential A'_α satisfies the Lorentz gauge condition. Is the function Λ unique? Explain your answer.
(b) Find the differential equation for the gauge function Λ for which the transformed vector potential A'_α satisfies the Coulomb gauge condition. Is the function Λ unique? Explain your answer.

EXERCISE 2.23

The 4-vector potential of a plane electromagnetic wave traveling in the $\hat{\mathbf{k}}$ direction is

$$A^\alpha = a^\alpha \exp(ik^\alpha r_\alpha) \tag{2.249}$$

where

$$a^\alpha = \left(\frac{\phi}{c}, \mathbf{a}\right) \tag{2.250}$$

and

$$k^\alpha = \left(\frac{\omega}{c}, \mathbf{k}\right) \tag{2.251}$$

are constants, and

$$k_\alpha k^\alpha = 0 \tag{2.252}$$

(a) What is the condition on a^α in the Lorentz gauge?

(b) What is the condition on a^α in the Coulomb gauge?

(c) If we write the vector potential in the Lorentz gauge in the form

$$A^\alpha = \left(\frac{\Phi_0}{c}, \mathbf{A}_0\right) \exp(ik^\alpha r_\alpha) \tag{2.253}$$

show that the function Λ in the gauge transformation

$$A'^\alpha = A^\alpha - \partial^\alpha \Lambda \tag{2.254}$$

that transforms the vector potential of the wave in the Lorentz gauge into the vector potential of the wave in the Coulomb gauge is

$$\Lambda = -i\frac{\Phi_0}{\omega} \exp(ik^\alpha r_\alpha) \tag{2.255}$$

2.4.2 Symmetric Stress Tensor for the Electromagnetic Field

In three-dimensional notation, the momentum conservation law for the electromagnetic field is expressed in terms of the Maxwell stress tensor $T_{ij}^{(M)}$, and the energy conservation law (Poynting's theorem) is expressed separately. In 4-vector notation, the energy and momentum conservation laws are combined into a single 4-vector statement. The energy and momentum of the electromagnetic field are combined into the 4-vector momentum, and the 4-vector momentum conservation law is expressed in terms of the symmetric stress tensor $T^{\alpha\beta}$, which includes the Maxwell stress tensor as its spacelike components.

To derive the 4-vector momentum conservation law for the electromagnetic field, we begin with the identity

$$\partial_\gamma \mathcal{D} - \partial_\alpha \mathcal{D}\delta^\alpha{}_\gamma = 0 \tag{2.256}$$

We then substitute for the Lagrangian density $\mathcal{D}$ and use the Maxwell equations (and some other identities) to simplify the result, and from this we obtain the conservation law. The algebra is a bit tedious and devoid of physical motivation, but the result is simple and elegant.

For the free electromagnetic field, without sources, the Lagrangian density (2.167) is

$$\mathcal{D} = \tfrac{1}{2}K_1 F^{\mu\nu} F_{\mu\nu} \tag{2.257}$$

if we ignore the Proca term. The 4-vector gradient of the Lagrangian density is

$$\partial_\gamma \mathcal{D} = K_1 F^{\alpha\beta} (\partial_\alpha \partial_\gamma A_\beta - \partial_\beta \partial_\gamma A_\alpha) \tag{2.258}$$

Swapping α and β in the second term, recognizing that $F^{\alpha\beta}$ is antisymmetric, we get

$$\partial_\gamma \mathcal{D} = 2K_1 F^{\alpha\beta} \partial_\alpha \partial_\gamma A_\beta \tag{2.259}$$

But we also know that

$$F^{\alpha\beta} \partial_\alpha \partial_\beta A_\gamma = 0 \tag{2.260}$$

since $F^{\alpha\beta}$ is antisymmetric whereas $\partial_\alpha \partial_\beta A_\gamma$ is symmetric in α and β. Adding this to (2.259), we get

$$\partial_\gamma \mathcal{D} = 2K_1 F^{\alpha\beta} \partial_\alpha F_{\gamma\beta} = 2K_1 [\partial_\alpha (F^{\alpha\beta} F_{\gamma\beta}) - F_{\gamma\beta} \partial_\alpha F^{\alpha\beta}] \tag{2.261}$$

But the second term in the brackets vanishes due to the Maxwell equation (in the absence of sources). This leaves us with the result

$$\partial_\gamma \mathcal{D} = 2K_1 \partial_\alpha (F^{\alpha\beta} F_{\gamma\beta}) \tag{2.262}$$

But we also have, from (2.257),

$$\partial_\alpha \mathcal{D} = \tfrac{1}{2} K_1 \partial_\alpha (F^{\mu\nu} F_{\mu\nu}) \tag{2.263}$$

Substituting (2.262) and (2.263) into (2.256), we obtain the tensor expression

$$\partial_\alpha T^\alpha{}_\beta = 0 \tag{2.264}$$

where the stress tensor is

$$T^\alpha{}_\beta = -2K_1 \left(-F^{\alpha\nu} F_{\beta\nu} + \tfrac{1}{4} F^{\mu\nu} F_{\mu\nu} \delta^\alpha{}_\beta\right) \tag{2.265}$$

We see immediately that the trace of the stress tensor vanishes:

$$T^\alpha{}_\alpha = -2K_1 \left(-F^{\alpha\nu} F_{\alpha\nu} + \tfrac{1}{4} F^{\mu\nu} F_{\mu\nu} \delta^\alpha{}_\alpha\right) = 0 \tag{2.266}$$

since $\delta^\alpha{}_\alpha = 4$.

To show that (2.264), the 4-vector divergence of the stress tensor, represents a conservation law, we integrate over all Minkowski space between the hypersurfaces $r^0 = ct_a$ and $r^0 = ct_b$ and use the divergence theorem. Since $T^\alpha{}_\beta$ is presumed to vanish at infinity, the only parts of the integrals that survive are those over the hypersurfaces $r^0 = ct_a$ and $r^0 = ct_b$, which are integrals over all space at times t_a and t_b. We therefore get

$$\int T^0{}_\beta (t_a) \, d^3\mathbf{r} = \int T^0{}_\beta (t_b) \, d^3\mathbf{r} = \text{constant} \tag{2.267}$$

for $\beta = 0 \ldots 3$.

Since the integral of the stress tensor over all space is conserved for the free field, it must be related to the energy and the momentum of the field. Specifically, we see that

$$T^0{}_0 = -2K_1 \left(-F^{0\nu} F_{0\nu} + \frac{1}{4} F^{\mu\nu} F_{\mu\nu}\right) = -K_1 \left(B^2 + \frac{E^2}{c^2}\right) \tag{2.268}$$

If we choose

$$K_1 = -\frac{1}{2\mu_0} \tag{2.269}$$

we get

$$T^0_{\ 0} = \frac{\varepsilon_0}{2} E^2 + \frac{B^2}{2\mu_0} = \mathcal{U} \tag{2.270}$$

where $\mathcal{U}$ is the familiar energy density of the electromagnetic field. In the same way, we find that

$$T^0_{\ 1} = -T^1_{\ 0} = -\frac{1}{\mu_0} F^{0\nu} F_{1\nu} = -cg_x \tag{2.271}$$

where $\mathbf{g} = \varepsilon_0 \mathbf{E} \times \mathbf{B}$ is the momentum density of the field. The rest of the stress tensor is just the three-dimensional Maxwell stress tensor

$$T^i_{\ j} = T^j_{\ i} = -T^{(M)}_{ij} = \varepsilon_0 E_i E_j + \frac{1}{\mu_0} B_i B_j - \mathcal{U}\delta_{ij} \tag{2.272}$$

The complete stress tensor is then

$$T^\alpha_{\ \beta} = \begin{bmatrix} \mathcal{U} & -cg_x & -cg_y & -cg_z \\ cg_x & & & \\ cg_y & & \left(-T^{(M)}_{ij}\right) & \\ cg_z & & & \end{bmatrix} \tag{2.273}$$

If we raise or lower one of the indices of the stress tensor we get the contravariant and covariant forms

$$T^{\alpha\beta} = \begin{bmatrix} \mathcal{U} & cg_x & cg_y & cg_z \\ cg_x & & & \\ cg_y & & \left(T^{(M)}_{ij}\right) & \\ cg_z & & & \end{bmatrix} = T^{\beta\alpha} \tag{2.274}$$

$$T_{\alpha\beta} = \begin{bmatrix} \mathcal{U} & -cg_x & -cg_y & -cg_z \\ -cg_x & & & \\ -cg_y & & \left(T^{(M)}_{ij}\right) & \\ -cg_z & & & \end{bmatrix} = T_{\beta\alpha} \tag{2.275}$$

We see that the stress tensor is symmetric, as defined in Chapter 1. That is, $T^{\alpha\beta} = T^{\beta\alpha}$ and $T_{\alpha\beta} = T_{\beta\alpha}$, but $T^\alpha_{\ \beta} \neq T^\beta_{\ \alpha}$ in general. Actually, this symmetry is not accidental. It happens because we added the vanishing term (2.260) to make the expression more symmetric. In fact, we could have added an arbitrary amount of the divergence-free tensor $F^{\alpha\beta} \partial_\beta A_\gamma$. This would have changed the elements in $T^{\alpha\beta}$ but would not have affected the conservation law (2.264). However, the choice we made not only makes the stress tensor symmetric, it also

1. makes T^{00} the energy density $\mathcal{U}$ of the electromagnetic field,
2. makes $T^{01} \ldots T^{03}$ the momentum density $c\mathbf{g}$ of the electromagnetic field, and
3. satisfies automatically the conservation of angular momentum of the field.

We can see this last point in the following way. By analogy to the angular momentum tensor (2.85) for a system of particles, we see that the angular momentum tensor of the electromagnetic field is

$$M^{\alpha\beta} = \int \left(r^\alpha g^\beta - r^\beta g^\alpha\right) d^3\mathbf{r} \tag{2.276}$$

The covariant generalization of the angular momentum stress tensor is then

$$M^{\alpha\beta\gamma} = r^\alpha T^{\beta\gamma} - r^\beta T^{\alpha\gamma} \tag{2.277}$$

Conservation of angular momentum requires that the divergence of this tensor vanish, where the divergence is

$$\partial_\gamma M^{\alpha\beta\gamma} = r^\alpha \partial_\gamma T^{\beta\gamma} + T^{\beta\gamma} \delta_\gamma^{\,\alpha} - r^\beta \partial_\gamma T^{\alpha\gamma} - T^{\alpha\gamma} \delta_\gamma^{\,\beta} \tag{2.278}$$

But the first and third terms vanish by (2.264), and we see that

$$\partial_\gamma M^{\alpha\beta\gamma} = T^{\beta\alpha} - T^{\alpha\beta} = 0 \tag{2.279}$$

when $T^{\alpha\beta}$ is symmetric, which it is.

The conservation law (2.264) changes when we include sources in the Maxwell equation for the field. If we go back to (2.261) and use the Maxwell equation with sources to evaluate the second term, we get in place of (2.262) the expression

$$\partial_\gamma \mathcal{D} = -\frac{1}{\mu_0} \partial_\alpha (F^{\alpha\beta} F_{\gamma\beta}) + F_{\gamma\beta} J^\beta \tag{2.280}$$

Combining this with (2.263), as before, we get the result

$$\partial_\alpha T^\alpha_{\,\beta} + F_{\beta\alpha} J^\alpha = 0 \tag{2.281}$$

The four components of this equation express the conservation of energy and momentum of the field. From the timelike component ($\beta = 0$) we get Poynting's theorem, and from the spacelike components we get the momentum conservation law, as discussed in the Prologue.

EXERCISE 2.24

Consider the attraction between two bar magnets, as illustrated in Figure 2.19. Since the components of the symmetric stress tensor are quadratic in the fields, we may focus our attention on the region between the poles, where the fields are strongest, and ignore the region outside the poles. In the region between the poles, we assume that the magnetic induction has the uniform value B in the $\hat{x}$ direction.

(a) Show that in the region between the poles, the symmetric stress tensor is

$$T^{\alpha\beta} = \mathcal{U} \begin{bmatrix} 1 & 0 & 0 & 0 \\ 0 & -1 & 0 & 0 \\ 0 & 0 & 1 & 0 \\ 0 & 0 & 0 & 1 \end{bmatrix} \tag{2.282}$$

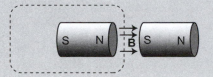

Figure 2.19 Attraction between bar magnets.

where the energy density is

$$\mathcal{U} = \frac{B^2}{2\mu_0}$$ (2.283)

(b) Integrate the $\hat{\mathbf{x}}$ component of the momentum conservation law over the volume indicated by the dotted line, and apply the divergence theorem to the spacelike terms to show that the total force on the bar magnet (the rate of change of the total momentum) is

$$F = \mathcal{U}A$$ (2.284)

in the $\hat{\mathbf{x}}$ direction, where A is the area of the pole faces.

(c) The total energy in the magnetic field is

$$\mathcal{W} = \mathcal{U}AL$$ (2.285)

where L is the distance between the pole faces. Use arguments based on the conservation of energy as the magnets are moved farther apart to find the force of attraction between them.

EXERCISE 2.25

The Proca Lagrangian density for the electromagnetic field is given by the expression

$$\mathcal{D} = \frac{1}{2\mu_0}\left[-\frac{1}{2}F^{\alpha\beta}F_{\alpha\beta} + \mu_\gamma^2 A^\alpha A_\alpha\right]$$ (2.286)

where the second term has the effect of giving the electromagnetic field a rest mass. Including the mass term, the wave equation for a real field in the absence of sources is

$$\partial_\alpha F^{\alpha\beta} + \mu_\gamma^2 A^\beta = 0$$ (2.287)

while the Lorentz gauge condition remains

$$\partial_\alpha A^\alpha = 0$$ (2.288)

Show that the symmetric stress tensor for the electromagnetic field including the mass term has the form

$$T^\alpha{}_\beta = \frac{1}{\mu_0}\left[-F^{\alpha\nu}F_{\beta\nu} + \mu_\gamma^2 A^\alpha A_\beta - \frac{1}{2}\left(-\frac{1}{2}F^{\mu\nu}F_{\mu\nu} + \mu_\gamma^2 A^\nu A_\nu\right)\delta^\alpha{}_\beta\right]$$ (2.289)

Show that the energy density is given by the formula

$$\mathcal{U} = \frac{\varepsilon_0 E^2}{2} + \frac{B^2}{2\mu_0} + \mu_\gamma^2\left(\frac{\varepsilon_0 \Phi^2}{2} + \frac{A^2}{2\mu_0}\right)$$ (2.290)

EXERCISE 2.26

In a previous exercise we have shown that the Lagrangian density

$$\mathcal{D} = -\frac{1}{4\mu_0} D(F^{\alpha\beta} F_{\alpha\beta}) - A_\alpha J^\alpha \tag{2.291}$$

where $D(x)$ is some function of x, leads to the covariant, but nonlinear, equations of motion

$$\partial_\alpha [D'(F^{\mu\nu} F_{\mu\nu}) F^{\alpha\beta}] = \mu_0 J^\beta \tag{2.292}$$

in which $D'(x) = dD/dx$. To recover the Maxwell equations in the limit of weak fields, it is necessary that $D'(x) \xrightarrow[x \to 0]{} 1$.

(a) Show that the stress tensor for the field is

$$T^\alpha{}_\beta = -\frac{1}{\mu_0} \left[D'(F^{\mu\nu} F_{\mu\nu}) F^{\alpha\gamma} F_{\beta\gamma} - \frac{1}{4} D(F^{\mu\nu} F_{\mu\nu}) \delta^\alpha{}_\beta \right] \tag{2.293}$$

and that it satisfies the conservation law

$$\partial_\alpha T^\alpha{}_\beta + F_{\beta\alpha} J^\alpha = 0 \tag{2.294}$$

(b) Born and Infeld propose the function

$$D(F^{\alpha\beta} F_{\alpha\beta}) = 4 \frac{E_0^2}{c^2} \left(\sqrt{1 + \frac{c^2}{2E_0^2} F^{\alpha\beta} F_{\alpha\beta}} - 1 \right) \tag{2.295}$$

where E_0 is a constant. We have shown in a previous exercise that for a point charge q at the origin, the electric field is then

$$E = \frac{q}{4\pi \varepsilon_0 r^2} \frac{1}{\sqrt{1 + \left(\dfrac{q}{4\pi \varepsilon_0 r^2 E_0} \right)^2}} \tag{2.296}$$

Show that the total energy in the electric field has the finite value

$$\mathcal{W} = K \sqrt{\frac{E_0 q^3}{4\pi \varepsilon_0}} \tag{2.297}$$

where the constant is, after integrating twice by parts,

$$K = \frac{2}{3} \int_0^\infty \frac{dr}{\sqrt{1 + r^4}} = \frac{2}{3} F\left(\frac{\pi}{2} \Big| \frac{1}{2} \right) = 1.236 \tag{2.298}$$

in which $F(\phi|m)$ is an elliptic integral of the first kind. Born and Infeld equate this energy to the rest energy of the particle.

EXERCISE 2.27

Show that the symmetric stress tensor for a free Lorentz scalar field is

$$T^{\alpha}{}_{\beta} = 2K_4\left[\partial^{\alpha}\phi\partial_{\beta}\phi - \tfrac{1}{2}\left(\partial^{\nu}\phi\partial_{\nu}\phi - \mu_{\phi}^2\phi^2\right)\delta^{\alpha}{}_{\beta}\right] \tag{2.299}$$

for some constants K_4 and μ_{ϕ}. Show that the energy density is

$$T^{00} = K_4\left[\frac{1}{c^2}\left(\frac{\partial\phi}{\partial t}\right)^2 + (\nabla\phi)^2 + \mu_{\phi}^2\phi^2\right] \tag{2.300}$$

Since the energy in the field must be positive, we now see that the constant K_4 must be positive. Previously, we found that a particle in a scalar potential has the Hamiltonian (2.156), where the scalar field of a stationary charge is given by (2.204). Do like charges attract or repel? *Hint:* Since the Hamiltonian is not explicitly time dependent, the total energy is conserved. What happens to the energy of the moving particle as it approaches the origin?

EXERCISE 2.28

(a) It is shown in Chapter 1 that the quantities

$$F^{\alpha\beta}F_{\alpha\beta} = 2\left(B^2 - \frac{E^2}{c^2}\right) \tag{2.301}$$

and

$$F^{\alpha\beta}\mathcal{F}_{\alpha\beta} = -\frac{4}{c}\mathbf{E}\cdot\mathbf{B} \tag{2.302}$$

are Lorentz scalars; that is, they are invariant under Lorentz transformations. Since the generic properties of a light wave must be independent of the reference frame in which the light wave is observed, the ratio of the electric and magnetic fields must be independent of the reference frame. Therefore, it follows that the invariant (2.301) must vanish, making

$$E = cB \tag{2.303}$$

in a light wave. In the same way, since $\mathbf{E}$ and $\mathbf{B}$ depend on the reference frame in which the light wave is viewed but the product (2.302) does not, it follows that the product must vanish, making

$$\mathbf{E}\cdot\mathbf{B} = 0 \tag{2.304}$$

in a light wave. Likewise, by symmetry, for a plane wave the energy flow must be in the direction of the wave vector, normal to the wave fronts, so the Poynting vector must be in this direction. From the definition

$$\mathbf{S} = \frac{1}{\mu_0}\mathbf{E}\times\mathbf{B} \tag{2.305}$$

of the Poynting vector, we see that $\mathbf{E}$ and $\mathbf{B}$ must be orthogonal to the wave vector $\mathbf{k}$ as well as to each other. These same results can be obtained in a more pedestrian fashion by solving the wave equation, as shown in the Prologue. Show that if

$\mathbf{S} = S\hat{\mathbf{x}}$ is the Poynting vector of a light wave traveling in the $\hat{\mathbf{x}}$ direction with intensity S, the symmetric stress tensor is

$$T^{\alpha\beta} = \begin{bmatrix} \mathcal{U} & \mathcal{U} & 0 & 0 \\ \mathcal{U} & \mathcal{U} & 0 & 0 \\ 0 & 0 & 0 & 0 \\ 0 & 0 & 0 & 0 \end{bmatrix} \tag{2.306}$$

where

$$\mathcal{U} = S/c \tag{2.307}$$

is the energy density in the light wave.

(b) Show that if light reflects off a surface with the reflection coefficient R and we ignore interference between the incident and reflected waves, the stress tensor just outside the surface is

$$T^{\alpha\beta} = \begin{bmatrix} (1+R)\mathcal{U} & (1-R)\mathcal{U} & 0 & 0 \\ (1-R)\mathcal{U} & (1+R)\mathcal{U} & 0 & 0 \\ 0 & 0 & 0 & 0 \\ 0 & 0 & 0 & 0 \end{bmatrix} \tag{2.308}$$

(c) Consider a light wave incident on a conductor. When the light wave scatters off the electrons in the conductor, the momentum of the photons is transferred to the electrons, dragging them in the direction of the light wave until a space charge develops and creates an electric field that stops their motion. This is called the photon-drag effect. We limit the following discussion to the case when the intensity of the incident light is constant and ignore variations transverse to the surface normal. We further assume that the photon-drag effect is small, so the electrons are not moved far from their quiescent positions. The density of electrons in the conductor is then n_e everywhere. We further assume that the incident light interacts only with the conduction electrons in the solid. Using the imaginary volume shown in Figure 2.20, apply the conservation law (2.281) to show that the total voltage induced in the direction normal to the surface of the conductor is

$$\Delta V = \int_0^\infty E_x \, dx = -\frac{S}{cqn_e}(1+R) \tag{2.309}$$

where q is the electron charge. For a laser intensity $S = 10^{13}$ W/m^2 incident on silver ($n_e = 6 \times 10^{28}$/m^3, $R = .99$ at a wavelength of 1 μm) what is the induced voltage? In most experiments the conductor is a doped semiconductor with a much smaller electron density, so the voltage is much larger.

Vacuum   Conductor

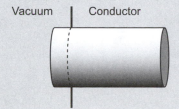

Figure 2.20 Imaginary volume for the photon-drag effect.

BIBLIOGRAPHY

The derivation of the relativistic equations of motion of the particles and fields forms part of almost all advanced texts on classical electrodynamics, notably

J. D. Jackson, *Classical Electrodynamics,* 3rd edition, John Wiley & Sons, New York (1999),

L. D. Landau and E. M. Lifshitz, *The Classical Theory of Fields,* 4th edition, Pergamon Press, Oxford (1975),

F. E. Low, *Classical Field Theory,* John Wiley & Sons, New York (1997).

In addition, several monographs have been written on the subject. These include

A. O. Barut, *Electrodynamics and Classical Theory of Fields and Particles,* Macmillan Company, New York (1964)

F. Rohrlich, *Classical Charged Particles,* Addison-Wesley Publishing Company, Reading, MA (1990).

3 Time-Independent Electromagnetic Fields

3.1 ELECTROSTATICS

The history of electrostatics is long and colorful. It began at least 2600 years ago with parlor tricks involving cat's fur and amber, passed through dramatic (and sometimes dangerous) experiments with Leyden jars and lightning, and evolved to electron microscopes and van de Graaff generators by the end of the second millenium A.D. The history of electrostatics is filled with some of the most famous names in mathematics and physics, including experimenters such as Coulomb, Cavendish, and Franklin, and mathematicians such as Gauss (who also invented the telegraph), Poisson, Laplace, and Green. The mathematical phase of electrostatics began in the late 18th century with Priestley's discovery of the inverse square law for the force between charged bodies, and the discoveries that followed led ultimately to the development of the Maxwell equations, on which classical electrodynamics is based today. Historically, of course, Coulomb's law came first and the Maxwell equations followed from it and some other discoveries, but in the preceding chapter we derived the Maxwell equations directly from Einstein's postulates and Hamilton's principle. We now reverse the historical order, beginning with the Maxwell equations and concluding with Coulomb's law.

3.1.1 Coulomb's Law

In the completely static case there is no motion of the charges, so the current $\mathbf{J}$ and the magnetic induction $\mathbf{B}$ both vanish. In this case, the Coulomb and Lorentz gauges are identical. The vector potential is at most a constant that we can ignore, and the scalar potential satisfies the Poisson equation

$$\nabla^2 \Phi = -\frac{\rho}{\varepsilon_0} \tag{3.1}$$

If the charge density ρ vanishes, this is called the Laplace equation.

A useful and interesting theorem, called the mean-value theorem, states that in a region of space in which there are no charges, the potential Φ_0 at the center of a sphere is just the average of the potential over the surface of the sphere. To prove this, we imagine a spherical surface S of radius R centered on point P. In the absence of sources, the potential satisfies the Laplace equation

$$\nabla^2 \Phi = 0 \qquad (3.2)$$

If we integrate this over the volume inside S and use the divergence theorem, we get

$$\oint_S \nabla \Phi \cdot \hat{\mathbf{n}} \, dS = R^2 \oint_S \frac{\partial \Phi}{\partial r} \, d\Omega = R^2 \frac{d}{dR} \oint_S \Phi \, d\Omega = 0 \qquad (3.3)$$

where $\hat{\mathbf{n}}$ is a unit vector normal to the surface dS and $d\Omega$ is a differential of solid angle. Therefore $\oint_S \Phi \, d\Omega = \text{constant}$, and since the potential is continuous at P when no charges are present, we see that

$$\Phi_0 = \frac{1}{4\pi R^2} \oint_S \Phi \, dS \qquad (3.4)$$

thus proving the theorem.

It follows from the mean-value theorem that if $\Phi > \Phi_0$ over some portions of the sphere, then necessarily $\Phi < \Phi_0$ over other portions of the sphere. Therefore, the potential Φ can have neither a maximum nor a minimum in a region where the charge density vanishes. In particular, if a charge-free region is bounded by an equipotential surface, the potential must be a constant throughout the region and the electric field vanishes. This explains why electrical conductors, which in equilibrium come to a uniform potential, can be used to shield an enclosed region of space from electric fields.

As a further consequence of the mean-value theorem, we note that any static arrangement of unrestrained charges is unstable, or at most metastable, since none of the charges finds itself in a local minimum of the potential energy due to the other charges. This is called Earnshaw's theorem. Earnshaw's theorem poses a serious difficulty for classical physics, since it predicts that stable matter held together by electrostatic forces is impossible, which is somewhat at variance with everyday experience. Even stable cyclic motions, such as an electron orbiting around a proton, do not save classical physics, since the accelerating electron must radiate all its energy in a few nanoseconds and fall into the proton. Quantum mechanics, of course, saves the day, but that is outside the scope of this discourse.

We now proceed to Coulomb's law. For a point charge q at position $\mathbf{r}_0$, the charge density is

$$\rho = q \delta (\mathbf{r} - \mathbf{r}_0) \qquad (3.5)$$

If we require that the potential vanish at infinity, the solution to the Poisson equation (3.1) is then

$$\Phi = \frac{q}{4\pi \varepsilon_0} \frac{1}{|\mathbf{r} - \mathbf{r}_0|} \qquad (3.6)$$

Since the potentials satisfy the principle of superposition, the potential due to an assembly of charges is the algebraic sum of the potentials of the individual charges. Thus, we may write the solution as a sum or integral over the charges of the form

$$\Phi = \frac{1}{4\pi \varepsilon_0} \sum_i \frac{q_i}{|\mathbf{r} - \mathbf{r}_i|} = \frac{1}{4\pi \varepsilon_0} \int \frac{\rho(\mathbf{r}') \, d^3 \mathbf{r}'}{|\mathbf{r} - \mathbf{r}'|} \qquad (3.7)$$

The electric field is the negative gradient of the potential,

$$\mathbf{E} = -\nabla \Phi \tag{3.8}$$

To find the field of a point charge q at position $\mathbf{r}_0$, we note first that

$$\frac{\partial}{\partial \mathbf{r}} \frac{1}{|\mathbf{r} - \mathbf{r}_0|} = -\frac{\mathbf{r} - \mathbf{r}_0}{|\mathbf{r} - \mathbf{r}_0|^3} \tag{3.9}$$

Then, from (3.6) and (3.8) we see that the electric field is

$$\mathbf{E} = \frac{q}{4\pi\varepsilon_0} \frac{\mathbf{r} - \mathbf{r}_0}{|\mathbf{r} - \mathbf{r}_0|^3} \tag{3.10}$$

Since the fields likewise satisfy the principle of superposition, the field due to an assembly of charges may be expressed as a sum or integral over the sources of the form

$$\mathbf{E} = \frac{1}{4\pi\varepsilon_0} \sum_i q_i \frac{\mathbf{r} - \mathbf{r}_i}{|\mathbf{r} - \mathbf{r}_i|^3} = \frac{1}{4\pi\varepsilon_0} \int \frac{\mathbf{r} - \mathbf{r}'}{|\mathbf{r} - \mathbf{r}'|^3} \rho \, d^3\mathbf{r}' \tag{3.11}$$

In Chapter 2 we found that the force on a charge q, defined as the rate of change of the momentum, is

$$\mathbf{F} = \frac{d\mathbf{p}}{dt} = q(\mathbf{E} + \mathbf{v} \times \mathbf{B}) = q\mathbf{E} \tag{3.12}$$

in the electrostatic case. Combining this with (3.10), we see that the force $\mathbf{F}$ on the charge q due to another charge q' is

$$\mathbf{F} = \frac{qq'}{4\pi\varepsilon_0} \frac{\mathbf{r} - \mathbf{r}'}{|\mathbf{r} - \mathbf{r}'|^3} = -\mathbf{F}' \tag{3.13}$$

where $\mathbf{F}'$ is the force on q'. This is known as Coulomb's law. Historically, it was the starting point of electrostatics, rather than a special case of a more general law. Note that in the electrostatic case Newton's third law of motion (the action is equal and opposite to the reaction) is valid. This is not true in the more general case, of course, and the difference is made up by the rate of change of the momentum in the field. That Newton's third law should be valid in electrostatics is not surprising, since, as observed in Chapter 1, the validity of the third law rests on Newton's (incorrect) assumption that forces propagate with infinite velocity. In the electrostatic case, this is effectively true.

EXERCISE 3.1

Consider the space charge–limited flow of current in a planar vacuum diode, as illustrated in Figure 3.1. We assume that the cathode represents an unlimited supply of electrons at zero velocity that are accelerated by the electric field toward the anode. The electrons in the anode–cathode gap form a space charge density

$$\rho = \frac{J}{v} \tag{3.14}$$

where J is the current density and v the electron velocity, that modifies the field as shown in Figure 3.1 and repels the electrons leaving the cathode. The current becomes limited

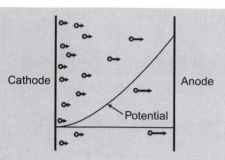

Figure 3.1 Space-charge-limited current in a diode.

when the electric field at the surface of the cathode vanishes. For nonrelativistic motion of the electrons, show that the space charge–limited current in the diode is given by the Child–Langmuir law

$$I = KV^{3/2} \tag{3.15}$$

where V is the potential of the anode relative to the cathode. The quantity

$$K = \frac{4\varepsilon_0 A}{9d^2} \sqrt{\frac{2e}{m}} \tag{3.16}$$

where A is the area of the cathode, d the anode–cathode separation, $e = |q|$ the magnitude of the electron charge, and m the electron mass, is called the perveance of the diode. The SI unit of perveance is the perv P, or more commonly the microperv μP.

EXERCISE 3.2

Consider a solid conducting sphere of radius r_0 at potential Φ_0, surrounded by a concentric, spherical conducting shell of radius R_0 at the same potential, with no charges in the region between the spheres. According to the Laplace equation, the field in the region between the two spheres must vanish everywhere, and so the charge on the inner sphere must also vanish, by Gauss's law. However, this is not true if the photon has mass.

(a) Show that if the photon has mass m_γ, then Gauss's law becomes

$$\oint_S \mathbf{E} \cdot \hat{\mathbf{n}} \, dS = \frac{Q}{\varepsilon_0} - \mu_\gamma^2 \int_V \Phi \, dV \tag{3.17}$$

where S is a closed surface bounding the volume V, $\mathbf{E}$ the electric field, $\hat{\mathbf{n}}$ a unit vector normal to dS in the outward direction, $Q = \int_V \rho \, dV$ the total charge inside S, and $\mu_\gamma = m_\gamma c/\hbar$ the inverse scale length, in which c is the speed of light and $\hbar$ Planck's constant divided by 2π.

(b) Show that in this case, the potential between the spheres is

$$\Phi = \Phi_0 \frac{r_0}{r} \left[K_+ e^{\mu_\gamma (r - r_0)} + K_- e^{-\mu_\gamma (r - r_0)} \right] \tag{3.18}$$

where the constants satisfy the boundary conditions

$$K_+ + K_- = 1 \tag{3.19}$$

$$\frac{r_0}{R_0}\left[K_+ e^{\mu_\gamma(R_0-r_0)} + K_- e^{-\mu_\gamma(R_0-r_0)}\right] = 1 \tag{3.20}$$

at $r = r_0$ and $r = R_0$, respectively. Note that for a nonvanishing photon mass, the potential has a minimum in the charge-free region between the spheres. Conversely, inside the inner sphere the potential is a constant and the field vanishes, but the charge density does not vanish.

(c) Show that the electric field at the surface of the inner sphere is

$$E_r(r_0) = -\left.\frac{d\Phi}{dr}\right|_{r_0} = \frac{\Phi_0}{r_0}[1 - \mu_\gamma r_0(2K_+ - 1)] \tag{3.21}$$

(d) Show that to lowest order in the photon mass, the charge on the inner sphere is

$$\frac{Q}{4\pi\varepsilon_0 r_0\Phi_0} = 1 - \mu_0 r_0(2K_+ - 1) + \frac{1}{3}\mu_0^2 r_0^2 \approx \frac{\mu_\gamma^2}{6}\left(2R_0^2 - R_0 r_0 + r_0^2\right) \tag{3.22}$$

(e) In an experiment to measure the photon mass (a modern version of Cavendish's experiment to test the $1/r^2$ force law), the outer sphere is alternately charged to a voltage $\Phi_0 = V_0$ at a frequency ω_0. The inner sphere is connected to the outer one by a wire that keeps the two spheres at the same potential, and the charge on the inner sphere is determined by measuring the current in the wire. Provided that the quasistatic result (3.22) may be used at this frequency, the peak current in the wire is

$$I_0 \approx \omega_0 Q(V_0) = \tfrac{2}{3}\pi\varepsilon_0 r_0\omega_0 V_0\mu_\gamma^2\left(2R_0^2 - R_0 r_0 + r_0^2\right) \tag{3.23}$$

The spheres are constructed with radii $r_0 = 1$ m and $R_0 = 2$ m, and the RF voltage used to charge the outer sphere is $V_0 = 10$ kV at the frequency $\omega_0 = 10 \times 10^6$ radians/s. Tests of the equipment show that current can be measured down to 1 pA, but no current is detected. What is the upper bound on the photon mass implied by this negative result?

3.1.2 Energy in Electrostatic Fields

As shown in Chapter 2, the electromagnetic field has an energy density

$$\mathcal{U} = \frac{\varepsilon_0}{2}E^2 + \frac{1}{2\mu_0}B^2 \tag{3.24}$$

where the second term vanishes in the electrostatic case. The total energy in the field is then

$$W = \frac{\varepsilon_0}{2}\int E^2\,d^3\mathbf{r} = -\frac{\varepsilon_0}{2}\int \mathbf{E}\cdot\nabla\Phi\,d^3\mathbf{r} \tag{3.25}$$

where the integral extends over all space. Rearranging this, we get

$$W = -\frac{\varepsilon_0}{2}\int [\nabla\cdot(\Phi\mathbf{E}) - \Phi\nabla\cdot\mathbf{E}]\,d^3\mathbf{r} \tag{3.26}$$

But by the divergence theorem, the first term may be converted to a surface integral that vanishes because the field and the potential both vanish at infinity. To evaluate the second term, we use (3.7) and Gauss's law to get

$$W = \frac{1}{2} \int \rho(\mathbf{r})\Phi(\mathbf{r})\, d^3\mathbf{r} = \frac{1}{2} \iint \frac{\rho(\mathbf{r})\rho(\mathbf{r}')}{4\pi\varepsilon_0|\mathbf{r} - \mathbf{r}'|}\, d^3\mathbf{r}\, d^3\mathbf{r}' \tag{3.27}$$

That is, the total energy is half the potential energy of each of the charges in the field of all the other charges. We see from the last form of the integral that the factor $\frac{1}{2}$ appears because the integral "double counts" the interaction of each charge with the other charges.

For point charges we run into some difficulty, because the potential Φ is infinite at the position of the point charge. That is, point charges have infinite self-energy in the electric field. For now, we ignore the (infinite) self-energy of the particles and calculate the energy of interaction in the following way. If we begin with a charge q_1 at position $\mathbf{r}_1$ and bring up another charge q_2 from infinity to position $\mathbf{r}_2$, the amount of work done on charge q_2 is just the potential energy of charge q_2 in the field of charge q_1,

$$W_2 = \frac{q_1 q_2}{4\pi\varepsilon_0\, |\mathbf{r}_1 - \mathbf{r}_2|} \tag{3.28}$$

If we now bring up, in succession, charges q_3, q_4, and so on, the work done on each charge is

$$W_i = \sum_{j<i} \frac{q_i q_j}{4\pi\varepsilon_0|\mathbf{r}_i - \mathbf{r}_j|} \tag{3.29}$$

where the sum is over the previous charges $j < i$, and the total work done is

$$W = \sum_i \sum_{j<i} \frac{q_i q_j}{4\pi\varepsilon_0|\mathbf{r}_i - \mathbf{r}_j|} = \frac{1}{2} \sum_i \sum_{j\neq i} \frac{q_i q_j}{4\pi\varepsilon_0|\mathbf{r}_i - \mathbf{r}_j|} \tag{3.30}$$

when we "double count" the interactions. Therefore, the total energy of the assembly of point charges, relative to their energy at infinite separation, may be expressed in the form

$$W = \frac{1}{2} \sum_i q_i \Phi_i \tag{3.31}$$

where the potential at position $\mathbf{r}_i$,

$$\Phi_i = \sum_{j\neq i} \frac{q_j}{4\pi\varepsilon_0|\mathbf{r}_i - \mathbf{r}_j|} \tag{3.32}$$

is due to all the other charges. These expressions are analogous to (3.27), except that they discard the (infinite) self-energy of the point charges.

3.1.3 Multipole Moments

For localized charge distributions, such as atoms and nuclei, the field at a point far from the charges may be expanded in inverse powers of the distance from the charges, and the coefficients in this expansion are called the multipole moments of the charge distribution. The interaction of the charge distribution with externally applied fields can also be expressed in terms of the same multipole moments, making them convenient for a variety of different problems. For example, the hydration of ions in water is the result of the interaction

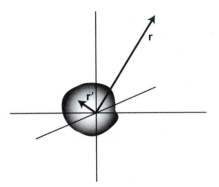

Figure 3.2 Potential surrounding a charge distribution.

between the charge on the ion and the dipole moments of the water molecules. Also, the observation of a finite electric quadrupole moment for a deuteron shows that it is prolate, or egg-shaped, rather than spherically symmetric.

As we have seen, the potential at point $\mathbf{r}$ due to a charge distribution $\rho(\mathbf{r}')$ is

$$\Phi(\mathbf{r}) = \frac{1}{4\pi\varepsilon_0} \int \frac{\rho(\mathbf{r}') \, d^3\mathbf{r}'}{|\mathbf{r} - \mathbf{r}'|} \tag{3.33}$$

If the source is localized near the origin and the field point $\mathbf{r}$ is a large distance away, as shown in Figure 3.2, it is useful to Taylor expand the reciprocal of the radius

$$f(\mathbf{r}') = \frac{1}{|\mathbf{r} - \mathbf{r}'|} = \frac{1}{\sqrt{(x - x')^2 + (y - y')^2 + (z - z')^2}} \tag{3.34}$$

as a function of $\mathbf{r}'$ for $r' \ll r$. The Taylor series for a function of three variables may be expressed in the form

$$f(\mathbf{r}') = f\Big|_{\mathbf{r}'=0} + \sum_{i=1}^{3} r_i' \frac{\partial f}{\partial r_i}\Big|_{\mathbf{r}'=0} + \frac{1}{2} \sum_{i,j=1}^{3} r_i' r_j' \frac{\partial^2 f}{\partial r_i' \partial r_j'}\Big|_{\mathbf{r}'=0} + \cdots \tag{3.35}$$

If we differentiate (3.34) we find, after some algebra, that

$$f(\mathbf{r}')\Big|_{\mathbf{r}'=0} = \frac{1}{r} \tag{3.36}$$

$$\frac{\partial f}{\partial r_i'}\Big|_{\mathbf{r}'=0} = \frac{r_i}{r^3} \tag{3.37}$$

$$\frac{\partial^2 f}{\partial r_i' \partial r_j'}\Big|_{\mathbf{r}'=0} = \frac{3r_i r_j - r^2 \delta_{ij}}{r^5} \tag{3.38}$$

and so on. Therefore, at large distances from the localized charge distribution the potential is

$$\Phi(\mathbf{r}) = \int \frac{1}{r} \rho(\mathbf{r}') \, d^3\mathbf{r}' + \int \sum_{i=1}^{3} \frac{r_i r_i'}{r^3} \rho(\mathbf{r}') \, d^3\mathbf{r}'$$

$$+ \int \frac{1}{2} \sum_{i,j=1}^{3} \frac{3r_i r_j - r^2 \delta_{ij}}{r^5} r_i' r_j' \rho(\mathbf{r}') \, d^3\mathbf{r}' + \cdots \tag{3.39}$$

But the summations and all factors of $\mathbf{r}$ may be moved outside the integrals over $\mathbf{r}'$. We also note that for $i = j$ the last term vanishes since

$$\sum_{i=1}^{3} (3r_i r_i - r^2) = 0 \tag{3.40}$$

for any $\mathbf{r}$. Therefore, in the third term of (3.39) we may add to the factor $r_i' r_j'$ inside the integral anything multiplied by δ_{ij}. Specifically, we add $\frac{1}{3} r'^2 \delta_{ij}$ to symmetrize the dependence of this term on $\mathbf{r}$ and $\mathbf{r}'$. We then find that the potential is given by the series

$$\Phi(\mathbf{r}) = \frac{1}{4\pi\varepsilon_0} \left[\frac{Q}{r} + \sum_{i=1}^{3} \frac{Q_i r_i}{r^3} + \frac{1}{6} \sum_{i,j=1}^{3} \frac{Q_{ij}(3r_i r_j - r^2 \delta_{ij})}{r^5} + \cdots \right] \tag{3.41}$$

where the multipole moments of the charge distribution are

$$Q = \int \rho(\mathbf{r}') d^3\mathbf{r}' \tag{3.42}$$

$$Q_i = \int r_i' \rho(\mathbf{r}') d^3\mathbf{r}' = p_i \tag{3.43}$$

$$Q_{ij} = \int (3r_i' r_j' - r'^2 \delta_{ij}) \rho(\mathbf{r}') d^3\mathbf{r}' \tag{3.44}$$

and so on. The first term, the monopole moment, is just the total charge. The second term is called the dipole moment $\mathbf{p}$, and the third is the quadrupole moment. Of the nine components of the quadrupole moment, only five are independent. Symmetry ($Q_{ij} = Q_{ji}$) reduces the number of independent components to six, and since the trace vanishes,

$$\sum_{i=1}^{3} Q_{ii} = \sum_{i=1}^{3} \int (3r_i' r_i' - r'^2) \rho(\mathbf{r}') d^3\mathbf{r}' = 0 \tag{3.45}$$

by (3.40), we see that the third diagonal component is dependent on the other two. Beyond the quadrupole moment the multipole expansion (3.41) becomes increasingly cumbersome.

In addition to using a multipole expansion to calculate the potential outside a localized charge distribution, the multipole expansion can be used to compute the energy $W = \int \rho(\mathbf{r}) \, \Phi(\mathbf{r}) \, d^3\mathbf{r}$ of a localized charge distribution in an externally applied potential Φ. Near the origin, the potential Φ may be Taylor expanded in the form

$$\Phi(\mathbf{r}) = \Phi \Big|_{\mathbf{r}=0} + \sum_{i=1}^{3} r_i \frac{\partial \Phi}{\partial r_i} \Big|_{\mathbf{r}=0} + \frac{1}{2} \sum_{i,j=1}^{3} r_i r_j \frac{\partial^2 \Phi}{\partial r_i \, \partial r_j} \Big|_{\mathbf{r}=0} + \cdots \tag{3.46}$$

But the externally applied potential (the potential in the absence of the local charge distribution) satisfies the Laplace equation

$$\sum_{i=1}^{3} \frac{\partial^2 \Phi}{\partial r_i \, \partial r_i} = 0 \tag{3.47}$$

so in the third term of (3.46), we may add to the factor $r_i r_j$ anything multiplied by δ_{ij} and the sum is unchanged. Specifically, we add $\frac{1}{3} r^2 \delta_{ij}$, in which case the energy of the charge distribution in the externally applied field becomes

$$W = \int \Phi(\mathbf{r})\rho(\mathbf{r})\, d^3\mathbf{r} = Q\Phi\Big|_{\mathbf{r}=0} + \sum_{i=1}^{3} Q_i \frac{\partial \Phi}{\partial r_i}\Big|_{\mathbf{r}=0} + \frac{1}{6}\sum_{i,j=1}^{3} Q_{ij} \frac{\partial^2 \Phi}{\partial r_i\, \partial r_j}\Big|_{\mathbf{r}=0} + \cdots$$

(3.48)

In terms of the applied electric field $\mathbf{E} = -\nabla\Phi$, this becomes

$$W = Q\Phi(0) - \mathbf{p}\cdot\mathbf{E}(0) - \frac{1}{6}\sum_{i,j=1}^{3} Q_{ij} \frac{\partial E_j}{\partial r_i}\Big|_{\mathbf{r}=0} + \cdots$$

(3.49)

We see from this that the total charge Q interacts with the potential at the origin, the dipole moment $\mathbf{p}$ interacts with the electric field, and the quadrupole moment Q_{ij} interacts with the gradient of the field. The net force on a rigid charge distribution, such as a quantum-mechanical atom or molecule, is found from the gradient of the energy. Thus,

$$F_k = -\frac{\partial W}{\partial r_k} = QE_k(0) + \sum_{i=1}^{3} p_i \frac{\partial E_i}{\partial r_k}\Big|_{\mathbf{r}=0} + \frac{1}{6}\sum_{i,j=1}^{3} Q_{ij} \frac{\partial^2 E_j}{\partial r_i\, \partial r_k}\Big|_{\mathbf{r}=0} + \cdots$$

(3.50)

The net force on a localized charge distribution is proportional to the total charge times the electric field, the dipole moment times the derivative of the electric field, the quadrupole moment times the second derivative of the electric field, and so on. If the charge distribution is not perfectly rigid, so that a dipole or higher order moment can be induced by the external field, then the situation becomes more complex. This is discussed in Chapters 6 and 7.

Some examples can make these notions clearer. Consider the interaction of two dipoles, $\mathbf{p}_1$ and $\mathbf{p}_2$, the first at the origin and the second at point $\mathbf{r}$, as shown in Figure 3.3. The potential at point $\mathbf{r}$ due to the dipole $\mathbf{p}_1$ at the origin is

$$\Phi(\mathbf{r}_2) = \frac{1}{4\pi\varepsilon_0} \frac{\mathbf{p}_1\cdot\mathbf{r}}{r^3}$$

(3.51)

and the electric field is

$$\mathbf{E} = -\nabla\Phi = \frac{1}{4\pi\varepsilon_0}\left(3\frac{\mathbf{p}_1\cdot\mathbf{r}}{r^5}\mathbf{r} - \frac{\mathbf{p}_1}{r^3}\right)$$

(3.52)

The energy of interaction is then

$$W = -\mathbf{E}\cdot\mathbf{p}_2 = \frac{1}{4\pi\varepsilon_0}\left[\frac{\mathbf{p}_1\cdot\mathbf{p}_2}{r^3} - 3\frac{(\mathbf{p}_1\cdot\mathbf{r})(\mathbf{p}_2\cdot\mathbf{r})}{r^5}\right]$$

(3.53)

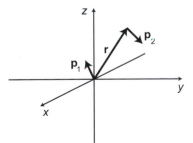

Figure 3.3 Interaction between two dipoles.

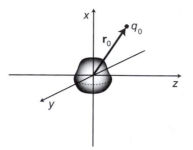

Figure 3.4 Interaction of a charge distribution with a point charge.

The energy is lowest when the dipoles are both oriented in the same direction parallel or antiparallel to the vector $\mathbf{r}$ that separates them.

As another example, we consider the interaction of a charge distribution located near the origin with a point charge q_0 located at position $\mathbf{r}_0$, as shown in Figure 3.4. We assume in the following that the total charge and dipole moment of the localized charge distribution vanish, so the quadrupole components are the first nonvanishing moments. If we make the distribution symmetric about the x axis and use the fact that the trace of Q_{ij} vanishes, then the quadrupole moments are

$$Q_{11} = \int (3x^2 - r^2)\rho(\mathbf{r})\, d^3\mathbf{r} \tag{3.54}$$

$$Q_{22} = Q_{33} = -\tfrac{1}{2}Q_{11} \tag{3.55}$$

$$Q_{ij} = 0, \qquad \text{for } i \neq j \tag{3.56}$$

For any other orientation of the charge distribution, the quadrupole moment matrix can be found by a rotation of coordinates, if desired. However, this is not necessary, since the energy clearly depends only on the angle θ between the axis of the spheroid and the radius vector $\mathbf{r}_0$. From (3.48) we see that the dominant term in the interaction energy (the quadrupole term) is

$$W = -\frac{1}{6}\sum_{i,j=1}^{3} Q_{ij}\frac{\partial E_j}{\partial r_i}\bigg|_{\mathbf{r}=0} = -\frac{Q_{11}}{6}\left(\frac{\partial E_x}{\partial x} - \frac{1}{2}\frac{\partial E_y}{\partial y} - \frac{1}{2}\frac{\partial E_z}{\partial z}\right) \tag{3.57}$$

But

$$\mathbf{E} = \frac{q_0}{4\pi\varepsilon_0}\frac{\mathbf{r} - \mathbf{r}_0}{|\mathbf{r} - \mathbf{r}_0|^3} \tag{3.58}$$

so

$$\frac{\partial E_i}{\partial r_i} = -\frac{q_0}{4\pi\varepsilon_0}\left(\frac{3r_{0i}^2}{r_0^5} - \frac{1}{r_0^3}\right) \tag{3.59}$$

and the interaction energy is

$$W = \frac{Q_{11}q_0}{16\pi\varepsilon_0}\frac{3x_0^2 - r_0^2}{r_0^5} = \frac{Q_{11}q_0}{4\pi\varepsilon_0}\frac{3\cos^2\theta - 1}{r_0^3} \tag{3.60}$$

To make this example more concrete, we consider an atomic system consisting of a positively charged nucleus at the origin (effectively a point charge) surrounded by a negative charge distribution created by the electrons. If the negative charge distribution of the electrons is prolate (egg shaped), then the quadrupole moment of the atom is negative, $Q_{11} < 0$. In this case the lowest energy occurs for $\theta = 0$ and $\theta = \pi$, so the atom tends to line up parallel to the electric field of the point charge. If the charge distribution is oblate (pancake shaped), then $Q_{11} > 0$. In this case the lowest energy occurs for $\theta = \pi/2$, and the atom tends to align its symmetry axis normal to the field.

EXERCISE 3.3

Molecular interactions are frequently interpreted in terms of multipoles.

(a) What is the energy of interaction between a dipole and a point charge?
(b) It is found experimentally (in molecular-beam experiments) that the bond energy between a water molecule ($p = 6.2 \times 10^{-30}$ C-m) and a Mg^+ ion is 1.7×10^{-19} J, with a bond length of 2.0×10^{-10} m. How does this compare to the charge–dipole interaction? Draw a graph of the interaction energy as a function of the bond length including the expected short-range repulsion, and use this to explain the difference between the charge–dipole interaction energy and the measured bond energy.

EXERCISE 3.4

Consider a molecule at the the origin with axial symmetry about the x axis having no net charge or dipole moment, but a quadrupole moment Q_{11}. The molecule interacts with a charge q_0 at point $\mathbf{r}_0 = (r_0, \theta_0, \phi_0)$.

(a) Show that the force on the quadrupole is

$$\mathbf{F} = -\frac{3q_0 Q_{11}}{16\pi \varepsilon_0 r_0^4}[(1 + 5\cos^2 \theta_0)\hat{\mathbf{r}}_0 - 2\cos \theta_0 \hat{\mathbf{x}}] \tag{3.61}$$

(b) Show that the torque on the quadrupole is

$$\tau = \frac{3q_0 Q_{11}}{8\pi \varepsilon_0 r_0^3} \cos \theta_0 \sin \theta_0 \tag{3.62}$$

EXERCISE 3.5

When a spherically symmetric atom is placed in an electric field, the charge distribution distorts to form a dipole moment. In the linear approximation, the induced dipole moment is

$$\mathbf{p} = \alpha \mathbf{E} \tag{3.63}$$

where α is the atomic polarizability.

(a) Show that relative to infinity (where the externally applied field is assumed to vanish), the energy of the atom in the electric field $\mathbf{E}$ is

$$\mathcal{W} = -\tfrac{1}{2}\varepsilon_0 E^2 V \tag{3.64}$$

where $\frac{1}{2}\varepsilon_0 E^2$ is the energy density of the electric field and

$$V = \frac{\alpha}{\varepsilon_0} \tag{3.65}$$

is a volume that is typically of the order of the volume of the atom. Thus, neutral atoms are drawn into regions where the electric field is highest.

(b) The atom can be represented crudely by a nucleus surrounded by a rigid, uniform, spherical distribution of negative charge of radius a, such that the nucleus can be pulled off center by the externally applied electric field against the restoring force of the spherical charge distribution itself. Compute the restoring electric field felt by the nucleus inside the uniform negative charge distribution, and show that for this simple atomic model the polarizability is

$$\alpha = 4\pi \varepsilon_0 a^3 \tag{3.66}$$

EXERCISE 3.6

The long-range interaction between atoms and molecules is dominated by the electrostatic interactions between their multipole moments. In the case of neutral, spherically symmetric atoms or molecules, the interaction is called the van der Waals attraction and has an r^{-6} dependence on the separation r between the atomic centers. Classically, this is interpreted as the interaction between a fluctuating dipole moment in one atom that induces a dipole moment in the other atom, and vice versa. Although the average fluctuating dipole moment of a spherically symmetric atom vanishes, the mean square fluctuation does not.

(a) Show that the instantaneous electric field of dipole 1 (located at the origin) is

$$\mathbf{E}_1(\mathbf{r}) = -\nabla \Phi_1 = \frac{1}{4\pi \varepsilon_0 r^3} [3(\mathbf{p}_1 \cdot \hat{\mathbf{r}})\hat{\mathbf{r}} - \mathbf{p}_1] \tag{3.67}$$

The dipole moment induced in atom 2 (located at $\mathbf{r}$) is then $\mathbf{p}_2 = \alpha_2 \mathbf{E}_1(\mathbf{r})$, where α_2 is the molecular polarizability.

(b) The energy of the induced dipole in the electric field of atom 1 is $\mathcal{W} = -\mathbf{p}_2 \cdot \mathbf{E}_1(\mathbf{r})$. Show that the instantaneous energy of the dipole–dipole interaction is

$$\mathcal{W} = \frac{-\alpha_2}{(4\pi \varepsilon_0)^2 r^6} [\mathbf{p}_1 \cdot \mathbf{p}_1 + 3(\mathbf{p}_1 \cdot \hat{\mathbf{r}})^2] \tag{3.68}$$

(c) Average this over a time that is long compared with the fluctuations of $\mathbf{p}_1$, and add the effects of the fluctuating dipole moment of atom 2 on atom 1 to show that the total interaction energy is

$$\langle \mathcal{W} \rangle_{\text{total}} = -\frac{\alpha_2 \langle p_1^2 \rangle + \alpha_1 \langle p_2^2 \rangle}{8\pi^2 \varepsilon_0^2 r^6} \tag{3.69}$$

where the brackets indicate a time average. This is negative, showing that the van der Waals interaction is attractive.

3.2 BOUNDARY-VALUE PROBLEMS WITH CONDUCTORS

3.2.1 Boundary Conditions and Uniqueness Theorems

Expressions (3.7) and (3.11) for the scalar potential and the electric field are perfectly general in the static approximation, provided that the fields and potentials vanish at infinity. If we know the charge distribution ρ and can evaluate the integrals, we can determine the field. In an ideal insulator, the charges are fixed and immobile. We can therefore establish any charge distribution we choose and specify it ahead of time. We all know about placing a surface charge on amber by rubbing it with cat's fur, and one may equally imagine creating a volume charge distribution by ion implantation or other means. In any event, the charge distribution is fixed. The charge density in an ideal insulator may therefore be regarded as known.

In an ideal conductor, on the other hand, the charges are free to move about under the influence of even the smallest field, so we do not know the charge distribution a priori. But in electrostatics the charges are at rest, so the fields in an ideal conductor must vanish. Since the field in the conductor is the gradient of the potential, the potential in a conductor must everywhere be the same. Thus, we can specify the potential of a conductor, or the total charge, but not the charge distribution. We do know a few things about the charge distribution, however. Specifically, all the charge in a conductor must reside on the surface of the conductor. This follows immediately from the Poisson equation (3.1). If the potential is constant in some volume, the charge density vanishes there.

In response to an external field, the charges on the conductor arrange themselves over the surface to cancel out the external field throughout the volume of the conductor, as indicated in Figure 3.5. Outside the conductor, the component of the electric field $\mathbf{E}$ parallel to the surface must vanish, for the conductor is an equipotential and therefore the gradient of the potential must vanish in the direction parallel to the surface. The field in the $\hat{\mathbf{n}}$ direction normal to the surface is related to the local surface charge density σ. To determine the relation, we construct an infinitesimal Gaussian surface in the shape of a pillbox, with the flat surfaces parallel and very close to the (locally flat) surface of the conductor, as shown in Figure 3.6. Applying Gauss's law to the pillbox, we see that

$$\oint_S \mathbf{E} \cdot \hat{\mathbf{n}} \, dS = \frac{\sigma A}{\varepsilon_0} \tag{3.70}$$

where S is the surface of the pillbox, A is the (infinitesimal) area of the top of the pillbox, and the surface charge density σ is nearly constant inside the pillbox. But the contributions

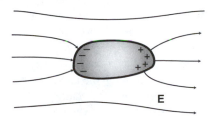

Figure 3.5 Surface charge distribution on a conductor.

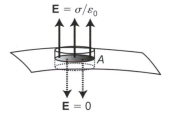

Figure 3.6 Local surface charge density on a conductor.

to the surface integral due to the sides of the pillbox may be ignored if the top and bottom are very close together, and the contribution due to the portion of the pillbox inside the conductor vanishes because the field vanishes there. The flux out of the pillbox is therefore just the flux $\int \mathbf{E} \cdot \hat{\mathbf{n}} \, dS = \mathbf{E} \cdot \hat{\mathbf{n}} A$ out the top of the pillbox, so the field at the surface is

$$\mathbf{E} = -\frac{\partial \Phi}{\partial n} \hat{\mathbf{n}} = \frac{\sigma}{\varepsilon_0} \hat{\mathbf{n}} \tag{3.71}$$

where $\partial \Phi / \partial n = \hat{\mathbf{n}} \cdot \nabla \Phi$ is the gradient of the potential in the direction normal to the surface.

But the charges are mobile, so the question arises whether any solution we obtain in the presence of conductors is unique. To be specific, we consider the problem of determining the potential $\Phi(\mathbf{r})$ throughout a volume V, where V is bounded by one or several conducting surfaces S_i and, perhaps, infinity. Inside the volume V, the potential satisfies the Poisson equation (3.1) for some prescribed charge distribution $\rho(\mathbf{r})$, and we specify either the total charge or the potential of each of the conducting surfaces. To see if our solution is unique, we assume for the moment that there exist two solutions to this problem, which we call Φ_1 and Φ_2, and consider the integral

$$I(\Phi_1, \Phi_2) = \int_V [\nabla(\Phi_2 - \Phi_1)]^2 \, d^3\mathbf{r} = \int_V \nabla(\Phi_2 - \Phi_1) \cdot \nabla(\Phi_2 - \Phi_1) \, d^3\mathbf{r} \tag{3.72}$$

If we integrate this by parts, we get

$$I(\Phi_1, \Phi_2) = \sum_i \int_{S_i} (\Phi_2 - \Phi_1) \nabla(\Phi_2 - \Phi_1) \cdot \hat{\mathbf{n}} \, dS - \int_V (\Phi_2 - \Phi_1) \nabla^2(\Phi_2 - \Phi_1) \, d^3\mathbf{r} \tag{3.73}$$

where $\hat{\mathbf{n}}$ is a unit vector normal to the surface element dS. But $\nabla^2(\Phi_2 - \Phi_1) = 0$, since both solutions satisfy the Poisson equation with the same charge density. Therefore, the second term vanishes. Moreover, on each of the conducting surfaces S_i the potential is a constant, so the factor $(\Phi_2 - \Phi_1)$ is at most a constant, which may be moved outside the integral. In view of Gauss's law, then, the integral is

$$I(\Phi_1, \Phi_2) = -\frac{1}{\varepsilon_0} \sum_i [\Phi_2(S_i) - \Phi_1(S_i)][Q_2(S_i) - Q_1(S_i)] \tag{3.74}$$

where $\Phi(S_i)$ is the potential and $Q(S_i)$ the charge on the surface S_i. The integral $I(\Phi_1, \Phi_2)$ therefore vanishes, proving the solution unique, for two types of boundary conditions on the conductors. If the total charge $Q(S_i) = Q_i$ on the ith conductor is specified as a boundary condition, then the second factor vanishes. If the potential $\Phi(S_i) = \Phi_i$ is specified, then the first factor vanishes. One might imagine fixing the potential physically by connecting a battery between ground and the conductor in question. The charge is then unknown, being determined by the current that flows through the battery, but the potential is fixed. Either of these two boundary conditions, specifying the charge or the potential of a conductor, may be used for any of the various conductors in a given problem, and the solution is still unique. Of course, if both boundary conditions are applied to any one conductor, the problem is overconstrained, since either one is sufficient to define a unique solution.

3.2.2 Energy and Capacitance

We are all familiar with the idea of a capacitor, which in an idealized configuration consists of two parallel, conducting plates in close proximity. As charge is moved from one plate to the other, the voltage difference between the plates increases linearly, with a coefficient of proportionality called the capacitance (or, more precisely, the reciprocal of the capacitance). The energy in the electric field is then expressible in terms of either the voltage or the charge and the capacitance. In this section we extend these ideas to an arbitrary number of conductors.

Consider, for a moment, a set of conductors each with a specified charge Q_i. Since the equations of electrostatics and the boundary conditions are linear, the potential Φ_i must be a linear function of all the charges Q_j,

$$\Phi_i = \sum_j K_{ij} Q_j \tag{3.75}$$

where the constants K_{ij} depend on the geometry of the conductors. If we have a single conductor charged to a positive potential, the potential at all points in the surrounding space is also positive. If we place other, uncharged conductors near the first, their potential must also be positive. From this we conclude that the coefficients K_{ij} are all positive,

$$K_{ij} > 0 \tag{3.76}$$

From (3.27), the total energy in the field is

$$W = \frac{1}{2} \sum_i Q_i \Phi_i = \frac{1}{2} \sum_{i,j} K_{ij} Q_i Q_j \tag{3.77}$$

If we bring up charge δQ_i and add it to the ith conductor, the change in the energy is

$$\delta W = \sum_i \Phi_i \, \delta Q_i = \sum_{i,j} K_{ij} Q_j \, \delta Q_i \tag{3.78}$$

But if we differentiate (3.77), we see that

$$\delta W = \frac{1}{2} \sum_{i,j} K_{ij} (Q_i \, \delta Q_j + Q_j \, \delta Q_i) = \frac{1}{2} \sum_{i,j} (K_{ij} + K_{ji}) Q_j \, \delta Q_i \tag{3.79}$$

Comparing these expressions, we conclude that the matrix of coefficients is symmetric,

$$K_{ij} = K_{ji} \tag{3.80}$$

Since the energy is positive definite, that is,

$$W = \frac{1}{2} \sum_{i,j} K_{ij} Q_i Q_j > 0 \tag{3.81}$$

unless all the charges vanish, the matrix K_{ij} must have an inverse. We call this inverse matrix the capacitance matrix, and the coefficients satisfy

$$\sum_i C_{ij} K_{ik} = \delta_{jk} \tag{3.82}$$

Since K_{ij} is symmetric, it follows that the capacitance matrix is symmetric. If we multiply (3.75) by C_{ik} and sum over i, we find that

$$Q_k = \sum_i C_{ik}\Phi_i \tag{3.83}$$

That is, the charge on the kth conductor is proportional to the mutual capacitance and the potential of all the other conductors in the system. Substituting (3.83) into (3.81) and using (3.82), we find that the total energy in a system of conductors is

$$W = \frac{1}{2}\sum_{i,j} C_{ij}\Phi_i\Phi_j \tag{3.84}$$

In a conventional capacitor, consisting of two parallel, conducting plates, the two plates have equal and opposite charges, $Q_1 = -Q_2 = Q$, so the energy is

$$W = \frac{1}{2}K_{11}Q_1^2 + \frac{1}{2}K_{22}Q_2^2 + K_{12}Q_1 Q_2 = \frac{Q^2}{2C} \tag{3.85}$$

where the conventional capacitance of the pair of plates is

$$C = \frac{1}{K_{11} + K_{22} - 2K_{12}} \tag{3.86}$$

The capacitance C is always positive because the energy is positive. When the plates are very close together, the potential on plate 1, relative to infinity, due to the charge on plate 2 is nearly as large as the potential due to the same charge on plate 1 itself. Referring to (3.75), we see that this implies that for two plates close together, K_{12} is nearly as large as $K_{11} = K_{22}$, so the denominator of (3.86) nearly vanishes. This makes the capacitance large, which is why we construct capacitors this way.

Since the energy density in the field is $\frac{1}{2}\varepsilon_0 E^2$, the work done *on* the field in displacing the surface of a conductor is

$$\delta W = -\tfrac{1}{2}\varepsilon_0 E^2\, dS\, \delta n = -dF\, \delta n \tag{3.87}$$

where W is the energy in the field, dS is the surface area, $\delta n = \hat{\mathbf{n}} \cdot \delta\mathbf{r}$ is the component of the displacement $\delta\mathbf{r}$ in the direction $\hat{\mathbf{n}}$ normal to the surface out of the conductor, as shown in Figure 3.7, and dF is the force exerted by the field on the surface dS. The total force on a conductor is therefore

$$\mathbf{F} = \frac{1}{2}\varepsilon_0 \int_S E^2\, \hat{\mathbf{n}}\, dS = \frac{1}{2}\int_S \sigma \mathbf{E}\, dS \tag{3.88}$$

where we have used (3.71) to express the force in terms of the surface charge density σ. The factor $\frac{1}{2}$ in the last expression is at first surprising but may be understood as follows.

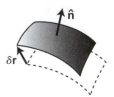

Figure 3.7 Force on a conductor in an electric field.

The field near the surface is the sum of the fields of the external charges and the surface charge. But the fields of the external charges and the surface charge cancel immediately inside the conductor and add immediately outside it, so the field due just to the external charges is half the total field immediately outside the conductor. This is the field that exerts a force on the surface charge.

For a set of conductors, the force may be computed from the change in the field energy when a conductor is displaced. For a generalized displacement δs, which might be a translation or a rotation, the work done *on* the field when the *charges Q_i are fixed* is

$$\delta W = \frac{1}{2} \sum_{i,j} Q_i Q_j \frac{dK_{ij}}{ds} \delta s = -F \delta s \tag{3.89}$$

where F is the generalized force on the conductor. When the *potentials Φ_i are fixed,* the change δW in the energy of the field is

$$\delta W = \sum_{i,j} K_{ij} Q_i \delta Q_j + \frac{1}{2} \sum_{i,j} Q_i Q_j \delta K_{ij} \tag{3.90}$$

where we have taken advantage of the symmetry of K_{ij} to simplify the first term. Comparing this expression with (3.89), we see that the second term is just the work done on the field by the displacement of the conductors. Since the charges on the conductors are the same, the forces (3.88) on the conductors are the same. We therefore identify the first term as the work done by the batteries as they move the charges to keep the potentials constant. But from (3.83) we see that

$$\delta Q_j = \sum_k \Phi_k \delta C_{jk} \tag{3.91}$$

and from (3.82) we see that

$$\sum_j (C_{jk} \delta K_{ij} + K_{ij} \delta C_{jk}) = 0 \tag{3.92}$$

Combining these results, we find that the work done by the batteries is

$$\sum_{i,j} K_{ij} Q_i \delta Q_j = \sum_{i,j,k} Q_i \Phi_k K_{ij} \delta C_{jk} = -\sum_{i,j,k} C_{jk} \Phi_k Q_i \delta K_{ij} = -\sum_{i,j} Q_i Q_j \delta K_{ij} \tag{3.93}$$

If we compare this with (3.89), we see that the work done by the batteries is exactly twice the work done by the displacement and has the opposite sign. Therefore, the change in the field energy due to a displacement with the potentials fixed has the same magnitude but the opposite sign compared with the change in the field energy for a displacement with the charges fixed. The force on each conductor is, of course, the same whether the potential is held fixed or the charge is held fixed. For example, if two spheres with charges Q and $-Q$ are displaced toward one another, the energy in the electrostatic field decreases because the charges attract one another. On the other hand, if a battery connects the two spheres while they are displaced, the field energy increases because the charge on the spheres increases as the mutual capacitance increases. The increased charge makes the energy in the electric field grow. The interaction between the two spheres is still attractive, of course, independent of whether the battery is connected.

EXERCISE 3.7

Using arguments about the energy in the field and arguments similar to that used to show that $K_{ij} > 0$, prove that

$$C_{ii} > 0 \tag{3.94}$$

and

$$C_{i \neq j} < 0 \tag{3.95}$$

EXERCISE 3.8

Consider the case of two spherical conductors of radii a_1 and a_2 whose centers are separated by the distance $d \gg a$, where $a = O(a_1, a_2)$.

(a) Estimate the coefficients K_{ij} to lowest order in a/d, and give arguments to show that this estimate is good to $O((a/d)^3)$.

(b) Invert the matrix to show that

$$C_{ii} = 4\pi \varepsilon_0 \frac{a_i d^2}{d^2 - a_1 a_2} \tag{3.96}$$

and

$$C_{i \neq j} = -4\pi \varepsilon_0 \frac{a_1 a_2 d}{d^2 - a_1 a_2} \tag{3.97}$$

(c) Show that the energy in the field is

$$W \approx \frac{1}{4\pi \varepsilon_0} \left(\frac{Q_1^2}{2a_1} + \frac{Q_2^2}{2a_2} + \frac{Q_1 Q_2}{d} \right) \tag{3.98}$$

(d) Show that if one sphere is given charge Q and the other charge $-Q$, the effective capacitance is

$$\frac{1}{C} = \frac{2W}{Q^2} \approx \frac{1}{4\pi \varepsilon_0} \left(\frac{1}{a_1} + \frac{1}{a_2} - \frac{2}{d} \right) \tag{3.99}$$

This shows that compared with the capacitance of two isolated spheres, the correction is $O(a/d)$.

EXERCISE 3.9

An old-fashioned radio is tuned by means of a variable capacitor whose capacitance varies through $10 \leq C \leq 100$ pF as the shaft is rotated through $0 \leq \theta \leq \pi$.

(a) The potential across the capacitor is maintained at $\Phi = 100$ V. Show that the torque on the shaft is

$$\tau = \frac{\Phi^2}{2} \frac{dC}{d\theta} \tag{3.100}$$

What is the value of the torque?

(b) The capacitor is rotated to $\theta = \pi$ ($C = 100$ pF) and charged to $\Phi = 100$ V, and the voltage source is disconnected. The capacitor is then rotated to $\theta = 0$ ($C = 10$ pF). What is the voltage on the capacitor now? What is the energy in the electric field?

(c) The capacitor is rotated to $\theta = \pi$ ($C = 100$ pF) and charged to $\Phi = 100$ V. The capacitor is then rotated to $\theta = 0$ ($C = 10$ pF) with the voltage source connected. What is the energy in the electric field now?

3.2.3 Method of Images

A priori, it is somewhat remarkable that the charges on a conductor can be arranged to cancel precisely the field of an arbitrary arrangement of charges external to the conductor. But, of course, the charges simply keep moving until the fields are exactly canceled. To get a feel for the way the charges arrange themselves on a conductor, it is worthwhile to examine some simple cases.

The method of images lends itself to simple situations with a high degree of symmetry. What we do is place a small number of charges inside the conductor to represent the effect of the surface charge. This does not describe the field inside the conductor, which vanishes, but it does describe the field outside the conductor. The simplest case is that of a point charge in front of an infinite, plane conductor, as shown in Figure 3.8. We assume that the surface of the conductor is at $x = 0$ and that the potential vanishes there. The charge q is located at point $\mathbf{x}_0$ on the axis. As indicated in Figure 3.8, the field at the surface of the conductor is normal to the conductor, so the symmetry of the situation is irresistible. If we place a charge $-q$ at point $-\mathbf{x}_0$, we get the field distribution indicated by the solid lines outside the conductor and by the dotted lines inside the conductor. This is called the image charge, owing to the obvious analogy with the image formed by a plane mirror. The potential at point $\mathbf{r}$ is then simply

$$\Phi(\mathbf{r}) = \frac{q}{4\pi\varepsilon_0}\left(\frac{1}{|\mathbf{r} - \mathbf{x}_0|} - \frac{1}{|\mathbf{r} + \mathbf{x}_0|}\right) \tag{3.101}$$

Clearly, this satisfies the boundary condition $\Phi = 0$ at the surface of the conductor.

We can compute the actual surface charge density on the conductor from the electric field at the surface of the conductor using (3.71). We get

$$\sigma = \varepsilon_0 \mathbf{E} \cdot \hat{\mathbf{n}} = -\frac{q}{2\pi}\frac{x_0}{|\mathbf{r} - \mathbf{x}_0|^3} = -\frac{q}{2\pi}\frac{x_0}{\left(r^2 + x_0^2\right)^{3/2}} \tag{3.102}$$

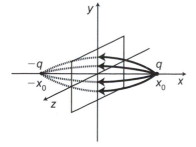

Figure 3.8 Point charge in front of a plane conductor, and image charge.

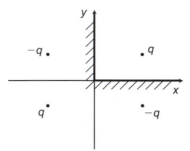

Figure 3.9 Corner between two planar conductors.

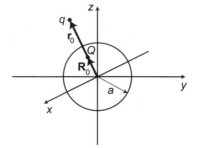

Figure 3.10 Point charge near a grounded, conducting sphere.

It is not surprising that the charge is spread out over an area whose radius is on the order of x_0. This surface charge distribution gives the same field in the space outside the conductor as the image charge at $-\mathbf{x}_0$.

Other highly symmetric problems likewise lend themselves to the method of images. For example, the corner between two planar conductors shown in Figure 3.9 is easily solved if the angle between the conductors is π/n, for any integer n. In this case the problem has $2n$-fold symmetry and the field is correctly given by the $2n$-fold symmetric charge distribution shown in Figure 3.9 for the case of $n = 2$.

The case of a point charge q near a grounded, conducting sphere of radius a is shown in Figure 3.10. Because of the field of the point charge, the charge on the surface of the sphere is attracted toward the side of the sphere facing the point charge. The electric field in the region outside the sphere can be represented by the field of a point charge Q, the image charge, placed off center inside the conductor as shown. According to Gauss's law, applied to a surface surrounding the charged sphere, the total charge on the surface of the conductor is just equal to the image charge Q. To prove that the image charge correctly describes the field outside the sphere, we must show that the surface of the sphere is an equipotential. To do this, we place the center of the sphere at the origin and the point charge q at position $\mathbf{r}_0 = r_0\hat{\mathbf{n}}_0$, where $\hat{\mathbf{n}}_0$ is a unit vector. By symmetry, the image charge Q must lie somewhere on the line from the center of the sphere to the point charge q, at a position inside the sphere that we call $\mathbf{R}_0 = R_0\hat{\mathbf{n}}_0$. The potential on the surface of the sphere ($\mathbf{r} = a\hat{\mathbf{n}}$, where $\hat{\mathbf{n}}$ is a unit vector) is then

$$\Phi(\mathbf{r}) = \frac{1}{4\pi\varepsilon_0}\left(\frac{q}{|a\hat{\mathbf{n}} - r_0\hat{\mathbf{n}}_0|} + \frac{Q}{|a\hat{\mathbf{n}} - R_0\hat{\mathbf{n}}_0|}\right) = 0 \tag{3.103}$$

since the sphere is grounded. We therefore require that

$$q\left|\hat{\mathbf{n}} - \frac{R_0}{a}\hat{\mathbf{n}}_0\right| + Q\left|\hat{\mathbf{n}} - \frac{r_0}{a}\hat{\mathbf{n}}_0\right| = 0 \tag{3.104}$$

for all $\hat{\mathbf{n}}$. Using the law of cosines, we get

$$q^2\left[1 + \left(\frac{R_0}{a}\right)^2 - 2\left(\frac{R_0}{a}\right)\cos\theta\right] = Q^2\left[1 + \left(\frac{r_0}{a}\right)^2 - 2\left(\frac{r_0}{a}\right)\cos\theta\right] \tag{3.105}$$

for all θ, where $\theta = \arccos(\hat{\mathbf{n}} \cdot \hat{\mathbf{n}}_0)$ is the angle between the source and observation positions. Solving this, we find that (3.103) is true if

$$r_0 R_0 = a^2 \tag{3.106}$$

and

$$Q = -q\frac{a}{r_0} \tag{3.107}$$

As the charge approaches the surface of the sphere ($r_0 \to a$), the image charge approaches the surface from the other side and $Q \to -q$, as expected. As the charge recedes to a distance that is large compared to the radius of the sphere, the image charge Q approaches the center of the sphere and vanishes.

EXERCISE 3.10

Consider an isolated, conducting sphere of radius a, centered at the origin, with no net charge on it. A point charge q is brought up from infinity to the point $r_0 > a$. Show that the potential of the sphere is

$$\Phi = \frac{q}{4\pi\varepsilon_0 r_0} \tag{3.108}$$

just as though the sphere were a point at the origin.

3.2.4 Separation of Variables

For certain symmetric situations it is possible to express the solution as a product of functions

$$\Phi(x, y, z) = F(q_1)G(q_2)H(q_3) \tag{3.109}$$

where $q_1 \ldots q_3$ are curvilinear coordinates that reflect the symmetry of the problem. Since each function has only a single argument q_i, it satisfies an ordinary differential equation instead of a partial differential equation, which greatly simplifies life. This is called the method of separation of variables. When one or more of the curvilinear coordinates has a finite range (such as $0 \leq \phi \leq 2\pi$ for an azimuthal angle), or when the boundaries of the problem establish a finite range for one or more of the coordinates, it is generally found that the solution can be represented as a sum of eigenfunctions of the differential equation for that coordinate. This gives us more possibilities for finding a solution.

Unfortunately, not all problems can be analyzed by separation of variables, since the method demands a high degree of symmetry. In the first place, only 11 curvilinear coordinate systems have been found for which the Laplace operator itself is separable. Fortunately these include some of our favorites, such as Cartesian, cylindrical, and spherical polar coordinates. In the second place, the boundary conditions must also simplify in the curvilinear coordinate system to make the method useful. These ideas are best illustrated by some examples, and in this section we discuss an example using Cartesian coordinates. In the following sections we take up spheroidal and spherical polar coordinates.

As our first example we consider the problem illustrated in Figure 3.11, consisting of a semi-infinite rectangular box formed from conducting surfaces. The long sides of the box

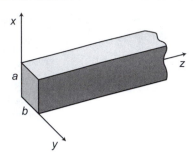

Figure 3.11 Rectangular conducting box.

are held at ground potential, while the end of the box is held at potential $\Phi_0(x, y)$. The clear choice in this example is to use Cartesian coordinates, in which case the Laplace equation is simply

$$\frac{d^2\Phi}{dx^2} + \frac{d^2\Phi}{dy^2} + \frac{d^2\Phi}{dz^2} = 0 \tag{3.110}$$

We assume that the solution can be expressed in the form

$$\Phi(x, y, z) = X(x)Y(y)Z(z) \tag{3.111}$$

If we substitute this into (3.110) and divide by $\Phi = XYZ$, we get

$$\frac{1}{X}\frac{d^2X}{dx^2} + \frac{1}{Y}\frac{d^2Y}{dy^2} + \frac{1}{Z}\frac{d^2Z}{dz^2} = 0 \tag{3.112}$$

But the first term depends only on x, the second only on y, and the third only on z. This can be true only if each term is equal to a constant, so we get the separate equations

$$\frac{d^2X}{dx^2} + k_x^2 X = 0 \tag{3.113}$$

$$\frac{d^2Y}{dy^2} + k_y^2 Y = 0 \tag{3.114}$$

$$\frac{d^2Z}{dz^2} - k_z^2 Z = 0 \tag{3.115}$$

where the separation constants satisfy the relation

$$k_z^2 = k_x^2 + k_y^2 \tag{3.116}$$

To satisfy the boundary conditions, it is sufficient that

$$X(0) = X(a) = 0 \tag{3.117}$$

$$Y(0) = Y(b) = 0 \tag{3.118}$$

and

$$Z \xrightarrow[z \to \infty]{} 0 \tag{3.119}$$

and over the end of the box

$$\Phi(x, y, 0) = \Phi_0(x, y) \tag{3.120}$$

The solutions to (3.113) and (3.114) with boundary conditions (3.117) and (3.118) are

$$X(x) = \sin(k_x x) \tag{3.121}$$

$$Y(y) = \sin(k_y y) \tag{3.122}$$

where

$$k_x a = m\pi, \qquad \text{for } m = 1, 2, 3, \ldots \tag{3.123}$$

$$k_y b = n\pi, \qquad \text{for } n = 1, 2, 3, \ldots \tag{3.124}$$

The solution to (3.115) that satisfies the boundary condition (3.119) at infinity is

$$Z(z) = e^{-k_z z} \tag{3.125}$$

so the solution in three dimensions is of the form

$$\Phi(x, y, z) \sim \sin\left(m\pi \frac{x}{a}\right) \sin\left(n\pi \frac{y}{b}\right) e^{-\pi z \sqrt{(m^2/a^2)+(n^2/b^2)}}, \qquad \text{for } m, n = 1, 2, 3, \ldots \tag{3.126}$$

It remains to satisfy the boundary condition (3.120) at $z = 0$. To do this, we use a sum of the solutions (3.126) of the form

$$\Phi(x, y, z) = \sum_{m,n=1}^{\infty} c_{mn} \sin\left(m\pi \frac{x}{a}\right) \sin\left(n\pi \frac{y}{b}\right) e^{-\pi z \sqrt{(m^2/a^2)+(n^2/b^2)}} \tag{3.127}$$

The boundary condition at $z = 0$ is then

$$\Phi_0(x, y) = \sum_{m,n=1}^{\infty} c_{mn} \sin\left(m\pi \frac{x}{a}\right) \sin\left(n\pi \frac{y}{b}\right) \tag{3.128}$$

To determine the coefficients c_{mn}, we exploit the orthogonality of the sines over the interval $0 \leq x \leq a$ and $0 \leq y \leq b$:

$$\int_0^a \sin\left(m\pi \frac{x}{a}\right) \sin\left(p\pi \frac{x}{a}\right) dx = \frac{1}{2} a \delta_{mp} \tag{3.129}$$

$$\int_0^b \sin\left(n\pi \frac{y}{b}\right) \sin\left(q\pi \frac{y}{b}\right) dy = \frac{1}{2} b \delta_{nq} \tag{3.130}$$

Thus, if we multiply (3.128) by

$$\sin\left(p\pi \frac{x}{a}\right) \sin\left(q\pi \frac{y}{b}\right)$$

and integrate over x and y, we find that

$$c_{pq} = \frac{4}{ab} \int_0^a dx \int_0^b dy \sin\left(p\pi \frac{x}{a}\right) \sin\left(q\pi \frac{y}{b}\right) \Phi_0(x, y) \tag{3.131}$$

That is, the coefficients c_{mn} are just the Fourier coefficients of the boundary condition $\Phi_0(x, y)$. For example, if $\Phi_0(x, y) = \text{constant} = V_0$, we find that

$$c_{mn} = \frac{4V_0}{\pi^2 mn} [1 - \cos(m\pi)] [1 - \cos(n\pi)]$$

$$= \frac{16V_0}{\pi^2 mn} \sin^2\left(\frac{1}{2} m\pi\right) \sin^2\left(\frac{1}{2} n\pi\right) \tag{3.132}$$

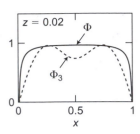

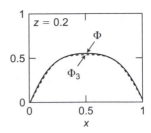

Figure 3.12 Convergence of the solution for a rectangular box.

The convergence of (3.127) to the correct solution is indicated in Figure 3.12, where the potential Φ at the midplane $\left(y = \frac{1}{2}b\right)$ is compared with the approximate solution

$$\Phi_3(x, y, z) = \sum_{m,n=1}^{3} c_{mn} \sin\left(m\pi \frac{x}{a}\right) \sin\left(n\pi \frac{y}{b}\right) e^{-\pi z \sqrt{(m^2/a^2)+(n^2/b^2)}} \qquad (3.133)$$

for $z = 0.02a$ and $z = 0.2a$. As shown there, the first few terms give a relatively poor representation of the potential at $z = 0.02a$ but an excellent representation at $z = 0.2a$. Since the decay rate $\pi\sqrt{(m^2/a^2) + (n^2/b^2)}$ is faster for the higher harmonics (larger m, n), the higher harmonics become relatively less important at larger z. Thus, the solution converges faster as we get farther from the end of the box.

EXERCISE 3.11

The potential $\Phi(x, y)$ in the region $y > 0$ is bounded at $y = 0$ by the periodic function

$$\Phi(x, 0) = \Phi_0, \qquad \text{for } |x - mL| < \frac{1}{2}a \qquad (3.134)$$

$$\Phi(x, 0) = -\Phi_0, \qquad \text{for } |x - (m + \frac{1}{2})L| < \frac{1}{2}a \qquad (3.135)$$

$$\Phi(x, 0) = 0, \qquad \text{otherwise} \qquad (3.136)$$

where $m = \cdots -2, -1, 0, 1, 2, \ldots$, as shown in Figure 3.13, and vanishes as $y \to \infty$. Show that the potential in the region $y > 0$ is

$$\Phi(x, y) = \sum_{m=1}^{\infty} c_m \cos\left(m2\pi \frac{x}{L}\right) e^{-m2\pi(y/L)} \qquad (3.137)$$

where

$$c_m = \frac{4\Phi_0}{\pi m} \sin^2\left(\frac{m\pi}{2}\right) \sin\left(m\pi \frac{a}{L}\right) \qquad (3.138)$$

Figure 3.13 Periodic potential on the boundary.

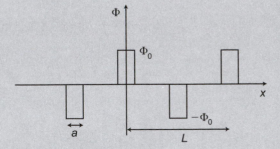

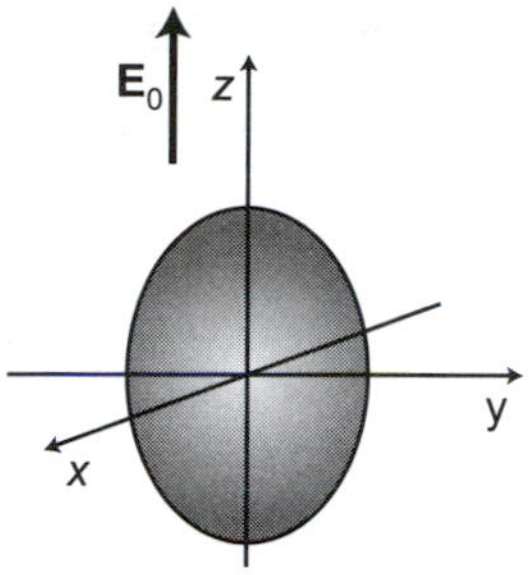

Figure 3.14 Prolate spheroidal conductor.

3.2.5 Spheroidal Coordinates

As a second example of the method of separation of variables, we consider a problem that can be solved using curvilinear coordinates. Specifically, we seek the potential Φ surrounding a prolate spheroidal conductor immersed in the otherwise uniform electric field $\mathbf{E}_0$ directed along the axis of symmetry, as shown in Figure 3.14. The potential must satisfy the Laplace equation subject to the boundary conditions

$$\Phi \to 0 \qquad \text{on the surface of the spheroid} \tag{3.139}$$

and

$$\Phi \to -E_0 z \qquad \text{far from the spheroid} \tag{3.140}$$

where $\hat{\mathbf{z}}$ is the axis of symmetry and the direction of the electric field. When the spheroid is very long and slender, like a needle, the electric field at the tip can be extremely large.

The equation for the surface of a prolate spheroid symmetric about the z axis is

$$\frac{z^2}{a^2} + \frac{r^2}{b^2} = 1 \tag{3.141}$$

where a and b are the major and minor semiaxes of the spheroid ($a > b$) and

$$r^2 = x^2 + y^2 \tag{3.142}$$

is the radial distance from the axis of symmetry. The foci of the spheroid are located on the z axis at the points

$$z^2 = c^2 = a^2 - b^2 \tag{3.143}$$

To find the potential surrounding the spheroid, it is convenient to change to a coordinate system in which one family of coordinate surfaces consists of spheroids confocal with the conductor, with the other families of coordinate surfaces orthogonal to these surfaces everywhere. To do this, we begin with the equation

$$\frac{z^2}{c^2\eta^2} + \frac{r^2}{c^2(\eta^2 - 1)} = 1 \tag{3.144}$$

Since the difference between the denominators of (3.144) is always $c^2\eta^2 - c^2(\eta^2 - 1) = c^2$, this describes a family of surfaces confocal with the original spheroid, with η as a parameter. Given any values of r and z, this quadratic equation in η^2 has two roots, which we call ρ^2 and ζ^2. The quantities ρ and ζ are then the spheroidal coordinates of the point (r, z). For $1 < \rho = \eta < \infty$, both terms in (3.144) are positive, so the surfaces $\rho = \text{constant}$ are

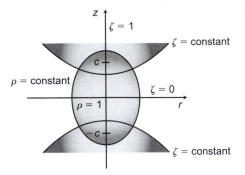

Figure 3.15 Prolate spheroidal coordinates.

spheroids confocal to the original surface. The surface of the original spheroid corresponds to $\rho = a/c$. For $-1 < \zeta = \eta < 1$ the first term in (3.144) is positive and the second term negative, so the surfaces $\zeta = $ constant correspond to hyperboloids of revolution of two sheets, which are also confocal with the original surface, as shown in Figure 3.15. The third coordinate is the azimuthal angle ϕ about the z axis, for which the coordinate surfaces are planes through the z axis. It can be shown that these three families of coordinate surfaces are everywhere orthogonal to one another.

Given the spheroidal coordinates (ρ, ζ), we can solve for r and z in the following way. Consider the expression

$$\left(\frac{z^2}{c^2 \eta^2} + \frac{r^2}{c^2(\eta^2 - 1)} - 1 \right) \eta^2 (\eta^2 - 1) \tag{3.145}$$

This expression is quadratic in η^2, and the coefficient of η^4 is -1. Moreover, if η is one of the coordinates ρ or ζ corresponding to (r, z), then the expression vanishes. Consequently (3.145) must be identically equal to the expression

$$-(\eta^2 - \rho^2)(\eta^2 - \zeta^2) \tag{3.146}$$

If we equate expressions (3.145) and (3.146), and set $\eta^2 = 0$ or $\eta^2 = 1$, we get

$$z = c\rho\zeta \tag{3.147}$$

and

$$r = c\sqrt{(\rho^2 - 1)(1 - \zeta^2)} \tag{3.148}$$

Then,

$$x = r\cos\phi = c\sqrt{(\rho^2 - 1)(1 - \zeta^2)}\cos\phi \tag{3.149}$$

and

$$y = r\sin\phi = c\sqrt{(\rho^2 - 1)(1 - \zeta^2)}\sin\phi \tag{3.150}$$

In terms of the polar coordinates R and θ, we see that

$$R^2 = r^2 + z^2 = c^2(\rho^2 + \zeta^2 - 1) \xrightarrow[\rho \to \infty]{} c^2\rho^2 \tag{3.151}$$

and

$$\cos^2\theta = \frac{z^2}{z^2 + r^2} = \frac{\rho^2\zeta^2}{(\rho^2 + \zeta^2 - 1)} \xrightarrow[\rho \to \infty]{} \zeta^2 \tag{3.152}$$

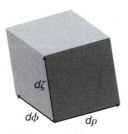

Figure 3.16 Rectangular volume in spheroidal coordinate system.

Therefore, far from the origin, the spheroidal coordinates $(\rho c, \zeta, \phi)$ are analogous to the spherical polar coordinates $(R, \cos\theta, \phi)$. At the origin, $\zeta = 0$ and $\rho = 1$.

Transforming the Laplace equation to spheroidal coordinates (or any other curvilinear coordinates) is inevitably a tedious process. We proceed in the following way. If we consider a small rectangular volume formed by the coordinates (ρ, ζ, ϕ) as shown in Figure 3.16, we find by differentiating (3.147) and (3.148) that the increments dx, dy, and dz along the edges are

$$dx = c\sqrt{\frac{1-\zeta^2}{\rho^2-1}}\,\rho\cos\phi\,d\rho + c\sqrt{\frac{\rho^2-1}{1-\zeta^2}}\,\zeta\cos\phi\,d\zeta - c\sqrt{(\rho^2-1)(1-\zeta^2)}\sin\phi\,d\phi \tag{3.153}$$

$$dy = c\sqrt{\frac{1-\zeta^2}{\rho^2-1}}\,\rho\sin\phi\,d\rho + c\sqrt{\frac{\rho^2-1}{1-\zeta^2}}\,\zeta\sin\phi\,d\zeta + c\sqrt{(\rho^2-1)(1-\zeta^2)}\cos\phi\,d\phi \tag{3.154}$$

and

$$dz = c\zeta\,d\rho + c\rho\,d\zeta \tag{3.155}$$

Summing the squares of all these terms, we find that the increment of length from one corner of the rectangular volume to the diagonal corner (the metric in the new coordinate system) is

$$dl^2 = dx^2 + dy^2 + dz^2 = h_\rho^2\,d\rho^2 + h_\zeta^2\,d\zeta^2 + h_\phi^2\,d\phi^2 \tag{3.156}$$

where

$$h_\rho = c\sqrt{\frac{\rho^2-\zeta^2}{\rho^2-1}} \tag{3.157}$$

$$h_\zeta = c\sqrt{\frac{\rho^2-\zeta^2}{1-\zeta^2}} \tag{3.158}$$

$$h_\phi = c\sqrt{(\rho^2-1)(1-\zeta^2)} \tag{3.159}$$

The increments of length corresponding to increments of the curvilinear coordinates are evidently

$$dl_\rho = h_\rho\,d\rho \tag{3.160}$$

$$dl_\zeta = h_\zeta\,d\zeta \tag{3.161}$$

$$dl_\phi = h_\phi\,d\phi \tag{3.162}$$

To express the Laplace equation in spheroidal coordinates, we recognize that the Laplace equation is equivalent, by the divergence theorem, to the statement

$$\oint_S \frac{\partial \Phi}{\partial n} \, dS = 0 \tag{3.163}$$

where $\partial \Phi / \partial n$ is the derivative normal to the surface element dS in the direction outward from the volume enclosed by S. To apply this to the infinitesimal rectangular volume shown in Figure 3.16 we note that the electric flux through the face normal to the coordinate ρ is

$$\frac{h_\zeta \, d\zeta \, h_\phi \, d\phi}{h_\rho} \frac{\partial \Phi}{\partial \rho}$$

so the net electric flux outward from the volume through this face and the opposite face is

$$\frac{\partial}{\partial \rho} \left(\frac{h_\zeta h_\phi}{h_\rho} \frac{\partial \Phi}{\partial \rho} \right) d\rho \, d\zeta \, d\phi$$

Doing the same calculation for the other faces and adding, we obtain the Laplace equation in the curvilinear coordinate system

$$\left[\frac{\partial}{\partial \rho} \left(\frac{h_\zeta h_\phi}{h_\rho} \frac{\partial \Phi}{\partial \rho} \right) + \frac{\partial}{\partial \zeta} \left(\frac{h_\rho h_\phi}{h_\zeta} \frac{\partial \Phi}{\partial \zeta} \right) + \frac{\partial}{\partial \phi} \left(\frac{h_\rho h_\zeta}{h_\phi} \frac{\partial \Phi}{\partial \phi} \right) \right] d\rho \, d\zeta \, d\phi = 0 \tag{3.164}$$

In the present case, when we substitute for h_ρ, h_ζ, and h_ϕ we find that the Laplace equation in prolate spheroidal coordinates is

$$\frac{\partial}{\partial \rho} \left[(\rho^2 - 1) \frac{\partial \Phi}{\partial \rho} \right] + \frac{\partial}{\partial \zeta} \left[(1 - \zeta^2) \frac{\partial \Phi}{\partial \zeta} \right] + \frac{\rho^2 - \zeta^2}{(\rho^2 - 1)(1 - \zeta^2)} \frac{\partial^2 \Phi}{\partial \phi^2} = 0 \tag{3.165}$$

For azimuthally symmetric problems, such as the one we used to begin this discussion, the dependence of Φ on ϕ vanishes. In this case, we assume a separable solution of the form

$$\Phi = R(\rho) Z(\zeta) \tag{3.166}$$

substitute into (3.165), and divide by Φ, as we did in the previous section. When we do this, we find that the functions $R(\rho)$ and $Z(\zeta)$ must satisfy the ordinary differential equations

$$\frac{d}{d\rho} \left[(1 - \rho^2) \frac{dR}{d\rho} \right] + l(l+1) R = 0, \qquad 1 \le \rho < \infty \tag{3.167}$$

and

$$\frac{d}{d\zeta} \left[(1 - \zeta^2) \frac{dZ}{d\zeta} \right] + l(l+1) Z = 0, \qquad -1 \le \zeta \le 1 \tag{3.168}$$

where we have written the separation constant in the form $l(l+1)$ for convenience in the following discussion.

The linear, homogeneous, second-order, ordinary differential equation (3.168) is called the Legendre equation, and it has been studied to death. It is actually an example of what are called Sturm–Liouville problems, and many of the properties of its solutions are common to other Sturm–Liouville problems that arise in physics. The end points $\zeta = \pm 1$, where the coefficient $(1 - \zeta^2)$ vanishes, are called singular points. If the solution is going to diverge, this is where it will happen. Since the Legendre equation is of second order, it has, for each value of l, two linearly independent solutions, which we call $P_l(\zeta)$ and $Q_l(\zeta)$. All other solutions can be formed from a linear combination of these two solutions. For integer values of l, it is conventional to choose for $P_l(\zeta)$ the solution that is regular at the end points $\zeta = \pm 1$, so $Q_l(\zeta)$ is the solution that diverges there.

To find the solution of (3.168), we assume a power series of the form

$$Z(\zeta) = \sum_{m=0}^{\infty} a_m \zeta^m \tag{3.169}$$

If we substitute this into (3.168), we get the power series

$$\sum_{m=0}^{\infty} \{(m+2)(m+1)\, a_{m+2} + [l(l+1) - m(m+1)]\, a_m\}\zeta^m = 0 \tag{3.170}$$

But if this is to be valid for all $-1 \le \zeta \le 1$, each term must vanish identically and we get the recursion formula

$$(m+2)(m+1)a_{m+2} = [m(m+1) - l(l+1)]a_m \tag{3.171}$$

We are free to choose either $a_0 = 0$ or $a_1 = 0$, in which case the solution consists of just odd or even powers of ζ. One choice corresponds to $P_l(\zeta)$, the other to the linearly independent solution $Q_l(\zeta)$. However, the sum diverges at the end points $\zeta = \pm 1$ unless the series truncates. This is possible for integer values of l, in which case the series truncates at $m = l$.

The regular solutions to the Legendre equation obtained in this way are called the Legendre polynomials $P_l(\zeta)$. Since the Legendre equation is homogeneous, the solutions are defined only up to a constant factor. It is conventional to choose the normalization

$$P_l(1) = 1 \tag{3.172}$$

The first few Legendre polynomials are then

$$P_0(\zeta) = 1 \tag{3.173}$$

$$P_1(\zeta) = \zeta \tag{3.174}$$

$$P_2(\zeta) = \tfrac{1}{2}(3\zeta^2 - 1) \tag{3.175}$$

$$P_3(\zeta) = \tfrac{1}{2}(5\zeta^3 - 3\zeta) \tag{3.176}$$

and so on, as shown in Figure 3.17. A general formula for the Legendre polynomials is

$$P_l(\zeta) = \frac{1}{2^l l!} \frac{d^l}{d\zeta^l}(\zeta^2 - 1)^l \tag{3.177}$$

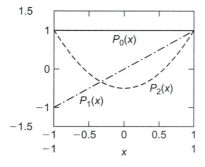

Figure 3.17 Legendre polynomials.

which is called the Rodrigues formula. To prove this, we let

$$z = k(1 - \zeta^2)^l \tag{3.178}$$

for some constant k, so that

$$(1 - \zeta^2)\frac{dz}{d\zeta} + 2l\zeta z = 0 \tag{3.179}$$

If we differentiate this expression $l + 1$ times, we get

$$\frac{d}{d\zeta}\left[(1 - \zeta^2)\frac{d^{l+1}z}{d\zeta^{l+1}}\right] + l(l + 1)\frac{d^l z}{d\zeta^l} = 0 \tag{3.180}$$

This is just the Legendre equation, which proves that $d^l z/d\zeta^l$ is a solution. Since the Legendre equation is homogeneous, we can set $k = 1/2^l l!$ to normalize the solution and get (3.177).

The Legendre polynomials for different values of l have the property that they are orthogonal over the interval $-1 \leq \zeta \leq 1$. That is,

$$\int_{-1}^{1} P_l(\zeta) P_{l'}(\zeta) \, d\zeta = \frac{2}{2l + 1}\delta_{ll'} \tag{3.181}$$

This is easily proved by substituting the Rodrigues formula and integrating repeatedly by parts. Since the powers of ζ form a complete set of functions on the interval $-1 \leq \zeta \leq 1$, the Legendre polynomials do so as well. We may therefore expand any function $f(\zeta)$ in the form

$$f(\zeta) = \sum_{l=0}^{\infty} a_l P_l(\zeta) \tag{3.182}$$

where the coefficients are

$$a_l = \frac{2l + 1}{2} \int_{-1}^{1} f(\zeta) P_l(\zeta) \, d\zeta \tag{3.183}$$

To solve the Legendre equation (3.167) for the interval $1 \leq \rho < \infty$, we note that the singular point $\rho = 1$ corresponds to the origin in space. Therefore, to represent the potential in regions of space that include the origin, we must construct the solution from products of the regular solutions, which are the Legendre polynomials $P_l(\rho)$. However, for regions of space that exclude the origin, such as the space outside a conducting ellipsoid, we can include the linearly independent functions $Q_l(\rho)$. These divergent functions are called Legendre functions of the second kind.

For $1 \leq \rho < \infty$, however, the power series (3.169) does not converge unless it truncates, so we must find the functions $Q_l(\rho)$ by different means. Using a standard technique, we assume that

$$R_l(\rho) = P_l(\rho)S_l(\rho) \tag{3.184}$$

is the required solution for some function $S_l(\rho)$. If we substitute this into (3.167), we get

$$(\rho^2 - 1)\left(P_l\frac{d^2 S_l}{d\rho^2} + 2\frac{dP_l}{d\rho}\frac{dS_l}{d\rho} + S_l\frac{d^2 P_l}{d\rho^2}\right) + 2\rho\left(P_l\frac{dS_l}{d\rho} + S_l\frac{dP_l}{d\rho}\right) = 0 \tag{3.185}$$

But $P_l(\rho)$ is a solution of the Legendre equation, so

$$(\rho^2 - 1)\frac{d^2 P_l}{d\rho^2} + 2\rho\frac{d P_l}{d\rho} - l(l+1)P_l = 0 \qquad (3.186)$$

If we multiply this by $S_l(\rho)$, subtract from (3.185), and rearrange the result, we get

$$(\rho^2 - 1)\left(2\frac{d P_l}{d\rho}\frac{d S_l}{d\rho} + P_l\frac{d^2 S_l}{d\rho^2}\right) + 2\rho P_l\frac{d S_l}{d\rho} = 0 \qquad (3.187)$$

If, now, we multiply this by P_l and rearrange the result, we obtain the differential equation

$$\frac{d}{d\rho}\left[(\rho^2 - 1)P_l^2\frac{d S_l}{d\rho}\right] = 0 \qquad (3.188)$$

for the function $S_l(\rho)$. Upon integrating this twice, we get

$$S_l(\rho) = a\int\frac{d\rho'}{(\rho'^2 - 1)P_l^2(\rho')} \qquad (3.189)$$

for some constant a. Choosing a convenient normalization, we arrive at the result

$$Q_l(\rho) = P_l(\rho)\int_\rho^\infty\frac{d\rho'}{(\rho'^2 - 1)P_l^2(\rho')} \qquad (3.190)$$

It is sufficient for our purposes to find just the first few functions $Q_l(\rho)$. For $l = 0$, the Legendre function of the second kind is

$$Q_0(\rho) = \operatorname{arctanh}\frac{1}{\rho} \xrightarrow[\rho\to\infty]{} \frac{1}{\rho} \qquad (3.191)$$

For $l = 1$,

$$Q_1(\rho) = \rho\left(\operatorname{arctanh}\frac{1}{\rho} - \frac{1}{\rho}\right) \xrightarrow[\rho\to\infty]{} \frac{1}{3\rho^2} \qquad (3.192)$$

Although they diverge at the origin ($\rho = 1$), the associated Legendre functions vanish at infinity, so they can be used to construct functions in the region $\rho > 1$.

Returning to our original problem of a conducting spheroid in a uniform field, we look for a solution of the Laplace equation (3.165) subject to the boundary conditions (3.139) and (3.140). In prolate spheroidal coordinates these become

$$\Phi(\rho, \zeta) = 0, \qquad\qquad \text{for } \rho = \frac{a}{c} \qquad (3.193)$$

$$\Phi(\rho, \zeta) \to -cE_0\rho\zeta, \qquad \text{for } \rho \to \infty \qquad (3.194)$$

when we use (3.147). In terms of the Legendre polynomials, the field at infinity can be represented by the product $-cE_0 P_1(\rho)P_1(\zeta)$. To make the potential vanish on the surface of the conductor, we add to this the proper amount of the product $Q_1(\rho)P_1(\zeta)$, which vanishes at infinity, and get

$$\Phi(\rho, \zeta) = E_0 c\rho\zeta\left[\frac{\operatorname{arctanh}\dfrac{1}{\rho} - \dfrac{1}{\rho}}{\operatorname{arctanh}\dfrac{c}{a} - \dfrac{c}{a}} - 1\right] \qquad (3.195)$$

Thus, after making a substantial investment to develop the machinery for problems of this type we obtain the solution almost trivially. But, of course, this is the reason we introduce curvilinear coordinates. Besides, we need the Legendre polynomials for spherical polar coordinates in the following section.

At the surface of the conductor the electric field is normal to the surface, so

$$E = -\frac{\partial \Phi}{\partial n} = -\frac{1}{h_\rho}\frac{\partial \Phi}{\partial \rho} \tag{3.196}$$

Evaluating the derivative at the surface of the conductor, we get

$$E = E_0 \zeta \sqrt{\frac{a^2 - c^2}{a^2 - c^2\zeta^2}} \left[1 + \frac{c^2}{\frac{a}{c}(a^2 - c^2)\left(\text{arctanh}\,\frac{c}{a} - \frac{c}{a}\right)} \right] \tag{3.197}$$

For a long, slender spheroid ($a/b \gg 1$), the enhancement of the electric field at the tip can be very large. To express this in other terms, we note that at the tip of the spheroid ($\zeta = 1$), the radius of curvature of the surface is

$$R = \frac{b^2}{a} \tag{3.198}$$

For a sharp needle ($R/a \ll 1$) we may use the approximation $\text{arctanh}\,x \approx -\frac{1}{2}\ln[\frac{1}{2}(1 - x)]$ to show that the electric field at the tip is

$$E \approx E_0 \frac{2a/R}{\ln(4a/R)} \tag{3.199}$$

For example, if we begin with a metallic wire 2 cm long ($a = 1$ cm), etch the tip to a radius $R = 1$ μm, and place the needle in an electric field $E_0 = 5 \times 10^6$ V/m, the surface electric field at the tip is $E \approx 10^{10}$ V/m. This is easily achieved in the laboratory, and the electric field at the tip is comparable to that found in atoms and molecules. This field enhancement at a sharp tip explains the operation of a lightning rod. Before the lightning strikes, the electric field at the tip of the rod is enhanced to the point where the surrounding air is ionized. This provides a high-conductivity region that extends upward in the direction of the electric field. This allows some of the charge to leak off through the tip of the rod and attracts the lightning when the strike occurs.

In summary, we have solved a complex problem by the following series of steps. First, we introduced curvilinear coordinates that reflect the symmetry of the problem. Second, we found the Laplace equation in the curvilinear coordinate system. Third, we used separation of variables to reduce the partial differential equation to ordinary differential equations. Fourth, we found the eigenfunctions and associated functions of the ordinary differential equations. Fifth, and this was the easy part, we constructed the solution from the eigenfunctions and associated functions. This same method can be used for other problems with this and other curvilinear coordinate systems. In particular we would mention ellipsoidal and paraboloidal coordinates, and our favorite, spherical polar coordinates, to which we turn in the next section.

EXERCISE 3.12

Consider a prolate spheroidal conductor in a uniform field E_0 aligned with the symmetry axis. Show that the perturbation of the field at large distances from the conductor is equivalent to a dipole of strength

$$p = \frac{\frac{4}{3}\pi c^3 \varepsilon_0 E_0}{\text{arctanh}\,\dfrac{c}{a} - \dfrac{c}{a}} \tag{3.200}$$

where a and b are the major and minor semiaxes of the spheroid and $c^2 = a^2 - b^2$.

EXERCISE 3.13

A prolate spheroidal conductor is charged to potential Φ_0 relative to infinity.

(a) Show that the potential in the region outside the spheroid is

$$\Phi(\rho, \zeta) = \Phi_0 \frac{\text{arctanh}\,\dfrac{1}{\rho}}{\text{arctanh}\,\dfrac{c}{a}} \tag{3.201}$$

where a and b are the major and minor semiaxes of the spheroid and $c^2 = a^2 - b^2$.

(b) Show that the electric field at the tip of the spheroid ($\zeta = 1$) is

$$E = \frac{\Phi_0 c}{b^2} \frac{1}{\text{arctanh}\,\dfrac{c}{a}} \tag{3.202}$$

Show that in the limit of a sphere ($b = a$) the electric field is

$$E = \frac{\Phi_0}{a} \tag{3.203}$$

as expected, and that in the limit of a sharp needle

$$E = \frac{2\Phi_0}{R \ln(4a/R)} \tag{3.204}$$

where R is the radius of curvature at the tip.

3.2.6 Spherical Harmonics

Arguably the most useful curvilinear coordinate system is spherical polar coordinates, (r, θ, ϕ). As illustrated in Figure 3.18, the Cartesian coordinates are

$$x = r \sin\theta \cos\phi \tag{3.205}$$

$$y = r \sin\theta \sin\phi \tag{3.206}$$

$$z = r \cos\theta \tag{3.207}$$

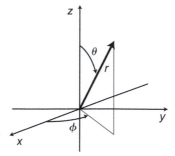

Figure 3.18 Spherical polar coordinates.

in which r is the radius from the origin, θ the angle from the z axis, and ϕ the azimuthal angle about the z axis. Differentiating these formulas we find that

$$dx = \sin\theta\cos\phi\,dr + r\cos\theta\cos\phi\,d\theta - r\sin\theta\sin\phi\,d\phi \tag{3.208}$$

$$dy = \sin\theta\sin\phi\,dr + r\cos\theta\sin\phi\,d\theta + r\sin\theta\cos\phi\,d\phi \tag{3.209}$$

$$dz = \cos\theta\,dr - r\sin\theta\,d\theta \tag{3.210}$$

Following the arguments used in the previous section, we sum the squares of these terms to get the metric

$$dl^2 = dx^2 + dy^2 + dz^2 = h_r^2\,dr^2 + h_\theta^2\,d\theta^2 + h_\phi^2\,d\phi^2 \tag{3.211}$$

where

$$h_r = 1 \tag{3.212}$$

$$h_\theta = r \tag{3.213}$$

and

$$h_\phi = r\sin\theta \tag{3.214}$$

From this we find that the Laplace equation in spherical polar coordinates is

$$\frac{\partial}{\partial r}\left(r^2\frac{\partial\Phi}{\partial r}\right) + \frac{1}{\sin\theta}\frac{\partial}{\partial\theta}\left(\sin\theta\frac{\partial\Phi}{\partial\theta}\right) + \frac{1}{\sin^2\theta}\frac{\partial^2\Phi}{\partial\phi^2} = 0 \tag{3.215}$$

which can also be written

$$r\frac{\partial^2}{\partial r^2}(r\Phi) + \frac{1}{\sin\theta}\frac{\partial}{\partial\theta}\left(\sin\theta\frac{\partial\Phi}{\partial\theta}\right) + \frac{1}{\sin^2\theta}\frac{\partial^2\Phi}{\partial\phi^2} = 0 \tag{3.216}$$

If we assume a separable solution of the form

$$\Phi(r,\theta,\phi) = \frac{R(r)}{r}U(\cos\theta)F(\phi) \tag{3.217}$$

substitute into (3.216), and divide by Φ, we obtain

$$(1 - u^2)\left\{\frac{r^2}{R}\frac{d^2R}{dr^2} + \frac{1}{U}\frac{d}{du}\left[(1-u^2)\frac{dU}{du}\right]\right\} + \frac{1}{F}\frac{d^2F}{d\phi^2} = 0 \tag{3.218}$$

where

$$u = \cos\theta \tag{3.219}$$

But the last term is a function of ϕ only, while the rest of the expression is independent of ϕ. Therefore, the third term can be at most a constant, which we call $-m^2$, so that

$$\frac{d^2 F}{d\phi^2} = -m^2 F \tag{3.220}$$

The solution to (3.220) is $e^{im\phi}$, so if the potential Φ is to be continuous, m must be an integer.

Continuing the argument, we substitute this into (3.218) and get

$$\frac{r^2}{R}\frac{d^2 R}{dr^2} + \frac{1}{U}\frac{d}{du}\left[(1-u^2)\frac{dU}{du}\right] - \frac{m^2}{1-u^2} = 0 \tag{3.221}$$

But the first term depends only on r and the rest depends only on u, so each can be at most a constant. If we call this constant $l(l+1)$, we obtain

$$\frac{d^2 R}{dr^2} = l(l+1)\frac{R}{r^2} \tag{3.222}$$

and

$$\frac{d}{du}\left[(1-u^2)\frac{dU}{du}\right] + \left[l(l+1) - \frac{m^2}{1-u^2}\right]U = 0 \tag{3.223}$$

The solution to (3.222) is

$$\frac{R(r)}{r} = Ar^l + \frac{B}{r^{l+1}} \tag{3.224}$$

where A and B are constants but l is not determined.

In the azimuthally symmetric case ($m = 0$), the remaining equation (3.223) becomes the Legendre equation, which is discussed at length in the previous section. As shown there, the solution diverges at the singular points $u = \pm 1$ unless l is an integer. For $m \neq 0$, (3.221) is called the generalized Legendre equation and its solutions are found in the same manner as those of the Legendre equation. We begin with the expression

$$z = (1 - u^2)^l \tag{3.225}$$

which satisfies the differential equation

$$(1 - u^2)\frac{dz}{du} + 2luz = 0 \tag{3.226}$$

If we differentiate this equation $l + m + 1$ times and multiply by $(1 - u^2)^{m/2}$, we get

$$(1-u^2)^{1+m/2}\frac{d^{l+m+2}z}{du^{l+m+2}} - (m+1)u(1-u^2)^{m/2}\frac{d^{l+m+1}z}{du^{l+m+1}}$$

$$+ (l+m+1)(l-m)(1-u^2)^{m/2}\frac{d^{l+m}z}{du^{l+m}} = 0 \tag{3.227}$$

We now let

$$v = (1-u^2)^{m/2}\frac{d^{l+m}z}{du^{l+m}} \tag{3.228}$$

so that

$$\frac{dv}{du} = (1 - u^2)^{m/2} \frac{d^{l+m+1} z}{du^{l+m+1}} - mu(1 - u^2)^{m/2-1} \frac{d^{l+m} z}{du^{l+m}} \tag{3.229}$$

Then (3.227) becomes

$$\frac{d}{du}\left[(1 - u^2)\frac{dv}{du}\right] + \left[l(l + 1) - \frac{m^2}{1 - u^2}\right] v = 0 \tag{3.230}$$

which is the generalized Legendre equation for v. With the proper standardization, (3.228) becomes the Rodrigues formula

$$P_l^m(u) = \frac{(-1)^m}{2^l l!}(1 - u^2)^{m/2}\frac{d^{l+m}}{du^{l+m}}(u^2 - 1)^l \tag{3.231}$$

for the associated Legendre polynomials $P_l^m(u)$. Clearly, $P_l^m(u)$ vanishes for $m > l$. It is not hard to show that

$$P_l^{-m}(u) = (-1)^m \frac{(l - m)!}{(l + m)!} P_l^m(u) \tag{3.232}$$

so m has the integer values $-l \le m \le l$.

The first few associated Legendre polynomials are

$$P_0^0(u) = P_0(u) = 1 \tag{3.233}$$

$$P_1^0(u) = P_1(u) = u \tag{3.234}$$

$$P_1^1(u) = -\sqrt{1 - u^2} \tag{3.235}$$

$$P_2^0(u) = P_2(u) = \tfrac{1}{2}(3u^2 - 1) \tag{3.236}$$

$$P_2^1(u) = -3u\sqrt{1 - u^2} \tag{3.237}$$

$$P_2^2(u) = 3(1 - u^2) \tag{3.238}$$

and so on for $m \ge 0$, as shown in Figure 3.19.

As noted in the previous section, the Legendre polynomials form a complete, orthogonal set of functions for the interval $-1 \le u \le 1$, so they can be used to represent an azimuthally symmetric function $f(\theta)$ over the interval $0 \le \theta \le \pi$. To represent functions $f(\theta, \phi)$ that are not azimuthally symmetric, we use the associated Legendre polynomials in products of the form $P_l^m(\cos\theta)e^{im\phi}$. For this purpose we introduce the spherical harmonics

$$Y_{lm}(\theta, \phi) = \sqrt{\frac{2l + 1}{4\pi}\frac{(l - m)!}{(l + m)!}} P_l^m(\cos\theta)e^{im\phi} \tag{3.239}$$

where $-l \le m \le l$. From (3.239) and (3.232) we see that

$$Y_{lm}^*(\theta, \phi) = (-1)^m Y_{l,-m}(\theta, \phi) \tag{3.240}$$

The spherical harmonics have the orthogonality property

$$\int_0^{2\pi} d\phi \int_0^{\pi} \sin\theta \, d\theta \, Y_{lm}^*(\theta, \phi)Y_{l'm'}(\theta, \phi) = \delta_{ll'}\delta_{mm'} \tag{3.241}$$

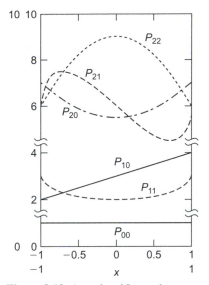

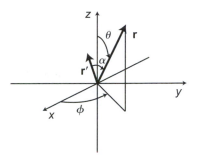

Figure 3.19 Associated Legendre polynomials.

Figure 3.20 Difference between radius vectors.

They may therefore be used to represent any function on the unit sphere in the form

$$f(\theta, \phi) = \sum_{l=0}^{\infty} \sum_{m=-l}^{l} a_{lm} Y_{lm}(\theta, \phi) \tag{3.242}$$

The coefficients a_{lm} are found by multiplying this expression by $Y_{l'm'}^*(\theta, \phi)$, integrating over the unit sphere, and using the orthogonality property (3.241). We get

$$a_{lm} = \int_{0}^{2\pi} d\phi \int_{0}^{\pi} \sin\theta \, d\theta \, Y_{lm}^*(\theta, \phi) f(\theta, \phi) \tag{3.243}$$

A particularly important function of (θ, ϕ) is the function $P_l(\cos\alpha)$, where α is the angle between the unit vectors $\mathbf{r}$ and $\mathbf{r}'$, as shown in Figure 3.20. In this case, (θ, ϕ) are the polar coordinates of the unit vector $\hat{\mathbf{r}}$. The polar coordinates (θ', ϕ') of the unit vector $\hat{\mathbf{r}}'$ play the role of parameters. The expansion of this function in terms of spherical harmonics is therefore

$$P_l(\cos\alpha) = \sum_{l'=0}^{\infty} \sum_{m'=-l'}^{l'} \sum_{l''=0}^{\infty} \sum_{m''=-l''}^{l''} a_{l'm'l''m''} Y_{l'm'}(\theta', \phi') Y_{l''m''}(\theta, \phi) \tag{3.244}$$

or equivalently, in view of (3.240),

$$P_l(\cos\alpha) = \sum_{l'=0}^{\infty} \sum_{m'=-l'}^{l'} \sum_{l''=0}^{\infty} \sum_{m''=-l''}^{l''} a_{l'm'l''m''} Y_{l'm'}^*(\theta', \phi') Y_{l''m''}(\theta, \phi) \tag{3.245}$$

Finding the coefficients $a_{l'm'l''m''}$ is a bit complicated, although the result (3.250) is remarkably simple. We begin by noting that the sum (3.245) must be real, and symmetric in (θ, ϕ)

and (θ', ϕ'). Therefore the sums over l'' and m'' collapse and we are left with

$$P_l(\cos\alpha) = \sum_{l'=0}^{\infty} \sum_{m'=-l'}^{l'} a_{l'm'} Y_{l'm'}^*(\theta', \phi') Y_{l'm'}(\theta, \phi) \tag{3.246}$$

If we now rotate coordinates so that the vector $\mathbf{r}'$ lies along the new z axis, then $\alpha = \theta$ and the only term that survives is $Y_{l0}(\theta, \phi)$. But it can be shown that a rotation of coordinates doesn't mix harmonics of different l. Therefore, the only terms that survive in (3.246) are those for $l' = l$, and we are left with

$$P_l(\cos\alpha) = \sum_{m=-l}^{l} a_{lm} Y_{lm}^*(\theta', \phi') Y_{lm}(\theta, \phi) \tag{3.247}$$

To determine the coefficients a_{lm} we first set $\mathbf{r}' = \mathbf{r}$ [that is, $(\theta', \phi') = (\theta, \phi)$], so that $\alpha = 0$ and $P_l(\cos\alpha) = 1$. We then integrate over all solid angles using the orthogonailty relation (3.241) and get

$$4\pi = \sum_{m=-l}^{l} a_{lm} \tag{3.248}$$

Next we take the square of (3.247) and integrate over all (θ', ϕ') and (θ, ϕ). On the left-hand side we can rotate the coordinate system to place $\mathbf{r}'$ along the z axis, so $\alpha = \theta$. We then evaluate the integral over (θ, ϕ) using (3.181), and the integral over (θ', ϕ') yields 4π. To evaluate the integrals on the right-hand side we use the orthogonality relation (3.241) and get

$$\frac{(4\pi)^2}{2l+1} = \sum_{m=-l}^{l} |a_{lm}^2| \tag{3.249}$$

But there are $(2l + 1)$ terms in the sums, so the only way to satisfy both (3.248) and (3.249) is to make all the coefficients a_m the same, that is, $a_{lm} = 4\pi/(2l + 1)$. Thus, we finally arrive at the desired result

$$P_l(\cos\alpha) = \frac{4\pi}{2l+1} \sum_{m=-l}^{l} Y_{lm}^*(\theta', \phi') Y_{lm}(\theta, \phi) \tag{3.250}$$

This is called the addition theorem for spherical harmonics.

Since the products $(Ar^l + Br^{-l-1}) Y_{lm}(\theta, \phi)$ satisfy the Laplace equation, we can represent any solution to the Laplace equation by a series of the form

$$\Phi(r, \theta, \phi) = \sum_{l=0}^{\infty} \sum_{m=-l}^{l} \left(A_{lm} r^l + \frac{B_{lm}}{r^{l+1}} \right) Y_{lm}(\theta, \phi) \tag{3.251}$$

A particularly important example is the expansion of the function $1/|\mathbf{r} - \mathbf{r}'|$ of two vectors in terms of their polar coordinates. Clearly this function depends only on the angle α between the vectors, so we can write

$$\frac{1}{|\mathbf{r} - \mathbf{r}'|} = \sum_{l=0}^{\infty} R_l(r, r') P_l(\cos\alpha) = \sum_{l=0}^{\infty} \sum_{m=-l}^{l} R_l(r, r') \frac{4\pi}{2l+1} Y_{lm}^*(\theta', \phi') Y_{lm}(\theta, \phi) \tag{3.252}$$

But the function $1/|\mathbf{r} - \mathbf{r}'|$ satisfies the Laplace equation $\nabla^2(1/|\mathbf{r} - \mathbf{r}'|) = 0$ as well as $\nabla'^2(1/|\mathbf{r} - \mathbf{r}'|) = 0$, so from (3.224) we see that the function $R_l(r, r')$ must be of the form

$$R_l(r, r') = \left(A_l r^l + \frac{B_l}{r^{l+1}} \right) \left(A_l' r'^l + \frac{B_l'}{r'^{l+1}} \right) \tag{3.253}$$

for some constants A_l, B_l, A_l', and B_l'. To determine the constants we consider two points on the z axis. Then all the functions Y_{lm} vanish except $l = 0$, for which $Y_{l0} = \sqrt{(2l+1)/4\pi}$, and (3.252) becomes

$$\frac{1}{|\mathbf{r} - \mathbf{r}'|} = \frac{1}{|r - r'|} = \sum_{l=0}^{\infty} \left(A_l r^l + \frac{B_l}{r^{l+1}} \right) \left(A_l' r'^l + \frac{B_l'}{r'^{l+1}} \right) \tag{3.254}$$

But for $r' < r$ we can use the Taylor series expansion

$$\frac{1}{|r - r'|} = \frac{1}{r\sqrt{1 - 2\dfrac{r'}{r} + \dfrac{r'^2}{r^2}}} = \sum_{l=0}^{\infty} \frac{r'^l}{r^{l+1}} \tag{3.255}$$

Comparing these expressions we see that $A_l' = 1$, $B_l' = 0$, $A_l = 0$, and $B_l = 1$. We therefore obtain the general expansion

$$\frac{1}{|\mathbf{r} - \mathbf{r}'|} = \sum_{l=0}^{\infty} \sum_{m=-l}^{l} \frac{4\pi}{2l+1} \frac{r_<^l}{r_>^{l+1}} Y_{lm}^*(\theta', \phi') Y_{lm}(\theta, \phi) \tag{3.256}$$

where $r_>$ is the larger and $r_<$ the smaller of r and r'.

We can use this result to describe the potential of a localized charge distribution $\rho(\mathbf{r}')$ in terms of multipole moments. If we substitute (3.256) into (3.33) we immediately find that the potential outside the charge distribution is

$$\Phi(\mathbf{r}) = \frac{1}{4\pi\varepsilon_0} \sum_{l=0}^{\infty} \sum_{m=-l}^{l} \frac{4\pi}{2l+1} \frac{q_{lm}}{r^{l+1}} Y_{lm}(\theta, \phi) \tag{3.257}$$

where the (l, m) multipole moment of the charge distribution is

$$q_{lm} = \int \rho(\mathbf{r}') r'^l Y_{lm}^*(\theta', \phi') \, d^3\mathbf{r}' \tag{3.258}$$

From (3.240) we see that

$$q_{l,-m} = (-1)^m q_{lm}^*$$

The leading term $(l = 0)$ in the expansion (3.257) is called the monopole moment,

$$q_{00} = \sqrt{\frac{1}{4\pi}} \int \rho(\mathbf{r}') \, d^3\mathbf{r}' = \sqrt{\frac{1}{4\pi}} \, Q \tag{3.259}$$

where Q is the total charge. The next term $(l = 1)$ in the expansion is called the dipole moment. It has three components,

$$q_{11} = -\sqrt{\frac{3}{8\pi}} \int \rho(\mathbf{r}') r' \sin\theta' (\cos\phi' - i\sin\phi') \, d^3\mathbf{r}' = -\sqrt{\frac{3}{8\pi}} (p_x - ip_y) \tag{3.260}$$

$$q_{10} = \sqrt{\frac{3}{4\pi}} \int \rho(\mathbf{r}') r' \cos\theta' \, d^3\mathbf{r}' = \sqrt{\frac{3}{4\pi}} \, p_z \tag{3.261}$$

$$q_{1,-1} = -q_{11}^* = \sqrt{\frac{3}{8\pi}} (p_x + ip_y) \tag{3.262}$$

where the Cartesian components p_i of the dipole moment are defined by (3.43). The third term ($l = 2$) in the expansion is called the quadrupole moment. It has five components,

$$q_{22} = \sqrt{\frac{15}{32\pi}} \int \rho(\mathbf{r}')r'^2 \sin^2 \theta' (\cos \phi' - i \sin \phi') \, d^3\mathbf{r}'$$

$$= \frac{1}{3}\sqrt{\frac{15}{32\pi}}(Q_{xx} - Q_{yy} - i2Q_{xy}) \tag{3.263}$$

$$q_{21} = -\sqrt{\frac{15}{8\pi}} \int \rho(\mathbf{r}')r'^2 \cos \theta' \sin \theta' (\cos \phi' - i \sin \phi') \, d^3\mathbf{r}'$$

$$= -\frac{1}{3}\sqrt{\frac{15}{8\pi}}(Q_{xz} - i Q_{yz}) \tag{3.264}$$

$$q_{20} = \sqrt{\frac{5}{16\pi}} \int \rho(\mathbf{r}')r'^2 (3\cos^2 \theta' - 1) \, d^3\mathbf{r}' = \frac{1}{2}\sqrt{\frac{5}{4\pi}} Q_{zz} \tag{3.265}$$

$$q_{2,-1} = -q_{21}^* = \frac{1}{3}\sqrt{\frac{15}{8\pi}}(Q_{xz} + i Q_{yz}) \tag{3.266}$$

and

$$q_{2,-2} = q_{22}^* = \frac{1}{3}\sqrt{\frac{15}{32\pi}}(Q_{xx} - Q_{yy} + i2Q_{xy}) \tag{3.267}$$

where the Cartesian components Q_{ij} of the quadrupole moment are defined by (3.44). Thus, we can switch between the Cartesian and spherical representations of the multipole moments at our convenience. Higher order multipoles become increasingly complex and are seldom used.

EXERCISE 3.14

(a) Using Rodrigues' formula, show that

$$\frac{dP_{l+1}}{dx} - \frac{dP_{l-1}}{dx} = (2l + 1)P_l \tag{3.268}$$

(b) Using this and the differential equation for the Legendre polynomials, derive the recurrence relation

$$(l + 1)P_{l+1} - (2l + 1)x P_l + l P_{l-1} = 0 \tag{3.269}$$

EXERCISE 3.15

A conducting sphere of radius R_0 with potential $\Phi = 0$ is placed at the origin in a uniform electric field

$$\Phi = -E_0 z = -E_0 r P_1(\cos \theta) \tag{3.270}$$

By expanding the potential in spherical harmonics and applying the boundary condition $\Phi = 0$ at the surface of the sphere, show that the potential outside the sphere is that of the external field plus a dipole at the origin:

$$\Phi = \frac{\mathbf{p}_0 \cdot \mathbf{r}}{r^3} - E_0 z \tag{3.271}$$

What is the magnitude and direction of the dipole moment $\mathbf{p}_0$?

3.2.7 Variational Methods

In Chapter 2 it is shown that the Maxwell equations for the electromagnetic field of prescribed charges can be derived from Hamilton's principle in the form

$$\delta S = \delta \int_a^b \mathcal{L} \, dt = 0 \tag{3.272}$$

where S is the action and the Lagrangian

$$\mathcal{L} = \int \mathcal{D} \, d^3\mathbf{r} = -\frac{1}{4\mu_0} \int F^{\alpha\beta} F_{\alpha\beta} \, d^3\mathbf{r} - \int A_\alpha J^\alpha \, d^3\mathbf{r} \tag{3.273}$$

is constructed from the Lagrangian density $\mathcal{D}$ of the field $F^{\alpha\beta}$ and its interaction with the current density J^α. In the electrostatic case the Lagrangian is time independent, so to satisfy (3.272) the Lagrangian itself must be a minimum. But in 3-vector notation, the Lagrangian density is

$$F^{\alpha\beta} F_{\alpha\beta} = 2\left(B^2 - \frac{E^2}{c^2} \right) \tag{3.274}$$

$$A_\alpha J^\alpha = \rho \Phi - \mathbf{A} \cdot \mathbf{J} \tag{3.275}$$

where $\mathbf{B} = \nabla \times \mathbf{A}$ is the magnetic induction, $\mathbf{E} = -\nabla \Phi$ the electric field, ρ the charge density, and $\mathbf{J}$ the current density. Substituting into (3.273), we find that the fields satisfy the variational principle

$$\delta\left(\frac{\varepsilon_0}{2} \int E^2 \, dV - \int \rho \Phi \, dV - \frac{1}{2\mu_0} \int B^2 \, dV + \int \mathbf{J} \cdot \mathbf{A} \, dV \right) = 0 \tag{3.276}$$

where the integrals are over all space. But the variations of $\mathbf{B}$ and $\mathbf{E}$ are independent, so the fields independently satisfy the variational principles

$$\delta\left(\frac{\varepsilon_0}{2} \int E^2 \, dV - \int \rho \Phi \, dV \right) = 0 \tag{3.277}$$

and

$$\delta\left(\frac{1}{2\mu_0} \int B^2 \, dV - \int \mathbf{J} \cdot \mathbf{A} \, dV \right) = 0 \tag{3.278}$$

For the electric field, this says that the energy of the field minus the potential energy of the charges in it is a minimum.

This principle can be used to find an approximate solution to problems in electrostatics when an exact solution is not necessary or is not possible. In practice, a trial function Φ

is used for the potential and parameters in the function Φ are varied to minimize the Lagrangian. In this process, inspired guesses about the general form of the potential are most useful. Even when the trial function is not a very good representation of the potential, however, it is often possible to get satisfactory estimates of the global features of the solution such as the energy in the field, the capacitance, and so on.

In applying the variational principle (3.277) to conductors, it is necessary that the trial function satisfy the boundary conditions. Thus, if the boundary condition requires that the potential on a given conductor be, say, Φ_0, then the trial function must satisfy this condition exactly. In this case, the charge on the conductor is not known. However, the potential on the conductor is not varied in the calculation, so the second term in (3.276) vanishes and the unknown charge does not appear in the calculation. If the total charge Q on a conductor is specified, rather than the potential, then we divide the sources into the (known) volume charge density ρ and the (unknown) surface charge density σ, so that (3.277) becomes

$$\delta\left(\frac{\varepsilon_0}{2} \int E^2 \, dV - \int \rho\Phi \, dV - \oint_S \sigma\Phi \, dS\right) = 0 \tag{3.279}$$

But the surface charge density is related to the electric field at the surface by (3.71), so this becomes

$$\delta\left(\frac{\varepsilon_0}{2} \int E^2 \, dV - \int \rho\Phi \, dV - \varepsilon_0 \oint_S \Phi \frac{\partial\Phi}{\partial n} \, dS\right) = 0 \tag{3.280}$$

where $\partial\Phi/\partial n = \hat{\mathbf{n}} \cdot \nabla\Phi$ is the gradient of the potential in the $\hat{\mathbf{n}}$ direction normal to the surface out of the volume V, that is, into the conductor. The functional in parentheses must then be minimized subject to the condition

$$\varepsilon_0 \oint_S \frac{\partial\Phi}{\partial n} \, dS = Q \tag{3.281}$$

on the conductor for which the total charge is specified. This can be accomplished using the method of Lagrange multipliers.

As an example of the variational method, we consider the field of a periodic charge distribution in the plane $y = 0$ for which the surface charge density alternates between $\sigma = \sigma_0$ and $\sigma = -\sigma_0$ with a period a, as indicated in Figure 3.21. We assume a potential of the form

$$\Phi = \Phi_0 e^{-k|y|} \sin\left(2\pi \frac{x}{a}\right) \tag{3.282}$$

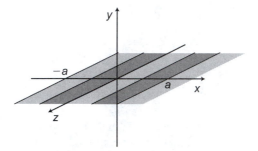

Figure 3.21 Potential due to a periodic charge distribution.

for some constants Φ_0 and k. The electric field is then

$$\mathbf{E} = \Phi_0 e^{-k|y|} \left[k\hat{\mathbf{y}} \sin\left(2\pi \frac{x}{a}\right) - \frac{2\pi \hat{\mathbf{x}}}{a} \cos\left(2\pi \frac{x}{a}\right) \right] \tag{3.283}$$

Evaluating the integrals over one period of the charge distribution, we get

$$\int_0^a dx \int_{-\infty}^{\infty} dy \frac{\varepsilon_0}{2} E^2 = \frac{\pi}{2} \varepsilon_0 \Phi_0^2 \left(\frac{ka}{2\pi} + \frac{2\pi}{ka} \right) \tag{3.284}$$

$$\int_0^a dx \, \sigma \Phi = \frac{2a}{\pi} \sigma_0 \Phi_0 \tag{3.285}$$

Substituting into (3.277) and computing the variation, we get

$$\left[\pi \varepsilon_0 \Phi_0 \left(\frac{ka}{2\pi} + \frac{2\pi}{ka} \right) - \frac{2a}{\pi} \sigma_0 \right] \delta\Phi_0 + \left[\frac{a}{4} \varepsilon_0 \Phi_0^2 \left(1 - \frac{4\pi^2}{k^2 a^2} \right) \right] \delta k = 0 \tag{3.286}$$

Since Φ_0 and k are independent, each of the quantities in square brackets must vanish identically, and we obtain the results

$$k = \frac{2\pi}{a} \tag{3.287}$$

and

$$\Phi_0 = \frac{\sigma_0 a}{\pi^2 \varepsilon_0} \tag{3.288}$$

A more organized way of picking trial functions is to expand the potential in a finite series of orthogonal functions and determine the coefficients by using the variational principle (3.276). This is called the Ritz method. As a trivial example, we consider the potential between two grounded planes at $x = 0$ and $x = a$ when the volume between them is uniformly filled with the charge density ρ. We expand the potential in a Fourier series of the form

$$\Phi = c_1 \sin\left(\pi \frac{x}{a}\right) + c_2 \sin\left(2\pi \frac{x}{a}\right) + c_3 \sin\left(3\pi \frac{x}{a}\right) \tag{3.289}$$

which satisfies the boundary conditions at $x = 0$ and $x = a$. To determine the coefficients c_n, we substitute into (3.276) and evaluate the integrals to get

$$\frac{\varepsilon_0}{2} \int_0^a E^2 \, dx - \int_0^a \rho \Phi \, dx = \frac{\pi^2 \varepsilon_0}{4a} (c_1^2 + 2^2 c_2^2 + 3^2 c_3^2) - \frac{2a\rho}{\pi} \left(\frac{c_1}{1} + 0 + \frac{c_3}{3} \right) \tag{3.290}$$

where we have taken advantage of the orthogonality of the cosine functions on the interval $0 < x < a$ to evaluate the first integral. The variation of this expression is

$$\left(\frac{\pi^2 \varepsilon_0}{2a} c_1 - \frac{2a\rho}{\pi} \right) \delta c_1 + \left(\frac{\pi^2 \varepsilon_0}{2a} 2^2 c_2 \right) \delta c_2 + \left(\frac{\pi^2 \varepsilon_0}{2a} 3^2 c_3 - \frac{2a\rho}{\pi} \right) \delta c_3 = 0 \tag{3.291}$$

Since all the coefficients are independent, we see by induction that

$$C_n = \begin{cases} \dfrac{4a^2 \rho}{\pi^3 \varepsilon_0 n^3}, & n = \text{odd} \\[2mm] 0, & n = \text{even} \end{cases} \tag{3.292}$$

Of course, the exact solution is

$$\Phi = \frac{\rho}{2\varepsilon_0} x(a - x) \tag{3.293}$$

It is easily shown that the coefficients (3.292) obtained by the Ritz method are just the coefficients of the Fourier expansion of (3.293).

EXERCISE 3.16

Consider once again the Cavendish experiment to measure the photon mass. In this experiment a solid conducting sphere of radius r_0 and a concentric conducting sphere of radius R_0 are both at potential Φ_0, and the charge on the inner sphere is measured.

(a) Show that if the photon has a mass $m_\gamma = \hbar\mu_\gamma/c$, corresponding to the scale length μ_γ^{-1}, the field satisfies the variational principle

$$\delta\left[\frac{\varepsilon_0}{2}\int E^2\, dV + \frac{\varepsilon_0\mu_\gamma^2}{2}\int \Phi^2\, dV - \int \rho\Phi\, dV\right] = 0 \tag{3.294}$$

(b) As a simple trial function, use the expression

$$\Phi = \Phi_0 + K(r - r_0)(R_0 - r) \tag{3.295}$$

which satisfies the boundary conditions and has one free parameter K. Substitute this into the variational principle (3.294) and show that to lowest order in μ_γ^2

$$K \approx -\mu_\gamma^2\Phi_0\frac{I_2}{I_1} \tag{3.296}$$

where the necessary integrals

$$I_1 = \int_{r_0}^{R_0} [2r - (R_0 + r_0)]^2\, dV = -4I_2 + (R_0 - r_0)^2 V \tag{3.297}$$

and

$$I_2 = -\int_{r_0}^{R_0} (r - r_0)(r - R_0)\, dV \tag{3.298}$$

are trivial but tedious to evaluate, and $V = \frac{4}{3}\pi(R_0^3 - r_0^3)$ is the volume between the concentric spheres.

(c) Using Gauss's law in the form (3.17), which includes the effect of photon mass, show that the charge on the inner sphere is

$$\frac{Q}{4\pi\varepsilon_0 r_0\Phi_0} = \mu_\gamma^2\left[\frac{r_0^2}{3} + r_0(R_0 - r_0)\frac{I_2}{I_1}\right] \tag{3.299}$$

When the integrals are evaluated, it is found that the approximate solution (3.299) agrees with the exact solution (3.22) very well for $R_0/r_0 \approx 1$, and is about 13 percent low at $R_0/r_0 \approx 1.5$.

EXERCISE 3.17

Consider a volume bounded by a square of length L in the x and y directions, and extending to $\pm\infty$ in the z direction, filled with a uniform charge density ρ. The boundaries of the volume are maintained at the potential $\Phi = 0$. Expand the potential in the interior in a Fourier series of the form

$$\Phi_N = \sum_{m,n=1}^{N} c_{mn} \sin\left(m\pi\frac{x}{L}\right) \sin\left(n\pi\frac{y}{L}\right) \qquad (3.300)$$

for some integer N, and use the Ritz method to show that the coefficients of the series are

$$c_{mn} = \frac{16L^2\rho}{\pi^4} \frac{1}{mn(m^2 + n^2)}, \qquad n = \text{odd} \qquad (3.301)$$

$$c_n = 0, \qquad\qquad\qquad n = \text{even} \qquad (3.302)$$

Sum the series, and show that the potential in the center of the square is

$$\Phi\left(\frac{1}{2}L, \frac{1}{2}L\right) = -\frac{16}{\pi^4}\frac{\rho L^2}{\varepsilon_0} \sum_{(\text{odd})m,n=1}^{N} \frac{(-1)^{(m+n)/2}}{mn(m^2 + n^2)} \xrightarrow{N\to\infty} 0.07367\frac{\rho L^2}{\varepsilon_0} \qquad (3.303)$$

The series may be summed on a computer, or the first few terms (up to $N = 3$) may be evaluated by hand to obtain a satisfactory approximation ($\Phi \approx 0.07219\rho L^2/\varepsilon_0$).

3.2.8 Numerical Methods

With the advent of fast computers on every desktop, the need for analytical solutions is diminishing rapidly. Still, there are times when nothing else is convenient. For example, electric fields as high as 10^{10} V/m are conveniently achieved in the laboratory at the tips of very fine needles. To compute the field near such a tip by numerical methods, it is necessary to use a grid that, at least locally around the tip, has a mesh size smaller than the tip radius. This is typically less than a micrometer and frequently as small as nanometers. Numerical computations become impractical when the overall scale of the experiment is on the order of centimeters, many orders of magnitude larger than the grid size, although varying the mesh size in different regions of the problem helps. Thus, analytical calculations of the field at the tip of a sharp needle, as discussed in an earlier section, are still very useful, if not absolutely necessary.

Having defended the need for analytical solutions to electrostatic problems, it is time to admit that numerical solutions also have their place. Numerical solutions of the Laplace equation are particularly simple to obtain by a technique called the relaxation method. From the mean-value theorem (3.4) we see that the potential at the center of a sphere is just the average of the potential over the surface of the sphere. If we fill the region to be analyzed with a cubic lattice, the potential at any point is approximately the average of the potential at the six (four, in two dimensions) nearest neighbors,

$$\Phi_{l,m,n} = \frac{1}{6}(\Phi_{l-1,m,n} + \Phi_{l+1,m,n} + \Phi_{l,m-1,n} + \Phi_{l,m+1,n} + \Phi_{l,m,n-1} + \Phi_{l,m,n+1}) \qquad (3.304)$$

To use the relaxation method, the computation begins with correct values of $\Phi_{l,m,n}$ on the boundaries and a guess at the potential $\Phi_{l,m,n}$ everywhere in the interior. Usually a constant is a satisfactory guess, since convergence is typically rapid and stable. The potential is then computed at each lattice point in succession using (3.304) and the process iterated until it converges. In two dimensions, for simple geometries, this can often be carried out using a commercial spreadsheet program.

A similar technique can be used when space charge is present. For a sphere of radius h filled with charge density ρ, the potential at the center, relative to the surface, is

$$\Delta\Phi = \frac{\rho h^2}{6\varepsilon_0} \tag{3.305}$$

This must be added to the potential computed from (3.304). In two dimensions the correction due to the charge density is

$$\Delta\Phi = \frac{\rho h^2}{4\varepsilon_0} \tag{3.306}$$

and the algorithm for the relaxation method is

$$\Phi_{m,n} = \frac{1}{4}\left(\Phi_{m-1,n} + \Phi_{m+1,n} + \Phi_{m,n-1} + \Phi_{m,n+1} + \frac{\rho h^2}{\varepsilon_0}\right) \tag{3.307}$$

As an example, we consider a square area of length L on each side, filled with uniform charge density ρ. The edges of the square are all grounded, so $\Phi = 0$ there. The result of a numerical computation using a commercial spreadsheet program is shown in Figure 3.22. For this computation the square region (including the boundaries) is spanned by an 11- by 11-point grid, so $h = L/10$, and the potential is normalized to the value $\Phi_0 = \rho L^2/\varepsilon_0$. The computed potential at the center normalized this way is $\Phi = 0.0731$. The correct answer is $\Phi = 0.0737$, which can be obtained by expanding the solution in a Fourier series. Thus, an excellent approximation is obtained even for a rather coarse grid.

While these simple methods are sufficient for many problems, more elaborate techniques have been developed for difficult problems. The extension to a rectangular (rather than square) grid is simple, but for many problems curvilinear grids are appropriate. In addition, more sophisticated algorithms that average over more points surrounding the central point can be used to provide more accurate results with coarser grids. Or, to be more precise, the solutions converge to the exact answer more rapidly as the grid size is reduced. The interested reader is referred to one of the many books available on the numerical integration of partial differential equations.

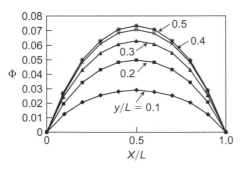

Figure 3.22 Potential in a square region filled with charge.

EXERCISE 3.18

Prove the mean-value theorem for two dimensions, and derive the algorithm (3.307) for solving the Poisson equation in two dimensions. Do the same for one dimension.

EXERCISE 3.19

Use a commercial spreadsheet program to solve the problem discussed in the text of a square of length L filled with uniform charge density ρ, with the boundaries held at $\Phi = 0$. Examine the convergence of the solution to the correct result as the grid size becomes smaller. Show by "numerical experiments" that the error (compared with the exact solution) remaining at the center of the grid after the iterations have converged is proportional to h^2, where h is the grid spacing. Be sure that the iterations have converged. You can test this convergence by examining the solution obtained by starting with $\Phi = 0$ everywhere, which relaxes to the true solution from below, with that obtained by starting with $\Phi = 1$ everywhere (except, of course, on the boundaries), which relaxes to the true solution from above.

3.2.9 Green Functions

Expression (3.7) represents the potential $\Phi(\mathbf{r})$ as an integral over the charge distribution $\rho(\mathbf{r}')$ with a weighting function that in this case is simply $1/4\pi\varepsilon_0|\mathbf{r} - \mathbf{r}'|$. Provided that the integrals can be evaluated, this provides an explicit solution to the problem of finding the electrostatic field of a known charge distribution. The weighting function is actually the potential of a point charge, and the formula simply sums up the contributions of all the charges in the distribution by taking advantage of the superposition principle of electromagnetic fields. The convenience of this representation, both conceptually and computationally, is obvious, but the formula works only for the case when the charge distribution is known in advance and the boundary condition is that the potential vanishes at infinity. Green functions offer a powerful way to generalize this approach to include conductors and other boundary conditions. In fact, Green functions of various types can be used to find the solutions to a wide variety of differential equations, provided only that the equations are linear. We use Green functions frequently in the following chapters.

We begin the discussion by deriving a couple of identities that we need. To do this we start with the divergence theorem, which states that for any vector field in a volume V enclosed by a surface S, we can convert a volume integral of the divergence of the vector to a surface integral of the form

$$\int_V \nabla' \cdot \mathbf{A}(\mathbf{r}')\,dV' = \oint_S \mathbf{A}(\mathbf{r}') \cdot \hat{\mathbf{n}}'\,dS' \tag{3.308}$$

where $\hat{\mathbf{n}}'$ is a unit vector pointing outward from volume V normal to the surface element dS', as shown in Figure 3.23. The surface S may be composed of several surfaces, as indicated in the figure. If we let

$$\mathbf{A}(\mathbf{r}') = \phi(\mathbf{r}')\nabla'\psi(\mathbf{r}') \tag{3.309}$$

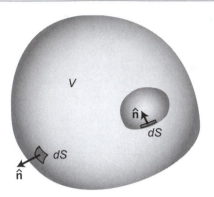

Figure 3.23 Volume V enclosed by the surface S.

for any two scalar functions $\phi(\mathbf{r})$ and $\psi(\mathbf{r})$, and substitute into (3.308), we get

$$\int_V [\nabla'\phi(\mathbf{r}') \cdot \nabla'\psi(\mathbf{r}') + \phi(\mathbf{r}')\nabla'^2\psi(\mathbf{r}')]\, dV' = \oint_S \phi(\mathbf{r}')\frac{\partial \psi(\mathbf{r}')}{\partial n'}\, dS' \qquad (3.310)$$

where $\partial\psi/\partial n' = \nabla'\psi \cdot \hat{\mathbf{n}}'$ is the derivative of the function ψ in the $\hat{\mathbf{n}}'$ direction out of volume V normal to the surface. If we now interchange $\phi(\mathbf{r}')$ and $\psi(\mathbf{r}')$ and subtract the result from (3.310), the symmetric terms $\nabla'\phi(\mathbf{r}') \cdot \nabla'\psi(\mathbf{r}')$ cancel out and we obtain the result

$$\int_V [\phi(\mathbf{r}')\nabla'^2\psi(\mathbf{r}') - \psi(\mathbf{r}')\nabla'^2\phi(\mathbf{r}')]\, dV' = \oint_S \left[\phi(\mathbf{r}')\frac{\partial \psi(\mathbf{r}')}{\partial n'} - \psi(\mathbf{r}')\frac{\partial \phi(\mathbf{r}')}{\partial n'}\right] dS'$$

$$(3.311)$$

This is called Green's theorem.

The trick to finding an explicit (albeit formal) solution to our problem lies in the choice of the function $\psi(\mathbf{r}')$. Specifically, we seek a solution to the Poisson equation

$$\nabla^2\Phi(\mathbf{r}) = -\frac{\rho(\mathbf{r})}{\varepsilon_0} \qquad (3.312)$$

subject to either Dirichlet boundary conditions,

$$\Phi(\mathbf{r}) = \Phi_D(\mathbf{r}), \qquad \text{for } \mathbf{r} \text{ on } S \qquad (3.313)$$

where $\Phi_D(\mathbf{r})$ is some function specified over the surface S, or Neumann boundary conditions,

$$\frac{\partial \Phi(\mathbf{r})}{\partial n} = \Phi'_N(\mathbf{r}), \qquad \text{for } \mathbf{r} \text{ on } S \qquad (3.314)$$

where $\partial\Phi(\mathbf{r})/\partial n$ is the derivative of Φ in the direction normal to the surface (out of the volume, into the surface), and $\Phi'_N(\mathbf{r})$ is some function specified over the surface S. For surfaces that correspond to conductors in electrostatics, Dirichlet boundary conditions have the simple form $\Phi_D = \text{constant}$ over the surface. For conductors on which the total charge Q is specified, the boundary condition is $\int \Phi'_N\, dS = Q/\varepsilon_0$ (in choosing the sign, remember that Φ'_N is the derivative in the direction out of the volume V and, therefore, into the conductor). It is shown by (3.74) that for either type of boundary condition, the solution is unique. A mixture of the two types of boundary conditions is also possible, although the solution is overconstrained if both types of boundary conditions are applied to the same surface. Thus, Dirichlet boundary conditions, in which Φ_D can vary over the surface rather than being merely a constant, and Neumann boundary conditions, in which Φ'_N is specified everywhere rather than just its integral over the surface, are actually more general than the boundary conditions that are ordinarily encountered in electrostatics. However, the more general boundary conditions are encountered in problems of thermal conduction, diffusion, and other problems described by the Poisson equation.

The trick we use to solve the boundary-value problem is to choose for the function $\psi(\mathbf{r}')$ the solution of the differential equation

$$\nabla'^2 G(\mathbf{r}', \mathbf{r}) = -\delta(\mathbf{r}' - \mathbf{r}) \tag{3.315}$$

together with boundary conditions that are discussed shortly. This is called the Green function for the problem. If we substitute $\Phi(\mathbf{r}')$ for $\phi(\mathbf{r}')$ and $G(\mathbf{r}', \mathbf{r})$ for $\psi(\mathbf{r}')$ in Green's theorem, we get

$$\int_V [\Phi(\mathbf{r}')\nabla'^2 G(\mathbf{r}', \mathbf{r}) - G(\mathbf{r}', \mathbf{r})\nabla'^2 \Phi(\mathbf{r}')]\, dV'$$

$$= \oint_S \left[\Phi(\mathbf{r}')\frac{\partial G(\mathbf{r}', \mathbf{r})}{\partial n'} - G(\mathbf{r}', \mathbf{r})\frac{\partial \Phi(\mathbf{r}')}{\partial n'} \right] dS' \tag{3.316}$$

If we now use (3.312) and (3.315) to evaluate the left-hand side, we find that

$$\Phi(\mathbf{r}) = \frac{1}{\varepsilon_0} \int_V \rho(\mathbf{r}')G(\mathbf{r}', \mathbf{r})\, dV' + \oint_S \left[G(\mathbf{r}', \mathbf{r})\frac{\partial \Phi(\mathbf{r}')}{\partial n'} - \Phi(\mathbf{r}')\frac{\partial G(\mathbf{r}', \mathbf{r})}{\partial n'} \right] dS' \tag{3.317}$$

At least in a formal sense, this remarkable result provides the solution to our boundary-value problem in the form of an integral over the source term $\rho(\mathbf{r}')/\varepsilon_0$ with a weighting function given by the Green function $G(\mathbf{r}', \mathbf{r})$, together with integrals over the boundaries. This represents a generalization of the formula (3.7) that we developed at the beginning of this chapter for the potential due to a charge distribution $\rho(\mathbf{r})$ with the Dirichlet boundary condition

$$\Phi = 0, \qquad \text{at } r = \infty \tag{3.318}$$

In this simple case, the Green function satisfies (3.315) with the boundary condition

$$G(\mathbf{r}', \mathbf{r}) = 0, \qquad \text{at } r' = \infty \tag{3.319}$$

which has the solution

$$G(\mathbf{r}', \mathbf{r}) = \frac{1}{4\pi\, |\mathbf{r}' - \mathbf{r}|} \tag{3.320}$$

The Green function is evidently the potential due to a point charge $q = \varepsilon_0$ at position $\mathbf{r}$. Since the Green function vanishes at infinity, the second term in (3.317) vanishes and the potential is simply

$$\Phi(\mathbf{r}) = \frac{1}{4\pi\varepsilon_0} \int \frac{\rho(\mathbf{r}')}{|\mathbf{r} - \mathbf{r}'|}\, d^3\mathbf{r}' \tag{3.321}$$

Clearly, in more complex cases the problem is in determining the Green function $G(\mathbf{r}', \mathbf{r})$. Nevertheless, the use of Green functions is a powerful technique that is not limited to electrostatics, or even to the Poisson equation. It can be applied to any linear differential equation, such as the wave equation, in one, two, three, or even more dimensions, and we have many occasions to use it in the following chapters. For example, with the same boundary conditions we have just used (Φ vanishes at infinity), the Green function for the wave equation is just a spherical wave expanding away from the unit impulse at point $\mathbf{r}'$ at time t'. The complete solution to the wave equation with sources is then the superposition of spherical waves radiated by the sources at all points in space and time. For a particle that radiates as it moves, the solution is the superposition of waves emitted along the trajectory of the particle. For the diffraction of a wave emerging from an aperture in otherwise empty space, there is no source term corresponding to the volume integral in (3.317). The wave in the region beyond the aperture is described by a surface integral that corresponds to the second term in (3.317). The surface integral represents the superposition of spherical waves emanating from the wave in the aperture. This is the essence of Huygens' construction for the propagation of waves in space.

Getting back to electrostatics, when the problem involves Dirichlet boundary conditions we choose the Green function that satisfies the boundary condition

$$G(\mathbf{r}', \mathbf{r}) = 0, \qquad \text{for } \mathbf{r}' \text{ on } S \tag{3.322}$$

In a physical sense, then, the Green function $G(\mathbf{r}', \mathbf{r})$ defined by (3.315) for the Dirichlet problem is the potential at field position $\mathbf{r}'$ due to a point charge $q = \varepsilon_0$ located at the source position $\mathbf{r}$, when all the conductors (and the boundary at infinity) are at ground potential. Using (3.313) and (3.322) on the right-hand side of (3.317), we obtain

$$\Phi(\mathbf{r}) = \frac{1}{\varepsilon_0} \int_V \rho(\mathbf{r}') G(\mathbf{r}', \mathbf{r}) \, dV' - \oint_S \Phi_D(\mathbf{r}') \frac{\partial G(\mathbf{r}', \mathbf{r})}{\partial n'} \, dS' \tag{3.323}$$

Provided that the integrals can be evaluated, this provides a solution to the problem in the form of integrals over the known charge density $\rho(\mathbf{r}')$ inside the volume V and over the potential $\Phi_D(\mathbf{r}')$ on the surface. It is not necessary to know the charge distributions on the conductors, only the potentials.

When the problem involves Neumann boundary conditions, it would be nice to choose the Green function that satisfies the boundary condition

$$\frac{\partial G(\mathbf{r}', \mathbf{r})}{\partial n'} = 0, \qquad \text{for } \mathbf{r}' \text{ on } S \tag{3.324}$$

However, if we integrate (3.315) over the volume inside S, which includes the point $\mathbf{r}$, and apply the divergence theorem, we see that

$$\oint_S \nabla' G(\mathbf{r}', \mathbf{r}) \cdot \hat{\mathbf{n}}' \, dS' = -1 \tag{3.325}$$

where $\hat{\mathbf{n}}'$ is a unit vector pointing out of the volume and therefore into the surface of the conductor. Therefore, the simplest boundary condition we can apply to the Green function is

$$\frac{\partial G(\mathbf{r}', \mathbf{r})}{\partial n'} = -\frac{1}{S}, \qquad \text{for } \mathbf{r}' \text{ on } S \tag{3.326}$$

where S is the total area of the boundary surface, and it must be remembered that the derivative $\partial G(\mathbf{r}', \mathbf{r})/\partial n'$ is in the direction into the surface. In a physical sense, then, the Green function $G(\mathbf{r}', \mathbf{r})$ for the Neumann problem is the potential at the field position $\mathbf{r}'$ due to a point charge $q = 1/\varepsilon_0$ located at the source position $\mathbf{r}$, when all the flux originating from the point charge is distributed uniformly over the boundaries. If we now substitute (3.314) and (3.326) into (3.317), we get

$$\Phi(\mathbf{r}) = \frac{1}{S} \oint_S \Phi(\mathbf{r}') \, dS' + \frac{1}{\varepsilon_0} \int_V \rho(\mathbf{r}') G(\mathbf{r}', \mathbf{r}) \, dV' + \oint_S \Phi'_N(\mathbf{r}') G(\mathbf{r}', \mathbf{r}) \, dS' \quad (3.327)$$

Since the desired solution $\Phi(\mathbf{r})$ appears on both sides of this equation, it is, in some sense, not a formal solution of the problem but an integral equation for $\Phi(\mathbf{r})$. However, the solution $\Phi(\mathbf{r})$ for Neumann boundary conditions is defined only to within an additive constant, and this constant may be chosen to make the integral $\int \Phi \, dS' = 0$. With this additional constraint (3.327) becomes, at least formally, a solution to the Neumann problem.

It is not difficult to show that the Green function $G(\mathbf{r}', \mathbf{r})$ is a symmetric function of the variables $\mathbf{r}'$ and $\mathbf{r}$, or that we at least have the flexibility to choose a symmetric function. To show this we use Green's theorem with the functions $\phi(\mathbf{r}') = G(\mathbf{r}', \mathbf{r}_1)$ and $\psi(\mathbf{r}') = G(\mathbf{r}', \mathbf{r}_2)$. Then, using (3.315) to evaluate the left-hand side, we get

$$G(\mathbf{r}_1, \mathbf{r}_2) - G(\mathbf{r}_2, \mathbf{r}_1) = \oint_S \left[G(\mathbf{r}', \mathbf{r}_1) \frac{\partial G(\mathbf{r}', \mathbf{r}_2)}{\partial n'} - G(\mathbf{r}', \mathbf{r}_2) \frac{\partial G(\mathbf{r}', \mathbf{r}_1)}{\partial n'} \right] dS' \quad (3.328)$$

For Dirichlet boundary conditions, we use (3.322), and the right-hand side vanishes, proving that the Green function $G(\mathbf{r}_1, \mathbf{r}_2)$ is symmetric. For Neumann boundary conditions, we use (3.326) and get

$$G(\mathbf{r}_1, \mathbf{r}_2) - G(\mathbf{r}_2, \mathbf{r}_1) = \frac{1}{S} \oint_S [G(\mathbf{r}', \mathbf{r}_2) - G(\mathbf{r}', \mathbf{r}_1)] \, dS' \quad (3.329)$$

or

$$G(\mathbf{r}_1, \mathbf{r}_2) - \frac{1}{S} \oint_S G(\mathbf{r}', \mathbf{r}_2) \, dS' = G(\mathbf{r}_2, \mathbf{r}_1) - \frac{1}{S} \oint_S G(\mathbf{r}', \mathbf{r}_1) \, dS' \quad (3.330)$$

Clearly, the new function $G(\mathbf{r}_1, \mathbf{r}_2) - \frac{1}{S} \oint_S G(\mathbf{r}', \mathbf{r}_2) \, dS'$ is symmetric in $\mathbf{r}_1$ and $\mathbf{r}_2$. Since the second term (the integral) is a function only of $\mathbf{r}_2$, the new function satisfies both (3.315) and (3.326) if $G(\mathbf{r}_1, \mathbf{r}_2)$ does. Therefore, the function

$$G^{(S)}(\mathbf{r}_1, \mathbf{r}_2) = G(\mathbf{r}_1, \mathbf{r}_2) - \frac{1}{S} \oint_S G(\mathbf{r}', \mathbf{r}_2) \, dS' \quad (3.331)$$

is the desired symmetric Green function. In fact $G(\mathbf{r}', \mathbf{r})$, like $\Phi(\mathbf{r}')$, is defined only to within an additive constant when Neumann boundary conditions are applied. Therefore, if we fix $G(\mathbf{r}', \mathbf{r})$ with the additional condition that $\oint G(\mathbf{r}', \mathbf{r}) \, dS' = 0$, analogous to the condition used earlier to make $\Phi(\mathbf{r}')$ unique, the Green function is immediately symmetric.

We observed earlier that in a physical sense the Green function $G(\mathbf{r}', \mathbf{r})$ is the potential at the field position $\mathbf{r}'$ due to a point charge $q = \varepsilon_0$ located at the source position $\mathbf{r}$. The symmetry of the Green function with respect to its arguments shows that we may exchange the positions of the source and observer without changing the potential, as shown in Figure 3.24. This is certainly true for a single, isolated point charge with no boundaries except

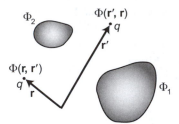

Figure 3.24 Green's reciprocity theorem.

at infinity, as we see from (3.320), for example. The fact that this remains true in the presence of grounded conductors or appropriately charged isolated conductors is a remarkable result. It is also central to the application of Green functions to the solution of inhomogeneous partial differential equations, for the way we use the Green function to construct the solution uses **r** and **r′** in the opposite sense. In (3.323), for example, we see that the first term (the volume integral) is the sum over all charges of the potential $G(\mathbf{r}', \mathbf{r})$ at the field position **r** due to a point charge $\rho \, dV'$ at the source position **r′**. On the other hand, the definition (3.315) of the Green function that was used to derive (3.323) makes $G(\mathbf{r}', \mathbf{r})$ the potential at **r′** due to a source at **r**, which reverses the roles of **r** and **r′**. The statement that the Green function is symmetric is called Green's reciprocity theorem.

As an example of a Green function when conductors are involved, we consider the potential due to a charge distribution $\rho(\mathbf{r})$ surrounding a grounded, conducting sphere of radius a. Using the method of images, we previously obtained the solution (3.103) for a point charge near a grounded sphere. From this we find that the Green function is

$$G(\mathbf{r}', \mathbf{r}) = \frac{1}{4\pi} \left(\frac{1}{|r'\hat{\mathbf{n}}' - r\hat{\mathbf{n}}|} + \frac{Q}{|r'\hat{\mathbf{n}}' - R\hat{\mathbf{n}}|} \right) \tag{3.332}$$

where Q and R are the charge and position of the image charge. The image charge and image position are given by (3.106) and (3.107). Using the law of cosines to evaluate the magnitudes of the denominators, we get

$$G(\mathbf{r}', \mathbf{r}) = \frac{1}{4\pi} \left(\frac{1}{\sqrt{r'^2 + r^2 - 2r'r\cos\theta}} - \frac{1}{\sqrt{\frac{r^2 r'^2}{a^2} + a^2 - 2r'r\cos\theta}} \right) \tag{3.333}$$

where θ is the angle between **r′** and **r**. Clearly, this is symmetric in the variables **r′** and **r**.

EXERCISE 3.20

The use of Green functions, eigenfunctions, and minimum principles is much more general than their application to the Poisson equation. For example, the Sturm–Liouville problem consists of the general linear, self-adjoint, homogeneous differential equation

$$\frac{d}{dx}\left[p(x)\frac{d\psi}{dx} \right] + [q(x) + \lambda r(x)]\psi = 0 \tag{3.334}$$

where $p(x) > 0$ and $q(x)$ and $r(x)$ are real, nonsingular functions of x over the interval $a < x < b$, together with boundary conditions on ψ at $x = a$ and $x = b$. This is a prototype for many linear equations in physics, such as the Helmholtz equation, the Legendre equation, and the Bessel equation. Solutions exist only for certain values of λ, which we call the eigenvalues λ_n, and the corresponding solutions are called the eigenfunctions ψ_n. For definiteness in the following, we apply the homogeneous boundary conditions

$$\alpha \frac{d\psi}{dx}(a) - \beta\psi(a) = 0 \tag{3.335}$$

$$\mu \frac{d\psi}{dx}(b) - \nu\psi(b) = 0 \tag{3.336}$$

where α, β, μ, and ν are constants, although periodic boundary conditions also work. It may be shown that the eigenvalues $\lambda_0 < \lambda_1 < \lambda_2 \cdots \infty$ are all real and distinct, and we assume for simplicity that they are all nondegenerate. It can be proved that the eigenfunctions ψ_n form a complete set on the interval $a < x < b$.

(a) Prove that the eigenfunctions corresponding to different eigenvalues are orthogonal with the weighting function $r(x)$, that is,

$$(\lambda_m - \lambda_n) \int_a^b r(x)\psi_m(x)\psi_n(x)\,dx = 0 \tag{3.337}$$

(b) Using the methods we developed in Chapter 2, show that $\psi(x)$ satisfies the variational principle

$$\int_a^b \left[p(x)\left(\frac{d\psi}{dx}\right)^2 - q(x)\psi^2 \right] dx = \text{minimum} \tag{3.338}$$

when $\psi(x)$ is varied subject to the auxiliary condition

$$\int_a^b r(x)\psi^2(x)\,dx = \text{constant} \tag{3.339}$$

Hint: If we use the method of Lagrange multipliers, this corresponds to varying $\psi(x)$ to minimize the functional

$$\delta \int_a^b \left[p(x)\left(\frac{d\psi}{dx}\right)^2 - q(x)\psi^2 - \lambda r(x)\psi^2 \right] dx = 0 \tag{3.340}$$

where λ is the Lagrange multiplier.

(c) Show that in one dimension, Green's theorem is

$$\int_a^b \left[\phi \frac{d}{dx}\left(p\frac{d\psi}{dx}\right) - \psi \frac{d}{dx}\left(p\frac{d\phi}{dx}\right) \right] dx = \left[p\left(\phi\frac{d\psi}{dx} - \psi\frac{d\phi}{dx}\right) \right]\Bigg|_a^b \tag{3.341}$$

The Green function satisfies the inhomogeneous differential equation

$$\frac{d}{dx}\left[p(x)\frac{d}{dx}G(x, x') \right] + [q(x) + \lambda r(x)]\,G(x, x') = -\delta(x - x') \tag{3.342}$$

and the homogeneous boundary conditions

$$\alpha \frac{dG}{dx}(a, x') - \beta G(a, x') = 0 \tag{3.343}$$

$$\mu \frac{dG}{dx}(b, x') - \nu G(b, x') = 0 \tag{3.344}$$

Use Green's theorem to show that the general solution to the inhomogeneous equation

$$\frac{d}{dx}\left[p(x)\frac{d\psi}{dx}\right] + [q(x) + \lambda r(x)]\,\psi = -\rho(x) \tag{3.345}$$

is given in terms of the Green function by

$$\psi(x) = \int_a^b G(x', x)\rho(x')\,dx' \tag{3.346}$$

(d) Use Green's theorem, derived earlier, to show that the Green function is symmetric,

$$G(x, x') = G(x', x) \tag{3.347}$$

(e) Show that the Green function may be represented by the eigenfunction expansion

$$G(x, x') = \sum_{n=0}^{\infty} \frac{\psi_n(x)\psi_n(x')}{(\lambda_n - \lambda)\langle \psi_n^2 \rangle} \tag{3.348}$$

where

$$\langle \psi_n^2 \rangle = \int_a^b r(x)\psi_n^2(x)\,dx \tag{3.349}$$

depends on the normalization of the eigenfunctions. *Hint:* Expand the Green function in a series of eigenfunctions, substitute into the differential equation for $G(x, x')$, and use the orthogonality of the eigenfunctions to determine the coefficients of the expansion.

3.3 MAGNETOSTATICS

Like electrostatics, magnetism has a long history that dates back into antiquity. The compass seems to have been discovered in China more than 4600 years ago, and it was in common use for navigation in the Indian Ocean by the third century A.D. The Chinese philosopher Kuopho speculated about the connection between magnetism and electricity in the fourth century A.D., and the effect of lightening on compasses supported this speculation. But proof of the connection awaited the development of the voltaic pile as a reliable laboratory-scale source of electric current, and this did not happen until about 200 years ago. Then the floodgates opened, and following Oersted's initial discovery in 1820, Ampere, Biot, Savart, Gauss, and others quickly filled in the details. By around the middle of the 19th century, our understanding of magnetostatics was essentially complete.

3.3.1 Biot–Savart Law

In the completely static case, when all the charges are motionless, there is no current and no magnetic field. However, if we ignore the fact that real currents arise from the motion of discrete charges, it is possible to imagine a steady current density $\mathbf{J}(\mathbf{r})$ arising from the motion of a continuous charge distribution. If there is no net accumulation of charge at any point, there is no time dependence in the charge density. From the continuity relation, then, we get

$$\nabla \cdot \mathbf{J} = -\frac{\partial \rho}{\partial t} = 0 \tag{3.350}$$

That is, the current density $\mathbf{J}$ is divergence free. The current flows along streamlines, and the streamlines form closed loops like the lines of magnetic induction $\mathbf{B}$. For real currents, consisting of discrete charges in motion, we may approximate this situation with some sort of microscopic average if the charge carriers are sufficiently small and numerous.

It is convenient to begin the discussion of magnetostatics with the vector potential $\mathbf{A}$, in terms of which the magnetic induction is

$$\mathbf{B} = \nabla \times \mathbf{A} \tag{3.351}$$

as discussed in Chapter 1. From the spacelike components of the 4-vector wave equation we get

$$\frac{1}{c^2} \frac{\partial^2 \mathbf{A}}{\partial t^2} - \nabla^2 \mathbf{A} = \mu_0 \mathbf{J} \tag{3.352}$$

which in the time-independent case becomes simply

$$\nabla^2 \mathbf{A} = -\mu_0 \mathbf{J} \tag{3.353}$$

The three orthogonal components of this vector expression are independent of one another, each with its own source, and they comprise three separate, scalar Poisson equations. If we require that it vanish at infinity, the vector potential, like the scalar potential, may be expressed as an integral over the source of the form

$$\mathbf{A}(\mathbf{r}) = \frac{\mu_0}{4\pi} \int \frac{\mathbf{J}(\mathbf{r}')\, d^3\mathbf{r}'}{|\mathbf{r} - \mathbf{r}'|} \tag{3.354}$$

To this we may add any function that satisfies the Laplace equation

$$\nabla^2 \mathbf{A}' = 0 \tag{3.355}$$

and use this freedom to satisfy whatever boundary conditions apply to the problem.

In addition, we may add to the solution (3.354) any function of the form $\nabla \Lambda$ without changing the magnetic field $\mathbf{B}$. As discussed in Chapter 2, this corresponds to a change of gauge. For most purposes we use this freedom to make

$$\nabla \cdot \mathbf{A} = 0 \tag{3.356}$$

In the time-independent case this satisfies both the Lorentz and Coulomb gauge conditions. In particular, the solution (3.354) satisfies the gauge condition (3.356). To see this, we differentiate (3.354) to find that

$$\nabla \cdot \mathbf{A}(\mathbf{r}) = \frac{\mu_0}{4\pi} \int \mathbf{J}(\mathbf{r}') \cdot \nabla \frac{1}{|\mathbf{r} - \mathbf{r}'|}\, d^3\mathbf{r}' = -\frac{\mu_0}{4\pi} \int \mathbf{J}(\mathbf{r}') \cdot \nabla' \frac{1}{|\mathbf{r} - \mathbf{r}'|}\, d^3\mathbf{r}' \tag{3.357}$$

Integrating once by parts gives

$$\nabla \cdot \mathbf{A}(\mathbf{r}) = -\frac{\mu_0}{4\pi} \left[\sum_{i=1}^{3} \int J_i(\mathbf{r}') \frac{1}{|\mathbf{r} - \mathbf{r}'|} \, d^2 r_{j \neq i} \Big|_{r_i = -\infty}^{r_i = \infty} - \int \frac{1}{|\mathbf{r} - \mathbf{r}'|} \nabla' \cdot \mathbf{J}(\mathbf{r}') \, d^3 r' \right]$$

(3.358)

The first term vanishes if the source vanishes at infinity, and since $\nabla' \cdot \mathbf{J}(\mathbf{r}')$ vanishes in the time-independent case, as shown by (3.350), we see that the solution (3.354) satisfies the gauge condition (3.356).

The spacelike components of the Maxwell equations in the magnetostatic case become

$$\nabla \times \mathbf{B} = \mu_0 \mathbf{J}$$

(3.359)

Using Stokes' theorem, we may write this in the form

$$\oint_C \mathbf{B}(\mathbf{r}) \cdot d\mathbf{l} = \mu_0 \int_S \mathbf{J}(\mathbf{r}) \cdot \hat{\mathbf{n}} \, dS = \mu_0 I$$

(3.360)

where C is a closed path in space bounding the surface S, $\hat{\mathbf{n}}$ is a unit vector normal to the surface element dS, and

$$I = \int_S \mathbf{J}(\mathbf{r}) \cdot \hat{\mathbf{n}} \, dS$$

(3.361)

is the total current flowing through the surface S, as shown in Figure 3.25. This is the form of the law that Ampere discovered, but it is correct only in the static limit. In the general case, when time-varying electric fields are also present, Ampere's law must be extended to include the "displacement currents" introduced by Maxwell.

From (3.354) we see that the magnetic induction in the time-independent case is

$$\mathbf{B}(\mathbf{r}) = \nabla \times \mathbf{A}(\mathbf{r}) = \frac{\mu_0}{4\pi} \int \nabla \times \left[\frac{\mathbf{J}(\mathbf{r}')}{|\mathbf{r} - \mathbf{r}'|} \right] d^3 r'$$

(3.362)

Working out the ith component of this equation, keeping in mind that $\mathbf{J}(\mathbf{r}')$ doesn't depend on $\mathbf{r}$ and using the fact that

$$\frac{\partial}{\partial r_i} \frac{1}{|\mathbf{r} - \mathbf{r}'|} = -\frac{r_i - r_i'}{|\mathbf{r} - \mathbf{r}'|^3}$$

(3.363)

gives the result

$$\mathbf{B}(\mathbf{r}) = \frac{\mu_0}{4\pi} \int \mathbf{J}(\mathbf{r}') \times \frac{\mathbf{r} - \mathbf{r}'}{|\mathbf{r} - \mathbf{r}'|^3} \, d^3 r'$$

(3.364)

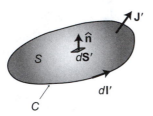

Figure 3.25 Current through a surface.

This is called the Biot–Savart law, and it occupies much the same position in magnetostatics that Coulomb's law occupies in electrostatics. Although we derive it here as a special case of the Maxwell equations, historically the Biot–Savart law came first.

As in electrostatics, conductors play an important role in magnetostatics. Specifically, conductors serve to confine the current density $\mathbf{J}$ to a localized region of space. When the current flows through conductors, the streamlines are confined inside the conductors. When the conductors have a bulky form, the current loops can follow any closed curves within the solid. The eddy currents in a copper bar immersed in a magnetic field are an example of current loops in a conductor that have only a limited relationship to the shape of the conductor. The surface currents of the Meissner effect that exclude the magnetic field from a superconductor provide another example. On the other hand, when the conductor has the form of a slender wire it is usually possible to ignore eddy current loops and to assume that the current in the wire flows through a stream tube that coincides with the surface of the wire. It necessarily follows that the wires must be closed loops in magnetostatics and that the total current flowing through the wire is the same at all points in the loop.

If the wire is sufficiently slender compared with the distance from the wire to the field point, then it is possible in (3.364) to approximate the radius $\mathbf{r} - \mathbf{r}'$ by the distance to the centerline of the wire and to assume that the direction of the current is parallel to the centerline. The total current, everywhere the same along the wire, is

$$I = d\hat{\mathbf{l}}' \cdot \int_S \mathbf{J}(\mathbf{r}') \, dS' \tag{3.365}$$

where $d\mathbf{l}'$ is a vector tangent to the centerline of the wire in the direction of the current, as shown in Figure 3.26, and the integral is over the cross section of the wire. The magnetic induction is then

$$\mathbf{B}(\mathbf{r}) = \frac{\mu_0 I}{4\pi} \oint d\mathbf{l}' \times \frac{\mathbf{r} - \mathbf{r}'}{|\mathbf{r} - \mathbf{r}'|^3} \tag{3.366}$$

where the integral is taken around the entire loop of wire. This expression is the customary form of the Biot–Savart law.

It is worth pointing out that the Biot–Savart law (3.364) [or (3.366)] is relativistically correct, at least within the magnetostatic approximation. It applies to currents that arise from charges in arbitrary motion, including relativistic velocities and acceleration caused by changes in the size and direction of the stream tubes through which the (steady) current is flowing. In all cases, far from the current, the strength of the field falls off as $1/r^2$ and the energy density in the field falls off as $1/r^4$. Thus, the energy in the field is mostly in the region near the source. It is shown in Chapter 10 that charges that experience acceleration radiate energy into the far field, yet steady currents do not. The fields from all the macroscopic currents in a stream tube, or circuit, add in just such a way that the far fields cancel out, rather than add constructively, when the macroscopic current is time independent.

Figure 3.26 Current flow in a slender wire.

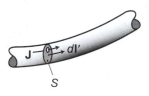

EXERCISE 3.21

Consider a semi-infinite solenoid, which we idealize as a cylindrical current sheet of radius a and azimuthal current σ per unit length, with one end at the origin and stretching along the negative z axis to $-\infty$.

(a) Use the Biot–Savart law to show that the magnetic induction on the axis is

$$B_z = \frac{\mu_0 \sigma}{2} \left[1 - \frac{z}{(a^2 + z^2)^{1/2}} \right] \tag{3.367}$$

(b) Use a Taylor-series expansion of the field near the axis together with $\nabla \cdot \mathbf{B} = 0$ and $\nabla \times \mathbf{B} = 0$ to show that the magnetic induction near the axis is

$$B_\rho \approx \frac{\mu_0 \sigma}{4} \frac{a^2 \rho}{(a^2 + z^2)^{3/2}} \tag{3.368}$$

$$B_z \approx \frac{\mu_0 \sigma}{2} \left[1 - \frac{z}{(a^2 + z^2)^{1/2}} - \frac{3}{4} \frac{a^2 z \rho^2}{(a^2 + z^2)^{5/2}} \right] \tag{3.369}$$

where $\rho = \sqrt{x^2 + y^2}$ is the radial distance from the z axis.

EXERCISE 3.22

Consider the spiral-wound, infinitely long solenoid illustrated in Figure 3.27.

(a) Beginning with the Biot–Savart law, show that the magnetic induction at the origin has the vector components

$$B_x = 0 \tag{3.370}$$

$$B_y = \frac{\mu_0 I}{L} \left[\frac{2\pi R}{L} K_0 \left(\frac{2\pi R}{L} \right) + K_1 \left(\frac{2\pi R}{L} \right) \right] \tag{3.371}$$

$$B_z = \frac{\mu_0 I}{L} \tag{3.372}$$

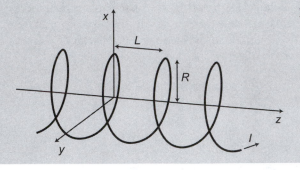

Figure 3.27 Spiral-wound solenoid.

where I is the current, R the radius, and L the pitch of the solenoid, and $K_0(x)$ and $K_1(x)$ are modified Bessel functions. *Hint:* you will need the integrals

$$\int_0^\infty \frac{x \sin(ax)\, dx}{(x^2 + b^2)^{3/2}} = a K_0(ab) \tag{3.373}$$

$$\int_0^\infty \frac{\cos(ax)\, dx}{(x^2 + b^2)^{3/2}} = \frac{a}{b} K_1(ab) \tag{3.374}$$

(b) From (3.372) we see that the axial component of the field is just that of an ideal solenoid (no pitch in the windings). Beginning with (3.371), show that the transverse component of the field has the expected value in the limits $L \to 0$ (ideal solenoid) and $L \to \infty$ (wire parallel to the z axis).

3.3.2 Forces and Energy

In electrostatics we found that the forces on conductors and the energy in the electric field can be described in terms of the charges on the conductors and a set of constants called the capacitance. We are also familiar with the idea that the energy in the magnetic field surrounding a single conductor can be described in terms of the current and a parameter we call the inductance. In this section we see how to generalize the notion of inductance to an arbitrary number of conductors, each with its own current. From this we can compute the total energy in the magnetic field, and from the energy we can compute the forces on the individual conductors.

We found in Chapter 2 that the force density on a 4-vector current, defined as the rate of change of the 4-vector momentum density, is

$$F_\beta = \mu \frac{dU_\beta}{dt} = J^\alpha F_{\beta\alpha} \tag{3.375}$$

where μ is the mass density and U_β is the 4-vector velocity. From the spacelike components, we see that the total force is

$$\mathbf{F} = \int [\rho(\mathbf{r})\mathbf{E}(\mathbf{r}) + \mathbf{J}(\mathbf{r}) \times \mathbf{B}(\mathbf{r})]\, d^3r = \int \mathbf{J}(\mathbf{r}) \times \mathbf{B}(\mathbf{r})\, d^3r \tag{3.376}$$

in the absence of electric fields. The force on a current I_1 in a loop interacting with the magnetic field of a current I_2 in another loop is

$$\mathbf{F}_{12} = I_1 \oint d\mathbf{l}_1 \times \mathbf{B}_2 = \frac{\mu_0 I_1 I_2}{4\pi} \oint \oint \frac{d\mathbf{l}_1 \times [d\mathbf{l}_2 \times (\mathbf{r}_1 - \mathbf{r}_2)]}{|\mathbf{r}_1 - \mathbf{r}_2|^3} \tag{3.377}$$

where we have used (3.366) to evaluate the magnetic field $\mathbf{B}_2(\mathbf{r})$ of the current I_2. This may be expressed in a more symmetric fashion by expanding the triple cross product to get

$$\mathbf{F}_{12} = \frac{\mu_0 I_1 I_2}{4\pi} \oint \oint \frac{[d\mathbf{l}_1 \cdot (\mathbf{r}_1 - \mathbf{r}_2)]\, d\mathbf{l}_2 - (d\mathbf{l}_1 \cdot d\mathbf{l}_2)(\mathbf{r}_1 - \mathbf{r}_2)}{|\mathbf{r}_1 - \mathbf{r}_2|^3} \tag{3.378}$$

But from (3.9) we see that

$$\frac{d\mathbf{l}_1 \cdot (\mathbf{r}_1 - \mathbf{r}_2)}{|\mathbf{r}_1 - \mathbf{r}_2|^3} = -d_1\left(\frac{1}{|\mathbf{r}_1 - \mathbf{r}_2|}\right) \tag{3.379}$$

along loop 1. That is, it is a perfect differential. Therefore, when we integrate the first term in (3.378) over $d\mathbf{l}_1$, we end up where we started and the integral vanishes. The force on loop 1 due to loop 2 is therefore

$$\mathbf{F}_{12} = -\frac{\mu_0 I_1 I_2}{4\pi} \oint \oint \frac{(d\mathbf{l}_1 \cdot d\mathbf{l}_2)(\mathbf{r}_1 - \mathbf{r}_2)}{|\mathbf{r}_1 - \mathbf{r}_2|^3} = -\mathbf{F}_{21} \tag{3.380}$$

This shows that the total force on one loop is identically equal and opposite the force on the other. This is somewhat remarkable, since the forces between the individual segments of the loops do not obey Newton's third law of motion. On the other hand, we might have anticipated this, since when the momentum of the particles is not conserved (when the action is not equal to the reaction), the difference is made up by the momentum of the fields. In the magnetostatic case, the momentum of the fields is constant.

As discussed in Chapter 2, the magnetic field possesses an energy density

$$\mathcal{U} = \frac{B^2}{2\mu_0} \tag{3.381}$$

In terms of the vector potential, then, the total energy in the field is

$$W = \frac{1}{2\mu_0} \int \mathbf{B} \cdot (\nabla \times \mathbf{A}) \, d^3\mathbf{r} = \frac{1}{2\mu_0} \sum_{i,j,k=1}^{3} \varepsilon_{ijk} \int B_k \frac{\partial A_j}{\partial r_i} \, d^3\mathbf{r} \tag{3.382}$$

If we integrate once by parts, we get

$$W = \frac{1}{2\mu_0} \sum_{i,j,k=1}^{3} \varepsilon_{ijk} \left[\int B_k A_j d^2 r_{l \neq i} \Big|_{r_i = -\infty}^{r_i = \infty} - \int A_j \frac{\partial B_k}{\partial r_i} \, d^3\mathbf{r} \right] \tag{3.383}$$

But the first term vanishes if the magnetic induction vanishes at infinity, and we are left with

$$W = \frac{1}{2\mu_0} \int \nabla \times \mathbf{B} \cdot \mathbf{A} \, d^3\mathbf{r} = \frac{1}{2} \int \mathbf{J} \cdot \mathbf{A} \, d^3\mathbf{r} \tag{3.384}$$

in the magnetostatic approximation, where we have used (3.359) to evaluate $\nabla \times \mathbf{B}$. This expression for the magnetostatic energy in terms of the current density and the vector potential is the magnetic equivalent to (3.27), which expresses the electrostatic energy in terms of the charge density and scalar potential.

Since $\nabla \cdot \mathbf{J} = 0$ in magnetostatics, the streamlines of the current density field $\mathbf{J}(\mathbf{r})$ form closed loops, as discussed earlier. Where bundles of nearby streamlines stay close together, they form what are called stream tubes. For example, the currents flowing through electrical wires form stream tubes defined by the surface of the wire. Within each stream tube, the total current flowing through the stream tube is the same at all points along the tube. If the current density in (3.354) is separated into a sum of currents

$$\mathbf{J}(\mathbf{r}) = \sum_m \mathbf{J}_m(\mathbf{r}) \tag{3.385}$$

in which each $\mathbf{J}_m(\mathbf{r})$ corresponds to a complete stream tube, then the vector potential can also be separated into a sum of the form

$$\mathbf{A}(\mathbf{r}) = \sum_m \mathbf{A}_m(\mathbf{r}) = \frac{\mu_0}{4\pi} \sum_m \int \frac{\mathbf{J}_m(\mathbf{r}')}{|\mathbf{r} - \mathbf{r}'|} \, d^3\mathbf{r}' \tag{3.386}$$

where $\mathbf{A}_m(\mathbf{r})$ is generated by the current $\mathbf{J}_m(\mathbf{r}')$. The energy in the magnetic field is then

$$\mathcal{W} = \frac{1}{2} \sum_{m,n} \int \mathbf{J}_m(\mathbf{r}) \cdot \mathbf{A}_n(\mathbf{r}) \, d^3\mathbf{r} = \frac{\mu_0}{8\pi} \sum_{m,n} \int\!\!\int \frac{\mathbf{J}_m(\mathbf{r}') \cdot \mathbf{J}_n(\mathbf{r})}{|\mathbf{r} - \mathbf{r}'|} \, d^3\mathbf{r} \, d^3\mathbf{r}' \quad (3.387)$$

If we assume that the current density $\mathbf{J}_m(\mathbf{r})$ is proportional to the total current I_m, then we can define the quantities

$$L_{mn} = \frac{\mu_0}{4\pi I_m I_n} \int\!\!\int \frac{\mathbf{J}_m(\mathbf{r}') \cdot \mathbf{J}_n(\mathbf{r})}{|\mathbf{r} - \mathbf{r}'|} d^3\mathbf{r} \, d^3\mathbf{r}' \quad (3.388)$$

For $m \neq n$, the quantity $L_{mn} = L_{nm}$ is called the mutual inductance of the circuits m and n. For $m = n$, it is called the self-inductance. The energy of the magnetic field is then

$$\mathcal{W} = \frac{1}{2} \sum_{m,n} I_m I_n L_{mn} \quad (3.389)$$

Clearly, this formula plays the same role in magnetostatics that (3.81) plays in electrostatics. In the simple case of a single conductor with a single current, the total energy due to the self-inductance is given by the familiar formula

$$\mathcal{W} = \tfrac{1}{2} L I^2 \quad (3.390)$$

The representation of the magnetic field energy in terms of inductances is particularly useful when the current densities $\mathbf{J}_m(\mathbf{r})$ flow through slender wires separated by distances that are large compared with the transverse dimensions of the wire. In this case, we may use the approximation

$$L_{mn} = \frac{\mu_0}{4\pi} \oint \oint \frac{d\mathbf{l}_m \cdot d\mathbf{l}_n}{|\mathbf{r}_m - \mathbf{r}_n|} = L_{nm}, \qquad \text{for } m \neq n \quad (3.391)$$

This formula is not valid for the self-inductance ($m = n$) because in this case the integral extends to small distances where the vector potential is not satisfactorily described by placing all the current on the centerline of the wire. In fact, (3.391) diverges for $m = n$.

If we consider a generalized displacement ds of the conductors, keeping the currents in the conductors fixed, we find that the change in the energy of the field is

$$\delta\mathcal{W} = \frac{1}{2} \sum_{m,n} I_m I_n \frac{dL_{mn}}{ds} \delta s \quad (3.392)$$

For example, if we translate only the ith conductor a distance $\delta\mathbf{r}$, the change in the magnetic field energy is

$$\delta\mathcal{W} = I_i \sum_{n \neq i} I_n \frac{dL_{in}}{d\mathbf{r}} \cdot \delta\mathbf{r} = -I_i \sum_{n \neq i} I_n \frac{\mu_0}{4\pi} \oint \oint \frac{(\mathbf{r}_i - \mathbf{r}_n) \cdot \delta\mathbf{r}}{|\mathbf{r}_i - \mathbf{r}_n|^3} d\mathbf{l}_i \cdot d\mathbf{l}_n \quad (3.393)$$

where we have used (3.9) and (3.391) to evaluate the derivative of the mutual inductance. If we compare this to (3.380), we see that

$$\delta\mathcal{W} = \sum_n \mathbf{F}_{in} \cdot \delta\mathbf{r} \quad (3.394)$$

But $\sum_n \mathbf{F}_{in} \cdot \delta\mathbf{r}$ is the work done by the field on the ith conductor. When this is positive, we might expect the energy in the field to decrease as it is consumed to do work on the

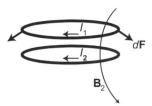

Figure 3.28 Force between two circular conductors.

conductor, so (3.394) would seem to have the wrong sign. The explanation is similar to that given for the displacement of charged conductors when the potentials are held fixed. That is, work must be done to maintain the currents in the conductors at the original values, and this work actually increases the energy of the field.

A concrete example can serve to illustrate this. We consider two circular conductors, as shown in Figure 3.28, carrying currents I_1 in the upper loop and I_2 in the lower loop, both in the same direction. The magnetic induction $\mathbf{B}_2$ of current I_2 intersects current I_1 as shown, and the total force $\mathbf{F} = \oint d\mathbf{F}$ on I_1 is therefore downward, toward I_2, by symmetry. If I_1 is displaced downward, the work done by the field is positive, so the energy of the field should decrease. However, the magnetic flux through the upper loop increases as the upper loop approaches the lower loop. Therefore, according to Faraday's law of induction and Lenz's law (which are, strictly speaking, outside the scope of magnetostatics), an emf is induced in the upper loop to oppose the increase in the flux. That is, a current is induced in the upper coil opposite the current I_1, and positive work must be done to maintain the current in this loop. The same is true in the lower loop, and the total work done by the external sources is just twice the work done by the field on the loop as it is displaced. This reverses the sign of the change in the field energy when the currents are held constant.

EXERCISE 3.23

The self-inductance of a current loop can be computed from the energy in the magnetic field. For a loop of thin wire, for which the radius a of the wire is small compared with a characteristic dimension of the loop, say $R = l/2\pi$, where l is the length of the wire, the energy in the field is found largely in the region near the wire. In this region, provided that the wire is not bent too sharply, the field can be estimated by treating the wire as an infinitely long straight conductor. At high frequencies, for which the inductance of a loop of wire is most often needed, the current is limited to the region near the surface of the wire by the skin effect, so the field energy inside the wire can be ignored in the following.

(a) Show that the energy density in the field near the wire is

$$\mathcal{U} = \frac{\mu_0 I^2}{8\pi^2 r^2} \tag{3.395}$$

where r is the radial distance from the wire.

(b) Integrate this from the surface of the wire out to some radius characteristic of the size of the loop to show that the self-inductance of the loop is approximately

$$L = \frac{\mu_0 l}{2\pi} \ln\left(\frac{\chi R}{a}\right) \tag{3.396}$$

where χ is a constant on the order of unity. For a circular loop of radius R, it is found that for $a/R \ll 1$ the constant is

$$\chi = \frac{8}{e^2} = 1.083 \tag{3.397}$$

(c) If a length of thin conductor is connected to a thick conductor, the inductance is dominated by the field energy around the thin conductor. For example, this is the case for a spark gap (with a filamentary discharge) connected to an otherwise low-inductance circuit. The inductance is then proportional to the length of the spark. Estimate the inductance of a spark gap with an arc channel 30 μm in diameter, a gap between the electrodes of 6 mm, and an outer diameter of 3 cm.

Although it is fundamentally incorrect to refer to the self-inductance of a segment of a circuit, this is often done and is frequently a good approximation. Thus, we can go down to the local electronics parts store and buy a 100-nH rf choke or a 40-mH smoothing choke and connect it to the rest of our circuit. Nevertheless, it is important to keep in mind that the inductance of the connections and the rest of the circuit can often be significant.

EXERCISE 3.24

A coaxial cable has an inner conductor with radius r and an outer conductor with radius R.

(a) By computing the energy in the magnetic field, show that the inductance per unit length is

$$L = \frac{\mu_0}{2\pi} \ln \frac{R}{r} \tag{3.398}$$

(b) By computing the energy in the electric field, show that the capacitance per unit length is

$$C = \frac{2\pi \varepsilon_0}{\ln(R/r)} \tag{3.399}$$

(c) What is the impedance $Z = \sqrt{L/C}$ of a coaxial cable for which $R/r = e$, where e is the base of natural logarithms?

3.3.3 Multipole Moments

Just like the scalar potential of a localized charge distribution, the vector potential of a localized current distribution can be expanded in terms of multipole moments of the current. The mathematical development of the expansion is similar to that of scalar multipole moments. Although the potential $\mathbf{A(r)}$ is a vector, each component has its own source, as shown by (3.354), and therefore its own multipole expansion. However, the components of the current $\mathbf{J}$ are related, in magnetostatics, by the conservation law $\nabla \cdot \mathbf{J} = 0$. This can be used to simplify the results.

We begin with the expression (3.354) for the vector potential as an integral over the source $\mathbf{J(r)}$. Provided that the source is localized around the origin and the observation point is a large distance away, we may use the Taylor expansion (3.35)–(3.38) of the function $1/|\mathbf{r} - \mathbf{r}'|$. If we substitute the Taylor expansion into (3.354), we get, after some algebra, the series

$$A_i(\mathbf{r}) = \frac{\mu_0}{4\pi} \left[\frac{1}{r} \int J_i(\mathbf{r}') \, d^3\mathbf{r}' + \frac{1}{r^3} \sum_{j=1}^{3} \int r_j r_j' J_i(\mathbf{r}') \, d^3\mathbf{r}' + O(r^{-3}) \right] \tag{3.400}$$

for the ith component of the vector $\mathbf{A}(\mathbf{r})$. This expression can be streamlined in several ways. To begin with, the first term vanishes because the current flows in loops. Since the current flows in loops, the net motion of the charges integrated over all space must vanish. To prove this we note that we may write

$$\int J_i(\mathbf{r}') \, d^3\mathbf{r}' = \sum_{j=1}^{3} \int J_j(\mathbf{r}') \frac{\partial r_i'}{\partial r_j'} \, d^3\mathbf{r}' \tag{3.401}$$

since $\partial r_i'/\partial r_j' = \delta_{ij}$. Integrating once by parts we find that

$$\int J_i(\mathbf{r}') \, d^3\mathbf{r}' = \sum_{j=1}^{3} \int J_j(\mathbf{r}') \, r_i' d^2 r_{k\neq j}' \Big|_{r_j'=-\infty}^{r_j'=\infty} - \sum_{j=1}^{3} \int r_i' \frac{\partial J_j(\mathbf{r}')}{\partial r_j'} \, d^3\mathbf{r}' = 0 \tag{3.402}$$

where the first term vanishes because the source vanishes at infinity, and the second term vanishes because $\nabla \cdot \mathbf{J} = 0$ for steady currents. Thus, the first term (the monopole moment) in the series (3.400) vanishes for magnetostatics. Note that this has nothing to do with the nonexistence of magnetic monopoles. Rather, (3.402) follows from the conservation of charge in a time-independent current distribution, that is, from $\nabla \cdot \mathbf{J} = 0$.

The second term in (3.400) can be simplified if we separate it into its symmetric and antisymmetric parts:

$$r_j' J_i = \tfrac{1}{2}(r_j' J_i + r_i' J_j) + \tfrac{1}{2}(r_j' J_i - r_i' J_j) \tag{3.403}$$

The integral over the symmetric part vanishes after integrating by parts, since

$$\sum_{j=1}^{3} r_j \int (r_j' J_i + r_i' J_j) \, d^3\mathbf{r}' = \sum_{j,k=1}^{3} r_j \int \left(r_j' \frac{\partial r_i'}{\partial r_k'} J_k + r_i' \frac{\partial r_j'}{\partial r_k'} J_k \right)$$

$$= \sum_{j,k=1}^{3} r_j \left[\int r_j' r_i' J_k \, d^2 r_{l\neq k}' \Big|_{r_k'=-\infty}^{r_k'=\infty} - \int r_j' r_i' \frac{\partial J_k}{\partial r_k'} \, d^3\mathbf{r}' \right]$$

$$= 0 \tag{3.404}$$

where the first term vanishes because the source vanishes at infinity and the second term vanishes because $\nabla \cdot \mathbf{J} = 0$, as before.

The remaining, antisymmetric part is conveniently expressed in terms of the magnetic dipole moment of the current distribution,

$$\mathbf{m} = \frac{1}{2} \int \mathbf{r}' \times \mathbf{J}(\mathbf{r}') \, d^3\mathbf{r}' \tag{3.405}$$

Note that the magnetic dipole moment is independent of the choice of origin, since $\int \mathbf{J}(\mathbf{r}') \, d^3\mathbf{r}' = 0$. To express the vector potential in terms of the magnetic dipole moment we begin with the cross product

$$(\mathbf{m} \times \mathbf{r})_i = \sum_{j,k,l,m=1}^{3} \varepsilon_{ijk} \varepsilon_{jlm} \frac{1}{2} \int r_k r'_l J_m \, d^3\mathbf{r} \tag{3.406}$$

But

$$\sum_{j=1}^{3} \varepsilon_{ijk} \varepsilon_{jlm} = \delta_{im}\delta_{kl} - \delta_{il}\delta_{km} \tag{3.407}$$

since $\varepsilon_{ijk}\varepsilon_{jlm}$ vanishes unless $(l, m) = (k, i)$ or $(l, m) = (i, k)$, so

$$(\mathbf{m} \times \mathbf{r})_i = \sum_{j=1}^{3} \frac{1}{2} \int (r_j r'_j J_i - r_j r'_i J_j) \, d^3\mathbf{r} \tag{3.408}$$

Therefore, we see from (3.403) and (3.404) that

$$\sum_{j=1}^{3} \int r_j r'_j J_i(\mathbf{r}') \, d^3\mathbf{r}' = (\mathbf{m} \times \mathbf{r})_i \tag{3.409}$$

If we substitute this into (3.400) we see that the vector potential $\mathbf{A}(\mathbf{r})$ is

$$\mathbf{A}(\mathbf{r}) = \frac{\mu_0}{4\pi} \frac{\mathbf{m} \times \mathbf{r}}{r^3} + O\left(\frac{1}{r^3}\right) \tag{3.410}$$

When the current distribution $\mathbf{J}(\mathbf{r}')$ has the form of a current flowing through a slender wire the magnetic dipole moment is

$$\mathbf{m} = \frac{1}{2} I \oint \mathbf{r}' \times d\mathbf{l}' \tag{3.411}$$

where I is the total current in the loop, $d\mathbf{l}$ is an increment of length along the loop in the direction of the current, and the integral is taken around the loop, as shown in Figure 3.29. But $\hat{\mathbf{n}} \, dS = \frac{1}{2}\mathbf{r}' \times d\mathbf{l}'$ represents an increment of area dS in the direction of the normal unit vector $\hat{\mathbf{n}}$, so the dipole moment may be written

$$\mathbf{m} = I\mathbf{S} \tag{3.412}$$

where

$$\mathbf{S} = \frac{1}{2} \oint \mathbf{r}' \times d\mathbf{l}' \tag{3.413}$$

is the total vector area of the loop, regardless of its shape.

If a localized current distribution is immersed in an externally generated magnetic field, the force and torque on the current distribution can be calculated using the multipole moments of the current. The procedure is similar to that used earlier to calculate the electrostatic forces on a localized charge distribution. Provided that the currents are localized near

Figure 3.29 Magnetic dipole moment of a current loop.

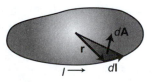

the origin, the externally generated magnetic induction may be expanded in a Taylor series of the form

$$B_i(\mathbf{r}') = B_i(0) + \sum_{j=1}^{3} r_j' \left. \frac{\partial B_i(\mathbf{r})}{\partial r_j} \right|_{\mathbf{r}=0} + \cdots \tag{3.414}$$

From (3.376), we see that the force on the current distribution is

$$F_i = \sum_{j,k=1}^{3} \varepsilon_{ijk} \int J_j B_k \, d^3\mathbf{r}'$$

$$= \sum_{j,k=1}^{3} \varepsilon_{ijk} B_k(0) \int J_j \, d^3\mathbf{r}' + \sum_{j,k,l=1}^{3} \varepsilon_{ijk} \left. \frac{\partial B_k(\mathbf{r})}{\partial r_l} \right|_{\mathbf{r}=0} \int r_l' J_j \, d^3\mathbf{r}' + \cdots \tag{3.415}$$

But the first term vanishes, as shown by (3.402). To simplify the second term, we assume that the external field is generated by currents that vanish in the region of the localized current distribution. In this region, then, $\nabla \times \mathbf{B} = 0$, or

$$\frac{\partial B_k}{\partial r_l} = \frac{\partial B_l}{\partial r_k} \tag{3.416}$$

To lowest order, therefore, we may write

$$F_i = \sum_{j,k,l=1}^{3} \varepsilon_{ijk} \frac{\partial}{\partial r_k} \int B_l(\mathbf{r}) r_l' J_j \, d^3\mathbf{r}' \bigg|_{\mathbf{r}=0} \tag{3.417}$$

Using (3.409), but with the substitution of $\mathbf{B}(\mathbf{r})$ for $\mathbf{r}$, we get

$$F_i = \sum_{j,k=1}^{3} \varepsilon_{ijk} \frac{\partial}{\partial r_k} (\mathbf{m} \times \mathbf{B})_j = \sum_{j,k,l,m=1}^{3} \varepsilon_{ijk} \varepsilon_{jlm} \frac{\partial}{\partial r_k} m_l B_m = \sum_{k=1}^{3} \frac{\partial}{\partial r_k} (m_k B_i - m_i B_k)$$

$$= \sum_{k=1}^{3} m_k \frac{\partial B_i}{\partial r_k} = \sum_{k=1}^{3} m_k \frac{\partial B_k}{\partial r_i} \tag{3.418}$$

since $\nabla \cdot \mathbf{B} = 0$ and $\nabla \times \mathbf{B} = 0$, as noted earlier. In vector notation, then, the force is

$$\mathbf{F} = (\mathbf{m} \cdot \nabla)\mathbf{B} = \nabla(\mathbf{m} \cdot \mathbf{B}) \tag{3.419}$$

Since the force on the current distribution is the gradient of a scalar, this scalar must be the negative of the potential energy. We therefore see that the energy of interaction between a magnetic dipole and a magnetic field may be written

$$\mathcal{W} = -\mathbf{m} \cdot \mathbf{B} \tag{3.420}$$

From (3.419) we see that the force on a localized current distribution depends not on the magnetic induction but on the gradient of the magnetic induction or, more precisely, on the gradient of the interaction of the magnetic moment with the magnetic induction. For example, in the Stern–Gerlach experiment (1922) to measure the magnetic dipole moment of silver atoms, a magnetic field was used to deflect a beam of silver atoms that was collimated by passing them through a small hole, as shown in Figure 3.30. To maximize the deflection, the magnetic field was designed to be as nonuniform as possible and had a strong gradient in the vertical direction. The atoms were deflected vertically, in the direction

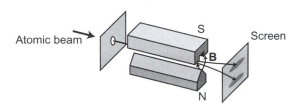

Figure 3.30 Stern–Gerlach experiment.

of the gradient, rather than horizontally, perpendicular to the magnetic field. Because of quantum mechanical space quantization, the angular momentum and the magnetic dipole moment of the silver atoms are quantized in the vertical direction. Specifically, the net magnetic dipole moment of a silver atom is just the intrinsic magnetic moment associated with the spin of its single valence electron, and the component in the direction of the field can have only two possible values,

$$\mathbf{m} \cdot \mathbf{B} = \pm \mu_B B \tag{3.421}$$

where the Bohr magneton $\mu_B = e\hbar/2m$ is the intrinsic magnetic dipole moment of an electron, in which $e = |q|$ and m are the charge and mass of the electron, and $\hbar$ is Planck's constant divided by 2π. This causes the atoms to be deflected into one of two discrete spots, corresponding to spin up and spin down. The Stern–Gerlach experiment demonstrated the existence of space quantization even before the development of wave mechanics.

Since the energy (3.420) of the magnetic dipole depends on the angle θ between the dipole moment and the magnetic induction, the field exerts a torque $\boldsymbol{\tau}$ on the dipole. But the energy is proportional to $\cos\theta$, so the torque (the gradient of the energy) must be proportional to $\sin\theta$. The torque is therefore

$$\boldsymbol{\tau} = \mathbf{m} \times \mathbf{B} \tag{3.422}$$

This shows that the torque acts to align the magnetic dipole moment with the magnetic induction. This is also evident from (3.420), since the energy decreases as the magnetic dipole moment aligns with the field.

An important phenomenon caused by the torque on a magnetic dipole in a magnetic field is nuclear magnetic resonance. In this case, the nuclear magnetic dipole moment is

$$\mathbf{m} = \gamma_N \mathbf{L} \tag{3.423}$$

where $\mathbf{L}$ is the angular momentum of the nucleus and γ_N is called the gyromagnetic ratio. The rate of change of the angular momentum is

$$\frac{d\mathbf{L}}{dt} = \boldsymbol{\tau} = \gamma_N \mathbf{L} \times \mathbf{B} \tag{3.424}$$

which is perpendicular to both $\mathbf{L}$ and $\mathbf{B}$, as shown in Figure 3.31. Clearly, the component of $\mathbf{L}$ in the direction parallel to $\mathbf{B}$ is constant, while the component normal to $\mathbf{B}$ rotates around $\mathbf{B}$ with the frequency

$$\omega_P = \gamma_N B \tag{3.425}$$

This is called Larmor precession.

Quantum mechanically, the energy of a magnetic dipole in a magnetic field is, from (3.420),

$$\mathcal{W} = -\mathbf{m} \cdot \mathbf{B} = -\hbar \gamma_N B m_J, \qquad m_J = -J, -(J-1), \ldots, J \tag{3.426}$$

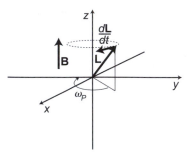

Figure 3.31 Torque on a magnetic dipole.

where m_J is the magnetic quantum number and J is the total angular momentum quantum number. When the nuclei in a magnetic field undergo transitions between angular momentum states, the energy change is

$$\Delta W = \pm \hbar \omega = \pm \hbar \gamma_N B \tag{3.427}$$

Thus, the frequency predicted from quantum mechanics is the same as the frequency predicted by the classical theory.

To measure the frequency at which the angular momentum precesses about the **B** direction, a small magnetic field is introduced that oscillates at a frequency near ω_P. As the frequency of the oscillating field is tuned through the precession frequency, the field suddenly interacts resonantly with the nuclear precession when $\omega = \omega_P$ and the effect of the interaction can be detected. To create magnetic resonance images of nonuniform materials, such as human organs, a nonuniform magnetic field is used. This makes it possible to correlate the magnetic resonance signal with the region of space where the magnetic induction has the resonant value.

EXERCISE 3.25

The vector potential of a magnetic dipole **m** is given by (3.410). Show that the magnetic induction is

$$\mathbf{B}(\mathbf{r}) = \frac{\mu_0}{4\pi} \left(\frac{3\mathbf{m} \cdot \mathbf{r}}{r^5} \mathbf{r} - \frac{\mathbf{m}}{r^3} \right) \tag{3.428}$$

EXERCISE 3.26

Consider a uniform shell of charge spinning with the angular velocity $\boldsymbol{\omega}$. Show that the magnetic dipole moment is

$$\mathbf{m} = \tfrac{1}{3} q R^2 \boldsymbol{\omega} \tag{3.429}$$

where q is the total charge and R the radius of the sphere.

3.3.4 Magnetic Scalar Potential

Consider the solid angle Ω subtended by loop C when it is viewed from point $\mathbf{r}$, as shown in Figure 3.32. If we displace the observation point by the amount $\delta \mathbf{r}$, this is equivalent to displacing the loop by the amount $-\delta \mathbf{r}$. The change in the solid angle due to the

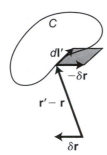

Figure 3.32 Solid angle subtended by a loop.

segment $d\mathbf{l}'$ of the loop at the point $\mathbf{r}'$ (the shaded region in Figure 3.32) is proportional to the component of the vector area $-\delta\mathbf{r} \times d\mathbf{l}'$ in the $\hat{\mathbf{R}}$ direction, where

$$\mathbf{R} = \mathbf{r}' - \mathbf{r} \tag{3.430}$$

That is,

$$d\delta\Omega = \frac{(-\delta\mathbf{r} \times d\mathbf{l}') \cdot \hat{\mathbf{R}}}{R^2} = \frac{[d\mathbf{l}' \times (\mathbf{r} - \mathbf{r}')] \cdot \delta\mathbf{r}}{|\mathbf{r} - \mathbf{r}'|^3} \tag{3.431}$$

after we rearrange the cross and dot products using a vector identity. The total change in the subtended solid angle is therefore

$$\delta\Omega = \delta\mathbf{r} \cdot \oint \frac{d\mathbf{l}' \times (\mathbf{r} - \mathbf{r}')}{|\mathbf{r} - \mathbf{r}'|^3} = \delta\mathbf{r} \cdot \nabla\Omega \tag{3.432}$$

where $\nabla\Omega$ is the gradient of the solid angle

$$\nabla\Omega = \oint \frac{d\mathbf{l}' \times (\mathbf{r} - \mathbf{r}')}{|\mathbf{r} - \mathbf{r}'|^3} \tag{3.433}$$

If we compare this with the Biot–Savart law (3.366), we see that we can represent the magnetic induction $\mathbf{B}$ as the gradient of a magnetic scalar potential

$$\mathbf{B} = -\nabla\Phi_M \tag{3.434}$$

where

$$\Phi_M = -\frac{\mu_0 I}{4\pi}\Omega \tag{3.435}$$

Since we can superpose the magnetic fields of *many* current loops, we can use the magnetic scalar potential to represent the magnetic field of an arbitrary current distribution. However, this representation of the magnetic induction is possible only in magnetostatics, where the Biot–Savart law is valid. Furthermore, the use of a scalar potential is limited to regions in which the current density vanishes. To see this, we note that if (3.434) is true, then

$$\nabla \times \mathbf{B} = -\nabla \times \nabla\Phi_M = 0 \tag{3.436}$$

identically, by a vector identity. But from Ampere's law we know that this is true only if $\mathbf{J} = 0$. Nevertheless, the magnetic scalar potential is useful in a restricted class of problems because of its simplicity, for if we substitute (3.434) into the Gauss law $\nabla \cdot \mathbf{B} = 0$, we get the Laplace equation

$$\nabla^2\Phi_M = 0 \tag{3.437}$$

This makes it possible for us to apply to magnetic problems much of the mathematical machinery we have developed to solve the Laplace equation in electrostatics. This is especially true, as we see later, in computing the magnetic induction created by magnetic materials. In this case, the current density vanishes and the magnetization of the material can be used to add a source term to (3.437). However, we do not generally encounter boundary-value problems in magnetostatics similar to those we encounter in electrostatics problems with conductors. An exception to this is provided by superconductors that exhibit the Meissner effect. For these materials, the magnetic field vanishes inside the superconductor. However, since $\nabla \cdot \mathbf{B} = 0$, the magnetic field must be tangent to the boundary rather than normal to it as in the electrostatic case. Thus, the boundary condition at the surface of the superconductor is

$$\nabla \Phi_M \cdot \hat{\mathbf{n}} = \frac{\partial \Phi_M}{\partial n} = 0 \qquad (3.438)$$

where $\hat{\mathbf{n}}$ is a unit vector normal to the surface, rather than $\nabla \Phi \times \hat{\mathbf{n}} = 0$, as in electrostatics. The boundary of the superconductor is not a magnetic equipotential surface.

Even when the current density vanishes outside some source region, care must be exercised when using the magnetic scalar potential. In the first place, the force on a charged particle is not $-\nabla \Phi_M$, so Φ_M cannot be interpreted as an energy. Moreover, the integral form (3.360) of Ampere's law says that for any contour C through which the current I flows,

$$\oint_C \nabla \Phi_M \cdot d\mathbf{l} = \oint_C d\Phi_M = -\mu_0 I \neq 0 \qquad (3.439)$$

That is, if we go all the way around the contour C, we do not end up at the same potential (the potential is multiple valued) unless the total current I passing through the surface S bounded by the contour vanishes. If the source is limited to a finite volume, we can completely exclude this region from the region in which we define the potential Φ_M and stretch the surface S around the source so that the current vanishes everywhere on S. The potential Φ_M is then single valued everywhere, provided that the region in which it is defined is simply connected. That is, it must be possible to distort any path between the points A and B into any other path between these points without passing through the source region. For example, if the current has the form of a wire extending to infinity, we must exclude the source as shown in Figure 3.33. Then, the region in which Φ_M is defined is simply connected, and we cannot form a contour that encloses the wire. If the source forms a loop, as also shown in Figure 3.33, then the excluded volume must be simply connected, as indicated. Otherwise, we could form a contour passing through the loop and enclosing a nonvanishing net current.

As an example of the use of the magnetic scalar potential, we consider the magnetic induction from a uniform spherical shell of charge rotating about the z axis, as illustrated

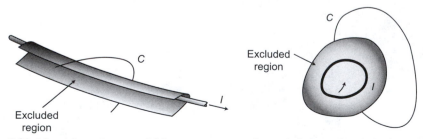

Figure 3.33 Magnetic scalar potential for a current extending to infinity or forming a loop.

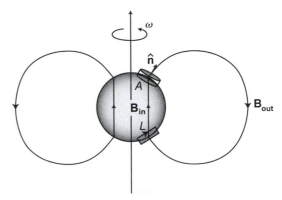

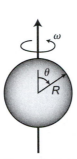

Figure 3.34 Rotating
spherical shell of charge.

Figure 3.35 Boundary conditions for a rotating
spherical shell of charge.

in Figure 3.34. This is sometimes used as a model for a classical electron, as discussed in
Chapter 11. At the angle θ from the z axis the surface current density is

$$\sigma(\theta) = \frac{q\omega \sin\theta}{4\pi R} \tag{3.440}$$

where q is the total charge on the sphere, ω the frequency of rotation, and R the radius of
the sphere. Outside the sphere the magnetic scalar potential can be represented by the
expansion

$$\Phi_{\text{out}} = \sum_{m=0}^{\infty} \frac{a_m}{r^{m+1}} P_m(\cos\theta) \tag{3.441}$$

while inside the sphere it can be represented by the expansion

$$\Phi_{\text{in}} = \sum_{m=1}^{\infty} b_m r^m P_m(\cos\theta) \tag{3.442}$$

for some coefficients a_m and b_m.

At the surface, of course, the potentials Φ_{out} and Φ_{in} do not match since they are not
defined within the surface, where the current density is nonvanishing. However, the mag-
netic induction **B** goes through the surface, and this can be used to establish the matching
conditions. We begin by constructing an imaginary pillbox at the surface, as shown in Figure 3.35, and applying Gauss's law. For a sufficiently small pillbox the magnetic field is
uniform over the flat surfaces. In the limit when the sides of the pillbox become vanishingly small we can ignore the flux through the sides, and Gauss's law becomes

$$\oint_S \mathbf{B} \cdot d\mathbf{S} = 0 = \int_A (\mathbf{B}_{\text{out}} - \mathbf{B}_{\text{in}}) \cdot \hat{\mathbf{n}} \, dA \tag{3.443}$$

where A is the area of the faces of the pillbox and $\hat{\mathbf{n}}$ is a unit vector normal to the surface
of the sphere. From this we get the boundary condition

$$(\nabla\Phi_{\text{out}} - \nabla\Phi_{\text{in}}) \cdot \hat{\mathbf{n}} = \frac{\partial\Phi_{\text{out}}}{\partial r} - \frac{\partial\Phi_{\text{in}}}{\partial r} = 0 \tag{3.444}$$

at the surface $r = R$. In the same way, we construct a small imaginary rectangular loop at
the surface as shown in Figure 3.35 and apply Ampere's law. The current through the loop

is $I = \sigma L$, where the current density σ is given by (3.440) and L is the length of the loop. In the limit when the ends of the loop become vanishingly small, their contribution can be ignored and we obtain from Ampere's law the result

$$\oint_C \mathbf{B} \cdot d\mathbf{l} = \int_L (\mathbf{B}_{\text{out}} - \mathbf{B}_{\text{in}}) \cdot \hat{\boldsymbol{\theta}} \, dl = \mu_0 I = \mu_0 \int_L \sigma \, dl \tag{3.445}$$

From this we get the boundary condition

$$(\nabla \Phi_{\text{out}} - \nabla \Phi_{\text{in}}) \cdot \hat{\boldsymbol{\theta}} = \frac{1}{R} \left(\frac{\partial \Phi_{\text{out}}}{\partial \theta} - \frac{\partial \Phi_{\text{in}}}{\partial \theta} \right) = \mu_0 \frac{q\omega \sin\theta}{4\pi R} \tag{3.446}$$

at $r = R$, where $\hat{\boldsymbol{\theta}}$ is a unit vector tangent to the surface of the sphere in the direction of increasing θ. When we substitute (3.441) and (3.442) into (3.444) and require that the boundary condition be satisfied for all θ, we find that $a_m = 0$ and

$$b_m = -\frac{m+1}{m} \frac{a_m}{R^{2m+1}} \tag{3.447}$$

for $m \geq 1$. When we substitute (3.441) and (3.442) into (3.446) and use (3.447), we find that

$$\sum_{m=1}^{\infty} \frac{2m+1}{m} \frac{a_m}{R^{m+1}} \frac{dP_m(\cos\theta)}{d\theta} = \frac{\mu_0 q\omega}{4\pi} \sin\theta = -\frac{\mu_0 q\omega}{4\pi} \frac{dP_1(\cos\theta)}{d\theta} \tag{3.448}$$

The only term that survives is $m = 1$, so the magnetic scalar potential outside the sphere is

$$\Phi_{\text{out}} = -\frac{\mu_0 \omega q R^2}{12\pi r^2} P_1(\cos\theta) \tag{3.449}$$

Thus, the magnetic field outside a rotating sphere of charge is identically that of a magnetic dipole located at the center of the sphere. No higher magnetic moments appear.

EXERCISE 3.27

Beginning with (3.449), show that the magnetic field outside a spinning spherical shell of charge is

$$\mathbf{B}(\mathbf{r}) = -\nabla \Phi_M(\mathbf{r}) = \frac{\mu_0}{4\pi} \left(\frac{3\mathbf{m} \cdot \mathbf{r}}{r^5} \mathbf{r} - \frac{\mathbf{m}}{r^3} \right) \tag{3.450}$$

where the magnetic dipole moment of the spinning sphere is

$$\mathbf{m} = \tfrac{1}{3} q R^2 \boldsymbol{\omega} \tag{3.451}$$

EXERCISE 3.28

Below the critical magnetic field H_c, type I superconductors exhibit the Meissner effect, in which all magnetic fields are excluded from the volume of the superconductor. Consider a superconducting sphere of radius a placed in an otherwise uniform magnetic induction $\mathbf{B}_0$.

Since the magnetic field is excluded from the sphere, the magnetic induction **B** is tangent to the sphere at its surface as shown in Figure 3.36.

(a) Show that the total magnetic induction outside the sphere is

$$\mathbf{B}(\mathbf{r}) = \mathbf{B}_0 + \frac{\mu_0}{4\pi}\left(\frac{3\mathbf{m}\cdot\mathbf{r}}{r^5}\mathbf{r} - \frac{\mathbf{m}}{r^3}\right) \tag{3.452}$$

where the induced dipole moment is

$$\mathbf{m} = -\frac{2\pi a^3}{\mu_0}\mathbf{B}_0 \tag{3.453}$$

(b) Show that the energy required to place the sphere in the magnetic field is

$$\mathcal{W} = \frac{\pi a^3 B_0^2}{\mu_0} \tag{3.454}$$

It is somewhat surprising that this is $\frac{3}{2}$ the energy of the uniform field in the volume $\frac{4}{3}\pi a^3$ of the sphere, and $\frac{1}{2}$ the energy of a simple dipole placed in the magnetic field.

Hint: The Maxwell stress tensor (see Chapter 2) shows that the magnetic induction exerts a pressure on the surface of the sphere. Compute the work $\delta\mathcal{W}$ done on the field when the radius of the sphere expands from r to $r + \delta r$.

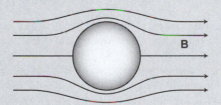

Figure 3.36 Magnetic field excluded from a superconducting sphere.

EXERCISE 3.29

The superconducting sphere of Exercise 3.28 is levitated by placing it in the nonuniform field above the end of a solenoid as shown in Figure 3.37. As discussed in an earlier exercise, the components of the field near the axis are

$$B_\rho \approx \frac{\mu_0\sigma}{4}\frac{R^2\rho}{(R^2+z^2)^{3/2}} \tag{3.455}$$

$$B_z \approx \frac{\mu_0\sigma}{2}\left[1 - \frac{z}{(R^2+z^2)^{1/2}} - \frac{3}{4}\frac{R^2 z\rho^2}{(R^2+z^2)^{5/2}}\right] \tag{3.456}$$

where σ is the azimuthal current per unit length of the solenoid, R the radius of the solenoid, z the height above the end of the solenoid, and $\rho = \sqrt{x^2 + y^2}$ the radial distance from the (vertical) z axis. Provided that the radius a of the sphere is sufficiently small ($a/R \ll 1$) that the magnetic field may be regarded as locally uniform, the energy of the

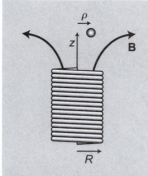

Figure 3.37 Superconducting sphere levitated above a solenoid.

sphere in the field is

$$W \approx \frac{\pi a^3}{\mu_0} B^2 \tag{3.457}$$

as discussed earlier. Show that the position of the sphere levitated above the magnet is stable against small horizontal displacements for

$$\frac{R}{z} > \sqrt{24} \tag{3.458}$$

BIBLIOGRAPHY

Electrostatics and magnetostatics are prominent parts of all texts on electrodynamics and mathematical physics, especially the older texts. A large (sometimes overwhelming) variety of techniques and solved problems can be found in

J. D. Jackson, *Classical Electrodynamics,* 3rd edition, John Wiley & Sons, New York (1999),

J. H. Jeans, *The Mathematical Theory of Electricity and Magnetism,* 5th edition, Cambridge University Press, Cambridge, U.K. (1960),

W. R. Smythe, *Static and Dynamic Electricity,* 3rd edition, Hemisphere Publishing Corporation, New York (1989).

An exhaustive discussion of analytical techniques useful in many branches of physics can be found in

P. M. Morse and H. Feshbach, *Methods of Theoretical Physics,* McGraw-Hill Book Company, New York (1953).

For a deeper introduction to the quantum mechanical aspects of atoms and solids, the reader is referred to

D. Bohm, *Quantum Theory,* Prentice Hall, Englewood Cliffs, NJ (1951),

C. Kittel, *Introduction to Solid State Physics,* 7th edition, John Wiley & Sons, New York (1996),

R. W. Robinett, *Quantum Mechanics: Classical Results, Modern Systems, and Visualized Examples,* Oxford University Press, Oxford (1997).

There are many books about the numerical solution of partial differential equations and other problems encountered in physics, but a good place to look for both methods and actual computer code (available on disk in BASIC, C, FORTRAN, and Pascal) is

W. H. Press, S. A. Teukolsky, W. T. Vetterling, and B. P. Flannery, *Numerical Recipes, The Art of Scientific Computing,* 2nd edition, Cambridge University Press, Cambridge, U.K. (1992).

4   Electromagnetic Waves

One of Maxwell's most important contributions to science was his discovery of wavelike solutions of the equations of electromagnetism. This discovery identified the nature of light as electromagnetic waves and made possible not only the advances by Hertz, Tesla, and Marconi in radio communication but also the advances of Planck, Einstein, and others in our understanding of the interaction of light with matter.

But electromagnetic waves also provide a conceptual framework of much broader importance. Since the Maxwell equations are linear, it is often useful to decompose more general electromagnetic fields into waves by means of Fourier series. As one of the most important applications of this technique, we develop the canonical equations of motion for the electromagnetic fields. These are the starting point for the quantum theory of radiation.

4.1   PLANE WAVES

4.1.1   Electric and Magnetic Fields in Plane Waves

We begin by summarizing very quickly the properties of plane, electromagnetic waves, and introduce some notation that is needed in the following sections. Although it is not covariant, the Coulomb gauge is convenient for describing the free radiation field, and we adopt it here. We recall from Chapter 2 that the scalar potential Φ in the Coulomb gauge is just the instantaneous electrostatic potential and can be ignored far from the sources. The electromagnetic field is then completely described by the 3-vector potential $\mathbf{A}(\mathbf{x}, t)$.

In the absence of sources, the vector potential satisfies the wave equation

$$\partial^\alpha \partial_\alpha \mathbf{A} = \frac{1}{c^2} \frac{\partial^2 \mathbf{A}}{\partial t^2} - \nabla^2 \mathbf{A} = 0 \tag{4.1}$$

and the Coulomb gauge condition

$$\nabla \cdot \mathbf{A} = 0 \tag{4.2}$$

To describe a plane wave propagating in the $\hat{\mathbf{n}}$ direction, we introduce the null (or lightlike) 4-vector

$$\hat{n}^\alpha = (1, \hat{\mathbf{n}}) \tag{4.3}$$

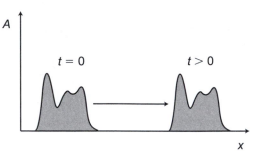

Figure 4.1 Propagation of a wave.

The solution of the wave equation is then

$$\mathbf{A} = \mathbf{A}(\xi) \tag{4.4}$$

for some vector function $\mathbf{A}(\xi)$, where the argument is

$$\xi = \hat{n}^{\alpha} r_{\alpha} = ct - \hat{\mathbf{n}} \cdot \mathbf{r} \tag{4.5}$$

The function $\mathbf{A}(\xi)$ need not be harmonic (sinusoidal) nor even periodic. Whatever the shape of the pulse at $t = 0$, it retains this shape as it propagates at the velocity c, as shown in Figure 4.1.

From the Coulomb gauge condition we see that

$$\hat{\mathbf{n}} \cdot \frac{d\mathbf{A}}{d\xi} = 0 \tag{4.6}$$

so the component $\mathbf{A} \cdot \hat{\mathbf{n}}$ of the vector potential in the direction of propagation of the wave is at most a constant. Since a constant added to the vector potential contributes nothing to the fields or to the equations of motion, we ignore it. This makes the vector potential $\mathbf{A}$ transverse to the direction of propagation.

In terms of the vector potential $\mathbf{A}(\xi)$, the electric and magnetic fields are

$$\mathbf{E} = -\frac{\partial \mathbf{A}}{\partial t} = -c\frac{d\mathbf{A}}{d\xi} \tag{4.7}$$

and

$$\mathbf{B} = \nabla \times \mathbf{A} = -\hat{\mathbf{n}} \times \frac{d\mathbf{A}}{d\xi} = \frac{1}{c}\hat{\mathbf{n}} \times \mathbf{E} \tag{4.8}$$

Thus, the electric and magnetic fields in a plane wave are transverse to the direction of propagation of the wave. We see, moreover, that

$$|\mathbf{B}| = \frac{1}{c}|\mathbf{E}| \tag{4.9}$$

and

$$\mathbf{E} \cdot \mathbf{B} = \mathbf{E} \cdot \hat{\mathbf{n}} = \mathbf{B} \cdot \hat{\mathbf{n}} = 0 \tag{4.10}$$

That is, the electric and magnetic fields remain in phase with one another at all times, and the vectors $\mathbf{E}$, $\mathbf{B}$, and $\hat{\mathbf{n}}$ form a right-handed orthogonal set, as shown in Figure 4.2.

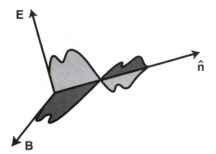

Figure 4.2 Electric and magnetic fields in a plane wave.

The energy density, momentum density, and Poynting vector of a plane wave are

$$\mathcal{U} = \frac{\varepsilon_0}{2}E^2 + \frac{1}{2\mu_0}B^2 = \varepsilon_0 E^2 = \frac{1}{\mu_0}B^2 = \frac{1}{\mu_0}\left(\frac{d\mathbf{A}}{d\xi}\right)^2 \tag{4.11}$$

$$\mathbf{g} = \varepsilon_0 \mathbf{E} \times \mathbf{B} = \frac{\mathcal{U}(\xi)}{c}\hat{\mathbf{n}} \tag{4.12}$$

and

$$\mathbf{S} = \frac{1}{\mu_0}\mathbf{E} \times \mathbf{B} = c\mathcal{U}(\xi)\hat{\mathbf{n}} \tag{4.13}$$

From (4.12) we see that in any infinitesimal volume dV the momentum and the energy form the 4-vector momentum

$$dp^\alpha = \frac{\mathcal{U}}{c}(1, \hat{\mathbf{n}})\, dV \tag{4.14}$$

where $(1, \hat{\mathbf{n}})$ is a null vector. This is analogous to the expression

$$p^\alpha = \frac{\mathcal{E}}{c}(1, \beta\hat{\mathbf{n}})$$

for a particle, but in this case the particle is traveling at the speed of light. From (4.13) we see that the energy in the wave flows at the speed of light. We find in Chapter 7 that in dispersive media the energy flows at the group velocity, rather than the phase velocity, whenever these two velocities are different. This must be true, since the energy travels with the pulse.

For a wave traveling in the $\hat{\mathbf{x}}$ direction, the Maxwell stress tensor is

$$T_{ij}^{(M)} = -\varepsilon_0 E_i E_j - \frac{1}{\mu_0}B_i B_j + \frac{1}{2}\left(\varepsilon_0 E^2 + \frac{1}{\mu_0}B^2\right)\delta_{ij} = \mathcal{U}\delta_{1i}\delta_{1j} \tag{4.15}$$

irrespective of the orientation of $\mathbf{E}$ and $\mathbf{B}$ around the direction of the wave. In dyadic notation, this is $\overset{\leftrightarrow}{\mathbf{T}}^{(M)} = \mathcal{U}\hat{\mathbf{x}}\hat{\mathbf{x}}$, where $\hat{\mathbf{x}}$ is a unit vector, so more generally for a wave traveling in the $\hat{\mathbf{n}}$ direction, the Maxwell stress tensor is

$$\overset{\leftrightarrow}{\mathbf{T}}^{(M)} = \mathcal{U}\hat{\mathbf{n}}\hat{\mathbf{n}} \tag{4.16}$$

The stress tensor is independent of the polarization of the wave. Since the stress tensor has components only in the direction of propagation of the wave, when a wave falls on a

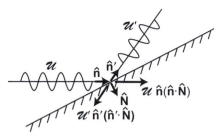

Figure 4.3 Pressure exerted by a wave reflecting off a surface.

surface it pushes the surface in the direction of propagation of the wave, not in the direction normal to the surface. Of course, if part of the wave is reflected, the reflected wave also pushes on the surface, as shown in Figure 4.3. For a perfectly reflected wave the resulting total force, called the radiation pressure, is normal to the surface.

For a wave traveling in the $\hat{\mathbf{x}}$ direction, the symmetric stress tensor has the particularly simple form

$$T^{\alpha\beta} = \begin{bmatrix} \mathcal{U} & c\mathbf{g} \\ c\mathbf{g} & \overset{\leftrightarrow}{\mathbf{T}}^{(M)} \end{bmatrix} = \begin{bmatrix} \mathcal{U} & \mathcal{U} & 0 & 0 \\ \mathcal{U} & \mathcal{U} & 0 & 0 \\ 0 & 0 & 0 & 0 \\ 0 & 0 & 0 & 0 \end{bmatrix} \tag{4.17}$$

The generalization to a wave traveling in an arbitrary direction is clear from the dyadic expression (4.16), and just involves the direction cosines of $\hat{\mathbf{n}}$.

For a wave with a simple harmonic time dependence, the vector potential is

$$\mathbf{A}(\mathbf{r}, t) = \mathbf{a}_1(\mathbf{r}) \cos(\omega t + \phi) + \mathbf{a}_2 \sin(\omega t + \phi) \tag{4.18}$$

where the constants ω and ϕ are the frequency and phase, respectively. When we use this expression, the wave equation (4.1) leads to the Helmholtz equation

$$\nabla^2 \mathbf{a} + k^2 \mathbf{a} = 0 \tag{4.19}$$

where

$$k^2 = \frac{\omega^2}{c^2} \tag{4.20}$$

is the square of the wave number. In problems involving fields with a harmonic time dependence, it is often convenient to express the solution (4.18) to the wave equation using complex variables in the form

$$\mathbf{A}(\mathbf{r}, t) = \mathrm{Re}(\mathbf{a}e^{-i\omega t}) = \tfrac{1}{2}(\mathbf{a}e^{-i\omega t} + \mathbf{a}^* e^{i\omega t}) \tag{4.21}$$

In this expression $\mathbf{a}$ is a complex vector into which the phase ϕ has been absorbed. So long as we are performing only linear operations on $\mathbf{A}$, we can use the complex expression for the field throughout the calculation and then take the real part at the end. Of course, it is not generally possible to use this technique in nonlinear problems, but quadratic quantities are easily computed in the following way. For two fields of the form

$$\mathbf{A} = \mathbf{A}_0 e^{-i\omega t} \tag{4.22}$$

and

$$\mathbf{B} = \mathbf{B}_0 e^{-i\omega t} \tag{4.23}$$

we see that

$$\text{Re}(\mathbf{A}) \cdot \text{Re}(\mathbf{B}) = \tfrac{1}{4}(\mathbf{A}_0 e^{-i\omega t} + \mathbf{A}_0^* e^{i\omega t}) \cdot (\mathbf{B}_0 e^{-i\omega t} + \mathbf{B}_0^* e^{i\omega t})$$

$$= \tfrac{1}{4}(\mathbf{A}_0 \cdot \mathbf{B}_0^* + \mathbf{A}_0^* \cdot \mathbf{B}_0) + \tfrac{1}{4}\mathbf{A}_0 \cdot \mathbf{B}_0 e^{-2i\omega t} + \tfrac{1}{4}\mathbf{A}_0^* \cdot \mathbf{B}_0^* e^{2i\omega t} \quad (4.24)$$

If we require only the average value of the product over one period of the wave, which is generally the case, we see that the last two terms vanish and we are left with

$$\langle \text{Re}(\mathbf{A}) \cdot \text{Re}(\mathbf{B}) \rangle = \tfrac{1}{2}\text{Re}(\mathbf{A}_0 \cdot \mathbf{B}_0^*) = \tfrac{1}{2}\text{Re}(\mathbf{A}_0^* \cdot \mathbf{B}_0) \quad (4.25)$$

where $\langle \ \rangle$ indicates the average over a complete cycle. In the same way we find that

$$\langle \text{Re}(\mathbf{A}) \times \text{Re}(\mathbf{B}) \rangle = \tfrac{1}{2}\text{Re}(\mathbf{A}_0 \times \mathbf{B}_0^*) = \tfrac{1}{2}\text{Re}(\mathbf{A}_0^* \times \mathbf{B}_0) \quad (4.26)$$

For a harmonic plane wave traveling in the $\hat{\mathbf{n}}$ direction, the solution of the Helmholtz equation (4.19) is $\mathbf{a} = \mathbf{A}_0 \exp(i\mathbf{k} \cdot \mathbf{r})$ and the complete solution to the wave equation is

$$\mathbf{A} = \text{Re}(\mathbf{A}_0 e^{-ik^\alpha r_\alpha}) \quad (4.27)$$

where

$$k^\alpha = k\hat{n}^\alpha = \left(\frac{\omega}{c}, k\hat{\mathbf{n}} \right) \quad (4.28)$$

is the wave 4-vector. The complex electric and magnetic fields are

$$\mathbf{E} = -\frac{\partial \mathbf{A}}{\partial t} = i\omega\mathbf{A} = i\omega\mathbf{A}_0 e^{-ik^\alpha r_\alpha} \quad (4.29)$$

and

$$\mathbf{B} = \nabla \times \mathbf{A} = i\mathbf{k} \times \mathbf{A} = i\mathbf{k} \times \mathbf{A}_0 e^{-ik^\alpha r_\alpha} \quad (4.30)$$

and from (4.25) and (4.26) we see that the average energy density is

$$\langle \mathcal{U} \rangle = \frac{\varepsilon_0 \omega^2}{2} \mathbf{A}_0 \cdot \mathbf{A}_0^* \quad (4.31)$$

If we represent the electric and magnetic fields by expressions of the form

$$\mathbf{E} = \text{Re}(\mathbf{E}_0 e^{-ik^\alpha r_\alpha}) \quad (4.32)$$

and

$$\mathbf{B} = \text{Re}(\mathbf{B}_0 e^{-ik^\alpha r_\alpha}) \quad (4.33)$$

then the energy density is

$$\langle \mathcal{U} \rangle = \frac{\varepsilon_0}{2} \mathbf{E}_0 \cdot \mathbf{E}_0^* = \frac{1}{2\mu_0} \mathbf{B}_0 \cdot \mathbf{B}_0^* \quad (4.34)$$

The polarization of a plane electromagnetic wave is defined by the direction of the electric field. For the following discussion, we introduce the basis vectors shown in Figure 4.4, consisting of two unit vectors $\hat{\mathbf{e}}_1$ and $\hat{\mathbf{e}}_2$ normal to each other and to the direction of propagation $\hat{\mathbf{n}}$, such that

$$\hat{\mathbf{e}}_1 \cdot \hat{\mathbf{e}}_2 = 0 \quad (4.35)$$

$$\hat{\mathbf{e}}_1 \times \hat{\mathbf{e}}_2 = \hat{\mathbf{n}} \quad (4.36)$$

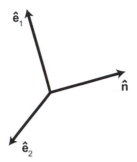

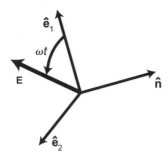

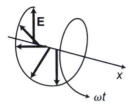

Figure 4.4 Coordinate system for linearly polarized waves.

Figure 4.5 Electric field of an elliptically polarized wave.

Figure 4.6 Positive helicity.

In terms of these unit vectors, we may write the electric field in the form

$$\mathbf{E} = (\hat{\mathbf{e}}_1 E_1 + \hat{\mathbf{e}}_2 E_2)e^{-ik^\alpha r_\alpha} \tag{4.37}$$

for some complex constants E_1 and E_2. If E_1 and E_2 have the same complex phase ϕ, so that

$$E_1 = |E_1|e^{i\phi} \tag{4.38}$$

$$E_2 = |E_2|e^{i\phi} \tag{4.39}$$

then the sum (4.37) is a linearly polarized wave

$$\text{Re}(\mathbf{E}) = (\hat{\mathbf{e}}_1|E_1| + \hat{\mathbf{e}}_2|E_2|)\cos(\mathbf{k} \cdot \mathbf{r} - \omega t + \phi) \tag{4.40}$$

where $(\hat{\mathbf{e}}_1|E_1| + \hat{\mathbf{e}}_2|E_2|)$ is a constant (real) vector. If the phases of E_1 and E_2 are different, so that

$$E_1 = |E_1|e^{i\phi_1} \tag{4.41}$$

and

$$E_2 = |E_2|e^{i\phi_2} \tag{4.42}$$

then the sum is an elliptically polarized wave

$$\text{Re}(\mathbf{E}) = \hat{\mathbf{e}}_1|E_1|\cos(\mathbf{k} \cdot \mathbf{r} - \omega t + \phi_1) + \hat{\mathbf{e}}_2|E_2|\cos(\mathbf{k} \cdot \mathbf{r} - \omega t + \phi_2) \tag{4.43}$$

As illustrated in Figure 4.5, the electric field rotates in the plane of $\hat{\mathbf{e}}_1$ and $\hat{\mathbf{e}}_2$. When $\sin(\phi_2 - \phi_1) > 0$, the direction of rotation is counterclockwise, as shown. The wave is then called left polarized, or is said to have positive helicity, as illustrated in Figure 4.6. Otherwise, the direction of rotation is clockwise and the wave is right polarized (negative helicity). When $E_1 = \pm i E_2$, we get circularly polarized waves.

Certain materials, such as quartz, have the property that light polarized in a particular direction, called the optical axis, travels at a velocity different from that of light polarized in a direction perpendicular to the optical axis. In fact, ordinary cellophane has this property, which is called birefringence. When light that is linearly polarized at an angle oblique to the optical axis propagates through a birefringent material, the component polarized along the optical axis travels at a different velocity and accumulates a different phase than the component polarized perpendicular to the optical axis. When the wave emerges from the material, the two components have a phase difference, so the light that entered the

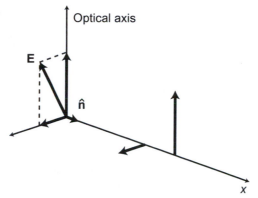

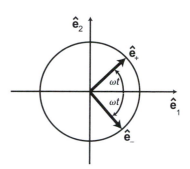

Figure 4.7 Conversion of a linearly polarized wave to elliptical polarization.

Figure 4.8 Rotation of circularly polarized vectors.

material linearly polarized emerges from it elliptically polarized, as shown in Figure 4.7. A so-called quarter-wave plate is designed to produce a 90-degree (quarter-wave) phase shift between the two components. It is used to convert linearly polarized light to circular polarization, or vice versa.

It is often convenient to decompose a linearly or elliptically polarized wave into circularly polarized waves. Instead of using the basis vectors $\hat{\mathbf{e}}_1$ and $\hat{\mathbf{e}}_2$, we represent the electric field with the left- and right-hand circularly polarized unit vectors

$$\hat{\mathbf{e}}_\pm = \frac{1}{\sqrt{2}}(\hat{\mathbf{e}}_1 \pm i\hat{\mathbf{e}}_2) \tag{4.44}$$

which have the orthogonality properties

$$\hat{\mathbf{e}}_\pm \cdot \hat{\mathbf{e}}_\pm^* = 1 \tag{4.45}$$

$$\hat{\mathbf{e}}_\pm \cdot \hat{\mathbf{e}}_\mp^* = 0 \tag{4.46}$$

The cross products have the convenient form

$$\hat{\mathbf{n}} \times \hat{\mathbf{e}}_\pm = \mp i\hat{\mathbf{e}}_\pm \tag{4.47}$$

$$\hat{\mathbf{e}}_\pm \times \hat{\mathbf{e}}_\mp = \mp i\hat{\mathbf{n}} \tag{4.48}$$

which we use later in the canonical description of the radiation field. That the vectors $\hat{\mathbf{e}}_\pm$ are left- and right-hand circularly polarized is evident when we examine the real parts, which are

$$\text{Re}\{\hat{\mathbf{e}}_\pm \exp[i(\mathbf{k} \cdot \mathbf{r} - \omega t)]\} = \frac{1}{\sqrt{2}}[\hat{\mathbf{e}}_1 \cos(\mathbf{k} \cdot \mathbf{r} - \omega t) \mp \hat{\mathbf{e}}_2 \sin(\mathbf{k} \cdot \mathbf{r} - \omega t)] \tag{4.49}$$

These vectors rotate counterclockwise (+) or clockwise (−) with increasing time t (at a fixed field position $\mathbf{r}$), as shown in Figure 4.8. We note in this connection that if $\hat{\mathbf{e}}_{\hat{\mathbf{n}}(\pm)}$ is a circularly polarized vector oriented about the direction of propagation $\hat{\mathbf{n}}$, then it is related to the circularly polarized vector $\hat{\mathbf{e}}_{-\hat{\mathbf{n}}(\pm)}$ oriented in the opposite direction by

$$\hat{\mathbf{e}}_{\hat{\mathbf{n}}(\pm)} = \hat{\mathbf{e}}_{-\hat{\mathbf{n}}(\pm)}^* = \hat{\mathbf{e}}_{-\hat{\mathbf{n}}(\mp)} \tag{4.50}$$

We can see this by considering (4.44) and rotating $\hat{\mathbf{n}}$ about the direction $\hat{\mathbf{e}}_1$, in which case $\hat{\mathbf{e}}_2 \to -\hat{\mathbf{e}}_2$. It is also obvious, since when referred to the opposite direction of propagation, a left-hand (or right-hand) circularly polarized vector reverses its helicity.

EXERCISE 4.1

For a plane wave traveling along the x axis, boost the components of the symmetric stress 4-tensor into a frame moving along the x axis with velocity $v = \beta c$. *Do not do this the hard way by performing a Lorentz transformation of the tensor!* There are several almost trivial ways to solve this problem.

EXERCISE 4.2

As we have seen in previous exercises, the Lorentz scalar field ϕ in the absence of sources satisfies the wave equation

$$\partial^\alpha \partial_\alpha \phi + \mu^2 \phi = 0 \tag{4.51}$$

where μ is the Proca mass constant. The symmetric stress tensor is given by the expression

$$T^{\alpha\beta} = 2K_2[\partial^\alpha \phi \partial^\beta \phi - \tfrac{1}{2}(\partial^\nu \phi \partial_\nu \phi - \mu^2 \phi^2)g^{\alpha\beta}] \tag{4.52}$$

for some constant K_2 that is arbitrary (and defines the system of units) except that the sign must be positive if two like particles attract and negative if they repel. Find the stress tensor for a plane wave of the 4-scalar field ϕ traveling in the $\hat{\mathbf{x}}$ direction with frequency ω, and show that when averaged over one cycle of the wave, it has the form

$$\langle T^{\alpha\beta} \rangle = \langle \mathcal{U} \rangle \begin{bmatrix} 1 & \beta_g & 0 & 0 \\ \beta_g & \beta_g^2 & 0 & 0 \\ 0 & 0 & 0 & 0 \\ 0 & 0 & 0 & 0 \end{bmatrix} \tag{4.53}$$

where $\langle \mathcal{U} \rangle$ is the energy density and β_g is the *group* velocity of the wave. *Hint:* Use a real representation for the wave, or, if you use a complex representation, remember that averages of quadratic forms involving the real part of the wave or linear functions of the wave are given by the expression

$$\langle \mathrm{Re}(f)\mathrm{Re}(g) \rangle = \tfrac{1}{2}\mathrm{Re}(fg^*) \tag{4.54}$$

EXERCISE 4.3

A complex vector may be represented in terms of linear or circularly polarized basis vectors as

$$\mathbf{E} = \hat{\mathbf{e}}_1 E_1 + \hat{\mathbf{e}}_2 E_2 = \hat{\mathbf{e}}_+ E_+ + \hat{\mathbf{e}}_- E_- \tag{4.55}$$

Use the orthogonality relations to show that the equations that transform between these representations are

$$E_\pm = \frac{1}{\sqrt{2}}(E_1 \mp i E_2) \qquad (4.56)$$

$$E_{1,2} = \frac{1}{\sqrt{2}}[(1, i)E_+ + (1, -i)E_-] \qquad (4.57)$$

EXERCISE 4.4

An elliptically polarized plane wave may be represented in terms of circularly polarized basis vectors as

$$\mathbf{E} = (\hat{\mathbf{e}}_+ E_+ + \hat{\mathbf{e}}_- E_-)e^{i(\mathbf{k}\cdot\mathbf{r}-\omega t)} \qquad (4.58)$$

where

$$E_\pm = a_\pm e^{i\alpha_\pm} \qquad (4.59)$$

for real $a_\pm$ and $\alpha_\pm$. Draw a diagram showing the real part of $\mathbf{E}$ in the $\hat{\mathbf{e}}_1$–$\hat{\mathbf{e}}_2$ plane as a sum of two vectors rotating in opposite directions. Indicate the position of the vectors at $\mathbf{k} \cdot \mathbf{r} - \omega t = 0$ and show that the tilt θ and eccentricity (ratio $a_>/a_<$ of the major and minor axes) χ of the ellipse are

$$\chi = \frac{a_>}{a_<} = \left| \frac{a_- + a_+}{a_- - a_+} \right| \qquad (4.60)$$

$$\theta = \tfrac{1}{2}(\alpha_- - \alpha_+) \qquad (4.61)$$

as shown in Figure 4.9.

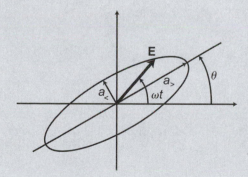

Figure 4.9 Tilt and eccentricity of an elliptically polarized wave.

EXERCISE 4.5

In quantum mechanics it is shown that photons are Bosons (particles with integer spin). Classically we can show that a circularly polarized electromagnetic wave packet propagating in the direction $\hat{\mathbf{n}}$ with the frequency ω and total energy $\mathcal{W} = \hbar\omega$ has the intrinsic angular momentum

$$\mathbf{L}_s = \pm\hbar\hat{\mathbf{n}} \tag{4.62}$$

(a) The angular momentum of an electromagnetic field is

$$\mathbf{L} = \varepsilon_0 \int \mathbf{r} \times (\mathbf{E} \times \mathbf{B})\, d^3\mathbf{r} \tag{4.63}$$

Show that in terms of the vector potential $\mathbf{A}$, this may be expressed as

$$L_i = \varepsilon_0 \sum_{j,k=1}^{3} \varepsilon_{ijk} \int E_j A_k \, d^3\mathbf{r} - \varepsilon_0 \sum_{j,k,l=1}^{3} \varepsilon_{ijk} \int r_j A_l \frac{\partial E_l}{\partial r_k} \, d^3\mathbf{r} \tag{4.64}$$

provided that the fields vanish at infinity. The first term is independent of the origin of the coordinates and is called the intrinsic angular momentum, or spin, of the field. The second term is the angular momentum due to the motion of the center of momentum relative to the origin of coordinates.

(b) For a nearly plane, circularly polarized wave, the complex vector potential may be represented by the approximate expression

$$\mathbf{A} \approx a(\mathbf{r}, t)\hat{\mathbf{e}}_{\pm} e^{-ik^{\alpha}r_{\alpha}} \tag{4.65}$$

for some slowly varying complex function $a(\mathbf{r}, t)$. Show that the spin angular momentum averaged over a period of the wave is

$$\langle \mathbf{L}_s \rangle = \varepsilon_0 \int \langle \mathbf{E} \times \mathbf{A} \rangle \, d^3\mathbf{r} \approx \pm\frac{1}{2}\varepsilon_0\omega\hat{\mathbf{n}} \int aa^* \, d^3\mathbf{r} \tag{4.66}$$

(c) Show that the average energy in the wave is

$$\langle \mathcal{W} \rangle = \frac{1}{2}\varepsilon_0\omega^2 \int aa^* \, d^3\mathbf{r} \tag{4.67}$$

Combine (4.66) and (4.67) to complete the proof.

4.1.2 Charged Particle in a Plane Wave

The simplest example of the interaction of light with matter is the motion of a charged particle in a plane electromagnetic wave. Although the motions are quite trivial when the fields are weak, they become more complex when the fields are strong and give rise to a variety of nonlinear phenomena, some of which are discussed in later chapters. With the development in recent years of high-power lasers, it is possible to observe these nonlinear effects, and nowadays they are interesting not just for themselves but also for a variety of applications in the laboratory.

When an electromagnetic field is incident on a charged particle, the electric field accelerates the particle in the direction of polarization, transverse to the direction of propagation of the field. However, when the transverse velocity becomes relativistic, the interaction of this motion with the transverse magnetic field of the incident wave produces a longitudinal acceleration that causes the particle to execute a "figure-eight" motion. In fact, a sufficiently intense optical pulse can accelerate the particle in the direction of propagation to an average longitudinal velocity that is also relativistic.

As a result of the acceleration, the particle emits radiation at the frequency of the incident field. Since the radiation emitted in weak incident fields is proportional to the intensity of the incident radiation, this is called Thomson scattering of the incident radiation. But as the optical fields become more intense and the particle motions become relativistic, the Thomson-scattered radiation becomes nonlinear and harmonics of the incident frequency appear. These effects are discussed in detail in Chapter 10, and the effects of the radiation on the motion of the particle itself (called the radiation reaction) are discussed in Chapter 11. In the following discussion, however, we ignore the effects of the radiation emitted. Unfortunately, the analysis of even this simple example is somewhat complicated and the results are given only in parametric form, with the 4-vector trajectory $r^\alpha = (ct, \mathbf{r})$ expressed in terms of an independent parameter ξ that is proportional to the proper time of the particle rather than the real time. However, this is sufficient for what we need later. Along the way we determine the parameter that distinguishes weak optical fields (nonrelativistic motions) from intense ones (relativistic motions).

To compute the trajectory, we consider a linearly polarized plane wave propagating in the $\hat{\mathbf{n}}$ direction, described by the 4-vector potential A^α. As shown earlier, for a wave linearly polarized in the $\hat{\mathbf{e}}$ direction, the 4-vector potential in the Coulomb gauge has the form

$$A^\alpha = \hat{e}^\alpha A(\xi) \tag{4.68}$$

where ξ is given by (4.5) and we define the polarization 4-vector

$$\hat{e}^\alpha = (0, \hat{\mathbf{e}}) \tag{4.69}$$

The equation of motion for the 4-vector momentum $p^\alpha = (\mathcal{E}/c, \mathbf{p})$ is

$$\frac{dp^\alpha}{d\tau} = \frac{q}{m}(\partial^\alpha A^\beta - \partial^\beta A^\alpha) p_\beta \tag{4.70}$$

The trick to solving this equation is to form the inner product with the (null) propagation 4-vector $\hat{n}^\alpha = (1, \hat{\mathbf{n}})$, defined by (4.3). When we do this, we find that

$$\hat{n}_\alpha \frac{dp^\alpha}{d\tau} = \frac{q}{m}(\hat{n}_\alpha \partial^\alpha A^\beta - \partial^\beta \hat{n}_\alpha A^\alpha) p_\beta = \frac{q}{m}\left[\hat{n}^\alpha \hat{n}_\alpha \frac{dA^\beta}{d\xi} - \partial^\beta(\hat{n}_\alpha \hat{e}^\alpha A)\right] = 0 \tag{4.71}$$

since $\hat{n}^\alpha \hat{n}_\alpha$ and $\hat{n}^\alpha \hat{e}_\alpha$ both vanish. From this we obtain the useful result

$$\hat{n}_\alpha p^\alpha = \frac{\mathcal{E}}{c} - \hat{\mathbf{n}} \cdot \mathbf{p} = \text{constant} = \frac{\mathcal{E}_0}{c} \tag{4.72}$$

Then from (4.5) and (4.72) we see that

$$\frac{d\xi}{d\tau} = \hat{n}^\alpha \frac{dr_\alpha}{d\tau} = \frac{\hat{n}^\alpha p_\alpha}{m} = \text{constant} = \frac{\mathcal{E}_0}{mc} \tag{4.73}$$

Thus, the parameter ξ is just a constant times the proper time of the particle. This simplifies things enormously, since we can now express everything in terms of the independent variable ξ.

The timelike component of the equation of motion (4.70) is

$$\frac{dp^0}{d\tau} = \frac{1}{c}\frac{d\mathcal{E}}{d\tau} = \frac{q}{m}(\partial^0 A^\beta - \partial^\beta A^0) p_\beta = \frac{q}{mc}\hat{e}^\beta p_\beta \frac{\partial A}{\partial t} = \frac{q}{m}\hat{e}^\beta p_\beta \frac{dA}{d\xi} \tag{4.74}$$

when we use (4.68) and (4.69). If we change from the proper time τ to the interval ξ, using (4.73), we get

$$\frac{d\mathcal{E}}{d\xi} = -\frac{qc^2}{\mathcal{E}_0}\hat{\mathbf{e}}\cdot\mathbf{p}\frac{dA}{d\xi} \tag{4.75}$$

To evaluate the momentum $\hat{\mathbf{e}}\cdot\mathbf{p}$ in the direction of polarization of the incident wave, we recall that the vector potential is invariant in this direction, so the canonical momentum is conserved. Then

$$\mathbf{p}\cdot\hat{\mathbf{e}} + q\mathbf{A}\cdot\hat{\mathbf{e}} = \mathbf{p}\cdot\hat{\mathbf{e}} + qA(\xi) = \text{constant} = P_0 \tag{4.76}$$

This gives the transverse momentum $\hat{\mathbf{e}}\cdot\mathbf{p}$ in terms of the vector potential $A(\xi)$. Equation (4.75) now becomes

$$\frac{d\mathcal{E}}{d\xi} = \frac{q^2c^2}{\mathcal{E}_0}\left[A(\xi) - \frac{P_0}{q}\right]\frac{dA}{d\xi} \tag{4.77}$$

This gives us the energy $\mathcal{E}$ in the form of an integral of the known function $A(\xi)$, and with this we immediately get the longitudinal momentum $\hat{\mathbf{n}}\cdot\mathbf{p}$ from (4.72).

To compute the 4-vector position from the 4-vector momentum, we use (4.73) to get

$$\frac{dr^\alpha}{d\xi} = \frac{p^\alpha}{m}\frac{d\tau}{d\xi} = \frac{p^\alpha c}{\mathcal{E}_0} \tag{4.78}$$

The longitudinal and transverse positions are then

$$\hat{\mathbf{n}}\cdot\mathbf{r} = \frac{c}{\mathcal{E}_0}\int\hat{\mathbf{n}}\cdot\mathbf{p}\,d\xi = \int\left(\frac{\mathcal{E}}{\mathcal{E}_0} - 1\right)d\xi \tag{4.79}$$

$$\hat{\mathbf{e}}\cdot\mathbf{r} = \frac{c}{\mathcal{E}_0}\int\hat{\mathbf{e}}\cdot\mathbf{p}\,d\xi = -\frac{qc}{\mathcal{E}_0}\int\left[A(\xi) - \frac{P_0}{q}\right]d\xi \tag{4.80}$$

and the time is

$$ct = \xi + \hat{\mathbf{n}}\cdot\mathbf{r} = \xi + \int\left(\frac{\mathcal{E}}{\mathcal{E}_0} - 1\right)d\xi \tag{4.81}$$

We consider two cases. In the first case, the charged particle is exposed to a cw, periodic (but not necessarily harmonic) wave. In the second case we examine a charged particle that is initially at rest and is exposed to a pulse of radiation.

Case 1

When the wave is periodic, we can evaluate the constants $\mathcal{E}_0$ and P_0 in terms of the average (with respect to ξ over one period) values $\langle\mathcal{E}\rangle$ and $\langle A\rangle$ of the energy and field. If we examine the motion in a frame of reference in which the particle returns periodically to its original position, the average momentum is

$$P_0 = q\langle A\rangle \tag{4.82}$$

and

$$\mathcal{E}_0 = \langle\mathcal{E}\rangle \tag{4.83}$$

The energy, found by integrating (4.77), is then

$$\mathcal{E}(\xi) - \langle\mathcal{E}\rangle = \frac{q^2c^2}{2\langle\mathcal{E}\rangle}[A^2(\xi) - \langle A^2\rangle] \tag{4.84}$$

The longitudinal and transverse positions are

$$\hat{\mathbf{n}} \cdot \mathbf{r} = \frac{q^2 c^2}{2\langle \mathcal{E} \rangle^2} \int [A^2(\xi) - \langle A^2 \rangle] \, d\xi \tag{4.85}$$

and

$$\hat{\mathbf{e}} \cdot \mathbf{r} = \frac{c}{\langle \mathcal{E} \rangle} \int \hat{\mathbf{e}} \cdot \mathbf{p} \, d\xi = -\frac{qc}{\langle \mathcal{E} \rangle} \int [A(\xi) - \langle A \rangle] \, d\xi \tag{4.86}$$

and the time is

$$ct = \xi + \hat{\mathbf{n}} \cdot \mathbf{r} = \xi + \frac{q^2 c^2}{2\langle \mathcal{E} \rangle^2} \int [A^2(\xi) - \langle A^2 \rangle] \, d\xi \tag{4.87}$$

Finally, we can find the average energy from the fact that

$$\mathcal{E}^2 = p^2 c^2 + m^2 c^4 = (\hat{\mathbf{n}} \cdot \mathbf{p})^2 c^2 + (\hat{\mathbf{e}} \cdot \mathbf{p})^2 c^2 + m^2 c^4 \tag{4.88}$$

If we substitute for $\hat{\mathbf{n}} \cdot \mathbf{p}$ and $\hat{\mathbf{e}} \cdot \mathbf{p}$ and compute the average, we get

$$\mathcal{E}_0^2 = \langle \mathcal{E} \rangle^2 = q^2 c^2 [\langle A^2 \rangle - \langle A \rangle^2] + m^2 c^4 \tag{4.89}$$

We see from (4.89) that the motion of the charged particle is relativistic when the dimensionless parameter

$$\frac{q^2 \langle A^2 \rangle}{m^2 c^2} = O(1) \tag{4.90}$$

This is to be expected, since the momentum is on the order of qA, and when this is comparable to mc, the motion is relativistic. This parameter also describes the degree to which the longitudinal motions are significant, for we see from (4.85), (4.86), and (4.89) that

$$\frac{\langle (\hat{\mathbf{n}} \cdot \mathbf{r})^2 \rangle}{\langle (\hat{\mathbf{e}} \cdot \mathbf{r})^2 \rangle} = O\left(\frac{q^2 \langle A^2 \rangle}{q^2 \langle A^2 \rangle + m^2 c^2} \right) \tag{4.91}$$

When the vector potential is small, the motion is nonrelativistic and the longitudinal motions are negligible compared with the transverse motions. When the motion is relativistic, the longitudinal motions are comparable in magnitude to the transverse motions and the amplitude of each is on the order of the optical wavelength.

Relativistic motion can occur near the focus of modern high-power pulsed lasers. In a harmonic field $A_0 \exp(ik^\alpha r_\alpha)$, the rms vector potential is

$$\langle A^2 \rangle = \frac{1}{2} A_0^2 = \frac{\mu_0 \langle S \rangle}{\omega k} = \frac{\mu_0}{4\pi^2 c} \lambda^2 \langle S \rangle \tag{4.92}$$

by (4.13) and (4.31), where $\langle S \rangle$ is the average (over one cycle) intensity of the wave. However, the intensity to which an optical beam can be focused depends on the wavelength of the light and the f-number of the optical system. The area of the focused spot is on the order of $\lambda^2 f^2$, so the relativistic parameter is

$$\frac{q\sqrt{\langle A^2 \rangle}}{mc} = O\left(\sqrt{\frac{\mu_0 q^2}{4\pi^2 m^2 c^3} \frac{P}{f^2}} \right) \tag{4.93}$$

where P is the laser power and $f = D/L$ the f-number, in which D is the diameter and L the focal length of the optical system. It is interesting that the wavelength cancels out from this expression. High-power Nd:glass lasers achieve peak power $P > 10^{13}$ W in picosecond

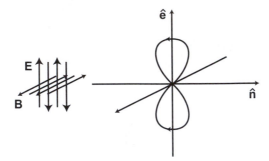

Figure 4.10 Motion of a charge in a linearly polarized plane wave.

pulses, and a typical focusing system might be $f \sim 10$. This corresponds to

$$\frac{q\sqrt{\langle A^2 \rangle}}{mc} = O(2) \tag{4.94}$$

so relativistic, nonlinear effects can be quite strong. As a result, the particle moves in a "figure-eight" orbit, as pictured in Figure 4.10, to which may be added a uniform motion that depends on the frame of reference. We note in passing that the rms electric and magnetic fields in a laser pulse of this intensity are

$$E_{\text{rms}} = \sqrt{\mu_0 c \langle S \rangle} = O\left(\sqrt{\frac{\mu_0 c P}{\lambda^2 f^2}}\right) = O(6 \times 10^{12} \text{ V/m}) \tag{4.95}$$

$$B_{\text{rms}} = \frac{E_{\text{rms}}}{c} = O(2 \times 10^4 \text{ T}) \tag{4.96}$$

These are the strongest electric and magnetic fields produced in the laboratory and they are stronger than the fields typically experienced by electrons orbiting around the lighter nuclei $[Z < O(10)]$. The amplitude of the transverse oscillations of an electron in a field of this intensity is on the order of 300 nm, which is three orders of magnitude larger than the size of an atom.

Case 2

When an optical pulse is incident on a particle initially at rest, the particle is accelerated both transversely and in the direction of propagation of the pulse. In this case the constants of the motion are

$$\mathcal{E}_0 = mc^2 \tag{4.97}$$

and

$$P_0 = 0 \tag{4.98}$$

so the energy is

$$\mathcal{E} = \frac{q^2 A^2(\xi)}{2m} + mc^2 \tag{4.99}$$

The longitudinal momentum is then

$$\hat{\mathbf{n}} \cdot \mathbf{p} = \frac{\mathcal{E} - \mathcal{E}_0}{c} = \frac{q^2 A^2(\xi)}{2mc} \tag{4.100}$$

E

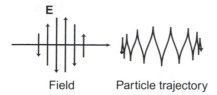

Field   Particle trajectory

Figure 4.11 Motion of a charge starting at rest.

This is always in the direction of the incident pulse and can be relativistic if $qA/mc > O(1)$, as discussed earlier. For an optical pulse the vector potential vanishes periodically, as do the longitudinal and transverse momenta. Thus, the particle comes instantaneously to rest at the ends of the transverse motions. Overall, the particle is accelerated in the direction of the pulse in a series of steps at twice the optical frequency and comes completely to rest again when the pulse has passed. This is illustrated in Figure 4.11. Experimentally, this effect is used to form short, intense bursts of x-rays for a variety of purposes, including pulse radiography on a picosecond time scale. A high-power laser pulse incident on a solid surface forms a plasma just above the surface, and the laser then accelerates the electrons into the surface at relativistic velocities. The x-rays are created by Bremsstrahlung when the electrons are deflected by collisions with the nuclei in the solid.

EXERCISE 4.6

Consider a charged particle initially at rest illuminated by the linearly polarized harmonic plane wave

$$A(\xi) = \sqrt{2} A_{\text{rms}}(\xi) \sin(k\xi) \qquad (4.101)$$

where the rms amplitude $A_{\text{rms}}(\xi)$ is slowly varying compared with $\sin(k\xi)$.

(a) Show that during the pulse, the average (with respect to ξ over one period) energy is

$$\langle \mathcal{E} \rangle (\xi) = mc^2 \sqrt{1 + a^2(\xi)} \qquad (4.102)$$

where

$$a(\xi) = \frac{q A_{\text{rms}}(\xi)}{mc} \qquad (4.103)$$

(b) Show that the total longitudinal displacement of the particle during the pulse is

$$\hat{\mathbf{n}} \cdot \Delta \mathbf{r} \approx \frac{1}{2} \int_{-\infty}^{\infty} a^2(\xi) \, d\xi \qquad (4.104)$$

A high-power tabletop Nd:glass laser produces 4 J of output in 40 ps at a wavelength of 1 µm. If this pulse is focused to a spot 10 µm in diameter incident on an electron, what is the dimensionless vector potential a at the focus? What is the total longitudinal displacement of the electron after the pulse has passed?

EXERCISE 4.7

For a charged particle in the left- or right-hand circularly polarized, cw harmonic plane wave

$$\mathbf{A}(\xi) = \mathrm{Re}(\hat{\mathbf{e}}_{\pm} A_0 e^{-ik\xi}) \tag{4.105}$$

where A_0 is a constant, show that the energy is

$$\mathcal{E} = \langle \mathcal{E} \rangle = mc^2 \sqrt{1 + a^2} \tag{4.106}$$

where

$$a = \frac{q A_0}{mc} \tag{4.107}$$

Show that in some frame of reference the longitudinal position is constant, so the particle moves in a circle described by the equations

$$k(\mathbf{r} - \langle \mathbf{r} \rangle) = \mathrm{Re}\left[\pm i \hat{\mathbf{e}}_{\pm} \frac{a}{\sqrt{1 + a^2}} e^{-ik(ct - \hat{\mathbf{n}} \cdot \bar{\mathbf{r}})}\right] \tag{4.108}$$

where $\langle \mathbf{r} \rangle$ is the center of the circle.

EXERCISE 4.8

Consider a charged particle that is illuminated by a harmonic plane wave that starts at $t = 0$ and then rises to an intensity that remains constant.

(a) Show that when the incident wave reaches a steady state, the motions are like those illustrated in Figure 4.11, with cusps at the ends of the transverse motions.
(b) Show that in a coordinate system in which the mean motion vanishes, the particle executes a "figure-eight" motion.

4.2 CANONICAL EQUATIONS OF AN ELECTROMAGNETIC FIELD

Both the classical and the quantum theories of radiation begin with the decomposition of the electromagnetic field into eigenmodes of a rectangular volume of space. This seems natural enough in the quantum theory of radiation, since the eigenmodes form a representation of the photons that comprise the radiation field. But it is also true in the classical theory, where the analogy between the oscillations of the eigenmmodes and those of classical harmonic oscillators is used to write the equations of motion in canonical form. The classical equipartition theorem of statistical mechanics can then be used to describe the radiation field in thermodynamic equilibrium. As is now well known, the classical result, called the Rayleigh–Jeans law, is in sharp disagreement with experiment. The failure of the classical theory, the so-called ultraviolet catastrophe, lies not in the mathematical description of the fields as a collection of modes, but in the use of the classical partition function. Quantum mechanically, as discovered by Planck, only those modes are thermally excited for which $\hbar\omega/k_B T \leq O(1)$, where $\hbar$ is Planck's constant, ω the frequency of the mode, k_B Boltzmann's constant, and T the temperature. Because of its usefulness for both the classical and quantum theories of radiation, we now develop the canonical form of the equations of motion for the fields of a charged particle.

4.2.1 Fourier Decomposition of the Field

The Fourier decomposition of the electromagnetic field in a rectangular volume V is conveniently accomplished by imposing periodic boundary conditions on the field in the volume. That is, we assume that the value of the field at one end of the volume is the same as that at the other end. The field in all space is then a repetition of the field in the volume under consideration. Provided that the results we obtain by this method are independent of the size of the volume we consider, we may regard the results as valid for free space. This same mathematical approach is also used for the wave functions in the quantum theory of a free-electron gas in a metal or the neutrons in a neutron star, for example, and in the theory of lattice vibrations (phonons) in crystals.

We consider the rectangular volume shown in Figure 4.12, whose sides are of length L_x, L_y, and L_z. To satisfy the periodic boundary conditions, the fields $\Phi(\mathbf{r}, t)$ and $\mathbf{A}(\mathbf{r}, t)$ in this volume may be represented by the Fourier series

$$\Phi(\mathbf{r}, t) = \sum_{\mathbf{n}} \Phi_{\mathbf{n}}(t) e^{i\mathbf{k_n} \cdot \mathbf{r}} \tag{4.109}$$

$$\mathbf{A}(\mathbf{r}, t) = \sum_{\mathbf{n}} \mathbf{A}_{\mathbf{n}}(t) e^{i\mathbf{k_n} \cdot \mathbf{r}} \tag{4.110}$$

in which the vector subscript is

$$\mathbf{n} = (n_x, n_y, n_z) \tag{4.111}$$

where $n_{x,y,z} = \ldots, -2, -1, 0, 1, 2, \ldots$ are integers and the wave vector is

$$\mathbf{k_n} = \left(\frac{2\pi n_x}{L_x}, \frac{2\pi n_y}{L_y}, \frac{2\pi n_z}{L_z} \right) \tag{4.112}$$

Since the fields Φ and $\mathbf{A}$ are real, we see from (4.109) and (4.110) that

$$\Phi_{\mathbf{n}}(t) = \Phi_{-\mathbf{n}}^*(t) \tag{4.113}$$

$$\mathbf{A}_{\mathbf{n}}(t) = \mathbf{A}_{-\mathbf{n}}^*(t) \tag{4.114}$$

The vectors $\mathbf{n}$ occupy the integer points of n-space, as shown in Figure 4.13, and corresponding to each point there is a wave vector $\mathbf{k_n}$ in k-space. The eigenfunctions satisfy

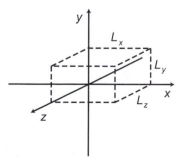

Figure 4.12 Vector potential field in a rectangular volume.

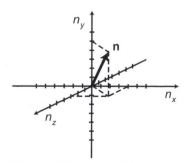

Figure 4.13 n-space of the eigenmodes of a rectangular volume.

the orthogonality relation

$$\int e^{i(\mathbf{k_n}-\mathbf{k_{n'}})\cdot\mathbf{r}} dV = V\delta_{\mathbf{nn'}} \tag{4.115}$$

where

$$V = L_x L_y L_z \tag{4.116}$$

is the volume of the region under consideration. Since we generally sum over many eigenvectors, it is useful to convert the sums over $\mathbf{n}$ to integrals over k-space. The volume of k-space corresponding to an infinitesimal volume of n-space is

$$d^3\mathbf{k} = \frac{2\pi}{L_x}\frac{2\pi}{L_y}\frac{2\pi}{L_z} d^3\mathbf{n} = \frac{(2\pi)^3}{V} d^3\mathbf{n} \tag{4.117}$$

Since each integer point in n-space occupies a unit volume, the number of eigenvalues in an infinitesimal region of k-space is

$$dN = d^3\mathbf{n} = \frac{V}{(2\pi)^3} d^3\mathbf{k} = \frac{V}{(2\pi)^3} k^2\, dk\, d\Omega \tag{4.118}$$

where $d\Omega$ is an infinitesimal solid angle in k-space.

In the Coulomb gauge we impose the condition

$$\nabla \cdot \mathbf{A} = 0 = \sum_{\mathbf{n}} i\mathbf{k_n} \cdot \mathbf{A_n}(t) e^{i\mathbf{k_n}\cdot\mathbf{r}} \tag{4.119}$$

for all $\mathbf{r}$, so for each $\mathbf{n}$ we have

$$\mathbf{k_n} \cdot \mathbf{A_n}(t) = 0 \tag{4.120}$$

That is, the vector potential waves are transverse in the Coulomb gauge, as we have seen previously. There are two possible orientations (polarizations) of the vector $\mathbf{A_n}$, which may be represented by the linearly polarized unit vectors $\hat{\mathbf{e}}_{\mathbf{n}(1,2)}$ or the circularly polarized unit vectors $\hat{\mathbf{e}}_{\mathbf{n}(\pm)}$, where the vector subscript $\mathbf{n}$ is needed to orient the basis set. Thus, the complete expansion of the vector potential $\mathbf{A}$ is

$$\mathbf{A}(\mathbf{r}, t) = \sum_{\mathbf{n}, p} \hat{\mathbf{e}}_{\mathbf{n}p} A_{\mathbf{n}p}(t) e^{i\mathbf{k_n}\cdot\mathbf{r}} \tag{4.121}$$

where $p = 1, 2$ or $p = \pm$, depending on the choice of basis vectors. In the following we use circularly polarized basis vectors, for which $\hat{\mathbf{e}}_{\mathbf{n}p}^* = \hat{\mathbf{e}}_{-\mathbf{n}p}$, as noted earlier. From this and the reality condition (4.114), we see that

$$A_{\mathbf{n}p} = A_{-\mathbf{n}p}^* \tag{4.122}$$

To find the equations of motion of the coefficients $\Phi_{\mathbf{n}}(t)$ and $A_{\mathbf{n}p}(t)$, we begin with the Maxwell equations for the fields. In the Coulomb gauge these are

$$\nabla^2\Phi = -\frac{\rho}{\varepsilon_0} \tag{4.123}$$

and

$$\frac{1}{c^2}\frac{\partial^2\mathbf{A}}{\partial t^2} - \nabla^2\mathbf{A} = \mu_0\mathbf{J} - \frac{1}{c^2}\nabla\frac{\partial\Phi}{\partial t} \tag{4.124}$$

where the terms on the right-hand side of (4.124) form the so-called transverse current, as discussed in Chapter 2. For a point charge q at position $\mathbf{r}_0(t)$, the charge density and current are

$$\rho = q\delta(\mathbf{r} - \mathbf{r}_0) \tag{4.125}$$

and

$$\mathbf{J} = q\mathbf{v}_0\delta(\mathbf{r} - \mathbf{r}_0) \tag{4.126}$$

where

$$\mathbf{v}_0 = \frac{d\mathbf{r}_0}{dt} \tag{4.127}$$

is the velocity of the particle. If we substitute the expansion (4.109) into the equation of motion (4.123), with (4.125) for the charge density, multiply by $e^{-i\mathbf{k}_{n'}\cdot\mathbf{r}}$, and integrate over the volume V using the orthogonality relation (4.115), we get

$$\Phi_{\mathbf{n}} = \frac{q}{\varepsilon_0}\frac{e^{-i\mathbf{k}_{\mathbf{n}}\cdot\mathbf{r}_0}}{k_{\mathbf{n}}^2 V} \tag{4.128}$$

This is just the Fourier expansion of the instantaneous Coulomb potential of a charged particle in a periodic box or, equivalently, the Coulomb potential of a periodic array of charged particles. It is worth noting that since the potential is periodic, the electric field is also periodic, so the net flux out of the volume V must vanish. According to Gauss's law, then, the total charge in the volume must also vanish. The field represented by (4.128) is therefore the field in volume V when the point charge q is balanced by an equal and opposite charge uniformly distributed throughout V. Since this uniform charge does not radiate, it does not affect the fields in which we are interested. The need for this "background charge" is also evident if we view Φ as the field of a periodic array of charges q. Without the background charge, the potential of the infinite array would diverge.

If we substitute (4.128), together with the expansion (4.121), into (4.124), use (4.126) for the current density of a point charge, multiply by $\hat{\mathbf{e}}_{\mathbf{n}'p'}^* e^{-i\mathbf{k}_{n'}\cdot\mathbf{r}}$, integrate over V using the orthogonality relation (4.115), and take advantage of the fact that $\mathbf{k}_{\mathbf{n}} \cdot \hat{\mathbf{e}}_{\mathbf{n}p}^* = 0$, we get the equation of motion

$$\frac{d^2 A_{\mathbf{n}p}}{dt^2} + \omega_{\mathbf{n}}^2 A_{\mathbf{n}p} = \frac{q}{\varepsilon_0 V}\mathbf{v}_0 \cdot \hat{\mathbf{e}}_{\mathbf{n}p}^* e^{-i\mathbf{k}_{\mathbf{n}}\cdot\mathbf{r}_0} \tag{4.129}$$

where

$$\omega_{\mathbf{n}} = k_{\mathbf{n}}c \tag{4.130}$$

is the frequency of the $\mathbf{n}$th eigenmode. By means of this Fourier decomposition we have converted the partial differential equation (4.124) into an infinite set of ordinary differential equations. These equations of motion are identical to those of ordinary forced harmonic oscillators, except that they are complex valued. The forcing function is just the appropriately phased component of the transverse current $q\mathbf{v}_0$. In the absence of the forcing term the solutions have the time dependence $e^{-i\omega_{\mathbf{n}}t}$, so the eigenmodes are traveling waves $A_{\mathbf{n}p} \propto e^{i(\mathbf{k}_{\mathbf{n}}\cdot\mathbf{r}-\omega_{\mathbf{n}}t)}$ that propagate in the $\hat{\mathbf{n}}$ direction. The modes that are excited by the current $q\mathbf{v}_0$ represent radiation emitted by the point charge in the $\hat{\mathbf{n}}$ direction.

The electric and magnetic fields are given by the expressions

$$\mathbf{E} = -\nabla\Phi - \frac{\partial\mathbf{A}}{\partial t} = \sum_{\mathbf{n},p} \hat{\mathbf{e}}_{\mathbf{n}p} \frac{dA_{\mathbf{n}p}}{dt} e^{i\mathbf{k_n}\cdot\mathbf{r}} - \nabla\Phi \tag{4.131}$$

and

$$\mathbf{B} = \nabla\times\mathbf{A} = \sum_{\mathbf{n},p} pk_{\mathbf{n}}\hat{\mathbf{e}}_{\mathbf{n}p} A_{\mathbf{n}p} e^{i\mathbf{k_n}\cdot\mathbf{r}} \tag{4.132}$$

in which we have used the property (4.47) of the circularly polarized vectors. The total energy in the field is found by integrating the energy density over the volume V. In the Coulomb gauge the field Φ is instantaneous and moves with the particle q. Thus, the total energy in the field Φ remains constant (actually, it diverges near the particle), and we ignore it from now on. The energy in the vector potential field is

$$W = \int_V dV \left[\frac{\varepsilon_0}{2}\mathbf{E}\cdot\mathbf{E}^* + \frac{1}{2\mu_0}\mathbf{B}\cdot\mathbf{B}^* \right] \tag{4.133}$$

since the fields are real. If we insert (4.131) and (4.132) into (4.133) and integrate over V using the orthogonality relations (4.45), (4.46), and (4.115), we find that the total energy in the field is

$$W = \frac{\varepsilon_0 V}{2} \sum_{\mathbf{n},p} \left(\frac{dA_{\mathbf{n}p}}{dt}\frac{dA_{\mathbf{n}p}^*}{dt} + \omega_{\mathbf{n}}^2 A_{\mathbf{n}p} A_{\mathbf{n}p}^* \right) \tag{4.134}$$

EXERCISE 4.9

Show that the energy in the Coulomb field (4.113) and (4.128) is

$$W = \frac{q^2}{4\pi^2\varepsilon_0} \int_0^\infty dk \tag{4.135}$$

This diverges in the limit $k \to \infty$, which is equivalent in some sense to integrating the energy density over all space and encountering the divergence in the limit $r \to 0$. However, the total energy is explicitly time independent.

4.2.2 Spontaneous Emission by a Harmonic Oscillator

As an illustration of the use of this formalism, we consider the radiation from an oscillating electric charge. This is the classical analogue of emission by a quantum-mechanical harmonic oscillator, and the method we use to calculate the emission is similar to the time-dependent perturbation theory (Fermi's "golden rule") used in quantum mechanics. In the limit of large quantum numbers, the classical and quantum calculations give the same answer, as the correspondence principle says they must.

In the classical case, the position of the charge q is

$$\mathbf{r}_0 = \mathbf{a}_0 \sin(\omega_0 t) \tag{4.136}$$

where $\mathbf{a}_0$ is the amplitude of the oscillation. We restrict ourselves to the nonrelativistic limit, so the velocity

$$\mathbf{v}_0 = \omega_0 \mathbf{a}_0 \cos(\omega_0 t) \tag{4.137}$$

is small compared with the speed of light. In this limit, we see that $\mathbf{k_n} \cdot \mathbf{r}_0 \ll 1$ for $ck_\mathbf{n} \approx \omega_0$ and the factor $e^{-i\mathbf{k_n} \cdot \mathbf{r}_0}$ in the equation of motion (4.129) is essentially unity.

To solve the equation of motion (4.129), we change the independent variable to t', multiply by the integrating factor $\sin[\omega_\mathbf{n}(t - t')]$, and integrate from $t' = 0$ to $t' = t$. After integrating by parts and using the initial condition

$$\frac{d A_{\mathbf{n}p}}{dt'} = A_{\mathbf{n}p} = 0 \qquad \text{at } t' = 0 \tag{4.138}$$

we find that the solution to the equation of motion (4.129) in the nonrelativistic limit is

$$A_{\mathbf{n}p}(t) = \frac{q}{\varepsilon_0 V \omega_\mathbf{n}} \int_0^t dt' \sin[\omega_\mathbf{n}(t - t')]\mathbf{v}_0(t') \cdot \hat{\mathbf{e}}_{\mathbf{n}p}^*$$

$$= \frac{q\omega_0 \mathbf{a}_0 \cdot \hat{\mathbf{e}}_{\mathbf{n}p}^*}{i4\varepsilon_0 V \omega_\mathbf{n}} \int_0^t dt' \left[e^{i\omega_\mathbf{n}(t-t')} - e^{-i\omega_\mathbf{n}(t-t')} \right] (e^{i\omega_0 t'} + e^{-i\omega_0 t'}) \tag{4.139}$$

If we expand the products and integrate, the terms whose exponents contain $\pm(\omega_0 + \omega_\mathbf{n})t'$ can be ignored because the integrals are much smaller than those of the near-resonant terms. This leaves us with

$$A_{\mathbf{n}p}(t) = \frac{q\omega_0 \mathbf{a}_0 \cdot \hat{\mathbf{e}}_{\mathbf{n}p}^*}{4\varepsilon_0 V \omega_\mathbf{n}(\omega_\mathbf{n} - \omega_0)} (e^{i\omega_0 t} + e^{-i\omega_0 t} - e^{i\omega_\mathbf{n} t} - e^{-i\omega_\mathbf{n} t}) \tag{4.140}$$

From (4.134) we see that the energy in mode $(\mathbf{n}, p)$ is

$$\mathcal{W}_{\mathbf{n}p} = \frac{\varepsilon_0 V}{2} \left(\frac{d A_{\mathbf{n}p}}{dt} \frac{d A_{\mathbf{n}p}^*}{dt} + \omega_\mathbf{n}^2 A_{\mathbf{n}p} A_{\mathbf{n}p}^* \right) \tag{4.141}$$

But when averaged over a period of the oscillation, the square of the amplitude is

$$\omega_\mathbf{n}^2 \langle A_{\mathbf{n}p}^* A_{\mathbf{n}p} \rangle \approx \frac{1}{8} \left(\frac{q\omega_0}{\varepsilon_0 V} \right)^2 \frac{|\mathbf{a}_0 \cdot \hat{\mathbf{e}}_{\mathbf{n}p}|^2}{(\omega_\mathbf{n} - \omega_0)^2} \left[-e^{i(\omega_\mathbf{n} - \omega_0)t} + 2 - e^{-i(\omega_\mathbf{n} - \omega_0)t} \right]$$

$$= \frac{1}{8} \left(\frac{q\omega_0}{\varepsilon_0 V} \right)^2 |\mathbf{a}_0 \cdot \hat{\mathbf{e}}_{\mathbf{n}p}^*|^2 \left\{ \frac{\sin\left[\frac{1}{2}(\omega_\mathbf{n} - \omega_0)t \right]}{\frac{1}{2}(\omega_\mathbf{n} - \omega_0)} \right\}^2 \tag{4.142}$$

Likewise, if we differentiate (4.140) and average over a period of the oscillation, we find that

$$\left\langle \frac{d A_{\mathbf{n}p}^*}{dt} \frac{d A_{\mathbf{n}p}}{dt} \right\rangle \approx \frac{1}{8} \left(\frac{q\omega_0^2}{\varepsilon_0 V \omega_\mathbf{n}} \right)^2 \frac{|\mathbf{a}_0 \cdot \hat{\mathbf{e}}_{\mathbf{n}p}|^2}{(\omega_\mathbf{n} - \omega_0)^2} \left[-e^{i(\omega_\mathbf{n} - \omega_0)t} + 2 - e^{-i(\omega_\mathbf{n} - \omega_0)t} + \frac{(\omega_\mathbf{n} - \omega_0)^2}{\omega_0 \omega_\mathbf{n}} \right]$$

$$\approx \frac{1}{8} \left(\frac{q\omega_0}{\varepsilon_0 V} \right)^2 |\mathbf{a}_0 \cdot \hat{\mathbf{e}}_{\mathbf{n}p}^*|^2 \left\{ \frac{\sin\left[\frac{1}{2}(\omega_\mathbf{n} - \omega_0)t \right]}{\frac{1}{2}(\omega_\mathbf{n} - \omega_0)} \right\}^2 \tag{4.143}$$

for near-resonant modes. Thus, the average energy in the mode $(\mathbf{n}, p)$ is

$$\langle \mathcal{W}_{\mathbf{n}p} \rangle = \frac{1}{8} \frac{q^2 \omega_0^2}{\varepsilon_0 V} |\mathbf{a}_0 \cdot \hat{\mathbf{e}}_{\mathbf{n}p}^*|^2 \left\{ \frac{\sin\left[\frac{1}{2}(\omega_\mathbf{n} - \omega_0)t \right]}{\frac{1}{2}(\omega_\mathbf{n} - \omega_0)} \right\}^2 \tag{4.144}$$

This shows something surprising, that for near-resonant modes at short times $[|(\omega_n - \omega_0)t|$ $\ll 1]$ the energy in each mode increases quadratically with time. However, the number of modes that are "near resonant" decreases as t^{-1}, so the total energy radiated into all modes increases linearly with time. The power radiated is therefore constant, as we would expect.

To compute the total energy radiated and find the total power explicitly, we sum over all modes $(\mathbf{n}, p)$. Summing first over the two polarizations, we find from the definition of the circularly polarized vectors that

$$\sum_p |\mathbf{a}_0 \cdot \hat{\mathbf{e}}_{np}^*|^2 = \sum_p \left(\mathbf{a}_0 \cdot \frac{\hat{\mathbf{e}}_{n1} + ip\hat{\mathbf{e}}_{n2}}{\sqrt{2}}\right)\left(\mathbf{a}_0 \cdot \frac{\hat{\mathbf{e}}_{n1} - ip\hat{\mathbf{e}}_{n2}}{\sqrt{2}}\right)$$

$$= \frac{1}{2}\sum_p [(\hat{\mathbf{a}}_0 \cdot \hat{\mathbf{e}}_{n1})^2 + (\mathbf{a}_0 \cdot \hat{\mathbf{e}}_{n2})^2] = \mathbf{a}_0 \cdot \mathbf{a}_0 - (\mathbf{a}_0 \cdot \hat{\mathbf{k}}_n)^2 \qquad (4.145)$$

since the unit vectors $(\hat{\mathbf{e}}_{n1}, \hat{\mathbf{e}}_{n2}, \hat{\mathbf{k}}_n)$ are orthogonal. The factor $1/2$ vanishes because the sum over p consists of two identical terms. In terms of the angle $\theta = \arccos(\hat{\mathbf{a}}_0 \cdot \hat{\mathbf{k}}_n)$ between the oscillator axis and the direction of emission, this may be expressed as

$$\sum_p |\mathbf{a}_0 \cdot \hat{\mathbf{e}}_{np}^*|^2 = a_0^2(1 - \cos^2\theta) = a_0^2 \sin^2\theta \qquad (4.146)$$

This radiation pattern, shown in Figure 4.14, is characteristic of an oscillating dipole. The radiation vanishes along the axis of the oscillation but is azimuthally symmetric about this axis.

To carry out the sum over $\mathbf{n}$, it is convenient to convert the sum to an integral in k-space, as discussed earlier. Using (4.118), we get

$$\langle\mathcal{W}\rangle = \frac{V}{(2\pi)^3}\int_{4\pi} d\Omega \int_0^\infty k^2\, dk \sum_p \langle\mathcal{W}_{\mathbf{n}p}\rangle$$

$$= \int_{4\pi} \frac{d\langle\mathcal{W}\rangle}{d\Omega}\, d\Omega \qquad (4.147)$$

The angular fluence (the energy radiated per unit solid angle $d\Omega$) is

$$\frac{d\langle\mathcal{W}\rangle}{d\Omega} = \frac{q^2\omega_0^2 a_0^2}{64\pi^3\varepsilon_0 c^3}\sin^2\theta \int_0^\infty \omega^2 d\omega \left\{\frac{\sin[\frac{1}{2}(\omega - \omega_0)t]}{\frac{1}{2}(\omega - \omega_0)}\right\}^2 \qquad (4.148)$$

where $\omega^2 d\omega = c^3 k^2 dk$. But this integral is dominated by the region near resonance, $|(\omega - \omega_0)t| < O(1)$, so for $\omega_0 t \gg 1$ we may set $\omega^2 \approx \omega_0^2$ and extend the lower limit of

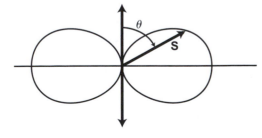

Figure 4.14 Radiation from an oscillating dipole.

integration to $-\infty$. Evaluating the integral with the help of the formula

$$\int_{-\infty}^{\infty} \frac{\sin^2 x}{x^2}\, dx = \pi \tag{4.149}$$

we get

$$\frac{d\langle \mathcal{W} \rangle}{d\Omega} = \frac{q^2 \omega_0^4 a_0^2 \sin^2 \theta}{32\pi^2 \varepsilon_0 c^3} t = \frac{d\langle \mathcal{P} \rangle}{d\Omega} t \tag{4.150}$$

where the power radiated per unit solid angle $d\Omega$ is

$$\frac{d\langle \mathcal{P} \rangle}{d\Omega} = \frac{q^2 \omega_0^4 a_0^2}{32\pi^2 \varepsilon_0 c^3} \sin^2 \theta \tag{4.151}$$

The integral over $d\Omega$ is just

$$\int_{4\pi} d\Omega \sin^2 \theta = 2\pi \int_0^\pi (1 - \cos^2 \theta)\, d\cos\theta = \frac{8\pi}{3} \tag{4.152}$$

so the average total power radiated by the oscillator is

$$\langle \mathcal{P} \rangle = \frac{q^2 \omega_0^4 a_0^2}{12\pi \varepsilon_0 c^3} \tag{4.153}$$

We note in passing that the energy per unit frequency interval emitted by an oscillator up to time t can be identified from (4.148) as

$$\frac{d^2 \langle \mathcal{W} \rangle}{d\omega\, d\Omega} = \frac{q^2 \omega_0^2 a_0^2 \omega^2 t^2}{64\pi^3 \varepsilon_0 c^3} \left\{ \frac{\sin[\frac{1}{2}(\omega - \omega_0)t]}{\frac{1}{2}(\omega - \omega_0)t} \right\}^2 \tag{4.154}$$

This is shown in Figure 4.15, where we see that the half-width of the spectrum at the first minimum is $\Delta\omega = 2\pi/t$. Thus, (4.154) shows that the spectral intensity at line center increases quadratically with time, but the spectrum gets narrower as the oscillator continues to radiate.

The quantum-mechanical expression for the spontaneous emission from a dipole-allowed transition is

$$\mathcal{P}_{m \to n} = \frac{q^2 \omega_0^4}{3\pi \varepsilon_0 c^3} |\langle m|\mathbf{r}|n \rangle|^2 \tag{4.155}$$

where $\langle m|\mathbf{r}|n \rangle$ is the dipole moment matrix element between the initial and final states. The quantum and classical results agree if we identify the dipole matrix element with the

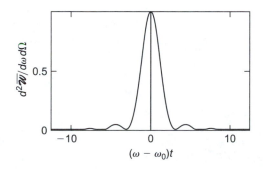

Figure 4.15 Spectrum of a pulsed oscillator.

classical amplitude of the oscillation; that is,

$$|\langle m|\mathbf{r}|n\rangle| \sim \frac{a_0^2}{4} \tag{4.156}$$

In fact, for a transition $m \to n = m - 1$, the dipole matrix element of a harmonic oscillator is

$$|\langle m|\mathbf{r}|m-1\rangle|^2 = \frac{m\hbar\omega}{2k} = \frac{1}{4}\left(\frac{E_m + E_{m-1}}{k}\right) = \frac{1}{4}\left(\frac{a_m^2 + a_{m-1}^2}{2}\right) \tag{4.157}$$

where $E_m = (m + \frac{1}{2})\hbar\omega_0$ is the energy of the mth vibrational level, k the spring constant of the oscillator, and

$$a_m = \sqrt{2E_m/k} \tag{4.158}$$

the classical turning point of a harmonic oscillator with the energy E_m. Thus, the quantum-mechanical amplitude of the oscillator is just the average of the classical turning points of the upper and lower states, which approaches the classical amplitude in the correspondence limit.

Not only are the classical and quantum-mechanical results (4.153) and (4.155) similar for the spontaneous emission, the quantum-mechanical analysis of the transition probability is also similar to the classical analysis. According to first-order time-dependent perturbation theory, the transition probability is

$$W_{a\to b} \approx \frac{1}{\hbar^2}\left|\int_0^t e^{i\omega_{ab}t'} V_{ab}\, dt'\right|^2 \tag{4.159}$$

where V_{ab} is the matrix element connecting the initial and final states a and b, and $\hbar\omega_{ab} = E_a - E_b$ is the energy difference between the states. If the interaction is not explicitly time dependent, the transition probability is

$$W_{a\to b} \approx \frac{|V_{ab}|^2}{\hbar^2}\left|\int_0^t e^{i\omega_{ab}t'} dt'\right|^2 = \frac{|V_{ab}|^2}{\hbar^2}\left[\frac{\sin\left(\frac{1}{2}\omega_{ab}t\right)}{\frac{1}{2}\omega_{ab}}\right]^2 \tag{4.160}$$

For transitions to a continuum of final states, the total transition probability is the integral of this probability over all final states. Provided that the matrix element is slowly varying near resonance, the result is

$$W_{\text{tot}} = \frac{|V_{ab}|^2}{\hbar^2}\int_0^\infty dE_b\rho(E_b)\left[\frac{\sin\left(\frac{1}{2}\omega_{ab}t\right)}{\frac{1}{2}\omega_{ab}}\right]^2 \tag{4.161}$$

where $\rho(E_b)$ is the density of final states at the energy E_b. In the case of radiative transitions, the density of final states is, by (4.118), $\rho \propto k^2 \propto \omega_0^2$, where $\hbar\omega_0$ is the difference between the atomic energy levels. If we compare (4.161) with (4.148), the correspondence between the quantum-mechanical and classical calculations is evident.

EXERCISE 4.10

Consider a plane, harmonic wave incident on a charged particle. In the nonrelativistic approximation, as discussed earlier, the particle oscillates at the frequency of the incident wave and therefore radiates at the same frequency. Show that the power radiated into the solid angle $d\Omega$ is proportional to the average incident intensity $\langle S \rangle$ and may therefore be described as scattering of the incident radiation,

$$\frac{d\langle\mathcal{P}\rangle}{d\Omega} = \frac{d\sigma_T}{d\Omega}\langle S\rangle \tag{4.162}$$

with the differential scattering cross section

$$\frac{d\sigma_T}{d\Omega} = r_c^2 \sin^2\theta \tag{4.163}$$

where θ is the angle between the direction of polarization of the incident light and the direction into which the light is scattered. The quantity

$$r_c = \frac{q^2}{4\pi\varepsilon_0 mc^2} \tag{4.164}$$

is called the classical radius of the particle, and for an electron it has the value $r_c = 2.82 \times 10^{-15}$ m. This is called Thomson scattering.

EXERCISE 4.11

In beta decay of a nucleus, an electron and an antineutrino are ejected from the nucleus along with some accompanying radiation. The radiation can be simulated by thinking of the electron as starting from rest and being instantaneously accelerated to some constant final velocity $\mathbf{v} = \boldsymbol{\beta}c$. Beginning with (4.129), use this model to show that the angular spectral fluence in the direction $\hat{\mathbf{n}}$ is

$$\frac{d^2\mathcal{W}}{d\omega\,d\Omega} = \frac{q^2}{16\pi^3\varepsilon_0 c}\left|\frac{\hat{\mathbf{n}}\times\boldsymbol{\beta}}{1-\hat{\mathbf{n}}\cdot\boldsymbol{\beta}}\right|^2 \tag{4.165}$$

where q is the electron charge. *Hint:* The integral that appears oscillates at infinity, but for present purposes it can be evaluated by using the device

$$\int_0^\infty e^{-i\omega t}\,dt = \lim_{\varepsilon\to 0}\int_0^\infty e^{-i\omega t-\varepsilon t}\,dt = \lim_{\varepsilon\to 0}\left(\frac{1}{\varepsilon+i\omega}\right) = \frac{1}{i\omega} \tag{4.166}$$

4.2.3 Canonical Equations of the Electromagnetic Field

Up to this point it has been convenient to use complex coefficients for the expansion of the field, recognizing that since the vector potential $\mathbf{A}(t)$ is real, the coefficients must satisfy the condition (4.122). Sometimes, however, it is more convenient to work with real variables that we can construct from combinations of the complex variables. We therefore introduce the real quantities

$$P_{\mathbf{n}p}(t) = \omega_{\mathbf{n}}\sqrt{\frac{\varepsilon_0 V}{2}}[A_{\mathbf{n}p}(t) + A^*_{\mathbf{n}p}(t)] \tag{4.167}$$

$$Q_{\mathbf{n}p}(t) = i\sqrt{\frac{\varepsilon_0 V}{2}}[A_{\mathbf{n}p}(t) - A^*_{\mathbf{n}p}(t)] \tag{4.168}$$

in terms of which the Fourier components of the field are

$$A_{\mathbf{n}p} = \frac{P_{\mathbf{n}p} - i\omega_{\mathbf{n}} Q_{\mathbf{n}p}}{\omega_{\mathbf{n}} \sqrt{2\varepsilon_0 V}} \tag{4.169}$$

In the absence of sources, the new variables satisfy the second-order differential equations

$$\frac{d^2 P_{\mathbf{n}p}}{dt^2} + \omega_{\mathbf{n}}^2 P_{\mathbf{n}p} = 0 \tag{4.170}$$

$$\frac{d^2 Q_{\mathbf{n}p}}{dt^2} + \omega_{\mathbf{n}}^2 Q_{\mathbf{n}p} = 0 \tag{4.171}$$

Although $P_{\mathbf{n}p}$ is not the ordinary momentum of the field, we can show that it is the canonical momentum conjugate to the generalized coordinate $Q_{\mathbf{n}p}$. To do this we introduce the function

$$\mathcal{H}_f = \frac{1}{2} \sum_{\mathbf{n},p} \left(P_{\mathbf{n}p}^2 + \omega_{\mathbf{n}}^2 Q_{\mathbf{n}p}^2 \right) \tag{4.172}$$

and show first that it leads to the canonical equations of motion and second that it represents the energy of the field. This proves that $\mathcal{H}_f$ is the Hamiltonian, so that $Q_{\mathbf{n}p}$ and $P_{\mathbf{n}p}$ are the generalized coordinate and the conjugate momentum. Evaluating the derivatives, we get

$$\frac{d P_{\mathbf{n}p}}{dt} = -\frac{\partial \mathcal{H}_f}{\partial Q_{\mathbf{n}p}} = -\omega_{\mathbf{n}}^2 Q_{\mathbf{n}p} \tag{4.173}$$

$$\frac{d Q_{\mathbf{n}p}}{dt} = \frac{\partial \mathcal{H}_f}{\partial P_{\mathbf{n}p}} = P_{\mathbf{n}p} \tag{4.174}$$

Differentiating and combining these, we recover (4.170) and (4.171), proving that (4.173) and (4.174) are the canonical equations of motion. Returning to (4.134), we substitute (4.169) and use the equations of motion (4.173) and (4.174) to show that the total energy in the field is

$$W_f = \frac{1}{4} \sum_{\mathbf{n},p} \left[\frac{1}{\omega_{\mathbf{n}}^2} \left(\frac{d P_{\mathbf{n}p}}{dt} \right)^2 + \left(\frac{d Q_{\mathbf{n}p}}{dt} \right)^2 + P_{\mathbf{n}p}^2 + \omega_{\mathbf{n}}^2 Q_{\mathbf{n}p}^2 \right]$$

$$= \frac{1}{2} \sum_{\mathbf{n},p} \left(P_{\mathbf{n}p}^2 + \omega_{\mathbf{n}}^2 Q_{\mathbf{n}p}^2 \right) = \mathcal{H}_f \tag{4.175}$$

This completes the proof.

When sources are present, the equation of motion for the complex components $A_{\mathbf{n}p}$ of the vector potential field is given by (4.129). To find the canonical equations of motion of the field with sources we note that the total Hamiltonian for a particle and a field is just the Hamiltonian $\mathcal{H}_m$ for the material particles in the field plus the Hamiltonian $\mathcal{H}_f$ for the field. For a set of charged particles, the Hamiltonian is

$$\mathcal{H}_m = \sum_i \left[\sqrt{(\mathbf{P}_i - q_i \mathbf{A}_i)^2 c^2 + m_i^2 c^4} + q_i \Phi_i \right]$$

$$\approx \sum_i \left[m_i c^2 + \frac{(\mathbf{P}_i - q_i \mathbf{A}_i)^2}{2m_i} + q_i \Phi_i \right] \tag{4.176}$$

in the nonrelativistic limit, where m_i is the mass, q_i the charge, and $\mathbf{P}_i$ the canonical momentum of the ith particle, and $\mathbf{A}_i = \mathbf{A}(\mathbf{r}_i)$ and $\Phi_i = \Phi(\mathbf{r}_i)$ are the potentials at position $\mathbf{r}_i$ of the particle. The total Hamiltonian is then

$$\mathcal{H} = \mathcal{H}_m + \mathcal{H}_f \tag{4.177}$$

and the canonical equations of motion for the fields are

$$\frac{dP_{\mathbf{n}p}}{dt} = -\frac{\partial \mathcal{H}}{\partial Q_{\mathbf{n}p}} = \sum_i q_i \mathbf{v}_i \cdot \frac{\partial \mathbf{A}_i}{\partial Q_{\mathbf{n}p}} - \omega_{\mathbf{n}}^2 Q_{\mathbf{n}p} \tag{4.178}$$

$$\frac{dQ_{\mathbf{n}p}}{dt} = \frac{\partial \mathcal{H}}{\partial P_{\mathbf{n}p}} = -\sum_i q_i \mathbf{v}_i \cdot \frac{\partial \mathbf{A}_i}{\partial P_{\mathbf{n}p}} + P_{\mathbf{n}p} \tag{4.179}$$

To evaluate the derivatives $\partial A_i / \partial Q_{\mathbf{n}p}$ and $\partial A_i / \partial P_{\mathbf{n}p}$ we use (4.121), which we write in the explicitly real form

$$A_i = \frac{1}{2} \sum_{\mathbf{n},p} (A_{\mathbf{n}p} \hat{\mathbf{e}}_{\mathbf{n}p} e^{i\mathbf{k}_i \cdot \mathbf{r}_i} + A_{\mathbf{n}p}^* \hat{\mathbf{e}}_{\mathbf{n}p}^* e^{-i\mathbf{k}_i \cdot \mathbf{r}_i}) \tag{4.180}$$

If we substitute (4.169) for $A_{\mathbf{n}p}$ and $A_{\mathbf{n}p}^*$, and take the indicated derivatives, we get the canonical equations of motion

$$\frac{dP_{\mathbf{n}p}}{dt} = -\omega_{\mathbf{n}}^2 Q_{\mathbf{n}p} - \frac{i}{\sqrt{2\varepsilon_0 V}} \sum_i q_i \mathbf{v}_i \cdot \frac{\hat{\mathbf{e}}_{\mathbf{n}p} e^{i\mathbf{k}_{\mathbf{n}} \cdot \mathbf{r}_i} - \hat{\mathbf{e}}_{\mathbf{n}p}^* e^{-i\mathbf{k}_{\mathbf{n}} \cdot \mathbf{r}_i}}{2} \tag{4.181}$$

$$\frac{dQ_{\mathbf{n}p}}{dt} = P_{\mathbf{n}p} - \frac{1}{\sqrt{2\varepsilon_0 V}} \sum_i q_i \mathbf{v}_i \cdot \frac{\hat{\mathbf{e}}_{\mathbf{n}p} e^{i\mathbf{k}_{\mathbf{n}} \cdot \mathbf{r}_i} + \hat{\mathbf{e}}_{\mathbf{n}p}^* e^{-i\mathbf{k}_{\mathbf{n}} \cdot \mathbf{r}_i}}{2\omega_{\mathbf{n}}} \tag{4.182}$$

for the fields when sources are included.

EXERCISE 4.12

In the Fourier decomposition of a free electromagnetic field, a photon is an eigenfunction of the field with energy $\mathcal{W} = \hbar\omega$. The total angular momentum of the field is

$$\mathbf{L} = \varepsilon_0 \int_V \mathbf{r} \times (\mathbf{E} \times \mathbf{B}) \, dV \tag{4.183}$$

where $\mathbf{r}$ is the radius vector from some chosen origin about which the angular momentum is calculated.

(a) Show that the total angular momentum of the field is

$$\mathbf{L} = \mathbf{l} + \mathbf{s} \tag{4.184}$$

where the "orbital" angular momentum

$$\mathbf{l} = \varepsilon_0 \sum_{l=1}^{3} \int_V E_l (\mathbf{r} \times \nabla) A_l \, dV \tag{4.185}$$

depends on the choice of the origin and the "intrinsic" angular momentum

$$\mathbf{s} = \varepsilon_0 \int_V \mathbf{E} \times \mathbf{A} \, dV \tag{4.186}$$

does not. *Hint:* Use the identity $\sum_k \varepsilon_{ijk}\varepsilon_{klm} = \delta_{il}\delta_{jm} - \delta_{im}\delta_{jl}$ and the fact that $\nabla \cdot \mathbf{E} = 0$ to show that the kth component of the momentum density is

$$\varepsilon_0 (\mathbf{E} \times \mathbf{B})_k = \varepsilon_0 \sum_{l=1}^{3} \left(E_l \frac{\partial A_l}{\partial r_k} - \frac{\partial E_l A_k}{\partial r_l} \right) \tag{4.187}$$

and the ith component of the angular momentum density is

$$\varepsilon_0 [\mathbf{r} \times (\mathbf{E} \times \mathbf{B})]_i = \varepsilon_0 \sum_{l=1}^{3} E_l (\mathbf{r} \times \nabla)_i A_l + \varepsilon_0 (\mathbf{E} \times \mathbf{A})_i$$

$$- \varepsilon_0 \sum_{l=1}^{3} \frac{\partial}{\partial r_l} \left(\sum_{j,k=1}^{3} \varepsilon_{ijk} r_j A_k E_l \right) \tag{4.188}$$

Then take advantage of the periodic boundary conditions to evaluate the integral over V of the last term in (4.188).

(b) Show that the intrinsic angular momentum (spin) of the field is

$$\mathbf{s} = p \frac{\mathcal{W}}{\omega} \hat{\mathbf{k}} = p\hbar \hat{\mathbf{k}} \tag{4.189}$$

for a (real) photon

$$\mathbf{A} = a \hat{\mathbf{e}}_{\hat{\mathbf{k}}p} e^{i(\mathbf{k}\cdot\mathbf{r}-\omega t)} + a^* \hat{\mathbf{e}}_{\hat{\mathbf{k}}p}^* e^{-i(\mathbf{k}\cdot\mathbf{r}-\omega t)} \tag{4.190}$$

traveling in the $\hat{\mathbf{k}}$ direction with polarization $p = \pm$.

EXERCISE 4.13

(a) The wave equation for a Lorentz scalar field ϕ in the presence of a point source q at $\mathbf{r}_0(t)$ is

$$\partial^\alpha \partial_\alpha \phi + \mu^2 \phi = -\frac{q}{2K_2\gamma} \delta^3(\mathbf{r} - \mathbf{r}_0) \tag{4.191}$$

If we decompose this field into the eigenfunctions of a rectangular volume $V = L_x L_y L_z$ with periodic boundary conditions

$$\phi = \sum_{\mathbf{n}} \phi_{\mathbf{n}}(t) e^{i\mathbf{k_n}\cdot\mathbf{r}} \tag{4.192}$$

show that the equation of motion for the component $\phi_{\mathbf{n}}(t)$ is

$$\frac{1}{c^2} \frac{d^2\phi_{\mathbf{n}}}{dt^2} + (k_{\mathbf{n}}^2 + \mu^2)\phi_{\mathbf{n}} = -\frac{q}{2K_2\gamma V} e^{-i\mathbf{k_n}\cdot\mathbf{r}_0(t)} \tag{4.193}$$

(b) The symmetric stress tensor for the scalar field is

$$T^{\alpha\beta} = 2K_2 \left[\partial^\alpha \phi \, \partial^\beta \phi - \frac{1}{2} (\partial^\nu \phi \, \partial_\nu \phi - \mu^2 \phi^2) g^{\alpha\beta} \right] \tag{4.194}$$

Show that the total energy of the field in the volume V is

$$W = K_2 V \sum_{\mathbf{n}} \left[\frac{1}{c^2} \frac{d\phi_{\mathbf{n}}}{dt} \frac{d\phi_{\mathbf{n}}^*}{dt} + \left(k_{\mathbf{n}}^2 + \mu^2 \right) \phi_{\mathbf{n}} \phi_{\mathbf{n}}^* \right] \tag{4.195}$$

(c) We define the generalized coordinates and momenta of the field as

$$Q_{\mathbf{n}} = \frac{i}{c} \sqrt{K_2 V} (\phi_{\mathbf{n}} - \phi_{\mathbf{n}}^*) \tag{4.196}$$

$$P_{\mathbf{n}} = \frac{\omega_{\mathbf{n}}}{c} \sqrt{K_2 V} (\phi_{\mathbf{n}} + \phi_{\mathbf{n}}^*) \tag{4.197}$$

where

$$\frac{\omega_{\mathbf{n}}^2}{c^2} = k_{\mathbf{n}}^2 + \mu^2 \tag{4.198}$$

Show that they are the canonical variables and that the Hamiltonian for the free field is

$$\mathcal{H}_\phi = \frac{1}{2} \sum_{\mathbf{n}} \left(P_{\mathbf{n}}^2 + \omega_{\mathbf{n}}^2 Q_{\mathbf{n}}^2 \right) \tag{4.199}$$

4.2.4 Blackbody Radiation and the Einstein Coefficients

The canonical formulation of the radiation field was used by Rayleigh (1900) and Jeans (1905) to describe the spectrum of radiation in thermal equilibrium at temperature T. The argument goes as follows. According to the equipartition theorem, at thermal equilibrium each of the oscillators that appears in the canonical description of the electromagnetic field should contain energy equal to $\frac{1}{2} k_B T$ for each degree of freedom ($Q_{\mathbf{n}p}$ or $P_{\mathbf{n}p}$) that enters into the Hamiltonian, where k_B is Boltzmann's constant. Therefore, the total energy density of the blackbody radiation field should be

$$\mathcal{U} = \frac{W}{V} = \sum_{\mathbf{n}, p} \frac{k_B T}{V} \tag{4.200}$$

Unfortunately, this expression diverges, since there are an infinite number of eigenvectors $\mathbf{n}$. However, if we look at the spectrum of the radiation, we get

$$\frac{d\mathcal{U}}{d\omega} = \frac{k_B T}{V} \frac{dN}{d\omega} \tag{4.201}$$

where dN is the number of eigenmodes in the frequency range $d\omega$. But from (4.118) we find that if we include both polarizations, the spectral density of eigenstates is

$$\frac{dN}{d\omega} = \frac{k^2 V}{\pi^2} \frac{dk}{d\omega} = \frac{V}{\pi^2} \frac{\omega^2}{c^3} \tag{4.202}$$

so the spectrum of blackbody radiation should be

$$\frac{d\mathcal{U}}{d\omega} = \frac{\omega^2 k_B T}{\pi^2 c^3} \tag{4.203}$$

This is called the Rayleigh–Jeans law, and in fact it is actually a good approximation to the spectrum at low frequencies. The failure to describe the spectrum at high frequencies is the "ultraviolet catastrophe." It led directly to the discovery of quantum mechanics.

The problem, of course, lies in the assumption that the average energy in each oscillator is $k_B T$. The correct expression was discovered by Planck, who proposed that the energy in an individual oscillator could only be an integer multiple of the quantum $\hbar \omega_{\mathbf{n}}$, where $\hbar$ is now known as Planck's constant. He then assigned the relative statistical likelihood $\exp(-n\hbar\omega_{\mathbf{n}}/k_B T)$ to the state possessing n quanta and obtained the expression

$$\mathcal{W}_{\mathbf{n}p} = \frac{\hbar\omega_{\mathbf{n}}}{\exp(\hbar\omega_{\mathbf{n}}/k_B T) - 1} \tag{4.204}$$

for the average energy in the oscillator $(\mathbf{n}, p)$. When this expression is used in (4.200), the total energy density in the blackbody field is

$$\mathcal{U} = \frac{\pi^2}{15} \frac{k_B^4 T^4}{\hbar^3 c^3} \tag{4.205}$$

which is finite. The spectrum is

$$\frac{d\mathcal{U}}{d\omega} = \frac{\mathcal{W}(\omega)}{V} \frac{dN}{d\omega} = \frac{\hbar\omega^3/\pi^2 c^3}{\exp(\hbar\omega/k_B T) - 1} \tag{4.206}$$

or in terms of the wavelength,

$$\frac{d\mathcal{U}}{d\lambda} = \frac{d\mathcal{U}}{d\omega}\left|\frac{d\omega}{d\lambda}\right| = \frac{16\pi^2\hbar c/\lambda^5}{\exp(2\pi\hbar c/k_B T\lambda) - 1} \tag{4.207}$$

This is shown in Figure 4.16, where the Rayleigh–Jeans law is also shown for comparison. We see that the classical approximation is valid for frequencies $\hbar\omega/k_B T \ll 1$. The Planck spectrum peaks at the wavelength for which $2\pi\hbar c/k_B T\lambda = 4.965$, that is

$$\lambda = \frac{1}{4.965} \frac{2\pi\hbar c}{k_B T} = \frac{2.898 \times 10^{-3} \, (\text{m-K})}{T \, (\text{K})} \tag{4.208}$$

This is known as Wein's displacement law and was derived by him (apart from the value of the constant) using thermodynamic arguments in 1893. For a blackbody at room temperature (300 K), the wavelength of the peak of the spectrum is around 10 μm, which is in the infrared. For the surface of the sun at about 6000 K the peak is around 500 nm, near the middle of the visible spectrum, and for the cosmic background radiation at 2.7 K the peak is near 1 mm.

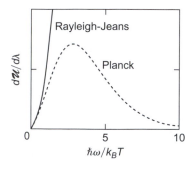

Figure 4.16 Blackbody spectrum.

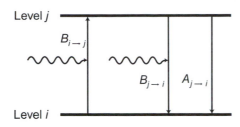

Level j

$B_{i \to j}$

$B_{j \to i}$ $A_{j \to i}$

Level i

Figure 4.17 Absorption, stimulated emission, and spontaneous emission.

When an oscillator (classical or quantum mechanical) is placed in a radiation field, it both emits radiation into the field and absorbs radiation from it. Using a very clever thermodynamic argument, Einstein (who else?) derived the relationship between these processes. The argument that he put forward in 1917 goes as follows. We consider an atom in a blackbody radiation field at temperature T and suppose that the atom can be in any one of several states that we indicate by the subscript i. Einstein recognized that the populations N_i and N_j of states i and j are coupled by the following three processes, as indicated in Figure 4.17:

1. The atom can undergo a transition from the higher energy state i to the lower energy state j by spontaneously emitting a photon with the energy

$$\mathcal{E}_{ij} = \hbar\omega_{ij} = \mathcal{E}_i - \mathcal{E}_j \tag{4.209}$$

where ω_{ij} is the frequency of the photon and $\mathcal{E}_i$ ($\mathcal{E}_j$) is the energy of state i (j). The probability per unit time of this occurring is described by the constant $A_{i \to j}$.

2. The atom can undergo a transition from state j to state i by absorbing a photon of the same energy. The probability of this occurring is proportional to the number of photons with the correct energy present in the blackbody radiation field, and we denote it by the product $B_{j \to i}[d\mathcal{U}(\omega_{ij})/d\omega]$.

3. The atom can be stimulated by the blackbody field to emit a photon of the same energy and undergo a transition from state i to state j with the probability $B_{i \to j}[d\mathcal{U}(\omega_{ij})/d\omega]$.

The constants $A_{i \to j}$, $B_{i \to j}$, and $B_{j \to i}$ are called the Einstein coefficients. According to the principle of detailed balance at thermal equilibrium, the rates of the forward and backward transitions between the states i and j must be equal, so that

$$N_i \left[A_{i \to j} + B_{i \to j} \frac{d\mathcal{U}(\omega_{ij})}{d\omega} \right] = N_j B_{j \to i} \frac{d\mathcal{U}(\omega_{ij})}{d\omega} \tag{4.210}$$

But the equilibrium populations of the states i and j are related by

$$\frac{N_i}{N_j} = e^{-\mathcal{E}_{ij}/k_B T} \tag{4.211}$$

Substituting this into (4.210), we find that

$$\frac{d\mathcal{U}(\omega_{ij})}{d\omega} = \frac{A_{i \to j}}{B_{j \to i} e^{\hbar\omega_{ij}/k_B T} - B_{i \to j}} \tag{4.212}$$

Comparing this with (4.206), we see that the two equations can be reconciled only if the coefficients satisfy the relations

$$B_{i \to j} = B_{j \to i} \tag{4.213}$$

and

$$\frac{A_{i \to j}}{B_{i \to j}} = \frac{\hbar \omega^3}{\pi^2 c^3} \tag{4.214}$$

These are called the Einstein relations.

The absorption of photons is often described by an absorption cross section, but here we must be careful. The absorption is spread over a (generally narrow) range of frequencies called the width of the absorption line, and the Einstein coefficient represents a sum over all the frequencies absorbed. Clearly, the linewidth must be narrow so that we may approximate the energy density of the field at all the absorbed frequencies by $d\mathcal{U}(\omega_{ij})/d\omega$ in Einstein's analysis. The integral of the absorption cross section over the width of the line is then

$$\int \sigma_{j \to i} \frac{d\Phi}{d\omega} \, d\omega \approx \frac{d\Phi(\omega_{ij})}{d\omega} \int \sigma_{j \to i} \, d\omega = B_{j \to i} \frac{d\mathcal{U}(\omega_{ij})}{d\omega} \tag{4.215}$$

where the flux of photons in a blackbody radiation field is

$$\frac{d\Phi}{d\omega} = \frac{c}{\hbar \omega} \frac{d\mathcal{U}}{d\omega} \tag{4.216}$$

If we combine these formulas, we get the relation

$$\int \sigma_{i \to j} \, d\omega = \frac{\pi^2 c^2}{\omega^2} A_{i \to j} \tag{4.217}$$

for the integrated cross section in terms of the Einstein coefficient. In elementary quantum mechanics, this formula is generally inverted to calculate the spontaneous emission rate from the absorption cross section, since the absorption cross section is easily computed semiclassically, that is, without quantizing the radiation field. In the present (classical) case, we have already computed the spontaneous emission from an oscillator, so we can use (4.217) to calculate the absorption cross section.

A few additional remarks are in order here. In the first place, the flux of photons in the blackbody radiation field is isotropic, so the Einstein coefficient $B_{i \to j}$ and cross section $\sigma_{i \to j}$ are averaged over all orientations of the atom. If we try to use (4.217) for the absorption of photons from an incident plane wave, this must be kept in mind. In quantum mechanics the orientation of an atom is described by the magnetic quantum number $m_J = -J, \ldots, J$, where J is the total angular momentum quantum number, so (4.217) represents an average over all initial and final magnetic quantum numbers. This being the case, we need to allow for the degeneracy d_i and d_j of the upper and lower levels. At equilibrium the populations are related by

$$\frac{N_i}{N_j} = \frac{d_i}{d_j} e^{-\mathcal{E}_{ij}/k_B T} \tag{4.218}$$

Following this through the arguments used previously, we obtain for the Einstein coefficients the relations

$$d_i B_{i \to j} = d_j B_{j \to i} \tag{4.219}$$

and

$$\frac{A_{i \to j}}{B_{i \to j}} = \frac{\hbar \omega^3}{\pi^2 c^3} \tag{4.220}$$

and for the cross sections,

$$\int \sigma_{i \to j} \, d\omega = \frac{\pi^2 c^2}{\omega^2} A_{i \to j} \tag{4.221}$$

and

$$\int \sigma_{j \to i} \, d\omega = \frac{d_i}{d_j} \int \sigma_{i \to j} \, d\omega = \frac{d_i}{d_j} \frac{\pi^2 c^2}{\omega^2} A_{i \to j} \tag{4.222}$$

EXERCISE 4.14

In quantum theory the radiation field is quantized and a photon has the 4-vector momentum

$$p^\alpha = \hbar k^\alpha \tag{4.223}$$

where $\hbar$ is Planck's constant divided by 2π. When the Proca term is included, the wave vector satisfies the dispersion relation

$$k^\alpha k_\alpha = \mu_\gamma^2 \tag{4.224}$$

for some constant μ_γ. But for a particle, the magnitude of the 4-vector momentum is

$$p^\alpha p_\alpha = m^2 c^2 \tag{4.225}$$

which suggests that the photon has a rest mass

$$m_\gamma = \frac{\hbar \mu_\gamma}{c} \tag{4.226}$$

which is independent of the frequency. In quantum statistical mechanics it is shown that the equilibrium number of photons of wave number $\mathbf{k}$ in each polarization at the temperature T is

$$n_{\mathbf{k}} = \frac{1}{\exp\left(\dfrac{\hbar \omega_{\mathbf{k}}}{k_B T}\right) - 1} \tag{4.227}$$

in which $\omega_{\mathbf{k}}$ is the frequency corresponding to the wave vector $\mathbf{k}$ and k_B is Boltzmann's constant. Count up the number of photons in a volume V, including both polarizations, and show that in the limit

$$\frac{m_\gamma c^2}{k_B T} \ll 1 \tag{4.228}$$

the total rest mass of photons in a blackbody radiation field is

$$\frac{m}{V} = \frac{2.404}{\pi^2} \left(\frac{k_B T}{\hbar c}\right)^3 m_\gamma \tag{4.229}$$

To evaluate the sum, you need the integral

$$\int_0^\infty \frac{x^2 dx}{e^x - 1} = \sum_{n=1}^\infty \frac{2}{n^3} = 2.404 \tag{4.230}$$

(it doesn't have an analytical value). Experimental data place an upper bound on the photon mass of $m_\gamma < 7 \times 10^{-52}$ kg. Within this upper bound, is the possible mass of the cosmic background radiation at 2.7 K enough to explain the "missing mass of the universe," that is, the mass required to close the universe, which corresponds to a density $m/V = 6 \times 10^{-27}$ kg/m^3 (about four hydrogen atoms per cubic meter)?

EXERCISE 4.15

In the free-electron theory of metals, the electron wave functions are plane waves whose eigenvalues are found by applying periodic boundary conditions at the boundaries of a rectangular volume

$$V = L_x L_y L_z \tag{4.231}$$

where L_x, L_y, and L_z are the lengths of the volume in the $\hat{\mathbf{x}}$, $\hat{\mathbf{y}}$, and $\hat{\mathbf{z}}$ directions. The allowed momentum vectors are then given by the (relativistically correct) de Broglie relation

$$\mathbf{p_n} = \hbar \mathbf{k_n} = \hbar \left(\frac{2\pi n_x}{L_x}, \frac{2\pi n_y}{L_y}, \frac{2\pi n_z}{L_z} \right) \tag{4.232}$$

for integer values of n_x, n_y, and n_z. Since electrons have half-integral spin, they are Fermions and obey the Pauli exclusion principle. Therefore, unlike photons, which have integral spin and are Bosons, only two electrons (corresponding to spin up and spin down) are allowed in each momentum state. At zero temperature the electrons settle into the lowest available energy states and fill the available levels up to some maximum value of the momentum p_F.

(a) If the number of electrons in the volume V is N, count up the number of available levels with $p < p_F$ to show that

$$p_F = \hbar (3\pi^2 n)^{1/3} \tag{4.233}$$

where $n = N/V$ is the density of electrons. Using the nonrelativistic expression

$$\mathcal{E} = \frac{p^2}{2m} \tag{4.234}$$

show that the Fermi energy $\mathcal{E}_F = p_F^2/2m$ is

$$\mathcal{E}_F = \frac{\hbar^2}{2m} (3\pi^2 n)^{2/3} \tag{4.235}$$

(b) Neutrons are also Fermions, and they obey the Pauli exclusion principle. Neutron stars therefore provide another example of a Fermion gas at low temperature, but in this case the Fermi energy is relativistic. Using the relativistic expression,

$$\mathcal{E}^2 = p^2 c^2 + m^2 c^4 \tag{4.236}$$

show that the Fermi energy (including the rest energy) is

$$\mathcal{E}_F = mc^2 \sqrt{1 + \left(3\pi^2 \frac{\hbar^3 n}{m^3 c^3}\right)^{2/3}} = mc^2 \sqrt{1 + \left(\frac{3}{8\pi}\lambda_C^3 n\right)^{2/3}} \tag{4.237}$$

where

$$\lambda_C = \frac{h}{mc} \tag{4.238}$$

is the Compton wavelength of a neutron, and $\lambda_C^3 n$ is the number of particles in a cubic Compton wavelength. In a neutron star, this is typically on the order of unity.

EXERCISE 4.16

Show that the integrated cross section for stimulated emission of a classical oscillator is

$$\int \sigma_{i \to j}\, d\omega = \frac{\pi q^2}{6\varepsilon_0 mc} \tag{4.239}$$

when averaged over all orientations of the oscillator.

4.3 WAVES IN PLASMAS

When electromagnetic waves travel through a plasma, a variety of new phenomena can occur. This is particularly true of a plasma imbedded in a magnetic field, like the van Allen belts in the earth's magnetosphere, for example. In the following discussion, however, we restrict ourselves to just a few of the simplest phenomena associated with electromagnetic waves in plasmas. These phenomena are important because plasmas are found in so many places from stellar coronas to fluorescent lamps, and especially in metals. Although the properties of free electrons in metals are modified by quantum effects and the presence of the lattice ions, the plasma effects are readily identifiable and serve to explain, at least qualitatively, many of the optical and electrical properties of metals, and even those of dielectrics at sufficiently high freqencies. There are two types of waves in simple plasmas, those that are transversely polarized and those that are longitudinally polarized. We discuss them in that order.

4.3.1 Transverse Electromagnetic Waves

The first case we consider is the propagation of a circularly polarized electromagnetic wave through a collisionless plasma. To be specific, we assume that the plasma is, on average, electrically neutral, consisting of an average number density n_0 of light particles (electrons) of charge q and mass m, and an average density n_0/Z of heavy particles (ions) of charge $-qZ$ that neutralize the plasma. We ignore the motions of the heavy particles in the following discussion.

We return to first principles, using the Maxwell equations in their vacuum form, and describe the wave by a 4-vector potential $A^\alpha = (\Phi/c, \mathbf{A})$. In the Coulomb gauge

$\nabla \cdot \mathbf{A} = 0$, so for plane waves the longitudinal component of the vector potential vanishes. We therefore represent the wave by the expression

$$A^{\alpha} = \text{Re}\left[\left(\frac{\Phi_0}{c}, \hat{\mathbf{e}}_{\pm} A_0\right) e^{-ik^{\alpha} r_{\alpha}}\right] \tag{4.240}$$

where $\hat{\mathbf{e}}_{\pm}$ is a circularly polarized basis vector oriented transverse to the direction of propagation of the wave vector $k^{\beta} = (\omega, \mathbf{k})$, as discussed earlier, and Φ_0 and $\mathbf{A}_0$ are constants. In general the wave vector k^{β} is not a null vector, which means that the velocity of propagation is not c. From the wave equation

$$\partial^{\beta} \partial_{\beta} \mathbf{A} = \mu_0 \mathbf{J} \tag{4.241}$$

we get

$$-k^{\beta} k_{\beta} \mathbf{A} = \left(k^2 - \frac{\omega^2}{c^2}\right) \mathbf{A} = \mu_0 \mathbf{J} \tag{4.242}$$

and from the Poisson equation

$$\nabla^2 \Phi = -\frac{\rho}{\varepsilon_0} \tag{4.243}$$

we get

$$k^2 \Phi = \frac{\rho}{\varepsilon_0} \tag{4.244}$$

To find $\mathbf{J}$ and ρ, we must determine the electron velocity $\mathbf{v} = \mathbf{p}/\gamma m$, where $\mathbf{p}$ is the electron momentum. This is found from the canonical equations of motion using the Hamiltonian

$$\mathcal{H} = \sqrt{(\mathbf{P} - e\mathbf{A})^2 + m^2 c^4} + q\Phi \tag{4.245}$$

where $\mathbf{P} = \mathbf{p} + q\mathbf{A}$ is the canonical momentum.

We look first for a solution with $\Phi = 0$. This implies that ρ vanishes everywhere, so the velocity in the longitudinal direction, which we take to be the $\hat{\mathbf{z}}$ direction, is at most a constant. As discussed earlier, when a plane wave is incident on an isolated charged particle, the particle is accelerated in the longitudinal direction by the interaction with the magnetic field. However, when a plane wave is incident on a plasma, space-charge forces prevent the electrons from moving very far, so we assume in the following that $\mathbf{p}_z = 0$. In this case the fields (other than the space-charge fields) are transverse to the direction of propagation. To find the current density $\mathbf{J}$, we note that the fields are invariant in the transverse direction, so the transverse canonical momentum of a particle in the field is conserved. The kinetic momentum in the transverse direction is therefore

$$\mathbf{p}_t = -q\mathbf{A} + \text{constant} = -q\mathbf{A} \tag{4.246}$$

in a coordinate system where the average transverse momentum vanishes. The total energy of the particles is then

$$\mathcal{E} = \gamma^2 m^2 c^4 = p^2 c^2 + m^2 c^4 = m^2 c^4 (1 + a^2) \tag{4.247}$$

where the dimensionless vector potential is

$$a^2 = \frac{q^2 A_0^2}{m^2 c^2} \tag{4.248}$$

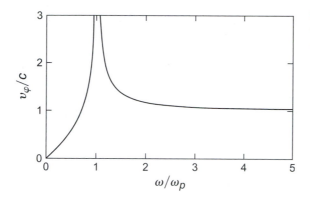

Figure 4.18 Dispersion relation for transverse electromagnetic waves in a plasma.

The current density is

$$\mathbf{J} = \frac{n_0 q}{\gamma m}\mathbf{p} = -\frac{n_0 q^2}{m\sqrt{1+a^2}}\mathbf{A} \tag{4.249}$$

Substituting this into the wave equation (4.68), we get the dispersion relation

$$\omega^2 - k^2 c^2 = \frac{\omega_p^2}{\sqrt{1+a^2}} \tag{4.250}$$

where the plasma frequency is

$$\omega_p^2 = \frac{n_0 q^2}{\varepsilon_0 m} \tag{4.251}$$

This is illustrated in Figure 4.18. Several interesting phenomena are embodied in the dispersion relation (4.250). For frequencies above the effective plasma frequency $\omega_p/\sqrt[4]{1+a^2}$, the propagation constant k is real and the waves propagate through the plasma with a phase velocity $v_\phi = \omega/k > c$. The group velocity, of course, is smaller than c. We also note that for an optical beam with finite transverse dimensions, the amplitude of the field is ordinarily greatest near the center of the beam. Near the center, then, the effective plasma frequency is smallest and the phase velocity

$$v_\phi^2 = \frac{\omega^2}{k^2} = c^2 + \frac{\omega_p^2}{k^2\sqrt{1+a^2}} \tag{4.252}$$

is slowest. This phenomenon allows the wave fronts near the edge to travel faster than those near the center, which causes nonlinear "self-focusing" of optical beams at high intensity.

At the effective plasma frequency the phase velocity becomes infinite, and below this frequency the propagation constant becomes imaginary. That is, the amplitude of the waves decays with distance at the rate

$$\alpha = \frac{1}{c}\sqrt{\frac{\omega_p^2}{\sqrt{1+a^2}} - \omega^2} \tag{4.253}$$

Since the plasma is collisionless, there are no dissipative processes to remove energy from the wave, so the decay is not a damping process. Rather, the wave is reflected by the plasma when the frequency is below the effective plasma frequency. This phenomenon is responsible for the reflection of low-frequency radio waves off the ionosphere, for example. In metals it is also responsible for the reflection of light in the infrared and visible parts of the spectrum. The plasma frequency for most metals is in the ultraviolet, and for frequencies above the plasma frequency metals become transparent, or would become transparent if it weren't for absorption processes caused by inner-shell electrons in the atoms. This is the origin of the ultraviolet transparency observed in alkali metals in the spectral region just above the plasma frequency, where inner-shell transitions are absent.

EXERCISE 4.17

What is the relation (phase and amplitude) between $\mathbf{B}$ and $\mathbf{E}$ in a transverse electromagnetic wave passing through a neutral, collisionless plasma at frequencies above, at, and below the plasma frequency?

EXERCISE 4.18

Consider the propagation of a circularly polarized wave

$$\mathbf{E}' = \mathrm{Re}[E'\hat{\mathbf{e}}_{\pm}e^{-ik^{\alpha}r_{\alpha}}] \tag{4.254}$$

$$\mathbf{B}' = \mathrm{Re}[B'\hat{\mathbf{e}}_{\pm}e^{-ik^{\alpha}r_{\alpha}}] \tag{4.255}$$

in which E' and B' are constants and the wave vector is

$$k^{\alpha} = \left(\frac{\omega}{c}, k\hat{\mathbf{n}}\right) \tag{4.256}$$

through a neutral, collisionless plasma in the direction parallel to the constant magnetic field

$$\mathbf{B}_0 = B_0\hat{\mathbf{n}} \tag{4.257}$$

(a) Use the Maxwell equations to show that

$$E' = \pm i\frac{\omega}{k}B' \tag{4.258}$$

and

$$\mathrm{Re}\left[\left(k^2 - \frac{\omega^2}{c^2}\right)B'\hat{\mathbf{e}}_{\pm}e^{-ik^{\alpha}r_{\alpha}}\right] = \pm\mu_0 k\mathbf{J} \tag{4.259}$$

where the current is

$$\mathbf{J} = n_0 q\mathbf{v} \tag{4.260}$$

in which n_0 is the density, q the charge, and $\mathbf{v}$ the velocity of the electrons.

(b) Linearize the equation of motion

$$\frac{d\mathbf{p}}{dt} = \frac{\partial\mathbf{p}}{\partial t} + (\mathbf{v}\cdot\nabla)\mathbf{p} = q[\mathbf{E}' + \mathbf{v}\times(\mathbf{B}' + \mathbf{B}_0)] \tag{4.261}$$

of the electrons to show that to lowest order in the amplitude of the wave, the current is

$$\mathbf{J} = \mp \frac{n_0 q^2}{m} \frac{\omega}{k} \frac{1}{\omega \mp \omega_c} \mathrm{Re}[B' \hat{\mathbf{e}}_\pm e^{-ik^\alpha r_\alpha}] \qquad (4.262)$$

where

$$\omega_c = -\frac{q B_0}{m} \qquad (4.263)$$

is the electron cyclotron frequency. The negative sign is introduced in (4.263) because (negatively charged) electrons are left circularly polarized about the direction of the longitudinal magnetic field. This choice makes the cyclotron frequency ω_c positive if B_0 is positive. *Hint:* Since both the circularly polarized wave and the longitudinal magnetic field give rise to circular motions of the electrons around the direction of the field, we can assume that the transverse momentum has the form

$$\mathbf{p} = \mathrm{Re}[p' \hat{\mathbf{e}}_\pm e^{-ik^\alpha r_\alpha}] \qquad (4.264)$$

for some constant p'. To lowest order in the amplitude of the wave, the electron transverse motions are nonrelativistic and longitudinal motions can be ignored.

(c) Substitute this into the result of (a) to derive the dispersion relation

$$\omega^2 - k^2 c^2 = \omega_p^2 \frac{\omega}{\omega \mp \omega_c} \qquad (4.265)$$

where

$$\omega_p^2 = \frac{n_0 q^2}{\varepsilon_0 m} \qquad (4.266)$$

is the plasma frequency. This is illustrated in Figure 4.19 for the case $\omega_c/\omega_p = 5$.

(d) Show that the phase velocity diverges at the frequencies

$$\omega = \sqrt{\omega_p^2 + \tfrac{1}{4}\omega_c^2} \pm \tfrac{1}{2}\omega_c \qquad (4.267)$$

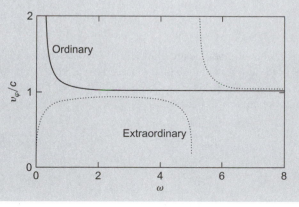

Figure 4.19 Dispersion relation for circularly polarized waves in a magnetized plasma.

In the literature, waves that rotate in the same direction as the electrons are sometimes called extraordinary waves and those that rotate opposite the electrons are called ordinary waves. For a magnetic field in the direction of propagation $(B_0 > 0)$, left-hand circularly polarized waves (positive helicity) are extraordinary waves and right-hand circularly polarized waves (negative helicity) are ordinary waves. The upper sign in (4.265) and (4.267) corresponds to extraordinary waves and the lower sign to ordinary waves.

4.3.2 Longitudinal Electrostatic Waves

Next we look for wavelike solutions in which the vector potential $\mathbf{A}$ vanishes and the scalar potential Φ has the form of a plane wave. The transverse electric and magnetic fields therefore vanish, and only the longitudinal electric field remains. In this case we don't need the wave equation; we can find the scalar potential just from the Poisson equation (4.243). To find the net charge density $\rho = q(n - n_0)$, we use the continuity relation

$$\frac{\partial \rho}{\partial t} = -\nabla \cdot \mathbf{J} = -\nabla \cdot \left(\frac{nq\mathbf{p}}{\gamma m} \right) \tag{4.268}$$

and to find the momentum in the longitudinal direction, we use the canonical equation of motion

$$\frac{d\mathbf{p}}{dt} = -q\nabla\Phi \tag{4.269}$$

where the total time derivative of the function $f(\mathbf{r}, t)$ is

$$\frac{df}{dt} = \frac{\partial f}{\partial t} + \mathbf{v} \cdot \nabla f \tag{4.270}$$

To solve these equations, we linearize (4.268) and (4.269), which also restricts us to the nonrelativistic limit. That is, we assume that $n \approx n_0$ and $\gamma \approx 1$ in (4.268), and ignore the second term in (4.270). We then get

$$\frac{\partial \rho}{\partial t} = -\frac{n_0 q}{m} \nabla \cdot \mathbf{p} \tag{4.271}$$

$$\frac{\partial \mathbf{p}}{\partial t} = -q\nabla\Phi \tag{4.272}$$

Since (4.244), (4.271), and (4.272) are all linear, it is convenient to work with the complex expressions for ρ, $\mathbf{p}$, and Φ, and then take the real part at the end, although we do not need to do this last step. If ρ and $\mathbf{p}$, like Φ, depend on time and distance only through the factor $e^{-ik^\alpha r_\alpha}$, then these equations become

$$\frac{\partial \rho}{\partial t} = -i\omega\rho = -i\frac{n_0 q}{m} \mathbf{k} \cdot \mathbf{p} \tag{4.273}$$

and

$$\frac{\partial \mathbf{p}}{\partial t} = -i\omega\mathbf{p} = -iq\mathbf{k}\Phi \tag{4.274}$$

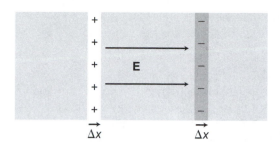

Figure 4.20 Electrostatic oscillations in a plasma.

Substituting back into (4.244), we find that

$$\Phi = \frac{n_0 q^2}{\varepsilon_0 m \omega^2} \Phi \tag{4.275}$$

That is,

$$\omega^2 = \frac{n_0 q^2}{\varepsilon_0 m} = \omega_p^2 \tag{4.276}$$

independent of the wavelength.

The existence of a characteristic frequency for electrostatic oscillations in a plasma was first discovered by Tonks and Langmuir in 1929. A physical picture of these oscillations is illustrated in Figure 4.20. Conceptually, we take the electrons in the volume between two parallel planes and translate them a distance Δx normal to the planes. This leaves behind a region of positive charge due to the ions, and creates a region of negative charge where the electron density is doubled. The field in the intervening region is

$$E = -\frac{n_0 q \, \Delta x}{\varepsilon_0} \tag{4.277}$$

The force restoring the electrons to their original position is then

$$F = qE = -\frac{n_0 q^2 \Delta x}{\varepsilon_0} = m \frac{d^2 \Delta x}{dt^2} \tag{4.278}$$

But this is just the equation for a simple harmonic oscillator. We therefore see that the electrons oscillate back and forth in the longitudinal direction at the plasma frequency ω_p given by (4.276).

Since the frequency of a plasma wave is independent of the wave number k, the phase velocity $v_\phi = \omega/k$ can have any value. For the same reason, the group velocity $v_g = d\omega/dk$ vanishes. Aside from the effects of thermal motions, classical plasma waves do not propagate.

We note in passing that the energy in the electric field is just equal to the electron kinetic energy, when averaged over a cycle of the oscillation. This is easy to see from (4.277) and (4.278), but it is also clear from the fact that the electrons and the field form a closed, conservative system. Since the field is at its maximum when the electrons are stopped at the limits of their motion and vanishes when the electrons are at their maximum velocity, the energy of the system simply oscillates back and forth between field energy and kinetic energy.

In metals, the electron density is high and quantum effects are important. Quantized electrostatic oscillations of the conduction electrons in a metal are called plasmons, and the quantum of energy is

$$\mathcal{E} = \hbar\omega_p \tag{4.279}$$

The quanta of typical metals range from $\mathcal{E} = 3.45$ eV in rubidium to $\mathcal{E} = 18.4$ eV in beryllium. The quantization of the plasmon energy can be observed experimentally when multikilovolt electron beams pass through metallic foils and excite plasma oscillations. The energy-loss spectra show multiple peaks spaced at the plasmon energy. Other quantum effects are also noticeable. Due to the exclusion principle, the electrons that oscillate into a region that is already filled by other electrons must occupy higher energy levels lying above the Fermi level. This is equivalent, in a sense, to increasing the restoring force on the electrons, and it has the effect of increasing the plasma frequency. The result is described approximately by the dispersion relation

$$\omega^2 \approx \omega_p^2 + \tfrac{3}{5}v_F^2 k^2 \tag{4.280}$$

where

$$v_F = \frac{\hbar}{m}(3\pi^2 n_0)^{1/3} \tag{4.281}$$

is the Fermi velocity, as discussed earlier. Clearly, the effect is most important for short-wavelength (small-phase-velocity) plasmons. For a plasmon with the phase velocity $v_\phi = \omega_p/k = c$ the correction is only $\Delta\omega/\omega \approx (3/10)(v_F^2/c^2) \approx 2 \times 10^{-5}$ in aluminum.

Another quantum effect that is generally more important than this arises from the band structure of the electron energy levels. The band structure of crystalline solids modifies the relation between the electron momentum and its wave number. An elegant way to represent this effect is called the effective mass. By analogy with the discussion of de Broglie waves in Chapter 2, we see that the electron momentum $p = \hbar k$ can be related to the group velocity by an effective mass m^*, where

$$p = \hbar k = m^* v_g = m^* \frac{d\omega}{dk} = \frac{m^*}{\hbar}\frac{d\mathcal{E}}{dk} \tag{4.282}$$

and $\mathcal{E} = \hbar\omega$ is the electron energy. If we differentiate this expression with respect to k, we get

$$\frac{1}{m^*} = \frac{1}{\hbar^2}\frac{d^2\mathcal{E}}{dk^2} \tag{4.283}$$

so the effective mass is related to the curvature of the function $\mathcal{E}(\mathbf{k})$. Due to the periodic nature of the crystal lattice, the dispersion relation $\mathcal{E}(\mathbf{k})$ for the electrons becomes rather complicated. At the edges of the so-called Brillouin zones in $\mathbf{k}$-space, gaps appear in the spectrum of allowed energy levels as indicated in Figure 4.21. These gaps separate the allowed energy levels into bands. If the electron energy is not too close to a band gap, the dispersion relation remains approximately

$$\mathcal{E} \approx \frac{\hbar^2 k^2}{2m} \tag{4.284}$$

so the effective mass for electrons in these levels is just the mass of a free electron. In metals, where the conduction band is only partially filled, the Fermi level lies somewhere in the middle of the conduction band, far from a band edge, and the effective mass is very nearly

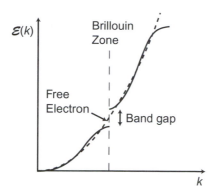

Figure 4.21 Band structure of the electron energy in a periodic lattice.

that of a free electron. In semiconductors, however, the Fermi level lies in or near the band gap between the valence and conduction bands, where the dispersion relation is distorted by the presence of the band gap. In semiconductors, therefore, the effective mass is typically much smaller than the electron mass, and may even be negative.

EXERCISE 4.19

When the thermal motions of the electrons in a plasma are important, the dispersion relation for the plasma waves becomes more complex and allows for both propagation and damping of the waves. In the simplest case of a nonrelativistic plasma in which the electron–electron collision frequency is very high and the electron–ion collision frequency can be ignored, the thermal motions contribute just an isotropic pressure term to the equations of motion. In the following we ignore heat conduction, viscosity, and other dissipative processes.

(a) Consider a group of $N_e = \text{constant} = n_e V$ electrons occupying a (macroscopically very small) volume V, where n_e is the electron density. Show that the net pressure force on this volume of electrons is $-V\nabla p_e$, where $p_e = n_e k_B T_e$ is the electron pressure, k_B Boltzmann's constant, and T_e the electron temperature. Show that the equation of motion for the mean velocity $\mathbf{v}$ of the electrons is

$$n_e m \frac{d\mathbf{v}}{dt} = -\nabla p_e - n_e q \nabla \Phi \tag{4.285}$$

where m is the electron mass and Φ the scalar potential, and the total time derivative is $d/dt = \partial/\partial t + \mathbf{v} \cdot \nabla$.

(b) Consider the internal energy $\frac{3}{2} N_e k_B T_e$ of this same group of electrons and the work $dW = p_e dV$ they do on the surrounding electrons as they expand the volume they occupy. Use the first law of thermodynamics (conservation of energy) to show that

$$n_e \frac{d}{dt}\left(\frac{3}{2} k_B T_e\right) = k_B T_e \frac{dn_e}{dt} \tag{4.286}$$

or, equivalently, that

$$\frac{dp_e}{dt} = \frac{5}{3} k_B T_e \frac{dn_e}{dt} \tag{4.287}$$

(c) Linearize these equations together with the continuity equation and the Poisson equation for small perturbations from the static conditions $n_e = n_0$, $T_e = T_0$, $\mathbf{v} = 0$, and $\Phi = 0$. Show that for longitudinal waves of the form $e^{ik^\alpha r_\alpha} = e^{i(\mathbf{k}\cdot\mathbf{r}-\omega t)}$, the dispersion relation (4.276) becomes

$$\omega^2 = \omega_p^2 + \frac{5}{3}\frac{k_B T_0}{m}k^2 = \omega_p^2\left(1 + \frac{5}{3}h^2 k^2\right) \tag{4.288}$$

where $\omega_p = \sqrt{n_0 q^2/\varepsilon_0 m}$ is the plasma frequency and

$$h = \sqrt{\frac{\varepsilon_0 k_B T_0}{n_0 q^2}} \tag{4.289}$$

is called the Debye shielding distance. This is roughly the size of a space-charge region over which the potential energy is comparable to the thermal energy. We see from (4.288) that the waves move from the plasma regime (dominated by electrostatics) to the acoustic regime (dominated by pressure) when the wavelength is comparable to the Debye shielding distance [$hk = O(1)$].

(d) Show that the phase velocity $v_\phi = \omega/k$ and the group velocity $v_g = d\omega/dk$ are given by

$$v_\phi^2 = \frac{5}{3}h^2\omega_p^2\frac{\omega^2}{\omega^2 - \omega_p^2} \tag{4.290}$$

and

$$v_g^2 = \frac{5}{3}h^2\omega_p^2\frac{\omega^2 - \omega_p^2}{\omega^2} \tag{4.291}$$

Both the phase velocity and the group velocity are imaginary for $\omega < \omega_p$, so waves cannot propagate below the plasma frequency. This is also evident from (4.288).

BIBLIOGRAPHY

The properties of electromagnetic waves in vacuum and the motions of particles in electromagnetic waves are discussed by

L. D. Landau and E. M. Lifshitz, *The Classical Theory of Fields,* 4th edition, Pergamon Press, Oxford (1975).

The quantum-mechanical aspects of radiative processes in atoms are treated in some depth by most intermediate-level textbooks, including

R. M. Eisberg and R. Resnick, *Quantum Physics of Atoms, Molecules, Solids, Nuclei, and Particles,* John Wiley and Sons, New York (1985),

K. Gottfried, *Quantum Mechanics,* Volume I, 6th printing, Benjamin/Cummings Publishing Company, Reading MA (1979).

Plasma physics is of interest for both technical and astrophysical applications. Brief and very readable introductions are found in

S. Chandrasekhar, *Plasma Physics,* University of Chicago Press, Chicago (1960),

L. Spitzer, Jr., *Physics of Fully Ionized Gases,* Interscience Publishers, New York (1956).

The physics of the free-electron gas and the band theory of crystalline solids are explained very clearly by

C. Kittel, *Introduction to Solid State Physics,* 7th edition, John Wiley & Sons, New York (1996).

5 Fourier Techniques and Virtual Quanta

Extending beyond the physical importance of electromagnetic waves themselves is the mathematical importance of the concept of waves, as it is used in the description of electromagnetic fields other than light. For example, when a charged particle passes an atom or a nucleus, it exerts a brief force on the atom or nucleus that can push the constituents of the target into more highly excited states. But the electromagnetic pulse from the passing particle can be resolved into electromagnetic waves, which we call virtual quanta, and the excitation of the target computed from the effects of the waves. The importance of these notions actually extends well beyond electrodynamics, and a corpus of mathematical tools based on plane waves has now been highly developed. Called Fourier analysis, these techniques are so powerful and broadly useful that physicists now describe many phenomena in terms of waves. For example, pulses of all types are described in terms of the waves of which they are composed, and the propagation of the pulses is computed from the phase shift of the waves. In addition, statistical fluctuations of all kinds are described in terms of their spectra. This chapter is dedicated to the application of Fourier analysis to electrodynamics, but we see the application to other phenomena as we go along.

5.1 FOURIER TRANSFORMATION

5.1.1 Fourier's Theorem

One of the most powerful and useful theorems in mathematical analysis is Fourier's inversion theorem. Simply stated, this theorem allows us to decompose a broad variety of arbitrary functions into waves and then to reconstruct the original function from the waves. Specifically, Fourier's theorem states that the function $f(t)$ may be represented as an integral of the form

$$f(t) = \frac{1}{\sqrt{2\pi}} \int_{-\infty}^{\infty} d\omega\, \tilde{f}(\omega) e^{-i\omega t} \tag{5.1}$$

where the Fourier transform is defined by the integral

$$\tilde{f}(\omega) = \frac{1}{\sqrt{2\pi}} \int_{-\infty}^{\infty} dt\, f(t) e^{i\omega t} \tag{5.2}$$

The first expression says that the time-dependent function $f(t)$ can be constructed from the sum of oscillating functions of the type $e^{-i\omega t}$ with the weighting function $\tilde{f}(\omega)$, which we call the Fourier transform, and the second expression is a formula for how to compute the weighting function.

Note carefully the conventions used in (5.1) and (5.2), since they are not universal! In particular, the Fourier transform (5.2) uses $+i\omega t$ in the exponent, whereas the inversion (5.1) uses $-i\omega t$. Also, the normalization factor $1/\sqrt{2\pi}$ appears symmetrically in front of the integral in both the transform and its inversion, and there is no factor of 2π in the exponent.

The inversion theorem is easy to prove for functions $f(t)$ that have a finite (but not necessarily continuous) derivative everywhere and are absolutely integrable (that is, the integral of $|f(t)|$ is finite) so that the Fourier transform exists. We begin by observing that we may write the left-hand side of (5.1) in the form

$$f(t) = \lim_{\varepsilon \to 0} \frac{1}{2\sqrt{\pi \varepsilon}} \int_{-\infty}^{\infty} dt' \, f(t) e^{-(t'-t)^2/4\varepsilon} \tag{5.3}$$

since $f(t)$ can be moved outside the integral over t' and the value of the integral together with the other factors is unity for any value of ε. But the right-hand side of (5.1) may be written

$$\frac{1}{\sqrt{2\pi}} \int_{-\infty}^{\infty} d\omega \, \tilde{f}(\omega) e^{-i\omega t} = \lim_{\varepsilon \to 0} \frac{1}{\sqrt{2\pi}} \int_{-\infty}^{\infty} d\omega \, \tilde{f}(\omega) e^{-i\omega t} e^{-\varepsilon \omega^2}$$

$$= \lim_{\varepsilon \to 0} \frac{1}{2\pi} \int_{-\infty}^{\infty} d\omega \int_{-\infty}^{\infty} dt' \, f(t') e^{i\omega(t'-t)-\varepsilon \omega^2} \tag{5.4}$$

If we complete the square in the exponent and invert the order of integration, we get

$$\frac{1}{\sqrt{2\pi}} \int_{-\infty}^{\infty} d\omega \, \tilde{f}(\omega) e^{-i\omega t} = \lim_{\varepsilon \to 0} \frac{1}{2\sqrt{\pi \varepsilon}} \int_{-\infty}^{\infty} dt' \, f(t') e^{-(t'-t)^2/4\varepsilon} \tag{5.5}$$

Therefore, provided that $f'(t) \leq f'_{\max}$ everywhere, as stipulated in the beginning, we see that

$$\left| f(t) - \frac{1}{\sqrt{2\pi}} \int_{-\infty}^{\infty} d\omega \, \tilde{f}(\omega) e^{-i\omega t} \right| = \lim_{\varepsilon \to 0} \left| \frac{1}{2\sqrt{\pi \varepsilon}} \int_{-\infty}^{\infty} dt' \, [f(t) - f(t')] e^{-(t'-t)^2/4\varepsilon} \right|$$

$$\leq \lim_{\varepsilon \to 0} \left| \frac{f'_{\max}}{2\sqrt{\pi \varepsilon}} \right| \left| \int_{-\infty}^{\infty} dt' |t - t'| e^{-(t'-t)^2/4\varepsilon} \right|$$

$$= \lim_{\varepsilon \to 0} \left| \frac{f'_{\max}}{2\sqrt{\pi \varepsilon}} \right| 4\varepsilon = 0 \tag{5.6}$$

Actually, the Fourier inversion theorem remains true for a broader class of functions than has been assumed here, including functions with a finite number of finite discontinuities, and so on. In fact, it can be extended to "generalized functions" such as the Dirac δ-function, which are defined only by some limit process. It seems that the class of functions for which Fourier's theorem is valid is broader than can be shown by any single proof, but the simple proof used here is sufficient for most of the cases that arise in physics. In other cases, some of which are encountered later, it is necessary to exercise caution. It is often possible to avoid singularities in the integral by displacing the path of integration in the complex ω-plane, and we have occasion to do this in later sections.

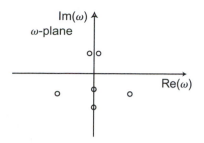

Figure 5.1 Singularities of the Fourier transform of a real function.

If the function $f(t)$ is real, then we see from (5.2) that its Fourier transform has the symmetry property

$$\tilde{f}^*(\omega) = \frac{1}{\sqrt{2\pi}} \int_{-\infty}^{\infty} dt \, f(t) e^{-i\omega^* t} = \tilde{f}(-\omega^*) \tag{5.7}$$

In particular, if $f(t)$ is real, then the singularities of $\tilde{f}(\omega)$ are distributed symmetrically about the imaginary axis in the complex ω-plane as shown in Figure 5.1.

As an example of Fourier's theorem, we consider the convolution of the functions $f(t)$ and $g(t)$

$$\int_{-\infty}^{\infty} dt \, f(\tau - t) g(t) = \frac{1}{\sqrt{2\pi}} \int_{-\infty}^{\infty} dt \, g(t) \int_{-\infty}^{\infty} d\omega \, \tilde{f}(\omega) e^{-i\omega(\tau - t)}$$

$$= \frac{1}{\sqrt{2\pi}} \int_{-\infty}^{\infty} d\omega \, \tilde{f}(\omega) e^{-i\omega\tau} \int_{-\infty}^{\infty} dt \, g(t) e^{i\omega t}$$

$$= \int_{-\infty}^{\infty} d\omega \, \tilde{f}(\omega) \tilde{g}(\omega) e^{-i\omega\tau} \tag{5.8}$$

This is called the Faltung (German for "folding") theorem.

As a special case of the Faltung theorem, we take $\tau = 0$ and $g(t) = f^*(-t)$, so that

$$\tilde{g}(\omega) = \frac{1}{\sqrt{2\pi}} \int_{-\infty}^{\infty} dt \, f^*(t) e^{-i\omega t} = \tilde{f}^*(\omega^*) \tag{5.9}$$

We then get

$$\int_{-\infty}^{\infty} dt \, f(t) f^*(t) = \int_{-\infty}^{\infty} d\omega \, \tilde{f}(\omega) \tilde{f}^*(\omega^*) \tag{5.10}$$

This is called Parseval's theorem. When the function $f(t)$ is real and the integral in the complex ω-plane is along the real axis ($\omega = \omega^*$), we may use the symmetry property (5.7) to eliminate the integral over negative frequencies, for then we have $\tilde{f}(\omega)\tilde{f}^*(\omega^*) = \tilde{f}(\omega)\tilde{f}(-\omega)$, which is symmetric. We therefore get

$$\int_{-\infty}^{\infty} dt \, f^2(t) = 2 \int_{0}^{\infty} d\omega \, |\tilde{f}(\omega)|^2 \tag{5.11}$$

This shows that in the real world, where we are concerned with real frequencies ω and real functions like the field $E(t)$, when we compute quantities like the total energy $\int |\tilde{E}^2(\omega)| \, d\omega$, we don't have to worry about negative frequencies, which are hard to imagine experimentally.

EXERCISE 5.1

The Fourier inversion formula may be regarded as the limit

$$f(t) = \frac{1}{\sqrt{2\pi}} \int_{-\infty}^{\infty} \tilde{f}(\omega) e^{-i\omega t} \, d\omega = \lim_{\omega_0 \to \infty} \frac{1}{\sqrt{2\pi}} \int_{-\omega_0}^{\omega_0} \tilde{f}(\omega) e^{-i\omega t} \, d\omega \qquad (5.12)$$

(a) Substitute for $\tilde{f}(\omega)$ to derive Fourier's single-integral formula

$$f(t) = \lim_{\omega_0 \to \infty} \frac{1}{\pi} \int_{-\infty}^{\infty} f(t') \frac{\sin[\omega_0(t - t')]}{(t - t')} \, dt' \qquad (5.13)$$

In a sense, then, the function $\sin[\omega_0(t - t')]/(t - t')$ has the properties of a Dirac δ-function, at least in the limit $\omega_0 \to \infty$, and illustrates the intimate connection between Fourier transforms and δ-functions. From this formula we get the representation

$$\delta(x) = \lim_{\omega_0 \to \infty} \frac{1}{\pi} \frac{\sin[\omega_0 x]}{x} \qquad (5.14)$$

Sketch this function for a few increasing values of ω_0 to see how it converges.

(b) Consider the function

$$f(t) = 0, \qquad \text{for } t < 0 \quad \text{or} \quad t > 1$$
$$f(t) = 1, \qquad \text{for } 0 < t < 1 \qquad (5.15)$$

as shown in Figure 5.2. Show that

$$f(t) = \lim_{\omega_0 \to \infty} \left[1 - \frac{1}{\pi} \int_{\omega_0 t}^{\infty} \frac{\sin \xi}{\xi} \, d\xi - \frac{1}{\pi} \int_{\omega_0(1-t)}^{\infty} \frac{\sin \xi}{\xi} \, d\xi \right] \qquad (5.16)$$

Evaluate this for $t = 0, 1$ to show that

$$f(0) = f(1) = \tfrac{1}{2} \qquad (5.17)$$

That is, Fourier's integral formula is defined even where the original function is not.

(c) The convergence of Fourier's single integral formula displays some interesting behavior around discontinuities, known as Gibbs' phenomenon. Consider times t

Figure 5.2 Gibbs' phenomenon for $\omega_0 = 100$.

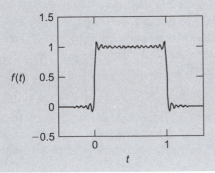

near the discontinuity at the origin, that is, $t = O(1/\omega_0)$. In the limit $\omega_0 \to \infty$, the second integral in (5.16) vanishes but the first integral oscillates. Show that

$$f(t) = \lim_{\omega_0 \to \infty} \left[\frac{1}{2} + \frac{1}{\pi} \int_0^{\omega_0 t} \frac{\sin \xi}{\xi} d\xi \right] \to \begin{cases} 1 & \text{for } t > 0 \\ 0 & \text{for } t < 0 \end{cases} \quad (5.18)$$

Show that the function overshoots to a peak value

$$f_{\max} = f\left(\frac{\pi}{\omega_0} \right) = 1.089 \quad (5.19)$$

at $\xi = \pi = \omega_0 t$. The overshoot is essentially independent of ω_0, but the peak moves toward the discontinuity as $\omega_0 \to \infty$. Thus, the oscillations do not diminish in the limit $\omega_0 \to \infty$ but collapse around the discontinuity, as shown in Figure 5.2. Similar considerations apply at the discontinuity at $t = 1$.

Gibbs' phenomenon is also observed in the summing of Fourier series

$$f(t) = \lim_{N \to \infty} \sum_{n=0}^{N} [a_n \sin(n\omega t) + b_n \cos(n\omega t)] \quad (5.20)$$

The connection with Fourier transforms is apparent when we compare this formula to (5.12). From the oscillations observed in Figure 5.2, even for $\omega_0 = 100$, it is apparent that Gibbs' phenomenon is quite persistent. This illustrates the dangers that lurk in the evaluation of Fourier transforms and Fourier series.

5.1.2 Asymptotic Behavior of Fourier Transforms

It is frequently important to understand the asymptotic behavior of a Fourier transform in the limit $\omega \to \infty$. For example, it may be necessary to understand the behavior of the spectrum from some physical event in the limit of high frequencies. Also, whenever we need to evaluate a Fourier transform by contour integration in the complex plane we need to know that the Fourier transform vanishes at infinity. This is the subject of the Riemann–Lebesgue lemma, which states that if a function $f(t)$ is absolutely integrable and continuous everywhere, then its Fourier transform vanishes at high frequencies:

$$\lim_{\omega \to \infty} |\tilde{f}(\omega)| = 0 \quad (5.21)$$

To prove the lemma, we note that $\exp(i\pi) = -1$, so we may write

$$\tilde{f}(\omega) = -\frac{1}{\sqrt{2\pi}} \int_{-\infty}^{\infty} f(t) e^{i\omega(t+(\pi/\omega))} \, dt = -\frac{1}{\sqrt{2\pi}} \int_{-\infty}^{\infty} f\left(t - \frac{\pi}{\omega} \right) e^{i\omega t} dt \quad (5.22)$$

We therefore see that

$$|\tilde{f}(\omega)| = \frac{1}{2\sqrt{2\pi}} \left| \int_{-\infty}^{\infty} \left[f(t) - f\left(t - \frac{\pi}{\omega} \right) \right] e^{i\omega t} \, dt \right|$$

$$\leq \frac{1}{2\sqrt{2\pi}} \int_{-\infty}^{\infty} \left| f(t) - f\left(t - \frac{\pi}{\omega} \right) \right| dt \quad (5.23)$$

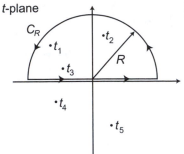

Figure 5.3 Fourier transform of a smooth function.

But the right-hand side vanishes as $\omega \to \infty$, since the function $f(t)$ is continuous by hypothesis, which proves the lemma. The lemma is actually valid for a function $f(t)$ that has a finite number of finite discontinuities. Between the singularities, the integral vanishes as shown. Around a singularity, $|f(t) - f(t - \pi/\omega)|$ approaches the size of the discontinuity as $\omega \to \infty$. Therefore, the integral vanishes as we allow the interval around the singularity to shrink.

How fast $\tilde{f}(\omega) \to 0$ depends on the function $f(t)$. If $f(t)$ is a smooth function, having at most a finite number of singularities in the complex plane with none on the real axis, then the Fourier transform vanishes exponentially. Otherwise, $\tilde{f}(\omega) \to 0$ algebraically.

To demonstrate the exponential asymptotic decay of the Fourier transform of a smooth function, we assume that $f(t)$ has a finite number of singularities in the complex t-plane, as shown in Figure 5.3, with none on the real axis and no branch points. To evaluate the transform for $\omega > 0$, we use the contour shown in Figure 5.3, consisting of a path along the real axis and a semicircle enclosing all the singularities in the upper half plane. The Fourier transform is then

$$\tilde{f}(\omega) = \frac{1}{\sqrt{2\pi}} \lim_{R \to \infty} \int_{-R}^{R} dt \, f(t) e^{i\omega t}$$

$$= \lim_{R \to \infty} \oint dt \, f(t) e^{i\omega t} - \lim_{R \to \infty} \int_{C_R} dt \, f(t) e^{i\omega t} \qquad (5.24)$$

where C_R is the semicircular contour. But for $\omega > 0$,

$$|f(t)e^{i\omega t}| = |f(t)|e^{-\omega y} \xrightarrow[R \to \infty]{} 0 \qquad (5.25)$$

where we have written $t = x + iy$, so the integral around the semicircular contour vanishes for large R. Therefore, by Cauchy's residue theorem, the Fourier transform is

$$\tilde{f}(\omega) = \frac{1}{\sqrt{2\pi}} \oint dt \, f(t) e^{i\omega t} = i\sqrt{2\pi} \sum_n \text{residue}[f(t_n)e^{i\omega t_n}] \qquad (5.26)$$

where the sum is to be taken over all the poles of $f(t)$ in the upper half plane. But

$$|\text{residue}[f(t_n)e^{i\omega t_n}]| = |\text{residue}[f(t_n)]|e^{-\omega y_n} \qquad (5.27)$$

for real frequencies, so

$$|\tilde{f}(\omega)| \leq \sqrt{2\pi} \sum_n |\text{residue}[f(t_n)]|e^{-\omega y_n} \xrightarrow[\omega \to \infty]{} \sqrt{2\pi} \, |\text{residue}[f(t_{min})]|e^{-\omega y_{min}} \qquad (5.28)$$

where t_{min} is the pole closest to the real axis. For $\omega < 0$, a similar result is obtained by closing the contour in the lower half plane. Thus, for smooth functions (functions with no singularities on the real axis) the Fourier transform vanishes exponentially as $\omega \to \pm\infty$.

For functions with discontinuities, or discontinuities in the nth derivative, the Fourier transforms vanish algebraically as $\omega \to \infty$. For example, for the square function

$$f(t) = \begin{cases} 1 & \text{for } |t| < 1 \\ 0 & \text{otherwise} \end{cases} \tag{5.29}$$

which has discontinuities at $t = -1$ and at $t = 1$, the Fourier transform is

$$\tilde{f}(\omega) = \sqrt{\frac{2}{\pi}} \frac{\sin(\omega)}{\omega} \tag{5.30}$$

which vanishes like ω^{-1}. In the same way, for the triangle function

$$f(t) = \begin{cases} 1 - |t| & \text{for } |t| < 1 \\ 0 & \text{otherwise} \end{cases} \tag{5.31}$$

which has discontinuities in the first derivative, the Fourier transform is

$$\tilde{f}(\omega) = i\sqrt{\frac{2}{\pi}} \frac{\sin(\omega)}{\omega^2} \tag{5.32}$$

which vanishes like ω^{-2}, and so on.

In the general case, we consider a function $f(t)$ that is absolutely integrable and whose first $n + 1$ derivatives are everywhere finite and continuous, except that the nth derivative has (and possibly higher derivatives have) a finite number of discontinuities at times $t = t_k$. We let

$$\Delta f^{(n)}(t_k) = \lim_{\varepsilon \to 0} \left[f^{(n)}(t_k + \varepsilon) - f^{(n)}(t_k - \varepsilon) \right] \tag{5.33}$$

be the discontinuity in the derivative $f^{(n)}(t)$ at $t = t_k$. Integrating by parts $n + 1$ times, we get

$$\tilde{f}(\omega) = \frac{1}{\sqrt{2\pi}} \int_{-\infty}^{\infty} f(t) e^{i\omega t} \, dt$$

$$= \frac{1}{\sqrt{2\pi}} \left[-\left(\frac{i}{\omega}\right) \sum_{m=0}^{n} \left(\frac{i}{\omega}\right)^m f^{(m)}(t) e^{i\omega t} \Big|_{t=-\infty}^{t=\infty} + \left(\frac{i}{\omega}\right)^{n+1} \int_{-\infty}^{\infty} f^{(n+1)}(t) e^{i\omega t} \, dt \right] \tag{5.34}$$

But if $f^{(m)}(t)$ is absolutely integrable, it must vanish at $t = \pm\infty$, so the first term vanishes. To evaluate the second term, we recognize that if $f^{(n)}(t)$ has a discontinuity at $t = t_k$, then $f^{(n+1)}(t)$ has a δ-function there. Where $f^{(n+1)}(t)$ is continuous, we can use the arguments of the Reimann–Lebesgue lemma to show that the integral vanishes in the limit $\omega \to \infty$, which leaves the integrals over the δ-functions. The Fourier transform therefore has the asymptotic behavior

$$\tilde{f}(\omega) \xrightarrow[\omega \to \infty]{} \frac{1}{\sqrt{2\pi}} \left(\frac{i}{\omega}\right)^{n+1} \sum_k \Delta f^{(n)}(t_k) e^{i\omega t_k} \tag{5.35}$$

The simple examples given here are seen to be special cases of this result. We have occasion to use (5.35) in later chapters.

EXERCISE 5.2

Nature abhors discontinuities, and the mathematical discontinuities we introduce to simplify our life are generally smoothed out in reality. Thus, at sufficiently high frequencies the algebraic decay that characterizes the Fourier transforms of discontinuous functions gives way to the exponential decay characteristic of smooth functions. For example, the wings of a pressure-broadened spectroscopic line fall off algebraically because the emitted wave is discontinuously interrupted by collisions. However, the collisions that cause pressure broadening are really of short but finite duration, so the far wings of the line must fall off exponentially.

(a) Consider the function

$$f(t, \tau) = \frac{1}{2}\left[\operatorname{erf}\left(\frac{t+1}{\tau}\right) - \operatorname{erf}\left(\frac{t-1}{\tau}\right)\right] \tag{5.36}$$

where the error function is

$$\operatorname{erf}(x) = \frac{2}{\sqrt{\pi}}\int_0^x e^{-t^2}\,dt \tag{5.37}$$

As illustrated in Figure 5.4, for $\tau \ll 1$ this resembles the unit square function (5.29) but is actually smooth and continuous on a fine scale. Show that the Fourier transform is

$$\tilde{f}(\omega, \tau) = \sqrt{\frac{2}{\pi}}\frac{\sin(\omega)}{\omega}e^{-(1/4)\omega^2\tau^2} \tag{5.38}$$

which decays algebraically for $\omega\tau \ll 1$ and exponentially for $\omega\tau \gg 1$.

(b) Consider a molecule that emits, between collisions, a wave of the form

$$E = e^{-i\omega_0 t}f\left(\frac{2t}{\Delta t_c}, \frac{2\tau_c}{\Delta t_c}\right) \tag{5.39}$$

where ω_0 is the oscillation frequency, Δt_c the time between collisions, and $\tau_c \ll \Delta t_c$ the duration of a collision. The spectral intensity is proportional to the

Figure 5.4 Square function with soft edges.

square of the Fourier transform. Show that

$$\tilde{E}^2 = \frac{\Delta t_c^2}{4\pi^2} \left\{ \frac{\sin[\frac{1}{2}(\omega - \omega_0)\Delta t_c]}{\frac{1}{2}(\omega - \omega_0)\Delta t_c} \right\}^2 e^{-(1/2)(\omega - \omega_0)^2 \tau_c^2} \tag{5.40}$$

For $(\omega - \omega_0)\Delta t_c \gg 1$, the sinusoidal oscillations average out to $1/2$ when averaged over a statistical distribution of collision times Δt_c.

(c) For CO at atmospheric pressure, the mean time between collisions is about $\Delta t_c \approx 4 \times 10^{-10}$ s and the duration of a collision is on the order of $\tau \sim 10^{-13}$ s. What is the relative intensity (compared to line center) at the point $(\omega - \omega_0)\tau_c \approx 1$, where the line shape changes from algebraic to exponential? Is this likely to be observed?

5.1.3 δ-Functions

The Dirac δ-function is defined by its property that

$$\int_a^b f(x)\,\delta(x - x_0)\,dx = \begin{cases} f(x_0) & \text{if } a < x_0 < b \\ 0 & \text{otherwise} \end{cases} \tag{5.41}$$

In this way, it is really an operator that picks out the value of $f(x)$ at $x = x_0$ and not actually a function at all, although it can be envisioned as a function whose value is zero everywhere except at x_0, where it is infinite, with a unit area under the peak. In this operational sense, the Fourier transform and its inverse form a δ-function. From (5.1) and (5.2), we may write

$$f(t) = \frac{1}{2\pi} \int_{-\infty}^{\infty} dt'\, f(t') \int_{-\infty}^{\infty} d\omega\, e^{-i\omega(t - t')} \tag{5.42}$$

so in some sense we see that

$$\int_{-\infty}^{\infty} e^{-i\omega t}\, d\omega = 2\pi\,\delta(t) \tag{5.43}$$

This expression is analogous to the orthogonality relation discussed in Chapter 4 for the functions $e^{i\mathbf{k}\cdot\mathbf{x}}$ in a volume V. Integrals like (5.43) clearly have meaning only if they are defined by some sort of limit process, and functions defined in this way are called "generalized functions." The limit may be taken in many ways, and all equivalent limits define the same generalized function. One way in which the limit in (5.43) may *not* be taken is by allowing the limits of integration to approach infinity, since in this case the integral oscillates. A valid definition of the δ-function is provided by the limit process

$$\int_{-\infty}^{\infty} e^{-i\omega t}\, d\omega = \lim_{\varepsilon \to 0} \int_{-\infty}^{\infty} e^{-i\omega t} e^{-\varepsilon \omega^2}\, d\omega = \lim_{\varepsilon \to 0} \sqrt{\frac{\pi}{\varepsilon}} e^{-t^2/4\varepsilon} \tag{5.44}$$

From this we obtain the definition of the δ-function

$$\delta(t) = \lim_{\varepsilon \to 0} \sqrt{\frac{1}{4\pi\varepsilon}} e^{-t^2/4\varepsilon} = \lim_{\beta \to \infty} \sqrt{\frac{\beta}{\pi}} e^{-\beta t^2} \tag{5.45}$$

This definition is not unique, and an equivalent definition of the generalized function is provided by the limit

$$\delta(t) = \lim_{\varepsilon \to 0} \left[\frac{\varepsilon}{\pi(\varepsilon^2 + t^2)} \right] \tag{5.46}$$

Both (5.45) and (5.46) show that the Dirac δ-function may correctly be thought of as the limit of a narrow function whose width vanishes and whose peak approaches infinity in a way that preserves the area under the peak. These more precise definitions are occasionally necessary to fix the value of integrals like (5.41) when $f(x)$ is pathological or when x_0 is at one of the limits of the integral, and we use them in later chapters.

The derivative of the δ-function can also be defined by means of a limit process such as

$$\delta'(x) = \lim_{\varepsilon \to 0} \frac{d}{dx} \left[\frac{\varepsilon}{\pi(x^2 + \varepsilon^2)} \right] \tag{5.47}$$

for example. The derivative $\delta'(x)$ has the property

$$\int_a^b f(x)\delta'(x - x_0)\, dx = f(x)\delta(x - x_0) \Big|_a^b - \int_a^b f'(x)\delta(x - x_0)\, dx$$

$$= \begin{cases} -f'(x_0) & \text{for } a < x_0 < b \\ 0 & \text{otherwise} \end{cases} \tag{5.48}$$

EXERCISE 5.3

Define the integral

$$2\pi\delta(t) = \int_{-\infty}^{\infty} d\omega\, e^{-i\omega t} \tag{5.49}$$

by a limit process using $e^{-\varepsilon|\omega|}$, and show that the δ-function may be defined by the expression

$$\delta(t) = \lim_{\varepsilon \to 0} \left[\frac{\varepsilon}{\pi(\varepsilon^2 + t^2)} \right] \tag{5.50}$$

5.1.4 Autocorrelation Functions and the Wiener–Khintchine Theorem

As an illustration of the use and significance of Fourier transforms, we examine the relationship between filters and autocorrelators. To be specific, we consider an optical pulse described by the electric field $E(t)$. From the discussion in Chapter 4, we see that the intensity (the Poynting vector) incident on a detector is

$$S(t) = \varepsilon_0 c E^2(t) \tag{5.51}$$

so the total fluence (energy per unit area) incident on the detector is

$$\mathcal{J} = \int_{-\infty}^{\infty} S(t)\, dt = \varepsilon_0 c \int_{-\infty}^{\infty} E^2(t)\, dt = 2\varepsilon_0 c \int_0^{\infty} |\tilde{E}(\omega)|^2\, d\omega \tag{5.52}$$

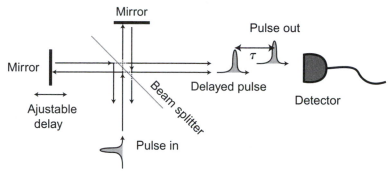

Figure 5.5 Autocorrelation measurement.

by (5.11). We call the quantity

$$\frac{d\mathcal{J}}{d\omega} = 2\varepsilon_0 c |\tilde{E}(\omega)|^2 \tag{5.53}$$

the spectral fluence. It describes the optical fluence per unit frequency interval. If we place a narrow-bandpass filter in front of a detector and measure the energy transmitted from an incident pulse, what we measure is the spectral fluence of the pulse.

Another measurement we can perform on an optical pulse is autocorrelation. As illustrated in Figure 5.5, we use a Michelson interferometer to split the pulse into two parts, delay one part relative to the other, and then recombine them into a single beam as shown. In the following we assume that the beam splitter has no losses and splits the beam 50/50. Since it divides the energy into two equal pieces, the beam splitter reduces the amplitude of each beam by $1/\sqrt{2}$. After passing through the beam splitter twice, the amplitude of the electric field in each pulse is then

$$E_S(t) = \tfrac{1}{2} E(t) \tag{5.54}$$

If we direct the combined beam, which now consists of two pulses, onto the detector, the fluence is

$$\mathcal{J} = \varepsilon_0 c \int_{-\infty}^{\infty} [E_S(t) + E_S(t-\tau)]^2 \, dt = \frac{1}{2}\left[\varepsilon_0 c \int_{-\infty}^{\infty} E^2(t) \, dt + C_1(\tau)\right] \tag{5.55}$$

where the first-order autocorrelation function is defined by

$$C_1(\tau) = \varepsilon_0 c \int_{-\infty}^{\infty} E(t) E(t-\tau) \, dt = C_1(-\tau) \tag{5.56}$$

This is also called the amplitude, or one-photon, autocorrelation function. It is a measure of the overlap between the pulse and its delayed counterpart, and its value depends on the relative delay of the two pulses in the interferometer.

If we use the Faltung theorem (5.8) with $f(t) = E(-t) = g(-t)$, we see that

$$\int_{-\infty}^{\infty} E(t) E(t-\tau) \, dt = \int_{-\infty}^{\infty} |\tilde{E}(\omega)|^2 e^{i\omega\tau} \, d\omega \tag{5.57}$$

and from the symmetry property (5.7), we find that

$$\int_{-\infty}^{\infty} E(t)E(t-\tau)\,dt = 2\int_{0}^{\infty} |\tilde{E}(\omega)|^2 \cos(\omega\tau)\,d\omega \tag{5.58}$$

Comparing this with (5.53), we obtain the relation

$$\mathcal{C}_1(\tau) = \int_{0}^{\infty} \frac{d\mathcal{J}}{d\omega} \cos(\omega\tau)\,d\omega \tag{5.59}$$

That is, the first-order autocorrelation function is just the Fourier (cosine) transform of the spectral fluence. If we Fourier transform (5.57) with respect to τ, using the orthogonality relation (5.43) and the even symmetry of $\mathcal{C}_1(\tau)$, we find that

$$\frac{d\mathcal{J}}{d\omega} = \frac{1}{\pi} \int_{0}^{\infty} \mathcal{C}_1(\tau) \cos(\omega\tau)\,d\tau \tag{5.60}$$

That is, the spectral fluence is just the Fourier (cosine) transform of the amplitude autocorrelation function.

Relations (5.59) and (5.60) are called the Weiner–Khintchine theorem. Simply put, they say that the information contained in the first-order autocorrelation function is equivalent to that contained in the spectral fluence, since a measurement of either one can be converted, by means of a Fourier transform, to the other. This shows the fundamental equivalence of a filter and an autocorrelator, which is also evident in the physical similarity of the autocorrelator shown in Figure 5.5 to a Michelson interferometer. We can also understand the equivalence between a filter and an autocorrelator by examining a typical interference filter constructed from a stack of dielectric layers, as shown in Figure 5.6. Each interface between layers reflects part of the incident beam, and the emerging beam is a sum of beams that have been delayed relative to one another by the separate reflections. Thus, the filter comprises, in a sense, a set of autocorrelators. The same is true of a reflection grating, which can be used as a spectral filter. As shown in Figure 5.7, the filtering comes from the interference between waves reflected from different rulings that arrive at the detector with different phases. Thus, the grating is just an autocorrelator with a large number of separate delays. Absorption filters are more subtle, but the same principles apply. It is shown in Chapter 7 that the absorption coefficient and the index of refraction of the material in a filter are not completely independent. They both depend on the response of the medium to an external stimulus, and this response must satisfy the requirements of causality. The relation between the absorption coefficient and the refractive index is

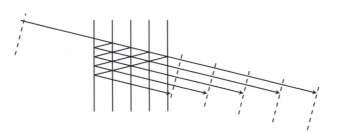

Figure 5.6 Interference filter.

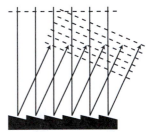

Figure 5.7 Reflection grating.

expressed mathematically by the Kramers–Kronig relations. Thus, the phase shift in the filter, which is related to an autocorrelation measurement, and the absorption, which is related to the spectral measurement, are not independent of one another.

As an example of the use of the Wiener–Khintchine theorem in statistical mechanics, we consider a random voltage signal $V(t)$. Although the voltage at any time t is completely random, the voltage a short time τ later bears some relation to the voltage at the time t. If we average over a large number of observations of the voltage at time t and that at time $t + \tau$, the correlation between these measurements has the form

$$\langle V(t)V(t + \tau) \rangle = \langle V^2 \rangle e^{-\tau/\tau_0} \tag{5.61}$$

where the brackets $\langle \; \rangle$ indicate an average over a large number of observations. The quantity τ_0 is called the correlation time, and the correlation between $V(t)$ and $V(t + \tau)$ vanishes for $\tau \gg \tau_0$. This implies that the autocorrelation function has the form

$$\mathcal{C}_1(\tau) = \int_{-\infty}^{\infty} V(t)V(t + \tau)\,dt = \langle V^2 \rangle \Delta t\, e^{-\tau/\tau_0} \tag{5.62}$$

where Δt is the period (presumed to be large compared with τ_0) over which the observations are made. The time Δt is equivalent to the pulse length in the example of an optical pulse discussed earlier, and $\langle V^2 \rangle \Delta t$ (the rms voltage times the time) is equivalent to the pulse energy, or fluence. By the Wiener–Khintchine theorem, then, the noise spectrum of this random voltage is

$$\frac{d\mathcal{J}}{d\omega} = \frac{1}{\pi} \int_{0}^{\infty} \mathcal{C}_1(\tau) \cos(\omega\tau)\,d\tau = \frac{1}{\pi} \langle V^2 \rangle \Delta t \frac{\tau_0}{1 + \omega^2 \tau_0^2} \tag{5.63}$$

This spectrum is flat ("white noise") out to frequencies for which $\omega\tau_0 \sim 1$, and it falls off algebraically at higher frequencies.

For optical signals, the autocorrelation function provides a classical definition of the coherence properties of the light. Lasers, for example, ordinarily have a finite spectral bandwidth that reflects the fact that the optical wave is not a pure, single-frequency harmonic wave. The amplitude and phase have a noise component in them. Characteristically, the autocorrelation function of a cw laser beam or a long pulse has a broad noise component with a "coherence spike." The width of the coherence spike is a measure of the coherence time of the optical signal.

Before leaving the topic of autocorrelation methods, it is important to point out that most optical autocorrelation experiments actually measure the *intensity* autocorrelation function, not the electric field *amplitude* autocorrelation function. Since the intensity is quadratic in the electric field, the intensity autocorrelation function is the second-order, or two-photon, autocorrelation function

$$\mathcal{C}_2(\tau) = (\varepsilon_0 c)^2 \int_{-\infty}^{\infty} E^2(t)E^2(t - \tau)\,dt = \mathcal{C}_2(-\tau) \tag{5.64}$$

For example, in a typical measurement of a very short laser pulse, the pulse is split as shown in Figure 5.8 and then passed through a two-photon second-harmonic generation system consisting of a nonlinear crystal. As discussed in Chapter 8, the polarization of the crystal is nonlinear in the electric field, so the polarization oscillates not only at the fundamental but also at the harmonics of the incident light. The amplitude of the second-harmonic

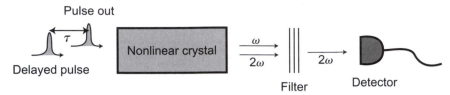

Figure 5.8 Two-photon autocorrelation measurement.

component of the polarization is proportional to the square of the incident field, that is, to the incident intensity. But the intensity of the radiation produced at the second harmonic is proportional to the square of the amplitude of the polarization, so the second-harmonic intensity is proportional to the square of the incident intensity. In quantum-mechanical terms, the probability of generating a second-harmonic photon from two first-harmonic photons is proportional to the probability of finding two photons at the same place at the same time. Since the photon density is proportional to the intensity, the two-photon probability is proportional to the square of the incident intensity. The energy of the second-harmonic pulse is therefore proportional to the two-photon autocorrelation function C_2, and a measurement of the energy of the second-harmonic radiation can be used to determine the two-photon autocorrelation function.

In interpreting two-photon autocorrelation measurements, it is important to remember that the second-order autocorrelation function is *not* related to the spectral intensity by the Wiener–Khintchine theorem. The difference between C_1 and C_2 is illustrated by comparing the autocorrelation functions for a pulse that is "chirped." A chirped pulse has a frequency that increases (for positive chirp) throughout the pulse, like the sound made by a bird. By passing such a pulse through a dispersive system, the pulse can be compressed or stretched, depending on the sign of the dispersion. If, for example, the system has dispersion such that the group velocity is faster for higher frequencies, then the high-frequency portions of the pulse near the back catch up to the low-frequency portions at the front and the pulse becomes shorter. In an optimally compressed pulse, the dispersion places all the frequency components on top of one another and the chirp vanishes. This is discussed in detail in the next section. Clearly, the width of the intensity autocorrelation function is shorter after the pulse is compressed. On the other hand, the spectral fluence is unchanged by dispersion because dispersion is a linear process, so by the Wiener–Khintchine theorem, the first-order autocorrelation function is unchanged by the compression process. The width of the first-order autocorrelation function must therefore be comparable to the length of the compressed (unchirped) pulse, so the first-order autocorrelation function of a chirped pulse must be shorter than the second-order correlation function, reflecting not the actual length of the pulse but the length of the pulse after it is optimally compressed.

To illustrate these ideas with a simple example, we consider the double pulse shown in Figure 5.9. The first pulse (which appears earlier, at smaller times) has a lower frequency and the second (which appears at later times) has a higher frequency. When the two-photon autocorrelation is measured, the result consists of peaks at three different delay times. The first appears when the leading pulse overlaps the trailing pulse. The second peak occurs when the two leading pulses overlap and the two trailing pulses overlap, and the third peak

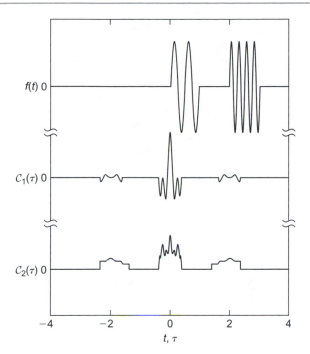

Figure 5.9 Autocorrelation of a double pulse.

occurs when the trailing pulse overlaps the leading pulse. On the other hand, when the one-photon (amplitude) autocorrelation is measured, the peak corresponding to the overlap of the leading pulse with the trailing pulse (nearly) vanishes due to cancellation because the pulses have different frequencies. Likewise the third peak in the autocorrelation (nearly) vanishes (as it must by symmetry), leaving only the central peak. Thus, the first-order (amplitude) correlation function, in which only the central peak is large, is effectively narrower than the second-order (intensity) correlation function, which has three separate peaks. On the other hand, we can compress the double pulse before we perform the auto-correlation by passing it through a dispersive section. If the dispersion has the correct value, the second pulse (at higher frequency) catches up to the first pulse at the end of the dispersive section and overlaps it. Since dispersion is a linear effect, no frequency components are altered and the spectral fluence is unchanged. Therefore, according to the Weiner–Kintchine theorem, the first-order autocorrelation C_1 must be the same as before. However, the second-order autocorrelation has changed and now reflects the length of the compressed pulse. This simple example also shows that *group-velocity* dispersion, rather than *phase-velocity* dispersion, is most important in pulse compression.

It is worth pointing out that the two-photon (intensity) autocorrelation function is positive definite, whereas the one-photon (amplitude) autocorrelation function oscillates at the frequency of the wave. This makes it easier in practice to measure the two-photon autocorrelation than the one-photon autocorrelation, since fewer data points are required. Put another way, if the experimental resolution is limited, the two-photon autocorrelation does not wash out by cancellation of the positive and negative oscillations whereas the one-photon autocorrelation does.

EXERCISE 5.4

A pulse consists of two narrow spikes separated by the interval T, so we can represent the electric field by the function

$$E = E_0\big[\delta\big(t + \tfrac{1}{2}T\big) + \delta\big(t - \tfrac{1}{2}T\big)\big] \tag{5.65}$$

(a) What is the spectral amplitude (the Fourier transform) of the pulse?
(b) Show that the spectral intensity is

$$\frac{d\mathcal{J}}{d\omega} = \frac{4\varepsilon_0 c E_0^2}{\pi}\cos^2\left(\frac{\omega T}{2}\right) \tag{5.66}$$

(c) Use the Weiner–Khintchine theorem to show that the amplitude autocorrelation function is

$$\mathcal{C}_1(\tau) = \varepsilon_0 c E_0^2[\delta(T + \tau) + \delta(T - \tau) + 2\delta(\tau)] \tag{5.67}$$

(d) Use the properties of the δ-function discussed in Chapter 1 to derive (5.67) directly from the definition of the amplitude autocorrelation function.
(e) Sketch what you would expect the intensity (two-photon) autocorrelation function to look like. Indicate the relative size of the peaks.

EXERCISE 5.5

A Pockels cell (a type of optical switch) is used to chop a pulse of duration τ_0 from the beam of a laser whose frequency is ω_0. Thus, the field transmitted by the Pockels cell may be represented by the expressions

$$E = \begin{cases} E_0 \cos(\omega_0 t) & \text{for } -\dfrac{\tau_0}{2} < t < \dfrac{\tau_0}{2} \\ 0 & \text{otherwise} \end{cases} \tag{5.68}$$

(a) Show that for $\omega_0\tau_0 \gg 1$, the spectral fluence of the transmitted pulse is

$$\frac{d\mathcal{J}}{d\omega} \approx \frac{E_0^2\tau_0^2}{4\pi\mu_0 c}\left\{\frac{\sin[(\omega - \omega_0)\tau_0/2]}{(\omega - \omega_0)\tau_0/2}\right\}^2 \tag{5.69}$$

Sketch this result.
(b) Show that for $\omega_0\tau_0 \gg 1$, the amplitude autocorrelation function is

$$\mathcal{C}_1 \approx \tfrac{1}{2}\varepsilon_0 c E_0^2(\tau_0 - |\tau|)\cos(\omega_0\tau) \quad \text{for } -\tau_0 < \tau < \tau_0$$
$$= 0 \qquad\qquad\qquad\qquad\qquad \text{otherwise} \tag{5.70}$$

Sketch this result.
(c) Show that for $\omega_0\tau_0 \gg 1$, the intensity autocorrelation function is

$$\mathcal{C}_2 \approx \tfrac{1}{4}\varepsilon_0^2 c^2 E_0^4(\tau_0 - |\tau|)\cos^2(\omega_0\tau) \quad \text{for } -\tau_0 < \tau < \tau_0$$
$$= 0 \qquad\qquad\qquad\qquad\qquad\quad \text{otherwise} \tag{5.71}$$

Sketch this result.

EXERCISE 5.6

As shown in Figure 5.10, the spectral fluence of the Vanderbilt University free-electron laser is dominated by a peak that is roughly Gaussian in frequency space. Experimental data are frequently quoted, for convenience, as the full width at the half-maximum points (FWHM). The observed bandwidth (FWHM) is $\Delta\omega_{1/2}/\omega_0 \approx 0.9\%$ at a center wavelength $\lambda_0 = 2\pi c/\omega_0 \approx 3.2$ μm. The two-photon autocorrelation function shown in Figure 5.10 was measured by two-photon absorption in a semiconductor crystal. The signal is proportional to the transmitted energy, so the autocorrelation appears as a decrease in the intensity.

(a) Based on the observed spectrum, which is approximately Gaussian in frequency space, use the Weiner–Khintchine theorem to compute the one-photon (amplitude) autocorrelation function of the optical pulse. Show that the one-photon autocorrelation function has the form of a fast oscillation with a Gaussian envelope, and that the width (FWHM) of the envelope is

$$\Delta\tau_{1/2} = \frac{8\ln 2}{\Delta\omega_{1/2}} \tag{5.72}$$

(b) Sketch your result for the envelope of the one-photon autocorrelation function and compare it qualitatively with the two-photon (intensity) autocorrelation function shown in Figure 5.10. How does the predicted width of the envelope of the one-photon autocorrelation function compare with the observed width (FWHM) $\Delta\tau_0' \approx 1.4$ ps of the two-photon autocorrelation function?

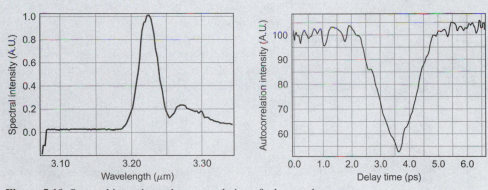

Figure 5.10 Spectral intensity and autocorrelation of a laser pulse.

EXERCISE 5.7

The noise voltage in a resonant circuit is random except that when averaged over a large number of observations, the voltage at a time τ later is found to have the autocorrelation

$$\langle V(t)V(t+\tau)\rangle = \langle V^2\rangle\cos(\omega_0\tau)e^{-\tau/\tau_0} \tag{5.73}$$

where $\langle V^2 \rangle$ is the rms noise voltage and ω_0 the resonant frequency of the circuit. Use the Weiner–Khintchine theorem to show that the spectral intensity of the noise in the circuit has the form

$$S(\omega) = \frac{\langle V^2 \rangle}{2\pi} \Delta t \left[\frac{\tau_0}{1 + (\omega + \omega_0)^2 \tau_0^2} + \frac{\tau_0}{1 + (\omega - \omega_0)^2 \tau_0^2} \right] \tag{5.74}$$

where Δt is the period over which the observations are taken. Sketch your result.

5.1.5 Pulse Compression

Optical pulse compression is used to produce the most powerful laser pulses that have been developed so far, and the compression of electromagnetic and acoustic pulses is of great importance in signal processing. Moreover, the spreading of quantum-mechanical (and other) wave packets by dispersion is just the inverse of pulse compression and has relevance to many other branches of science.

To see how optical pulse compression works, we examine the compression of a pulse whose spectrum is narrow and centered on the frequency ω_0. A pulse of this type is conveniently described by the electric field

$$E(t) = E_0(t)e^{-i\omega_0 t} \tag{5.75}$$

where the envelope $E_0(t)$ is a slowly varying function of the time. The rapid variation at the optical frequency ω_0 is placed in the exponential function, but phase variations in $E_0(t)$ can be used to describe departures of the local frequency in the pulse from the central frequency ω_0. When the pulse enters a dispersive region, as shown in Figure 5.11, the Fourier component of the pulse at a frequency ω is delayed by an amount $\tau(\omega)$ that depends on the frequency. Specifically, we consider two harmonic waves of the form $e^{-i\omega t}$, one at frequency ω and the other at frequency $\omega + d\omega$, that enter the dispersive region simultaneously, with a certain phase front of one wave coincident with a phase front of the other wave. When the phase front of the wave at ω emerges from the dispersive region, the phase front of the wave at $\omega + d\omega$ has been delayed by the relative time $d\tau$. Note that the delay of the *phase front* depends on the *phase* velocity of the wave through the dispersive region. Compared with the first wave, the phase of the wave at $\omega + d\omega$ is shifted by the amount

$$d\phi = \omega \, d\tau = \frac{d\tau}{d\omega} \omega \, d\omega \tag{5.76}$$

The positive sign follows because the phase of the wave is proportional to $-i\omega\tau$. Integrating this expression from the frequency ω_0 to ω, we see that the phase shift relative to a

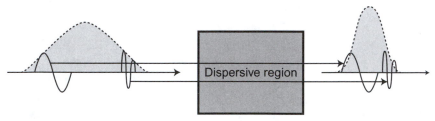

Figure 5.11 Dispersion of a pulse.

wave at the frequency ω_0 is

$$\Delta\phi = \int_{\omega_0}^{\omega} d\phi = \int_{\omega_0}^{\omega} \frac{d\tau}{d\omega}\,\omega\,d\omega \tag{5.77}$$

But the pulse $E(t)$ entering the dispersive region can be decomposed into harmonic waves by means of its Fourier transform, so

$$E_0(t) = \frac{1}{\sqrt{2\pi}} \int_{-\infty}^{\infty} d\omega\,\tilde{E}(\omega)e^{-i(\omega-\omega_0)t} \tag{5.78}$$

where

$$\tilde{E}(\omega) = \frac{1}{\sqrt{2\pi}} \int_{-\infty}^{\infty} dt\,E_0(t)e^{i(\omega-\omega_0)t} \tag{5.79}$$

Since the phase of each Fourier component is shifted by the relative amount $\Delta\phi$, the wave emerging from the dispersive region is

$$E_0'(t') = \frac{1}{\sqrt{2\pi}} \int_{-\infty}^{\infty} d\omega\,\tilde{E}(\omega)e^{-i(\omega-\omega_0)t'}e^{i\Delta\phi}$$

$$= \frac{1}{2\pi} \int_{-\infty}^{\infty} dt\,E_0(t) \int_{-\infty}^{\infty} d\omega\,e^{i(\omega-\omega_0)(t-t')}e^{i\Delta\phi} \tag{5.80}$$

where the overall delay $\tau(\omega_0)$ of the pulse as it passes through the dispersive system has been absorbed into time t'.

To complete the solution, we must evaluate the integral over ω so that the emerging pulse $E_0'(t')$ is represented as an integral over the entering pulse $E_0(t)$ with a kernel that depends on the time difference $t - t'$. Since the spectrum of $E(t)$ is narrowly centered about the frequency ω_0, we may expand the delay in a Taylor series of the form

$$\frac{d\tau}{d\omega} = \frac{d\tau}{d\omega}\bigg|_{\omega_0} + \frac{d^2\tau}{d\omega^2}\bigg|_{\omega_0}(\omega-\omega_0) + O[(\omega-\omega_0)^2] \tag{5.81}$$

Substituting into (5.77) and integrating, we see that the phase of the emerging wave is

$$\Delta\phi = \tau_0(\omega-\omega_0) - \frac{1}{2\mu}(\omega-\omega_0)^2 + O[(\omega-\omega_0)^3] \tag{5.82}$$

where

$$\tau_0 = \omega_0 \frac{d\tau}{d\omega}\bigg|_{\omega_0} \tag{5.83}$$

$$\frac{1}{\mu} = -\frac{d\tau}{d\omega}\bigg|_{\omega_0} - \omega_0 \frac{d^2\tau}{d\omega^2}\bigg|_{\omega_0} = -\frac{d}{d\omega}\left(\omega\frac{d\tau}{d\omega}\right)\bigg|_{\omega_0} \tag{5.84}$$

If we substitute into (5.80) and change to the variable

$$t'' = t' - \tau_0 \tag{5.85}$$

we get

$$E_0'(t'' + \tau_0) = \frac{1}{2\pi} \int_{-\infty}^{\infty} dt\,E_0(t) \int_{-\infty}^{\infty} d\omega\,e^{i[(\omega-\omega_0)(t-t'')-(1/2\mu)(\omega-\omega_0)^2]} \tag{5.86}$$

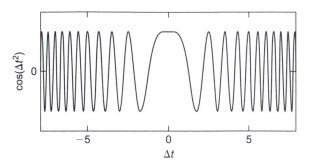

Figure 5.12 Real part of the kernel for dispersion of a pulse.

If we now absorb the time shift τ_0 into the function $E_0''(t'') = E_0'(t'' + \tau_0)$, complete the square in the exponent, and integrate over ω, we obtain the result

$$E_0''(t'') = \sqrt{\frac{\mu}{i2\pi}} \int_{-\infty}^{\infty} dt\, E_0(t) e^{i(\mu/2)(t-t'')^2} \tag{5.87}$$

This formula expresses the slowly varying part of the compressed pulse in the form of an integral over the initial pulse. For a real function $E_0(t)$, the kernel

$$\exp\left[i\frac{\mu}{2}(t-t'')^2\right]$$

weights the contributions from the initial pulse $E_0(t)$ most heavily for times near the time of interest, that is, for $(\mu/2)(t-t'')^2 \ll 1$. Note that the kernel does not actually vanish for increasing $|t - t''|$, but the contributions from times far from t'' vanish by cancellation, since the kernel oscillates with increasing frequency as $|t - t''|$ increases, as indicated in Figure 5.12. This rapidly oscillating behavior is not too difficult to deal with in analytical calculations, but it can present serious difficulties in numerical simulations.

In terms of (5.87), we can understand pulse compression in the following way. As we see from Figure 5.12, the dominant contribution to the integral comes from regions where the frequency of oscillation of the exponential function vanishes. This is called the point of stationary phase. When the initial pulse includes a frequency shift $\Delta\omega(t)$, the phase of the envelope function $E_0(t)$ varies locally like $-i\Delta\omega t$. When this linear term combines with the quadratic variation $(\mu/2)(t - t'')^2$ in the exponent, the point of stationary phase shifts to time $t = t'' + (\Delta\omega/\mu)$ and the contribution from the initial pulse at this time t becomes heavily weighted in the emerging pulse at time t''. In this way, sections of the initial pulse are shifted around in time according to their local frequency by the effects of dispersion. This is the mechanism of pulse compression.

It is interesting to note that while the delay τ of the phase front in (5.76) depends on the phase velocity in the dispersive section, the compressed pulse (5.87) depends only on derivatives of the delay, and only through the quantity μ, not τ_0. Since the group velocity depends on the derivative $v_g = d\omega/dk$, the dependence of the kernel on the derivative of τ reflects the importance of *group-velocity* dispersion. To see this more explicitly, we consider the case when the dispersion is caused by propagation through a length L of some medium whose index of refraction varies with the frequency. The phase delay, which depends on the phase velocity $v_\phi = \omega/k$, is then

$$\tau = \frac{L}{v_\phi} = L\frac{k}{\omega} \tag{5.88}$$

From (5.84) we find that

$$\frac{1}{\mu} = \frac{d}{d\omega}\left(\frac{L}{v_\phi} - \frac{L}{v_g}\right) \tag{5.89}$$

We see that even materials in which the phase velocity is a decreasing function of the frequency do not necessarily have the effect of delaying the high-frequency components of the pulse relative to the low-frequency components. Thus, whereas phase-velocity dispersion is necessarily negative in a transparent medium, as we see in Chapter 7, group-velocity dispersion can have either sign and may outweigh phase-velocity dispersion.

The compression of chirped pulses has become very important in high-power laser technology. The fundamental issue associated with the creation of extremely high peak power pulses is damage to the optical components, especially the optical coatings on the surfaces of windows and other laser components, at the extremely high fields in the laser beam. To avoid this problem, the optical pulse is first formed at a low power, generally by a technique called mode locking, with a pulse length of the order of tens of femtoseconds. This pulse is then stretched by passing it through a dispersive region, usually a pair of gratings arranged as shown in Figure 5.13. In addition to stretching the pulse, the dispersion delays certain frequency components relative to others and introduces a chirp in the pulse. The stretched, chirped pulse is then amplified until the total pulse energy is of the order of a few Joules, after which the pulse is recompressed by means of another grating pair. This final compression lowers the total pulse energy, due to unavoidable losses, but it increases the peak power and produces extremely short pulses. In practice, the pulse may be stretched and then recompressed by a factor on the order of 10^5. In this way pulses have been produced with duration shorter than 20 fs (about eight optical cycles at a wavelength of 0.8 μm) and peak power in excess of 100 TW (10^{14} W). As they pass through the amplification stages, the pulses are stretched to about a nanosecond, a value limited by the size of available gratings. Of course, the bandwidth of the amplifying medium must be broad enough to amplify all the frequency components of the chirped pulse. While the stretching process makes the pulse longer, it does not narrow the spectrum but merely converts the bandwidth of the shorter pulse to a chirp. The most widely used medium in so-called chirped-pulse amplifiers is Ti-doped sapphire, which has a useable bandwidth from 680 to 1100 nm.

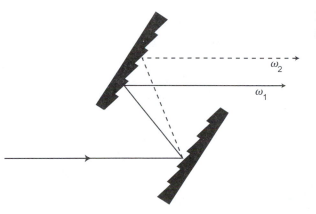

Figure 5.13 Dispersion by a grating pair.

A similar approach is used in modern chirped-pulse radars. To obtain the signal-to-noise advantages of long-pulse radars together with the range resolution of short-pulse radars and to avoid the technological problems of producing and radiating high-power short pulses, a long radar pulse is produced with chirp. When the reflected pulse is received, it is compressed in a dispersive delay line to improve the distance resolution. This compression is accomplished by converting the incoming signal to surface acoustic waves on a solid medium, which are then detected by an array of tuned filters. More recently, fast digital signal processing is used to extract the range information with greater accuracy and flexibility. Of course, bats trying to echolocate their dinner have been using chirped pulses for millions of years. In fact, bats use a complex waveform consisting of constant-frequency and negatively chirped segments. The range to the prey is extracted by a sophisticated system of acoustic delay lines and sharply tuned spiral ganglion cells in the bat's brain that allow the bat to filter and autocorrelate the echo with the original sound. In addition, the system has been miniaturized for installation in very small aircraft and simplified for operation by unsophisticated users. The existence of well-fed bats is proof of the success of this highly evolved system. Despite the explosive growth in recent years in the theory and application of filters and autocorrelators in signal processing, the most sophisticated signal-processing systems we have conceived remain primitive compared with those that have evolved in nature.

In closing this section, we note that de Broglie waves suffer dispersion even in vacuum, as discussed in Chapter 2. This causes initially localized wave packets to spread out after some time. Since pulse spreading is just pulse compression in reverse, we see that the methods of this section can be used to compute the spread of quantum-mechanical wave packets.

EXERCISE 5.8

In a linearly chirped pulse, the frequency is

$$\omega = \omega_0 + \omega' t \tag{5.90}$$

so a linearly chirped Gaussian pulse has the form

$$E(t) = E_0(t)e^{-i\omega_0 t} = e^{-t^2/\sigma^2}e^{-i[\omega_0 t + (\omega'/2)t^2]} \tag{5.91}$$

The envelope function is therefore

$$E_0(t) = e^{-[(1/\sigma^2)+i(\omega'/2)]t^2} \tag{5.92}$$

(a) Show that if a simple (not chirped) Gaussian pulse

$$E_0(t) = e^{-t^2/\sigma_0^2} \tag{5.93}$$

enters a dispersive region, the pulse that emerges is stretched and linearly chirped, and that

$$\omega' = -\frac{\mu}{1 + \dfrac{\mu^2\sigma_0^4}{4}} \tag{5.94}$$

$$\sigma^2 = \sigma_0^2\left(1 + \frac{4}{\mu^2\sigma_0^4}\right) \tag{5.95}$$

(b) Conversely, if this pulse is inserted into a region with the opposite dispersion, then the original pulse is recovered. From this show that the optimum dispersion, which compresses a chirped pulse to the minimum length, is

$$\mu = \omega' \left(1 + \frac{4}{\omega'^2 \sigma^4} \right) \tag{5.96}$$

Note that the original pulse of length σ_0 is the minimum-length pulse. It cannot be further compressed because it has no chirp.

EXERCISE 5.9

Consider a relativistic particle of mass m and energy $E_0 = \hbar\omega_0 = \gamma mc^2$, where ω_0 is a characteristic frequency and $\hbar$ is Planck's constant divided by 2π. As the particle passes the origin along the x axis, its wave function may be expressed by

$$\psi(t) = e^{-i\omega_0 t} \psi_0(t) \tag{5.97}$$

As shown in Chapter 2, the dispersion relation for the wave function of the particle is

$$\omega^2 - k^2 c^2 = \frac{m^2 c^4}{\hbar^2} \tag{5.98}$$

Because of this dispersion, when the particle reaches point L, the envelope of the wave function has the form

$$\psi_L(t) = \sqrt{\frac{\mu}{i 2\pi}} \int_{-\infty}^{\infty} dt' \, \psi_0(t') e^{i(\mu/2)(t'-t)^2} \tag{5.99}$$

by analogy with the preceding discussion. Show that the dispersion parameter is

$$\frac{1}{\mu} = \frac{\hbar L}{mc^3} \frac{2\gamma^2 - 1}{\gamma^2 (\gamma^2 - 1)^{3/2}} \xrightarrow[\gamma \to \infty]{} \frac{2\hbar L}{\gamma^3 mc^3}$$

$$\xrightarrow[\beta \to 0]{} \frac{\hbar L}{\beta^3 mc^3} \tag{5.100}$$

where $\gamma = 1/\sqrt{1 - \beta^2}$. Thus, the dispersion vanishes as the particle velocity approaches c and increases as the particle takes longer to reach the point L. When the initial wave function has a square envelope, the wave function at L can be expressed in terms of Fresnel integrals. As shown in Chapter 9, this is equivalent to the diffraction of an optical wave passing through a slit.

5.2 METHOD OF VIRTUAL QUANTA

When a charged particle moves at an ultrarelativistic velocity ($\gamma \gg 1$), its electric and magnetic fields are Lorentz contracted in the direction of motion as described in Chapter 1. As a result, the electric fields, which are in the radial direction, are essentially transverse to

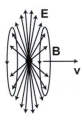

Figure 5.14 Fields of an ultrarelativistic particle.

the direction of motion, as illustrated in Figure 5.14. The magnetic fields are always transverse to the direction of motion, as shown there. If the motion is along the x axis, the $\hat{\mathbf{y}}$ components of the electric and magnetic fields at point $(x, y, 0)$ are

$$E_y = \frac{\gamma q}{4\pi\varepsilon_0} \frac{y}{[y^2 + \gamma^2(x - \beta ct)^2]^{3/2}} \approx E_y(x - ct) \tag{5.101}$$

and

$$B_z = \frac{\gamma q}{4\pi\varepsilon_0} \frac{v}{c^2} \frac{y}{[y^2 + \gamma^2(x - \beta ct)^2]^{3/2}} \approx \frac{E_y(x - ct)}{c} \tag{5.102}$$

where $\beta c \approx c$ is the velocity and βct the position of the charged particle along the x axis at time t. There is also a small component of the electric field in the $\hat{\mathbf{x}}$ direction, but this is of order $1/\gamma$ compared with the transverse field. Since the electric and magnetic fields are essentially orthogonal to the direction of motion and to each other, are in the ratio $|\mathbf{B}| \approx |\mathbf{E}/c|$, and move at nearly the speed of light, they are essentially the same as an optical field. Thus, the response of a target particle at point $(x, y, 0)$ to the field of the incident particle is essentially the same as it would be to an optical field. If we Fourier resolve the field into its frequency components and the interaction is linear, then we can compute the response of the target to the incident field from its response to optical fields of the same frequency. This response might be known from optical excitation measurements, for example. The requirement of linearity generally means that the incident particle must not be significantly deflected by the interaction and the target particle should not be disturbed very much. The requirements on the target are generally fulfilled if the incident particle is traveling at an ultrarelativistic velocity, as we see from the following line of reasoning. Since the only characteristic length in the problem is the impact parameter $b = y$, the duration of the interaction is on the order of

$$\tau_b = \frac{b}{\gamma c} \tag{5.103}$$

But the maximum distance the target can move is on the order of $c\tau_b = b/\gamma \ll b$, so the movement is small compared with the impact parameter. Shifting to the rest frame of the incident particle proves the same conclusion with respect to the deflection of the incident particle.

The equivalence between the field of a relativistic incident particle and an optical pulse is known as the Weizsaeker–Williams approximation, and the Fourier components into which the field is resolved are called "virtual quanta," or sometimes "equivalent photons." For many purposes the Fourier decomposition of (5.101) and (5.102) is enough to solve the problem at hand, but in the following we take a slightly different approach. By Fourier resolving the field in three dimensions, we can discuss the scattering of the field in much more detail.

5.2.1 Fourier Decomposition of the Field of a Relativistic Charge

We begin the discussion by resolving the field of a stationary charge into its Fourier components and then boosting these fields into a coordinate system moving at a highly relativistic velocity. The Fourier components of the field of the particle at rest are longitudinal waves, since the electric field is in the radial direction. However, when the waves are boosted into an ultrarelativistic frame, the wave vectors shift in the direction of the boost, whereas the components of the field-strength tensor shift to the directions transverse to the boost, as discussed in Chapter 1. As a result of the boost, then, the Fourier components of the fields become transverse waves, like optical waves. This has to be regarded as one of the mathematical miracles of special relativity.

The electrostatic field of a charge q at the origin of its own rest frame is described by the Maxwell equations

$$\nabla \cdot \mathbf{E} = \frac{\rho}{\varepsilon_0} = \frac{q}{\varepsilon_0}\delta(\mathbf{r}) \tag{5.104}$$

and

$$\nabla \times \mathbf{E} = 0 \tag{5.105}$$

Of course, the magnetic field vanishes in the particle rest frame. If we expand the electric field in terms of its Fourier components,

$$\mathbf{E}(\mathbf{r}) = \frac{1}{(2\pi)^{3/2}} \int_{-\infty}^{\infty} d^3k \, e^{-i\mathbf{k}\cdot\mathbf{r}} \tilde{\mathbf{E}}(\mathbf{k}) \tag{5.106}$$

and substitute this expansion into (5.105), we find that

$$\mathbf{k} \times \tilde{\mathbf{E}} = 0 \tag{5.107}$$

That is, the Fourier components of the field of a stationary charge are longitudinal waves. It is therefore convenient to write

$$\tilde{\mathbf{E}}(\mathbf{k}) = \hat{\mathbf{k}}\tilde{E}(\mathbf{k}) \tag{5.108}$$

where $\hat{\mathbf{k}} = \mathbf{k}/k$ is a unit vector. If we substitute this result into (5.104), multiply by $e^{i\mathbf{k}'\cdot\mathbf{r}}$, and integrate over all space, we find that

$$\frac{-i}{(2\pi)^{3/2}} \int_{-\infty}^{\infty} d^3k \, k \tilde{E}(\mathbf{k}) \int_{-\infty}^{\infty} d^3r \, e^{i(\mathbf{k}'-\mathbf{k})\cdot\mathbf{r}} = \frac{q}{\varepsilon_0} \tag{5.109}$$

But the integral over d^3r is just a representation of the δ-function, specifically $(2\pi)^3\delta(\mathbf{k}' - \mathbf{k})$, so the Fourier transform of the electrostatic field of a point charge is

$$\tilde{\mathbf{E}}(\mathbf{k}) = \frac{i}{(2\pi)^{3/2}} \frac{q\hat{\mathbf{k}}}{\varepsilon_0 k} \tag{5.110}$$

If the charge is placed in a reference frame K' moving along the x axis with the velocity βc, the field in the laboratory frame K is found by Lorentz transforming the wave 4-vector $k'^\alpha = (0, \mathbf{k}')$ and the (Lorentz tensor) fields $\mathbf{E}'$ and $\mathbf{B}' = 0$ from the rest frame back into the laboratory frame. For $\gamma \gg 1$, the wave vector $k^\alpha = (\omega/c, \mathbf{k})$ is

$$k^0 = \frac{\omega}{c} = \beta\gamma k'_x = O(\gamma) \tag{5.111}$$

$$k^1 = k_x = \gamma k'_x = \frac{\omega}{\beta c} = O(\gamma) \tag{5.112}$$

$$k^2 = k_y = k'_y = O(1) \tag{5.113}$$

$$k^3 = k_z = k'_z = O(1) \tag{5.114}$$

In the ultrarelativistic limit ($\gamma \gg 1$, $\beta \approx 1$) we see that k_y, $k_z = O(k/\gamma)$, so the wave vectors in the laboratory frame are shifted into a narrow cone around the forward direction. The magnitude (squared) of the wave vector is the Lorentz scalar

$$k^\alpha k_\alpha = \frac{\omega^2}{c^2} - k^2 = -\left(\frac{\omega^2}{\beta^2\gamma^2 c^2} + k_y^2 + k_z^2\right) = O(1) \tag{5.115}$$

The electric field components in the laboratory frame are

$$\tilde{E}_x = \tilde{E}'_x = \frac{i}{(2\pi)^{3/2}} \frac{q}{\varepsilon_0} \frac{\dfrac{\omega}{\beta\gamma c}}{\dfrac{\omega^2}{\beta^2\gamma^2 c^2} + k_y^2 + k_z^2} = O(1) \tag{5.116}$$

$$\tilde{E}_y = \gamma\tilde{E}'_y = \frac{i}{(2\pi)^{3/2}} \frac{q}{\varepsilon_0} \frac{\gamma k_y}{\dfrac{\omega^2}{\beta^2\gamma^2 c^2} + k_y^2 + k_z^2} = O(\gamma) \tag{5.117}$$

$$\tilde{E}_z = \gamma\tilde{E}'_z = \frac{i}{(2\pi)^{3/2}} \frac{q}{\varepsilon_0} \frac{\gamma k_z}{\dfrac{\omega^2}{\beta^2\gamma^2 c^2} + k_y^2 + k_z^2} = O(\gamma) \tag{5.118}$$

Thus, the electric field lies in the plane of the wave vector $\mathbf{k}$ and the direction of motion $\hat{\mathbf{x}}$. From (5.115) we find that

$$\frac{\omega^2}{k^2 c^2} = 1 + O\left(\frac{1}{\gamma^2}\right) \tag{5.119}$$

That is, the waves travel at nearly the speed of light. Moreover, we see that

$$\frac{\tilde{\mathbf{E}} \cdot \mathbf{k}}{\tilde{E} k} = O\left(\frac{1}{\gamma}\right) \tag{5.120}$$

so the electric field is essentially transverse to the direction of propagation. Moreover, we see that since $\mathbf{E} \cdot \mathbf{B}$ and $E^2 - c^2 B^2$ are Lorentz invariants,

$$\tilde{\mathbf{E}} \cdot \tilde{\mathbf{B}} = \tilde{\mathbf{E}}' \cdot \tilde{\mathbf{B}}' = 0 \tag{5.121}$$

and

$$\frac{\tilde{E}^2 - c^2\tilde{B}^2}{\tilde{E}^2} = \frac{\tilde{E}'^2}{\tilde{E}^2} = O\left(\frac{1}{\gamma^2}\right) \tag{5.122}$$

That is, the electric and magnetic fields are perpendicular to one another in the laboratory frame K and have essentially the correct ratio for a light wave. Finally, we note that since $\tilde{B}_x = \tilde{B}'_x = 0$ for a boost in the $\hat{\mathbf{x}}$ direction, the magnetic field is perpendicular to the direction of motion $\hat{\mathbf{x}}$ as well as to the electric field $\tilde{\mathbf{E}}$. It is therefore perpendicular to the wave vector $\mathbf{k}$. Thus, for $\gamma \gg 1$ the waves into which we have decomposed the field of the moving point charge have all the essential properties of light waves.

In the laboratory frame, the electric field of the moving point charge is

$$\mathbf{E}(\mathbf{r}, t) = \frac{1}{(2\pi)^{3/2}} \int_{-\infty}^{\infty} d^3\mathbf{k}' e^{-ik^\alpha r_\alpha} \tilde{\mathbf{E}}(\mathbf{k}) \tag{5.123}$$

where the Fourier transform field $\tilde{\mathbf{E}}(\mathbf{k})$ is given by (5.116)–(5.118). Note that the integral is over $\mathbf{k}'$, not $\mathbf{k}$, since this is just a dummy index for the sum and the original Fourier transform (5.106) was over wave vectors in the particle rest frame. From (5.112)–(5.114) we see that $d^3\mathbf{k}' = d^3\mathbf{k}/\gamma$, so (5.123) can be written as an integral over wave vectors in the laboratory frame in the form

$$\mathbf{E}(\mathbf{r}, t) = \frac{1}{(2\pi)^{3/2}\gamma} \int_{-\infty}^{\infty} d^3\mathbf{k}\, e^{-ik^\alpha r_\alpha} \tilde{\mathbf{E}}(\mathbf{k}) \tag{5.124}$$

with a corresponding expression for $\mathbf{B}(\mathbf{r}, t)$. This is the required Fourier decomposition of the field.

EXERCISE 5.10

Consider the motion of a particle of charge Q along the x axis at a highly relativistic velocity ($\gamma \gg 1$). Starting with the electric field of a moving charge discussed in Chapter 1, show that the electric field at point b on the y axis has the Fourier components

$$\tilde{E}_y(\omega) = \frac{1}{(2\pi)^{3/2}} \frac{Q}{\varepsilon_0 bc} \frac{\omega b}{\gamma c} K_1\left(\frac{\omega b}{\gamma c}\right) \tag{5.125}$$

in frequency space, where $K_1(x)$ is a modified Bessel function. *Hint:* you need the integral

$$\int_{-\infty}^{\infty} \frac{e^{i\omega t}\, dt}{(t^2 + \tau^2)^{3/2}} = 2\frac{\omega}{\tau} K_1(\omega\tau) \tag{5.126}$$

5.2.2 Bremsstrahlung

When an electron (or other charged particle) passes through ordinary matter, the incident particle is deflected by collisions with the electrons and the nuclei that make up the material. The acceleration experienced by the incident particle in the collisions causes the particle to emit what is called "Bremsstrahlung." This is German for "braking radiation," so called because of its association with the slowing down of fast particles as they travel through matter, even though the radiation is actually associated principally with the deflection of the particles rather than their slowing down. To use the method of virtual quanta to calculate the Bremsstrahlung emitted by an electron in a collision with a nucleus, we invert the problem and consider a heavy particle of mass M and charge Q incident on a light particle of mass m

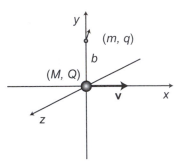

Figure 5.15 Heavy particle incident on a light particle.

and charge q, as shown in Figure 5.15. The heavy particle is presumed to move uniformly along the x axis, and the light particle, initially at point b on the y axis, recoils due to the interaction and emits the radiation. To find the Bremsstrahlung in the laboratory frame, we have only to Lorentz transform back into the rest frame of the heavy particle.

To calculate the radiation emitted by the electron in the collision, we think of the electron as responding to a pulse of light equivalent to the field of the incident heavy particle. The radiation emitted is called Thomson scattering of the incident radiation. As shown in Chapter 10, the differential cross section for Thomson scattering is

$$\frac{d\sigma_T}{d\Omega} = r_c^2 \sin^2 \psi \tag{5.127}$$

where $d\Omega$ is the element of solid angle into which the radiation is scattered and ψ is the angle between the direction of polarization of the incident light and the direction into which the light is scattered. That is, the scattered radiation has the pattern of dipole radiation in the rest frame of the incident electron. The quantity

$$r_c = \frac{q^2}{4\pi \varepsilon_0 mc^2} \tag{5.128}$$

is called the classical radius of the particle, and for an electron it has the value $r_c = 2.82 \times 10^{-15}$ m. The Thomson cross section (5.127) is independent of the intensity of the incident radiation and independent of the frequency provided that the particle motions remain nonrelativistic and the Compton effect can be ignored. This is valid at low frequencies and low intensities. In the present context the low-frequency limit includes the x-ray region and the low-intensity limit is valid for sufficiently large impact parameters $b/r_c Z \gg 1$, where Z is the atomic number of the nucleus.

The quantity in which we are interested is the angular spectral fluence of the scattered radiation. This is the total energy dW scattered per unit solid angle $d\Omega$ per unit frequency interval $d\omega$. To find it, we need to compute the total spectral fluence incident on the particle and multiply by the Thomson scattering cross section. We proceed as follows.

In the rest frame of the light particle, the electric field of the incident particle is given by (5.124). To carry out the integral over $\mathbf{k}$, we note that in the ultrarelativistic limit

$$k_x \approx \frac{\omega}{c} \tag{5.129}$$

In this same limit we may ignore the longitudinal electric field $\tilde{E}_x$ compared with the transverse components of $\tilde{\mathbf{E}}$, so for a target particle at point b on the y axis we need only the component $\tilde{E}_y$, given by (5.117). Substituting into (5.124), we get

$$E_y(t) = \frac{i}{(2\pi)^3}\frac{Q}{\varepsilon_0 c}\int_{-\infty}^{\infty} d\omega \int_{-\infty}^{\infty} dk_y \int_{-\infty}^{\infty} dk_z \frac{k_y}{\frac{\omega^2}{\gamma^2 c^2} + k_y^2 + k_z^2} e^{ik_y b} e^{-i\omega t} \quad (5.130)$$

for $\beta \approx 1$. We integrate first over k_z, using the formula

$$\int_{-\infty}^{\infty} \frac{dx}{a^2 + x^2} = \frac{\pi}{a} \quad (5.131)$$

and then over k_y, using the formula

$$\int_{-\infty}^{\infty} \frac{e^{i\xi x} x\, dx}{(a^2 + x^2)^{1/2}} = i2aK_1(a\xi) \quad (5.132)$$

where $K_1(x)$ is a modified Bessel function of first order. This gives us the result

$$E_y(t) = \frac{1}{(2\pi)^{1/2}}\int_{-\infty}^{\infty} d\omega\, \tilde{E}_y(\omega) e^{-i\omega t} \quad (5.133)$$

in which

$$\tilde{E}_y(\omega) = \frac{1}{(2\pi)^{3/2}}\frac{Q}{\varepsilon_0 c}\frac{1}{b}\frac{\omega b}{\gamma c}K_1\!\left(\frac{\omega b}{\gamma c}\right) \quad (5.134)$$

is the Fourier transform of the incident field at the point b.

Since the field at point b on the y axis has the character of an incident light pulse, we may attribute to it the total incident fluence (energy per unit area)

$$\mathcal{J}_b = \hat{\mathbf{x}} \cdot \int_{-\infty}^{\infty} dt\, \frac{1}{\mu_0}\mathbf{E}\times\mathbf{B} = \varepsilon_0 c\int_{-\infty}^{\infty} dt\, E^2 = \varepsilon_0 c\int_{-\infty}^{\infty} d\omega\, \tilde{E}(\omega)\,\tilde{E}^*(\omega) = \int_0^{\infty} d\omega\, \frac{d\mathcal{J}_b}{d\omega} \quad (5.135)$$

where the last step (limiting the integral to positive frequencies only) is possible because the field is real. The spectral fluence (fluence per unit frequency interval) incident on the particle is therefore

$$\frac{d\mathcal{J}_b}{d\omega} = 2\varepsilon_0 c\tilde{E}(\omega)\tilde{E}^*(\omega) = \frac{1}{4\pi^3}\frac{Q^2}{\varepsilon_0 c}\frac{1}{b^2}\left(\frac{\omega b}{\gamma c}\right)^2 K_1^2\!\left(\frac{\omega b}{\gamma c}\right) \quad (5.136)$$

This is illustrated in Figure 5.16, where we see that the spectrum has the characteristic width

$$\omega_b = \frac{\gamma c}{b} = \frac{1}{\tau_b} \quad (5.137)$$

where $\tau_b = b/\gamma c$ is the duration of the collision, as discussed earlier. The angular spectral fluence (spectral fluence per unit solid angle) of the radiation scattered by the target particle is therefore

$$\frac{d^2\mathcal{W}}{d\omega\, d\Omega} = \frac{d\mathcal{J}_b}{d\omega}\frac{d\sigma_T}{d\Omega} = \frac{1}{4\pi^3}\frac{Q^2}{\varepsilon_0 c}\frac{r_c^2}{b^2}\left(\frac{\omega b}{\gamma c}\right)^2 K_1^2\!\left(\frac{\omega b}{\gamma c}\right)\sin^2\psi \quad (5.138)$$

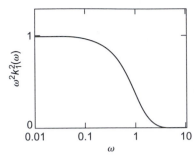

Figure 5.16 Spectral fluence of virtual photons.

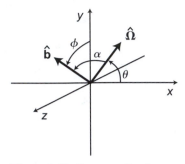

Figure 5.17 Geometry for the integration over all impact parameters.

To calculate the total radiation emitted when a charged particle passes through matter we must integrate this result over all impact parameters b and all azimuthal angles ϕ, as indicated in Figure 5.17. The required integral is therefore

$$\int_0^\infty b\,db \int_0^{2\pi} d\phi \frac{d^2\mathcal{W}}{d\omega\,d\Omega} = \frac{1}{4\pi^3} \frac{Q^2 r_c^2}{\varepsilon_0 c} \int_0^{2\pi} \sin^2 \psi \, d\phi \int_0^\infty \frac{db}{b} \left(\frac{\omega b}{\gamma c}\right)^2 K_1^2\left(\frac{\omega b}{\gamma c}\right) \quad (5.139)$$

where ψ is now the angle between the unit vector $\hat{\mathbf{b}}$ in the direction of the impact parameter and the unit vector $\hat{\mathbf{\Omega}}$ in the direction of observation. Without loss of generality we may place the direction of observation at the angle θ in the x–y plane. But $\cos\psi = \hat{\mathbf{\Omega}} \cdot \hat{\mathbf{b}} = \sin\theta \cos\phi$, so

$$\sin^2 \psi = 1 - \cos^2 \psi = 1 - \sin^2 \theta \cos^2 \phi \quad (5.140)$$

and the integral over azimuthal angles is just

$$\int_0^{2\pi} \sin^2 \psi \, d\phi = 2\pi \left(1 - \frac{1}{2}\sin^2\theta\right) \quad (5.141)$$

To evaluate the integral over the impact parameter b, we note that the spectral fluence vanishes as $b \to \infty$, so the integral is well behaved there. In fact, as shown in Figure 5.16, the spectral fluence cuts off fairly quickly for $b > b_{\max}$, where

$$b_{\max} = \frac{\gamma c}{\omega} \quad (5.142)$$

However, as $b \to 0$ the spectral fluence diverges, so the integral has a logarithmic singularity there. We deal with this by truncating the integral in one of the following ways.

1. Implicit in the preceding analysis is the assumption that the motions of the target particle are nonrelativistic. This is necessary for the validity of the Thomson cross section (5.127). However, as shown in Chapter 10, the scattering cross section falls off when the particle motions become relativistic, so we can truncate the integral when this occurs. To estimate the point at which the target particle motions become relativistic, we note that the peak field experienced by the target particle is $E_{\max} = \gamma Q/4\pi\varepsilon_0 b^2$. The momentum is therefore of the order of $p_{\max} = q E_{\max} \tau_b$,

where τ_b is the collision duration. If the motion of the target particle remains nonrelativistic, then

$$\frac{p_{\max}}{mc} = \frac{Q}{q}\frac{r_c}{b} \ll 1 \tag{5.143}$$

The minimum impact parameter is therefore

$$b_{\min} = \frac{Q}{q}r_c \tag{5.144}$$

2. At very small impact parameters, the angular momentum γmcb becomes smaller than $\hbar$. Quantum-mechanical considerations suggest that we cannot distinguish angular momentum this small, so the minimum value of the impact parameter is

$$\gamma mcb_{\min} = O(\hbar) \tag{5.145}$$

For Bremsstrahlung created by the deflection of an electron by a nucleus of atomic number $Z = Q/q$, the crossover between these limits occurs at

$$\gamma Z = \frac{1}{\alpha} \tag{5.146}$$

where $\alpha = q_e^2/4\pi\varepsilon_0\hbar c = 1/137$ is the so-called fine structure constant.

We can now evaluate the integral over b from $b_{\min}$ to $b_{\max}$. Since they enter only logarithmically, the exact limits are not critical and we get

$$\int_0^\infty b\,db \int_0^{2\pi} d\phi \frac{d^2\mathcal{W}}{d\omega\,d\Omega} = \frac{Q^2 r_c^2}{2\pi^2\varepsilon_0 c}\left(1 - \frac{1}{2}\sin^2\theta\right)\ln\left(\frac{b_{\max}}{b_{\min}}\right) \tag{5.147}$$

where $b_{\max}$ is found from (5.142) and $b_{\min}$ must be determined from consideration (1) or (2). Clearly, this is valid only when $b_{\max} \gg b_{\min}$, that is, when the argument of the logarithm is large compared with unity.

The limits of validity of this result depend on several effects that we have not yet considered. At very high frequencies, a limit is established by the quantum recoil of the target particle, called the Compton effect. The momentum of the emitted photon is $p_k = \hbar k = \hbar\omega/c$, so the recoil velocity can be neglected when $p_k < O(mc)$ or

$$\frac{\hbar\omega}{mc^2} \ll 1 \tag{5.148}$$

When the quantum-mechanical recoil is taken into account, the scattering cross section is reduced as discussed in Chapter 10.

At low frequencies, (5.147) again fails due to altogether different effects. In this limit Bremsstrahlung is dominated by large impact parameters, and these are affected by screening. Particularly in high-Z materials, the production of Bremsstrahlung is dominated by collisions with the nuclei, since these produce much larger deflections than collisions with the electrons. At large impact parameters the nuclear charge is screened by the electrons, which reduces the effective field of the nucleus.

EXERCISE 5.11

Lorentz transform the scattered radiation into the rest frame of the heavy particle to find the angular spectral fluence of Bremsstrahlung from a light particle of charge q and mass m traveling in the $\hat{\mathbf{x}}$ direction at velocity βc, incident on a heavy particle of charge Q. You may proceed as follows.

(a) By Lorentz transforming the energy and the wave vector of radiation scattered in the $\hat{\mathbf{n}}$ direction, show that

$$\frac{d^2\mathcal{W}}{d\omega\,d\Omega} = \frac{1}{\gamma^2(1-\beta\hat{\mathbf{n}}\cdot\hat{\mathbf{x}})^2}\frac{d^2\mathcal{W}'}{d\omega'd\Omega'} \approx \frac{4\gamma^2}{(1+\gamma^2\theta^2)^2}\frac{d^2\mathcal{W}'}{d\omega'd\Omega'} \qquad (5.149)$$

in the ultrarelativistic limit, where primed quantities refer to the moving coordinate system and unprimed quantities to the laboratory frame, and $\cos\theta = \hat{\mathbf{n}}\cdot\hat{\mathbf{x}}$. *Hint:* Since they represent radiation, $\mathcal{W}$ and ω are both the timelike components of null 4-vectors and therefore transform together.

(b) Use this result to transform (5.147) to the laboratory frame, and show that for a light particle incident on a heavy particle the angular spectral fluence of Bremsstrahlung is

$$\int_0^\infty b\,db \int_0^{2\pi} d\phi\,\frac{d^2\mathcal{W}}{d\omega\,d\Omega} = \frac{2Q^2r_c^2}{\pi^2\varepsilon_0 c}\frac{\gamma^2(1+\gamma^4\theta^4)}{(1+\gamma^2\theta^2)^4}\ln\left(\frac{b_{max}}{b_{min}}\right) \qquad (5.150)$$

in the ultrarelativistic limit.

(c) Show that the maximum impact parameter is

$$b_{max} = \frac{2\gamma^2 c}{\omega(1+\gamma^2\theta^2)} \qquad (5.151)$$

where ω is the frequency in the laboratory frame.

5.2.3 Excitation by a Fast Charged Particle

The concept of an optical fluence equivalent to the pulsed electric field of a passing particle can be used to compute the cross section for excitation of an atom or other system by a charged particle using known optical excitation cross sections. The idea was first applied by Fermi, who used x-ray absorption data to calculate the energy loss of high-energy particles due to ionization of the atoms in the material through which they are passing. The idea has also been applied to nuclear photodisintegration, pion production, and (in a quantum-mechanical version) electron–positron pair production. Since the incident particles pass the target at all azimuthal angles, the polarization of the virtual quanta (which is oriented radially away from the axis of the trajectory of the moving charge) is transverse to the direction of motion of the particle beam but otherwise random. The excitation probability therefore depends only on the polarization-averaged optical cross section.

Provided that the probability of excitation is small compared with unity, the probability of excitation of the target by an incident optical spectral fluence $d\mathcal{J}_b/d\omega$ is

$$P = \int_0^\infty \frac{\sigma(\omega)}{\hbar\omega}\frac{d\mathcal{J}_b}{d\omega}\,d\omega \qquad (5.152)$$

where $\sigma(\omega)$ is the optical excitation cross section and $(d\omega/\hbar\omega)(d\mathcal{J}_b/d\omega)$ is the number of photons per unit area in the frequency interval $d\omega$ near the frequency ω. In the semiclassical approximation, the electron-impact excitation cross section is

$$\sigma_e = \int_0^\infty db \, 2\pi b P_b \tag{5.153}$$

where P_b (assumed to be $\ll 1$) is the quantum-mechanical probability of excitation for a particle incident with the classical impact parameter b. If we use the equivalent optical fluence to compute the excitation probability, we find that the cross section is

$$\sigma_e = \int_0^\infty db \int_0^\infty d\omega \, \frac{\sigma(\omega)}{\hbar\omega} 2\pi b \frac{d\mathcal{J}_b}{d\omega} = \frac{q^2}{2\pi^2\varepsilon_0 c} \int_0^\infty d\omega \, \frac{\sigma(\omega)}{\hbar\omega} \int_0^\infty \frac{dx}{x} x^2 K_1^2(x) \tag{5.154}$$

where $x = \omega b/\gamma c$. Mathematically, the integral over x is well behaved at infinity, but diverges logarithmically at the lower limit, that is, for small impact parameters. In fact, as discussed in the previous section, the function $x^2 K_1^2(x)$ vanishes rapidly above $x = O(1)$ and is near unity below this. We may therefore use the approximation

$$\int \frac{dx}{x} x^2 K_1^2(x) \approx \ln\left(\frac{b_{\max}}{b_{\min}}\right) \tag{5.155}$$

where the exact value of the limits is not critical, since they enter only logarithmically. Based on the cutoff of $x^2 K_1^2(x)$ near $x = 1$, we may use

$$b_{\max} = \frac{\gamma c}{\omega} \tag{5.156}$$

for the upper limit unless physical considerations restrict it to a smaller value. The lower limit $b_{\min}$ must be determined by other considerations. Three possibilities suggest themselves.

1. When the impact parameter b is comparable to the size of the target, the fluence of virtual photons is not uniform over the target, so the concept of an optical fluence ceases to be meaningful.
2. For very small impact parameters, the equivalent fluence becomes enormous and the excitation probability predicted by (5.152) exceeds unity. Of course, this is impossible.
3. At very small impact parameters, the angular momentum $\gamma m c b$ becomes smaller than $\hbar$. Quantum-mechanical considerations suggest that we cannot distinguish angular momentum this small, as discussed earlier.

Which of these considerations is most restrictive depends on the circumstances. We consider two examples.

An important source of neutrons that must be considered in the shielding of high-energy particle accelerators is the (γ, n) reaction

$$\gamma + {}^A X \rightarrow {}^{A-1}X + n \tag{5.157}$$

in which a γ-ray photon interacts with a nucleus ${}^A X$ of mass number A to produce a photoneutron. Many materials exhibit a "giant resonance" in the cross section around 20 MeV. The corresponding electron-impact process is

$$e^- + {}^A X \rightarrow e^- + {}^{A-1}X + n \tag{5.158}$$

The electron-impact cross section rises rapidly above the giant resonance and then more slowly toward 100 MeV. At an energy of 50 MeV ($\gamma = 100$), the minimum impact parameter based on quantum considerations is

$$b_{min} = O\left(\frac{\hbar}{\gamma mc}\right) = O(4 \times 10^{-15} \text{ m}) \tag{5.159}$$

This is comparable to the size of a typical nucleus. At energies above about 50 MeV, the minimum impact parameter is established by the nuclear radius, $b_{min} = R$. The cross section is then

$$\sigma_e \approx \frac{q^2}{2\pi^2 \varepsilon_0 c} \int_0^\infty \frac{\sigma(\omega)}{\hbar\omega} \ln\left(\frac{\gamma c}{\omega_0 R}\right) d\omega \tag{5.160}$$

When the photonuclear cross section is narrowly confined near a giant resonance at ω_0, we may extract the other factors from the integral to get

$$\sigma_e \approx \frac{q^2}{2\pi^2 \varepsilon_0 c\hbar\omega_0} \ln\left(\frac{\gamma c}{\omega_0 R}\right) \int_0^\infty \sigma(\omega)\, d\omega \tag{5.161}$$

This exhibits a slow (logarithmic) variation of the cross section with electron energy. Below about 50 MeV, the minimum impact parameter is limited by quantum-mechanical considerations and the electron-impact excitation cross section is then

$$\sigma_e \approx \frac{q^2}{2\pi^2 \varepsilon_0 c\hbar\omega_0} \ln\left(\frac{\gamma^2 mc^2}{\hbar\omega_0}\right) \int_0^\infty \sigma(\omega)\, d\omega \tag{5.162}$$

As a second example, we consider the electron-impact ionization of an atom. The photoionization cross section peaks just above threshold and falls off at higher frequencies, so the electron-impact cross section is dominated by low-energy virtual photons that are produced at large impact parameters. The smallest impact parameter may be taken to be on the order of the size of an atom, $b_{min} = a_0$ (the Bohr radius). For dilute material we can use (5.156) for the upper limit of integration, so

$$\sigma_e \approx \frac{q^2}{2\pi^2 \varepsilon_0 c} \int_{\omega_0}^\infty d\omega \frac{\sigma(\omega)}{\hbar\omega} \ln\left(\frac{\gamma c}{a_0 \omega}\right) \tag{5.163}$$

where ω_0 is the photoionization threshold.

When the atom is part of a condensed phase, another effect, called the density effect, must be taken into account. Since the photoionization is dominated by low-energy virtual photons produced at large impact parameters, absorption of these photons by the material through which the electron is traveling reduces the optical fluence at large distances. For example, for $\gamma = 1000$, 200-eV virtual photons are numerous out to radii as large as $\gamma c/\omega = 1$ µm, but the absorption length for 200-eV photons in water, for example, is comparable. Moreover, the index of refraction for the low-energy photons is large enough to make the velocity of the incident particle superluminal at these frequencies. This alters the character of the virtual photons. The density effect becomes more pronounced as the electron energy gets higher and reduces the energy dependence of electron-impact ionization in condensed materials. The excitation and ionization of condensed matter by fast charged particles is discussed in detail from another point of view in Chapter 7, where the density effect appears in a natural way.

EXERCISE 5.12

The field of a bare magnetic monopole with no electric charge satisfies the Maxwell equations

$$\partial_\alpha F^{\alpha\beta} = 0 \tag{5.164}$$

and

$$\partial_\alpha \mathcal{F}^{\alpha\beta} = \mu_0 J_m{}^\beta \tag{5.165}$$

These may be obtained from the equations for an electric charge by exchanging

$$\mathcal{F}^{\alpha\beta} \leftrightarrow F^{\alpha\beta} \tag{5.166}$$

and

$$J^\beta \leftrightarrow J_m{}^\beta \tag{5.167}$$

where the transformation $\mathcal{F}^{\alpha\beta} \leftrightarrow F^{\alpha\beta}$ is accomplished most easily by exchanging the electric and magnetic fields as described in Chapter 1.

(a) Exchange the fields in the formulas for the fields of an electric charge to show that the electric field of a magnetic monopole may be decomposed into virtual quanta of the form

$$\mathbf{E}(\mathbf{r}, t) = \frac{1}{(2\pi)^{3/2}} \int_{-\infty}^{\infty} d^3\mathbf{k}' e^{-ik^\alpha r_\alpha} \tilde{\mathbf{E}}(\mathbf{k}) \tag{5.168}$$

where

$$\tilde{E}_x = 0 \tag{5.169}$$

$$\tilde{E}_y = \frac{i}{(2\pi)^{3/2}} \frac{q_m}{\varepsilon_0} \frac{\beta\gamma k_z}{\dfrac{\omega^2}{\beta^2\gamma^2 c^2} + k_y^2 + k_z^2} \tag{5.170}$$

$$\tilde{E}_z = \frac{-i}{(2\pi)^{3/2}} \frac{q_m}{\varepsilon_0} \frac{\beta\gamma k_y}{\dfrac{\omega^2}{\beta^2\gamma^2 c^2} + k_y^2 + k_z^2} \tag{5.171}$$

Thus, from (5.135) we see that the spectral fluence of virtual quanta from a magnetic monopole differs from that of an electrically charged particle only by the ratio $(q_m/q)^2$.

(b) Using the Dirac charge quantization rule from Chapter 2, show that the cross section for excitation by passage of an ultrarelativistic magnetic monopole through matter is larger than that of an electrically charged particle by the factor

$$\frac{\sigma_m}{\sigma_e} = \frac{n^2}{4\alpha^2} \tag{5.172}$$

where the dimensionless quantity α is

$$\alpha = \frac{q^2}{4\pi\varepsilon_0 \hbar c} \tag{5.173}$$

and n is an integer. When the charge q is the electron charge, $\alpha = 1/137$ is called the fine-structure constant. This shows that the effect of a magnetic monopole passing through matter should be about 10^4 times as great as that of an electron or proton. If magnetic monopoles exist, they should be relatively easy to observe.

5.2.4 Transition Radiation

When a charged particle crosses the interface from a vacuum into a conductor or dielectric, it gives off radiation as shown in Figure 5.18. This is called transition radiation, and it is equivalent to specular reflection of the incident virtual quanta at the interface. Clearly, this description of the phenomenon is valid only at frequencies where the conductor or dielectric is a reflector. At x-ray wavelengths the same ideas apply, except that the incident quanta are Bragg reflected off the crystal planes, as illustrated in Figure 5.19. The radiation in this case is called parametric x-rays, and the angular spectrum is modified by the Bragg condition. Here we limit our discussion to the transition radiation created when a charged particle enters a perfect conductor. When the medium is a dielectric, the same approach is valid except that the Fresnel reflection coefficient is used to calculate the fraction of the incident virtual quanta that is reflected as transition radiation.

For a perfect conductor, the entire radiation field is reflected. The total energy in the field is

$$W = \int_{-\infty}^{\infty} d^3\mathbf{r} \left(\frac{\varepsilon_0}{2}\mathbf{E} \cdot \mathbf{E} + \frac{1}{2\mu_0}\mathbf{B} \cdot \mathbf{B} \right) \tag{5.174}$$

But in the ultrarelativistic limit the fields are very nearly electromagnetic waves, so the electric and magnetic contributions to the energy are equal. We can therefore write

$$W = \varepsilon_0 \int_{-\infty}^{\infty} d^3\mathbf{r}\,\mathbf{E} \cdot \mathbf{E}^* \tag{5.175}$$

since the fields are real. If we use the Fourier inversion (5.124) to express the electric fields as integrals over $\mathbf{k}$ and $\mathbf{k}'$ we get

$$W = \frac{1}{(2\pi)^3} \frac{1}{\gamma^2} \int_{-\infty}^{\infty} d^3\mathbf{r} \int_{-\infty}^{\infty} d^3\mathbf{k} \int_{-\infty}^{\infty} d^3\mathbf{k}' \, \tilde{\mathbf{E}}(\mathbf{k}) \cdot \tilde{\mathbf{E}}^*(\mathbf{k}') e^{-i(k^\alpha - k'^\alpha)r_\alpha} \tag{5.176}$$

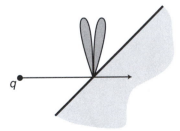

Figure 5.18 Transition radiation.

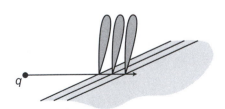

Figure 5.19 Parametric x-rays.

When we carry out the integral over $d^3\mathbf{r}$, we get a δ-function $(2\pi)^3\delta(\mathbf{k}' - \mathbf{k})$, and after we evaluate the integral over $\mathbf{k}'$ trivially, we find that the energy in the field is

$$W = \frac{\varepsilon_0}{\gamma^2}\int_{-\infty}^{\infty} d^3\mathbf{k}\,\tilde{\mathbf{E}}(\mathbf{k})\cdot\tilde{\mathbf{E}}^*(\mathbf{k}) = \frac{\varepsilon_0}{\gamma^2}\int_{-\infty}^{\infty} d^3\mathbf{k}[\tilde{E}_y(\mathbf{k})\tilde{E}_y^*(\mathbf{k}) + \tilde{E}_z(\mathbf{k})\tilde{E}_z^*(\mathbf{k})] \quad (5.177)$$

where we have ignored the longitudinal component of the field in the ultrarelativistic limit. Substituting (5.117) and (5.118) for the Fourier transforms we get

$$W = \frac{q^2}{8\pi\varepsilon_0}\int_{-\infty}^{\infty} d^3\mathbf{k}\,\frac{k_y^2 + k_z^2}{\left(\dfrac{\omega^2}{\gamma^2 c^2} + k_y^2 + k_z^2\right)^2} \quad (5.178)$$

In the ultrarelativistic limit, we see from (5.112) that the frequency is

$$\omega \approx k_x c \approx kc \quad (5.179)$$

Before they are reflected, all the waves in the virtual quantum field are traveling to the right, since they are fixed to the particle. Waves for which $k_x < 0$ correspond to negative frequencies and still travel to the right. Therefore, we can avoid the negative frequencies by noting that the integrand is even, so (5.177) becomes

$$W = \frac{\gamma^4 q^2}{4\pi^3\varepsilon_0}\int_{k_x>0} d^3\mathbf{k}\,\frac{k_y^2 + k_z^2}{\left[k^2 + \gamma^2\left(k_y^2 + k_z^2\right)\right]^2} \quad (5.180)$$

The direction $\hat{\mathbf{k}}$ is the direction of propagation of the waves. If we change to polar coordinates in $\mathbf{k}$-space, so that

$$d^3\mathbf{k} = k^2 dk\, d\Omega \quad (5.181)$$

in which $d\Omega$ is an element of solid angle, we may write the energy in the field in the form

$$W = \frac{\gamma^4 q^2}{4\pi^3\varepsilon_0 c}\int_0^{\infty} d\omega\int_{2\pi} d\Omega\,\frac{\theta^2}{(1 + \gamma^2\theta^2)^2} \quad (5.182)$$

where

$$\sin\theta = \frac{\sqrt{k_y^2 + k_z^2}}{k} \approx \theta = O(1/\gamma) \ll 1 \quad (5.183)$$

is the angle between the direction of propagation of the wave and the direction of motion of the charged particle. We immediately identify the angular spectral fluence of the virtual quanta as

$$\frac{d^2W}{d\omega\,d\Omega} = \frac{\gamma^4 q^2}{4\pi^3\varepsilon_0 c}\frac{\theta^2}{(1 + \gamma^2\theta^2)^2} \quad (5.184)$$

The angular spectral fluence of transition radiation is just the specular reflection of the virtual quanta from the face of the conductor, as illustrated in Figure 5.18. Note that the angular fluence of the virtual quanta vanishes in the direction of motion of the charge ($\theta = 0$).

The spectrum of transition radiation represented by (5.184) is independent of frequency and diverges when it is integrated over all frequencies. As noted previously, this is related to the fact that the energy in the field of a point charge diverges when integrated over all space due to the singularity at the position of the charge. In the case of transition radiation, this is not a problem since metals are poor reflectors at high frequencies, so transition radiation diminishes in the ultraviolet. To integrate (5.184) over all solid angles, we note that

$$d\Omega = 2\pi \sin\theta \, d\theta \approx 2\pi\theta \, d\theta \qquad (5.185)$$

at angles $\theta = O(1/\gamma) \ll 1$ where the integrand is large. The total spectral intensity is then

$$\frac{dW}{d\omega} = \frac{\gamma^4 q^2}{2\pi^2 \varepsilon_0 c} \int_0^{\pi/2} d\theta \, \frac{\theta^3}{(1+\gamma^2\theta^2)^2} \approx \frac{q^2}{2\pi^2 \varepsilon_0 c} \ln\gamma \qquad (5.186)$$

This shows that the total spectral intensity of transition radiation increases only slowly with particle energy for $\gamma \gg 1$.

It is interesting to examine the spot size over which the transition radiation from a single particle is radiated. It varies with the frequency of the radiation, for we see from Figure 5.16 that the fluence of virtual photons extends to

$$\frac{\omega b}{\gamma c} = \frac{2\pi b}{\gamma \lambda} = O(1) \qquad (5.187)$$

where $\lambda = 2\pi/\omega$ is the wavelength of the radiation. For example, for 50-MeV electrons ($\gamma = 100$), the radius out to which visible photons ($\lambda \approx 0.5 \, \mu$m) are radiated is $b \approx 8 \, \mu$m. But the diffraction angle for an aperture of diameter $2b$ is $\theta = O(\lambda/2b)$, so on the basis of diffraction effects we would expect the angular extent of virtual quanta to be $\theta = O(\pi/\gamma)$, independent of the wavelength. Comparing this with the angular distribution predicted by (5.184), we see that the angular fluence of transition radiation (and of the virtual quanta themselves) is roughly diffraction limited.

EXERCISE 5.13

For a single charged particle at the origin moving in the $\hat{\mathbf{x}}$ direction at a highly relativistic velocity, the electric field surrounding the particle can be represented by an equivalent radiation field. As shown by (5.124), this representation has the form

$$\mathbf{E}(\mathbf{r}, t) = \frac{1}{(2\pi)^{3/2}\gamma} \int_{-\infty}^{\infty} d^3\mathbf{k} \, e^{-ik^\alpha r_\alpha} \tilde{\mathbf{E}}(\mathbf{k}) \qquad (5.188)$$

where the wave 4-vector is $k^\alpha = (\omega/c, \mathbf{k})$ and $\omega^2 \approx k^2 c^2$ in the ultrarelativistic limit. Consider a bunch of N identical particles traveling parallel to one another at positions $\delta\mathbf{r}_i$ relative to the center of the bunch.

(a) Show that the total electric field of the bunch is

$$\mathbf{E}(\mathbf{r}, t) = \frac{1}{(2\pi)^{3/2}\gamma} \int_{-\infty}^{\infty} d^3\mathbf{k} \, e^{-ik^\alpha r_\alpha} \tilde{\mathbf{E}}(\mathbf{k}) \sum_i e^{i\mathbf{k}\cdot\delta\mathbf{r}_i} \qquad (5.189)$$

(b) Show that the spectral fluence of the virtual photons for the bunch of N particles is related to that of a single particle by the expression

$$\left(\frac{d^2\mathcal{W}}{d\omega\, d\Omega}\right)_N = \left(\frac{d^2\mathcal{W}}{d\omega\, d\Omega}\right)_1 \left|\sum_i e^{i\mathbf{k}\cdot\delta\mathbf{r}_i}\right|^2 \qquad (5.190)$$

If the particles are distributed randomly, then $e^{i\mathbf{k}\cdot\delta\mathbf{r}_i}$ is a random unit vector in the complex plane and the sum of N such vectors is $O(\sqrt{N})$ in magnitude. The additional factor is then

$$\left|\sum_i e^{i\mathbf{k}\cdot\delta\mathbf{r}_i}\right|^2 \approx N \qquad (5.191)$$

for $N \gg 1$, so the total spectral fluence is just the sum of the individual fluences. However, if the particles are very close together ($\mathbf{k}\cdot\delta\mathbf{r}_i \ll 1$), then the fields add coherently ($e^{i\mathbf{k}\cdot\delta\mathbf{r}_i} \approx 1$) and

$$\left|\sum_i e^{i\mathbf{k}\cdot\delta\mathbf{r}_i}\right|^2 \approx N^2 \qquad (5.192)$$

When the bunch contains a large number of particles, $N \gg 1$, coherent effects substantially increase the spectral fluence for wavelengths larger than the bunch size. A measurement of the spectrum therefore provides an estimate (equivalent to an autocorrelation) of the bunch size.

BIBLIOGRAPHY

Most textbooks on complex variables and applied mathematics include discussions of Fourier transforms. Good examples include

G. F. Carrier, M. Krook, and C. E. Pearson, *Functions of a Complex Variable,* McGraw-Hill Book Company, New York (1966),

P. M. Morse and H. Feshbach, *Methods of Theoretical Physics,* McGraw-Hill Book Company, New York (1953).

A brief but excellent introduction to Fourier transforms and δ-functions is found in

M. J. Lighthill, *Introduction to Fourier Analysis and Generalized Functions,* Cambridge University Press, Cambridge, U.K. (1958).

The use of Fourier transforms and the Weiner–Khintchine theorem in statistical mechanics is explained in a very thorough fashion by

R. K. Pathria, *Statistical Mechanics,* 2nd edition, Butterworth-Heinemann, Oxford (1966).

The method of virtual quanta, or equivalent photons, and examples of its use are discussed by

J. D. Jackson, *Classical Electrodynamics,* 3rd edition, John Wiley & Sons, New York (1999),

F. E. Low, *Classical Field Theory,* John Wiley & Sons, New York (1997).

6 Macroscopic Materials

Aside from the discussion of boundary-value problems involving conductors, our attention has focused largely on the electrodynamics of point charges in vacuum. Within the realm of classical electrodynamics, this provides a correct description of the universe and all that is in it, since rigid, extended particles are inconsistent with the principles of relativity. However, this description is useful only for problems that involve just a small number of particles. On the other hand, even the tiniest macroscopic object consists of enormous numbers of particles, and the large variety of interesting phenomena that arise from macroscopic collections of particles must be described by a theory that smoothes out the microscopic complexity of matter. It is the purpose of this chapter to develop such a theory and to explore some of the electromagnetic phenomena that arise from large collections of particles acting together.

Most of the macroscopic phenomena that arise from the interaction of electric and magnetic fields with matter, including the refraction and scattering of visible light, occur on a length scale large enough that the material may be treated as continuous. A few phenomena, such as x-ray scattering, occur on a much smaller length scale, and the microscopic nonuniformity of matter is critical. Unfortunately, we do not have time to discuss these fascinating phenomena here. In this chapter we develop the general formalism for the electrodynamics of continuous materials. We include just enough discussion of the microscopic nature of solids, liquids, gases, and plasmas to motivate the macroscopic description.

We discuss in a general way both linear and nonlinear phenomena. Most macroscopic phenomena are linear in the strength of the electric and magnetic fields, and linear phenomena and the powerful techniques that have been developed to deal with them are discussed in depth in Chapter 7. But ferromagnetism and ferroelectricity are familiar examples of nonlinear phenomena, and in recent years, with the development of high-power lasers, many interesting nonlinear optical phenomena have been discovered, some of which are achieving scientific and even commercial importance. These include harmonic generation, optical phase conjugation, and solitons, to name just a few. In this chapter we discuss ferromagnetism, in part because of its intrinsic interest, but also to illustrate the concepts of energy and dissipation in nonlinear macroscopic materials. The broader class of phenomena that compose nonlinear optics is deferred to Chapter 8.

6.1 POLARIZATION AND MAGNETIZATION

6.1.1 The Macroscopic Form of the Maxwell Equations

Our discussion begins with the Maxwell equations, which we modify to represent the effects of the macroscopic medium in a simple way. At the microscopic level, matter consists of electrons and nuclei in prodigious numbers. Even in a classical description the electrons and nuclei are constantly moving, and on a microscopic scale the electric and magnetic fields are fluctuating rapidly in time and space. To describe what we would refer to as macroscopic phenomena, we need a set of equations that average over these fluctuations. The kind of average that we need depends on the phenomena that we are trying to describe and on the relation between the length and time scales of the macroscopic phenomena and those of the microscopic fluctuations. We find that for a wide variety of phenomena it is sufficient to perform a spatial average over the microscopic fluctuations, and that it is neither possible nor necessary to perform a temporal average.

Since macroscopic materials are composed of atoms, which are in turn composed of smaller entities, the shortest length scale on which averaging might be useful must be large compared with the spacing of the atoms. In common table salt, for example, the lattice spacing is $a = 5.6 \times 10^{-10}$ m. Clearly, phenomena that involve length scales of this order or smaller cannot be described by averages over many atoms. For optical waves, a wavelength of this order of magnitude corresponds to a photon energy of $h\nu = hc/\lambda \approx 2$ keV, which is in the x-ray region of the spectrum. We expect, therefore, that electromagnetic phenomena of longer wavelengths can be described by the averaged properties of matter but electromagnetic phenomena of shorter wavelengths cannot. For example, it is well known that x-rays are scattered by the microscopic structure of crystalline solids, whereas visible light is not. In the following we average our electric and magnetic fields over a small volume that nevertheless includes a large number of atoms or molecules. This excludes short-wavelength optical phenomena such as x-ray scattering.

The microscopic electromagnetic fields also fluctuate in time due to the motions of the electrons and nuclei. Electronic motions are relatively fast, while nuclear motions are much slower. In general, electronic motions are associated with optical phenomena in the visible and ultraviolet part of the spectrum, nuclear motions with optical phenomena in the infrared and microwave regions, and below. The electrons orbit the nuclei with various periods, depending on the quantum level in which they are found, but we might characterize the slowest electronic motions by the absorption and emission of visible light. For a wavelength $\lambda \approx 6 \times 10^{-7}$ m, which corresponds to red light, the lower end of the visible spectrum, the period is $T = \lambda/c \approx 2 \times 10^{-15}$ s. The fastest nuclear motions correspond to molecular vibrations and to optical phonons in crystals. The highest frequencies for these motions lie in the near infrared, with wavelengths extending down to $\lambda \approx 3 \times 10^{-6}$ m, which corresponds to a period $\lambda/c \approx 10^{-14}$ s. Molecular rotations in liquids and gases are much slower, with characteristic frequencies in the microwave region (thus the usefulness of microwave ovens), corresponding to periods on the order of 10^{-10} s. Acoustic phonons in solids (sound waves, classically speaking) extend to arbitrarily low frequencies, depending on the size of the sample. Finally, the existence of persistent phenomena, such as ferromagnetism and ferroelectricity, shows that time averaging is just not a useful concept, in general. Fortunately, time averaging is not necessary for constructing a simple theory of electrodynamics in macroscopic media. To deal with the complex time dependence of matter in the presence of electromagnetic fields, we use Fourier techniques. This takes the Maxwell equations out of the time domain and into the frequency domain, where life is simpler. For most purposes this provides all the information that is needed, and it is not necessary to go back to the time domain. This is fortunate, since the properties of the material arising from all the different time scales discussed here generally make it impossible to invert the Fourier transform.

In summary, then, to construct a macroscopic theory of electromagnetic phenomena, it is necessary to average the properties of matter over a microscopic region of space that is large compared to an atom but small compared with the phenomena of interest. This limits the macroscopic theory to ultraviolet and longer wavelengths. On the other hand, it is not generally possible to average over time. The effects of time-dependent phenomena in the microscopic behavior of matter, such as the absorption and dispersion of light, show up at frequencies even below the microwave region and continue to appear at frequencies up to the limits of spatial averaging in the ultraviolet region of the spectrum. For linear and almost linear phenomena, we deal with these effects by transforming to the frequency domain. This is the subject of Chapters 7 and 8. In this chapter we establish the basic principles.

On the microscopic level, the electric field **E** and magnetic induction **B** are described by the Maxwell equations

$$\nabla \cdot \mathbf{E} = \frac{\rho}{\varepsilon_0} \tag{6.1}$$

$$\nabla \cdot \mathbf{B} = 0 \tag{6.2}$$

$$\nabla \times \mathbf{E} + \frac{\partial \mathbf{B}}{\partial t} = 0 \tag{6.3}$$

$$\nabla \times \mathbf{B} - \frac{1}{c^2} \frac{\partial \mathbf{E}}{\partial t} = \mu_0 \mathbf{J} \tag{6.4}$$

where the charge density ρ and current density **J** of the electrons and nuclei are rapidly varying on a microscopic length scale. If we average the fields and the charge and current densities over a volume that is large compared with the microscopic fluctuations but small compared with the macroscopic dimensions important in the problem being considered, the Maxwell equations for the average quantities are

$$\nabla \cdot \langle \mathbf{E} \rangle = \frac{\langle \rho \rangle}{\varepsilon_0} \tag{6.5}$$

$$\nabla \cdot \langle \mathbf{B} \rangle = 0 \tag{6.6}$$

$$\nabla \times \langle \mathbf{E} \rangle + \frac{\partial \langle \mathbf{B} \rangle}{\partial t} = 0 \tag{6.7}$$

$$\nabla \times \langle \mathbf{B} \rangle - \frac{1}{c^2} \frac{\partial \langle \mathbf{E} \rangle}{\partial t} = \mu_0 \langle \mathbf{J} \rangle \tag{6.8}$$

where the brackets $\langle \ \rangle$ are used to indicate an average over some volume that is microscopically large but macroscopically small.

The key step in the development of the macroscopic form of the Maxwell equations is the division of the charges and currents into what are generally called "free" and "bound" charges and currents. For the moment we confine the discussion to a reference frame in which the macroscopic medium is at rest. Free charges, in this reference frame, are those that can move around the system over macroscopic distances. For example, the conduction electrons in a metal and the positive and negative ions in an electrolyte are free charges. Bound charges are those that are confined near a particular site in the solid, liquid or gas. The site itself can move according to the macroscopic motions of the material, but the bound charge moves with the site. We therefore write

$$\langle \rho \rangle = \rho_f + \rho_b \tag{6.9}$$

where ρ_f and ρ_b represent the microscopic average free and bound charges, respectively.

It might seem that since the atoms and molecules that make up a macroscopic material are each individually neutral, they cannot contribute significantly to a macroscopic charge density, but this is not true. In many molecules the positive and negative charges are not symmetrically distributed, so the molecule possesses an intrinsic dipole moment. In all cases, an external electric field will draw the positive and negative charges somewhat apart to form an induced dipole moment. If the molecular dipole moments are randomly oriented, of course, a small volume containing many molecules will contain no net macroscopic dipole moment. But this changes if an external field, or some other mechanism such as strain in the crystal, causes the dipoles to align themselves in some preferred direction, at least on average. If the net dipole moment per unit volume is uniform, the bound charge density ρ_b still vanishes. If it is not uniform, however, a net bound-charge density results.

A physical picture of the origin of the volume charge ρ_b in a nonuniform medium is indicated in Figure 6.1. As shown there, a uniform medium has no net volume charge, since the positive charge on one molecule is canceled out by the negative charge of its neighbors. But in a nonuniform medium, the cancelation is not complete and a net charge density appears. In the figure on the right, there is a net negative charge density across the region where the negative tails of the molecules face one another.

With these considerations in mind, we represent the average microscopic bound charge density as the divergence

$$\rho_b = -\nabla \cdot \mathbf{P} \tag{6.10}$$

of some vector field $\mathbf{P}$, called the polarization, which we must relate to the dipole moment of the molecules in the material. Actually, the definition of the vector $\mathbf{P}$ is arbitrary to the extent that we may add to it the curl $\nabla \times \Lambda$ of any vector field Λ without changing the bound charge density, since $\nabla \cdot (\nabla \times \Lambda) = 0$. We can find the total dipole moment of the bound charge by integrating (6.10) over some volume V bounded by the surface S that

Uniform polarization

Nonuniform polarization

Figure 6.1 Volume charge arising from the divergence of the polarization.

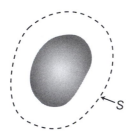

Figure 6.2 Surface surrounding a macroscopic body.

completely surrounds the macroscopic medium, as shown in Figure 6.2. After an integration by parts, we see that the total dipole moment defined in Chapter 3 is

$$\int_V \rho_b r_i \, d^3\mathbf{r} = -\sum_{j=1}^{3} \int_V r_i \frac{\partial P_j}{\partial r_j} \, d^3\mathbf{r} = -\sum_{j=1}^{3} \int r_i P_j \, d^2 r_{k \neq j} \Big|_{r_j \text{ on } S}^{r_j \text{ on } S} + \int_V P_i \, d^3\mathbf{r} \quad (6.11)$$

in which the first term on the right vanishes, since $\mathbf{P} = 0$ outside the material. But the molecules in the material are individually neutral, so the total dipole moment is just the sum of the molecular dipole moments $\mathbf{p}_n$, and we are left with

$$\int_V \rho_b \mathbf{r} \, d^3\mathbf{r} = \int_V \mathbf{P} \, d^3\mathbf{r} = \sum_V \mathbf{p}_n \quad (6.12)$$

where $\mathbf{p}_n$ is the dipole moment of the nth molecule and the sum is over all the molecules in the volume V. We therefore identify the vector $\mathbf{P}$ with the (microscopic average) molecular dipole moment per unit volume.

Outside the volume of the material, the vector $\mathbf{P}$ must vanish. Therefore, if we integrate the bound charge over the volume V of Figure 6.2 and use the divergence theorem, we get

$$\int_V \rho_b \, d^3\mathbf{r} = -\int_V \nabla \cdot \mathbf{P} \, d^3\mathbf{r} = -\int_S \mathbf{P} \cdot d\mathbf{S} = 0 \quad (6.13)$$

since $\mathbf{P} = 0$ on the surface S. Therefore, the bound charges are necessarily neutral, overall. Any net charge on the body must be included in the free charge.

The current density may be handled in a similar fashion. We divide the microscopic average current density into bound and free components,

$$\langle \mathbf{J} \rangle = \mathbf{J}_b + \mathbf{J}_f \quad (6.14)$$

Again it might be thought that currents localized within neutral molecules could not contribute to a significant macroscopic current density, but again this is wrong. A physical picture of macroscopic volume currents arising from microscopic bound currents is illustrated in Figure 6.3. As shown there, the microscopic currents due to the microscopic magnetic dipoles in a uniform medium cancel out. However, when the alignment or density of the dipoles is not uniform, the leftover currents contribute to an average volume current. In the figure on the right the net volume current flows upward and out of the page.

We write the bound-current density in the form

$$\mathbf{J}_b = \frac{\partial \mathbf{P}}{\partial t} + \nabla \times \mathbf{M} \quad (6.15)$$

for some vector field $\mathbf{M}$ that we call the magnetization. Actually, $\mathbf{M}$ is not unique, since we may add to it the gradient $\nabla \Lambda$ of any scalar field Λ without changing the bound current density, because $\nabla \times \nabla \Lambda = 0$. We identify $\mathbf{M}$ with the magnetic dipole moment of the molecules in the following way.

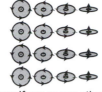

Figure 6.3 Volume current arising from the curl of the magnetization.

Uniform magnetization   Nonuniform magnetization

The ith component of the total magnetic dipole moment of a medium is found by integrating over the volume V shown in Figure 6.2 to get

$$\frac{1}{2} \sum_{j,k=1}^{3} \varepsilon_{ijk} \int_{V} r_j J_k \, d^3\mathbf{r}$$

$$= \frac{1}{2} \sum_{j,k=1}^{3} \varepsilon_{ijk} \int_{V} r_j \frac{\partial P_k}{\partial t} d^3\mathbf{r} + \frac{1}{2} \sum_{j,k,l,m=1}^{3} \varepsilon_{ijk}\varepsilon_{klm} \int_{V} r_j \frac{\partial M_m}{\partial r_l} d^3\mathbf{r}$$

$$= \frac{1}{2} \frac{\partial}{\partial t} \sum_{j,k=1}^{3} \varepsilon_{ijk} \int_{V} r_j P_k \, d^3\mathbf{r}$$

$$+ \frac{1}{2} \sum_{j,l,m=1}^{3} (\delta_{il}\delta_{jm} - \delta_{im}\delta_{jl}) \left[\int r_j M_m \, d^2 r_{n \neq l} \Big|_{r_l \text{ on } S}^{r_l \text{ on } S} - \delta_{jl} \int_V M_m \, d^3\mathbf{r} \right] \quad (6.16)$$

But the integral over $d^2 r_{n \neq l}$ vanishes because $\mathbf{M} = 0$ outside the medium, and we are left with

$$\frac{1}{2} \int_{V} \mathbf{r} \times \mathbf{J}_b \, d^3\mathbf{r} = \frac{1}{2} \frac{\partial}{\partial t} \int_{V} \mathbf{r} \times \mathbf{P} \, d^3\mathbf{r} + \int_{V} \mathbf{M} \, d^3\mathbf{r} \quad (6.17)$$

The first term on the right is the magnetic dipole moment due to the motion of the bound charges as the polarization changes with time. In a steady-state situation, this term vanishes and we are left with

$$\frac{1}{2} \int_{V} \mathbf{r} \times \mathbf{J}_b \, d^3\mathbf{r} = \int_{V} \mathbf{M} \, d^3\mathbf{r} = \sum_{V} \mathbf{m}_n \quad (6.18)$$

where $\mathbf{m}_n$ is the magnetic dipole moment of the nth molecule in the volume V. We therefore identify the vector $\mathbf{M}$ with the (microscopic average) molecular magnetic dipole moment per unit volume due to bound currents.

The macroscopic bound currents arising from the magnetization always flow in closed loops. To see this, we note that the divergence of the magnetization current is

$$\nabla \cdot \mathbf{J}_b(\text{magnetization}) = \nabla \cdot (\nabla \times \mathbf{M}) = 0 \quad (6.19)$$

by a vector identity. This being the case, the magnetization current cannot contribute to any change in the bound charge distribution. Macroscopic bound currents associated with changes in the bound charge are described by the first term in (6.15). If we take the divergence of (6.15) and use (6.10), we obtain the continuity relation, which shows that bound charge is conserved.

In terms of the polarization $\mathbf{P}$, the macroscopic form of Gauss's law becomes

$$\nabla \cdot \mathbf{D} = \rho_f \quad (6.20)$$

where the displacement vector **D** is defined by the relation

$$\mathbf{D} = \varepsilon_0 \langle \mathbf{E} \rangle + \mathbf{P} \tag{6.21}$$

In terms of the magnetization **M**, the macroscopic form of the Maxwell–Ampere law becomes

$$\nabla \times \mathbf{H} - \frac{\partial \mathbf{D}}{\partial t} = \mathbf{J}_f \tag{6.22}$$

where the magnetic field **H** is defined by the relation

$$\mathbf{H} = \frac{1}{\mu_0} \langle \mathbf{B} \rangle - \mathbf{M} \tag{6.23}$$

The homogeneous Maxwell equations remain

$$\nabla \cdot \langle \mathbf{B} \rangle = 0 \tag{6.24}$$

and

$$\nabla \times \langle \mathbf{E} \rangle + \frac{\partial \langle \mathbf{B} \rangle}{\partial t} = 0 \tag{6.25}$$

In what follows, we leave out the brackets $\langle \, \rangle$ for simplicity and generally ignore the subscript f with the understanding that the charge density ρ and current density **J** refer to the free charge and free current.

EXERCISE 6.1

Consider a volume V bounded by a surface S filled with a polarization $\mathbf{P}(\mathbf{r}')$ that depends on the position $\mathbf{r}'$. Assume that there are no free charges. As shown in Chapter 3, the potential of a point dipole at $\mathbf{r}'$ is

$$\Phi(\mathbf{r}) = \frac{1}{4\pi\varepsilon_0} \frac{\mathbf{p} \cdot \mathbf{R}}{R^3} \tag{6.26}$$

where $\mathbf{R} = \mathbf{r} - \mathbf{r}'$, so we can find the potential outside V from the integral

$$\Phi(\mathbf{r}) = \frac{1}{4\pi\varepsilon_0} \int_V \frac{\mathbf{P} \cdot \mathbf{R}}{R^3} dV \tag{6.27}$$

(a) Show that

$$\nabla'\left(\frac{1}{R}\right) = \frac{\mathbf{R}}{R^3} \tag{6.28}$$

(b) Use this result together with the divergence theorem to show that the potential may be expressed in the form

$$\Phi(\mathbf{r}) = \frac{1}{4\pi\varepsilon_0} \int_V \frac{\rho_b}{R} dV + \frac{1}{4\pi\varepsilon_0} \oint_S \frac{\sigma_b}{R} dS \tag{6.29}$$

where the bound volume and surface-charge densities are respectively

$$\rho_b(\mathbf{r}') = -\nabla' \cdot \mathbf{P}(\mathbf{r}') \tag{6.30}$$

$$\sigma_b(\mathbf{r}') = \mathbf{P}(\mathbf{r}') \cdot \hat{\mathbf{n}} \tag{6.31}$$

and $\hat{\mathbf{n}}$ is a unit vector outward normal to the surface S.

(c) Formula (6.29) looks like Coulomb's law. Does it apply inside V? Explain your answer.

(d) Consider a cylinder whose aspect ratio (ratio of the length L to the diameter D) is on the order of unity, filled with a uniform polarization parallel to its axis. Indicate the position and sign of the bound charge density. Sketch the equipotential surfaces. Sketch the lines of force of the electric field $\mathbf{E} = -\nabla\Phi$ inside and outside the cylinder.

(e) Sketch the lines of force of the displacement $\mathbf{D} = \varepsilon_0\mathbf{E} + \mathbf{P}$ inside and outside the cylinder. What is the approximate value of $\mathbf{D}$ at the very center of the cylinder for a very long cylinder $(L/D \gg 1)$, and for a very short cylinder $(L/D \ll 1)$? Explain your answers.

EXERCISE 6.2

Consider a volume V bounded by the surface S filled with a magnetization $\mathbf{M}(\mathbf{r}')$ that depends on the position $\mathbf{r}'$. Assume that there are no free currents. As shown in Chapter 3, the vector potential of a point magnetic dipole at $\mathbf{r}'$ is

$$\mathbf{A}(\mathbf{r}) = \frac{\mu_0}{4\pi} \frac{\mathbf{m} \times \mathbf{R}}{R^3} \tag{6.32}$$

where $\mathbf{R} = \mathbf{r} - \mathbf{r}'$, so we can find the potential outside V from the integral

$$\mathbf{A}(\mathbf{r}) = \frac{\mu_0}{4\pi} \int_V \frac{\mathbf{M} \times \mathbf{R}}{R^3} dV \tag{6.33}$$

(a) Show that

$$\nabla'\left(\frac{1}{R}\right) = \frac{\mathbf{R}}{R^3} \tag{6.34}$$

(b) Use this result together with the divergence theorem to show that the vector potential may be expressed in the form

$$\mathbf{A}(\mathbf{r}) = \frac{\mu_0}{4\pi} \int_V \frac{\mathbf{J}_b}{R} dV + \frac{\mu_0}{4\pi} \oint_S \frac{\mathbf{K}_b}{R} dS \tag{6.35}$$

where the bound volume and surface-current densities are respectively

$$\mathbf{J}_b(\mathbf{r}') = \nabla' \times \mathbf{M}(\mathbf{r}') \tag{6.36}$$

$$\mathbf{K}_b(\mathbf{r}') = \mathbf{M}(\mathbf{r}') \times \hat{\mathbf{n}} \tag{6.37}$$

and $\hat{\mathbf{n}}$ is a unit vector outward normal to the surface S.

(c) Formula (6.35) looks like a vector version of Coulomb's law. Does it apply inside V? Explain your answer.

(d) Consider a cylinder whose aspect ratio (ratio of the length L to the diameter D) is on the order of unity, filled with a uniform polarization parallel to its axis. Indicate the position and sign of the bound-current density. Sketch the lines of force of the vector potential $\mathbf{A}$. Sketch the lines of force of the magnetic induction $\mathbf{B} = \nabla \times \mathbf{A}$ inside and outside the cylinder.

(e) Sketch the lines of force of the magnetic field $\mathbf{H} = (1/\mu_0)(\mathbf{B} - \mathbf{M})$ inside and outside the cylinder. What is the approximate value of $\mathbf{H}$ at the very center of the cylinder for a very long cylinder $(L/D \gg 1)$, and for a very short cylinder $(L/D \ll 1)$? Explain your answers.

6.1.2 The Constitutive Relations

To close the set of equations (6.20)–(6.25), we need to know how the polarization $\mathbf{P}$ and magnetization $\mathbf{M}$ depend on the electric field $\mathbf{E}$ and magnetic induction $\mathbf{B}$ or, equivalently, the magnetic field $\mathbf{H}$. That is, we need to know the relations

$$\mathbf{P} = \mathbf{P}(\mathbf{E}, \mathbf{H}) \tag{6.38}$$

and

$$\mathbf{M} = \mathbf{M}(\mathbf{E}, \mathbf{H}) \tag{6.39}$$

which are called the constitutive relations. In general, the constitutive relations are nonlinear, as exemplified by the saturation of magnetization in high magnetic fields and by nonlinear optical phenomena in intense laser fields. The polarization and magnetization may also depend not merely on the instantaneous values of $\mathbf{E}$ and $\mathbf{B}$ (or $\mathbf{H}$) but on their entire history. This is obviously true for ferromagnetism, since the magnetization of a ferromagnetic sample may date back to events that occurred in prehistoric times. But the dependence of the instantaneous polarization and magnetization on the fields at earlier times is also responsible for the dispersion of electromagnetic waves passing through optical media and other materials. Later on we discuss some simple models for the polarization and magnetization of macroscopic media, but in the following sections we discuss their general properties, including relativistic covarianced and energy conservation.

We note in passing that (for historical reasons) it is conventional to express the magnetization $\mathbf{M}$ as a function of the magnetic field $\mathbf{H}$ rather than the magnetic induction $\mathbf{B}$. This harkens back to the fact that experimentally it is usually the current $\mathbf{J}_f$ that is known, and this is related directly to $\mathbf{H}$ through (6.22). Although this is confusing in some circumstances, it is convenient in others. In any event, it doesn't change the validity of anything. It changes only the form of the equations, since $\mathbf{H}$ and $\mathbf{B}$ are related by (6.23).

When the macroscopic medium is free to move, the bound sources $\mathbf{P}$ and $\mathbf{M}$ must be transformed into the laboratory reference frame. To accomplish this, we combine the polarization and magnetization into the antisymmetric 4-tensor

$$M^{\alpha\beta} = \begin{bmatrix} 0 & -cP_x & -cP_y & -cP_z \\ cP_x & 0 & M_z & -M_y \\ cP_y & -M_z & 0 & M_x \\ cP_z & M_y & -M_x & 0 \end{bmatrix} \tag{6.40}$$

Note the change in sign of $\mathbf{M}$ in $M^{\alpha\beta}$ compared with $\mathbf{B}$ in $F^{\alpha\beta}$ (see Chapter 1). This is related to the appearance of $\mathbf{M}$ with a minus sign in (6.23). To show that $M^{\alpha\beta}$ is indeed a tensor, we form the contraction with ∂_α and get

$$\partial_\alpha M^{\alpha 0} = -\sum_{i=1}^{3} \frac{\partial P_i}{\partial r_i} = c\rho_b = J_b^0 \tag{6.41}$$

$$\partial_\alpha M^{\alpha 1} = \frac{\partial P_x}{\partial t} + \frac{\partial M_z}{\partial y} - \frac{\partial M_y}{\partial z} = (J_b)_x = J_b^1 \tag{6.42}$$

and so on. Since the right-hand side is the 4-vector bound current density $J_b^\alpha = (c\rho_b, \mathbf{J}_b)$, which is a known vector, we see from the quotient rule (Chapter 1) that $M^{\alpha\beta}$ is a tensor.

To compute the polarization and magnetization of a moving medium, such as an electron beam in a microwave amplifier tube, we first transform the electric and magnetic fields into the coordinate system where the medium is at rest. We then compute the magnetization and polarization using the constitutive relations (6.38) and (6.39) in the rest frame and transform back into the coordinate system in which the body (or fluid) is moving by Lorentz transformation of the 4-tensor $M^{\alpha\beta}$.

Alternatively, we can form the 4-vector generalizations of the constitutive relations by looking for the appropriate expressions involving the field-strength tensor $F^{\alpha\beta}$ (or its dual $\mathcal{F}^{\alpha\beta} = \frac{1}{2}\varepsilon^{\alpha\beta\gamma\delta}F_{\gamma\delta}$) and the 4-vector velocity U^α. This has practical advantages because it eliminates the Lorentz transformations back and forth into the rest frame of the material, even if the results are less elegant and symmetrical. We begin by noting that the contraction

$$M^{\alpha\beta}U_\beta = \gamma c^2 \left(\frac{\mathbf{v} \cdot \mathbf{P}}{c}, \mathbf{P} - \frac{\mathbf{v} \times \mathbf{M}}{c^2} \right) \xrightarrow[|\mathbf{v}| \to 0]{} c^2(0, \mathbf{P}) \tag{6.43}$$

reduces to the polarization in the rest frame of the medium and forms the covariant generalization of the 3-vector polarization. Likewise, the contraction

$$F^{\alpha\beta}U_\beta = \gamma \left(\frac{\mathbf{v} \cdot \mathbf{E}}{c}, \mathbf{E} + \mathbf{v} \times \mathbf{B} \right) \xrightarrow[|\mathbf{v}| \to 0]{} (0, \mathbf{E}) \tag{6.44}$$

reduces to the electric field in the rest frame of the medium. For a linear, isotropic dielectric, the 3-vector constitutive relation in the rest frame of the material is

$$\mathbf{P}' = \varepsilon_0 \chi_e \mathbf{E}' \tag{6.45}$$

where the constant χ_e is called the dielectric susceptibility. The covariant generalization of this is clearly

$$M^{\alpha\beta}U_\beta = \frac{\chi_e}{\mu_0} F^{\alpha\beta}U_\beta \tag{6.46}$$

which has as its components the 3-vector expressions

$$\mathbf{v} \cdot \mathbf{P} = \varepsilon_0 \chi_e \mathbf{v} \cdot \mathbf{E} \tag{6.47}$$

and

$$\mathbf{P} - \frac{\mathbf{v} \times \mathbf{M}}{c^2} = \varepsilon_0 \chi_e (\mathbf{E} + \mathbf{v} \times \mathbf{B}) \tag{6.48}$$

Note the appearance of the magnetization $\mathbf{M}$ in (6.48). The timelike component (6.47) is redundant, since it can be derived from the spacelike component (6.48). The second relation is the relativistically correct, although not manifestly covariant, 3-vector generalization of the linear constitutive relation (6.45). When the constitutive relation is more complicated than the simple relation (6.45), the right-hand side of (6.48) becomes a more complex function of $F^{\alpha\beta}U_\beta$. For example, in a nonlinear medium we might expect to see scalar expressions of the form $F^{\alpha\beta}U_\beta F_{\alpha\gamma}U^\gamma = \gamma^2[(\boldsymbol{\beta}\cdot\mathbf{E})^2 - (\mathbf{E}+\mathbf{v}\times\mathbf{B})^2]$, and for an anisotropic medium we would expect to find vector expressions of the form $\chi_{\alpha\mu}U_\nu F^{\mu\nu}$, where $\chi_{\alpha\beta}$ is a tensor susceptibility. Clearly, the requirements of covariance restrict the analytical form that nonlinearities and other complications can assume.

Magnetization is treated in a similar manner. For a linear, isotropic magnetic material, the constitutive relation in the rest frame is conventionally expressed in the form

$$\mathbf{M}' = \chi_m \mathbf{H}' \tag{6.49}$$

in which the constant χ_m is called the magnetic susceptibility. Using (6.23) we see that this may also be expressed

$$\mathbf{M}' = \frac{\chi_m \mathbf{B}'}{\mu_0(1+\chi_m)} \tag{6.50}$$

which is more convenient for the present purposes. As we recall from Chapter 1, the substitutions $\mathbf{E}/c \to \mathbf{B}$ and $\mathbf{B} \to -\mathbf{E}/c$ convert the field-strength tensor into its dual tensor, $F^{\alpha\beta} \to \mathcal{F}^{\alpha\beta}$. Similarly, with the substitutions $c\mathbf{P} \to -\mathbf{M}$ and $\mathbf{M} \to c\mathbf{P}$, we get $M^{\alpha\beta} \to \mathcal{M}^{\alpha\beta}$, where

$$\mathcal{M}^{\alpha\beta} = \frac{1}{2}\varepsilon^{\alpha\beta\gamma\delta}M_{\gamma\delta} = \begin{bmatrix} 0 & M_x & M_y & M_z \\ -M_x & 0 & cP_z & -cP_y \\ -M_y & -cP_z & 0 & cP_x \\ -M_z & cP_y & -cP_x & 0 \end{bmatrix} \tag{6.51}$$

is the dual of the magnetization tensor. The covariant generalization of the linear constitutive relation (6.50) is therefore

$$\mathcal{M}^{\alpha\beta}U_\beta = -\frac{\chi_m}{\mu_0(1+\chi_m)}\mathcal{F}^{\alpha\beta}U_\beta \tag{6.52}$$

This has the components

$$\mathbf{v}\cdot\mathbf{M} = \frac{\chi_m}{\mu_0(1+\chi_m)}\mathbf{v}\cdot\mathbf{B} \tag{6.53}$$

$$\mathbf{M}+\mathbf{v}\times\mathbf{P} = \frac{\chi_m}{\mu_0(1+\chi_m)}\left(\mathbf{B}-\frac{\mathbf{v}\times\mathbf{E}}{c^2}\right) \tag{6.54}$$

Once again, (6.53) is redundant and (6.54) is the relativistically correct 3-vector form of the linear constitutive relation (6.50). As remarked earlier in regard to the polarization, generalizations of (6.52) to nonlinear and anisotropic media are possible.

EXERCISE 6.3

In the rest frame of a linear, isotropic dielectric medium, the polarization and magnetization are given by the constitutive relations

$$\mathbf{P}' = \varepsilon_0\chi_e\mathbf{E}' \tag{6.55}$$

and

$$\mathbf{M}' = 0 \tag{6.56}$$

In any other reference frame these transform into a polarization and a magnetization according to the rules for the transformation of tensors, even though the magnetic susceptibility vanishes in the rest frame.

(a) Transform the electric and magnetic fields from the laboratory frame K to the rest frame K' of the medium, keeping only terms up to first order in the velocity $\mathbf{v}$ of the medium. Compute the polarization in the rest frame using (6.55) and transform back into the laboratory frame, keeping only terms up to first order in the velocity. Show that

$$\mathbf{P} \approx \mathbf{P}' \approx \varepsilon_0 \chi_e (\mathbf{E} + \mathbf{v} \times \mathbf{B}) \tag{6.57}$$

and

$$\mathbf{M} \approx \mathbf{v} \times \mathbf{P}' \approx \varepsilon_0 \chi_e \mathbf{v} \times \mathbf{E} \tag{6.58}$$

(b) How does this result compare with (6.48) and (6.54)?
(c) Give a physical explanation for the velocity-dependent terms in (6.57) and (6.58).

6.1.3 Boundary Conditions

The macroscopic volume charge density due to the bound charges is proportional to the divergence of the polarization, as expressed in (6.10). At the surface of an object the polarization has a discontinuity, and this is equivalent to a macroscopic surface-charge density. This, in turn, leads to a discontinuity in the electric field. To find the discontinuity in the component of the electric field normal to the surface we use the cylindrical Gaussian surface shown in Figure 6.4. For this surface, from (6.20) and the divergence theorem, we have

$$\oint_S \mathbf{D} \cdot d\mathbf{S} = \int_V \rho_f \, dV \tag{6.59}$$

Provided that the flat faces of the cylinder are close to the boundary of the object, we may ignore the integrals over the sides and the only contribution to the free charge is from the surface free-charge density σ_f, so we are left with

$$(\mathbf{D}_{\text{outside}} - \mathbf{D}_{\text{inside}}) \cdot \hat{\mathbf{n}} = \sigma_f \tag{6.60}$$

where $\hat{\mathbf{n}}$ is a unit vector normal to the surface pointing outward from the material. In the absence of surface free charge, the normal component of the displacement $\mathbf{D}$ is continuous

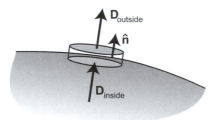

Figure 6.4 Gaussian surface for bound surface charge.

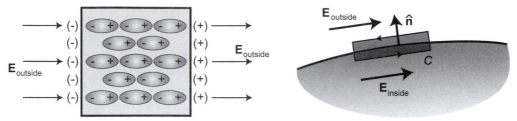

Figure 6.5 Effective surface charge due to polarization at the boundary.

Figure 6.6 Rectangular loop for tangential fields.

across the boundary. In terms of the electric field, the discontinuity is

$$(\mathbf{E}_{\text{outside}} - \mathbf{E}_{\text{inside}}) \cdot \hat{\mathbf{n}} = \frac{\mathbf{P} \cdot \hat{\mathbf{n}} + \sigma_f}{\varepsilon_0} = \frac{\sigma_b + \sigma_f}{\varepsilon_0} \tag{6.61}$$

where $\sigma_b = \mathbf{P} \cdot \hat{\mathbf{n}}$ is the effective bound surface-charge density. The physical significance of the effective surface bound charge is illustrated in Figure 6.5. If the medium is uniformly polarized, the displaced charges inside the medium are canceled by the nearby charges displaced in the opposite direction. At the boundary, however, the displaced charges are uncompensated and leave a net charge at the surface.

To determine the discontinuity in the tangential component of the electric field, we use the rectangular loop shown in Figure 6.6. From (6.25), using Stokes' theorem, we see that

$$\oint_C \mathbf{E} \cdot d\mathbf{l} = -\int_S \frac{\partial \mathbf{B}}{\partial t} \cdot d\mathbf{S} = 0 \tag{6.62}$$

where the right-hand side vanishes in the limit when the long sides of the loop approach the boundary of the object and the surface area inside the loop vanishes. Since this is true for any orientation of the loop in the plane of the surface, it follows that

$$(\mathbf{E}_{\text{outside}} - \mathbf{E}_{\text{inside}}) \times \hat{\mathbf{n}} = 0 \tag{6.63}$$

That is, the tangential component of the electric field $\mathbf{E}$ is continuous across the boundary.

As an illustration of these phenomena, we consider the simple dielectric-filled capacitor illustrated in Figure 6.7. As shown there, the field of the surface charge due to the displaced bound charges in the dielectric partially cancels the field of the charges on the plates of the capacitor. If the surface-charge density on the plates of the capacitor is σ, the electric field in the space between the plates and the dielectric is $E_{\text{out}} = \sigma/\varepsilon_0$. Across the surface of the dielectric the electric field changes due to the bound charge, and we find from (6.61) that the field inside the dielectric is

$$E_{\text{inside}} = E_{\text{outside}} - \frac{P}{\varepsilon_0} = \frac{\sigma - P}{\varepsilon_0} \tag{6.64}$$

In the absence of the dielectric, the voltage difference between the plates of the capacitor is $V = \sigma d/\varepsilon_0$ and the capacitance, defined as the total charge divided by the voltage, is $C = \varepsilon_0 A/d$, where A is the area and d the separation of the plates. When the dielectric is inserted into the space between the plates, the voltage difference is reduced to

$$V = \frac{\sigma d - (\sigma - P)t}{\varepsilon_0} \tag{6.65}$$

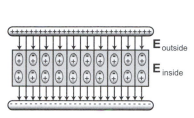

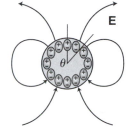

Figure 6.7 Simple dielectric capacitor.

Figure 6.8 Uniformly polarized dielectric sphere.

and the capacitance increases. Since the voltage drops while the charge on the plates of the capacitor remains the same, the energy of the capacitor decreases as the dielectric is inserted. Thus, the electric field does work on the dielectric, pulling it into the region between the plates.

As another example, we examine the electric field of a uniformly polarized sphere of radius R, as shown in Figure 6.8. The polarization **P** is assumed to be oriented in the $\hat{\mathbf{z}}$ direction. At the surface of the dielectric the polarization contributes a bound surface-charge density

$$\sigma_b = \hat{\mathbf{n}} \cdot \mathbf{P} = P \cos \theta \tag{6.66}$$

where $\hat{\mathbf{n}}$ is a unit vector normal to the surface of the sphere and θ the angle from the direction of polarization. This charge distribution creates a field outside and inside the sphere, as shown in the figure. To find this field we proceed as follows. We imagine that the material is composed of two equal and opposite, uniform charge distributions (the electrons and the nuclei), each one the size of the sphere. If the positive charge distribution is displaced by the amount δx relative to the negative charges, then a uniform polarization of the medium results. Outside the sphere, the electric field is the sum of the fields of the two spherical charge distributions. But the field of a spherical charge distribution is the same as the field of a point charge at the center of the sphere, so the field outside the dielectric is just the field of two point charges slightly displaced from one another. That is, the field outside the spherical dielectric is the field of a point dipole at the origin. From the discussion of multipoles in Chapter 3, with a little effort, we get

$$\mathbf{E}_{\text{outside}}(\mathbf{r}) = \frac{3\hat{\mathbf{r}}\,(\mathbf{p} \cdot \hat{\mathbf{r}}) - \mathbf{p}}{4\pi\varepsilon_0 r^3} \tag{6.67}$$

at point **r**, where

$$\mathbf{p} = \tfrac{4}{3}\pi R^3 \mathbf{P} \tag{6.68}$$

is the total dipole moment.

At the surface of the sphere the normal component of the field is

$$\mathbf{E}_{\text{outside}} \cdot \hat{\mathbf{n}} = \frac{\mathbf{p} \cdot \hat{\mathbf{n}}}{2\pi\varepsilon_0 R^3} = \frac{2}{3}\frac{P}{\varepsilon_0} \cos \theta \tag{6.69}$$

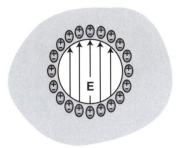

Figure 6.9 Spherical hole in a uniformly polarized medium.

If we apply Gauss's law to the charge distribution on the surface of the sphere, we see that the field just inside the sphere satisfies the boundary condition

$$\mathbf{E}_{\text{inside}} \cdot \hat{\mathbf{n}} = -\frac{1}{3}\frac{P}{\varepsilon_0}\cos\theta \tag{6.70}$$

where the negative sign arises because the unit vector $\hat{\mathbf{n}}$ points outward from the dielectric sphere. This is satisfied by the uniform field

$$\mathbf{E}_{\text{inside}} = -\frac{\mathbf{P}}{3\varepsilon_0} \tag{6.71}$$

so the field inside the sphere is just a uniform field pointed in the direction opposite the polarization.

For a spherical cavity in a material with a uniform polarization $\mathbf{P}$, illustrated in Figure 6.9, the bound surface charge density on the surface of the sphere is opposite that of a uniformly polarized sphere with the same polarization, since $\mathbf{P} \cdot \hat{\mathbf{n}}$ has the opposite sign. Therefore, the field inside the spherical cavity is just

$$\mathbf{E}_{\text{inside}} = \frac{\mathbf{P}}{3\varepsilon_0} \tag{6.72}$$

Alternatively, we can argue that if the field throughout a uniformly polarized infinite material vanishes, then the field of the same material with a spherical (or any other shape) cavity must be everywhere exactly opposite the field that would be found from a volume of similarly polarized material having the shape of the cavity.

We can find the boundary conditions on magnetic fields in a similar manner. If we use the Gaussian surface in Figure 6.4 together with (6.24) and the divergence theorem, we find that

$$(\mathbf{B}_{\text{outside}} - \mathbf{B}_{\text{inside}}) \cdot \hat{\mathbf{n}} = 0 \tag{6.73}$$

That is, the normal component of the magnetic induction $\mathbf{B}$ is continuous. Likewise, we may use the rectangular loop of Figure 6.6 together with (6.22) to show that

$$(\mathbf{H}_{\text{outside}} - \mathbf{H}_{\text{inside}}) \times \hat{\mathbf{n}} = -\mathbf{K}_f \tag{6.74}$$

where $\mathbf{K}_f$ is the free surface-current density. In the absence of free surface currents, the tangential component of the magnetic field $\mathbf{H}$ is continuous. In terms of the magnetic induction $\mathbf{B}$ the discontinuity is

$$(\mathbf{B}_{\text{outside}} - \mathbf{B}_{\text{inside}}) \times \hat{\mathbf{n}} = -\mu_0(\mathbf{M} \times \hat{\mathbf{n}} + \mathbf{K}_f) = -\mu_0(\mathbf{K}_b + \mathbf{K}_f) \tag{6.75}$$

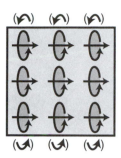

Figure 6.10 Effective surface current due to magnetization at the boundary.

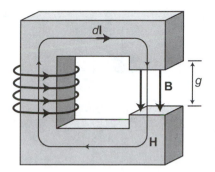

Figure 6.11 Simple magnetic circuit.

where $\mathbf{K}_b$ is the effective surface-current density of the bound charges. The physical origin of the effective surface current is illustrated in Figure 6.10. For a uniformly magnetized material, the bound currents are canceled by those of neighboring atoms. At the surface of the material, however, the current at the boundary is uncompensated.

These ideas are illustrated by the simple magnetic circuit shown in Figure 6.11, which consists of an iron core wrapped with wire and a small air gap. The magnetic field is largely confined to the iron core and the region of the gap. Inside the iron, the surface currents increase the magnetic induction by adding their current to that of the windings. Since the normal component of the magnetic induction B is continuous across the faces of the gap, the magnetic induction in the gap is the same as that in the core. On the other hand, the magnetic field H is quite different in the air gap and in the iron core. To find the magnetic induction, we apply Stokes' theorem to (6.22) and use the contour indicated in Figure 6.11. Then

$$\oint \mathbf{H} \cdot d\mathbf{l} = nI \tag{6.76}$$

where n is the number of turns and I is the current in the winding. To simplify matters we assume that in the gap the magnetic induction B is constant and inside the iron core the magnetic field H is constant. In the gap $B = \mu_0 H$, so the line integral is

$$HL + \frac{B}{\mu_0} g = nI \tag{6.77}$$

where L is the length of the core and g is the length of the gap. But the magnetic induction in the gap is the same as that in the core, where the induction is given by a curve like that in Figure 6.12. Therefore, to find the magnetic induction in the gap we rearrange (6.77) and solve the nonlinear equation

$$B(H) = \frac{\mu_0}{g}(nI - HL) \tag{6.78}$$

in which each side is a function of the magnetic field $\mathbf{H}$. This points out the usefulness of the historical convention of representing the magnetization $\mathbf{M}$ and magnetic induction $\mathbf{B} = \mu_0(\mathbf{M} + \mathbf{H})$ as functions of $\mathbf{H}$, rather than making $\mathbf{B}$ the independent variable.

For a nonlinear magnetic material such as iron, the solution must be obtained graphically, as indicated in Figure 6.13, even for the simple geometry we consider here. To deal

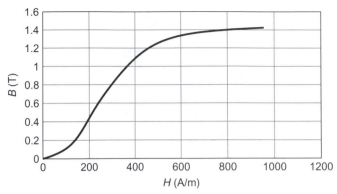

Figure 6.12 Magnetic induction in mild steel.

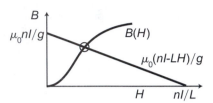

Figure 6.13 Graphical solution for the magnetic induction.

with a more complicated geometry, or to learn anything about the details of the field, such as the uniformity, it is necessary to resort to numerical computation. Fortunately, powerful computer codes now exist for this purpose, and they are widely available.

EXERCISE 6.4

In a linear, isotropic dielectric the polarization is related to the electric field in the medium by the simple relation

$$\mathbf{P} = \varepsilon_0 \chi_e \mathbf{E} \tag{6.79}$$

where χ_e is a constant called the dielectric susceptibility. Using the solution obtained above for a uniformly polarized dielectric sphere, show that the field of a dielectric sphere of radius R in the external uniform field $\mathbf{E}_0$ is

$$\mathbf{E}(\mathbf{r}) = \mathbf{E}_0 + \frac{\chi_e}{3 + \chi_e} \frac{R^3}{r^3} [3\hat{\mathbf{r}}(\mathbf{E}_0 \cdot \hat{\mathbf{r}}) - \mathbf{E}_0] \tag{6.80}$$

outside the sphere, and

$$\mathbf{E} = \frac{\mathbf{E}_0}{1 + \frac{1}{3}\chi_e} \tag{6.81}$$

inside the sphere.

EXERCISE 6.5

Your daughter is using a large nail wrapped with $n = 500$ turns of wire to make an electromagnet. The nail is made of mild steel (see Figure 6.12), and the wire carries a current $I = 1$ A. To estimate the magnetic induction at the ends of the nail we idealize the shape of the nail to that of a steel rod with a length $L = 15$ cm and a radius $R = 1.5$ mm. The ends are spherical. For simplicity, we assume that inside the nail the magnetic field is $H \approx$ constant and outside the ends of the nail (at least near the axis) the lines of force are radially outward normal to the surface of the nail. This means that $B \propto R^2/r^2$, where r is the distance from the center of curvature of the spherical ends of the nail.

(a) Using (6.76) for an appropriate contour, evaluate the integral and obtain a graphical solution for the magnetic induction B at the ends of the nail.

(b) Show that for any reasonable grade of steel, the magnetic induction at the ends of the nail is on the order of

$$B \approx \frac{\mu_0 n I}{2R} \tag{6.82}$$

EXERCISE 6.6

We observe that in the absence of free charges and free currents the Maxwell equations for electrostatics and magnetostatics become, respectively,

$$\nabla \cdot \mathbf{D} = 0 \tag{6.83}$$

$$\nabla \times \mathbf{E} = 0 \tag{6.84}$$

$$\mathbf{D} = \varepsilon_0 \mathbf{E} + \mathbf{P} \tag{6.85}$$

and

$$\nabla \cdot \mathbf{B} = 0 \tag{6.86}$$

$$\nabla \times \mathbf{H} = 0 \tag{6.87}$$

$$\mathbf{B} = \mu_0 \mathbf{H} + \mu_0 \mathbf{M} \tag{6.88}$$

The second set of equations is obtained from the first, and vice versa, by the substitutions $\mathbf{D} \leftrightarrow \mathbf{B}, \mathbf{E} \leftrightarrow \mathbf{H}, \mathbf{P} \leftrightarrow \mu_0 \mathbf{M}$, and $\varepsilon_0 \leftrightarrow \mu_0$.

(a) Exploit this analogy, together with the electrostatic results derived here, to find the magnetic induction $\mathbf{B}$ inside and outside a sphere of radius R filled with the uniform magnetization $\mathbf{M}$. What is the magnetic field $\mathbf{H}$? Sketch your result.

(b) What is the magnetic induction inside a cavity in a uniformly magnetized infinite material?

EXERCISE 6.7

Consider a cylinder of length L and radius R, filled with the uniform magnetization $\mathbf{M}$. If you believe in magnetic monopoles, you might argue that the magnetic field of the cylinder is caused by the surface magnetic monopole "charge" density $\sigma_m = \mathbf{M} \cdot \hat{\mathbf{n}}$ at the ends of the cylinder, where $\hat{\mathbf{n}}$ is a unit vector normal outward to the surface of the material, rather than by the surface current density $\mathbf{K}_b = \mathbf{M} \times \hat{\mathbf{n}}$ on the sides of the cylinder. Outside the material, the magnetic induction $\mathbf{B}$ that one would compute is the same either way. Inside the magnetic material, however, the magnetic induction is very different.

(a) For $L = O(R)$, sketch the lines of force of $\mathbf{B}$ and $\mathbf{H}$ inside and outside the material for magnetic monopoles and for magnetic dipoles.

(b) Consider a very long cylinder ($L \gg R$). Using the analogy with electrostatics, as appropriate, show that at the geometric center of the cylinder the fields are approximately as follows.
Magnetic dipoles:

$$\mathbf{B} \approx \mu_0 \mathbf{M} \tag{6.89}$$

$$\mathbf{H} \approx 0 \tag{6.90}$$

Magnetic monopoles:

$$\mathbf{B} \approx 0 \tag{6.91}$$

$$\mathbf{H} \approx -\mathbf{M} \tag{6.92}$$

(c) Show that for a very short cylinder the fields are approximately as follows.
Magnetic dipoles:

$$\mathbf{B} \approx 0 \tag{6.93}$$

$$\mathbf{H} \approx -\mathbf{M} \tag{6.94}$$

Magnetic monopoles:

$$\mathbf{B} \approx -\mu_0 \mathbf{M} \tag{6.95}$$

$$\mathbf{H} \approx -2\mathbf{M} \tag{6.96}$$

(d) Is it possible (at least in principle) to exploit this difference to detect monopoles by measuring the magnetic field in a cavity inside a magnetic material? If so, suggest an experiment to do this. What shape cavity would give the best results? If not, explain why not.

6.1.4 Magnetic Scalar Potential

Certain permanent magnetic materials such as samarium cobalt ($SmCo_5$) possess a very high residual magnetization $\mu_0 M \approx 1$ T but a very small magnetic susceptibility $\chi_m \ll 1$. That means that the magnetization $\mathbf{M}$ is fixed when the material is fabricated and is thereafter independent of the magnetic field $\mathbf{H}$, at least for magnetic fields smaller than the so-called coercivity H_c. This makes it relatively simple to compute the magnetic field using the magnetic scalar potential introduced in Chapter 3. The method works only for time-independent fields.

We begin with the Maxwell equations. When the free current density $\mathbf{J}_f$ vanishes, the time-independent Maxwell equations (6.22) and (6.24) become

$$\nabla \cdot \mathbf{B} = \mu_0 (\nabla \cdot \mathbf{H} + \nabla \cdot \mathbf{M}) = 0 \tag{6.97}$$

$$\nabla \times \mathbf{H} = 0 \tag{6.98}$$

Since the curl vanishes everywhere, the magnetic field may be derived from a scalar potential Φ_M,

$$\mathbf{H} = -\nabla \Phi_M \tag{6.99}$$

Substituting this into (6.97) we find that Φ_M satisfies the Poisson equation

$$\nabla^2 \Phi_M = -\rho_M \tag{6.100}$$

where the magnetic charge density is

$$\rho_M = -\nabla \cdot \mathbf{M} \tag{6.101}$$

Note that this magnetic charge density has nothing to do with magnetic monopoles. These equations are valid both inside and outside the magnetic material, since (6.97) and (6.98) apply everywhere. They allow us to apply the results and intuition we have gained from electrostatics to simple problems in magnetostatics, but they are useful only when we know the magnetization $\mathbf{M}$ ahead of time.

At the boundary of the magnetic material an effective surface magnetic charge density σ_M appears that is similar to the effective bound surface-charge density (6.61) due to the polarization. From the boundary condition (6.73) we see that

$$(\mathbf{H}_{\text{outside}} - \mathbf{H}_{\text{inside}}) \cdot \hat{\mathbf{n}} = \mathbf{M} \cdot \hat{\mathbf{n}} = \sigma_M \tag{6.102}$$

where $\hat{\mathbf{n}}$ is a unit vector outward normal to the surface. By analogy to Coulomb's law, then, the magnetic scalar potential due to the magnetization in the volume V bounded by the surface S is

$$\Phi_M(\mathbf{r}) = \frac{1}{4\pi} \int_V \frac{\rho_M \, dV'}{|\mathbf{r'} - \mathbf{r}|^2} + \frac{1}{4\pi} \oint_S \frac{\sigma_M \, dS'}{|\mathbf{r'} - \mathbf{r}|^2} \tag{6.103}$$

This is valid both inside and outside the magnetic material.

EXERCISE 6.8

Consider a cylinder of length L and radius R centered at the origin and aligned along the z axis, filled with the uniform magnetization $\mathbf{M} = M\hat{\mathbf{z}}$.

(a) Show that the magnetic scalar potential along the z axis is

$$\Phi_M = \tfrac{1}{2}M\left[\sqrt{\left(z - \tfrac{1}{2}L\right)^2 + R^2} - \sqrt{\left(z + \tfrac{1}{2}L\right)^2 + R^2} + 2z\right] \tag{6.104}$$

inside the cylinder ($-\tfrac{1}{2}L < z < \tfrac{1}{2}L$). What is the potential outside the cylinder?

(b) Show that the magnetic induction along the z axis is

$$\mathbf{B} = \frac{1}{2}\mu_0\mathbf{M}\left[\frac{z + \tfrac{1}{2}L}{\sqrt{\left(z + \tfrac{1}{2}L\right)^2 + R^2}} + \frac{z - \tfrac{1}{2}L}{\sqrt{\left(z - \tfrac{1}{2}L\right)^2 + R^2}}\right] \tag{6.105}$$

inside the cylinder. What is the induction outside the cylinder?

EXERCISE 6.9

The power of the magnetic scalar potential is that it allows us to use the techniques developed for electrostatics.

(a) Exploit the similarity between (6.100) and the corresponding equation in electrostatics to show that the magnetic scalar potential satisfies the variational principle

$$\delta\left(\frac{1}{2}\int H^2 \, dV - \int \rho_M \Phi_M \, dV\right) = 0 \tag{6.106}$$

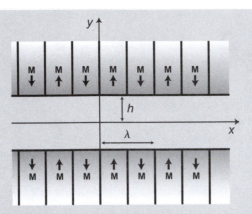

Figure 6.14 Periodic permanent-magnet structure.

(b) Consider the periodic permanent-magnet structure shown in Figure 6.14. The poles have a uniform magnetization **M** that alternates as shown, so the pole faces form a magnetic surface charge density M that alternates with the period λ. Based on the similarity of this to an electrostatics problem discussed in Chapter 3, we assume that the magnetic potential from the upper pole face can be approximated by the expression

$$\Phi_U = \Phi_0 \sin(kx)e^{-k|y-h|} \tag{6.107}$$

where $k = 2\pi/\lambda$, and h is the distance from the pole face to the axis of the structure. Use the variational principle (6.106) to evaluate Φ_0 for the upper pole face, and show that when both pole faces are included the magnetic inductance along the axis is

$$B_y = \frac{4}{\pi}\mu_0 M \sin(kx)e^{-kh} \tag{6.108}$$

Sketch the lines of **H** and **B**. Magnetic structures of this type are called wigglers or undulators. They are used in synchrotron radiation sources to deflect the electron beam back and forth, and create intense, nearly monochromatic radiation.

6.1.5 Conservation of Energy and Poynting's Theorem

Inevitably, when the subject of energy arises in connection with macroscopic materials, it brings with it the subject of thermodynamics. Electromagnetic energy is no exception to this rule. Unfortunately, a thorough discussion of thermodynamics in the presence of electromagnetic fields is beyond the scope of this discourse. Nevertheless, the concepts of work and energy are relevant and important, and they deserve at least some attention. The following discussion is brief and superficial, and to the extent that it deals with thermodynamics at all, the discussion ignores the connection between the polarization and magnetization on the one hand and other thermodynamic variables (such as temperature and stress) on the other hand. This excludes a variety of interesting effects such as piezoelectricity and magnetostriction, and for a discussion of these phenomena the reader must look elsewhere. But the extension of Poynting's theorem (the overall conservation of energy) to macroscopic systems (described by **D** and **H**) is addressed here, and we apply the results to some simple systems.

To extend Poynting's theorem to macroscopic materials we begin with the Maxwell equations in their macroscopic form (6.20)–(6.25). If we take the scalar product of (6.22) with **E** and of (6.25) with **H** and subtract, we get

$$\mathbf{E} \cdot (\nabla \times \mathbf{H}) - \mathbf{H} \cdot (\nabla \times \mathbf{E}) = \mathbf{E} \cdot \mathbf{J} + \mathbf{E} \cdot \frac{\partial \mathbf{D}}{\partial t} + \mathbf{H} \cdot \frac{\partial \mathbf{B}}{\partial t} \qquad (6.109)$$

We can simplify this with the vector identity

$$\nabla \cdot (\mathbf{a} \times \mathbf{b}) = \mathbf{b} \cdot (\nabla \times \mathbf{a}) - \mathbf{a} \cdot (\nabla \times \mathbf{b}) \qquad (6.110)$$

and the result is

$$\mathbf{E} \cdot \frac{\partial \mathbf{D}}{\partial t} + \mathbf{H} \cdot \frac{\partial \mathbf{B}}{\partial t} + \nabla \cdot (\mathbf{E} \times \mathbf{H}) + \mathbf{E} \cdot \mathbf{J} = 0 \qquad (6.111)$$

This is Poynting's theorem in macroscopic media.

We identify

$$\mathbf{S} = \mathbf{E} \times \mathbf{H} \qquad (6.112)$$

as the Poynting vector and recognize that it represents the flux of electromagnetic energy in the material. To see this, we consider an electromagnetic field at the surface of some material. As shown in Chapter 2, the energy flux normal to the surface in the vacuum is given by the vector

$$\mathbf{S} = \frac{1}{\mu_0} \mathbf{E} \times \mathbf{B} = \mathbf{E} \times \mathbf{H} \qquad (6.113)$$

since **M** = 0 in the vacuum. The normal component of this is

$$\mathbf{S} \cdot \hat{\mathbf{n}} = (\mathbf{E} \times \mathbf{H}) \cdot \hat{\mathbf{n}} = (\hat{\mathbf{n}} \times \mathbf{E}) \cdot \mathbf{H} = (\mathbf{H} \times \hat{\mathbf{n}}) \cdot \mathbf{E} \qquad (6.114)$$

which depends only on the tangential components of **E** and **H**. But energy cannot accumulate at the interface, so the energy flow immediately inside the surface must be the same as that outside the surface. Since the tangential components of **E** and **H** are continuous at the boundary, the correct expression for the energy flow must be (6.112) inside the material or outside it.

The last term in (6.111) represents the rate at which work is done by the electric field on the free charge in the material, where the total work is

$$\Delta W_f = \int \mathbf{E} \cdot \mathbf{J} \, dt \qquad (6.115)$$

In ordinary conductors this work is responsible for what is called Joule heating, that is, heating caused by the flow of current through a resistive material. The first two terms in (6.111) represent work done on the fields themselves and on the bound charge,

$$\Delta W_b = \int \mathbf{E} \cdot d\mathbf{D} + \int \mathbf{H} \cdot d\mathbf{B} \qquad (6.116)$$

To see this, we observe that the first term is just

$$\mathbf{E} \cdot d\mathbf{D} = \varepsilon_0 \mathbf{E} \cdot d\mathbf{E} + \mathbf{E} \cdot d\mathbf{P} = d\left(\tfrac{1}{2}\varepsilon_0 E^2\right) + \mathbf{E} \cdot \mathbf{J}_b \, dt \qquad (6.117)$$

in which $\frac{1}{2}\varepsilon_0 E^2$ is the energy density of the electric field, and $\mathbf{E} \cdot \mathbf{J}_b$ is the rate of doing work on the bound-charge current $\mathbf{J}_b = \partial \mathbf{P}/\partial t$. This work includes work done to increase the kinetic and potential energy of the bound charges, and work done to overcome dissipation. In the same way, the second term in (6.116) is

$$\mathbf{H} \cdot d\mathbf{B} = \frac{1}{\mu_0}\mathbf{B} \cdot d\mathbf{B} - \mathbf{M} \cdot \frac{\partial \mathbf{B}}{\partial t}\, dt = d\left(\frac{B^2}{2\mu_0}\right) + \mathbf{E} \cdot (\nabla \times \mathbf{M}) + \nabla \cdot (\mathbf{E} \times \mathbf{M}) \quad (6.118)$$

after we use Faraday's law and a vector identity. But the first term in (6.118) is the change in the magnetic field energy density and the second is the work done on the bound current $\mathbf{J}_b = \nabla \times \mathbf{M}$. This includes both the change of the kinetic and potential energy of the bound charges and the work done to overcome dissipation in the magnetization process. The last term in (6.118) is an energy-flow term left over from redefining the Poynting vector $\mathbf{S} = \mathbf{E} \times \mathbf{H}$ when we revert to a description explicitly in terms of the bound charges and currents.

The division of the electromagnetic work into free and bound components is somewhat arbitrary, as is the separation into free and bound charge itself. In plasmas, for example, the electrons can be regarded as free charge and described by a complex conductivity σ. Alternatively, they can be regarded as bound charge in the limit when the restoring force vanishes. In this case the electrons are described by a complex dielectric susceptibility χ_e. We come back to this point in Chapter 7 when we discuss the Lorentz–Drude model.

Even in the absence of free charge, it is difficult to separate in a completely general way the part of the work $\Delta\mathcal{W}_b$ done to increase the electromagnetic energy (nondissipative work) and the part done to increase the thermal energy (dissipative work). The polarization and magnetization of a material in an electromagnetic field are typically neither linear nor even unique functions of the instantaneous electric and magnetic fields. This is always true in the presence of dispersion and absorption, and even in quasistatic systems the polarization and magnetization are not unique functions of the electric and magnetic fields when hysteresis is important. In cyclic processes, however, the material returns periodically to (approximately) the same state at the end of each cycle, so the electromagnetic energy density is the same. The net energy that went into the material during the cycle is then thermal heating, or dissipation. In Chapter 7 we discuss electromagnetic energy and dissipation in linear, dispersive materials. For linear materials we can develop general expressions for the dissipation and the average electromagnetic field energy (including the energy of magnetization and polarization) for electromagnetic waves with a narrow spectrum (long pulses). The processes that occur throughout such pulses are, after all, nearly cyclic in this case. In the present section we discuss quasi-static processes, including the effects of hysteresis.

For most materials, specifically those that do not exhibit hysteresis, the polarization and magnetization in the quasi-static limit are unique functions of the electric and magnetic fields $\mathbf{E}$ (or $\mathbf{D}$) and $\mathbf{B}$ (or $\mathbf{H}$). In nonconducting materials, we can uniquely define the electromagnetic energy density $\mathcal{U}$ as the energy required to reach the point $(\mathbf{E}, \mathbf{B})$ adiabatically (that is, reversibly), since in this case the dissipation vanishes. We are then left with

$$\Delta\mathcal{U} = \Delta\mathcal{W}_b = \int_{\text{adiabatic}} \mathbf{E}\,(\mathbf{D}) \cdot d\mathbf{D} + \int_{\text{adiabatic}} \mathbf{H}\,(\mathbf{B}) \cdot d\mathbf{B} \quad (6.119)$$

When the state of polarization, and magnetization depends on other variables, such as temperature and pressure, then the path of integration must be more carefully defined. If, for example, the process occurs at constant temperature and volume, then the energy computed from the path integral is called the Helmholtz function. If the temperature and pressure are constant along the path, then the energy is called the Gibbs function, and so on. To simplify the discussion here, we ignore the dependence of the polarization and magnetization on other thermodynamic variables.

When hysteresis is important, the magnetization and polarization are no longer unique functions of the electromagnetic fields $\mathbf{E}$ and $\mathbf{B}$, even in the quasi-static limit. In ferromagnetic materials, for example, the magnetization $\mathbf{B}(\mathbf{H})$ follows different curves depending on the initial state of the system, as shown in Figure 6.15. As indicated there, the change $\int \mathbf{H} \cdot d\mathbf{B}$ in the total energy density is the area to the left of the magnetization curve. Depending on which curve we are following, the integral from one value of the magnetic induction B to another corresponds to a different change in the total energy density. As a result, for a given change in the magnetic induction, the change in the total energy density (electromagnetic energy plus thermal energy and possibly other forms of energy such as stress in a magnetostrictive system) does not have a uniquely defined value. Similar remarks apply to the polarization in ferroelectric materials.

However, if we have a cyclic process, such as the oscillating magnetic field in a transformer core, the electric and magnetic fields and the polarization and magnetization return to their original state at the end of the cycle, so the electromagnetic field energy must be the same as it was at the beginning of the cycle. Thus, the net flow of electromagnetic energy into the volume of the material, as represented by the Poynting vector, is dissipated entirely as heat, ΔQ. In Figure 6.16 the integral

$$\Delta Q = \Delta W_b = \oint H\, dB \tag{6.120}$$

is just the area inside the hysteresis curve. We therefore see that hysteresis introduces a new dissipative mechanism in the material, a sort of "magnetic friction" originating from the irreversible processes that occur in the magnetization of the iron. To minimize the losses in transformer cores, for example, it is desirable to select a type of steel for which the area inside the hysteresis curve is as small as possible.

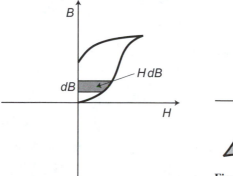

Figure 6.15 Energy density of a ferromagnetic material.

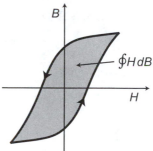

Figure 6.16 Cyclic energy dissipation due to hysteresis in a ferromagnetic material.

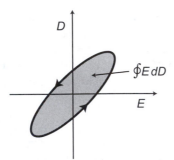

Figure 6.17 Cyclic energy dissipation due to phase lag in a dielectric.

Similar considerations apply to dissipative processes in the absorption of electromagnetic radiation by lossy dielectrics. This is easy to understand in the case of polar molecules in a liquid. As the molecules rotate to stay aligned with the alternating electric field they experience a sort of "friction," or "viscous drag," that damps their rotation. As the electric field completes one cycle, the displacement $D = \varepsilon_0 E + P$ follows a closed curve in the D versus E plane as shown in Figure 6.17. In the quasi-static case the polarization is a unique function of the electric field and the loop shrinks to a line. As in any adiabatic (reversible) process, the dissipation vanishes. In the general case, however, the polarization lags behind the electric field or, to put it another way, the electric field leads the displacement in phase. In the D versus E plane, the path around each cycle becomes an ellipse, as shown in Figure 6.17. The area inside the ellipse represents dissipation that will heat your dinner in the microwave oven tonight.

Finally, we note that if the polarization response is invariant under time reversal, then dissipation must be absent, and vice versa. This follows because under time reversal the integral $\oint \mathbf{E} \cdot d\mathbf{D}$ around one cycle changes sign, which contradicts the hypothesis. This conclusion also agrees with the thermodynamic notion of a reversible process, one in which there is no dissipation. Similar remarks apply to the magnetization.

EXERCISE 6.10

A choke is constructed from mild steel (see Figure 6.12) in the form of a toroid with a cross sectional area $A = 2 \text{ cm}^2$ and a circumference $C = 50 \text{ cm}$. The toroid is wrapped with $n = 200$ turns of wire.

(a) Estimate (roughly) how much energy is required to increase the current in the wire from $I = 0$ to $I = 1$ A. Sketch a magnetization diagram like Figure 6.12 and show how you made the estimate.

(b) Estimate (roughly) how much energy is required to increase the current from $I = 1$ A to $I = 2$ A. Show on a magnetization diagram how you made the estimate.

(c) Sketch the magnetization curve that describes how the magnetic induction behaves as the current is reduced from $I = 2$ A to $I = 0$. Indicate on the sketch the energy recovered from the choke during this process. What has become of the rest of the energy?

EXERCISE 6.11

In a polar liquid at high frequencies, the molecular dipoles do not have time to rotate in response to changes in the electric field. If we assume that the molecular rotations are viscously damped and ignore the effect of the moment of inertia, we can describe the behavior of the polarization by the linear differential equation

$$\frac{d\mathbf{P}}{dt} = \frac{\varepsilon_0 \chi_0 \mathbf{E} - \mathbf{P}}{\tau} \tag{6.121}$$

where $\chi_0 = \chi_e(0)$ is the dc susceptibility and τ the relaxation time.

(a) Show that for a wave of the form

$$\mathbf{E}(t) = \text{Re}(\mathbf{E}_0 e^{-i\omega t}) \tag{6.122}$$

the polarization is

$$\mathbf{P}(t) = \text{Re}(\mathbf{P}_0 e^{-i\omega t}) \tag{6.123}$$

where

$$\mathbf{P}_0 = \frac{\varepsilon_0 \chi_0}{1 - i\omega\tau} \mathbf{E}_0 \tag{6.124}$$

(b) Show that the average (over a complete cycle) rate of energy dissipation is

$$\left\langle \frac{dQ}{dt} \right\rangle = \left\langle \mathbf{E} \cdot \frac{d\mathbf{D}}{dt} \right\rangle = \varepsilon_0 \chi_0 \omega \frac{\omega\tau}{1 + \omega^2\tau^2} \langle E^2 \rangle \tag{6.125}$$

Hint: Recall that the average over a complete cycle is $\langle \mathbf{A} \cdot \mathbf{B} \rangle = \frac{1}{2}\text{Re}(\mathbf{A}_0^* \cdot \mathbf{B}_0)$.

EXERCISE 6.12

As shown in Chapter 3, the energy of an induced dipole $\mathbf{p} = \alpha\mathbf{E}$, where α is the atomic polarizability, in the electric field $\mathbf{E}$ is $\mathcal{W} = -\frac{1}{2}\alpha E^2$. Consider a dielectric material made up of such dipoles, for which the polarization is $P = \varepsilon_0 \chi_e E$.

(a) Assume that $\chi_e \ll 1$, so the electric field is only weakly perturbed by the dielectric. Show that the total energy of the dielectric material in the field is then

$$\Delta\mathcal{W} \approx -\frac{1}{2}\varepsilon_0 \chi_e E^2 V \tag{6.126}$$

where V is the volume of the dielectric material.

(b) Show that for a dielectric sphere the precise answer is

$$\Delta\mathcal{W} = -\frac{1}{2}\varepsilon_0 \frac{\chi_e}{1 + \frac{1}{3}\chi_e} E^2 V \xrightarrow[\chi_e \ll 1]{} -\frac{1}{2}\varepsilon_0 \chi_e E^2 V \tag{6.127}$$

Hint: The electric field of a uniformly polarized sphere in an external field is described by (6.80) and (6.81).

(c) Since the energy is quadratic in the electric field it is always negative, so in a nonuniform field the dielectric material is always attracted to the region of strongest electric field. When the electric field is provided by a focused laser beam, the dielectric material is attracted to the center of the laser spot. This is the principle behind the "optical tweezers" used to manipulate individual cells and cellular components for experiments in microbiology, for example. The force is very small, but to put it in context consider a laser beam with the parabolic intensity profile

$$S = S_0\left(1 - \frac{r^2}{R^2}\right) \tag{6.128}$$

where S_0 is the intensity at the center of the beam and R is the radius of the beam. Show that a small dielectric sphere near the center of the beam oscillates around the center with the frequency

$$\omega = \sqrt{\frac{\chi_e}{1 + \frac{1}{3}\chi_e} \frac{S_0}{c\rho R^2}} \tag{6.129}$$

where ρ is the density of the material. For a water droplet in a 1-mW laser beam focused by a microscope to a 1-μm radius, what is the frequency of oscillation?

(d) What is the relationship between the force of the optical tweezers and the "radiation pressure?"

EXERCISE 6.13

The amount of energy that can be stored in a dielectric is limited by breakdown of the dielectric at high electric fields. The amount of energy that can be stored in a magnetic material is limited by saturation at high magnetic fields. The amount of energy that can be stored in a flywheel is limited by the tensile strength of the material.

(a) For silicon oil, the susceptibility is $\chi_e \approx 2.5$ and the dielectric strength is $E \approx 15$ MV/m. What is the maximum electric energy density in the oil?

(b) For a mild steel, the average susceptibility is $\chi_m \approx 1600$ and the saturation field is $H \approx 700$ A/m. What is the maximum magnetic energy density in the steel?

(c) The energy density in a flywheel is largest at the outer rim of the flywheel, which we may view as a hoop rotating about its axis. The tensile strength of carbon fiber $\sigma_T \approx 3.5$ GPa. What is the maximum kinetic energy density in the carbon fiber?

6.2 PROPERTIES OF DIELECTRIC AND MAGNETIC MATERIALS

The dielectric and magnetic properties of materials compose a fascinating subject with a long history. Unfortunately, it would take us too far afield from the study of classical electrodynamics to go into the subject in depth, and in any event the properties of materials can be properly understood only with quantum mechanics. Nevertheless, a brief look at the physics underlying some of the macroscopic phenomena that are observed is useful in preparing us for the discussion in the following chapters.

6.2.1 Dielectric Materials

We take up first the phenomenon of polarization. Since the materials in which we are interested are composed of atoms and molecules, the polarization of the medium is related to the polarization of the individual atoms or molecules. In the case of crystalline solids, the unit cell takes on the role of a "molecule" in this discussion. The degree to which the atoms and molecules are polarized depends first of all on whether or not the atom or molecule possesses an intrinsic dipole moment, and secondarily on the degree to which it is influenced by its neighbors. Polar dielectrics tend to have a larger dielectric susceptibility than nonpolar dielectrics and a different temperature dependence. In general the interaction between neighboring molecules is small, but in certain solids it can be overwhelmingly important. We begin our discussion with nonpolar dielectrics. At first we ignore the interaction between neighboring molecules, and then we examine the effects of these interactions (which give us the Clausius–Mossotti equation and ferroelectrics). At the end of the section we take up polar dielectrics.

In the absence of an intrinsic dipole moment, the only dipole moment an atom or molecule can have is that due to distortion of the charge distribution by the electric field in which the atom finds itself. As a simple model we may imagine an atom as consisting of an electron bound to a nucleus. If the nucleus is regarded as fixed and the electron as bound to the nucleus like a spherically symmetric harmonic oscillator, then the "spring constant" of the electron is $k = m\omega_0^2$, where m is the mass and ω_0 the oscillation frequency of the electron. In a locally uniform, time-independent electric field $\mathbf{E}$ the dipole moment of the atom is

$$\mathbf{p} = \varepsilon_0 \alpha_{\mathrm{mol}} \mathbf{E} \tag{6.130}$$

where the molecular polarizability is

$$\alpha_{\mathrm{mol}} = \frac{Z_{\mathrm{eff}} q^2}{\varepsilon_0 m \omega_{\mathrm{eff}}^2} \tag{6.131}$$

in which q is the electron charge, Z_{eff} the number of electrons in the molecule that effectively take part in the polarization, and ω_{eff}^2 an average frequency. The polarization of the medium is

$$\mathbf{P} = N\mathbf{p} = \varepsilon_0 \chi_e \mathbf{E} \tag{6.132}$$

where N is the number density of atoms and $\mathbf{E}$ the electric field in the material. If the effect of neighboring atoms on the electric field experienced by each atom is negligible, then the dielectric susceptibility χ_e is

$$\chi_e = \frac{P}{\varepsilon_0 E} = \frac{N Z_{\mathrm{eff}} q^2}{\varepsilon_0 m \omega_{\mathrm{eff}}^2} \tag{6.133}$$

Of course, this picture totally ignores quantum mechanics and lumps the complexities of many-electron atoms into the undetermined parameters Z_{eff} and ω_{eff}. Nevertheless, for electronic motions, the frequency ω_{eff} is typically in the visible or ultraviolet region and (6.133) gives the correct order of magnitude for the static polarizability of nonpolar materials. In addition, the model does correctly suggest the limitations of the result (6.133). First of all, the derivation assumes that the electric field is time independent. This is reasonable when the frequency of the applied field is small compared with the frequency ω_0

of the harmonic oscillator, but we expect deviations when the frequency is comparable to or larger than the characteristic frequency. For atoms this occurs in the visible and ultraviolet regions, and gives rise to the phenomenon known as dispersion. When the material is composed of molecules, rather than atoms, the distortion of the molecular bonds by the electric field contributes to the molecular dipole moment. In this case dispersion sets in in the infrared region, at frequencies characteristic of molecular vibrations. The same is true in crystals, where phonon frequencies are important. Finally, we note the linear relationship (6.132) between the dipole moment and the electric field, which follows from the assumption of a linear spring. For an atom we might expect this linear relationship to break down when the displacement of the electron becomes comparable to the size of the atom, that is, when the electric field reaches a value of the order of

$$E_c = \frac{m\omega_0^2 a_0}{q} = O(10^9 \text{ V/m}) \tag{6.134}$$

where a_0 is the radius of the first Bohr orbit. In fact this is observed experimentally, and fields of this order and larger can be obtained at the focus of high-power lasers. For example, in the beam of a terawatt laser focused to a spot radius of the order of 10 μm, the peak electric field is on the order of 10^{12} V/m and nonlinear effects are very strong.

In the preceding discussion, we have ignored the effect of the surrounding atoms and molecules on the polarization of a given atom or molecule. In fact, the interaction between neighboring dipoles reinforces the polarization and enhances the dielectric susceptibility. The effect is negligible for gases but significant for liquids and solids. In the simplest case, we can estimate the effect by replacing all the surrounding molecules by a uniformly polarized material and consider the effect of the surrounding polarized material on the molecule of interest in the (otherwise empty) volume that it occupies. As shown in the previous section, the surrounding polarization creates the field

$$\mathbf{E} = \frac{\mathbf{P}}{3\varepsilon_0} \tag{6.135}$$

in the empty volume. Consequently, we see from (6.130) that the total polarization of the medium is

$$\mathbf{P} = N\mathbf{p} = N\varepsilon_0 \alpha_{\text{mol}} \left(\mathbf{E} + \frac{\mathbf{P}}{3\varepsilon_0} \right) \tag{6.136}$$

Solving this equation for $\mathbf{P}$, we find that the dielectric susceptibility is

$$\chi_e = \frac{N\alpha_{\text{mol}}}{1 - \frac{1}{3}N\alpha_{\text{mol}}} \tag{6.137}$$

or, if we solve (6.137) for α_{mol} we get

$$\alpha_{\text{mol}} = \frac{1}{N} \frac{\chi_e}{1 - \frac{1}{3}\chi_e} \tag{6.138}$$

This is called the Clausius–Mossotti relation. To the degree that the model is realistic, it should be possible to determine the molecular polarizability α_{mol} from measurements of the susceptibility at different densities and get the same result. For gases the factor $\frac{1}{3}\chi_e$ is negligible, of course, but for liquids and solids it can be significant. For example, for oxygen at its normal boiling point the molecular polarizability for the gas phase is found to be

$\alpha_{mol}(\text{gas}) = 1.97 \times 10^{-29}$ m^3, whereas $\alpha_{mol}(\text{liquid}) = 1.95 \times 10^{-29}$ m^3. This is a difference of only one percent, despite the fact that for the liquid the correction factor is $\frac{1}{3}\chi_e = 0.162$. That is, the effect of the surrounding polarization is substantial. Thus, the Clausius–Mossotti relation actually provides a pretty fair representation of the effect of density for many nonpolar gases and liquids.

Of course, this description of the density effect ignores the details of the fields of the nearest neighbors. For cubic crystals and amorphous solids the fields of the nearest neighbors cancel out, at least on the average, so the Clausius–Mossotti relation remains satisfactory. For other crystal symmetries, the nearest-neighbor fields do not average to the value assumed here and the factor $\frac{1}{3}$ in (6.138) is not correct.

In a few crystals, such as BaTiO$_3$, for example, the effect of the nearest neighbors is extremely strong and a "polarization catastrophe" results. Below the so-called Curie temperature ($T_C = 408$ K in BaTiO$_3$) the influence of nearby atoms is strong enough to cause the crystal to polarize itself spontaneously in an effect called ferroelectricity. At temperatures just above the Curie temperature the dielectric susceptibility can still be very large, of the order of 10^4. We can understand this from (6.137) if we think of a unit cell of the crystal as a molecule. The Curie temperature then corresponds to the point where $N\alpha_{mol} = 3\varepsilon_0$ and the denominator vanishes. At temperatures below the Curie temperature, the fields of the neighboring atoms are sufficient to cause the unit cells of the crystal to polarize spontaneously. Since local regions of the crystal tend to have different orientations of their spontaneous polarization, the overall macroscopic polarization ordinarily vanishes, on average. However, when an electric field is applied, the local regions all line up with the field and the net polarization can be enormous. When the electric field is removed, much of the polarization remains. Because of the similarity of this phenomenon to the residual magnetization in ferromagnets when the magnetic field is removed, the persistence of the polarization is called the ferroelectric effect. The phenomenon of ferroelectricity emphasizes once again that the relationship between the polarization and the electric field is not necessarily linear, nor even unique. It depends on the past history of the polarization of the medium.

Next we turn to macroscopic materials composed of polar molecules. When the molecules have an intrinsic dipole moment and the dipole moment is free to rotate, as in a liquid, an externally applied electric field causes the dipoles to line up parallel to the field and create a net polarization of the medium. The effective molecular polarizability of such a medium is generally larger than that of a nonpolar medium. In a liquid the rotational motions are damped by the surrounding molecules, so it is appropriate to consider the state of the system at thermal equilibrium. We recall from Chapter 3 that the energy of a permanent dipole $\mathbf{p}$ in an electric field $\mathbf{E}$ is $W = -\mathbf{p} \cdot \mathbf{E}$. At thermal equilibrium the number of dipoles with the energy W is $dn \propto e^{-W/k_B T}$, where k_B is Boltzmann's constant and T the temperature. The average dipole moment of a polar molecule in the electric field is therefore

$$\langle \mathbf{p} \cdot \hat{\mathbf{E}} \rangle = p \frac{\int_0^\pi e^{(Ep/k_B T)\cos\theta} \cos\theta \sin\theta \, d\theta}{\int_0^\pi e^{(Ep/k_B T)\cos\theta} \sin\theta \, d\theta} = p \left[\coth\left(\frac{Ep}{k_B T}\right) - \frac{k_B T}{Ep} \right] \quad (6.139)$$

where $\hat{\mathbf{E}}$ is a unit vector in the direction of the electric field and θ the angle between $\mathbf{p}$ and $\mathbf{E}$, and the brackets $\langle \, \rangle$ indicate an ensemble average. Under ordinary conditions the energy of interaction is small compared with the thermal energy ($Ep/k_B T \ll 1$), so we may expand (6.139) in powers of Ep/k_BT to show that the polarization depends linearly on the electric field. The susceptibility is

$$\chi_e = \frac{Np^2}{3\varepsilon_0 k_B T} \tag{6.140}$$

Thus, the dielectric susceptibility due to the intrinsic dipole moment is inversely proportional to the temperature due to thermal agitation of the dipoles in the field. On the other hand, the susceptibility (6.133) due to the induced dipole moment is independent of the temperature, so the total susceptibility consists of a term independent of the temperature and one inversely proportional to the temperature.

The classical model also suggests the limitations of (6.140). In the first place we assume thermal equilibrium, which implies that the electric field changes only slowly compared with the rotational relaxation time. The relaxation time in a liquid is bound to be longer than a rotational period of a free molecule, and we therefore expect dispersion to set in at microwave frequencies, which are characteristic of the relaxation time of typical liquids at room temperature. We also expect deviations from linearity when the interaction energy is comparable to or larger than the thermal energy $[Ep/k_BT \geq O(1)]$. For a typical dipole moment $[p = O(ea_0)$, where $e = |q_e|$ is the magnitude of the electronic charge and a_0 the Bohr radius] at room temperature, the critical electric field is

$$E_C = O\left(\frac{k_B T}{ea_0}\right) = O(10^9 \text{ V/m}) \tag{6.141}$$

The density effect is also important for polar liquids, but the Clausius–Mossotti relation does not work in this case. In fact, (6.138) would predict a polarization "catastrophe," since at normal temperatures the dielectric susceptibility of a polar liquid (such as water, for example) is greater than three. The Clausius–Mossotti equation applies only to nonpolar molecules because the field surrounding a polar molecule affects the orientation of the surrounding molecules and this causes a correlation between the orientations of nearby molecules in the liquid. The effect of the correlations has been analyzed by Onsager, who obtained the result

$$\chi_e = \tfrac{3}{4}\left(x + \sqrt{1 + \tfrac{2}{3}x + x^2} - 1\right) \tag{6.142}$$

where

$$x = \frac{Np^2}{3\varepsilon_0 k_B T} \tag{6.143}$$

6.2.2 Magnetic Materials

The magnetic susceptibilities of the elements span nearly five orders of magnitude, and unlike the dielectric susceptibility, the magnetic susceptibility can be either positive or negative. Materials for which the susceptibility is negative are called diamagnetic, and those for which the susceptibility is positive are called paramagnetic or ferromagnetic. As shown in Figure 6.18, the susceptibility varies from element to element in a regular way, depending on the electron configuration. With the notable exception of oxygen (which forms paramagnetic molecules O_2, as discussed later), elements with valence electrons in p-orbitals tend to be diamagnetic. This corresponds to elements on the right-hand side of the periodic table. Elements

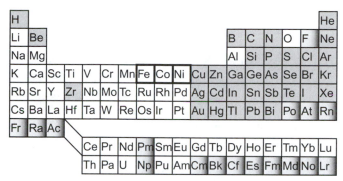

Figure 6.18 Magnetic susceptibility of the elements. Uniform shading indicates diamagnetic elements. No shading indicates paramagnetic elements. Fe, Co, and Ni are ferromagnetic. Graded shading indicates no data.

with valence electrons in s-orbitals tend to be weakly paramagnetic, while those with valence electrons in d- or f-orbitals tend to be strongly paramagnetic. Thus, the elements in columns I and II of the periodic table tend to be paramagnetic, while elements toward the center of the periodic table are strongly paramagnetic. At room temperature just three elements—iron, cobalt, and nickel—are ferromagnetic. There is, of course, an explanation for this regularity and for the exceptions. Although we do not have time to discuss all the details, it is not hard to understand the general principles. The magnetization of diamagnetic, paramagnetic, and ferromagnetic materials closely parallels the polarization of nonpolar, polar, and ferroelectric media, respectively, and we discuss the three phenomena in the same order.

Diamagnetism is observed in materials whose molecules have no intrinsic magnetic dipole moment. For atoms with valence electrons in p-orbitals, the spin magnetic moment of each valence electron cancels the magnetic moment due to its orbital angular momentum (remember that although the spin angular momentum is only $\frac{1}{2}\hbar$, the gyromagnetic ratio for electron spin is $g = 2$, so the spin magnetic moment is the same as the orbital magnetic moment of a p-orbital). The diamagnetic effect arises because as the magnetic field at the position of the molecule increases, the orbits adjust themselves in accordance with Lenz's law. If we imagine that each electron in its orbit is equivalent to a submicroscopic circuit carrying a tiny current, then Lenz's law tells us that the current in the circuit changes in a way that opposes the change in the magnetic flux through the circuit. Diamagnetism, then, corresponds to a reduction of the magnetic induction due to the response of the orbital magnetic moment. Diamagnetism is a small effect. In bismuth, for example, the magnetic susceptibility is $\chi_m = -1.6 \times 10^{-4}$. Actually, diamagnetism is present in all materials, but in most materials it is masked by paramagnetic or ferromagnetic effects, which are much larger.

We can describe diamagnetism classically by means of Larmor's theorem. We consider a set of N identical particles (same charge q and same mass m) in a uniform magnetic field $\mathbf{B}$. The particles are assumed to interact with each other and with some potential that is invariant under rotations about the axis of the magnetic field, such as a nuclear attraction. If we limit ourselves to the nonrelativistic case, the Lagrangian of the particles is

$$\mathcal{L} = \sum_{i=1}^{N} \left[\frac{1}{2} m v_i^2 + q \mathbf{A}\left(\mathbf{r}_i\right) \cdot \mathbf{v}_i \right] - \mathcal{W} \qquad (6.144)$$

where

$$A(\mathbf{r}) = \tfrac{1}{2}\mathbf{B} \times \mathbf{r} \tag{6.145}$$

is the vector potential of the uniform field $\mathbf{B}$ and $\mathcal{W}$ is a potential representing the interactions between the particles as well as any other potential that is invariant under rotations about an axis parallel to the field $\mathbf{B}$. If we change to a new coordinate system rotating at the frequency $\mathbf{\Omega}$ about an axis parallel to the field $\mathbf{B}$, the velocity of the particles in the new frame of reference is

$$\mathbf{v}'_i = \mathbf{v}_i - \mathbf{\Omega} \times \mathbf{r}'_i \tag{6.146}$$

If we substitute this into (6.144) we get

$$\mathcal{L} = \sum_{i=1}^{N} \left[\frac{1}{2}m v_i'^2 + m(\mathbf{\Omega} \times \mathbf{r}'_i) \cdot \mathbf{v}'_i + \frac{1}{2}q(\mathbf{B} \times \mathbf{r}'_i) \cdot \mathbf{v}'_i \right] - \mathcal{W} \tag{6.147}$$

to lowest order in Ω and B. If we choose the frequency of rotation to be

$$\mathbf{\Omega}_L = -\frac{q\mathbf{B}}{2m} \tag{6.148}$$

the effect of the magnetic induction vanishes in the rotating coordinate system. We therefore see that the motion of charges in a magnetic field is just the motion in the absence of the magnetic field plus a rotation of the entire system at the Larmor frequency Ω_L. This is Larmor's theorem. Note that the Larmor frequency Ω_L is *half* the cyclotron frequency of the particles in the same magnetic field, and the drection of the rotation vector $\mathbf{\Omega}_L$ is opposite the direction of the magnetic field $\mathbf{B}$ for positive charges.

The atomic magnetic moment created by this rotation of the charge distribution is

$$\boldsymbol{\mu} = \frac{1}{2} \int \mathbf{r} \times \mathbf{J} \, d^3\mathbf{r} = \frac{1}{2} \sum_i \int \mathbf{r}_i \times (\mathbf{\Omega}_L \times \mathbf{r}_i)\rho_i \, d^3\mathbf{r}_i \tag{6.149}$$

where $\mathbf{r}_i$ is the position and ρ_i the charge density of the ith electron. If we expand the cross product and evaluate the integrals for a spherically symmetric atom (or average over all orientations of the atom), we arrive at the formula

$$\boldsymbol{\mu} = -\frac{q}{4m} \sum_i \int \left[r_i^2 \mathbf{B} - (\mathbf{r}_i \cdot \mathbf{B})\mathbf{r}_i \right] \rho_i \, d^3\mathbf{r}_i = -\frac{Zq^2}{6m} \langle r^2 \rangle \mathbf{B} \tag{6.150}$$

where Z is the atomic number and $\langle r^2 \rangle$ the average radius of the electrons in the atom. Note that the magnetic moment points in the direction opposite the magnetic field, in agreement with Lenz's law. Provided that the magnetization is small ($\chi_m \ll 1$), the diamagnetic susceptibility is then

$$\chi_m = \frac{M}{H} \approx \frac{\mu_0 M}{B} = \frac{\mu_0 N \mu}{B} = -\frac{\mu_0 N Z q^2}{6m} \langle r^2 \rangle \tag{6.151}$$

This is called the Langevin formula. It is interesting to note that this classical formula agrees with the quantum-mechanical result, as must be true because the Larmor precessional frequency is the same for classical and quantum systems. Of course, $\langle r^2 \rangle$ must be calculated using quantum mechanics, but we can estimate it for simple systems. For example, in hydrogen $\langle r^2 \rangle = 3a_0^2$, exactly. In helium ($Z = 2$), the orbits are contracted by the (partially shielded) increased nuclear charge, so we use the estimate $\langle r^2 \rangle \approx a_0^2$. For comparison with experimental data, we note that magnetic susceptibilities are usually quoted per mole, in the form

$$\chi_M = \frac{N_A}{N}\chi_m = -\frac{\mu_0 N_A Z q^2}{6m}\langle r^2 \rangle \tag{6.152}$$

where $N_A = 6.02 \times 10^{26}$/kg-mole is Avogadro's number. Substituting into (6.152), we get $\chi_M(\text{He}) \approx -2 \times 10^{-8}$ m³/kg-mole. This may be compared with the experimental value $\chi_M(\text{He}) = -2.4 \times 10^{-8}$ m³/kg-mole.

Before leaving the subject of diamagnetism, and despite the good agreement obtained here for helium, it should be pointed out that in some sense it is impossible to estimate the diamagnetic susceptibility of a system in thermodynamic equilibrium on classical grounds. This is because the average motions of a particle in thermal equilibrium depend only on the parameter $\mathcal{E}/k_B T$, where $\mathcal{E}$ is the energy of the particle, k_B Boltzmann's constant, and T the temperature. Since the magnetic field doesn't change the energy of an orbiting particle, the equilibrium classical diamagnetic susceptibility should vanish. The procedure we have used works, of course, because quantum mechanics fixes the orbits of the electrons in the atom, independent of statistical mechanical considerations. The (rigid) electron configuration then rotates as we have described here (subject to the limitations imposed by surrounding atoms), since Larmor's theorem is valid in quantum mechanics as well. In metals, of course, the situation is more complicated, since the conduction electrons are free to move about the metal.

Paramagnetism is observed in materials whose atomic or molecular constituents possess a permanent magnetic dipole moment. Whereas diamagnetic materials have a negative susceptibility because the induced magnetization opposes the applied field, paramagnetic materials have a positive susceptibility. Under the influence of the magnetic field the dipole moments of the atoms and molecules line up with the magnetic field and create a relatively large magnetization. The situation is analogous to the dielectric susceptibility of polar substances, and we can make use of the results obtained there. The energy of a magnetic dipole $\boldsymbol{\mu}$ in the magnetic inductance $\mathbf{B}$ is $W = -\boldsymbol{\mu} \cdot \mathbf{B}$, which replaces the energy $W = -\mathbf{p} \cdot \mathbf{E}$ in dielectrics. Because we define the magnetic susceptibility in terms of $\mathbf{H}$ rather than $\mathbf{B}$, we replace χ_e by $\chi_m/(1 + \chi_m)$. The magnetic susceptibility (for $\chi_m \ll 1$) is therefore

$$\frac{\chi_m}{1 + \chi_m} \approx \chi_m = \frac{\mu_0 N \mu^2}{3k_B T} \tag{6.153}$$

provided that the interaction energy is small compared with the thermal energy ($\mu B/k_B T \ll 1$). The molar susceptibility is

$$\chi_M = \frac{\mu_0 N_A \mu^2}{3k_B T} \tag{6.154}$$

The inverse temperature dependence of the paramagnetic susceptibility was actually discovered experimentally, and is known as Curie's law (for Pierre Curie).

For example, the ground state configuration of the oxygen molecule O_2 is $^3\Sigma$. The orbital angular momentum of the Σ state vanishes, but the spin angular momentum of the triplet state (total spin quantum number $S = 1$) is $\mathbf{L}_S = \hbar\sqrt{S(S+1)}\,\hat{\mathbf{n}}$, where $\hat{\mathbf{n}}$ is a unit vector. The magnetic moment associated with the electron spin is oriented in the opposite direction and has the value

$$\boldsymbol{\mu}_{\text{spin}} = -2\sqrt{S(S+1)}\mu_B \hat{\mathbf{n}} \tag{6.155}$$

where $\mu_B = e\hbar/2m = 9.274 \times 10^{-24}$ J/T is the Bohr magneton, in which $e = |q_e|$ is the (positive) charge on an electron. If we substitute this into (6.154), we get $\chi_M(O_2) = 1.4 \times 10^{-4}$ m^3/kg-mole at a temperature $T = 90$ K. This is to be compared with the experimental value $\chi_M(O_2) = 0.97 \times 10^{-4}$ m^3/kg-mole at the same temperature.

In condensed matter, the magnetization of an atom or molecule is influenced by the magnetization of its neighbors. This is similar to the effect observed in dielectrics, but in magnetic materials the effect is ordinarily very small. The exception occurs in ferromagnets. Unlike ferroelectrics, where the polarization of the surrounding atoms is enough to cause spontaneous polarization of the medium, the corresponding magnetic effect in ferromagnetic materials is orders of magnitude too small. Instead, a quantum mechanical effect (called the exchange field, equivalent to a magnetic induction of the order of 10^3 T) links the spins of certain electrons and causes them to align parallel to one another. Although the individual interactions are small, the overall effect corresponds to a large magnetization. A completely magnetized crystal of iron has a magnetization $M = 1.7 \times 10^6$ A/m, which is equivalent to a magnetic induction $B = \mu_0 M = 2.2$ T.

Because of the exchange field, all the spins in the crystal line up spontaneously. However, the crystal subdivides itself into magnetic domains that are arranged in a way that minimizes the energy of the field. Moreover, ordinary polycrystalline material consists of many small regions that are separated by grain boundaries and oriented randomly. When a magnetic field is applied to a sample of ferromagnetic material that is initially unmagnetized, the first thing that happens is that the boundaries between the domains move to allow the more favorably oriented domains to grow at the expense of the others. This process is reversible, but at some point the domain boundaries bump up against grain boundaries and other crystal imperfections, and further growth in the magnetization occurs only by displacements of the domain boundaries across the imperfections. These displacements are irreversible in the sense that as the magnetic field is reduced the domain boundaries are pinned by the crystal imperfections, so the domains do not return to the original configuration. Domain growth saturates at very high fields, and further increase in the magnetization occurs only by rotation of the magnetization to crystal orientations that are otherwise energetically less favorable. This overall behavior is illustrated in Figure 6.19, where we have plotted the total magnetic induction $B = \mu_0(H + M)$ as a function of the applied field H.

When the magnetic field applied to an irreversibly magnetized sample is reduced, the magnetic induction follows the upper curve in Figure 6.20. When the applied field is reduced to zero, some residual magnetic induction remains. This is called the remnant field $B_r = \mu_0 M$, where M is the magnetization. This is what makes "permanent magnets"

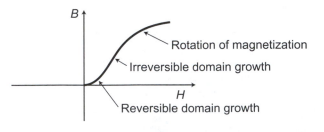

Figure 6.19 Magnetization of a ferromagnetic material.

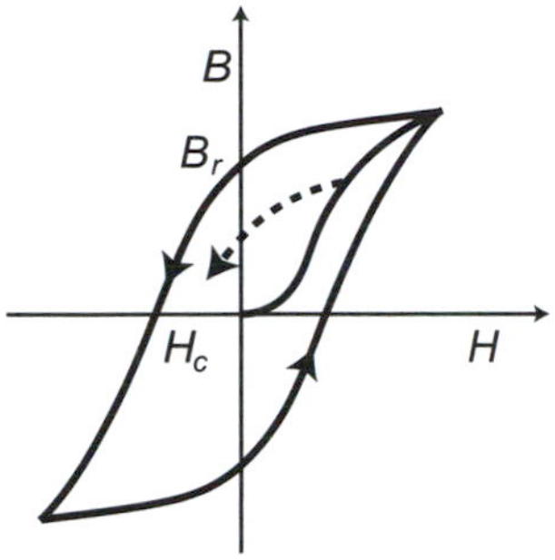

Figure 6.20 Hysteresis curve for a ferromagnetic material.

magnets. If the applied magnetic field is reversed, the magnetization continues along the upper curve into the left-hand quadrants of the diagram. At some point the magnetic induction vanishes. The magnetic field required to achieve this, starting with a fully magnetized sample, is called the coercivity (or coercive force), denoted H_c. This is what makes "permanent magnets" permanent, since the coercivity is a measure of the difficulty of demagnetizing a permanent magnet.

If we don't fully magnetize the sample on the first pass, when the magnetic field is reduced the magnetic induction follows the dotted curve in Figure 6.20. This illustrates the fact that the magnetization is not a unique function of the magnetic field. The magnetization depends on the history of the applied magnetic field.

Many types of magnetic material have been developed for special purposes. For permanent magnets, materials are selected for high remnant field and high coercivity. For transformer cores, on the other hand, the highest rate of change of the magnetic induction is desired, so a material with a low coercivity is used to place the operation of the transformer core on the slope of the hysteresis curve rather than on the flat top. For linear transformers, in which a sinusoidal input produces a sinusoidal output with no (or minimal) higher frequency components, the operation must be restricted to a linear portion of the hysteresis curve. Sometimes, however, nonlinear effects are desirable. For example, it may be desirable to produce harmonics of the input. In other cases, a choke can be built that has a large inductance until the current builds up to a large value that saturates the core. At this point the inductance of the choke drops abruptly, and the choke acts as a saturable "switch" to turn on the current quite suddenly. This technique has been used to form steeply rising (<20 ns) pulses of very high current (>1 kA) and voltage (>1 MV) for radar and other applications.

BIBLIOGRAPHY

An excellent, in-depth discussion of the electrodynamics of macroscopic media is given by
> L. D. Landau, E. M. Lifshitz, and L. B. Pitaevskii, *Electrodynamics of Continuous Media,* Pergamon Press, Oxford (1984).

The quantum-mechanical aspects of a broad variety of phenomena in solids, including polarization and magnetism are discussed in a very readable way by several authors, including
> N. W. Ashcroft and N. D. Mermin, *Solid State Physics,* Holt, Rinehart and Winston, New York (1976),
> C. Kittel, *Introduction to Solid State Physics,* 7th edition, John Wiley & Sons, New York (1996).

7

Linear, Dispersive Media

7.1 LINEAR MEDIA

Although nonlinear phenomena are of enormous interest these days, most common electromagnetic phenomena are well described by linear theory. The theory of linear media is mathematically quite elegant, for we can make powerful use of Fourier transforms and the theory of functions of a complex variable. In fact, we use the same mathematical approach in Chapter 9 to discuss nonlinear optical phenomena in the limit when the nonlinearity is a perturbation. The theorems that are proved in this way are of remarkable generality and are widely used to analyze the optical and electromagnetic properties of solids. We therefore take up the subject of linear media not only for the beauty of the mathematics and usefulness of the formulas, but also because the results serve to highlight the sometimes obscure and frequently surprising connections between various phenomena.

7.1.1 Waves in a Nondispersive Medium

In Chapter 7 it is shown that the Maxwell equations for a macroscopic material can be written in the form

$$\nabla \cdot \mathbf{D} = \rho_f \tag{7.1}$$

$$\nabla \cdot \langle \mathbf{B} \rangle = 0 \tag{7.2}$$

$$\nabla \times \langle \mathbf{E} \rangle + \frac{\partial \langle \mathbf{B} \rangle}{\partial t} = 0 \tag{7.3}$$

$$\nabla \times \mathbf{H} - \frac{\partial \mathbf{D}}{\partial t} = \mathbf{J}_f \tag{7.4}$$

where ρ_f and $\mathbf{J}_f$ are the free charge density and free current density, and the displacement vector $\mathbf{D}$ and the magnetic field $\mathbf{H}$ are defined by the relations

$$\mathbf{D} = \varepsilon_0 \langle \mathbf{E} \rangle + \mathbf{P} \tag{7.5}$$

and

$$\mathbf{H} = \frac{1}{\mu_0} \langle \mathbf{B} \rangle - \mathbf{M} \tag{7.6}$$

In the following we leave off the brackets $\langle \ \rangle$ for simplicity, with the understanding that the electric field and magnetic induction are really averages over a small region of space.

In the simplest approximation we might be tempted to assume that the polarization and magnetization are linearly proportional to the instantaneous electric field and magnetic induction. In a uniform, isotropic medium, then, we would write

$$\mathbf{P} = \varepsilon_0 \chi_e \mathbf{E} \tag{7.7}$$

$$\mathbf{M} = \chi_m \mathbf{H} \tag{7.8}$$

for some constant coefficients χ_e and χ_m, and

$$\mathbf{D} = \varepsilon_0(1 + \chi_e)\mathbf{E} = \varepsilon \mathbf{E} \tag{7.9}$$

$$\mathbf{B} = \mu_0(1 + \chi_m)\mathbf{H} = \mu \mathbf{H} \tag{7.10}$$

for some new, constant values of the permittivity and permeability. If we substitute these expressions into the Maxwell equations (7.1)–(7.4), leaving off the brackets $\langle \ \rangle$ for simplicity, we get

$$\nabla \cdot \mathbf{E} = \frac{\rho_f}{\varepsilon} \tag{7.11}$$

$$\nabla \cdot \mathbf{H} = 0 \tag{7.12}$$

$$\nabla \times \mathbf{E} + \mu \frac{\partial \mathbf{H}}{\partial t} = 0 \tag{7.13}$$

$$\nabla \times \mathbf{H} - \varepsilon \frac{\partial \mathbf{E}}{\partial t} = \mathbf{J}_f \tag{7.14}$$

These are equivalent to the Maxwell equations for a vacuum with the substitutions $\varepsilon_0 \to \varepsilon = \varepsilon_0(1 + \chi_e)$, $\mu_0 \to \mu = \mu_0(1 + \chi_m)$, and $c^2 \to 1/\mu\varepsilon$.

The simple equations (7.7)–(7.14) admit wavelike solutions that propagate with velocity $v = 1/\sqrt{\mu\varepsilon}$. However, it is well known that the velocity of electromagnetic waves in a medium depends on the frequency of the waves. This phenomenon is called dispersion. Since dispersion implies that ε and μ are not constants, it is not strictly possible to write the Maxwell equations in the form (7.11)–(7.14), even for linear media. However, at sufficiently low frequencies (which in some materials can extend to optical frequencies), the permeability and permittivity remain close to their dc values, and in these cases the approximations represented by (7.11)–(7.14) may be used.

7.1.2 Constitutive Relations in Dispersive Media

The origin of dispersion is to be found in the different time scales that characterize real media and make different phenomena important at different frequencies. The response of the medium to the applied fields is not instantaneous, for the electrons and nuclei have finite mass. Nor does the response vanish when the field is removed, for the medium takes a finite time to relax back to its quiescent state. The polarization and magnetization of the medium at time t therefore depend on the entire histories $\mathbf{E}(t')$ and $\mathbf{H}(t')$ of the fields at the point in question, and this must be represented by the constitutive relations. In the following discussion we ignore the dependence of the polarization on the magnetic field and the dependence of the magnetization on the electric field, and to simplify the notation we assume that the medium is isotropic. In this case, of course, the polarization is in the direction of the electric field and the magnetization is in the direction of the magnetic field. The extension to an anisotropic medium is straightforward, but the algebra is more complicated. We further assume that the medium is linear, but dispersive, so that the constitutive relations may be written in the form

$$\mathbf{P}(t) = \varepsilon_0 \int_{-\infty}^{\infty} \mathbf{E}(t')G_e(t - t')\,dt' \tag{7.15}$$

$$\mathbf{M}(t) = \int_{-\infty}^{\infty} \mathbf{H}(t')G_m(t - t')\,dt' \tag{7.16}$$

where $G_e(t - t')$ and $G_m(t - t')$ are called the response functions for the polarization and the magnetization, respectively. Physically, the response function describes the behavior of the system as a function of the time t following a unit impulse at time t'. The principle of causality says that the polarization at time t cannot depend on the electric field at later times $t' > t$. Therefore,

$$G(\tau) = 0 \qquad \text{for } \tau < 0 \tag{7.17}$$

for either the dielectric or the magnetic response function.

It is convenient for many problems to change from the time domain to the frequency domain. If we take the Fourier transform of the constitutive relations (7.15) and (7.16) using the Faltung theorem of Chapter 5, we get

$$\tilde{\mathbf{P}}(\omega) = \varepsilon_0 \chi_e(\omega)\tilde{\mathbf{E}}(\omega) \tag{7.18}$$

and

$$\tilde{\mathbf{M}}(\omega) = \chi_m(\omega)\tilde{\mathbf{H}}(\omega) \tag{7.19}$$

The functions $\chi_e(\omega)$ and $\chi_m(\omega)$ are called the dielectric and magnetic susceptibilities. In terms of the response function $G(\tau)$, the susceptibility is

$$\chi(\omega) = \int_{-\infty}^{\infty} G(\tau)e^{i\omega\tau}\,d\tau \tag{7.20}$$

Aside from a factor of $\sqrt{2\pi}$, the susceptibility is just the Fourier transform of the response function. Inverting the transform, we see that the response function is

$$G(\tau) = \frac{1}{2\pi} \int_{-\infty}^{\infty} \chi(\omega)e^{-i\omega\tau}\,d\omega \tag{7.21}$$

We discuss the analytical properties of the susceptibility $\chi(\omega)$ in the next section, but for now we note that since the response function $G(t)$ is real, the susceptibility defined by (7.20) has the symmetry property

$$\chi(-\omega) = \chi^*(\omega^*) \tag{7.22}$$

EXERCISE 7.1

For the following response functions, compute the complex susceptibility χ and plot the poles in the complex plane. Then invert the Fourier transform to confirm that the computed dielectric susceptibility is correct.

(a) Nondispersive medium; responds instantly to the applied impulse:

$$G(\tau) = \chi_0 \delta(\tau) \tag{7.23}$$

Show that

$$\chi(\omega) = \chi_0 \tag{7.24}$$

plot the singularities, and then invert the transform.

(b) Damped medium; responds to the applied impulse and then relaxes slowly back to its original state:

$$G(\tau) = \chi_0 H(\tau) \exp(-\alpha\tau) \tag{7.25}$$

where

$$\begin{aligned} H(\tau) &= 0, & \tau < 0 \\ H(\tau) &= 1, & \tau > 0 \end{aligned} \tag{7.26}$$

is the Heaviside unit step function. Show that

$$\chi(\omega) = \frac{\chi_0}{\alpha - i\omega} \tag{7.27}$$

plot the singularities, and then invert the transform.

(c) Conducting medium; responds to the applied impulse, but with no restoring force it does not relax back:

$$G(\tau) = \chi_0 H(\tau) \tag{7.28}$$

Show that

$$\chi(\omega) = \chi_0 \left[\pi \delta(\omega) + \frac{i}{\omega} \right] \tag{7.29}$$

plot the singularities, and then invert the transform.

(d) Resonant medium; responds to the applied impulse, after which it oscillates:

$$G(\tau) = \chi_0 H(\tau) \sin(\omega_0 \tau) \tag{7.30}$$

Show that

$$\chi(\omega) = \chi_0 \left[i\frac{\pi}{2}\delta(\omega - \omega_0) - i\frac{\pi}{2}\delta(\omega + \omega_0) - \frac{\omega_0}{\omega^2 - \omega_0^2} \right] \tag{7.31}$$

plot the singularities, and then invert the transform.

Hint: The integrals for (7.29) and (7.31) must be defined by an appropriate limit process. To evaluate the results without missing the singularity at $\omega = 0$, recall that

$$\lim_{\varepsilon \to 0} \left(\frac{\varepsilon}{\omega^2 + \varepsilon^2} \right) = \pi \delta(\omega) \tag{7.32}$$

To invert the transforms, except in the trivial case of (7.24), use contour integration and evaluate the integrals by computing the residues of the poles enclosed inside the contours. For $\tau < 0$, close the contour with a semicircle at infinity that encloses the upper half plane. On this contour, the factor $\exp(-i\omega\tau)$ vanishes exponentially for Im $\omega > 0$, making the integral over the semicircle vanish. For $\tau > 0$, close the contour with a semicircle enclosing the lower half plane so that the integral over the semicircle again vanishes.

7.1.3 Kramers–Kronig Relations

The Kramers–Kronig relations compose one of the most elegant and general theorems in physics, because they depend for their validity only on the principle of causality: The response cannot come before the stimulus. Based simply on this principle, the Kramers–Kronig relations describe the interdependence of the real and imaginary parts of the susceptibility $\chi(\omega)$. But as applied to wave propagation, the real part of the susceptibility describes essentially the index of refraction and the imaginary part the absorption coefficient of a medium. Thus, the Kramers–Kronig relations explain in the most fundamental and general terms, completely independent of the underlying physical mechanisms, the intimate connection between refraction and absorption. In fact, given one, the other follows immediately. We begin by establishing the domain in the complex ω (frequency) plane where the susceptibility $\chi(\omega)$ is analytic, and then we use this information to derive the Kramers–Kronig relations.

The domain in the complex frequency plane in which the susceptibility function is analytic is established just by the principle of causality (7.17), completely independent of the physical mechanism causing the dispersion. We assume, for the moment, that the response function $G(\tau)$ is absolutely integrable and has at most a finite number of finite discontinuities, so the Riemann–Lebesgue lemma assures us that

$$\chi(\omega) \xrightarrow[|\omega| \to \infty]{} 0 \tag{7.33}$$

In fact, conductors violate this assumption, but we deal with this later. For now we consider the integral $\oint \chi(\omega) e^{-i\omega\tau} d\omega$ around a contour enclosing the upper half of the complex ω plane, as shown in Figure 7.1. From Cauchy's residue theorem we know that

$$\oint \chi(\omega) e^{-i\omega\tau} d\omega = i2\pi \sum_n [e^{-i\omega_n\tau} \times \text{residue of } \chi(\omega_n)] \tag{7.34}$$

where ω_n is the nth pole of the function $\chi(\omega)$ inside the contour. Provided that $\chi(\omega)$ has only a finite number of poles, we can make the radius R large enough to enclose all the poles in the upper half plane. The contour integral consists of two parts, these being the integral along the real axis and the integral around the semicircle, so that

$$\lim_{R \to \infty} \oint \chi(\omega) e^{-i\omega\tau} d\omega = \int_{-\infty}^{\infty} \chi(\omega) e^{-i\omega\tau} d\omega + \lim_{R \to \infty} \int_R \chi(\omega) e^{-i\omega\tau} d\omega \tag{7.35}$$

But the first term on the right is just the response function $G(\tau)$, which vanishes for $\tau < 0$ to satisfy the requirements of causality. The magnitude of the second integrand is

$$|\chi(\omega) e^{-i\omega\tau}| < |\chi(\omega)| e^{y\tau} \tag{7.36}$$

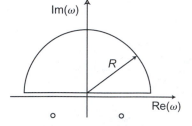

Figure 7.1 Contour enclosing the upper half plane.

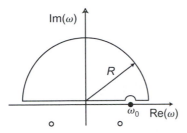

Figure 7.2 Contour for deriving the Kramers–Kronig relations.

where $\omega = x + iy$. For $\tau < 0$ this vanishes exponentially above the real axis, so the second term vanishes also. We see from this that

$$i2\pi \sum_n [e^{-i\omega_n \tau} \times \text{residue of } \chi(\omega_n)] = 0, \qquad \text{for } \tau < 0 \tag{7.37}$$

Therefore, if the response function $G(\tau)$ satisfies the principle of causality, there can be no poles of $\chi(\omega)$ in the upper half plane.

We next consider the function $\chi(\omega)/(\omega - \omega_0)$, where ω_0 is a point on the real axis, as shown in Figure 7.2. Like $\chi(\omega)$, it has no poles in the upper half plane, but it has a simple pole at ω_0. If we integrate $\chi(\omega)/(\omega - \omega_0)$ around the upper plane using the contour shown in Figure 7.2, the result vanishes by Cauchy's theorem. But the contour is composed of several segments, and we see that

$$\int_{-\infty}^{\omega_0 - \varepsilon} \frac{\chi(\omega)}{\omega - \omega_0} \, d\omega + \int_{\omega_0 + \varepsilon}^{\infty} \frac{\chi(\omega)}{\omega - \omega_0} \, d\omega + \int_\varepsilon \frac{\chi(\omega)}{\omega - \omega_0} \, d\omega + \lim_{R \to \infty} \int_R \frac{\chi(\omega)}{\omega - \omega_0} \, d\omega = 0 \tag{7.38}$$

The last term vanishes in the limit $R \to \infty$, since $\chi(\omega)$ vanishes due to the Riemann–Lebesgue lemma. In the limit $\varepsilon \to 0$, the first two terms become the principal part of the integral from $-\infty$ to ∞. The third term, the integral along the semicircle around the pole at ω_0, can be evaluated from half the residue at the pole, so

$$\lim_{\varepsilon \to 0} \int_\varepsilon \frac{\chi(\omega)}{\omega - \omega_0} \, d\omega = -i\pi \chi(\omega_0) \tag{7.39}$$

where the negative sign appears because we are integrating clockwise halfway around the pole. We therefore obtain the result

$$\chi(\omega_0) = \frac{1}{i\pi} P \int_{-\infty}^{\infty} \frac{\chi(\omega)}{\omega - \omega_0} \, d\omega \tag{7.40}$$

where P stands for principal part. The importance of this remarkable equation follows from the appearance of i on the right-hand side, for if we separate the real and imaginary parts of (7.40), we get the relations

$$\text{Re}\, \chi(\omega_0) = \frac{1}{\pi} P \int_{-\infty}^{\infty} \frac{\text{Im}\, \chi(\omega)}{\omega - \omega_0} \, d\omega \tag{7.41}$$

$$\text{Im}\, \chi(\omega_0) = -\frac{1}{\pi} P \int_{-\infty}^{\infty} \frac{\text{Re}\, \chi(\omega)}{\omega - \omega_0} \, d\omega \tag{7.42}$$

which were first obtained by Kramers (1927) and Kronig (1926). From them we see that if the susceptibility has a real part then it necessarily has an imaginary part, and vice versa.

Since dissipation is associated with the imaginary part of the susceptibility and dispersion with the real part, you can't, as they say, have one without the other.

These relations can be put in a more convenient form in the following way. If we multiply the integrand of (7.41) by $(\omega + \omega_0)/(\omega + \omega_0)$ we get

$$\text{Re } \chi(\omega_0) = \frac{1}{\pi} P \int_{-\infty}^{\infty} \frac{\omega \text{ Im } \chi(\omega)}{\omega^2 - \omega_0^2} d\omega + \frac{1}{\pi} P \int_{-\infty}^{\infty} \frac{\omega_0 \text{ Im } \chi(\omega)}{\omega^2 - \omega_0^2} d\omega \qquad (7.43)$$

But from (7.22) we see that Im $\chi(\omega)$ is an odd function of ω, so the second term vanishes. Since the integrand in first term is even, we may write it as an integral over positive frequencies only,

$$\text{Re } \chi(\omega_0) = \frac{2}{\pi} P \int_{0}^{\infty} \frac{\omega \text{ Im } \chi(\omega)}{\omega^2 - \omega_0^2} d\omega \qquad (7.44)$$

In the same way, we find that

$$\text{Im } \chi(\omega_0) = -\frac{2}{\pi} P \int_{0}^{\infty} \frac{\omega_0 \text{ Re } \chi(\omega)}{\omega^2 - \omega_0^2} d\omega \qquad (7.45)$$

Note carefully, when using the Kramers–Kronig relations in the form (7.41) and (7.42), that it is important to remember the symmetry of $\chi(\omega)$. If, for example, the imaginary part of $\chi(\omega)$ has a sharp peak at ω_α, corresponding to a sharp absorption at that frequency, then it has a sharp peak at $-\omega_\alpha$ also. This must be accounted for in the integral from $-\infty$ to ∞. For this reason it seems invariably to be safer and more convenient to use (7.44) and (7.45).

The Kramers–Kronig relations are widely used to compute the real part of the susceptibility when the imaginary part is known and vice versa. However, we can learn a few things just by inspection. For example, we see from (7.44) that if a medium has an imaginary component of the susceptibility at some frequency ω_α, then it must have a real component over a broad range of frequencies around ω_α. In the limit when Im $\chi(\omega) = \alpha\delta(\omega - \omega_\alpha)$, the real part is

$$\text{Re } \chi(\omega_0) = \frac{2}{\pi} \frac{\alpha\omega_\alpha}{\omega_\alpha^2 - \omega_0^2} \qquad (7.46)$$

as illustrated in Figure 7.3.

In the same fashion we may show that in a region where the imaginary part of the susceptibility is negligible, the real part must be an increasing function of the frequency. To show this, we assume that Im $\chi(\omega) = 0$ in the range $\omega_1 < \omega < \omega_2$, so that in this same range

$$\text{Re } \chi(\omega_0) = \frac{2}{\pi} \int_{0}^{\omega_1} \frac{\omega \text{ Im } \chi(\omega)}{\omega^2 - \omega_0^2} d\omega + \frac{2}{\pi} \int_{\omega_2}^{\infty} \frac{\omega \text{ Im } \chi(\omega)}{\omega^2 - \omega_0^2} d\omega \qquad (7.47)$$

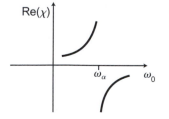

Figure 7.3 Real part of the susceptibility when the imaginary part is strongly peaked.

If we differentiate this with respect to ω_0 we get

$$\frac{d}{d\omega_0} \text{Re} \chi(\omega_0) = \frac{4\omega_0}{\pi} \int_0^{\omega_1} \frac{\omega \, \text{Im} \chi(\omega)}{\left(\omega^2 - \omega_0^2\right)^2} \, d\omega + \frac{4\omega_0}{\pi} \int_{\omega_2}^{\infty} \frac{\omega \, \text{Im} \chi(\omega)}{\left(\omega^2 - \omega_0^2\right)^2} \, d\omega \tag{7.48}$$

Both terms on the right are positive provided that $\text{Im} \chi(\omega) > 0$ everywhere. Since $\text{Im} \chi(\omega) > 0$ is associated with an optically absorbing medium, the real part of the susceptibility must be increasing with frequency except, perhaps, in a laser. But the real part of the susceptibility is associated with the index of refraction. We therefore see that in transparent regions of the spectrum, blue light must be refracted more than red light just to satisfy the requirements of causality. This is called normal dispersion. Anomalous dispersion, in which the index of refraction is a decreasing function of the frequency, can occur only in a spectral region where absorption is important.

Finally, we can use the asymptotic properties of the real and imaginary parts of the susceptibility to derive a sum rule. Quite generally, we would expect the response function $G(t)$ to be continuous for any reasonable physical model of the dielectric response but to have a discontinuity in the derivative at $t = 0$, where the impulse is applied. As shown in Chapter 5, the asymptotic behavior of the Fourier transform must then be

$$\chi(\omega) \xrightarrow[\omega \to \infty]{} -\frac{\Delta G'}{\omega^2} \tag{7.49}$$

where $\Delta G'$ is the (real) discontinuity in the derivative $G'(t) = dG/dt$ at $t = 0$. The imaginary part of the susceptibility must fall off even faster. If we use these properties in (7.44) we see that in the limit $\omega_0 \to \infty$ the right-hand side is dominated by frequencies for which $\omega/\omega_0 \ll 1$, so we may remove the factor $\omega_0^2 - \omega^2 \approx \omega_0^2$ from the integral. We then obtain the sum rule

$$\lim_{\omega_0 \to \infty} \left[\omega_0^2 \, \text{Re} \chi(\omega_0)\right] = -\frac{2}{\pi} \int_0^{\infty} \omega \, \text{Im} \chi(\omega) \, d\omega \tag{7.50}$$

As we see later, this relates the sum over all frequencies of the optical absorption properties of a substance (which are related to the imaginary part of the susceptibility) to the high-frequency limit of the refractive index (which is related to the real part of the susceptibility). Like the Kramers–Kronig relations from which it is derived, the sum rule depends for its validity only on the principle of causality.

EXERCISE 7.2

As shown by (7.29), it is characteristic of conductors that the dielectric susceptibility $\chi_e(\omega)$ has a simple pole and a δ-function on the real axis at the origin, so for sufficiently small ω (but $\omega \neq 0$) we may write

$$\chi_e \approx \frac{i\sigma_0}{\omega} + \cdots \tag{7.51}$$

where σ_0 is the dc conductivity. Thus, $\chi_e(\omega)$ has a singularity on the real axis, along with other singularities in the lower half plane. Rederive the Kramers–Kronig relations for this case, and show that

$$\operatorname{Re} \chi_e(\omega_0) = \pi \sigma_0 \, \delta(\omega_0) + \frac{2}{\pi} \mathrm{P} \int_0^\infty d\omega \, \frac{\omega \, \operatorname{Im} \chi_e(\omega)}{\omega^2 - \omega_0^2} \tag{7.52}$$

$$\operatorname{Im} \chi_e(\omega_0) = \frac{\sigma_0}{\omega_0} - \frac{2}{\pi} \mathrm{P} \int_0^\infty d\omega \, \frac{\omega_0 \, \operatorname{Re} \chi_e(\omega)}{\omega^2 - \omega_0^2} \tag{7.53}$$

Hint: To evaluate the residue at the origin, note that

$$\lim_{\omega \to 0} \left[\frac{1}{\omega - \omega_0} \right] = \lim_{\omega \to 0} \left[\frac{1}{\omega - \omega_0} \frac{\omega + \omega_0}{\omega + \omega_0} \right] = \lim_{\omega \to 0} \left[\frac{\omega_0}{\omega^2 - \omega_0^2} + i \frac{i\omega}{(i\omega)^2 + \omega_0^2} \right] \tag{7.54}$$

The second term, it will be recalled, is a representation of the δ-function.

7.1.4 Plane Waves in Dispersive Media

To deal with dispersive systems it is convenient to move from the time domain into the frequency domain by means of Fourier transforms. We begin by taking the Fourier transform of the Maxwell equations (7.1)–(7.4) using the rule $d/dt \to -i\omega$, to get

$$\nabla \cdot \tilde{\mathbf{D}} = \tilde{\rho}_f \tag{7.55}$$

$$\nabla \cdot \tilde{\mathbf{B}} = 0 \tag{7.56}$$

$$\nabla \times \tilde{\mathbf{E}} - i\omega \tilde{\mathbf{B}} = 0 \tag{7.57}$$

$$\nabla \times \tilde{\mathbf{H}} + i\omega \tilde{\mathbf{D}} = \tilde{\mathbf{J}}_f \tag{7.58}$$

If we substitute the linear relations (7.18) and (7.19) back into the Maxwell equations and assume that the medium is uniform (so we can ignore space and time derivatives of χ_e and χ_m), we get

$$\nabla \cdot \tilde{\mathbf{E}} = \frac{\tilde{\rho}_f}{\varepsilon} \tag{7.59}$$

$$\nabla \cdot \tilde{\mathbf{H}} = 0 \tag{7.60}$$

$$\nabla \times \tilde{\mathbf{E}} - i\omega\mu \tilde{\mathbf{H}} = 0 \tag{7.61}$$

$$\nabla \times \tilde{\mathbf{H}} + i\omega\varepsilon \tilde{\mathbf{E}} = \tilde{\mathbf{J}} \tag{7.62}$$

where the permittivity $\varepsilon(\omega)$ and permeability $\mu(\omega)$ are

$$\varepsilon(\omega) = \varepsilon_0[1 + \chi_e(\omega)] \tag{7.63}$$

and

$$\mu(\omega) = \mu_0[1 + \chi_m(\omega)] \tag{7.64}$$

To describe the propagation of waves in a dispersive medium, we combine the Maxwell equations into a single equation for $\tilde{\mathbf{H}}$ alone. We begin with the vector identity

$$\nabla \times (\nabla \times \tilde{\mathbf{H}}) = \nabla(\nabla \cdot \tilde{\mathbf{H}}) - \nabla^2 \tilde{\mathbf{H}} \tag{7.65}$$

and use (7.62) and (7.60) to evaluate $\nabla \times \tilde{\mathbf{H}}$ and $\nabla \cdot \tilde{\mathbf{H}}$. Then, using (7.61) to evaluate $\nabla \times \tilde{\mathbf{E}}$, we get the Helmholtz equation

$$\nabla^2 \tilde{\mathbf{H}} + n^2 \frac{\omega^2}{c^2} \tilde{\mathbf{H}} = -\nabla \times \tilde{\mathbf{J}} \tag{7.66}$$

in a uniform medium, where the (complex) index of refraction is

$$n^2(\omega) = \frac{\mu\varepsilon}{\mu_0\varepsilon_0} = [1 + \chi_e(\omega)][1 + \chi_m(\omega)] \tag{7.67}$$

The corresponding equation for $\tilde{\mathbf{E}}$ is found in a similar manner. The advantage of working in the frequency domain lies in the fact that in the frequency domain each Fourier component of the field has its own Helmholtz equation with its own source term. This is a consequence of the linearity of the constitutive relations (7.18) and (7.19). The difficulty lies in the fact that the Fourier transform cannot, in general, be simply inverted to find the field in the time domain. In the case of vacuum electromagnetic fields, the inversion is possible and the fields in the time domain are expressed as the retarded fields or the Lienard–Wiechert fields. In dispersive media the retarded time (the time for the fields to travel from the source point to the observation point), which depends on the velocity of propagation, is not the same for all frequencies.

In the absence of sources, the Helmholtz equation (7.66) has the plane-wave solution

$$\tilde{u}(\mathbf{r}, \omega) = \tilde{u}_0(\omega)e^{i\mathbf{k}\cdot\mathbf{r}} \tag{7.68}$$

where $\tilde{u}$ is any component of $\tilde{\mathbf{H}}(\omega)$ or $\tilde{\mathbf{E}}(\omega)$ and the wave vector satisfies the dispersion relation

$$\mathbf{k} \cdot \mathbf{k} = n^2\frac{\omega^2}{c^2} \tag{7.69}$$

Since the index of refraction is in general complex, the wave vector is also complex. For the moment we limit the discussion to true plane waves, for which

$$\mathbf{k} = k\hat{\mathbf{k}} \tag{7.70}$$

where $\hat{\mathbf{k}}$ is a real unit vector and

$$k = \frac{n}{c}\omega \tag{7.71}$$

is complex. Separating the real and imaginary parts of $k = k' + ik''$, we get

$$k' = \frac{n'}{c}\omega \tag{7.72}$$

$$k'' = \frac{n''}{c}\omega \tag{7.73}$$

where the complex index of refraction is

$$n = n' + in'' \tag{7.74}$$

for real n' and n''. The wave (7.68) then has the form

$$\tilde{u} = \tilde{u}_0 e^{(ik'-k'')\hat{\mathbf{k}}\cdot\mathbf{r}} \tag{7.75}$$

where $\hat{\mathbf{k}}$ is the direction of propagation. The intensity (the Poynting vector) of the wave is quadratic in the amplitude of the field, so the intensity varies as

$$S \propto e^{-2k''\hat{\mathbf{k}}\cdot\mathbf{r}} \tag{7.76}$$

Therefore, $2k'' = 2\omega n''/c$ is the decay constant for the intensity of the wave. It is proportional to the imaginary part of the index of refraction. In a typical case the absorption of the wave is small, so $k''/k' = n''/n' \ll 1$. In this case, the intensity decay in one radian of travel is

$$2\frac{k''}{k'} = 2\frac{n''}{n'} \approx \frac{\operatorname{Im}\chi_e}{1 + \operatorname{Re}\chi_e} + \frac{\operatorname{Im}\chi_m}{1 + \operatorname{Re}\chi_m} \ll 1 \tag{7.77}$$

and the phase velocity is

$$\frac{v_\phi^2}{c^2} = \frac{1}{n'^2} \approx \frac{1}{(1 + \operatorname{Re}\chi_e)(1 + \operatorname{Re}\chi_m)} \tag{7.78}$$

We see that the decay rate is associated with the imaginary parts of the dielectric and magnetic susceptibilities, while the phase velocity is associated with the real parts.

To find the fields themselves we proceed as in the vacuum case. From (7.59), in the absence of free charge, we see that

$$\nabla \cdot \tilde{\mathbf{E}} = i\mathbf{k} \cdot \tilde{\mathbf{E}} = 0 \tag{7.79}$$

and from (7.60) we get

$$\nabla \cdot \tilde{\mathbf{H}} = i\mathbf{k} \cdot \tilde{\mathbf{H}} = 0 \tag{7.80}$$

For true plane waves (real $\hat{\mathbf{k}}$) the fields are transverse to the direction of propagation even in dispersive media. From (7.61) we find that

$$\nabla \times \tilde{\mathbf{E}} = i\mathbf{k} \times \tilde{\mathbf{E}} = i\omega\mu\tilde{\mathbf{H}} \tag{7.81}$$

For true plane waves the electric field and the magnetic field are perpendicular to each other and to the direction of propagation. However, they are not necessarily in phase, for we see that

$$\tilde{\mathbf{H}} = \frac{\mathbf{k} \times \tilde{\mathbf{E}}}{\mu\omega} \tag{7.82}$$

where $\mu(\omega)$ and $\mathbf{k}$ may be complex.

When the wave vector is complex, we can separate it into its real and imaginary parts,

$$\mathbf{k} = \mathbf{k}' + i\mathbf{k}'' \tag{7.83}$$

where $\mathbf{k}'$ and $\mathbf{k}''$ are real. Provided that the wave vectors $\mathbf{k}'$ and $\mathbf{k}''$ are parallel, so that there are no spatial variations in the directions orthogonal to $\hat{\mathbf{k}}$, the waves described by (7.68) are true plane waves, although they may be exponentially decaying as they propagate. In some cases, which we encounter later in this chapter, the vectors $\mathbf{k}'$ and $\mathbf{k}''$ are not parallel even when the index of refraction is real. In these cases, when the index of refraction is real, we see from (7.69) that

$$\mathbf{k} \cdot \mathbf{k} = k'^2 - k''^2 + 2i\mathbf{k}' \cdot \mathbf{k}'' = \text{real} \tag{7.84}$$

Thus, the imaginary part of the wave vector is orthogonal to the real part ($\mathbf{k}' \cdot \mathbf{k}'' = 0$) for real $n(\omega)$. When $\mathbf{k}'$ and $\mathbf{k}''$ are not parallel, the wave has oscillatory behavior in one direction, which defines the phase fronts, and exponential decay in another direction. Although the phase fronts are planes, the waves are not true plane waves, since there are spatial variations in the directions parallel to the phase fronts. When the imaginary component of the wave vector is not parallel to the real component, the waves are no longer purely transverse. There is a component of the field in the direction normal to the phase fronts since, from (7.80),

$$\tilde{\mathbf{H}} \cdot \mathbf{k}' = -i\tilde{\mathbf{H}} \cdot \mathbf{k}'' \neq 0 \tag{7.85}$$

EXERCISE 7.3

Consider an optical medium that is transparent except for a finite number of very narrow absorption lines at wavelengths λ_i. Use the Kramers–Kronig relations to show that the index of refraction at a wavelength λ between the absorption lines is given by the expression

$$n^2 - 1 = \sum_i \frac{A_i \lambda^2}{\lambda^2 - \lambda_i^2} \tag{7.86}$$

for some constants A_i. This expression is frequently used as an analytical fit to the experimentally measured index of refraction of various substances. The coefficients A_i and λ_i are called the Sellmeier coefficients.

7.1.5 Phase Velocity and Group Velocity

Because of dispersion it is not possible, in general, to invert the Fourier transform analytically to find the propagation of the wave in the time domain. However, when the wave is composed of just a narrow band of frequencies, it is possible to approximate the effects of dispersion and invert the transform. The result demonstrates a remarkable property of pulses: They do not propagate at the velocity of the waves of which they are composed. We consider the propagation of a plane-wave pulse traveling in the $\hat{\mathbf{x}}$ direction, for which the Fourier transform is given by the solution (7.75) of the Helmholtz equation. Substituting this into Fourier's inversion theorem, we see that the pulse is represented by the integral

$$u(x, t) = \frac{1}{\sqrt{2\pi}} \int_{-\infty}^{\infty} \tilde{u}_0(\omega) e^{i(kx - \omega t)} \tag{7.87}$$

In the following we ignore the imaginary part of the wave vector (and the index of refraction) and assume that in the frequency range of interest k is real. Provided that the spectrum of $\tilde{u}(\omega)$ is confined to a narrow region around the frequency ω_0, we may expand the dispersion relation in the Taylor series

$$k(\omega) \approx k_0 + \frac{dk}{d\omega}(\omega - \omega_0) \tag{7.88}$$

so that

$$u(x, t) \approx \frac{1}{\sqrt{2\pi}} \int_{-\infty}^{\infty} \tilde{u}_0(\omega) e^{i\{[k_0 + (dk/d\omega)(\omega - \omega_0)]x - \omega t\}} \, d\omega$$

$$= e^{i[k_0 - (dk/d\omega)\omega_0]x} \frac{1}{\sqrt{2\pi}} \int_{-\infty}^{\infty} \tilde{u}_0(\omega) e^{i\omega[(dk/d\omega)x - t]} \, d\omega \tag{7.89}$$

The factor in front of the integral is just a phase factor and can be ignored. The remainder depends on the position and time only through the argument

$$\frac{dk}{d\omega} x - t = \frac{x}{v_g} - t \tag{7.90}$$

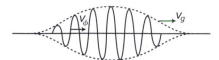

Figure 7.4 Waves traveling in a wave packet.

and therefore represents a pulse traveling in the $\hat{\mathbf{z}}$ direction with group velocity

$$v_g = \frac{d\omega}{dk} \tag{7.91}$$

This is a remarkable and important result, and it has some surprising implications. Therefore, a few comments are in order:

1. Since the phase velocity of the waves is different from (typically greater than) the group velocity of the pulse, the individual waves move through the pulse as indicated in Figure 7.4. If we place ourselves in a coordinate system moving at the group velocity, we see the waves form at the back of the pulse, move forward through it at velocity $v_\phi - v_g$, and then disappear by destructive interference at the front of the pulse. This nonsynchronous motion of the phase fronts is expressed in (7.89) by the phase factor in front of the integral.

2. Although the phase velocity of electromagnetic waves in matter is almost always less than c, the speed of light in vacuum, exceptions occur when the absorption is significant. In addition, the phase velocity can be greater than c in waveguides and, as discussed in Chapter 2, the phase velocity of matter waves in de Broglie's theory is always greater than the speed of light. However, in all these cases the group velocity is always less than c, as Einstein has taught us it must be. Since the influence of a pulse travels at the group velocity, causality is never violated.

The relationship between group velocity and phase velocity is illustrated by the propagation of light through BaF_2. This material is transparent in the near ultraviolet, visible, and infrared portions of the spectrum from about 0.2 to 11 μm. Over this region, the index of refraction is an increasing function of the frequency (a decreasing function of the wavelength). This is called normal dispersion, and it is always true in transparent materials, as shown by the Kramers–Kronig relations. In terms of the wavelength λ and the index of refraction, the group velocity is

$$v_g = \frac{c}{n}\left(1 + \frac{\lambda}{n}\frac{dn}{d\lambda}\right) = v_\phi\left(1 + \frac{\lambda}{n}\frac{dn}{d\lambda}\right) \tag{7.92}$$

The phase and group velocities are shown in Figure 7.5. Since $dn/d\lambda < 0$ in transparent media, the group velocity is always less than the phase velocity unless absorption is important. In so-called anomalous dispersion, the index of refraction increases locally with increasing wavelength. However, this occurs only within the region of absorption caused by some optical transition that is responsible for the anomalous dispersion.

When the dispersion in (7.88) is expanded to the next order in $(\omega - \omega_0)$,

$$k \approx k_0 + \left.\frac{dk}{d\omega}\right|_0 (\omega - \omega_0) + \frac{1}{2}\left.\frac{d^2k}{d\omega^2}\right|_0 (\omega - \omega_0)^2 \tag{7.93}$$

and the Fourier inversion (7.87) is evaluated, it is found that the wave packet spreads out in time. This is similar to the spreading and compression of pulses by a dispersive system

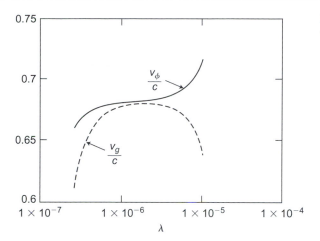

Figure 7.5 Phase and group velocity in BaF.

as discussed in Chapter 5. The connection is evident when we recognize that the phase delay of a wave traveling a distance L in a dispersive medium is

$$\tau = \frac{L}{v_\phi} = L\frac{k}{\omega} \tag{7.94}$$

In quantum mechanics this causes the wave function of particles to spread out in accordance with the uncertainty principle. For smooth pulses, the size of the wave packet initially increases quadratically with time and then increases linearly. Provided that the dispersion relation does not have some pathological behavior, we may estimate the time Δt for a wave packet to spread by an amount comparable to its own initial size Δx by recognizing that the spread of the pulse is caused by a spread in the group velocity. The spread in group velocity is

$$\Delta v_g = O\left(\frac{d^2\omega}{dk^2}\Delta k\right) \tag{7.95}$$

where Δk is the width of the spectrum in k space. But for a smooth pulse the width in k space satisfies the "uncertainty" relation $\Delta x \Delta k = O(1)$. Since the packet begins to spread appreciably when $\Delta v_g \Delta t = \Delta z$, we see that the time Δt for the wave packet to spread satisfies the rule

$$\frac{d^2\omega}{dk^2}\Delta t = O(\Delta x^2) \tag{7.96}$$

Finally, before we close this section, we need to point out that some of the remarks made earlier about group velocity and causality are not quite true, or at least are not stated precisely. It is true that the influence of a wave travels at a velocity (called the signal velocity) that is less than or equal to the velocity of light in a vacuum, and causality is not violated. However, the group velocity as defined here is not always less than the speed of light, and when that is the case the signal velocity is not the group velocity. The theory of information propagation and signal velocity in these cases is somewhat murky, and we do not go into it here. But the prediction and experimental observation of superluminal group velocity has attracted considerable attention and merits some discussion.

As shown by the integral (7.89), the envelope of the pulse propagates as

$$u(x, t) = u_0(n_g x - ct) \tag{7.97}$$

where $n_g = c\, dk/d\omega$ is called the group refractive index. But the phase refractive index is $n = ck/\omega$, so the group index is

$$n_g = c\frac{dk}{d\omega} = n + \omega\frac{dn}{d\omega} \tag{7.98}$$

Thus, if we can make $dn/d\omega$ sufficiently negative, we can make $n_g < 1$, which corresponds to a superluminal group velocity, or even negative. A negative group velocity means that the pulse emerges from the dispersive region before it enters. But the Kramers–Kronig relations remind us that $dn/d\omega > 0$ (normal dispersion) unless there is absorption present, in which case the pulse does not propagate a meaningful distance, or unless there is gain at some frequencies. Negative group velocity experiments exploit the latter possibility. In some very thoughtful and clever experiments, it has been possible to create gain in Cs vapor that makes $n \approx 1$ and $dn/d\omega < 0$ over a portion of the spectrum. A Raman gain mechanism (see Chapter 8) is used to create a pair of closely spaced gain lines at frequencies ω_1 and ω_2, as shown schematically in Figure 7.6. Between the gain lines the index of refraction decreases roughly linearly, so the group velocity is roughly constant over the entire frequency range. Since the linewidth of the optical pulse is comparable to the spectral width between the gain lines, this constant group velocity is necessary to assure that the optical pulse propagates with minimal distortion. When the group index is negative, the pulse propagates through the Cs cell as illustrated schematically in Figure 7.7. For negative group velocity, the pulse appears to emerge before it enters.

So, do we really violate either Einstein's postulates or the principle of causality in these experiments? Well, of course not. But what is going on to make it look as though we do? In fact, in the experiments done to date the optical pulses have been on the order of

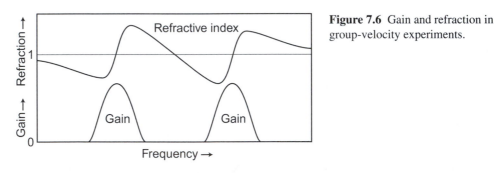

Figure 7.6 Gain and refraction in group-velocity experiments.

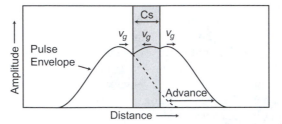

Figure 7.7 Pulse propagation through a superluminal region.

microseconds long overall, while the advance of the pulses has been less than a tenth of this. But this is not really the point. The point really is that while the peak of the optical pulse emerges from the Cs cell before we expect it to—or even before it enters the cell on the other side—the leading edge of the pulse has preceded both these events, and in fact has caused them. That is, the emerging pulse is not composed of the same energy as the entering pulse. It is actually composed largely of energy removed from the Cs vapor by the leading edge of the pulse as it is amplified, and the trailing edge of the pulse is attenuated to replace the energy in the Cs vapor. This makes the pulse appear to arrive too early, but in fact we have not actually transmitted energy at a superluminal velocity. That would violate Einstein's postulates. So what would happen if we had a pulse with a sharp leading edge? Would we violate Einstein's postulates or the principle of causality then? Well, this is where it gets murky. A sharp-edged pulse necessarily has a broad bandwidth, as discussed in Chapter 5, so the analysis we have done does not apply here. Nor is the experiment satisfactory for a sharp-edged pulse, because even the smooth, microsecond pulses already fill the entire spectral region between the gain lines, and a sharp-edged pulse extends into spectral regions that do not have superluminal group velocity. The pulse therefore inevitably emerges strongly distorted, and the situation becomes rather complicated. Fortunately, in some sense, we don't know how to make sharp-edged pulses so the question is not important from a practical point of view, at least not yet. But the question still remains in principle, and the answer still is not explicitly resolved.

EXERCISE 7.4

As discussed in Chapter 5, the spread of a pulse traveling a distance L through a dispersive medium is related to the compression of a pulse in a dispersive system with the delay

$$\tau = \frac{L}{v_\phi} = L\frac{k}{\omega} \tag{7.99}$$

where v_ϕ is the phase velocity. As shown there, an initially unchirped Gaussian pulse spreads according to the formula

$$\sigma^2 = \sigma_0^2 \left(1 + \frac{4}{\mu^2 \sigma_0^4}\right) \tag{7.100}$$

where

$$\frac{1}{\mu} = -\frac{d}{d\omega}\left[\omega \frac{d\tau}{d\omega}\right]_{\omega_0} \tag{7.101}$$

and ω_0 is the center frequency of the pulse.

(a) Show that the rms pulse length $\delta^2 = \frac{1}{2}\sigma^2$ spreads according to the formula

$$\delta^2 = \delta_0^2 \left(1 + \frac{t^2}{t_0^2}\right) \tag{7.102}$$

where the characteristic time for spreading is

$$\frac{1}{t_0} = \frac{1}{\delta_0^2}\left.\frac{d\omega}{dk}\right|_{\omega_0} \left|\frac{d}{d\omega}\left[\omega \frac{d}{d\omega}\left(\frac{k}{\omega}\right)\right]_{\omega_0}\right| \tag{7.103}$$

(b) As discussed in Chapter 2, the de Broglie wave 4-vector $k^\alpha = (\omega/c, \mathbf{k})$ for a particle of mass m is related to the particle momentum $p^\alpha = \gamma m(c, \mathbf{v})$ by

$$p^\alpha = \hbar k^\alpha \tag{7.104}$$

so the dispersion relation is

$$p^\alpha p_\alpha = \hbar^2 k^\alpha k_\alpha = m^2 c^2. \tag{7.105}$$

Show that a Gaussian wave packet spreads with the characteristic time

$$t_0 = \frac{\gamma m v^2 \delta_0^2}{\hbar(1 - \beta^4)} = \frac{\gamma m \Delta x_0^2}{\hbar(1 - \beta^4)} \tag{7.106}$$

where $\beta = v/c$ and $\Delta x_0^2 = v^2 \delta_0^2$ is the initial rms spatial spread of the wave packet. Note that the spreading of wave packets does not depend on the details of the quantum mechanical wave equation, but only on the dispersion relation, which de Broglie determined from more fundamental considerations.

EXERCISE 7.5

Consider a nonmagnetic material for which $\mathrm{Im}\,\chi_e = 0$ for $\omega_1 < \omega < \omega_2$, and $\mathrm{Im}\,\chi_e(\omega) > 0$ (the material is absorbing) everywhere else. In such a material the Kramers–Kronig relations show that

$$\frac{d}{d\omega}\,\mathrm{Re}\,\chi_e(\omega) > 0 \tag{7.107}$$

in the transparent region. This is called normal dispersion.

(a) Use the Kramers–Kronig relations to show that in the transparent region,

$$\frac{d}{d\omega}[\omega^2\,\mathrm{Re}\,\chi_e(\omega)] > 0 \tag{7.108}$$

(b) Use (7.107) and (7.108) to prove that the group index satisfies

$$n_g > 1 \tag{7.109}$$

That is, under conditions of normal dispersion in a transparent medium the group velocity $v_g = c/n_g$ is less than c.

7.1.6 Conservation of Energy in Dispersive Media

As shown in Chapter 6, Poynting's theorem in the presence of macroscopic materials becomes

$$\frac{\partial \mathcal{W}_b}{\partial t} + \nabla \cdot \mathbf{S} + \mathbf{E} \cdot \mathbf{J} = 0 \tag{7.110}$$

where the Poynting vector $\mathbf{S} = \mathbf{E} \times \mathbf{H}$ represents the flow of electromagnetic energy, and

$$\frac{\partial \mathcal{W}_b}{\partial t} = \mathbf{E} \cdot \frac{\partial \mathbf{D}}{\partial t} + \mathbf{H} \cdot \frac{\partial \mathbf{B}}{\partial t} \tag{7.111}$$

describes the work done on the bound charges and on the fields themselves. We ignore, for the moment, the Joule heating $\mathbf{E} \cdot \mathbf{J}$. For an electromagnetic pulse passing through the medium, the total energy that flows into the medium is

$$-\int_{-\infty}^{\infty} \nabla \cdot \mathbf{S} \, dt = \Delta \mathcal{W}_b = \int_{-\infty}^{\infty} \mathbf{E} \cdot \frac{\partial \mathbf{D}}{\partial t} \, dt + \int_{-\infty}^{\infty} \mathbf{H} \cdot \frac{\partial \mathbf{B}}{\partial t} \, dt \qquad (7.112)$$

In the absence of dispersion we may write

$$\mathbf{D} = \varepsilon_0 (1 + \chi_e) \mathbf{E} = \varepsilon \mathbf{E} \qquad (7.113)$$

$$\mathbf{H} = \frac{1}{\mu_0}(1 - \chi_m)\mathbf{B} = \frac{1}{\mu}\mathbf{B} \qquad (7.114)$$

for constant values of the permittivity ε and permeability μ. If we consider a medium that begins in a quiescent state, with no macroscopic electric or magnetic fields, and apply an electromagnetic pulse of finite duration, the total work done by turning on the fields is

$$\Delta \mathcal{W}_b = \int_0^{\mathbf{D}} \mathbf{E} \cdot d\mathbf{D} + \int_0^{\mathbf{B}} \mathbf{H} \cdot d\mathbf{B} = \frac{1}{2}\varepsilon E^2 + \frac{1}{2}\mu H^2 \qquad (7.115)$$

This work depends only on the final value of the fields and is independent of how the fields are turned on. This work may therefore be regarded as the energy in the fields, including the polarization and magnetization. When the fields return to zero at the end the pulse the total work vanishes, as does the flow of energy into and out of the material due to the Poynting vector. There is no change in the energy of the system, so the dissipation must vanish. In the absence of dispersion, then, there can be no dissipation other than the work $\mathbf{E} \cdot \mathbf{J}$ done on the free charges, which is called Joule heating.

When dispersion is included, things get more complicated. Among other things, the presence of dispersion implies dissipation, and vice versa. As shown by the Kramers–Kronig relations, you can't have one without the other. In the presence of dispersion the flow of energy represented by the Poynting vector does not go entirely into the field, but some of it is dissipated in the medium as heat, even without free charges. Since the response of the medium, including the dissipative effects, depends on the entire history of the electric and magnetic fields, it is not possible or meaningful to derive a general expression for the instantaneous rate at which energy is being dissipated in the medium. We therefore consider the total energy dissipated during an electromagnetic pulse of finite duration. By expressing the pulse in terms of its Fourier components we can compute the dissipation from the frequency-dependent permittivity and permeability.

The electric (as distinguished from magnetic) contribution to the work done on the fields and the bound charges, from the first term of (7.112), is

$$\int_{-\infty}^{\infty} \mathbf{E} \cdot \frac{\partial \mathbf{D}}{\partial t} \, dt = -\frac{1}{2\pi} \int_{-\infty}^{\infty} dt \int_{-\infty}^{\infty} d\omega \int_{-\infty}^{\infty} d\omega' \, \tilde{\mathbf{E}}(\omega') \cdot i\omega \tilde{\mathbf{D}}(\omega) e^{-i(\omega + \omega')t} \qquad (7.116)$$

where we have used the rule $\partial/\partial t \to -i\omega$ to evaluate the Fourier transform of $\partial \mathbf{D}/\partial t$. The integral over t produces a δ-function, and after integrating over ω' we get

$$\int_{-\infty}^{\infty} \mathbf{E} \cdot \frac{\partial \mathbf{D}}{\partial t} \, dt = -i \int_{-\infty}^{\infty} \omega \tilde{\mathbf{E}}(-\omega) \cdot \tilde{\mathbf{D}}(\omega) \, d\omega = -i \int_{-\infty}^{\infty} \omega \varepsilon(\omega) \tilde{\mathbf{E}}^*(\omega) \cdot \tilde{\mathbf{E}}(\omega) \, d\omega \qquad (7.117)$$

where we have used the symmetry relation $\mathbf{E}(-\omega) = \mathbf{E}^*(\omega)$, which is valid for real $\mathbf{E}(t)$ and real ω. But the integrand in (7.117) is antisymmetric, apart from the permittivity, so only the antisymmetric part of $\varepsilon(\omega)$ survives. From (7.22) we see that this is the imaginary part, and we are left with

$$\int_{-\infty}^{\infty} \mathbf{E} \cdot \frac{\partial \mathbf{D}}{\partial t}\, dt = \int_{-\infty}^{\infty} \omega \, \mathrm{Im}\, \varepsilon(\omega) \tilde{\mathbf{E}}^*(\omega) \cdot \tilde{\mathbf{E}}(\omega)\, d\omega \qquad (7.118)$$

Since the integrand is now symmetric in ω, we may write this as an integral over positive frequencies only, of the form

$$\int_{-\infty}^{\infty} \mathbf{E} \cdot \frac{\partial \mathbf{D}}{\partial t}\, dt = 2 \int_{0}^{\infty} \omega \, \mathrm{Im}\, \varepsilon(\omega) |\tilde{\mathbf{E}}(\omega)|^2\, d\omega \qquad (7.119)$$

In the same way we get

$$\int_{-\infty}^{\infty} \mathbf{H} \cdot \frac{\partial \mathbf{B}}{\partial t}\, dt = 2 \int_{0}^{\infty} \omega \, \mathrm{Im}\, \mu(\omega) |\tilde{\mathbf{H}}(\omega)|^2\, d\omega \qquad (7.120)$$

for the magnetic energy deposited in the medium. When we substitute these results back into (7.112), we find that

$$-\int_{-\infty}^{\infty} \nabla \cdot \mathbf{S}\, dt = 2 \int_{0}^{\infty} [\omega \, \mathrm{Im}\, \varepsilon(\omega) |\tilde{\mathbf{E}}(\omega)|^2 + \omega \, \mathrm{Im}\, \mu(\omega) |\tilde{\mathbf{H}}(\omega)|^2]\, d\omega \qquad (7.121)$$

This shows that after the pulse has come and gone, when the fields have returned to zero, there remains a net amount of energy deposited in the medium as heat. This dissipation depends only on the imaginary parts of the dielectric and magnetic susceptibilities. In the absence of dispersion, the permittivity and the susceptibility χ_e are constants. From the symmetry relation (7.22), then, we see that $\chi_e = \chi_e^*$, so ε must be real. In the same way, we see that in the absence of dispersion χ_m and μ must be real. Therefore, it follows from (7.121) that without dispersion there can be no dissipation. This agrees with our earlier result.

The fact that the integral in (7.121) is over frequency, rather than time, reflects the fact that the polarization and magnetization depend on the entire history of the electromagnetic pulse. It is not generally meaningful to try to identify the instantaneous rate at which energy is dissipated in the medium or the rate at which energy is added to the fields. For example, if a brief pulse is used to excite plasma oscillations in a time comparable to the mean collision time, the energy that appears initially in the oscillations eventually appears as heat after the collisions damp the oscillations. An exception to the general rule is provided by a long, smooth optical pulse traveling through a dispersive, dissipative medium. When the pulse is long enough for the distant history to be forgotten, the polarization and magnetization depend only on recent history. In a smooth pulse the recent history looks much like the present. This makes it possible to identify both the instantaneous rate at which energy is dissipated as well as the rate at which energy is added to the fields, at least when they are averaged over a period of the wave. The simplification relies on the fact that for a long, smooth pulse the spectrum consists of a narrow band of frequencies. This allows us to use a Taylor series expansion to approximate the permittivity and permeability over the important part of the spectrum, and evaluate the necessary integrals analytically.

When the pulse is long and smooth, so the spectrum is confined to a narrow region around the frequency ω_0, it is convenient to represent the electric field in the form

$$\mathbf{E}(t) = \tfrac{1}{2}[\mathbf{E}_0(t)e^{-i\omega_0 t} + \mathbf{E}_0^*(t)e^{i\omega_0 t}] \tag{7.122}$$

where $\mathbf{E}_0(t)$ is a slowly varying, complex function of the time. The Fourier transform of the electric field is

$$\tilde{\mathbf{E}}(\omega) = \tfrac{1}{2}[\tilde{\mathbf{E}}_0(\omega - \omega_0) + \tilde{\mathbf{E}}_0^*(-\omega - \omega_0)] \tag{7.123}$$

where the Fourier transform of the envelope function

$$\tilde{\mathbf{E}}_0(x) = \frac{1}{\sqrt{2\pi}} \int_{-\infty}^{\infty} \mathbf{E}_0(t)e^{ixt}\, dt \tag{7.124}$$

is narrowly confined to the region near $x = 0$.

The time derivative of the displacement vector can be similarly represented by the expression

$$\frac{\partial \mathbf{D}}{\partial t} = \frac{\partial}{\partial t} \frac{1}{\sqrt{2\pi}} \int_{-\infty}^{\infty} d\omega\, \varepsilon(\omega)\tilde{\mathbf{E}}(\omega) = \frac{1}{2}[\dot{\mathbf{D}}_0(t)e^{-i\omega_0 t} + \dot{\mathbf{D}}_0^* e^{i\omega_0 t}] \tag{7.125}$$

where

$$\dot{\mathbf{D}}_0(t) = \frac{-i}{\sqrt{2\pi}} \int_{-\infty}^{\infty} d\omega\, e^{-i(\omega - \omega_0)t}\omega\varepsilon(\omega)\tilde{\mathbf{E}}_0(\omega - \omega_0) \tag{7.126}$$

Since this integral is dominated by frequencies near ω_0, we Taylor expand the permittivity in the form

$$\omega\varepsilon(\omega) \approx (\omega\varepsilon)\Big|_{\omega_0} + \frac{d\omega\varepsilon}{d\omega}\Big|_{\omega_0}(\omega - \omega_0) \tag{7.127}$$

and get

$$\dot{\mathbf{D}}_0(t) = -i(\omega\varepsilon)\Big|_{\omega_0} \frac{1}{\sqrt{2\pi}} \int_{-\infty}^{\infty} dx\, e^{-ixt}\tilde{\mathbf{E}}_0(x) - i\frac{d\omega\varepsilon}{d\omega}\Big|_{\omega_0} \int_{-\infty}^{\infty} dx\, x e^{-ixt}\tilde{\mathbf{E}}_0(x)$$

$$= -i(\omega\varepsilon)\Big|_{\omega_0} \mathbf{E}_0(t) + \frac{d\omega\varepsilon}{d\omega}\Big|_{\omega_0} \frac{\partial \mathbf{E}_0(t)}{\partial t} \tag{7.128}$$

From (7.111) we see that the rate at which work is being done on the fields and the bound charges by the electric fields is $\mathbf{E} \cdot (\partial\mathbf{D}/\partial t)$. If we substitute (7.122) and (7.125) for $\mathbf{E}$ and $\partial\mathbf{D}/\partial t$, and average over one period of the fields, the terms proportional to $e^{\pm i2\omega_0 t}$ cancel out and we are left with

$$\left\langle \mathbf{E} \cdot \frac{\partial \mathbf{D}}{\partial t} \right\rangle = \frac{1}{4}[\mathbf{E}_0 \cdot \dot{\mathbf{D}}_0^* + \mathbf{E}_0^* \cdot \dot{\mathbf{D}}_0] \tag{7.129}$$

If we substitute (7.128) for $\dot{\mathbf{D}}_0$, collect terms involving the real and imaginary parts of ε, and ignore $\partial\mathbf{E}_0/\partial t$ compared with $\omega_0\mathbf{E}_0$, we find that

$$\left\langle \mathbf{E} \cdot \frac{\partial \mathbf{D}}{\partial t} \right\rangle = \frac{1}{2}\,\mathrm{Im}(\omega\varepsilon)\Big|_{\omega_0} \mathbf{E}_0^* \cdot \mathbf{E}_0 + \frac{1}{4}\,\mathrm{Re}\frac{d\omega\varepsilon}{d\omega}\Big|_{\omega_0} \frac{\partial}{\partial t}(\mathbf{E}_0 \cdot \mathbf{E}_0^*) \tag{7.130}$$

In terms of the real field $\mathbf{E}(t)$, we recall that when averaged over a complete cycle the result is

$$\langle \mathbf{E}(t) \cdot \mathbf{E}(t) \rangle = \tfrac{1}{2}\mathbf{E}_0^*(t) \cdot \mathbf{E}_0(t) \tag{7.131}$$

We may therefore write

$$\left\langle \mathbf{E} \cdot \frac{\partial \mathbf{D}}{\partial t} \right\rangle = \text{Im}(\omega\varepsilon)\Big|_{\omega_0} \langle E^2 \rangle + \frac{1}{2}\text{Re}\left(\frac{d\omega\varepsilon}{d\omega}\right)\Big|_{\omega_0} \left\langle \frac{\partial E^2}{\partial t} \right\rangle \tag{7.132}$$

In the same way, we find that the magnetic contribution to the total energy is

$$\left\langle \mathbf{H} \cdot \frac{\partial \mathbf{B}}{\partial t} \right\rangle = \text{Im}(\omega\mu)\Big|_{\omega_0} \langle H^2 \rangle + \frac{1}{2}\text{Re}\left(\frac{d\omega\mu}{d\omega}\right)\Big|_{\omega_0} \left\langle \frac{\partial H^2}{\partial t} \right\rangle \tag{7.133}$$

At each point in time, then, we obtain the conservation law

$$\langle \nabla \cdot \mathbf{S} \rangle + \left\langle \frac{dQ}{dt} \right\rangle + \frac{\partial \langle \mathcal{U} \rangle}{\partial t} = 0 \tag{7.134}$$

where the rate of dissipation averaged over a complete cycle is

$$\left\langle \frac{dQ}{dt} \right\rangle = \text{Im}(\omega\varepsilon)\big|_{\omega_0}\langle E^2 \rangle + \text{Im}(\omega\mu)\big|_{\omega_0}\langle H^2 \rangle \tag{7.135}$$

and the energy in the field, including the energy in polarization and magnetization of the medium averaged over a cycle, is

$$\langle \mathcal{U} \rangle = \frac{1}{2}\text{Re}\frac{d\omega\varepsilon}{d\omega}\Big|_{\omega_0}\langle E^2 \rangle + \frac{1}{2}\text{Re}\frac{d\omega\mu}{d\omega}\Big|_{\omega_0}\langle H^2 \rangle \tag{7.136}$$

It is interesting to compare expressions (7.135) for the dissipation and (7.136) for the field energy with our previous results. The rate of electrical dissipation could easily be derived from the earlier result (7.118) simply by recognizing that for a long pulse with a narrow spectrum, we can use the approximation $\omega\,\text{Im}\,\varepsilon(\omega) \approx \omega_0\,\text{Im}\,\varepsilon(\omega_0)$ over the spectral region of importance. Moving this factor outside the integral in (7.118), we get

$$\int_{-\infty}^{\infty} \mathbf{E} \cdot \frac{\partial \mathbf{D}}{\partial t}\,dt = \omega_0\,\text{Im}\,\varepsilon(\omega_0)\int_{-\infty}^{\infty} \tilde{\mathbf{E}}^*(\omega) \cdot \tilde{\mathbf{E}}(\omega)\,d\omega = \text{Im}(\omega\varepsilon)\Big|_{\omega_0}\int_{-\infty}^{\infty} \mathbf{E}(t) \cdot \mathbf{E}(t)\,dt \tag{7.137}$$

By converting the total dissipation to an integral over time in this way, we can identify the instantaneous rate of electrical dissipation when the pulse is long and smooth. Similar remarks apply to the magnetic dissipation, but not to either the electric or the magnetic energy density. Comparing (7.136) with (7.115), we see that (7.136) is not a trivial extension from (7.115). The correct expression, in which $\text{Re}(d\omega\varepsilon/d\omega)$ and $\text{Re}(d\omega\mu/d\omega)$ replace ε and μ, was first noted by Brillouin, in 1921. It is correct, of course, only for an electromagnetic field with a narrow spectrum, as is evident from the fact that it depends on the permittivity and the permeability evaluated around the frequency ω_0. Ordinarily, the difference between (7.115) (valid for constant ε and μ) and (7.136) (valid in the presence of dispersion) is small, for we see that

$$\frac{d\omega\varepsilon}{d\omega} = \varepsilon\left(1 + \frac{d\ln\varepsilon}{d\ln\omega}\right) \tag{7.138}$$

and ε is typically a slowly varying function of the frequency. In the example of BaF_2, discussed earlier, the correction is only about 1% in the visible part of the spectrum, rising to 20% near the absorption edge at 10 μm. However, the correction can be larger in the presence of anomalous dispersion, where the index of refraction is a strong function of the frequency. This is precisely the circumstance when the group velocity is significantly different from the phase velocity. We therefore recognize that the Brillouin correction to the energy density of a propagating wave is intimately connected with the fact that the energy in the electromagnetic wave travels at the group velocity (the velocity of the pulse) rather than at the phase velocity.

Another example where the Brillouin correction to the energy density is important is provided by the energy density in a plasma oscillation, as discussed in Chapter 4. As we see shortly, the dielectric susceptibility in a collisionless plasma at the frequency ω is

$$\chi_e = -\frac{\omega_p^2}{\omega^2} \tag{7.139}$$

where ω_p is the plasma frequency. In this case we find that

$$\frac{d\omega\varepsilon}{d\omega} = 1 + \frac{\omega_p^2}{\omega^2} \tag{7.140}$$

Since the frequency of plasma oscillations is $\omega = \omega_p$, independent of the wave number k, the group velocity $v_g = d\omega/dk$ vanishes and the energy density is

$$\langle \mathcal{U} \rangle = \varepsilon_0 \langle E^2 \rangle \tag{7.141}$$

The contribution from the magnetic field is absent because the magnetic field vanishes in a plasma oscillation. To see this we note that the electric field is longitudinal (normal to the phase fronts), so the curl vanishes. Then from Faraday's law,

$$\nabla \times \mathbf{E} + \frac{\partial \mathbf{B}}{\partial t} = 0 \tag{7.142}$$

it follows that the magnetic field vanishes identically. But the energy density $\mathcal{U}$ in (7.141) is a factor of two larger than the energy density of the vacuum electric field alone. The extra energy is the kinetic energy of the electrons as they oscillate. As shown in Chapter 4, the average electron kinetic energy is just equal to the average electric field energy. In this case the Brillouin formula applies even when the group velocity of the wave vanishes.

EXERCISE 7.6

Consider the propagation of a long optical pulse through a transparent (loss-free) but dispersive medium. In this case, the permittivity and permeability are purely real in the spectral region of interest and we can put together a heuristic derivation of the Brillouin correction to the energy density.

(a) Beginning with the Maxwell equations, show that for a long, smooth optical pulse in one dimension, the electric and magnetic fields are related by

$$\mu\langle H^2 \rangle = \varepsilon\langle E^2 \rangle \tag{7.143}$$

where the brackets indicate an average over one cycle.

(b) Show that

$$\frac{d \ln \omega n}{d \ln \omega} = \frac{v_\phi}{v_g} \tag{7.144}$$

(c) Starting with the definition of the Poynting vector, show that

$$|\langle \mathbf{S} \rangle| = \langle EH \rangle = v_\phi \left(\tfrac{1}{2} \varepsilon \langle E^2 \rangle + \tfrac{1}{2} \mu \langle H^2 \rangle \right) \tag{7.145}$$

and use the fact that the energy flows at the group velocity to determine the energy density in a transparent, dispersive medium.

(d) Starting with the Brillouin formula (7.136), show that the energy density in a transparent medium is

$$\langle \mathcal{U} \rangle = \left(\frac{1}{2} \varepsilon \langle E^2 \rangle + \frac{1}{2} \mu \langle H^2 \rangle \right) \frac{d \ln \omega n}{d \ln \omega} \tag{7.146}$$

Compare this with your answer to part (c).

EXERCISE 7.7

In a polar liquid at high frequencies, the molecular dipoles do not have time to rotate in response to changes in the electric field. If the molecular rotations are viscously damped, the polarization has the behavior

$$\frac{d\mathbf{P}}{dt} = \frac{\varepsilon_0 \chi_0 \mathbf{E} - \mathbf{P}}{\tau} \tag{7.147}$$

where $\chi_0 = \chi_e(0)$ is the dc susceptibility and τ is the relaxation time.

(a) Take the Fourier transform of this equation to show that the complex susceptibility is

$$\chi_e(\omega) = \frac{\chi_0}{1 - i\omega\tau} \tag{7.148}$$

(b) Use (7.135) to show that the average (over one cycle) rate of energy dissipation is

$$\left\langle \frac{dQ}{dt} \right\rangle = \varepsilon_0 \chi_0 \omega \frac{\omega\tau}{1 + \omega^2 \tau^2} \langle E^2 \rangle \tag{7.149}$$

Compare your answer with the result obtained in the exercises following Section 6.1.5.

(c) The intensity absorption coefficient is

$$\alpha = 2k'' \tag{7.150}$$

where the wave vector $k = k' + ik''$ satisfies the dispersion relation

$$k^2 c^2 = n^2 \omega^2 \tag{7.151}$$

For $\omega^2 \tau^2 \ll 1$, show that the intensity absorption length $L = 1/\alpha$ is

$$L = \frac{c\sqrt{1 + \chi_0}}{\omega^2 \tau \chi_0} \tag{7.152}$$

The measured value of the dc susceptibility of water at room temperature is $\chi_e(0) \approx 80$, and the absorption depth in water at 2.45 GHz, the frequency of a microwave oven, is $L = O(1 \text{ cm})$. Justify the approximation $\omega^2 \tau^2 \ll 1$, and estimate the relaxation time.

EXERCISE 7.8

Consider the propagation of a circularly polarized wave

$$\mathbf{E} = \text{Re}[E\hat{\mathbf{e}}_{\pm}e^{-ik^{\alpha}r_{\alpha}}] \tag{7.153}$$

in which E is a constant and the wave vector is

$$k^{\alpha} = \left(\frac{\omega}{c}, k\hat{\mathbf{n}}\right) \tag{7.154}$$

through a neutral, collisionless plasma in the direction parallel to the constant magnetic field

$$\mathbf{B}_0 = B_0\hat{\mathbf{n}} \tag{7.155}$$

(a) If we ignore the magnetic field of the wave compared to the constant magnetic field $\mathbf{B}_0$, the nonrelativistic equation of motion of the electrons is

$$m\frac{d\mathbf{v}}{dt} = q(\mathbf{E} + \mathbf{v} \times \mathbf{B}_0) \tag{7.156}$$

where m is the mass, q the charge, and $\mathbf{v}$ the velocity of the electron. Taking advantage of the properties of the circularly polarized vectors $\hat{\mathbf{e}}_{\pm}$, solve the equation of motion and find the macroscopic current density $\mathbf{J} = n_0 q\mathbf{v}$ of the electrons, where n_0 is the electron density. If we attribute this to a bound current density $d\mathbf{P}/dt$, where $\mathbf{P}$ is the polarization, show that the susceptibility is

$$\chi_e = -\frac{\omega_p^2}{\omega(\omega \mp \omega_c)} \tag{7.157}$$

in which

$$\omega_p^2 = \frac{n_0 q^2}{\varepsilon_0 m} \tag{7.158}$$

is the plasma frequency and

$$\omega_c = -\frac{qB_0}{m} \tag{7.159}$$

is the electron cyclotron frequency. The negative sign is introduced in (7.159) because for $B_0 > 0$ the motion of the (negatively charged) electrons corresponds to positive helicity about the direction of the longitudinal magnetic field. This choice of sign makes the cyclotron frequency ω_c positive if B_0 is positive. In the literature, waves that rotate in the same direction as the electrons are called extraordinary waves and those that rotate opposite the electrons are called ordinary waves. For a magnetic field in the direction of propagation ($B_0 > 0$), left-hand circularly polarized waves (positive helicity) are extraordinary waves and right-hand circularly polarized waves (negative helicity) are ordinary waves. The upper sign in (7.157) corresponds to extraordinary waves and the lower sign to ordinary waves.

(b) Using the Brillouin formula (7.136), show that the energy density is

$$\mathcal{U} = \frac{\varepsilon_0}{2}\left[1 + \frac{\omega_p^2}{(\omega \mp \omega_c)^2}\right]E^2 \tag{7.160}$$

For extraordinary waves (upper sign), this diverges at the cyclotron frequency.

7.1.7 Lorentz–Drude Model

A simple model of the polarizability developed by Lorentz and Drude illustrates the general principles discussed earlier. Their model describes the electrons in the medium as harmonically bound to the ions, which are fixed in space. In an electric field **E**, the displacement **r** of an electron then satisfies the equation

$$\frac{d^2\mathbf{r}}{dt^2} + \gamma \frac{d\mathbf{r}}{dt} + \omega_r^2 \mathbf{r} = \frac{q}{m}\mathbf{E} \tag{7.161}$$

in the nonrelativistic limit, where q is the charge on the electron, ω_r is the resonant frequency of the harmonically bound electron, and γ is a damping coefficient. The polarization per unit volume is just the dipole moment $q\mathbf{r}$ of an oscillator times the electron density n_e, that is, $\mathbf{P} = n_e q\mathbf{r}$. The polarization therefore satisfies the equation

$$\frac{d^2\mathbf{P}}{dt^2} + \gamma \frac{d\mathbf{P}}{dt} + \omega_r^2 \mathbf{P} = \varepsilon_0 \omega_p^2 \mathbf{E} \tag{7.162}$$

where

$$\omega_p = \sqrt{\frac{n_e q^2}{\varepsilon_0 m}} \tag{7.163}$$

is the plasma frequency. If we take the Fourier transform of (7.162) using the rule $d/dt \rightarrow -i\omega$ we get

$$\left(\omega_r^2 - \omega^2 - i\gamma\omega\right)\tilde{\mathbf{P}} = \varepsilon_0 \omega_p^2 \tilde{\mathbf{E}} \tag{7.164}$$

Comparing this with (7.18), we find that the dielectric susceptibility is

$$\chi_e(\omega) = \frac{\omega_p^2}{\omega_r^2 - \omega^2 - i\gamma\omega} \tag{7.165}$$

The poles of this expression are all located in the lower half-plane, as causality says they must be, at the points

$$\omega = -\frac{i\gamma}{2} \pm \sqrt{\omega_r^2 - \frac{\gamma^2}{4}} \tag{7.166}$$

As expected, these points are symmetrically distributed with respect to the imaginary axis. This is shown in Figure 7.8.

The physical connection between the real and imaginary parts of the susceptibility implied by the Kramers–Kronig relations is clear in the Lorentz–Drude model. Both the real and imaginary parts have their origin in the electron oscillations. As shown by (7.165), the

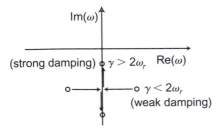

Figure 7.8 Poles of the susceptibility.

imaginary part of the susceptibility comes from the damping of the oscillators. This makes physical sense, since the imaginary part of the susceptibility is associated with dissipative processes. The real part of the susceptibility is associated with the index of refraction and the energy density. The index of refraction is associated with the electron oscillations because the electrons radiate as they oscillate. This radiation adds coherently to the incident fields with a phase that is related to that of the incident radiation through the electron dynamics. Depending on the relative phase, the fields radiated by the electrons can retard or advance the phase of the incident fields. Retardation corresponds to a phase velocity less than c and an index of refraction greater than unity. Advancement corresponds to a phase velocity greater than c. The index of refraction is also associated with the energy density. This includes both the kinetic energy of the electron motions and the potential energy in the oscillator "springs," as well as the usual energy in the electromagnetic field.

In the high-frequency limit, $\omega/\omega_r \gg 1$, the dielectric susceptibility of a collection of classical harmonic oscillators has the asymptotic behavior

$$\chi_e \xrightarrow[\omega \to \infty]{} -\frac{\omega_p^2}{\omega^2} = -\frac{n_e q^2}{\varepsilon_0 m} \frac{1}{\omega^2} \tag{7.167}$$

in agreement with the generally expected asymptotic behavior (7.49). The asymptotic value is independent of the resonant frequency and the damping coefficient of the harmonic oscillators, and is the same even when the absorption and dispersion are caused by a collection of oscillators of different frequencies and different damping coefficients. We may therefore write the sum rule (7.50) in the form

$$\int_0^\infty \omega \, \text{Im} \, \chi_e(\omega) \, d\omega = \frac{\pi q^2 n_e}{2\varepsilon_0 m} \tag{7.168}$$

This says that the total absorption, integrated over all frequencies, depends only on the electron density and not on the nature of the molecular or atomic substance of which the electrons are part. The sum rule (7.168) is useful as a consistency check on experimental measurements of the real and imaginary parts of the index of refraction of metals and optical materials, for example.

It is conventional to write the sum rule (7.168) in the form

$$\int_0^\infty f(\omega) \, d\omega = 1 \tag{7.169}$$

The quantity

$$f(\omega) = \frac{2}{\pi} \frac{\varepsilon_0 m}{n_e q^2} \omega \, \text{Im} \, \chi_e(\omega) \tag{7.170}$$

is called the oscillator strength per electron. It represents the absorption spectrum per electron normalized to the total absorption of a classical electron. The sum rule (7.169) is valid also for quantum mechanical systems. For electrons with transitions to several different final states it is common to speak of the oscillator strengths for the various transitions. The sum of the oscillator strengths for the various transitions is then unity.

In dielectric media, a phenomenon called anomalous dispersion occurs near a narrow absorption feature such as resonant absorption in a metal vapor. To describe this phenomenon in terms of the harmonic oscillator model, we assume that magnetic effects may be

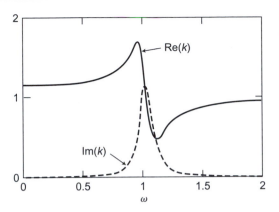

Figure 7.9 Resonant absorption and anomalous dispersion.

ignored, so that $\mu(\omega) = \mu_0$. We further assume that the harmonic oscillators are under-damped, so that

$$\gamma/\omega \ll 1 \tag{7.171}$$

From (7.165) we see that when the absorption per wavelength is small the real and imaginary parts of $\mu\varepsilon$ are

$$\text{Re}(\mu\varepsilon) = \mu_0\varepsilon_0\left[1 + \frac{\omega_p^2\left(\omega_r^2 - \omega^2\right)}{\left(\omega_r^2 - \omega^2\right)^2 + \gamma^2\omega^2}\right] = \frac{1}{\omega^2}(k'^2 - k''^2) \approx \frac{k'^2}{\omega^2} \approx \frac{n'^2}{c^2} \tag{7.172}$$

$$\text{Im}(\mu\varepsilon) = \mu_0\varepsilon_0\left[\frac{\omega_p^2\gamma\omega}{\left(\omega_r^2 - \omega^2\right)^2 + \gamma^2\omega^2}\right] = \frac{2k'k''}{\omega^2} \tag{7.173}$$

where $k = k' + ik''$ and $n = n' + in''$ are the complex wave number and index of refraction. The real and imaginary parts of the wave number are shown in Figure 7.9. We see that the absorption is strongly peaked about the resonant frequency, with a width $\Delta\omega = O(\gamma)$ and a maximum value

$$k'(\omega_r)k''(\omega_r) = \frac{\mu_0\varepsilon_0\omega_r\omega_p^2}{2\gamma} \tag{7.174}$$

Outside the region of absorption, the dispersion is normal; that is, the index of refraction increases with frequency everywhere except in the absorbing region. In this region the index of refraction dips abruptly, which is called anomalous dispersion.

In the limit $\omega_r \to 0$, the electrons are unbound and the dielectric becomes a conductor. The susceptibility then becomes

$$\chi_e \to -\frac{\omega_p^2}{\omega(\omega + i\gamma)} \tag{7.175}$$

But as shown in Chapter 6, the current density due to the "bound" electrons (all electrons that contribute to the dielectric susceptibility are regarded as "bound") is

$$\mathbf{J}_b = \frac{\partial\mathbf{P}}{\partial t} \tag{7.176}$$

for a nonmagnetic material. If we take the Fourier transform of this expression, we get

$$\tilde{\mathbf{J}}_b = -i\omega\tilde{\mathbf{P}} = -i\omega\varepsilon_0\chi_e\tilde{\mathbf{E}} \tag{7.177}$$

But we may equally well regard the current as arising from free electrons, in which case the current is described by Ohm's law,

$$\tilde{\mathbf{J}} = \sigma(\omega)\tilde{\mathbf{E}} \tag{7.178}$$

where $\sigma(\omega)$ is the (complex) frequency-dependent conductivity. Comparing (7.178) with (7.177), we find that the conductivity and the susceptibility are related by

$$\sigma(\omega) = -i\omega\varepsilon_0\chi_e(\omega) = \frac{i\varepsilon_0\omega_p^2}{\omega - i\gamma} \tag{7.179}$$

The factor $-i\omega$ appears because the current due to the bound electrons comes from the time derivative of the polarization. In the limit $\omega \to 0$,

$$\sigma \to \frac{\varepsilon_0\omega_p^2}{\gamma} = \sigma_0 \tag{7.180}$$

where σ_0 is the dc conductivity. In terms of the free-electron theory of metals the conductivity is $\sigma_0 = n_e q_e^2 \tau/m_e$, where τ is the mean time between collisions. We therefore see that $\gamma = 1/\tau$; that is, the damping rate is to be identified with the collision frequency. Using (7.180) we may express the frequency-dependent complex conductivity (7.179) in the form

$$\sigma(\omega) = \frac{\sigma_0}{1 - i\omega\tau} \tag{7.181}$$

In the limit of very low frequencies the conductivity approaches a finite, real value. In the limit of very high frequencies the conductivity becomes purely imaginary. Damping becomes unimportant in the high-frequency limit, since the amplitude of the oscillations diminishes due to inertia, and the electrons simply oscillate out of phase with the electric field.

The complex index of refraction of a nonmagnetic Drude conductor,

$$n = \sqrt{1 + \chi_e} = \sqrt{1 - \frac{\omega_p^2}{\omega^2 + i\omega\gamma}} \tag{7.182}$$

is illustrated in Figure 7.10. Typically for metals $\omega_p\tau \gg 1$, and Figure 7.10 is drawn for $\omega_p\tau = 100$. Ignoring, for the moment, the factor $i\omega\gamma$ compared with ω^2 in the denominator,

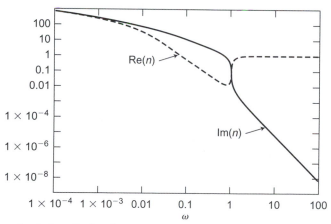

Figure 7.10 Complex index of refraction.

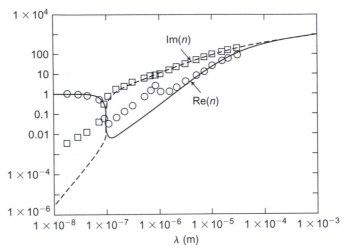

Figure 7.11 Complex index of refraction of aluminum.

we see that for $\omega/\omega_p < 1$ the index of refraction is essentially real, and for $\omega/\omega_p > 1$ it is essentially imaginary. The real and imaginary parts cross over at $\omega/\omega_p \approx 1$. In the limit $\omega/\omega_p \to \infty$ the susceptibility χ_e vanishes, as predicted by the Reimann–Lebesgue lemma, and the index of refraction approaches unity. The imaginary part $\mathrm{Im}\, n(\omega) \to 1/2\omega^3\tau$. In the opposite limit $\omega \to 0$ the index of refraction has the asymptotic behavior $n(\omega) \to \sqrt{\omega_p^2/\omega\gamma}\,(i)^{1/2} = \sqrt{\omega_p^2/2\omega\gamma}\,(1 + i)$. The real and imaginary parts converge to the same value.

The Drude model is plotted versus wavelength and compared with experimental data for aluminum in Figure 7.11. The solid curves correspond to the parameters $\omega_p = 1.98 \times 10^{16}$ radians/s and $\gamma = 9.8 \times 10^{13}$/s. Since aluminum has an electron density $n_e = 1.8 \times 10^{29}$/m^3 and a dc conductivity $\sigma_0 = 3.65 \times 10^7/\Omega$-m at room temperature, the straightforward application of (7.163) and (7.180) would predict the parameters $\omega_p = 2.4 \cdot 10^{16}$ radians/s and $\gamma = 1.4 \cdot 10^{13}$/s, which are somewhat at variance with the parameters inferred from optical measurements. With the "best fit" values of ω_p and γ, the agreement between the observed and calculated values of the real and imaginary parts of the complex index of refraction is good in the infrared ($\lambda > 10^{-6}$ m). At shorter wavelengths, the Drude model has the correct qualitative behavior, showing a sharp crossover of the real and imaginary parts at the plasma frequency ($\lambda \approx 10^{-7}$ m), but there are quantitative discrepencies. The disagreement at wavelengths below about 1.7×10^{-6} m has its origin in the excitation of conduction electrons to higher bands and, at still shorter wavelengths, to the excitation of inner-shell electrons.

For a plane wave traveling through a Drude conductor with vanishing magnetization, the dispersion relation becomes

$$k^2 = \mu\varepsilon\omega^2 = \frac{\omega^2}{c^2}(1 + \chi_e) = \frac{\omega^2}{c^2}\left[1 + \frac{\dfrac{i\sigma_0}{\varepsilon_0\omega}}{1 - \dfrac{i\omega}{\gamma}}\right] \tag{7.183}$$

In the low-frequency limit,

$$\frac{\varepsilon_0 \omega}{\sigma_0} \ll 1 \tag{7.184}$$

$$\frac{\omega}{\gamma} \ll 1 \tag{7.185}$$

the dispersion relation becomes

$$k^2 = i \mu_0 \sigma_0 \omega \tag{7.186}$$

so

$$k = \sqrt{\frac{\mu_0 \sigma_0 \omega}{2}} (1 + i) \tag{7.187}$$

We see, therefore, that the amplitude of the wave decays in a distance

$$\delta = \sqrt{\frac{2}{\mu_0 \sigma_0 \omega}} \tag{7.188}$$

This is called the skin depth, for if a wave is incident on a conductor, it penetrates into the conductor only to the depth δ. For example, in copper the electron density is $n_e = 8 \times 10^{28}/\text{m}^3$ and the dc conductivity is $\sigma_0 = 5.6 \times 10^7/\Omega\text{-m}$. Therefore, the plasma frequency is $\omega_p = \sqrt{n_e q^2/\varepsilon_0 m} = 1.6 \times 10^{16}$ radians/s and the damping frequency is $\gamma = \varepsilon_0 \omega_p^2/\sigma_0 = 4 \times 10^{13}$ radians/s. For ordinary 60-Hz ac current, the skin depth is $\delta \approx 1$ cm. However, at a frequency of 2.45 GHz, as in a microwave oven, the skin depth is $\delta \approx 1$ μm. Since the electromagnetic fields and the associated currents are confined near the surface, the dissipation in the resonant cavities of the magnetron (the source of microwave power) is confined to the surface as well.

Before closing this section, it is important to note that in the limit $\omega_r \to 0$, which is characteristic of conductors, the susceptibility develops a singularity on the real axis at the origin. This modifies the derivation of the Kramers–Kronig relations. As shown in the exercise following the previous section, the Kramers–Kronig relations take the form

$$\text{Re}\,\chi_e(\omega) = \pi \sigma_0 \delta(\omega) + \frac{2}{\pi} \text{P} \int_0^\infty d\omega' \frac{\omega' \, \text{Im}\,\chi_e(\omega')}{\omega'^2 - \omega^2} \tag{7.189}$$

$$\text{Im}\,\chi_e(\omega) = \frac{\sigma_0}{\omega} - \frac{2}{\pi} \text{P} \int_0^\infty d\omega' \frac{\omega \, \text{Re}\,\chi_e(\omega')}{\omega'^2 - \omega^2} \tag{7.190}$$

for conductors.

EXERCISE 7.9

(a) Starting with the differential equation

$$\frac{d^2 G}{d\tau^2} + \gamma \frac{dG}{dt} + \omega_r^2 G = \omega_p^2 \delta(\tau) \tag{7.191}$$

and boundary conditions

$$G(\tau) = 0 \qquad \text{for } \tau < 0 \tag{7.192}$$

and

$$G(\tau) \xrightarrow[\tau \to \infty]{} 0 \tag{7.193}$$

find the response function $G(\tau)$ for the Lorentz–Drude model.

(b) Starting with the susceptibility (7.165), use contour integration and the method of residues to invert the Fourier transform and find the response function $G(\tau)$. Compare your answer with that of part (a).

EXERCISE 7.10

In deriving the skin-depth formula (7.188), it was assumed that

$$\omega/\gamma \ll 1 \quad \text{(low-frequency regime)} \tag{7.194}$$

and

$$\sigma_0/\varepsilon_0\omega \gg 1 \quad \text{(high-conductivity regime)} \tag{7.195}$$

For a Nd:YAG laser ($\lambda = 1.06$ μm) incident on a copper mirror, show that one of these assumptions is not valid, and show that the skin depth is given by the formula

$$\delta = \sqrt{\frac{m}{\mu_0 n_e q^2}} = \frac{c}{\omega_p} \tag{7.196}$$

which is independent of frequency (within the limits of validity of the assumptions). Compute the skin depth for a Nd:YAG laser incident on a copper mirror. The effective electron density in copper is $n_e \approx 8 \times 10^{28}$ electrons/m^3. What is the range of wavelengths for which the formula is valid?

EXERCISE 7.11

In a collisionless plasma, electrostatic waves have the frequency $\omega_p = \sqrt{n_e q^2/\varepsilon_0 m}$, called the plasma frequency. The Lorentz–Drude model, for which the dielectric susceptibility is

$$\chi_e = \frac{\omega_p^2}{\omega_r^2 - \omega^2 - i\omega\gamma} \tag{7.197}$$

includes the collisionless plasma as a special case in which the resonant frequency ω_r and damping frequency γ both vanish.

(a) Beginning with the Maxwell equations for a plane, longitudinal wave in a medium with no free charges or currents, show that in the more general case ($\omega_r, \gamma \neq 0$) the frequency of electrostatic oscillations satisfies the dispersion relation

$$\omega^2 + i\gamma\omega = \omega_p^2 + \omega_r^2 \tag{7.198}$$

which has the solution

$$\omega = \sqrt{\omega_p^2 + \omega_r^2 - \tfrac{1}{4}\gamma^2} - i\tfrac{1}{2}\gamma \tag{7.199}$$

Hint: The magnetic field vanishes in an electrostatic wave.

(b) What is the phase velocity?

(c) What is the group velocity?

7.2 REFLECTION AND REFRACTION AT SURFACES

7.2.1 Boundary Conditions

When electromagnetic waves encounter a discontinuity in the properties of the material through which they are traveling, they are reflected and refracted. The details of the reflection and refraction depend on the direction of propagation and the polarization of the waves relative to the discontinuity, and many interesting phenomena can occur. For example, reflection can be enhanced or eliminated by films of one or many layers on the surface, and reflections can be used to polarize light in a desired direction or to convert linear polarization to circular polarization.

We consider a linearly polarized plane electromagnetic wave incident on a plane interface between two optical media, as shown in Figure 7.12. Elliptically polarized waves can be described by combining two linearly polarized waves with the appropriate phases. The incident wave therefore has the form

$$\mathbf{E}_i = \mathbf{e}_i e^{i(\mathbf{k}_i \cdot \mathbf{r} - \omega t)} \tag{7.200}$$

$$\mathbf{H}_i = \mathbf{h}_i e^{i(\mathbf{k}_i \cdot \mathbf{r} - \omega t)} \tag{7.201}$$

where the vectors $\mathbf{e}_i$ and $\mathbf{h}_i$ are constants. To simplify matters we assume that the waves are incident through a transparent (loss-free) medium, so the refractive index n_i is real. For a true plane wave the wave vector $\mathbf{k}_i$ is then also real, and the wave is transverse. Since the field vectors $\mathbf{E}_i$ and $\mathbf{H}_i$ are in phase, we may take both $\mathbf{e}_i$ and $\mathbf{h}_i$ to be real. The direction $\hat{\mathbf{e}}_i$ of the electric field defines the polarization of the incident wave. The directions $\hat{\mathbf{k}}_i$ and $\hat{\mathbf{n}}$ define the plane of incidence, where $\hat{\mathbf{n}}$ is a unit vector normal to the interface, as shown in Figure 7.12.

The reflected and transmitted waves can be expressed in the same form,

$$\mathbf{E}_r = \mathbf{e}_r e^{i(\mathbf{k}_r \cdot \mathbf{r} - \omega t)} \tag{7.202}$$

$$\mathbf{E}_t = \mathbf{e}_t e^{i(\mathbf{k}_t \cdot \mathbf{r} - \omega t)} \tag{7.203}$$

with corresponding expressions for the magnetic fields. From (7.81), Faraday's law, we see that the magnetic fields are related to the electric fields by

$$\mathbf{H}_v = \frac{\mathbf{k}_v \times \mathbf{E}_v}{\mu_v \omega} = \frac{\mathbf{k}_v \times \mathbf{e}_v}{\mu_v \omega} e^{i(\mathbf{k}_v \cdot \mathbf{r} - \omega t)} = \mathbf{h}_v e^{i(\mathbf{k}_v \cdot \mathbf{r} - \omega t)} \tag{7.204}$$

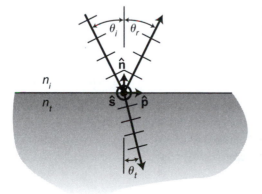

Figure 7.12 Reflection and refraction of a plane wave at an interface.

where the subscript v takes on the values $v = i$, r, t for the incident, reflected, and transmitted waves, respectively, and μ_v is the corresponding permeability.

We find in the following that the real and imaginary parts of the transmitted wave vector $\mathbf{k}_t = \mathbf{k}_t' + i\mathbf{k}_t''$ are not always parallel to one another. That is, the transmitted wave can exhibit oscillatory behavior in one direction and exponentially decaying behavior (imaginary wave vector) in another direction. In this case, the phase fronts are planes normal to the real part $\mathbf{k}_t'$ of the wave vector but the waves are not true plane waves, since they are not invariant in the planes of the phase fronts. By symmetry all three wave vectors $\mathbf{k}_v$ must lie entirely in the plane of incidence. That is, both the real and imaginary components of all the vectors $\mathbf{k}_v$ vanish identically in the direction normal to the plane of incidence. The waves are therefore invariant in the direction normal to the plane of incidence. From (7.69) we see that the incident and reflected wave vectors satisfy the dispersion relations

$$k_i = n_i \frac{\omega}{c} = k_r \tag{7.205}$$

where $n_i = n_r$ is the (real) refractive index on the incident side of the interface, and the transmitted wave vector satisfies the dispersion relation

$$\mathbf{k}_t \cdot \mathbf{k}_t = k_t'^2 - k_t''^2 + 2i\mathbf{k}_t' \cdot \mathbf{k}_t'' = n_t^2 \frac{\omega^2}{c^2} \tag{7.206}$$

where $n_t = n_t' + in_t''$ is the (complex) refractive index on the transmitted side of the interface. When the index of refraction on the transmitted side of the interface is real, we see that the real and imaginary components of the transmitted wave vector must be orthogonal to one another, that is, $\mathbf{k}_t' \cdot \mathbf{k}_t'' = 0$

At the interface the components of $\mathbf{D} = \varepsilon\mathbf{E}$ and $\mathbf{B}$ normal to the boundary are conserved, as are the components of $\mathbf{E}$ and $\mathbf{H} = \mathbf{B}/\mu$ parallel to the boundary. The boundary conditions on the parallel components are sufficient, since we can find the other components from the general relationships between $\mathbf{E}$, $\mathbf{H}$, and $\mathbf{k}$. The required boundary conditions are then

$$(\mathbf{E}_i + \mathbf{E}_r - \mathbf{E}_t) \cdot \hat{\mathbf{p}} = 0 \tag{7.207}$$

$$(\mathbf{H}_i + \mathbf{H}_r - \mathbf{H}_t) \cdot \hat{\mathbf{p}} = 0 \tag{7.208}$$

$$(\mathbf{E}_i + \mathbf{E}_r - \mathbf{E}_t) \cdot \hat{\mathbf{s}} = 0 \tag{7.209}$$

and

$$(\mathbf{H}_i + \mathbf{H}_r - \mathbf{H}_t) \cdot \hat{\mathbf{s}} = 0 \tag{7.210}$$

where $\hat{\mathbf{p}}$ is a unit vector lying in the intersection of the plane of incidence and the interface between the media, and $\hat{\mathbf{s}}$ is a unit vector perpendicular (German *senkrecht*) to the plane of incidence, as shown in Figure 7.12. In general we need only two of these boundary conditions, one on $\mathbf{E}$ and one on $\mathbf{H}$.

When we substitute (7.202) for the fields, we see that to satisfy the boundary conditions everywhere on the interface it is necessary that the components of the wave vectors match in the directions parallel to the interface. That is,

$$\mathbf{k}_i \cdot \hat{\mathbf{p}} = \mathbf{k}_r \cdot \hat{\mathbf{p}} = \mathbf{k}_t \cdot \hat{\mathbf{p}} \tag{7.211}$$

Considering only the incident and reflected waves for now, we see from (7.205) and (7.211) that the reflected wave vector is real and that

$$\theta_i = \theta_r \tag{7.212}$$

where $\sin \theta_i = \hat{\mathbf{k}}_i \cdot \hat{\mathbf{p}}$ and $\sin \theta_r = \hat{\mathbf{k}}_r \cdot \hat{\mathbf{p}}$. That is, the angle of incidence equals the angle of reflection. This is called the law of reflection. It is true even when the refractive indices n_i and n_t are both complex, provided only that the incident wave is a true plane wave.

With respect to the transmitted wave, we consider first the case when the index of refraction on the transmitted side of the interface is also real. We assume, for the moment, that the transmitted wave vector is real. Then from (7.211) we find that

$$n_i \sin \theta_i = n_t \sin \theta_t \tag{7.213}$$

where $\sin \theta_t = \hat{\mathbf{k}}_t \cdot \hat{\mathbf{p}}$. This is called Snell's law.

When the refractive index on the transmitted side of the interface is smaller than that on the incident side, $n_t < n_i$, we define the critical angle of incidence

$$\theta_c = \arcsin \frac{n_t}{n_i} \tag{7.214}$$

At this angle of incidence, Snell's law shows that the transmitted wave vector lies in the plane of the interface. For angles of incidence greater than the critical angle, there is no real solution of (7.213). Thus, when a wave propagates at a sufficiently large angle toward an interface with a medium of *smaller* index of refraction, the wave is totally reflected as shown in Figure 7.13. This is called total internal reflection, since it generally occurs when a wave propagating inside an optical element of refractive index $n > 1$ is incident on the surface of the element, outside which the refractive index is $n = 1$ (air). Experimentally, the determination of the critical angle provides a sensitive measure of the refractive index.

The fact that the wave is totally reflected by the interface does not mean that it doesn't extend into the less dense medium at all. Rather, there is an exponentially decaying wave, called the evanescent wave, that extends beyond the interface. The transmitted wave vector in this case is complex, with the imaginary part orthogonal to the real part ($\mathbf{k}_t' \cdot \mathbf{k}_t'' = 0$), as (7.206) shows that it must be. But from (7.211) we see that the imaginary part of $\mathbf{k}_t$ must lie in the plane of incidence orthogonal to $\hat{\mathbf{p}}$, so it must be normal to the interface. That is, there is no exponential decay of the wave along the surface. The real part of $\mathbf{k}_t$ therefore lies in the interface, parallel to $\hat{\mathbf{p}}$. From (7.211) it has the value

$$k_t' = k_i \sin \theta_i \tag{7.215}$$

Then from (7.205) and (7.206) we find that the imaginary part has the value

$$k_t''^2 = k_i^2 (\sin^2 \theta_i - \sin^2 \theta_c) \tag{7.216}$$

Figure 7.13 Total internal reflection.

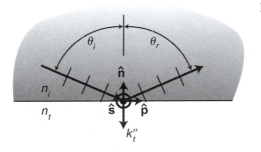

If we write $\mathbf{k}_t \cdot \hat{\mathbf{n}} = i/l$, where l is the length over which the evanescent wave decays, we see that

$$l = \frac{\lambda_i}{2\pi} \frac{1}{\sqrt{\sin^2 \theta_i - \sin^2 \theta_c}} = \frac{\lambda}{2\pi n_i} \frac{1}{\sqrt{\sin^2 \theta_i - \sin^2 \theta_c}} \qquad (7.217)$$

where $\lambda = 2\pi c/\omega$ is the wavelength in vacuum. For angles just above the critical angle, the evanescent wave extends many wavelengths beyond the interface. This might be expected, since for angles just below the critical angle the wave extends to infinity. Nevertheless, since the evanescent wave vanishes sufficiently far from the interface, it is clear that it carries no energy away from the interface. It is easy to show that when averaged over a cycle of the field, the component of the Poynting vector normal to the interface vanishes for a real index of refraction n_t. To demonstrate this, we note that the average (over one cycle) value of the Poynting vector is

$$\langle \mathbf{S}_t \rangle = \frac{1}{2} \operatorname{Re}(\mathbf{e}_t^* \times \mathbf{h}_t) = \frac{1}{2\mu_t} \operatorname{Re}\left[\mathbf{e}_t^* \times \left(\frac{\mathbf{k}_t \times \mathbf{e}_t}{\omega}\right)\right] \qquad (7.218)$$

where we have used (7.204) to evaluate $\mathbf{h}_t$. Expanding the triple cross product gives

$$\langle \mathbf{S}_t \cdot \hat{\mathbf{n}} \rangle = \frac{1}{2\mu_t \omega} \operatorname{Re}[(\mathbf{e}_t \cdot \mathbf{e}_t^*)(\mathbf{k}_t \cdot \hat{\mathbf{n}}) - (\mathbf{e}_t^* \cdot \mathbf{k}_t)(\mathbf{e}_t \cdot \hat{\mathbf{n}})] \qquad (7.219)$$

But the first term vanishes, since $\mathbf{e}_t \cdot \mathbf{e}_t^*$ is real and $\mathbf{k}_t^* \cdot \hat{\mathbf{n}}$ is pure imaginary, as we just proved. In the second term, the factor $\mathbf{e}_t^* \cdot \mathbf{k}_t$ does not vanish, in general, even though $\nabla \cdot \mathbf{D} = 0$ implies that $\mathbf{e}_t \cdot \mathbf{k}_t$ does. However, for waves polarized normal to the plane of incidence, called s-polarization, symmetry indicates that the normal component of $\mathbf{e}_t$ must vanish, so the second term also vanishes. For waves polarized in the plane of incidence, called p-polarization, we make the same arguments by evaluating the Poynting vector in terms of the magnetic field, which is now perpendicular to the plane of incidence. Therefore, the component of the Poynting vector carrying energy away from the interface vanishes in all cases. When the index of refraction n_t is complex, the normal component of the Poynting vector no longer vanishes because in this case it represents the flow of energy dissipated in the medium beyond the interface. In fact, the reflection of waves at the interface can be used to measure the absorption coefficient (the imaginary part of the index of refraction) of the substance into which the evanescent wave extends. This is useful for strongly absorbing materials.

Up to this point all the results have been based on matching the wave vectors to meet the boundary conditions. That is, we have used only the kinematic constraints. To proceed further we must use the boundary conditions on the electric and magnetic fields themselves. We consider separately the cases of s-polarization and p-polarization.

In the case of s-polarization the electric field is parallel to $\hat{\mathbf{s}}$, so the boundary condition (7.209) becomes

$$e_i + e_r = e_t \qquad (7.220)$$

The magnetic field has a component in the $\hat{\mathbf{p}}$ direction, and from (7.81), we see that

$$\mathbf{H}_v \cdot \hat{\mathbf{p}} = \frac{(\mathbf{k}_v \times \mathbf{E}_v) \cdot \hat{\mathbf{p}}}{\mu_v \omega} = \frac{E_v}{\mu_v \omega} \mathbf{k}_v \cdot \hat{\mathbf{n}} \qquad (7.221)$$

after rearranging the vector product, since $\hat{\mathbf{s}} \times \hat{\mathbf{p}} = \hat{\mathbf{n}}$. Substituting this into (7.208), we get

$$\frac{e_i - e_r}{\mu_i} \mathbf{k}_i \cdot \hat{\mathbf{n}} = \frac{e_t}{\mu_t} \mathbf{k}_t \cdot \hat{\mathbf{n}} \tag{7.222}$$

But from the dispersion relation (7.206), we see that

$$\mathbf{k}_t \cdot \mathbf{k}_t = n_t^2 \frac{\omega^2}{c^2} = (\mathbf{k}_t \cdot \hat{\mathbf{n}})^2 + (\mathbf{k}_t \cdot \hat{\mathbf{p}})^2 = (\mathbf{k}_t \cdot \hat{\mathbf{n}})^2 + n_i^2 \frac{\omega^2}{c^2} \sin^2 \theta_i \tag{7.223}$$

where we have used the kinematic constraint (7.211) and the dispersion relation (7.205) to evaluate $\mathbf{k}_t \cdot \hat{\mathbf{p}}$. Substituting this into (7.222), we get

$$\frac{e_i - e_r}{\mu_i} n_i \cos \theta_i = \frac{e_t}{\mu_t} \sqrt{n_t^2 - n_i^2 \sin^2 \theta_i} \tag{7.224}$$

Combining (7.220) and (7.224), we get

$$\frac{e_r}{e_i} = \frac{\mu_t \sqrt{n_i^2 - n_i^2 \sin^2 \theta_i} - \mu_i \sqrt{n_t^2 - n_i^2 \sin^2 \theta_i}}{\mu_t \sqrt{n_i^2 - n_i^2 \sin^2 \theta_i} + \mu_i \sqrt{n_t^2 - n_i^2 \sin^2 \theta_i}} \qquad \text{(s-polarization)} \tag{7.225}$$

$$\frac{e_t}{e_i} = \frac{2\mu_t \sqrt{n_i^2 - n_i^2 \sin^2 \theta_i}}{\mu_t \sqrt{n_i^2 - n_i^2 \sin^2 \theta_i} + \mu_i \sqrt{n_t^2 - n_i^2 \sin^2 \theta_i}} \qquad \text{(s-polarization)} \tag{7.226}$$

For p-polarized waves, the magnetic field is in the $\hat{\mathbf{s}}$ direction and the electric field has a component in the $\hat{\mathbf{p}}$ direction. It is most convenient, therefore, to use the boundary conditions (7.210) and (7.207). In the $\hat{\mathbf{s}}$ direction we get

$$h_i + h_r = h_t \tag{7.227}$$

The electric field components in the $\hat{\mathbf{p}}$ direction are found from the Maxwell equations, which for a wave of the form (7.203) become

$$\mathbf{e} \cdot \hat{\mathbf{p}} = -\frac{(\mathbf{k} \times \mathbf{h}) \cdot \hat{\mathbf{p}}}{\varepsilon \omega} = -\frac{h}{\varepsilon \omega} \mathbf{k} \cdot \hat{\mathbf{n}} \tag{7.228}$$

Substituting this into (7.207) we get

$$\frac{h_i - h_r}{\varepsilon_i} \mathbf{k}_i \cdot \hat{\mathbf{n}} = \frac{h_t}{\varepsilon_t} \mathbf{k}_t \cdot \hat{\mathbf{n}} \tag{7.229}$$

These equations are the same as the corresponding ones for s-polarization with the substitutions $e \to h$ and $\mu \to \varepsilon$. The reflected and transmitted fields are therefore

$$\frac{h_r}{h_i} = \frac{\varepsilon_t \sqrt{n_i^2 - n_i^2 \sin^2 \theta_i} - \varepsilon_i \sqrt{n_t^2 - n_i^2 \sin^2 \theta_i}}{\varepsilon_t \sqrt{n_i^2 - n_i^2 \sin^2 \theta_i} + \varepsilon_i \sqrt{n_t^2 - n_i^2 \sin^2 \theta_i}} \qquad \text{(p-polarization)} \tag{7.230}$$

$$\frac{h_t}{h_i} = \frac{2\varepsilon_t \sqrt{n_i^2 - n_i^2 \sin^2 \theta_i}}{\varepsilon_t \sqrt{n_i^2 - n_i^2 \sin^2 \theta_i} + \varepsilon_i \sqrt{n_t^2 - n_i^2 \sin^2 \theta_i}} \qquad \text{(p-polarization)} \tag{7.231}$$

These formulas [(7.225), (7.226), (7.230), and (7.231)] were first derived by Fresnel in 1823, in a slightly less general form, by assuming that light consists of transverse waves in an elastic medium (the ether). This was before Maxwell discovered that light is electro-magnetic waves. As they appear here, the formulas require that the index of refraction on the incident side be real (μ_i and ε_i both real), but they are valid for complex values of the index of refraction on the transmitted side (μ_t and ε_t both complex). In general, then, the electric and magnetic fields on the transmitted side are complex, which corresponds to both an amplitude change and a phase change.

EXERCISE 7.12

Consider an s-polarized wave that is totally internally reflected from the interface (in the x–y plane) between two dielectric media. Show that the flow of energy in the evanescent wave parallel to the interface is

$$\int_0^\infty \mathbf{S}_t \cdot \hat{\mathbf{p}}\, dz = \frac{\mu_i}{\pi \mu_t}\, \frac{\mu_t^2 \cos^2 \theta_i}{\mu_t^2 \cos^2 \theta_i + \mu_i^2 (\sin^2 \theta_i - \sin^2 \theta_c)}\, \frac{\lambda_i S_i \sin \theta_i}{\sqrt{\sin^2 \theta_i - \sin^2 \theta_c}} \tag{7.232}$$

where $\mathbf{S}_i$ and $\mathbf{S}_t$ are the Poynting vectors of the incident and transmitted waves. What is the flow of energy when the incident wave is p-polarized?

7.2.2 Dielectric Reflection

Since the incident and reflected waves travel in the same medium, their relative intensity is proportional to ee^* or hh^*. The reflectance is therefore

$$R = \frac{e_r e_r^*}{e_i e_i^*} = \frac{h_r h_r^*}{h_i h_i^*} \tag{7.233}$$

The most important dielectric materials are those that are transparent, in which the index of refraction n_t of the medium on the transmitted side of the interface is also real. Except in the region of total internal reflection, then, the fields are all real and we can ignore the com-plex conjugates. It is also generally true for dielectrics that $\mu_i = \mu_t = \mu_0$, and we assume this in the following. The reflectance is then

$$R = \left(\frac{\sqrt{n_i^2 - n_i^2 \sin^2 \theta_i} - \sqrt{n_t^2 - n_i^2 \sin^2 \theta_i}}{\sqrt{n_i^2 - n_i^2 \sin^2 \theta_i} + \sqrt{n_t^2 - n_i^2 \sin^2 \theta_i}} \right)^2 \qquad \text{(s-polarization)} \tag{7.234}$$

$$R = \left(\frac{n_i^2 \sqrt{n_i^2 - n_i^2 \sin^2 \theta_i} - n_i^2 \sqrt{n_t^2 - n_i^2 \sin^2 \theta_i}}{n_i^2 \sqrt{n_i^2 - n_i^2 \sin^2 \theta_i} + n_i^2 \sqrt{n_t^2 - n_i^2 \sin^2 \theta_i}} \right)^2 \qquad \text{(p-polarization)} \tag{7.235}$$

For waves incident normal to the interface, the distinction between s- and p-polarized waves disappears and we get the simple formula

$$R = \left(\frac{n_t - n_i}{n_t + n_i} \right)^2 \qquad \text{(normal incidence)} \tag{7.236}$$

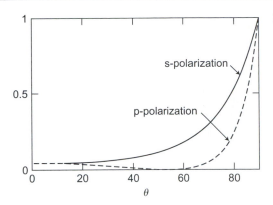

Figure 7.14 Reflectance of glass.

For glass ($n_t = 1.5$) in air ($n_i = 1$), the reflectance at normal incidence is 4 percent. At normal incidence, the reflectance is indifferent to whether the wave is passing from the medium of lower index to that of higher or the other way around. Thus, in passing through a slab of glass or other material, the light is reflected equally from both the front and back surfaces. At oblique angles of incidence, the reflection coefficients are equal for waves incident from opposite sides of the interface provided that the angles of incidence are related by Snell's law.

The reflectance for glass ($n_t = 1.5$) in air ($n_i = 1$) as a function of the angle of incidence is shown in Figure 7.14. As we see there, the reflectance of p-polarized light is less than that of s-polarized light except, of course, at normal incidence. In fact, the reflectance for p-polarized light vanishes at Brewster's angle,

$$\tan \theta_B = \frac{n_t}{n_i} \tag{7.237}$$

For glass in air this corresponds to $\theta_B = 56°$. At grazing incidence ($\theta_i \approx 90°$) the reflectance approaches unity for both polarizations.

Brewster's phenomenon has several important applications. First of all, it can be used to create linearly polarized light. By shining light of arbitrary polarization on a dielectric surface at Brewster's angle of incidence, we obtain a reflected beam of light completely polarized in the s-direction. Unfortunately, the reflectance for the s-polarized light is small at Brewster's angle (15 percent in the case of glass), so most of the light is lost. A more efficient but more complicated way to polarize light is to use a series of surfaces aligned at Brewster's angle. Each surface reflects some of the s-polarized light out of the beam but allows the p-polarized light to pass unattenuated. After many surfaces, the remaining light is almost completely p-polarized. Brewster's phenomenon is frequently used in lasers to prevent reflection losses at the windows at the ends of the lasers. Light of one polarization (the p-polarization) is unattenuated when it passes through the window, and it is this polarization that lases.

The disappearance of p-polarized light from the reflected beam has an interesting interpretation. At Brewster's angle of incidence the angle of refraction is

$$\sin \theta_t = \frac{n_i}{n_t} \sin \theta_B = \frac{n_i}{n_t} \frac{n_t}{\sqrt{n_i^2 + n_t^2}} = \cos \theta_B \tag{7.238}$$

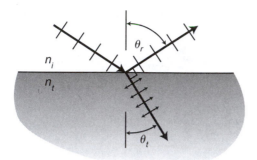

Figure 7.15 Brewster's phenomenon.

so $\theta_r + \theta_B = \frac{1}{2}\pi$. That is, the reflected and refracted rays are orthogonal, as shown in Figure 7.15. But the reflected wave is actually the radiation produced by all the microscopic dipoles in the medium. These dipoles are aligned with the electric field in the medium, which for p-polarized light is in the plane of incidence normal to the direction of propagation, as shown in Figure 7.15. Thus, at Brewster's angle the p-polarized reflection corresponds to dipole radiation emitted in the direction along the axis of the dipole, which vanishes as shown in Chapter 4.

Returning to (7.225), we see that if we avoid the region of total internal reflection ($n_t^2 - n_i^2 \sin^2 \theta_i > 0$), all the factors in the formula are real. We therefore see that for reflection from a dielectric with a larger refractive index $n_t/n_i > 1$ (we assume that $\mu_i = \mu_t = \mu_0$), the electric field of an s-polarized wave changes sign upon reflection; that is, the electric field experiences a phase reversal on reflection from a dielectric with a larger refractive index. For reflection from a dielectric with a smaller refractive index $n_t/n_i < 1$, the electric field of the reflected wave has the same phase as the incident field. The electric field of the transmitted wave is always in phase with the incident field. For p-polarized waves the situation is more complicated, since the electric fields of the incident and reflected waves are not parallel to one another. From (7.230) we see that the magnetic field, which is in the $\hat{s}$ direction for all three waves (incident, reflected, and transmitted), has behavior opposite that of the electric field of s-polarized waves. That is, the magnetic field experiences a phase reversal upon reflection from a medium with a lower refractive index but not upon reflection from a higher refractive index. However, near normal angles of incidence the reflected wave is traveling in roughly the opposite direction from the incident wave, so the electric field relative to the magnetic field has the opposite sign. Thus, the electric field of a p-polarized wave experiences a reversal whenever that of an s-polarized wave does, as must be the case. Near grazing incidence, however, reflection reverses the phase of a p-polarized wave relative to an s-polarized wave. For p-polarization, as for s-polarization, the transmitted wave always has the same phase as the incident wave.

In the region of total internal reflection, the phase change upon reflection is not limited to 0 or π. We consider s-polarization first. Since the second term in the numerator and denominator of (7.225) is now imaginary, the amplitude of the reflected wave, relative to that of the incident wave, has the form $e_r = [(a - ib)/(a + ib)]e_i$. The relative phase shift for s-polarized waves is therefore

$$\Delta\phi_s = -2\arctan\left(\frac{b}{a}\right) = -2\arctan\left(\frac{\mu_i}{\mu_t}\sqrt{\frac{n_i^2 \sin^2 \theta_i - n_t^2}{n_i^2 - n_i^2 \sin^2 \theta_i}}\right) \tag{7.239}$$

For p-polarized waves we know the phase shift of the reflected magnetic field, which is in the $\hat{s}$ direction. Since the angle of incidence is large, for total internal reflection, the reflected wave is traveling in roughly the same direction. The phase change in the electric field of p-polarized waves is therefore the same as that of the magnetic field. From (7.230) we find that

$$\Delta\phi_p = -2\arctan\left(\frac{\varepsilon_i}{\varepsilon_t}\sqrt{\frac{n_i^2\sin^2\theta_i - n_t^2}{n_i^2 - n_i^2\sin^2\theta_i}}\right) \tag{7.240}$$

Thus, the phase shifts are different for p-polarized and s-polarized incident waves. In general, then, when a linearly polarized wave is totally internally reflected from a dielectric interface, the s- and p-polarized components of the incident wave are phase shifted relative to one another. An incident wave that is linearly polarized is converted to an elliptically polarized wave upon reflection. This is the principle underlying the Fresnel rhomb, which uses total internal reflection to convert linearly polarized light into circularly polarized light, and vice versa, of course.

EXERCISE 7.13

A Fresnel rhomb is shown in Figure 7.16. Light incident normally on the entrance and exit faces is totally internally reflected twice inside the rhomb, so the s- and p-components are twice phase shifted relative to one another. For glass with a refractive index $n = 1.51$, find the angles of incidence for which the Fresnel rhomb converts linear polarization to circular polarization. There are two angles for which this is true.

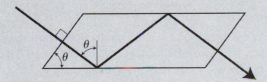

Figure 7.16 Fresnel rhomb.

EXERCISE 7.14

Show that for a lossy dielectric with a complex index of refraction $n_t = n_t' + in_t''$ the reflection coefficient at normal incidence is

$$R = \frac{(n_t' - n_i')^2 + n_t''^2}{(n_t' + n_i')^2 + n_t''^2} \tag{7.241}$$

EXERCISE 7.15

Light of wavelength λ in vacuum is incident normal to a substrate on which there is a coating of thickness d. If the refractive indices of the incident medium, the coating, and the substrate are n_i, n_c, and n_s (all real), show that the reflection coefficient is

$$R = \frac{n_1^2 + n_2^2 + 2n_1n_2 \cos \phi}{1 + n_1^2n_2^2 + 2n_1n_2 \cos \phi} \tag{7.242}$$

where

$$\phi = \frac{4\pi n_c d}{\lambda} \tag{7.243}$$

$$n_1 = \frac{n_i - n_c}{n_i + n_c} \tag{7.244}$$

$$n_2 = \frac{n_c - n_s}{n_c + n_s} \tag{7.245}$$

Solar cells made from Si ($n_s = 3.5$) are antireflection coated with SiO ($n_c = 1.45$) to increase their efficiency. What is the reflectance from uncoated Si, and what is the minimum reflectance from coated Si?

7.2.3 Metallic Reflection

The reflective properties of metals at low frequencies are dominated by the conduction electrons. For most metals the low-frequency region includes the infrared and parts of the visible spectrum, but at shorter wavelengths inter-band transitions introduce strong absorption bands. According to the Drude model of conductivity, discussed earlier, the dielectric susceptibility of a metal is

$$\chi_e = \frac{i\omega_p^2\tau^2}{\omega\tau(1 - i\omega\tau)} \tag{7.246}$$

where ω_p is the plasma frequency and τ is the collision time. In a typical metal, taking aluminum as an example, the plasma frequency is $\omega_p = 1.98 \times 10^{16}$ radians/s. This corresponds to a photon energy $\hbar\omega_p = 13$ eV, or a wavelength $\lambda_p = 2\pi c/\omega_p \approx 95$ nm, which lies in the far ultraviolet. The collision time is $\tau = 1.0 \times 10^{-14}$ s. This corresponds to a wavelength $\lambda_\tau = 2\pi c\tau = 19$ μm, in the far-infrared region of the spectrum. As is typical of most good metallic conductors, $\omega_p\tau \approx 200 \gg 1$.

At frequencies smaller than the plasma frequency, which correspond to wavelengths in the visible region of the spectrum and longer, the susceptibility becomes very large compared with unity. At microwave frequencies and below ($\omega\tau \ll 1$) the susceptibility is essentially pure imaginary, while for infrared and visible frequencies ($1 \ll \omega\tau \ll \omega_p\tau$) the susceptibility approaches a real, negative value. The origin of this behavior is easy to understand. At very low frequencies, less than the collision frequency, the electrons move in response to the electric field as though they were in a viscous medium. Their motion and the associated polarization therefore lag behind the field by 90°, which corresponds to the imaginary factor i in the dielectric susceptibility. At frequencies above the collision frequency, the electrons move as though they were completely free. They oscillate harmonically exactly 180° out of phase with the field, so the polarization is opposite the field and the susceptibility is negative (strictly speaking, of course, the electrons are negatively charged, so they oscillate in phase with the field, but the associated polarization is still opposite the field). It is interesting to examine how this behavior affects the reflectance of metals.

At frequencies that are small compared with the plasma frequency ($\omega/\omega_p \ll 1$), most metals are excellent reflectors. That is, the reflectance R is close to unity. To calculate the loss, we can calculate the reflectance and subtract it from unity, but the part we are looking for is small and the expansion is tedious. It is quicker and more instructive to calculate the Poynting vector just inside the surface of the metal and compare this with the Poynting vector incident on the surface. For this purpose, we use (7.219) and limit ourselves to waves incident normal to the surface. In this case, symmetry assures us that the transmitted wave, like the incident wave, is a transverse wave propagating in the direction normal to the surface, so both factors in the second term of (7.219) vanish. The quantity $\mathbf{e} \cdot \mathbf{e}^*$ is real, so the average Poynting vector is

$$\langle \mathbf{S}_v \cdot \hat{\mathbf{n}} \rangle = \frac{\mathbf{e}_v \cdot \mathbf{e}_v^*}{2\omega} \, \mathrm{Re}\left(\frac{\mathbf{k}_v \cdot \hat{\mathbf{n}}}{\mu_v} \right) \tag{7.247}$$

for both the incident and transmitted waves, when μ_v is allowed to be complex. From (7.223) we see that the wave vector is

$$\left(\frac{\mathbf{k}_v \cdot \hat{\mathbf{n}}}{\mu_v} \right)^2 = n_v^2 \frac{\omega^2}{c^2} \tag{7.248}$$

for a normally incident wave. The loss coefficient is therefore

$$L = \frac{\langle \mathbf{S}_t \cdot \hat{\mathbf{n}} \rangle}{\langle \mathbf{S}_i \cdot \hat{\mathbf{n}} \rangle} = \frac{e_t e_t^* \, \mathrm{Re}\sqrt{\mu_t \varepsilon_t}}{e_i e_i^* \, \mathrm{Re}\sqrt{\mu_i \varepsilon_i}} \tag{7.249}$$

Inside the metal we see from (7.226) that the electric field is

$$\frac{e_t}{e_i} = \frac{2\mu_t \sqrt{n_i^2}}{\mu_t \sqrt{n_i^2} + \mu_i \sqrt{n_t^2}} = \frac{2\mu_t \sqrt{\mu_i \varepsilon_i}}{\mu_t \sqrt{\mu_i \varepsilon_i} + \mu_i \sqrt{\mu_t \varepsilon_t}} \tag{7.250}$$

for a normally incident wave. For a typical metal in a vacuum, $\mu_i = \mu_t = \mu_0$, $\varepsilon_i = \varepsilon_0$, and $\omega_p \tau \gg 1$. For frequencies that are small compared with the plasma frequency ($\omega \ll \omega_p$), the susceptibility is large and we may ignore the first term in the denominator of (7.250), which leaves

$$\frac{e_t e_t^*}{e_i e_i^*} \approx 4 \sqrt{\frac{\varepsilon_i \varepsilon_i^*}{\varepsilon_t \varepsilon_t^*}} \approx \frac{4}{\sqrt{\chi_e \chi_e^*}} = \frac{4\omega\tau \sqrt{1 + \omega^2 \tau^2}}{\omega_p^2 \tau^2} \tag{7.251}$$

We also see that

$$\frac{\mathrm{Re}\sqrt{\mu_t \varepsilon_t}}{\mathrm{Re}\sqrt{\mu_i \varepsilon_i}} \approx \mathrm{Re}\sqrt{\chi_e} = \mathrm{Re}\sqrt{\frac{-\omega_p^2 \tau^2}{\omega\tau(\omega\tau + i)}} \tag{7.252}$$

At the very lowest frequencies, $\omega\tau \ll 1$, the susceptibility is pure imaginary, so that

$$\frac{\mathrm{Re}\sqrt{\mu_t \varepsilon_t}}{\mathrm{Re}\sqrt{\mu_i \varepsilon_i}} \approx \mathrm{Re}\sqrt{\frac{i\omega_p^2 \tau^2}{\omega\tau}} = \frac{\omega_p \tau}{\sqrt{2\omega\tau}} \tag{7.253}$$

The reflective loss coefficient is then

$$L \approx \frac{\sqrt{8\omega\tau}}{\omega_p \tau} = \sqrt{\frac{8\varepsilon_0 \omega}{\sigma_0}} \tag{7.254}$$

This is called the Hagen–Rubens relation. In this frequency region, the losses are controlled by the currents flowing near the surface due to the skin effect discussed earlier. Since the total current flowing in the surface is just enough to cancel out the magnetic field of the incident wave, it is proportional to the square root of the incident intensity. But the total dissipation is proportional to the current density squared times the volume, that is, to the total current squared divided by the penetration depth. Since the skin depth is inversely proportional to the square root of the frequency, as shown earlier, this accounts for the square root dependence of the reflection losses on frequency in (7.254).

At higher frequencies, $1 \ll \omega\tau \ll \omega_p\tau$, the susceptibility is almost purely negative real, so

$$\frac{\mathrm{Re}\sqrt{\mu_t \varepsilon_t}}{\mathrm{Re}\sqrt{\mu_i \varepsilon_i}} \approx \mathrm{Re}\sqrt{\frac{\omega_p^2}{\omega^2}\left(-1 + \frac{i}{\omega\tau}\right)} \approx \frac{\omega_p}{\omega}\left(\frac{1}{2\omega\tau}\right) \qquad (7.255)$$

The reflective loss coefficient is then

$$L \approx \frac{2}{\omega_p \tau} \qquad (7.256)$$

independent of frequency over the range $1 \ll \omega\tau \ll \omega_p\tau$. The fact that the losses are independent of the frequency follows from the fact that the skin depth is independent of frequency in this range, as shown by (7.196).

The Drude model provides a qualitatively correct description of the reflectance of many metals, including copper, silver, and gold, although it does not do so well for nickel, molybdenum, and some other metals. A comparison with silver is shown in Figure 7.17, where the solid line is a curve calculated from (7.249), with (7.250) and (7.246), using the parameters $\sigma_0 = 6.3 \times 10^7/\Omega$-m (the experimental value of the dc conductivity) and $n_e = 6 \times 10^{28}/\mathrm{m}^3$ (one conduction electron per atom). As shown there, the Drude model provides a reasonable description of the reflectance of silver at infrared wavelengths but underestimates the losses at shorter wavelengths.

As the optical frequency approaches the plasma frequency, the susceptibility drops to the order of unity and the approximations used in (7.251) and the subsequent equations are not valid. In this region the losses increase significantly. The plasma frequency in silver is

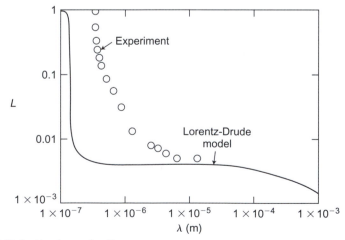

Figure 7.17 Reflection losses in silver.

$\omega_p = 1.4 \times 10^{16}$ radians/s, which corresponds to a wavelength $\lambda_p = 140$ nm, in the far ultraviolet. However, as illustrated in Figure 7.17, absorption at visible and ultraviolet wavelengths sets in much sooner than this due to electrons in the (filled) d-bands, which overlap the conduction band. This absorption is even stronger in copper and gold, and gives them their characteristic color.

At frequencies well above the collision frequency we can ignore the damping term in the susceptibility, so the refractive index due to conduction electrons becomes essentially real at frequencies above the plasma frequency. In the absence of other absorption processes the metal ceases to be strongly absorbing in this frequency range. This is the origin of the so-called ultraviolet transparency observed in alkali metals. In these metals inner-shell absorption processes are absent since the next lower electron bands are well below the conduction band. Immediately below the plasma frequency the susceptibility becomes real negative, so the refractive index is essentially pure imaginary. In the near absence of dissipative processes the metal becomes strongly reflecting.

EXERCISE 7.16

Beginning with the Drude expression for the dielectric susceptibility of a metal, show that the refractive index due to conduction electrons becomes essentially real for wavelengths shorter than

$$\lambda_0 = 2\pi c \sqrt{\frac{m_e}{\mu_0 n_e q_e^2}} \tag{7.257}$$

and pure imaginary for longer wavelengths. Compare this wavelength with the observed long-wavelength limit λ_0 of ultraviolet transparency in the alkali metals:

	n_e (one e^-/atom)	λ_0 (observed)
Li	4.6×10^{28}/m^3	155 nm
Na	2.5×10^{28}/m^3	210 nm
K	1.3×10^{28}/m^3	315 nm
Rb	1.1×10^{28}/m^3	340 nm
Cs	0.9×10^{28}/m^3	

7.2.4 Surface Waves

In addition to incident, reflected, and transmitted waves, surface waves, called polaritons, can travel along the surface of a metal in a vacuum. The waves are evanescent in both media, meaning that they decay exponentially in the direction normal to the interface in both the metal and the vacuum. To describe these waves, we assume that they have the form

$$\mathbf{E}_o = \mathbf{e}_o e^{i(\mathbf{k}_o \cdot \mathbf{r} - \omega t)} \tag{7.258}$$

$$\mathbf{E}_i = \mathbf{e}_i e^{i(\mathbf{k}_i \cdot \mathbf{r} - \omega t)} \tag{7.259}$$

outside and inside the metal, respectively. From Gauss's law we find that for each wave

$$\mathbf{k}_o \cdot \mathbf{e}_o = 0 = (\mathbf{k}_o \cdot \hat{\mathbf{p}})(\mathbf{e}_o \cdot \hat{\mathbf{p}}) + (\mathbf{k}_o \cdot \hat{\mathbf{n}})(\mathbf{e}_o \cdot \hat{\mathbf{n}}) \tag{7.260}$$

$$\mathbf{k}_i \cdot \mathbf{e}_i = 0 = (\mathbf{k}_i \cdot \hat{\mathbf{p}})(\mathbf{e}_i \cdot \hat{\mathbf{p}}) + (\mathbf{k}_i \cdot \hat{\mathbf{n}})(\mathbf{e}_i \cdot \hat{\mathbf{n}}) \tag{7.261}$$

From the rest of the Maxwell equations we get the dispersion relations

$$\mathbf{k}_o \cdot \mathbf{k}_o = \mu_0 \varepsilon_0 \omega^2 = (\mathbf{k}_o \cdot \hat{\mathbf{p}})^2 + (\mathbf{k}_o \cdot \hat{\mathbf{n}})^2 \tag{7.262}$$

$$\mathbf{k}_i \cdot \mathbf{k}_i = \mu_0 \varepsilon(\omega) \omega^2 = (\mathbf{k}_i \cdot \hat{\mathbf{p}})^2 + (\mathbf{k}_i \cdot \hat{\mathbf{n}})^2 \tag{7.263}$$

for the waves outside and inside the metal, where $\varepsilon(\omega)$ is the permittivity of the metal at the frequency ω and we have assumed that the metal is nonmagnetic. If we ignore the damping terms in the dielectric susceptibility (7.246), which we can do for frequencies $\omega\tau \gg 1$, the permittivity in the Drude model is

$$\frac{\varepsilon(\omega)}{\varepsilon_0} = 1 - \frac{\omega_p^2}{\omega^2} \tag{7.264}$$

which is purely real, and negative for $\omega < \omega_p$. Since the index of refraction is real, the real and imaginary components of the wave vectors must be orthogonal to each other. For a wave that is uniform along the surface of the conductor, the components $\mathbf{k}_o \cdot \hat{\mathbf{p}}$ and $\mathbf{k}_i \cdot \mathbf{p}$ parallel to the surface must be real, so the components $\mathbf{k}_o \cdot \hat{\mathbf{n}}$ and $\mathbf{k}_i \cdot \hat{\mathbf{n}}$ normal to the surface must be imaginary.

At the interface the fields must satisfy the usual boundary conditions. The kinetic boundary condition on the wave vectors is

$$\mathbf{k}_o \cdot \hat{\mathbf{p}} = \mathbf{k}_i \cdot \hat{\mathbf{p}} \tag{7.265}$$

while from the continuity of the tangential electric field and the normal electric displacement we get the boundary conditions

$$\mathbf{e}_o \cdot \hat{\mathbf{p}} = \mathbf{e}_i \cdot \hat{\mathbf{p}} \tag{7.266}$$

$$\mathbf{e}_o \cdot \hat{\mathbf{s}} = \mathbf{e}_i \cdot \hat{\mathbf{s}} \tag{7.267}$$

$$\varepsilon_0 \mathbf{e}_o \cdot \hat{\mathbf{n}} = \varepsilon(\omega) \mathbf{e}_i \cdot \hat{\mathbf{n}} \tag{7.268}$$

Since they don't appear in any of the other equations, we can ignore the components of the electric field in the $\hat{\mathbf{s}}$ direction. It is easy to show that the analogous waves in which $\mathbf{e}_o, \mathbf{e}_i \to \mathbf{h}_o, \mathbf{h}_i$ in the plane of the wave vectors, and the electric field is s-polarized, are not possible for nonmagnetic materials.

If we substitute the boundary conditions (7.265)–(7.268) into Gauss's laws (7.260) and (7.261) and then combine them, we find that

$$\varepsilon(\omega) \mathbf{k}_o \cdot \hat{\mathbf{n}} = \varepsilon_0 \mathbf{k}_i \cdot \hat{\mathbf{n}} \tag{7.269}$$

Since the fields must decay exponentially away from the interface in both directions, we see that a solution is possible only for Re $\varepsilon(\omega) < 0$, which corresponds to $\omega < \omega_p$. Similarly, if we substitute the kinematic boundary condition (7.265) into the dispersion relations (7.262) and (7.263) and combine them, we find that

$$(\mathbf{k}_o \cdot \hat{\mathbf{n}})^2 - (\mathbf{k}_i \cdot \hat{\mathbf{n}})^2 = \frac{\omega^2}{c^2}\left[1 - \frac{\varepsilon(\omega)}{\varepsilon_0}\right] \tag{7.270}$$

Combining these equations and using (7.262), we obtain the dispersion relation

$$(\mathbf{k}_o \cdot \hat{\mathbf{p}})^2 = \frac{\omega^2}{c^2}\frac{\varepsilon(\omega)}{\varepsilon_0 + \varepsilon(\omega)} \tag{7.271}$$

for the surface waves. But from the dispersion relation (7.262), we see that the waves

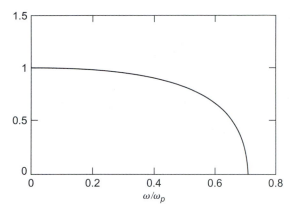

Figure 7.18 Dispersion relation for surface plasma waves.

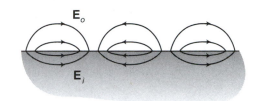

Figure 7.19 Fields of a surface plasma wave.

above the surface are evanescent [that is, $(\mathbf{k}_o \cdot \hat{\mathbf{n}})^2 < 0$] only if $(\mathbf{k}_o \cdot \hat{\mathbf{p}})^2 > \mu_0 \varepsilon_0 \omega^2 > 0$. Clearly, we get real values for $\mathbf{k}_o \cdot \hat{\mathbf{p}}$ only for $\varepsilon(\omega) > 0$ or $\varepsilon(\omega) < -\varepsilon_0$. The first case was excluded previously, so we see that we get evanescent waves only for $\varepsilon(\omega) < -\varepsilon_0$.

The phase velocity parallel to the surface is

$$\beta_\phi = \frac{v_\phi}{c} = \frac{\omega}{c\mathbf{k}_o \cdot \hat{\mathbf{p}}} = \sqrt{\frac{\varepsilon_0 + \varepsilon(\omega)}{\varepsilon(\omega)}} \tag{7.272}$$

which for the Drude model becomes

$$\beta_\phi = \sqrt{\frac{\omega_p^2 - 2\omega^2}{\omega_p^2 - \omega^2}}$$

This is real for $\omega/\omega_p < 1/\sqrt{2}$. The complete dispersion relation is shown in Figure 7.18. The phase velocity vanishes at $\omega/\omega_p = 1/\sqrt{2}$, in which case the waves are called surface plasmons. In the more general case they are called surface plasmon polaritons. The electric field pattern in a surface plasma wave is shown in Figure 7.19.

EXERCISE 7.17

At any given point $\mathbf{r}$ in a surface plasma wave, the electric field is given by (7.259), which we may write in the form

$$\mathbf{E} = \mathbf{E}_0 e^{-i\omega t} \tag{7.273}$$

for some constant complex vector $\mathbf{E}_0$.

(a) Show that for real (negative) $\varepsilon(\omega)$, the complex electric field inside the conductor is

$$\mathbf{E} = E_0 e^{-i\omega t} (\hat{\mathbf{n}} \pm i\hat{\mathbf{p}}\sqrt{-\varepsilon(\omega)/\varepsilon_0}) \tag{7.274}$$

where E_0 is a constant, and the choice of a positive sign for the second term is appropriate for a wave traveling in the $\hat{\mathbf{p}}$ direction.

(b) Show that if the electron motion is small compared with a wavelength (a necessary condition for the polarization to be a linear function of the field), the position of an electron in the field of a surface plasma wave is

$$\mathbf{r} = \frac{-qE_0}{m\omega^2} e^{-i\omega t} (\hat{\mathbf{n}} \pm i\hat{\mathbf{p}}\sqrt{-\varepsilon(\omega)/\varepsilon_0}) \qquad (7.275)$$

which corresponds to ellipses. Elliptical motions are characteristic of many types of surface waves, including water waves, where transverse motions and longitudinal motions are coupled by the presence of the surface. For a wave traveling in the $\hat{\mathbf{p}}$ direction, do the electrons circulate in the clockwise or the counterclockwise direction?

7.3 ENERGY LOSS BY FAST PARTICLES TRAVELING THROUGH MATTER

7.3.1 Ionization and Excitation

As high-energy charged particles travel through matter, they are scattered and lose energy due to collisions with the electrons and nuclei in the target material. The deflection of the incident particles, which is due largely to collisions with the nuclei, is essentially a problem in mechanics and we don't discuss it further here. However, the energy-loss processes provide interesting and useful examples of electrodynamics. The calculation of energy loss per unit distance, or "stopping power," as it is called, is an important problem, and has been associated historically with names like Bohr, Fermi, Bethe, and Landau.

The energy-loss processes comprise two mechanisms, excitation and ionization of the medium on the one hand, and radiative processes on the other. Besides transition radiation, which is associated with the surfaces of the target material, radiative processes include Bremsstrahlung and Cherenkov radiation. As discussed in Chapter 5, Bremsstrahlung is caused by the deflection of the incident particles in collisions with individual scattering centers, especially nuclei, in the target material. For low-energy particles Bremsstrahlung is small, but for very high energy particles the energy lost to Bremsstrahlung exceeds that lost to ionization and excitation of the target material, as shown by the elaborate calculations in Figure 7.20. For example, for electrons incident on a low-Z material such as

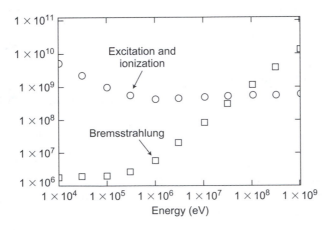

Figure 7.20 Stopping power of aluminum (eV/m).

aluminum, Bremsstrahlung is the dominant energy-loss mechanism for incident energies above 50 MeV. For electrons incident on a high-Z material such as lead, nuclear scattering is relatively more important and Bremsstrahlung remains the dominant energy-loss mechanism down to about 10 MeV. Although screening of the nuclear charge reduces the Bremsstrahlung in solid materials as discussed in Chapter 5, the basic mechanism of Bremsstrahlung can be described in terms of single-particle encounters and isn't discussed further here. Cherenkov radiation is produced by particles traveling at superluminal velocities [$v > c/n(\omega)$], but since the index of refraction $n(\omega)$ depends on the frequency ω, Cherenkov radiation appears only at those frequencies for which the velocity is superluminal. As an energy-loss mechanism, Cherenkov radiation is generally smaller than other processes. Nevertheless, the radiation is frequently useful and important in its own right, and is discussed in detail in Chapter 10. For now we focus our attention on electron excitation and ionization processes.

In the following we limit the discussion to the passage of electrons through matter and ignore energy loss by heavy particles. For electrons, the collisions with the nuclei are nearly elastic and contribute to the energy loss only through the production of Bremsstrahlung. But as mentioned earlier, we ignore Bremsstrahlung in the following. The electrons in the target are treated as a continuous medium, and the response of these electrons to the incident particle, including the ionization and excitation of the atoms, is conveniently described by the complex permittivity and permeability of the material. Since the energy loss due to excitation and ionization is dominated by high frequencies, we can evaluate the total energy loss from the known asymptotic behavior of the complex index of refraction or, equivalently, from the sum rule we derived earlier.

Since the theory involves the interaction of the field of the incident particle with the optical properties of the target material, it bears a resemblance to the method of virtual quanta discussed in Chapter 5. In the method of virtual quanta, the field of the incident particle is Fourier decomposed into waves called virtual quanta. As the incident particle passes an atom or other target, the excitation and ionization of the atom are found from the interaction of the virtual quanta with the target using the optical cross sections for the processes of interest. When the incident particle passes through condensed matter, the field of the particle (the virtual photons) is altered by the optical properties of the surrounding material, which gives rise to a "density effect." In the present approach, the field of the particle as it passes through the material is computed in a way that explicitly includes the effect of the complex index of refraction of the medium. Thus, the density effect appears naturally in the theory. The energy transmitted to the medium could, in principle, be computed by integrating over all the material affected by the fields of the incident particle. However, we are interested here only in the total energy lost by the incident particle, and we can compute this more simply by finding the field experienced by the incident particle as a result of the polarization that the particle induces in the medium. We then compute the energy lost by the particle due to its interaction with this polarization field. In most cases, the polarization of the medium is the polarization of the individual atoms and molecules induced by the incident particle. The energy lost by the incident particle is closely related to the excitation found by the method of virtual quanta, except that the density effect of the surrounding material is included. In other cases, such as the excitation of volume plasmons, a collective oscillation of the medium is excited, and this involves many electrons extending over many atoms. But as we have seen, collective oscillations of this type are also conveniently described in terms of the complex permittivity of the medium, making the present theory useful in this case as well.

In summary, then, the theory is developed as follows. As a particle with the charge q travels through a target material at the velocity $\mathbf{v}$, it polarizes the material in the region near the particle. This polarization creates an electric field $\mathbf{E}$ at the position of the particle, and the work

$$\frac{dW}{ds} = -\frac{q\mathbf{E} \cdot \mathbf{v}}{v} \tag{7.276}$$

that the particle does to move against this field is the energy lost by the particle in its interaction with the target.

We begin, therefore, by finding the field surrounding a fast particle as it moves through the target material. The Maxwell equations are

$$\nabla \cdot \mathbf{D} = q\delta(\mathbf{r} - \mathbf{v}t) \tag{7.277}$$

$$\nabla \cdot \mathbf{B} = 0 \tag{7.278}$$

$$\nabla \times \mathbf{E} + \frac{\partial \mathbf{B}}{\partial t} = 0 \tag{7.279}$$

and

$$\nabla \times \mathbf{H} - \frac{\partial \mathbf{D}}{\partial t} = q\mathbf{v}\delta(\mathbf{r} - \mathbf{v}t) \tag{7.280}$$

To solve these equations, we adopt the by now familiar approach of Fourier transforming everything in sight with respect to all the variables that can be found. Specifically, we define the Fourier transform of the function $f(\mathbf{r}, t)$ in space and time by

$$\tilde{f}_{\mathbf{k}}(\omega) = \frac{1}{(2\pi)^2} \int_{-\infty}^{\infty} dt \int_{-\infty}^{\infty} d^3\mathbf{r}\, e^{i(\omega t - \mathbf{k} \cdot \mathbf{r})} f(\mathbf{r}, t) \tag{7.281}$$

If we integrate first over space and then over time, we find that

$$\frac{1}{(2\pi)^2} \int_{-\infty}^{\infty} dt \int_{-\infty}^{\infty} d^3\mathbf{r}\, e^{i(\omega t - \mathbf{k} \cdot \mathbf{r})} \delta(\mathbf{r} - \mathbf{v}t) = \frac{1}{2\pi}\delta(\omega - \mathbf{k} \cdot \mathbf{v}) \tag{7.282}$$

The Maxwell equations then become

$$i\mathbf{k} \cdot \tilde{\mathbf{D}}_{\mathbf{k}} = \frac{q}{2\pi}\delta(\omega - \mathbf{k} \cdot \mathbf{v}) \tag{7.283}$$

$$i\mathbf{k} \cdot \tilde{\mathbf{B}}_{\mathbf{k}} = 0 \tag{7.284}$$

$$i\mathbf{k} \times \tilde{\mathbf{E}}_{\mathbf{k}} - i\omega\tilde{\mathbf{B}}_{\mathbf{k}} = 0 \tag{7.285}$$

$$i\mathbf{k} \times \tilde{\mathbf{H}}_{\mathbf{k}} + i\omega\tilde{\mathbf{D}}_{\mathbf{k}} = \frac{q\mathbf{v}}{2\pi}\delta(\omega - \mathbf{k} \cdot \mathbf{v}) \tag{7.286}$$

For a linear, dispersive medium the constitutive relations are

$$\tilde{\mathbf{D}}_{\mathbf{k}}(\omega) = \varepsilon(\omega)\tilde{\mathbf{E}}_{\mathbf{k}}(\omega) \tag{7.287}$$

$$\tilde{\mathbf{H}}_{\mathbf{k}}(\omega) = \frac{1}{\mu(\omega)}\tilde{\mathbf{B}}_{\mathbf{k}}(\omega) \tag{7.288}$$

so the Maxwell equations may finally be written

$$\varepsilon \mathbf{k} \cdot \tilde{\mathbf{E}}_\mathbf{k} = -i\frac{q}{2\pi}\delta(\omega - \mathbf{k} \cdot \mathbf{v}) \tag{7.289}$$

$$\mathbf{k} \cdot \tilde{\mathbf{B}}_\mathbf{k} = 0 \tag{7.290}$$

$$\mathbf{k} \times \tilde{\mathbf{E}}_\mathbf{k} = \omega\tilde{\mathbf{B}}_\mathbf{k} \tag{7.291}$$

and

$$\frac{1}{\mu}\mathbf{k} \times \tilde{\mathbf{B}}_\mathbf{k} + \omega\varepsilon\tilde{\mathbf{E}}_\mathbf{k} = -i\frac{q\mathbf{v}}{2\pi}\delta(\omega - \mathbf{k} \cdot \mathbf{v}) \tag{7.292}$$

We can now combine these equations to get a single equation for $\tilde{\mathbf{E}}_\mathbf{k}$. If we form the cross product of $\mathbf{k}$ with (7.291), expand the triple cross product on the left, and combine with (7.292), we get

$$\frac{1}{\mu\omega}[(\mathbf{k} \cdot \tilde{\mathbf{E}}_\mathbf{k})\mathbf{k} - k^2\tilde{\mathbf{E}}_\mathbf{k}] + \varepsilon\omega\tilde{\mathbf{E}}_\mathbf{k} = -i\frac{q\mathbf{v}}{2\pi}\delta(\omega - \mathbf{k} \cdot \mathbf{v}) \tag{7.293}$$

and if we now take the cross product of this with $\mathbf{k}$, we are left with

$$(k^2 - \mu\varepsilon\omega^2)\mathbf{k} \times \tilde{\mathbf{E}}_\mathbf{k} = i\frac{\mu\omega q}{2\pi}\mathbf{k} \times \mathbf{v}\delta(\omega - \mathbf{k} \cdot \mathbf{v}) \tag{7.294}$$

Finally, we can invert the Fourier transforms (7.289) and (7.294) with respect to ω to show that

$$\mathbf{k} \cdot \mathbf{E}_\mathbf{k}(t) = \frac{-iq}{(2\pi)^{3/2}}\frac{e^{-i\mathbf{k}\cdot\mathbf{v}t}}{\varepsilon(\mathbf{k} \cdot \mathbf{v})} \tag{7.295}$$

$$\mathbf{k} \times \mathbf{E}_\mathbf{k}(t) = \frac{iq}{(2\pi)^{3/2}}\frac{e^{-i\mathbf{k}\cdot\mathbf{v}t}}{\varepsilon(\mathbf{k} \cdot \mathbf{v})}\frac{n^2(\mathbf{k} \cdot \mathbf{v})(\mathbf{k} \cdot \mathbf{v})(\mathbf{k} \times \mathbf{v})}{k^2c^2 - n^2(\mathbf{k} \cdot \mathbf{v})(\mathbf{k} \cdot \mathbf{v})^2} \tag{7.296}$$

where the index of refraction is

$$n^2(\omega) = c^2\mu(\omega)\varepsilon(\omega) \tag{7.297}$$

These two equations provide the components of $\mathbf{E}_\mathbf{k}(t)$ in the directions parallel and perpendicular to the wave vector $\mathbf{k}$. We treat the nonrelativistic case first.

Comparing (7.295) and (7.296), we see that in the nonrelativistic limit the transverse component of the electric field is much smaller than the longitudinal component, so we can ignore it. We encounter the same behavior in the method of virtual quanta. In the nonrelativistic limit, the field of the particle is radial and nearly spherically symmetric, and the virtual quanta are essentially longitudinal waves. In the ultrarelativistic limit, on the other hand, the electric field flattens in the direction of motion due to Lorentz contraction and the virtual quanta become essentially transverse waves. But in the nonrelativistic limit, the electric field is essentially $\mathbf{E}_\mathbf{k} \approx (\mathbf{E}_\mathbf{k} \cdot \mathbf{k})\mathbf{k}/k^2$ and the total electric field in the material is

$$\mathbf{E} \approx \frac{-iq}{(2\pi)^3}\int_{-\infty}^{\infty}d^3\mathbf{k}\,\frac{\mathbf{k}}{k^2}\frac{e^{i\mathbf{k}\cdot(\mathbf{r}-\mathbf{v}t)}}{\varepsilon(\mathbf{k} \cdot \mathbf{v})} \tag{7.298}$$

At the position $\mathbf{r} = \mathbf{v}t$ of the particle, the exponent in (7.298) vanishes, so by symmetry we see that the total field lies along the direction of motion.

The total field (7.298) consists of two parts, these being the field of the incident particle and the field due to the polarization of the surrounding material. The field of the particle itself is given by the same expression (7.298), except that $\varepsilon(\mathbf{k} \cdot \mathbf{v})$ is replaced by ε_0. Since the self-field does no work on the particle, the energy lost by the incident particle per unit length, or stopping power, is

$$\frac{dW}{ds} = -\frac{q\mathbf{E} \cdot \mathbf{v}}{v} = \frac{iq^2}{(2\pi)^3 v} \int_{-\infty}^{\infty} d^3\mathbf{k} \, \frac{\mathbf{k} \cdot \mathbf{v}}{k^2} \left[\frac{1}{\varepsilon(\mathbf{k} \cdot \mathbf{v})} - \frac{1}{\varepsilon_0} \right] \tag{7.299}$$

Individually, the two terms of this integral diverge, since the self-field at the position of the particle diverges. However, the divergences occur at large k, where the difference is small. Therefore, we can evaluate the difference by evaluating the integrals inside some finite volume, such as the sphere $k < K$, and then letting $K \to \infty$. This still leaves a logarithmic divergence, but we will see how to deal with this shortly. Now, the term $\mathbf{k} \cdot \mathbf{v}/\varepsilon_0 k^2$ in the integrand is antisymmetric in the vector $\mathbf{k}$, so if we integrate over a symmetric volume, the integral of this term vanishes by symmetry. At the position of the particle, therefore, the contribution of the self-field vanishes in (7.299) and the rate of energy loss is given by the integral

$$\frac{dW}{ds} = -\frac{q\mathbf{E} \cdot \mathbf{v}}{v} = \frac{iq^2}{(2\pi)^3 v} \int_{-\infty}^{\infty} \frac{d^3\mathbf{k}}{k^2} \frac{\mathbf{k} \cdot \mathbf{v}}{\varepsilon(\mathbf{k} \cdot \mathbf{v})} \tag{7.300}$$

provided that we evaluate the integral correctly.

It is interesting to observe that while the work done on the incident particle by its self-field vanishes, the work done by the induced polarization does not. This is because the polarization is not an instantaneous effect. It lags behind the electric field due to the inertia and the damping of the electrons in the material, so the polarization field is asymmetric due to the motion of the particle. Because of this asymmetry, the induced electric field at the position of the incident particle doesn't entirely cancel out. The work done on the incident particle by the field computed from the complex polarizability is precisely the work done by the particle to excite and ionize the surrounding material.

Since the integrand in (7.300) is axially symmetric about the direction of motion, we change to the variables $\omega = k_z v$ along the direction of motion and k_t transverse to it, so that $k^2 = (\omega^2/v^2) + k_t^2$ and $d^3\mathbf{k} = 2\pi k_t dk_t (d\omega/v)$. The energy loss is then

$$\frac{dW}{ds} = \frac{iq^2}{4\pi^2} \int_{-\infty}^{\infty} \frac{\omega \, d\omega}{\varepsilon(\omega)} \int_0^{\infty} \frac{k_t \, dk_t}{\omega^2 + v^2 k_t^2} \tag{7.301}$$

Since the integral over k_t is logarithmically divergent, we truncate it at some limiting wave number k_0. We return to this shortly, but in the meantime (7.301) becomes

$$\frac{dW}{ds} = \frac{iq^2}{8\pi^2 v^2} \int_{-\infty}^{\infty} \frac{\omega \, d\omega}{\varepsilon(\omega)} \ln\left(1 + \frac{k_0^2 v^2}{\omega^2} \right) \tag{7.302}$$

Apart from the factor $1/\varepsilon(\omega)$, the integrand is antisymmetric in ω. But from the properties of $\varepsilon(\omega)$ we know that $\mathrm{Re}[1/\varepsilon(\omega)]$ is symmetric, while $\mathrm{Im}[1/\varepsilon(\omega)]$ is antisymmetric. We are therefore left with

$$\frac{dW}{ds} = \frac{-q^2}{4\pi^2 v^2} \int_0^{\infty} \omega \, d\omega \, \mathrm{Im}\left[\frac{1}{\varepsilon(\omega)} \right] \ln\left(1 + \frac{k_0^2 v^2}{\omega^2} \right) \tag{7.303}$$

As indicated by the logarithmic divergence at large k, the total rate of energy loss is dominated by high-energy "virtual quanta" of the field of the incident particle. For the most part, these high-energy quanta cause ionization of the atoms in the target material. The limiting wave number k_0 can therefore be estimated in the following way. Quantum mechanically, when a photon of wave number k is absorbed by an atom, it transfers to the atom the momentum $\hbar k$. But the momentum transferred cannot exceed the momentum $\gamma \beta mc$ of the incident electron, so the spectrum of the virtual quanta must be truncated at $\hbar k = \gamma \beta mc$. In this way we arrive at the criterion

$$\frac{k_0 v}{\omega} = \gamma \beta^2 \frac{mc^2}{\hbar \omega} \approx \frac{mv^2}{\hbar \omega} \qquad (7.304)$$

Because it enters only logarithmically, the exact value of k_0 is not critical. Then, since the logarithm is a slowly varying function of ω, we approximate it by some average value $\bar{\omega}$ and move it outside the integral. Moreover, the incident electron energy is generally large compared with the mean excitation energy $\hbar \bar{\omega}$, so $k_0 v / \bar{\omega} \gg 1$ and we can ignore the first term in the logarithm of (7.303).

To evaluate the remaining integral in (7.303) we take advantage of the asymptotic behavior of the permittivity $\varepsilon \xrightarrow[\omega \to \infty]{} \varepsilon_0$ to write

$$\mathrm{Im}\left(\frac{1}{\varepsilon}\right) \approx -\frac{\mathrm{Im}\,\varepsilon}{\varepsilon_0^2} \qquad (7.305)$$

We can then use the sum rule (7.168) to get

$$\frac{dW}{ds} \approx \frac{q^2}{2\pi^2 \varepsilon_0^2 v^2} \ln\left(\frac{k_0 v}{\bar{\omega}}\right) \int_0^\infty \omega \, d\omega \, \mathrm{Im}\,\varepsilon(\omega) = \frac{n_e q^4}{4\pi \varepsilon_0^2 mv^2} \ln\left(\frac{mv^2}{\hbar \bar{\omega}}\right) \qquad (7.306)$$

where n_e is the electron density in the target and $\bar{\omega}$ is some average value of the frequency. Since the photo-ionization cross section generally peaks somewhere not too far above threshold, we expect $\hbar \bar{\omega}$ to be of the order of some mean ionization energy for the electrons in the target material. The effective value of $\hbar \bar{\omega}$ is shown in Figure 7.21 as a function of the atomic number of the target material. It increases with Z, as would be expected. The simple

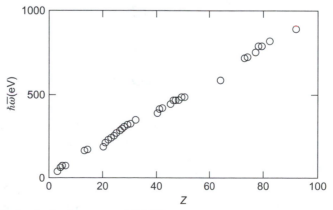

Figure 7.21 Effective ionization energy, $\hbar \bar{\omega}$ (eV).

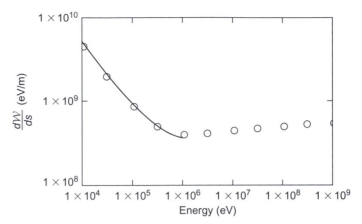

Figure 7.22 Collisional energy loss in aluminum.

result (7.306) is compared with more elaborate calculations (circles) for aluminum in Figure 7.22. The value $\hbar\bar{\omega} \approx 166$ eV (see Figure 7.21) is used in all the calculations shown. The agreement is remarkably good.

EXERCISE 7.18

For volume plasmons, the dielectric susceptibility is

$$\chi_e(\omega) = \frac{-\omega_p^2 \tau^2}{\omega\tau(\omega\tau + i)} \tag{7.307}$$

where ω_p is the plasma frequency and τ is the electron collision time. For many metals $\omega_p\tau \gg 1$, so the plasmon resonance is quite narrow.

(a) Beginning with (7.303), show that the rate of energy loss due to excitation of plasma waves is

$$\frac{dW}{ds} \approx \frac{q^2\omega_p^2}{4\pi\varepsilon_0 v^2} \ln\left(\frac{k_0 v}{\omega_p}\right) \tag{7.308}$$

In this case, we are describing the excitation of a collective effect involving many electrons, and the macroscopic permittivity ceases to have a clear meaning for lengths that are not large compared with the atomic spacing $a = O(n^{-1/3})$, where n is the number density of the atoms in the metal. We therefore truncate the integral over transverse wave numbers at

$$k_0 = n^{1/3} \tag{7.309}$$

It is interesting to note that while the energy-loss formula (7.303) depends explicitly on the imaginary part of the permittivity, which depends in (7.307) on the value of the collision time, the final result (7.308) is independent of τ. That is, the excitation of volume plasmons does not depend on dissipative processes in the medium.

(b) For a 30-k eV electron passing through a 10-μm-thick aluminum foil ($\hbar\omega_p = 15.3$ eV), approximately how many plasmons are excited?

7.3.2 Relativistic Limit and the Density Effect

At relativistic velocities, the transverse component of $\mathbf{E_k}$ becomes comparable to the longitudinal component and it must be included in the calculation of the total energy loss. In fact, at the zeros of the denominator of (7.296) (in the complex $\omega = \mathbf{k} \cdot \mathbf{v}$ plane) the expression for the transverse component of the field diverges. For a transparent material [real values of the index of refraction $n(\omega) = c\sqrt{\mu(\omega)\varepsilon(\omega)}$], this corresponds to the condition

$$k^2 c^2 = n^2 (\mathbf{k} \cdot \mathbf{v})^2 \tag{7.310}$$

That is, the component of the velocity in the direction of propagation is equal to the phase velocity of the wave. This is just the condition for creating Cherenkov radiation, which we discuss in detail in Chapter 10. For now, however, we focus on the ionization processes. These are dominated by high-frequency virtual quanta, $\hbar\omega \geq O(100 \, \text{eV})$ for all but the lightest elements, as indicated in Figure 7.21. Since the permittivity has the asymptotic behavior

$$\varepsilon(\omega) \xrightarrow[\omega \to \infty]{} \varepsilon_0 \left(1 - \frac{\omega_p^2}{\omega^2} \right) < \varepsilon_0 \tag{7.311}$$

where $\omega_p^2 = (n_e q^2 / \varepsilon_0 m)$ is the plasma frequency and n_e the electron density, the index of refraction is less than unity at frequencies above the plasma frequency. For most substances $\hbar\omega_p = O(10 \, \text{eV}) \ll \hbar\bar{\omega}$, so the incident electron velocity is not superluminal in the frequency region where most of the ionization takes place.

To find the total rate of energy loss when both the longitudinal and transverse components of the field are important, we expand the triple cross product $\mathbf{k} \times (\mathbf{k} \times \mathbf{E_k})$, rearrange the result, and take the dot product with $\mathbf{v}$. The total rate at which the particle loses energy to the field component $\mathbf{E_k}$ is then

$$-q\mathbf{E_k} \cdot \mathbf{v} = -\frac{q}{k^2} \left\{ (\mathbf{E_k} \cdot \mathbf{k})(\mathbf{k} \cdot \mathbf{v}) - [\mathbf{k} \times (\mathbf{k} \times \mathbf{E_k})] \cdot \mathbf{v} \right\} \tag{7.312}$$

where $\mathbf{k} \cdot \mathbf{E_k}$ and $\mathbf{k} \times \mathbf{E_k}$ are given by (7.295) and (7.296). Integrating this over all $\mathbf{k}$ gives the total energy loss. After a bit of algebra, we find that

$$\frac{dW}{ds} = \frac{iq^2}{(2\pi)^3 v} \int_{-\infty}^{\infty} d^3k \, \frac{\mathbf{k} \cdot \mathbf{v}}{\varepsilon(\mathbf{k} \cdot \mathbf{v})} \frac{c^2 - n^2 (\mathbf{k} \cdot \mathbf{v}) v^2}{k^2 c^2 - n^2 (\mathbf{k} \cdot \mathbf{v})(\mathbf{k} \cdot \mathbf{v})^2} \tag{7.313}$$

As before, the contribution due to the self-field of the electron vanishes by symmetry, so this expression represents the total energy loss due to the interaction of the electron with the target material. In the limit $v/c \ll 1$, we recover (7.300). Since the integrand is axially symmetric about the direction of motion, we change to the variables ω and k_t, as in the non-relativistic case, and get

$$\frac{dW}{ds} = \frac{iq^2}{(2\pi)^2} \int_0^{k_0} k_t \, dk_t \int_{-\infty}^{\infty} \omega \, d\omega \frac{1}{\varepsilon(\omega)} \frac{\frac{1}{v^2} - \frac{n^2(\omega)}{c^2}}{k_t^2 + \omega^2 \left[\frac{1}{v^2} - \frac{n^2(\omega)}{c^2} \right]} \tag{7.314}$$

Since the integral over k_t diverges logarithmically, we truncate it at k_0, as we did in the nonrelativistic limit.

This time we begin by evaluating first the integral over ω, using contour integration in the complex ω plane. To do this, we close the contour in the upper half plane as illustrated

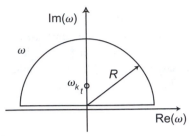

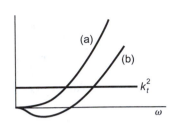

Figure 7.23 Contour for computing the stopping power.

Figure 7.24 Graphical solution for the location of the poles.

in Figure 7.23. As shown previously, the functions $\varepsilon(\omega)$ and $\mu(\omega)$ have no poles in the upper half plane, so the only poles are those at the zeros of the denominator, that is, at the solutions of the equation

$$g(\omega) = k_t^2 + \omega^2 \left[\frac{1}{v^2} - \frac{n^2(\omega)}{c^2} \right] = 0 \qquad (7.315)$$

In fact, there is usually only one such pole. To see this we begin by noting that the permittivity $\varepsilon(\omega)$ is real for $\omega = i\omega''$ on the imaginary axis, since Im $\varepsilon(\omega)$ is antisymmetric about the imaginary axis and therefore vanishes on the imaginary axis. The same is true of $\mu(\omega)$, and therefore of $n^2(\omega)$, so the poles occur on the imaginary axis for real k_t. Furthermore, we see from the Kramers–Kronig relation (7.44) that on the imaginary axis the dielectric susceptibility is

$$\chi_e(i\omega_0'') = \frac{2}{\pi} \int_0^\infty \frac{\omega \, \text{Im} \, \chi_e(\omega)}{\omega^2 + \omega_0''^2} \, d\omega \qquad (7.316)$$

where ω_0'' is real. The principal part of the integral is not required, since the denominator is nonvanishing along the path of integration (the positive real axis). Since Im $\varepsilon(\omega)$ is positive, we see from (7.316) that the susceptibility $\chi_e(\omega)$, and therefore the permittivity $\varepsilon(\omega)$, are decreasing functions of the frequency ω_0''. Similar remarks apply to the permeability $\mu(\omega)$, so $n^2(i\omega'')$ is a monotonically decreasing function of ω'' along the positive imaginary axis. Therefore, the function $\omega''^2\{(1/v^2) - [n^2(i\omega'')/c^2]\}$ increases monotonically from zero to infinity, and crosses the value k_t^2 just once for $k_t^2 > 0$. This is illustrated in Figure 7.24, where we see that for low velocities $v < c/n(0)$ [curve (a)] there is never more than one solution. Note, however, that for $v > c/n(0)$ [the case of Cherenkov radiation, curve (b)] there is a second solution ($\omega = 0$) for $k_t = 0$.

The integral we require is then

$$\int_{-\infty}^\infty \omega \, d\omega \, \frac{1}{\varepsilon(\omega)} \frac{\dfrac{1}{v^2} - \dfrac{n^2(\omega)}{c^2}}{k_t^2 + \omega^2 \left[\dfrac{1}{v^2} - \dfrac{n^2(\omega)}{c^2} \right]}$$

$$= \left[\oint d\omega - \int_R d\omega \right] \frac{\omega}{\varepsilon(\omega)} \frac{\dfrac{1}{v^2} - \dfrac{n^2(\omega)}{c^2}}{k_t^2 + \omega^2 \left[\dfrac{1}{v^2} - \dfrac{n^2(\omega)}{c^2} \right]} \qquad (7.317)$$

where $\int_R d\omega$ is the integral around the semicircle at the radius R. In the limit $R \to \infty$, this integral becomes

$$\int_R d\omega \frac{\omega}{\varepsilon(\omega)} \frac{\dfrac{1}{v^2} - \dfrac{n^2(\omega)}{c^2}}{k_t^2 + \omega^2 \left[\dfrac{1}{v^2} - \dfrac{n^2(\omega)}{c^2}\right]} \xrightarrow{R \to \infty} \frac{1}{\varepsilon_0} \int_R \frac{d\omega}{\omega} = \frac{i\pi}{\varepsilon_0} \qquad (7.318)$$

The first term in (7.317), the integral around the entire contour, is just $i2\pi$ times the residue at the pole ω_{k_t}, where ω_{k_t} is the solution of (7.315) corresponding to the transverse wave number k_t. The residue of a function having the form $f(\omega)/g(\omega)$ at a zero of the denominator $g(\omega)$ is just $f(\omega_{k_t})/[dg(\omega_{k_t})/d\omega]$, so we find that

$$\oint d\omega \frac{\omega}{\varepsilon(\omega)} \frac{\dfrac{1}{v^2} - \dfrac{n^2(\omega)}{c^2}}{k_t^2 + \omega^2 \left[\dfrac{1}{v^2} - \dfrac{n^2(\omega)}{c^2}\right]} = i2\pi \frac{\omega_{k_t}}{\varepsilon(\omega_{k_t})} \frac{\dfrac{1}{v^2} - \dfrac{n^2(\omega_{k_t})}{c^2}}{\dfrac{dg(\omega_{k_t})}{d\omega}} \qquad (7.319)$$

But from (7.315) we see that

$$\frac{1}{v^2} - \frac{n^2(\omega_{k_t})}{c^2} = -\frac{k_t^2}{\omega_{k_t}^2} \qquad (7.320)$$

and if we differentiate (7.315), we see that

$$2k_t + \frac{d}{d\omega}\left\{\omega_{k_t}^2 \left[\frac{1}{v^2} - \frac{n^2(\omega_{k_t})}{c^2}\right]\right\} \frac{d\omega_{k_t}}{dk_t} = 0 = 2k_t + \frac{dg(\omega_{k_t})}{d\omega} \frac{d\omega_{k_t}}{dk_t} \qquad (7.321)$$

Substituting these results back into (7.317) we get

$$\int_{-\infty}^{\infty} \omega \, d\omega \frac{1}{\varepsilon(\omega)} \frac{\dfrac{1}{v^2} - \dfrac{n^2(\omega)}{c^2}}{k_t^2 + \omega^2 \left[\dfrac{1}{v^2} - \dfrac{n^2(\omega)}{c^2}\right]} = i\pi \left[\frac{1}{\varepsilon(\omega_{k_t})} \frac{k_t}{\omega_{k_t}} \frac{d\omega_{k_t}}{dk_t} - \frac{1}{\varepsilon_0}\right] \qquad (7.322)$$

so the energy loss becomes

$$\frac{dW}{ds} = \frac{q^2}{4\pi\varepsilon_0} \int_0^{k_0} k_t \, dk_t - \frac{q^2}{4\pi} \int_0^{k_0} \frac{d\omega_{k_t}}{dk_t} \frac{k_t^2 \, dk_t}{\omega_{k_t}\varepsilon(\omega_{k_t})}$$

$$= \frac{q^2 k_0^2}{8\pi\varepsilon_0} - \frac{q^2}{4\pi} \int_{\omega_{k_t}(0)}^{\omega_{k_t}(k_0)} \frac{\omega_{k_t} \, d\omega_{k_t}}{\varepsilon(\omega_{k_t})} \left[\frac{1}{v^2} - \frac{n^2(\omega_{k_t})}{c^2}\right] \qquad (7.323)$$

To evaluate the remaining integral we assume that magnetic effects can be ignored, so that $\mu(\omega_{k_t}) = \mu_0$. Then the stopping power due to excitation and ionization becomes

$$\frac{dW}{ds} = \frac{q^2 k_0^2}{8\pi\varepsilon_0} - \frac{q^2}{4\pi} \int_{\omega_{k_t}(0)}^{\omega_{k_t}(k_0)} \frac{\omega_{k_t} \, d\omega_{k_t}}{\varepsilon(\omega_{k_t})} \left[\frac{1}{v^2} - \mu_0\varepsilon(\omega_{k_t})\right]$$

$$= \frac{q^2 k_0^2}{8\pi\varepsilon_0} - \frac{q^2}{4\pi\varepsilon_0 v^2} \int_{\omega_{k_t}(0)}^{\omega_{k_t}(k_0)} \omega_{k_t} \, d\omega_{k_t} \left[\frac{\varepsilon_0}{\varepsilon(\omega_{k_t})} - 1\right] + \frac{q^2}{4\pi\varepsilon_0\gamma^2 v^2} \int_{\omega_{k_t}(0)}^{\omega_{k_t}(k_0)} \omega_{k_t} \, d\omega_{k_t}$$

$$\qquad (7.324)$$

where we have used $(1/v^2) - (1/c^2) = 1/\gamma^2 v^2$.

To evaluate the upper limit of the integrals $\omega_{k_t}(k_0)$, we observe that large k_0 corresponds to large ω_{k_t}, so we can use the asymptotic expansion (7.311) for the permittivity. Then from (7.315) we get

$$\omega_{k_t}^2(k_0) = -\gamma^2 v^2 \left(k_0^2 + \frac{n_e q^2}{\varepsilon_0 m c^2} \right) \tag{7.325}$$

At the lower limit, corresponding to $k_t = 0$, we distinguish two cases. In the first case, corresponding to low velocities $v < c/n(0)$, the lower limit is $\omega_{k_t}(0) = 0$ [curve (a) in Figure 7.24]. In this case the energy loss is

$$\frac{dW}{ds} = \frac{q^2}{4\pi\varepsilon_0 v^2} \int_0^{\gamma v k_0} \omega'' \, d\omega'' \left[\frac{\varepsilon_0}{\varepsilon(i\omega'')} - 1 \right] - \frac{n_e q^4}{8\pi\varepsilon_0^2 m c^2} \tag{7.326}$$

where we have changed to the real variable $\omega'' = \omega_{k_t}/i$ and kept only the leading term of (7.325) in the upper limit of the integral. To evaluate the remaining integral we assume, as we did in the nonrelativistic case, that the integral is dominated by the high-energy virtual quanta. In this region $\varepsilon(\omega) \xrightarrow[\omega \to \infty]{} \varepsilon_0$, so

$$\frac{\varepsilon_0}{\varepsilon(\omega)} - 1 \approx -\chi_e(\omega) \approx -\frac{n_e q^2}{\varepsilon_0 m \omega^2} \tag{7.327}$$

Substituting this into (7.326) and using (7.304) for k_0, we get

$$\frac{dW}{ds} = \frac{n_e q^4}{4\pi\varepsilon_0^2 m c^2 \beta^2} \left[\ln\left(\frac{\gamma^2 \beta^2 m c^2}{\hbar\langle\omega\rangle} \right) - \frac{\beta^2}{2} \right] \tag{7.328}$$

where we have truncated the lower limit of the integral at $\langle\omega\rangle$ to avoid the logarithmic singularity introduced by the high-frequency approximation (7.327). In the limit $v/c \ll 1$, we recover the nonrelativistic result (7.306). The final result (7.328) is compared with the nonrelativistic approximation (7.306) as well as more elaborate calculations (circles) in Figure 7.25.

In the second case, $v > c/n(0)$, the particle is superluminal at low frequencies. This is the condition for Cherenkov radiation at these frequencies, although the production of

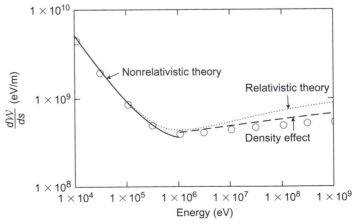

Figure 7.25 Collisional energy loss and density effect in aluminum.

Cherenkov radiation is not the important effect here. This is clear from the fact that the effect of this change is to reduce the energy loss. Rather, the important effect is the change in the equivalent photon distribution around the incident particle, which gives rise to the density effect, as described earlier. The lower limit of the integrals is $\omega_{k_t}(0) = i\omega_0''$ [curve (b) in Figure 7.24], where

$$n(i\omega_0'') = c/v \tag{7.329}$$

If this frequency is below the mean ionization frequency $\bar{\omega}$ we have to use the larger frequency. At ultrarelativistic velocities, however, the asymptotic expression (7.311) can be used for the permittivity, and we find that for $\gamma \gg 1$ the lower limit is

$$\omega_{k_t}^2(0) = -\gamma^2\omega_p^2 \tag{7.330}$$

At sufficiently high energy $\gamma\omega_p$ always exceeds $\bar{\omega}$, and the stopping power due to excitation and ionization is then

$$\frac{dW}{ds} = \frac{n_e q^4}{4\pi\varepsilon_0^2 mc^2}\left[\ln\left(\frac{\gamma mc^2}{\hbar\omega_p}\right) - \frac{1}{2}\right] \tag{7.331}$$

for $\gamma \gg 1$. Since the ultrarelativistic limit is dominated by high-energy equivalent photons, the atomic structure of the target material is unimportant. Only the total number density of electrons enters the result (through n_e and ω_p). This result is compared with more elaborate calculations in Figure 7.25. The difference between (7.331) and (7.328) is called the density effect. As shown in Figure 7.25, the density effect reduces the collisional energy loss at the highest energies. When this is taken into account, the agreement with the more eleborate calculations is quite good.

It should be reiterated before leaving this topic that the principal energy-loss mechanism for ultrarelativistic electrons passing through matter is not the ionization and excitation we have just calculated. Rather, as illustrated in Figure 7.20, it is Bremsstrahlung, due largely to scattering of the electrons by the nuclei. The contribution of excitation and ionization to the stopping power in the ultrarelativistic energy range is important mostly for the fact that this contribution is directly responsible for exciting the target material (to expose a photographic emulsion, for example) and for heating the material locally. The contribution of Bremsstrahlung to exciting and heating the material is important only to the extent that the Bremsstrahlung is absorbed locally by the medium.

BIBLIOGRAPHY

The interaction of electromagnetic fields with linear dispersive materials is discussed by
> J. D. Jackson, *Classical Electrodynamics,* 3rd edition, John Wiley & Sons, New York (1999),

and in great depth by
> L. D. Landau, E. M. Lifshitz, and L. B. Pitaevskii, *Electrodynamics of Continuous Media,* Pergamon Press, Oxford (1984).

The excitation of crystalline solids is discussed by
> J. N. Hodgson, *Optical Absorption and Dispersion in Solids,* Chapman & Hall, London (1970),
> H. Raether, *Excitation of Plasmons and Interband Transitions,* Springer-Verlag, Berlin (1980).

For an introduction to the quantum theory of solids, including the free-electron theory of metals and the band theory of solids, the reader is referred to

N. W. Ashcroft and N. D. Merman, *Solid State Physics,* Holt, Rinehart, and Winston, New York (1976),

C. K. Kittel, *Introduction to Solid State Physics,* 7th edition, John Wiley & Sons, New York (1996).

Useful data on the optical constants of solids can be found in

E. D. Palik, *Handbook of Optical Constants of Solids,* Academic Press, New York (1985).

Extensive computations of stopping power are discussed in

Stopping Powers for Electrons and Positrons, ICRU Report 37, International Commission on Radiation Units and Measurements, Washington, DC (1984).

8 Nonlinear Optics

The dielectric and magnetic response of macroscopic materials to applied electromagnetic fields is almost always close to linear. Ferroelectricity and ferromagnetism are egregious exceptions to this rule, but other examples of nonlinearity have been known for over a century. They include the Kerr effect, the Faraday effect, and a number of other electro-optic and magneto-optic effects. They are characterized by the fact that while they are exquisitely small effects on the microscopic scale, they are easily distinguished in the laboratory because they produce an observable effect that would otherwise be completely absent. The Kerr effect, for example, converts linearly polarized light to elliptical polarization in an otherwise anisotropic liquid medium. It is these "almost linear" effects in which we are interested here.

It is not surprising to find that the response of a material is linear in the field when the applied field is small compared with the fields that naturally occur in atoms and molecules. For example, the electric field at the first Bohr orbit of a hydrogen atom is $E_0 \approx 5 \times 10^{11}$ V/m, which is large compared with the fields with which we ordinarily deal in the laboratory. However, fields of this magnitude can be achieved at the tips of very fine needles, but more commonly and importantly nowadays are found near the focus of very intense laser beams. Modern high-power lasers (so-called "chirped-pulse amplifiers") produce pulses of light with a peak power of the order of 10^{13} W at a wavelength near 1 μm. Focused with an f/10 optical system to a spot 10 μm across, the beam has a peak intensity close to 10^{23} W/m^2. The electric field near the focus is on the order of 6×10^{12} V/m, which exceeds the electric field at the first Bohr orbit in a hydrogen atom by more than an order of magnitude. It's no wonder that nonlinear effects are important in such intense beams. But in actual fact, the very specific nature of the effects caused by nonlinear polarization (such as second-harmonic generation) makes them easy to detect even at much smaller fields. This, together with the cumulative effects of coherent interactions, makes it possible to observe nonlinear effects in optical fields many orders of magnitude smaller than atomic fields. We therefore use the small-field approximation for everything we do in this chapter, and expand the polarization as a power series in the amplitude of the field.

Prior to the development of lasers it was possible to observe nonlinear effects by using a dc electric or magnetic field to break the symmetry of an otherwise isotropic substance such as a liquid or a cubic crystal. This induces optical activity in the medium that can be readily observed and identified. For example, a strong electric field in a liquid such as nitrobenzene causes the molecules to be oriented preferentially in the direction of the electric field. The index of refraction of the material is then larger for light polarized parallel to the electric field. Light waves propagating through the liquid in this anisotropic state have their polarization altered in an observable fashion, as discovered by Kerr in 1876. In particular, light that is linearly polarized oblique to the dc field is converted to elliptically polarized light, since the component of polarization parallel to the dc field is retarded relative to the component polarized perpendicular to the applied field. The Kerr effect can be used to make a fast (microsecond), electrically operated optical shutter by placing the liquid between two crossed polarizers. Clearly, the Kerr effect is quadratic in the electric field, since the anisotropy induced in the liquid depends on the direction but not the sign of the electric field. A linear effect can be observed in certain crystals that lack inversion symmetry, so the sign of the electric field matters. This was first observed by Roentgen in quartz in 1883, but nowadays it is named for Pockels, who studied it extensively beginning in 1894. The Pockels effect, like the Kerr effect, can be used with crossed polarizers to form a fast electro-optic shutter called a Pockels cell. Pockels cells are commonly used to select short pulses from high-power laser systems.

Similarly, a magnetic field can be used to break the symmetry of a liquid or solid medium and create what are called magneto-optical effects. In the simplest case, the optical field is propagated parallel to the magnetic field. Due to the Larmor precession of the molecules in the magnetic field, the right-hand and left-hand circularly polarized components of a light wave at a frequency ω interact with the rotating molecules at the effective frequencies $\omega \pm \omega_L$, where ω_L is the Larmor frequency. As a result of dispersion, then, the different polarization components have slightly different indices of refraction, and this difference can be used to rotate the direction of polarization of linearly polarized light. The linear magneto-optic effect was discovered in 1845 by Faraday, after whom it is named. The quadratic effect was discovered by Kerr in 1901, but it is named for Cotton and Mouton, who studied it extensively beginning in 1905. However, magneto-optic effects caused by the optical fields themselves are generally small, and in the following discussion of nonlinear optics we ignore them.

It was not possible to observe nonlinear effects induced entirely by optical fields prior to the development of lasers. A simple comparison of laser and conventional (thermal) sources is enough to show why. Because of the importance of coherence, the useful figure of merit for comparing light sources is the spectral brilliance (energy $d\mathcal{W}$ per unit area dA, per unit solid angle $d\Omega$, per unit relative frequency interval $d\omega/\omega$, per unit time dt):

$$B_\omega = \frac{\omega \, d^4\mathcal{W}}{dA \, d\Omega \, d\omega \, dt} \tag{8.1}$$

In terms of the energy density $\mathcal{U}$, the spectral brilliance of a blackbody spectrum (see Chapter 4) is

$$B_\omega = \frac{\omega c}{4\pi} \frac{d\mathcal{U}}{d\omega} = \left[\frac{\hbar^4 \omega^4}{k_B^4 T^4} \frac{1}{\exp(\hbar\omega/k_B T)} \right] \frac{k_B^4 T^4}{4\pi^3 \hbar^3 c^2} \tag{8.2}$$

where k_B is Boltzmann's constant, T the temperature, $\hbar$ Planck's constant divided by 2π, and c the speed of light. The dimensionless quantity in square brackets has its maximum value 4.780 at $\hbar\omega/k_B T = 3.921$, so the peak spectral brilliance is

$$B_\omega = 4.780 \frac{k_B^4 T^4}{4\pi^3 \hbar^3 c^2} \tag{8.3}$$

At a temperature of 10^4 K, the peak spectral brilliance is $B_\omega \approx 1 \times 10^8$ W/m^2-steradian. On the other hand, the spectral brilliance of a laser beam in the lowest Gaussian mode (see Chapter 9) is

$$B_\omega = \frac{8\pi c}{\lambda^3} \frac{\mathcal{W}}{\Delta\omega\,\Delta t} \tag{8.4}$$

where $\mathcal{W}$ is the energy, λ the wavelength, $\Delta\omega$ the bandwidth, and Δt the pulse length of the laser beam. But for a Fourier transform limited pulse, $\Delta\omega\,\Delta t = O(1)$, so the peak spectral brilliance is

$$B_\omega = O\left(\frac{8\pi c}{\lambda^3} \mathcal{W}\right) \tag{8.5}$$

For a 1-J laser pulse at a wavelength $\lambda = 1$ µm the spectral brilliance is on the order of 8×10^{27} W/m^2-steradian. Compared with the spectral brilliance of a blackbody source, this is a difference of almost twenty orders of magnitude. Lasers are amazing devices.

As a result, the subject of nonlinear optics developed rapidly as soon as lasers were introduced. The first laser (a ruby laser) was built by Maiman in 1960. The first observation of a nonlinear optical effect (specifically, second-harmonic generation) was reported only a year later by Franken, who used a 1-J ruby laser. In the years since then, a wealth of nonlinear optical phenomena has been observed. Some of these phenomena, such as second-harmonic generation and parametric amplification, are now of great importance both scientifically and technically. Others, such as self-focusing, are often detrimental, causing distortion of laser beams and damage to optical components.

Broadly speaking, nonlinear optical effects may be divided into three categories, these being multiphoton effects (sum- and difference-frequency generation), single-frequency effects (nonlinear index of refraction), and Raman processes. At a fundamental level these phenomena are all related, of course, and much of the mathematical description is common to all nonlinear phenomena. Nevertheless, the distinction is useful pedagogically, and we discuss the three categories separately.

A simple example of sum-frequency generation is provided by the harmonics observed in nonlinear Thomson scattering. As discussed in Chapter 4, at sufficiently high intensity the motions of a charged particle in an optical field become relativistic and the scattered light contains harmonics of the incident light. These harmonics are generally weak except at enormous incident intensity. However, the scattered light can be substantially enhanced by coherent effects. As shown in Chapter 10, coherent scattering from multiple particles increases the radiation by the factor N compared with incoherent scattering, where N is roughly the number of particles in a cubic wavelength. As shown later, coherent effects can be produced by particles that are much farther apart if the coherence is maintained over a volume larger than a wavelength. Specifically, if the harmonic radiation travels at the same phase velocity as the incident (fundamental) radiation, then the radiation is coherent over a much larger volume and the intensity at the harmonic increases enormously. This is called phase matching. In a quantum-mechanical picture, sum-frequency or harmonic generation is viewed as the addition of two (or more) photons to generate a single photon of higher energy. Phase matching in this picture is the condition that the momentum of the photons, as well as the energy, be conserved. Thus, we have the simultaneous conditions

$$\hbar\omega_3 = \hbar\omega_1 + \hbar\omega_2 \tag{8.6}$$

and

$$\hbar\mathbf{k}_3 = \hbar\mathbf{k}_1 + \hbar\mathbf{k}_2 \tag{8.7}$$

for a process in which photons at the frequencies ω_1 (wave vector $\mathbf{k}_1$) and ω_2 ($\mathbf{k}_2$) combine to form a photon at the frequency ω_3 ($\mathbf{k}_3$). Phase matching in condensed matter requires great care, since the index of refraction of the material is ordinarily different at the fundamental and the harmonics because of dispersion.

The second category of nonlinear effects is the nonlinear index of refraction. The simplest example of an effect caused by a nonlinear index of refraction is the self-focusing of an intense wave passing through a plasma. As discussed in Chapter 4, for optical frequencies above the plasma frequency, relativistic effects decrease the phase velocity of a wave in a plasma at high intensity. This is equivalent to increasing the index of refraction, so an optical beam that is more intense in the center sees a higher index of refraction there, which focuses the beam as though it were passing through a positive lens. Since the focused beam has an even higher intensity than the unfocused beam, the effect becomes even stronger, and we see that self-focusing is an unstable (and frequently damaging) process. Other examples of index-of-refraction effects include phase-conjugate reflection, in which the rays of light are reflected directly back on themselves rather than specularly as from a mirror, and the formation and propagation of solitons. Solitons are pulses in which nonlinear effects conspire to cancel the effects of dispersion so that the pulse does not spread as it propagates. This makes solitons useful for high-speed data transmission over long optical fibers. Finally, it is worth noting that two-photon (and more generally multiphoton) absorption can occur at high intensity. This process can be viewed as a nonlinear enhancement of the imaginary part of the index of refraction.

The last category of nonlinear optical effects that we discuss here is Raman processes. In the Raman effect, an incident frequency ω combines with a natural resonant frequency ω_r of the material (typically a molecular rotation or vibration) to form sidebands at the frequencies $\omega \pm \omega_r$. In nonlinear Raman processes, phase matching occurs naturally, so the scattered radiation grows coherently. Ordinarily, coherent radiation at the lower frequency $\omega - \omega_r$ (called the Stokes line) is generated most strongly. Nonlinear Brillouin scattering (scattering involving phonon vibrations) is related to nonlinear Raman scattering and is important for many applications. Unfortunately, we don't have time to discuss it here.

We begin with a discussion of the basic properties of nonlinear polarization in the limit of (relatively) weak electric fields, and then discuss, in turn, multiphoton, single-frequency, and Raman processes.

8.1 NONLINEAR SUSCEPTIBILITY

8.1.1 Nonlinear Polarization

We have seen in Chapter 6 that the polarization of a material is caused by the microscopic displacement of the positive and negative charges in the material relative to one another. This displacement can result from distortion of the electron trajectories within the atoms and molecules or from the vibration and rotation of the molecules of which the material is composed. For small electric and magnetic fields the polarization is linear in the amplitude of the fields, but for larger fields the response of the material becomes nonlinear. We have seen, for example, that the response of free electrons in a plasma is nonlinear when relativistic effects are important, but relativistic effects can be ignored for bound electrons. This is evident from the fact that for electrons moving at near the speed of light, the amplitude of the oscillations is comparable to the wavelength of the incident optical field. Even for ultraviolet radiation the wavelength is large compared with atomic dimensions, and at higher frequencies (x-rays) we can no longer describe the material as a continuous medium. Therefore, other (nonrelativistic) effects dominate the nonlinear polarization of dielectric optical materials. For a liquid composed of polar molecules, the dielectric response is nonlinear when the energy of the dipoles in the field is comparable to the thermal energy. Nonpolar liquids exhibit a nonlinearity that results from the field-induced orientation of molecules that have a larger polarizability along one axes, as in the Kerr effect. In crystalline materials the dominant source of nonlinear effects is the nonlinear potential in which the charges (electrons and ions) move, and we begin our discussion with this type of nonlinearity.

Although the polarizability of a real material is correctly described only by quantum mechanics, we adopt here a classical model similar to the Lorentz–Drude model introduced in Chapter 7. In this model the charges are described as damped harmonic oscillators. To account for the nonlinearity, we add quadratic or higher-order terms to the restoring force. Provided we are not required actually to calculate the precise values of the optical constants, we can understand the physics of the polarization dynamics satisfactorily and more simply using this classical picture. The results obtained in this way are at least qualitatively correct, and the exact values of the optical constants must in any event be determined experimentally.

To simplify matters further, for the moment, we use a one-dimensional model to describe the dynamics of the polarization of the material in an electric field. It is easy to introduce the anisotropic features of a real crystal at the end. Following the arguments of Chapter 7, then, we see that the polarization P obeys the equation of motion

$$\frac{d^2 P}{dt^2} + \gamma \frac{dP}{dt} + \omega_r^2 P + K^{(2)} P^2 = \eta E \tag{8.8}$$

where ω_r is the resonant frequency and γ the damping coefficient of the oscillator, and $K^{(2)} P^2$ represents the effect of a nonlinear (quadratic) contribution to the restoring force. The parameter η is

$$\eta = \frac{Nq^2}{m} = \varepsilon_0 \omega_p^2 \tag{8.9}$$

where q is the charge, m the mass, and N the number density of the oscillators, and $\omega_p = \sqrt{Nq^2/\varepsilon_0 m}$ is the plasma frequency corresponding to the density of the oscillating charges.

Since we are interested in the limit when the electric field is small, we solve (8.8) by a perturbation technique. To carry out the perturbation expansion we treat η as a small parameter to indicate the weakness of the coupling to the electric field $E(t)$. At the end we set η equal to its correct value. In the limit when η is very small, we expand the solution as a power series in η having the form

$$P(t) = \eta P^{(1)}(t) + \eta^2 P^{(2)}(t) + \cdots \tag{8.10}$$

The Fourier transform of the polarization is then

$$\tilde{P}(\omega) = \eta \tilde{P}^{(1)}(\omega) + \eta^2 \tilde{P}^{(2)}(\omega) + \cdots \tag{8.11}$$

To compute the polarization, we substitute (8.10) into (8.8) and collect terms of like powers of η. To first order in η, we obtain the equation

$$\frac{d^2 P^{(1)}}{dt^2} + \gamma \frac{d P^{(1)}}{dt} + \omega_r^2 P^{(1)} = E \tag{8.12}$$

If we take the Fourier transform of this equation using the rule $d/dt \to -i\omega$, we find that the first-order polarization is

$$\eta \tilde{P}^{(1)}(\omega) = \varepsilon_0 \chi^{(1)}(\omega) \tilde{E}(\omega) \tag{8.13}$$

where the linear susceptibility is

$$\chi^{(1)}(\omega) = \frac{\eta}{\varepsilon_0} \frac{1}{\omega_r^2 - \omega^2 - i\gamma\omega} \tag{8.14}$$

(since the material is presumed to be nonmagnetic, we ignore the subscript e for the dielectric susceptibility).

To second order in the small parameter η, we obtain the equation

$$\frac{d^2 P^{(2)}}{dt^2} + \gamma \frac{d P^{(2)}}{dt} + \omega_r^2 P^{(2)} + K^{(2)} P^{(1)2} = 0 \tag{8.15}$$

If we take the Fourier transform of this and use (8.13) we find that

$$\left(\omega_r^2 - \omega^2 - i\gamma\omega\right) \tilde{P}^{(2)}(\omega) = -\frac{K^{(2)}}{\sqrt{2\pi}} \int_{-\infty}^{\infty} dt \, e^{i\omega t} P^{(1)2}(t)$$

$$= -\frac{\varepsilon_0^2 K^{(2)}}{(2\pi)^{3/2}\eta^2} \int_{-\infty}^{\infty} dt \int_{-\infty}^{\infty} d\omega' \int_{-\infty}^{\infty} d\omega'' e^{i(\omega-\omega'-\omega'')t} \chi^{(1)}(\omega')\chi^{(1)}(\omega'')\tilde{E}(\omega')\tilde{E}(\omega'') \tag{8.16}$$

But the integral over t produces the δ-function $2\pi\delta(\omega - \omega' - \omega'')$. Using (8.14), then, we find that the nonlinear polarization is

$$\eta^2 \tilde{P}^{(2)}(\omega) = \varepsilon_0 \int_{-\infty}^{\infty} d\omega' \int_{-\infty}^{\infty} d\omega'' \chi^{(2)}(\omega', \omega'')\tilde{E}(\omega')\tilde{E}(\omega'')\delta(\omega - \omega' - \omega'') \tag{8.17}$$

where the second-order susceptibility is

$$\chi^{(2)}(\omega', \omega'') = -\frac{\varepsilon_0^2 K^{(2)}}{\sqrt{2\pi}\eta} \chi^{(1)}(\omega' + \omega'')\chi^{(1)}(\omega')\chi^{(1)}(\omega'') \tag{8.18}$$

We see from (8.18) that the second-order susceptibility involves the product of three first-order susceptibilities. We therefore expect materials that have a large first-order susceptibility to have a relatively large second-order susceptibility as well, allowing, of course, for differences in the nonlinear coefficient $\eta^2 K^{(2)}$. In fact, this is observed to be approximately true experimentally and is called Miller's rule. Note that while the linear susceptibility $\chi^{(1)}(\omega)$, as defined by (8.14), is dimensionless, the second-order susceptibility $\chi^{(2)}(\omega', \omega'')$, as defined by (8.18), has the dimensions m/V in SI units.

At this point we could continue the perturbation expansion to higher orders and find the terms in the polarization proportional to the third and higher powers of the electric field. However, these higher-order terms are generally not important and we ignore them here. Even without a third-order nonlinearity, higher-order effects such as third-harmonic generation appear through the combination of fields generated at lower order in the perturbation.

The presence of the δ-function in (8.17) has some interesting consequences. It shows that if the field has components at the frequencies ω' and ω'', then the polarization has a component at the frequency $\omega = \omega' + \omega''$. In fact, since the electric field $\mathbf{E}(t)$ is real, its Fourier transform has the symmetry property

$$\tilde{\mathbf{E}}(-\omega) = \tilde{\mathbf{E}}^*(\omega^*) \tag{8.19}$$

in the complex frequency plane. For real frequencies, this shows that the field has Fourier components at the frequencies $\pm\omega$. The nonlinear polarization (8.17) therefore has components at all the sum and difference frequencies

$$\omega = \pm\omega' \pm \omega'' \tag{8.20}$$

In particular, if the (real) electric field has a component at only one positive (real) frequency, call it $\bar{\omega}$, then its Fourier transform has a component at $-\bar{\omega}$ and the Fourier transform of the polarization has components at $\omega = 0$, $\omega = \pm\bar{\omega}$ and $\omega = \pm 2\bar{\omega}$. The appearance of a dc component of the polarization in response to an optical field is called optical rectification. Physically it appears because the potential in a quadratic anharmonic oscillator is asymmetrical, as shown in Figure 8.1. Therefore, as the energy of the oscillator increases, its mean position is displaced from equilibrium, and this leads to a dc component of the polarization. The origin of the component at $2\bar{\omega}$ lies in the fact that when an anharmonic oscillator is driven by a sinusoidal force, the resulting motion is periodic but not sinusoidal. However, the periodic motion can be resolved into a Fourier series, and the various Fourier components of the motion of the oscillator radiate at the corresponding harmonic frequencies. With a restoring force consisting only of linear and quadratic terms only the second harmonic appears at lowest order in the perturbation, although higher harmonics can appear from combinations of lower harmonics as the lower harmonics grow sufficiently intense.

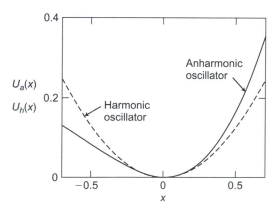

Figure 8.1 Potential energy of a quadratic anharmonic oscillator.

EXERCISE 8.1

Use (8.18) to estimate the second-order nonlinear susceptibility of a typical nonlinear optical material by arguing that when the charge on the anharmonic oscillator is displaced by an atomic dimension $d = 1/\sqrt[3]{N}$, where N is the density of oscillating charges, the nonlinear contribution to the restoring force should be of the same order of magnitude as the linear contribution. Show that the nonlinear susceptibility is then on the order of

$$\chi^{(2)} \approx \frac{\varepsilon_0 d^2}{\sqrt{2\pi} q} \frac{\omega_r^2}{\omega_p^2} (n^2 - 1)^3 \qquad (8.21)$$

where n is the index of refraction. For LiNbO$_3$, a commonly used nonlinear optical material, the index of refraction is $n \approx 2.3$. The material is transparent in the near infrared and visible but absorbs light in the ultraviolet, so we would estimate that the resonant frequency ω_r and the plasma frequency ω_p both lie in the ultraviolet and have comparable magnitudes. The valence electron density of common materials is typically on the order of $N \approx 10^{30}/m^3$. How does your estimate compare with the value $\chi^{(2)} \approx 10^{-11}$ m/V, which is roughly the average value of the nonlinear susceptibility for different crystal orientations of LiNbO$_3$?

8.1.2 Anisotropic Materials

In the real world, of course, the polarization and the electric field are three-dimensional vector quantities, and the one-dimensional equation of motion (8.8) must be replaced by a tensor expression that reflects the spatial symmetry of the material. In fact, a quadratic nonlinearity of the form (8.8) is not possible for an isotropic material or, for that matter, any material that is invariant under spatial inversion, since for a scalar $K^{(2)}$ (or any tensor $K^{(2)}_{i;jk}$ that is invariant under inversion) the quadratic term in (8.8) is invariant under inversion whereas the other terms change sign. Anisotropic materials are important in nonlinear optics for several reasons. In the first place, the second-order nonlinearity $K^{(2)}$ (but not the third-order nonlinearity $K^{(3)}$) vanishes identically in isotropic materials, as just mentioned, so anisotropy is a necessary condition for second-order processes to occur. In the second place, the birefringence that results from an anisotropic susceptibility can be used to compensate for dispersion and achieve phase matching. That is, the phase velocity of waves at different frequencies can be equalized by using polarization differences to offset dispersion.

The treatment of anisotropic materials is similar to the one-dimensional treatment discussed in the previous section. Using Cartesian tensor notation, for weak electric fields we can expand the Fourier transform of the polarization in a power series of the form

$$\tilde{P}_i(\omega) = \tilde{P}_i^{(1)}(\omega) + \tilde{P}_i^{(2)}(\omega) + \cdots \qquad (8.22)$$

Following the arguments of the previous section, we find that the linear and quadratic terms in the nonlinear polarization of an anisotropic material are sums over the contributions from components of the field in different directions of the form

$$\tilde{P}_i^{(1)} = \varepsilon_0 \sum_j \chi_{i;j}^{(1)} \tilde{E}_j \qquad (8.23)$$

$$\tilde{P}_i^{(2)}(\omega) = \varepsilon_0 \sum_{j,k} \int_{-\infty}^{\infty} d\omega' \int_{-\infty}^{\infty} d\omega'' \chi_{i;jk}^{(2)}(\omega', \omega'') \tilde{E}_j(\omega') \tilde{E}_k(\omega'') \delta(\omega - \omega' - \omega'') \qquad (8.24)$$

Although the first-order and second-order susceptibilities depend on several subscripts and several arguments, their complexity is reduced by symmetry requirements. We have already seen, for example, that the second-order nonlinearity vanishes in materials that possess inversion symmetry. Other spatial symmetry properties specific to certain crystalline materials can be used to reduce the number of independent tensor components of the susceptibilities, but this is outside the scope of the present discussion. Rather we focus our attention on some general symmetry properties.

Since the electric field $\mathbf{E}(t)$ and the polarization $\mathbf{P}(t)$ are real quantities, it follows from (8.23) and (8.24) that

$$\chi_{i;j}^{(1)*}(\omega) = \chi_{i;j}^{(1)}(-\omega^*) \tag{8.25}$$

$$\chi_{i;jk}^{(2)*}(\omega', \omega'') = \chi_{i;jk}^{(2)}(-\omega'^*, -\omega''^*) \tag{8.26}$$

and from the symmetry of (8.24) it follows that

$$\chi_{i;jk}^{(2)}(\omega', \omega'') = \chi_{i;kj}^{(2)}(\omega'', \omega') \tag{8.27}$$

That is, we can permute the indices j and k if we permute the frequencies ω' and ω'' with them. This is called intrinsic permutation symmetry. However, this symmetry does not, in general, extend to permutations with the first subscript even if we recognize that the first subscript is by definition associated with the sum frequency $\omega' + \omega''$. The semicolon is included in the subscript to emphasize this point. Actually, we see shortly that permutations of the first index are possible with some restrictions. Expressions analogous to (8.26) and (8.27) apply to higher-order susceptibilities.

Additional symmetries are possible when the effects of absorption and dispersion can be ignored. As observed in Chapter 6, time reversal symmetry is equivalent to the absence of dissipation. If the polarization response is invariant under time reversal we have

$$\chi_{i;j}^{(1)}(\omega) = \chi_{i;j}^{(1)}(-\omega) \tag{8.28}$$

$$\chi_{i;jk}^{(2)}(\omega', \omega'') = \chi_{i;jk}^{(2)}(-\omega', -\omega'') \tag{8.29}$$

Combining these with (8.25) and (8.26) we see that for real ω, ω', and ω'',

$$\chi_{i;j}^{(1)*}(\omega) = \chi_{i;j}^{(1)}(\omega) \tag{8.30}$$

and

$$\chi_{i;jk}^{(2)*}(\omega', \omega'') = \chi_{i;jk}^{(2)}(\omega', \omega'') \tag{8.31}$$

That is, for nondissipative materials the susceptibility is real for ω, ω', and ω'' on the real axis. In fact, we have already seen in Chapter 7 that the linear susceptibility is real (for real ω) in the absence of dissipation.

Additional permutation symmetries can be found by examining the flow of energy in nondissipative media. As discussed in Chapter 6, Poynting's theorem in a macroscopic material states that

$$\mathbf{E} \cdot \frac{\partial \mathbf{D}}{\partial t} + \mathbf{H} \cdot \frac{\partial \mathbf{B}}{\partial t} + \nabla \cdot (\mathbf{E} \times \mathbf{H}) + \mathbf{E} \cdot \mathbf{J} = 0 \tag{8.32}$$

where the Poynting vector $\mathbf{S} = \mathbf{E} \times \mathbf{H}$ represents the flow of electromagnetic energy and the term $\mathbf{E} \cdot \mathbf{J}$ in (8.32) describes the dissipation of energy by macroscopic currents

(Joule heating). In a nonconducting material we can ignore this last term. We now consider the passage of a finite optical pulse through an arbitrary volume V bounded by a surface S. If we integrate (8.32) over this volume from $t = -\infty$ to $t = \infty$ and use the divergence theorem, we get

$$\int_{-\infty}^{\infty} dt \int_V dV \left(\mathbf{E} \cdot \frac{\partial \mathbf{D}}{\partial t} + \mathbf{H} \cdot \frac{\partial \mathbf{B}}{\partial t} \right) + \int_{-\infty}^{\infty} dt \oint_S dS \, \hat{\mathbf{n}} \cdot (\mathbf{E} \times \mathbf{H}) = 0 \qquad (8.33)$$

where $\hat{\mathbf{n}}$ is a unit vector normal to the surface S. But in a nondissipative material the net flow of energy into or out of the volume must vanish after the pulse is completely gone, so the surface integral must vanish. In addition, for nonmagnetic materials we may write

$$\mathbf{H} = \frac{\mathbf{B}}{\mu_0} \qquad (8.34)$$

$$\mathbf{D} = \varepsilon_0 \mathbf{E} + \mathbf{P} \qquad (8.35)$$

so that

$$\int_V dV \int_{-\infty}^{\infty} \left(\varepsilon_0 \mathbf{E} \cdot \frac{\partial \mathbf{E}}{\partial t} + \mathbf{E} \cdot \frac{\partial \mathbf{P}}{\partial t} + \frac{\mathbf{B}}{\mu_0} \cdot \frac{\partial \mathbf{B}}{\partial t} \right) dt = 0 \qquad (8.36)$$

Since this is true for an arbitrary volume V, the integrand (that is, the integral over t) must vanish everywhere. But the first and last terms are perfect differentials, so they vanish when they are integrated over the duration of the pulse. We are therefore left with the result

$$\int_{-\infty}^{\infty} \mathbf{E} \cdot \frac{\partial \mathbf{P}}{\partial t} \, dt = 0 \qquad (8.37)$$

We can convert this to an integral over real ω in the usual way, and we find that

$$\int_{-\infty}^{\infty} \tilde{\mathbf{E}}^*(\omega) \cdot \tilde{\mathbf{P}}(\omega) \omega \, d\omega = 0 \qquad (8.38)$$

in a nondissipative material. The additional factor of ω comes from evaluating the Fourier transform of the time derivative in (8.37).

We can evaluate the Fourier transform of the polarization using (8.22) together with (8.23) and (8.24). When we do so, we find that

$$\sum_{i,j} \int_{-\infty}^{\infty} \omega \, d\omega \, \chi_{i;j}^{(1)}(\omega) \tilde{E}_i^*(\omega) \tilde{E}_j(\omega)$$

$$+ \sum_{i,j,k} \int_{-\infty}^{\infty} \omega \, d\omega \int_{-\infty}^{\infty} d\omega' \int_{-\infty}^{\infty} d\omega'' \chi_{i;jk}^{(2)}(\omega', \omega'') \tilde{E}_i^*(\omega) \tilde{E}_j(\omega') \tilde{E}_k(\omega'') \delta(\omega - \omega' - \omega'')$$

$$= 0 \qquad (8.39)$$

Since this is true for arbitrary $\tilde{E}_i(\omega)$, the two terms must vanish independently. But the electric fields are real and satisfy the symmetry property (8.19). Substituting this into (8.39), we obtain the separate results

$$\sum_{i,j} \int_{-\infty}^{\infty} \omega \, d\omega \, \chi_{i;j}^{(1)}(\omega) \tilde{E}_i(-\omega) \tilde{E}_j(\omega) = 0 \qquad (8.40)$$

$$\sum_{i,j,k} \int_{-\infty}^{\infty} \omega \, d\omega \int_{-\infty}^{\infty} d\omega' \int_{-\infty}^{\infty} d\omega'' \chi_{i;jk}^{(2)}(\omega', \omega'') \tilde{E}_i(-\omega) \tilde{E}_j(\omega') \tilde{E}_k(\omega'') \delta(\omega - \omega' - \omega'') = 0$$

$$(8.41)$$

To derive a permutation relation for the linear susceptibility we substitute the time-reversal symmetry relation (8.28) into (8.40) and make the change of variable $\omega \rightarrow -\omega$ to get

$$\sum_{i,j} \int_{-\infty}^{\infty} \omega \, d\omega \, \chi_{i;j}^{(1)}(\omega) \tilde{E}_i(\omega) \tilde{E}_j(-\omega) = 0 \tag{8.42}$$

If we exchange the dummy indices, i and j, and compare the result with (8.40), we obtain the permutation relation

$$\chi_{i;j}^{(1)}(\omega) = \chi_{j;i}^{(1)}(\omega) \tag{8.43}$$

That is, the order of the indices does not matter for nondissipative materials. Since the linear susceptibility tensor $\chi_{i;j}^{(1)}$ is symmetric for nondissipative materials, it is diagonal in some coordinate system. The axes of this coordinate system are called the principal axes of the material. Since it is diagonal in this coordinate system, the linear susceptibility tensor of a nondissipative material has only three independent components.

We can proceed in a similar fashion to derive a permutation relation for the second-order susceptibility. If we substitute the time-reversal symmetry relation (8.29) into (8.41) and make the changes $\omega' \rightarrow -\omega'$, $\omega'' \rightarrow -\omega''$, we get

$$\sum_{i,j,k} \int_{-\infty}^{\infty} \omega \, d\omega \int_{-\infty}^{\infty} d\omega' \int_{-\infty}^{\infty} d\omega'' \chi_{i;jk}^{(2)}(\omega', \omega'') \, \delta(\omega + \omega' + \omega'')$$

$$\times \tilde{E}_i(-\omega) \tilde{E}_j(-\omega') \tilde{E}_k(-\omega'') = 0 \tag{8.44}$$

For convenience we define the alternative form of the susceptibility

$$\psi_{i;jk}^{(2)}(\omega, \omega', \omega'') = \chi_{i;jk}^{(2)}(\omega', \omega'') \, \delta(\omega + \omega' + \omega'') \tag{8.45}$$

which is a function of three frequencies. Then (8.44) can be expressed

$$\sum_{i,j,k} \int_{-\infty}^{\infty} \omega \, d\omega \int_{-\infty}^{\infty} d\omega' \int_{-\infty}^{\infty} d\omega'' \psi_{i;jk}^{(2)}(\omega, \omega', \omega'') \tilde{E}_i(-\omega) \tilde{E}_j(-\omega') \tilde{E}_k(-\omega'') = 0 \tag{8.46}$$

Since the product $\tilde{E}_i(-\omega) \tilde{E}_j(-\omega') \tilde{E}_k(-\omega'')$ is invariant if the indices are permuted with the arguments, the function $\psi_{i;jk}^{(2)}(\omega, \omega', \omega'')$ must have the same property. From this we obtain the symmetry relations

$$\psi_{i;jk}^{(2)}(\omega, \omega', \omega'') = \psi_{j;ik}^{(2)}(\omega', \omega, \omega'') = \psi_{i;kj}^{(2)}(\omega, \omega'', \omega') \tag{8.47}$$

and so on. But the δ-function vanishes unless $\omega = -(\omega' + \omega'')$, so in terms of the original definition of the susceptibility these symmetry relations may be expressed

$$\chi_{i;jk}^{(2)}(\omega', \omega'') = \chi_{j;ik}^{(2)}(-\omega' - \omega'', \omega'') = \chi_{i;kj}^{(2)}(\omega'', \omega') \tag{8.48}$$

and so on. This is called full permutation symmetry, and is valid for nondissipative materials. Note that ω' and ω'' may be positive or negative as we are using them here.

In the case when dispersion is small, we can ignore the dependence of the susceptibility on the frequency. In this case the symmetry relations (8.48) are simply

$$\chi_{i;jk}^{(2)} = \chi_{j;ik}^{(2)} = \chi_{i;kj}^{(2)} \tag{8.49}$$

That is, all permutations of the indices are equivalent. Simply put, this says that the tendency of optical fields polarized in the directions j and k to produce a field polarized in the direction i is the same as the tendency of fields polarized in the i and j directions to produce a field polarized in the direction k, and so on. The relation (8.49) is called Kleinman symmetry, and it is valid when both absorption and dispersion can be ignored. At sufficiently low frequencies (typically well below the resonant frequencies and relaxation rates of the nonlinear material) dissipation and dispersion vanish. In this limit we expect Kleinman symmetry to apply quite generally. In general, the second-rank tensor $\chi^{(2)}_{i;jk}$ has 27 independent components. However, Kleinman symmetry reduces the total number of independent components to ten, these being

$$\chi^{(2)}_{1;11}, \ \chi^{(2)}_{2;22}, \ \chi^{(2)}_{3;33} \qquad \text{(three independent components)} \qquad (8.50)$$

$$\chi^{(2)}_{1;22}, \ \chi^{(2)}_{1;33}, \ \chi^{(2)}_{2;11}, \ \chi^{(2)}_{2;33}, \ \chi^{(2)}_{3;11}, \ \chi^{(2)}_{3;22} \qquad \text{(six independent components)} \qquad (8.51)$$

$$\chi^{(2)}_{1;23} \qquad \text{(one independent component)} \qquad (8.52)$$

All the other components can be found from these by permuting the indices.

A nonlinear medium consisting of anharmonic oscillators provides a good example of these general symmetries. For an anisotropic material, the tensor form of the equation of motion (8.8) is

$$\frac{d^2 P_i}{dt^2} + \sum_j \gamma_{i;j} \frac{d P_j}{dt} + \sum_j \omega^2_{i;j} P_j + \sum_{j,k} K^{(2)}_{i;jk} P_j P_k = \eta E_i \qquad (8.53)$$

where the sums are over $j, k, \ldots = 1 \ldots 3$. The off-diagonal coefficients $\gamma_{i;j}$, $\omega^2_{i;j}$ and $K^{(2)}_{i;jk}$ describe the how polarization in the i direction is coupled to polarization in the j and k directions. To solve the tensor equation of motion (8.53) we proceed as in the one-dimensional case. We assume a solution of the form

$$P_i(t) = \eta P^{(1)}_i(t) + \eta^2 P^{(2)}_i(t) + \cdots \qquad (8.54)$$

which has the Fourier transform

$$\tilde{P}_i(\omega) = \eta \tilde{P}^{(1)}_i(\omega) + \eta^2 \tilde{P}^{(2)}_i(\omega) + \cdots \qquad (8.55)$$

If we substitute (8.54) into the equation of motion and collect like powers of η, to lowest order we obtain the equation

$$\frac{d^2 P^{(1)}_i}{dt^2} + \sum_j \gamma_{i;j} \frac{d P^{(1)}_j}{dt} + \sum_j \omega^2_{i;j} P^{(1)}_j = E^{(1)}_i \qquad (8.56)$$

The Fourier transform of this equation is

$$\sum_j \kappa_{i;j} \tilde{P}^{(1)}_j = \tilde{E}_i \qquad (8.57)$$

where

$$\kappa_{i;j} = \omega^2_{i;j} - i\omega\gamma_{i;j} - \omega^2 \delta_{ij} \qquad (8.58)$$

This has the solution

$$\eta \tilde{P}^{(1)}_i = \varepsilon_0 \sum_j \chi^{(1)}_{i;j} \tilde{E}_j \qquad (8.59)$$

where the linear susceptibility tensor

$$\chi^{(1)}_{i;j} = \frac{\eta}{\varepsilon_0} \kappa^{-1}_{i;j} \qquad (8.60)$$

is the inverse of the tensor $\kappa_{i;j}$, that is,

$$\varepsilon_0 \sum_j \chi_{i;j}^{(1)} \kappa_{j;k} = \eta \delta_{ik} \tag{8.61}$$

Continuing the mathematical arguments as before, we find that the quadratic term in the polarization is

$$\eta^2 \tilde{P}_i^{(2)}(\omega) = \varepsilon_0 \sum_{j,k} \int_{-\infty}^{\infty} d\omega' \int_{-\infty}^{\infty} d\omega'' \chi_{i;jk}^{(2)}(\omega',\omega'') \tilde{E}_j(\omega') \tilde{E}_k(\omega'') \delta(\omega - \omega' - \omega'') \tag{8.62}$$

where the second-order susceptibility tensor is

$$\chi_{i;jk}^{(2)}(\omega',\omega'') = -\sum_{p,q,r} \frac{\varepsilon_0^2 K_{p;qr}^{(2)}}{\sqrt{2\pi}\,\eta} \chi_{i;p}^{(1)}(\omega'+\omega'') \chi_{q;j}^{(1)}(\omega') \chi_{r;k}^{(1)}(\omega'') \tag{8.63}$$

While the anharmonic oscillator model cannot be used to give quantitative results for the susceptibility, it has the correct qualitative behavior for the electronic contribution to both the linear and nonlinear susceptibility, as we have already seen for linear media in Chapter 7. In particular, it illustrates the symmetries discussed earlier. For example, we note that since the tensor $\kappa_{i;j}$ satisfies the reality condition (8.25), the linear susceptibility $\chi_{i;j}^{(1)}$ does as well. It follows from (8.63) and the reality of $K_{p;qr}^{(2)}$, then, that the nonlinear susceptibility $\chi_{i;jk}^{(2)}$ satisfies the reality condition (8.26). We observe further that the equation of motion (8.53) is invariant under time reversal when the damping factors $\gamma_{i;j}$ vanish, and in this case we see explicitly from (8.58), (8.60), and (8.63) that the susceptibilities $\chi_{i;j}^{(1)}$ and $\chi_{i;jk}^{(2)}$ are real. We also see from (8.63) that the nonlinear susceptibility explicitly exhibits the intrinsic permutation symmetry (8.27), but is not symmetric under permutations of the first index unless the linear susceptibility exhibits the symmetry $\chi_{i;j}^{(1)} = \chi_{j;i}^{(1)}$. However, we know from (8.43) that in the absence of dissipation, $\chi_{i;j}^{(1)}(\omega) = \chi_{j;i}^{(1)}(\omega)$. In this case, then, the nonlinear susceptibility (8.63) explicitly exhibits full permutation symmetry. Finally, we note that in the limit $\omega \to 0$ dissipation always vanishes, so $\chi_{i;j}^{(1)}(\omega) = \chi_{j;i}^{(1)}(\omega)$. This assures that $\kappa_{i;j}$ is symmetric, which implies in turn that the matrix $\omega_{i;j}$ is symmetric, that is, that $\omega_{i;j} = \omega_{j;i}$. Since this matrix is symmetric, there always exists a coordinate system in which $\omega_{i;j}$ is diagonal. In this coordinate system, called the principal axes of the material, motions along one axis are not coupled to motions along the other axes except, possibly, through velocity-dependent ($\gamma_{i;j}$) terms at higher frequencies.

Before leaving the properties of the dielectric susceptibility it should be pointed out that the notation and conventions used here are not universal, and the conventions used by various authors can be confusing. Most of the literature on nonlinear optics uses cgs units, in which the displacement is $\mathbf{D} = \mathbf{E} + 4\pi\mathbf{P}$, so the susceptibility (both linear and nonlinear) departs from that used in SI units by the factor $1/4\pi\varepsilon_0$. In addition most, but not all, authors define the susceptibility at the frequency ω in terms of the amplitude $\mathbf{P}$ of the polarization produced by an oscillating electric field of amplitude $\mathbf{E}$, rather than using the Fourier amplitude $\tilde{\mathbf{P}} = \sqrt{\frac{\pi}{2}}\mathbf{P}$ and $\tilde{\mathbf{E}} = \sqrt{\frac{\pi}{2}}\mathbf{E}$ as done here. For the linear susceptibility this does not matter, since in the expression $\mathbf{P} = \varepsilon_0 \chi^{(1)}\mathbf{E}$ both sides are linear in the amplitude. However, for higher-order susceptibilities the right-hand side is proportional to E^n, so the nth-order susceptibilities are related by the equivalence

$$\chi^{(n)} \text{ (wave amplitude)} = \left(\frac{\pi}{2}\right)^{(n-1)/2} \chi^{(n)} \text{ (Fourier amplitude)} \tag{8.64}$$

To confuse matters further, in much of the literature the amplitude of the electric field is defined to be $2\mathbf{E}$, and the amplitude of the polarization is $2\mathbf{P}$.

In addition, for tensor materials that possess the symmetry property $\chi^{(2)}_{i;jk} = \chi^{(2)}_{i;kj}$ the contracted notation d_{il} is widely used in place of $\chi^{(2)}_{i;jk}$. In this notation the subscript $l = 1 \ldots 6$ on d_{il} does not refer to individual Cartesian axes, but rather to certain combinations of axes. Specifically, the convention is

$$\chi^{(2)}_{i;11} = 2d_{i1} \tag{8.65}$$

$$\chi^{(2)}_{i;22} = 2d_{i2} \tag{8.66}$$

$$\chi^{(2)}_{i;33} = 2d_{i3} \tag{8.67}$$

$$\chi^{(2)}_{i;23} = \chi^{(2)}_{i;32} = 2d_{i4} \tag{8.68}$$

$$\chi^{(2)}_{i;31} = \chi^{(2)}_{i;13} = 2d_{i5} \tag{8.69}$$

$$\chi^{(2)}_{i;12} = \chi^{(2)}_{i;21} = 2d_{i6} \tag{8.70}$$

The symmetry $\chi^{(2)}_{i;jk} = \chi^{(2)}_{i;kj}$ is always valid for second-harmonic generation, by intrinsic perturbation symmetry, since $\omega_j = \omega_k$. It is also valid when $\chi^{(2)}_{i;jk}$ satisfies the Kleinman symmetry condition (8.49). In general d_{ij} has eighteen components. However, when Kleinman symmetry is valid, the number of independent elements of d_{ij}, like those of $\chi^{(2)}_{i;jk}$, is reduced to ten.

EXERCISE 8.2

We have seen that a second-order nonlinearity is possible only in materials that are not symmetric under inversion. However, third-order nonlinearity is possible even in isotropic materials. Show that in a one-dimensional material consisting of damped harmonic oscillators, the third-order susceptibility is related to the linear susceptibility by

$$\chi^{(3)}(\omega', \omega'', \omega''') = -\frac{\varepsilon_0^3 K^{(3)}}{2\pi\eta} \chi^{(1)}(\omega' + \omega'' + \omega''')\chi^{(1)}(\omega')\chi^{(1)}(\omega'')\chi^{(1)}(\omega''') \tag{8.71}$$

EXERCISE 8.3

Prove that the nonlinear coefficient $K^{(2)}_{i;jk}$ is symmetric, that is,

$$K^{(2)}_{i;jk} = K^{(2)}_{j;ik} = K^{(2)}_{i;kj} \tag{8.72}$$

and so on.

EXERCISE 8.4

Shortly after he observed second-harmonic generation, Franken was able to detect optical rectification. He did this by focusing a ruby laser into a sample of KDP configured as the dielectric in a parallel-plate capacitor and measuring the voltage. We can now estimate the size of the effect he saw.

(a) For a cw incident wave with the Fourier transform

$$\tilde{E}_i(\omega) = \bar{E}_i \delta(\omega - \bar{\omega}) + \bar{E}_i^* \delta(\omega + \bar{\omega}) \tag{8.73}$$

show that the Fourier transform of the nonlinear polarization is

$$\eta^2 \tilde{P}_i^{(2)}(\omega) = \varepsilon_0 \sum_{j,k} \left\{ \chi_{i;jk}^{(2)}(\bar{\omega}, \bar{\omega}) \bar{E}_j \bar{E}_k \delta(\omega - 2\bar{\omega}) + \chi_{i;jk}^{(2)*}(\bar{\omega}, \bar{\omega}) \bar{E}_j^* \bar{E}_k^* \delta(\omega + 2\bar{\omega}) \right.$$
$$\left. + \left[\chi_{i;jk}^{(2)}(\bar{\omega}, -\bar{\omega}) + \chi_{i;jk}^{(2)*}(\bar{\omega}, -\bar{\omega}) \right] \bar{E}_j \bar{E}_k^* \delta(\omega) \right\} \tag{8.74}$$

Hint: As discussed in Chapter 1,

$$\int_{-\infty}^{\infty} \delta(x - a)\delta(x - b)\, dx = \delta(a - b) \tag{8.75}$$

(b) Invert the Fourier transform to show that the dc component of the polarization is

$$\eta^2 \langle P_i \rangle = \sqrt{2\pi}\, \varepsilon_0 \sum_{j,k} \mathrm{Re}\, \chi_{i;jk}^{(2)}(\bar{\omega}, -\bar{\omega}) \langle E_j E_k \rangle \tag{8.76}$$

where $\langle \,\rangle$ indicates the average over one cycle.

(c) If the laser intensity and the medium are roughly uniform, the dc component of the electric field is $\langle E \rangle = -\eta^2 \langle P \rangle / \varepsilon_0$. For an incident laser intensity

$$\langle S \rangle = \varepsilon_0 n c \langle E^2 \rangle = 10^{12} \ \mathrm{W/m^2} \tag{8.77}$$

where $n = 1.5$ is the refractive index, and a nonlinear susceptibility $\chi^{(2)} = 10^{12}$ m/V, estimate the magnitude of the dc electric field. What is the amplitude of the optical electric field?

8.2 MULTIPHOTON PROCESSES

8.2.1 Coupled-Wave Equation

As an optical wave at the frequency ω propagates through a nonlinear material, it induces a polarization in the material at the original frequency ω and at the harmonics 2ω, 3ω, and so on. The polarization at the original frequency generates an optical field that is coherent with the incident wave and adds to it. Typically, this slows down the wave and we describe the effect by the refractive index of the material. When the induced polarization includes harmonics of the original frequency, it generates optical fields at the corresponding harmonics. These fields, in turn, propagate along with the original optical field either coherently or incoherently. When the propagation of the harmonics is coherent with the original field, the harmonics are strongly reinforced by the harmonic components of the polarization and can grow quite intense. In some cases, the intensity of the harmonic fields can become a significant fraction of that of the incident fundamental field. In this case, the incident field is depleted and converted to harmonic radiation.

If the incident field consists of two waves at the frequencies ω' and ω'', the induced polarization contains components at the two fundamentals and their harmonics, as before, and at the sum and difference frequencies $\omega = \pm\omega' \pm \omega''$, as indicated by (8.17). In principle, all these frequencies could grow in strength as the waves propagate, but experimentally the effects of dispersion can generally be compensated at only one frequency, so only one harmonic or one sum frequency remains coherent with the fundamental and grows significantly. The others can be ignored, which simplifies the analysis (and the experiments) significantly.

To describe the generation and propagation of waves at several frequencies in a nonlinear optical material, we begin, of course, with the Maxwell equations. To lowest order (as in a linear material), the waves at different frequencies propagate independently. They are coupled together only through the nonlinear susceptibility. To keep the problem tractable, we assume that this coupling is weak. As a result of the interaction, the amplitude and phase of the waves change slowly, on a length scale that is long compared with a wavelength. This secular change is described by the coupled-wave equation, which we now develop.

After Fourier transforming into the frequency domain, the Maxwell equations become

$$\nabla \cdot (\varepsilon_0 \tilde{\mathbf{E}} + \tilde{\mathbf{P}}) = \tilde{\rho}_f \tag{8.78}$$

$$\nabla \cdot \tilde{\mathbf{B}} = 0 \tag{8.79}$$

$$\nabla \times \tilde{\mathbf{E}} - i\omega\tilde{\mathbf{B}} = 0 \tag{8.80}$$

and

$$\nabla \times \left(\frac{\tilde{\mathbf{B}}}{\mu_0} - \tilde{\mathbf{M}} \right) + i\omega(\varepsilon_0 \tilde{\mathbf{E}} + \tilde{\mathbf{P}}) = \tilde{\mathbf{J}}_f \tag{8.81}$$

For nonconducting, nonmagnetic materials we ignore the free charge density $\tilde{\rho}_f$, the free current density $\tilde{\mathbf{J}}_f$, and the magnetization $\tilde{\mathbf{M}}$. To simplify the following discussion, we ignore the tensor nature of the material through which the waves are propagating. Although the nonlinear material must be anisotropic to possess a second-order nonlinearity, as discussed earlier, anisotropy is not intrinsic to the processes of interest here. In practice, the tensor properties of the material are actually important for achieving phase matching by careful handling of the polarization of the waves, and the effects of phase mismatch are discussed later. But for present purposes we assume that the polarization can be represented by the constitutive relation

$$\tilde{\mathbf{P}}(\omega) = \varepsilon_0 \chi^{(1)}(\omega)\tilde{\mathbf{E}}(\omega) + \eta^2\tilde{\mathbf{P}}^{(2)}(\omega) \tag{8.82}$$

where the nonlinear polarization is

$$\eta^2 \tilde{P}_i^{(2)}(\omega) = \varepsilon_0 \int_{-\infty}^{\infty} d\omega' \int_{-\infty}^{\infty} d\omega'' \chi^{(2)}(\omega', \omega'')\tilde{E}_i(\omega')\tilde{E}_i(\omega'')\delta(\omega - \omega' - \omega'') \tag{8.83}$$

If we begin with the identity

$$\nabla \times (\nabla \times \tilde{\mathbf{E}}) = \nabla(\nabla \cdot \tilde{\mathbf{E}}) - \nabla^2\tilde{\mathbf{E}} \tag{8.84}$$

and use (8.80), (8.78), and (8.81) together with the constitutive relation (8.82) to evaluate $\nabla \times \tilde{\mathbf{E}}$, $\nabla \cdot \tilde{\mathbf{E}}$, and $\nabla \times \tilde{\mathbf{B}}$, we obtain a Helmholtz equation with a nonlinear source term,

$$\nabla^2\tilde{\mathbf{E}} + k^2\tilde{\mathbf{E}} = -\frac{\eta^2}{\varepsilon_0 n^2(\omega)}\left[\nabla\left(\nabla \cdot \tilde{\mathbf{P}}^{(2)}\right) + k^2\tilde{\mathbf{P}}^{(2)}\right] \tag{8.85}$$

where the wave number and the index of refraction are

$$c^2 k^2(\omega) = n^2(\omega)\omega^2 \tag{8.86}$$

and

$$n^2(\omega) = 1 + \chi^{(1)}(\omega) \tag{8.87}$$

To lowest order in the small parameter η, the nonlinear terms vanish and we are left with the homogeneous Helmholtz equation. For plane waves linearly polarized in the $\hat{\mathbf{x}}$ direction and traveling in the $\hat{\mathbf{z}}$ direction, the solution is $\tilde{\mathbf{E}}(\omega) = \hat{\mathbf{x}}E_0(\omega)e^{ikz}$ for some function $E_0(\omega)$. For waves traveling in the positive $\hat{\mathbf{z}}$ direction, k is positive if ω is positive. If the nonlinear terms on the right-hand side of (8.85) are included, this is no longer an exact solution since the spectrum changes with distance, but if the nonlinearity is small an expression of this form should provide an approximation to the correct solution, at least locally. We therefore assume an approximate solution of the form

$$\tilde{\mathbf{E}}(\omega) \approx \hat{\mathbf{x}}E_\omega(z)e^{ikz} \tag{8.88}$$

$$\tilde{\mathbf{P}}^{(2)}(\omega) \approx \hat{\mathbf{x}}P_\omega(z)e^{ikz} \tag{8.89}$$

For transverse plane waves, of course, $\nabla \cdot \tilde{\mathbf{P}}^{(2)} = 0$ identically, and we ignore this term on the right-hand side of (8.85). To simplify the left-hand side of (8.85), we assume that $E_\omega(z)$ is a slowly varying function of z. By "slowly varying" it is meant that the variations are small in a wavelength, so that

$$\frac{dE_\omega}{dz} \ll kE_\omega \tag{8.90}$$

If we substitute (8.88) and (8.89) into (8.85) and evaluate the derivatives, using (8.90), the lowest-order surviving terms are

$$\frac{dE_\omega}{dz} = \frac{ik}{2\varepsilon_0 n^2}\eta^2 P_\omega \tag{8.91}$$

But from (8.83) we see that

$$\eta^2 P_\omega = \varepsilon_0 \int_{-\infty}^{\infty} d\omega' \int_{-\infty}^{\infty} d\omega'' \chi^{(2)}(\omega', \omega'')E_{\omega'}E_{\omega''}e^{i(k'+k''-k)z}\delta(\omega' + \omega'' - \omega) \tag{8.92}$$

Combining these two equations and using (8.86), we get

$$\frac{dE_\omega}{dz} = \frac{i\omega}{2nc} \int_{-\infty}^{\infty} d\omega' \int_{-\infty}^{\infty} d\omega'' \chi^{(2)}(\omega', \omega'')E_{\omega'}E_{\omega''}e^{i(k'+k''-k)z}\delta(\omega' + \omega'' - \omega) \tag{8.93}$$

This result is called the coupled-wave equation, since it describes the coupling of waves at frequencies ω' and ω'' to a wave at the frequency $\omega = \omega' + \omega''$, as indicated by the δ-function. We solve this equation for second-harmonic generation and for sum- and difference-frequency generation in the next two sections, but first we show in a general way that the coupled-wave equation conserves energy in a nondissipative material.

As discussed in Chapter 7, in the absence of dissipation the total fluence of a pulse in one dimension with intensity $S(t)$ is

$$\int_{-\infty}^{\infty} S \, dt = \varepsilon_0 c \int_{-\infty}^{\infty} d\omega \, n(\omega)E_\omega^* E_\omega = \varepsilon_0 c \int_{-\infty}^{\infty} d\omega \, n(\omega)E_{-\omega}E_\omega \tag{8.94}$$

since the field is real. If we differentiate this expression with respect to distance, using the fact that $n(-\omega) = n(\omega)$ in the absence of dissipation, evaluate the derivatives using the coupled-wave equation (8.93), and make the change of variable $\omega \rightarrow -\omega$, we find that the variation of the total fluence is

$$
\frac{d}{dz} \int_{-\infty}^{\infty} S \, dt = 2\varepsilon_0 c \int_{-\infty}^{\infty} d\omega \, n(\omega) E_{-\omega} \frac{dE_\omega}{dz}
$$

$$
= -i\varepsilon_0 \int_{-\infty}^{\infty} d\omega \int_{-\infty}^{\infty} d\omega' \int_{-\infty}^{\infty} d\omega'' \omega \chi^{(2)}(\omega', \omega'') E_\omega E_{\omega'} E_{\omega''} e^{i(k+k'+k'')z}
$$

$$
\times \, \delta(\omega + \omega' + \omega'') \tag{8.95}
$$

By symmetry we could equally well have written this with the factor $\omega' \chi^{(2)}(\omega, \omega'')$ or $\omega'' \chi^{(2)}(\omega', \omega)$ in the integrand. But for a nondissipative material we have the full permutation symmetry relation (8.48), which in the present (isotropic) case becomes simply

$$
\chi^{(2)}(\omega', \omega'') = \chi^{(2)}(\omega, \omega'') = \chi^{(2)}(\omega', \omega) \tag{8.96}
$$

for $\omega + \omega' + \omega'' = 0$. Using this symmetry relation and summing the ω, ω', and ω'' forms of (8.95) we get

$$
\frac{d}{dz} \int_{-\infty}^{\infty} S \, dt = -\frac{i\varepsilon_0}{3} \int_{-\infty}^{\infty} d\omega \int_{-\infty}^{\infty} d\omega' \int_{-\infty}^{\infty} d\omega'' \chi^{(2)}(\omega', \omega'') E_\omega E_{\omega'} E_{\omega''} e^{i(k+k'+k'')z}
$$

$$
\times \, (\omega + \omega' + \omega'') \delta(\omega + \omega' + \omega'') \tag{8.97}
$$

which obviously vanishes identically. This proves that the coupled-wave equation conserves energy in a nondissipative material. By summing over the ω, ω', and ω'' forms of the coupled-wave equation, we explicitly compute the variation of the total fluence of the waves at the three coupled frequencies. As shown by the result (8.97), when the waves at ω' and ω'' combine to create a wave at the sum frequency $\omega = \omega' + \omega''$, the intensities of the incident waves decrease to provide the energy for the new wave.

8.2.2 Second-Harmonic Generation

We can now apply the coupled-wave equation to some specific cases, and we choose for our first example that of second-harmonic generation, sometimes called frequency doubling. In this process, an optical beam at frequency $\bar{\omega}$ is incident on the surface of the nonlinear material, which we place at $z = 0$, as shown in Figure 8.2. As it propagates through the

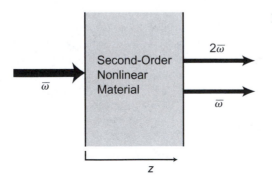

Figure 8.2 Second-harmonic generation.

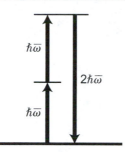

Figure 8.3 Addition of photons in second-harmonic generation.

nonlinear material, the wave creates a polarization in the material at the frequencies $\omega = 0$ and $\omega = 2\bar{\omega}$. We ignore the polarization at $\omega = 0$ for now, but the polarization at $2\bar{\omega}$ creates an optical field at the same frequency that propagates along with the original wave. If the new wave stays in phase with the polarization that is creating it, then it grows. In a quantum-mechanical view of this process, two photons from the incident wave combine through the nonlinear interaction and form a third photon whose energy and momentum are twice the energy and momentum of the incident photons, as described earlier by (8.6) and (8.7). This is indicated schematically in Figure 8.3. Eventually, the new wave at $2\bar{\omega}$, in concert with the original wave, produces polarization at the frequency $\omega = 3\bar{\omega}$, and so on until all the harmonics appear. Generally, however, these other harmonics are weak. Their growth is suppressed by the fact that they are not phase matched to the waves that create them, so they are not generated coherently.

For a cw beam the spectral amplitude consists of a series of spikes at the fundamental and all the harmonics. Since the electric field is real we may write its Fourier transform in the form

$$E_\omega(z) = \frac{1}{\sqrt{2\pi}} \int_{-\infty}^{\infty} e^{i\omega t} E(t)\, dt = \sum_{m=-\infty}^{\infty} E_m(z)\delta(\omega - \omega_m) \tag{8.98}$$

where

$$\omega_m = m\bar{\omega} \tag{8.99}$$

$$E_{-m} = E_m^* \tag{8.100}$$

We can now substitute (8.98) into the coupled-wave equation (8.93) and integrate over ω' and ω'' using the rule

$$\int_{-\infty}^{\infty} \delta(x - a)\delta(x - b)\, dx = \delta(a - b) \tag{8.101}$$

from Chapter 1. The result is

$$\sum_m \frac{dE_m}{dz}\delta(\omega - \omega_m) = \frac{i\omega}{2nc} \sum_{p,q} E_p E_q \chi^{(2)}(\omega_p, \omega_q) e^{i(k_p + k_q - k)z}\delta(\omega_p + \omega_q - \omega) \tag{8.102}$$

where

$$k_m = k(\omega_m) = \frac{n_m \omega_m}{c} \tag{8.103}$$

is the wave number at the frequency ω_m and $n_m = n(\omega_m)$ is the index of refraction at that frequency. To separate out the component at the frequency ω_m, we integrate (8.102) over the interval $\omega_m - \varepsilon < \omega < \omega_m + \varepsilon$ for some small $\varepsilon < \bar{\omega}$. When we do this, we get the coupled-amplitude equation

$$\frac{dE_m}{dz} = \frac{i\omega_m}{2n_m c} \sum_{p+q=m} \chi^{(2)}(\omega_p, \omega_q) E_p E_q e^{i(k_p + k_q - k_m)z} \tag{8.104}$$

where the sum is over all values of p and q that satisfy the condition $p + q = m$.

In general, the factor $e^{i(k_p + k_q - k_m)z}$ oscillates unless the phase velocities of the waves are matched. This suppresses the amplification of waves unless

$$k_p + k_q - k_m = 0 \tag{8.105}$$

which is called the phase-matching condition. Ordinarily, dispersion makes it impossible to satisfy the phase-matching condition even approximately except at two specific frequencies. We therefore assume from now on that the optical field is composed of just two waves, the incident wave at frequency $\omega_1 = \bar{\omega}$ and the second-harmonic wave at $\omega_2 = 2\bar{\omega}$. In the sum (8.104), then, the only term that survives for $m = 2$ is $p = q = 1$, and the amplitude of the second-harmonic wave satisfies the simpler equation

$$\frac{dE_2}{dz} = \frac{i\omega_2}{2n_2 c} \chi^{(2)}(\omega_1, \omega_1) E_1 E_1 e^{i(2k_1 - k_2)z} \tag{8.106}$$

In the same way we find that for the fundamental ($m = 1$), only the terms $(p, q) = (2, -1)$ and $(p, q) = (-1, 2)$ survive in (8.104). The Fourier amplitude of the fundamental therefore satisfies the equation

$$\frac{dE_1}{dz} = \frac{i\omega_1}{2n_1 c} \left[\chi^{(2)}(\omega_2, -\omega_1) + \chi^{(2)}(-\omega_1, \omega_2) \right] E_2 E_{-1} e^{i(k_2 + k_{-1} - k_1)z}$$

$$= \frac{i\omega_1}{n_1 c} \chi^{(2)}(\omega_1, \omega_1) E_2 E_1^* e^{i(k_2 - 2k_1)z} \tag{8.107}$$

after we use the reality condition (8.100) and the permutation symmetry (8.48).

The general solution to (8.106) and (8.107) is rather complicated, so we examine some special cases that illustrate features of the behavior to be expected more generally. We consider first the case when generation of the second harmonic is sufficiently weak that the fundamental is not depleted by conversion to the second harmonic. That is, we assume that $E_1 = $ constant. With no second-harmonic wave incident on the surface of the nonlinear material [$E_2(0) = 0$], the solution is simply

$$E_2(z) = \frac{i\omega_2}{2n_2 c} \chi^{(2)}(\omega_1, \omega_1) E_1^2 e^{i(1/2)(2k_1 - k_2)z} \frac{\sin\left[\frac{1}{2}(2k_1 - k_2)z\right]}{\frac{1}{2}(2k_1 - k_2)} \tag{8.108}$$

But the total fluence in a pulse is

$$\int_{-\infty}^{\infty} S \, dt = \frac{1}{2\pi} \int_{-\infty}^{\infty} dt \int_{-\infty}^{\infty} d\omega \int_{-\infty}^{\infty} d\omega' \varepsilon_0 c n(\omega) e^{-i(\omega + \omega')t} E_\omega E_{\omega'} \tag{8.109}$$

If we substitute (8.98) for E_ω and $E_{\omega'}$, carry out the integrals over ω and ω', and average over one cycle, we find that the average flux on the mth harmonic is

$$\langle S_m \rangle = \frac{\varepsilon_0 c n_m}{\pi} E_m E_m^* \tag{8.110}$$

The flux on the second harmonic is therefore

$$\langle S_2(z) \rangle = \frac{n_2 \varepsilon_0 c}{\pi} E_2 E_2^* = \frac{\varepsilon_0 \omega_2^2}{4\pi n_2 c} \chi^{(2)2}(\omega_1, \omega_1)(E_1 E_1^*)^2 \left\{ \frac{\sin\left[\frac{1}{2}(2k_1 - k_2)z\right]}{\frac{1}{2}(2k_1 - k_2)} \right\}^2 \quad (8.111)$$

From this we see that when the waves are not perfectly phase matched, the flux on the second harmonic oscillates with the spatial frequency $(2k_1 - k_2)$. We define the coherence length as the reciprocal of the wave-vector mismatch,

$$L_{\text{coherence}} = \frac{1}{|2k_1 - k_2|} = \frac{\bar{\lambda}_{\text{vacuum}}}{4\pi |n_2 - n_1|} \quad (8.112)$$

where $\bar{\lambda}_{\text{vacuum}}$ is the vacuum wavelength of the fundamental. This illustrates the importance of phase matching, since in this case, as shown by (8.111), the peak value of the second-harmonic flux is proportional to $L_{\text{coherence}}^2$. Unfortunately, the effects of dispersion limit the coherence length in real materials. For example, in the nonlinear crystal KH_2PO_4, commonly called KDP, the indices of refraction at $\bar{\lambda}_{\text{vacuum}} = 1$ μm are $n_1 = 1.50873$ and $n_2 = 1.52983$. The coherence length is therefore $L_{\text{coherence}} = 4$ μm, which is not enough to achieve significant conversion to the second harmonic. In actual fact the requirement is not that phase matching be perfect, but that the phase error be small over the characteristic distance required for conversion (or the length of the nonlinear material, if this is shorter). For significant conversion, then, it is necessary only that

$$\frac{L_{\text{coherence}}}{L_{\text{conversion}}} > O(1) \quad (8.113)$$

where

$$\frac{1}{L_{\text{conversion}}} = \left| \frac{\omega_1 \chi^{(2)}(\omega_1, \omega_1)}{n_1 c} E_1 \right| = \left(\frac{4\pi^3 \chi^{(2)2}(\omega_1, \omega_1)}{\varepsilon_0 c n_1^3 \bar{\lambda}_{\text{vacuum}}^2} \langle S_1 \rangle \right)^{1/2} \quad (8.114)$$

For KDP the nonlinear susceptibility is $\chi^{(2)} \approx 10^{-12}$ m/V, so at an incident intensity $\langle S \rangle = 10^{13}$ W/m^2 (approximately the damage limit) the conversion length is $L_{\text{conversion}} \approx 2$ mm. Since this is long compared with $L_{\text{coherence}}$, various tricks are used to improve the coherence length. The most popular is to take advantage of the birefringence of many crystals and orient the polarization of the waves at ω_1 and ω_2 in different directions to make the indices of refraction approximately the same. By this means, satisfactory phase matching can be achieved in many nonlinear materials.

When the waves are perfectly phase matched, the second harmonic flux grows in proportion to z^2 until the second harmonic becomes comparable to the fundamental. At this point, the depletion of the fundamental must be included in the analysis. For the case of perfect phase matching ($k_2 = 2k_1$), the coupled-amplitude equations (8.106) and (8.107) simplify to

$$\frac{dE_2}{dz} = \frac{i\omega_2}{2n_2 c} \chi^{(2)}(\omega_1, \omega_1) E_1 E_1 \quad (8.115)$$

$$\frac{dE_1}{dz} = \frac{i\omega_1}{n_1 c} \chi^{(2)}(\omega_1, \omega_1) E_2 E_1^* \quad (8.116)$$

If we differentiate (8.115) and evaluate dE_1/dz using (8.116), we obtain the equation

$$\frac{d^2u_2}{dz^2} = -2\kappa^2 u_1 u_1^* u_2 \tag{8.117}$$

where

$$u_1 = \frac{E_1}{|E_1(0)|} \tag{8.118}$$

$$u_2 = \frac{E_2}{|E_1(0)|} \tag{8.119}$$

and the conversion rate is

$$\kappa^2 = \frac{\omega_1^2}{n_1^2 c^2}\chi^{(2)^2}(\omega_1,\omega_1)\,|E_1(0)|^2 = \frac{\pi\mu_0\omega_1^2}{n_1^3 c}\chi^{(2)^2}(\omega_1,\omega_2)\langle S_1(0)\rangle \tag{8.120}$$

since $n_1 = n_2$ when the waves are phase matched. If there is no second-harmonic wave incident on the nonlinear crystal, then the initial conditions are

$$u_1(0) = 1 \tag{8.121}$$

$$u_2(0) = 0 \tag{8.122}$$

From the flux conservation law we find that

$$u_1 u_1^* + u_2 u_2^* = 1 \tag{8.123}$$

so the differential equation (8.117) becomes

$$\frac{d^2u_2}{dz^2} = -2\kappa^2(1 - u_2 u_2^*)u_2 \tag{8.124}$$

But $-2\kappa^2(1 - u_2 u_2^*)$ is always real, so the complex phase of u_2 is constant. Without loss of generality we may take u_2 as real, and (8.117) becomes simply

$$\frac{d^2u_2}{dz^2} = -2\kappa^2\big(1 - u_2^2\big)u_2 \tag{8.125}$$

The solution is

$$u_2 = \tanh\kappa z \xrightarrow[z\to\infty]{} 1 \tag{8.126}$$

as shown in Figure 8.4. For perfectly phase-matched waves, the fraction of the incident wave that is converted to the second harmonic asymptotically approaches unity.

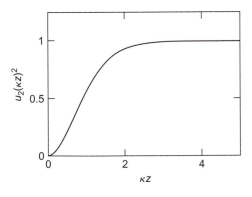

Figure 8.4 Conversion fraction in second-harmonic generation.

EXERCISE 8.5

Franken's first experiments on second-harmonic generation used a pulsed ruby laser ($\lambda_{vacuum} = 694$ nm) focused on quartz, which has a nonlinear susceptibility $\chi^{(2)} \approx 10^{-12}$ m/V. However, the experiments were not phase matched, since the index of refraction at the two wavelengths is $n(694$ nm$) = 1.55$, and $n(347$ nm$) = 1.58$ in quartz.

(a) What is the coherence length in Franken's experiments?

(b) For a fundamental intensity $\langle S_1 \rangle = 10^{11}$ W/m^2, what is the maximum second-harmonic intensity possible without phase matching?

(c) If the quartz crystal has a length of 1 cm, what second-harmonic intensity might be expected with perfect phase matching?

EXERCISE 8.6

The most common technique for achieving phase matching is to take advantage of the bire-fringence of the anisotropic crystals used for second-harmonic generation. As an example we consider what is called Type I phase matching in a uniaxial crystal. In a uniaxial crystal the optic axis, called the c axis, has a different polarizability than the other two orthogonal axes. Light polarized perpendicular to the c axis is said to have ordinary polarization and an ordinary index of refraction n_o. The other polarization, which has a component along the c axis, is called the extraordinary polarization and has an index of refraction $n_e(\theta)$, where θ is the angle between the direction of the wave vector **k** and the c axis. This is shown in Figure 8.5. Because they are orthogonal, the two polarizations propagate independently in the linear approximation, but they are coupled by the second-order susceptibility.

(a) Beginning with the Maxwell equations for a linear medium, compute the phase velocity of the extraordinary polarization and show that the extraordinary index of refraction is given by

$$\frac{1}{n_e^2(\theta)} = \frac{\sin^2 \theta}{n_c^2} + \frac{\cos^2 \theta}{n_o^2} \tag{8.127}$$

where the principal value n_c of the extraordinary polarization corresponds to polarization along the c axis. *Hint:* The electric field vector is not, in general, parallel to the displacement vector.

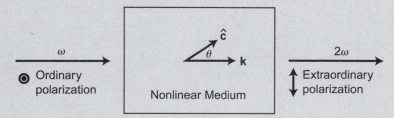

Figure 8.5 Phase matching in a uniaxial crystal.

(b) Show that if the index of refraction for polarization along the c axis is less than that for polarization normal to the c axis ($n_c < n_o$), the phase-matching angle for second-harmonic generation is

$$\sin^2 \theta = \frac{\dfrac{1}{n_o^2(\omega)} - \dfrac{1}{n_o^2(2\omega)}}{\dfrac{1}{n_c^2(2\omega)} - \dfrac{1}{n_o^2(2\omega)}} \qquad (8.128)$$

8.2.3 Sum-Frequency Generation

In the process of second-harmonic generation, the nonlinear material takes two photons from the first harmonic and creates a single photon at the second harmonic. It is also possible to take two photons of different frequencies and combine them to create a single photon with the same total energy. In a classical picture of this process, when the incident wave consists of components at two frequencies, call them $\bar{\omega}_1$ and $\bar{\omega}_2$, the polarization generated in the nonlinear material has components at the harmonics of the incident frequencies, as before, and at the sum and difference frequencies $\omega = \pm\bar{\omega}_1 \pm \bar{\omega}_2$. If any of these frequencies is phase matched to the incident waves, then a wave at that frequency is generated and amplified coherently.

The mathematical description of this process is similar to that of second-harmonic generation. Since the harmonics and the sum frequencies that are generated by the incident beams generate, in turn, further harmonics and sum frequencies, the complete field includes components at all these frequencies. When we compute the Fourier transform of this field, we find a narrow peak at each frequency. Therefore, we represent the Fourier transform of the optical field by the sum of δ-functions

$$E_\omega(z) = \sum_{m,n=-\infty}^{\infty} E_{mn}(z)\delta(\omega - \omega_{mn}) \qquad (8.129)$$

where

$$\omega_{mn} = m\bar{\omega}_1 + n\bar{\omega}_2 \qquad (8.130)$$

$$E_{-m,-n} = E_{mn}^* \qquad (8.131)$$

If we substitute (8.129) into the coupled-wave equation (8.93), integrate over ω' and ω'', and separate out the component at the frequency ω_{mn} as before, we get the coupled-amplitude equation

$$\frac{dE_{mn}}{dz} = i\frac{\omega_{mn}}{n_{mn}c} \sum_{\substack{p+r=m \\ q+s=n}} \chi^{(2)}(\omega_{pq}, \omega_{rs}) E_{pq} E_{rs} e^{i(k_{pq}+k_{rs}-k_{mn})z} \qquad (8.132)$$

where

$$k_{mn} = k(\omega_{mn}) = \frac{\omega_{mn}n_{mn}}{c} \qquad (8.133)$$

is the wave number at the frequency ω_{mn}, and $n_{mn} = n(\omega_{mn})$ is the index of refraction at that frequency. In (8.132) the sum is over all values of p, q, r, and s that satisfy the conditions $p + r = m$ and $q + s = n$.

As in second-harmonic generation, the oscillating factor $e^{i(k_{pq}+k_{rs}-k_{mn})z}$ suppresses the amplification unless the phase-matching condition

$$k_{pq} + k_{rs} - k_{mn} = 0 \tag{8.134}$$

is satisfied. Generally, it is possible to satisfy this condition at just one frequency. For sum-frequency generation we assume that the optical field is composed of just three waves, the incident waves at the frequencies $\bar{\omega}_1$ and $\bar{\omega}_2$, and the sum-frequency wave at $\omega_{11} = \bar{\omega}_1 + \bar{\omega}_2$. In the sum (8.132) the only term that survives for $(m, n) = (1, 1)$ is $p = s = 1$ and $q = r = 0$. The Fourier amplitude of the sum-frequency wave therefore satisfies the simpler equation

$$\frac{dE_{11}}{dz} = i\frac{\omega_{11}}{n_{11}c}\chi^{(2)}(\omega_{10}, \omega_{01})E_{10}E_{01}e^{i(k_{10}+k_{01}-k_{11})z} \tag{8.135}$$

In the same way we find that for the waves at $\omega_{10} = \omega_{11} + \omega_{0,-1}$ and $\omega_{01} = \omega_{11} + \omega_{-1,0}$ the Fourier amplitudes satisfy the equations

$$\frac{dE_{10}}{dz} = i\frac{\omega_{10}}{n_{10}c}\chi^{(2)}(\omega_{11}, \omega_{0,-1})E_{11}E_{0,-1}e^{i(k_{11}+k_{0,-1}-k_{10})z} \tag{8.136}$$

$$\frac{dE_{01}}{dz} = i\frac{\omega_{01}}{n_{01}c}\chi^{(2)}(\omega_{11}, \omega_{-1,0})E_{11}E_{-1,0}e^{i(k_{11}+k_{-1,0}-k_{01})z} \tag{8.137}$$

But $\omega_{0,-1} = -\omega_{01}$ and $\omega_{-1,0} = -\omega_{10}$, so if we use the symmetry relation (8.131) and the full permutation symmetry condition, these equations become

$$\frac{dE_{10}}{dz} = i\frac{\omega_{10}}{n_{10}c}\chi^{(2)}(\omega_{10}, \omega_{01})E_{11}E_{01}^*e^{i(k_{11}-k_{01}-k_{10})z} \tag{8.138}$$

$$\frac{dE_{01}}{dz} = i\frac{\omega_{01}}{n_{01}c}\chi^{(2)}(\omega_{10}, \omega_{01})E_{11}E_{10}^*e^{i(k_{11}-k_{10}-k_{01})z} \tag{8.139}$$

The three equations (8.135), (8.138), and (8.139) form a complete set from which we can find the amplitudes of the three waves. Before doing this, however, we can derive some useful and interesting relations between the intensities of the waves. If we compute the variation of the average intensity with distance using (8.110), we get

$$\left\langle\frac{dS_{11}}{dz}\right\rangle = -\frac{\varepsilon_0\omega_{11}}{\pi}\chi^{(2)}(\omega_{10}, \omega_{01})\,\text{Im}\left[E_{10}E_{01}E_{11}^*e^{i(k_{10}+k_{01}-k_{11})z}\right] \tag{8.140}$$

$$\left\langle\frac{dS_{10}}{dz}\right\rangle = \frac{\varepsilon_0\omega_{10}}{\pi}\chi^{(2)}(\omega_{10}, \omega_{01})\,\text{Im}\left[E_{10}E_{01}E_{11}^*e^{i(k_{10}+k_{01}-k_{11})z}\right] \tag{8.141}$$

and

$$\left\langle\frac{dS_{01}}{dz}\right\rangle = \frac{\varepsilon_0\omega_{01}}{\pi}\chi^{(2)}(\omega_{10}, \omega_{01})\,\text{Im}\left[E_{10}E_{01}E_{11}^*e^{i(k_{10}+k_{01}-k_{11})z}\right] \tag{8.142}$$

If we sum these three equations, we see that

$$\left\langle\frac{dS_{11}}{dz}\right\rangle + \left\langle\frac{dS_{10}}{dz}\right\rangle + \left\langle\frac{dS_{01}}{dz}\right\rangle$$

$$= \frac{\varepsilon_0}{\pi}\chi^{(2)}(\omega_{10}, \omega_{01})\text{Im}\left[E_{10}E_{01}E_{11}^*e^{i(k_{10}+k_{01}-k_{11})z}\right](\omega_{10} + \omega_{01} - \omega_{11}) = 0 \tag{8.143}$$

so energy is conserved, as expected. We get a more interesting result if we divide each equation by ω_{mn}, for then we find that

$$\frac{1}{\omega_{10}}\left\langle\frac{dS_{10}}{dz}\right\rangle = \frac{1}{\omega_{01}}\left\langle\frac{dS_{01}}{dz}\right\rangle = -\frac{1}{\omega_{11}}\left\langle\frac{dS_{11}}{dz}\right\rangle \qquad (8.144)$$

These are called the Manley–Rowe relations. They are valid for all sum- and difference-frequency generation processes in nondissipative materials, and were discovered in 1959, prior to the invention of the laser and the observation of nonlinear optical effects. Although we have derived them classically, the Manley–Rowe relations have an interesting quantum-mechanical interpretation. Since the energy of a photon is $\hbar\omega$, the quantity $\langle S_{mn}(z)\rangle/\omega_{mn}$ is proportional to the flux of photons of frequency ω_{mn} at the point z. The Manley–Rowe relations therefore state that the changes in the photon flux at the incident frequencies ω_{10} and ω_{01} are the same and are just the negative of the change in the photon flux at the sum frequency ω_{11}. The Manley–Rowe relations are equivalent to the statement that in sum-frequency generation, one photon at the frequency ω_{10} and one photon at the frequency ω_{01} are destroyed, and one photon at the sum frequency ω_{11} is created. This is illustrated schematically in Figure 8.6. Clearly, the Manley–Rowe relations (8.144) compose a stronger statement than the conservation of energy (8.143).

The general solution of the coupled-amplitude equations (8.135), (8.138), and (8.139) is even more complicated for sum-frequency generation than for second-harmonic generation, since there are two waves incident on the nonlinear material. We therefore limit our discussion to what is called upconversion. In this process, one of the incident waves (called the pump wave) is much stronger than the other and is, as a consequence of the Manley–Rowe relations, not significantly depleted. If we call this the wave at ω_{10}, then the weaker wave at ω_{01} is depleted and converted to a wave at the sum frequency ω_{11}, as indicated in Figure 8.7. Since it has a higher frequency but the same number of photons as the incident wave, the upconverted wave has more energy than the incident wave, part of which comes from the pump wave at ω_{10}. This is evident in a quantum-mechanical picture, as shown in Figure 8.6.

When the amplitude E_{10} of the pump is constant, we have only to deal with (8.135) and (8.139). To further simplify the mathematics, we limit our discussion to some special cases that illustrate features of the behavior we expect more generally. The first case

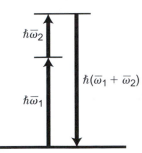

Figure 8.6 Addition of photons in sum-frequency generation.

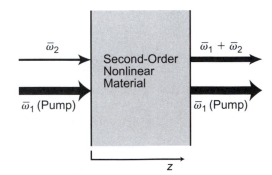

Figure 8.7 Sum-frequency generation (upconversion).

that we examine is one in which the efficiency of upconversion is small, so neither of the incident waves is depleted. We may then regard E_{10} and E_{01} as constants, and the solution of (8.135) is simply

$$E_{11} = i\frac{\omega_{11}}{n_{11}c}\chi^{(2)}(\omega_{10}, \omega_{01})E_{10}E_{01}e^{i(1/2)(k_{10}+k_{01}-k_{11})z}\frac{\sin\left[\frac{1}{2}(k_{10}+k_{01}-k_{11})z\right]}{\frac{1}{2}(k_{10}+k_{01}-k_{11})} \qquad (8.145)$$

The remarks made following (8.108) regarding phase matching in second-harmonic generation apply equally to sum-frequency generation.

In the case when the waves are phase matched, however, upconversion differs from second-harmonic generation. In the following discussion we assume that $k_{10} + k_{01} - k_{11} = 0$, so we must solve the equations

$$\frac{dE_{11}}{dz} = i\frac{\omega_{11}}{n_{11}c}\chi^{(2)}(\omega_{10}, \omega_{01})E_{10}E_{01} \qquad (8.146)$$

$$\frac{dE_{01}}{dz} = i\frac{\omega_{01}}{n_{01}c}\chi^{(2)}(\omega_{10}, \omega_{01})E_{11}E_{10}^* \qquad (8.147)$$

If we differentiate (8.146) with respect to z and use (8.147) to evaluate dE_{01}/dz we get

$$\frac{d^2u_{11}}{dz^2} = -\kappa^2 u_{11} \qquad (8.148)$$

where

$$\kappa^2 = \frac{\pi\mu_0\omega_{01}\omega_{11}}{n_{01}n_{10}n_{11}}\chi^{(2)2}(\omega_{10}, \omega_{01})\langle S_{10}\rangle \qquad (8.149)$$

is the growth rate, and where

$$u_{01} = \frac{E_{01}}{E_{01}(0)} \qquad (8.150)$$

$$u_{11} = \sqrt{\frac{n_{11}\omega_{01}}{n_{01}\omega_{11}}}\frac{E_{11}}{E_{01}(0)} \qquad (8.151)$$

are the dimensionless amplitudes. From the Manley–Rowe relations we get

$$u_{01}u_{01}^* + u_{11}u_{11}^* = 1 \qquad (8.152)$$

If there is no upconverted wave incident on the crystal, the initial conditions are

$$u_{01}(0) = 1 \qquad (8.153)$$

$$u_{11}(0) = 0 \qquad (8.154)$$

and the solution is

$$u_{01} = \cos\kappa z \qquad (8.155)$$

$$u_{11} = \sin\kappa z \qquad (8.156)$$

The intensity of the wave at the sum frequency is proportional to $u_{11}u_{11}^*$, which oscillates as indicated in Figure 8.8 even when the waves are phase matched. The maximum conversion occurs at the point $\kappa z = \pi/2$, after which the energy flows back into the incident wave. Although it would seem that second-harmonic generation is a degenerate

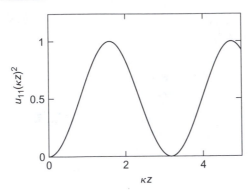

Figure 8.8 Upconversion fraction.

case of sum-frequency generation, and in some sense it is, second-harmonic generation asymptotically approaches complete conversion because the pump, which is also the wave being converted, is depleted and the conversion stops. In upconversion, on the other hand, the pump (the strong wave at the frequency ω_{10}) is not depleted, so conversion continues (back, now, to the incident frequency) even after all of the incident wave has been upconverted. Imperfect phase matching reduces the upconversion efficiency when (8.134) is not satisfied.

Sum-frequency generation now has many important applications. Besides being used to generate coherent optical pulses at wavelengths not otherwise obtainable, it can also be used to convert weak signals at long wavelengths to shorter wavelengths where they can be more easily detected. Fast, single-photon detectors are available in the visible part of the spectrum but not in the infrared. By mixing a weak infrared signal with an intense near-infrared or visible signal it is possible to convert the infrared photons to the visible part of the spectrum. Since the conversion rate κ depends on the intensity $\langle S_{10} \rangle$ of the visible signal, as shown by (8.149), the technique works best with picosecond and femtosecond laser pulses because the threshold intensity for optical damage is higher for shorter pulses.

In addition to sum-frequency generation, difference-frequency generation is also possible. We consider the case when a strong wave at the frequency ω_{11} is incident on a nonlinear material along with a weak wave at a lower frequency ω_{10}. As the waves propagate through the nonlinear material, the interaction creates a wave at the difference frequency $\omega_{01} = \omega_{11} - \omega_{10}$. When the waves are phase matched, the weaker waves grow stronger until they deplete the incident strong wave. This is illustrated schematically in Figure 8.9. For reasons that are somewhat obscure, difference-frequency generation is called a parametric process. In mechanics, a parametric oscillator is one that is driven not by a periodic force but by a periodic variation in one of the parameters, such as the spring constant. In the present case the spring constant (the dielectric susceptibility) is modulated at an optical frequency by the strong wave due to the nonlinear effect. Referring to Figure 8.10, the incident strong wave at frequency ω_{11} is called the pump, the incident weak wave at frequency ω_{10} is called the signal, and the remaining wave at frequency ω_{01} is called the idler. Growth of the signal at ω_{10} is called parametric amplification. Experimentally, either the signal or the idler (or both) can be used as the output.

Difference-frequency generation (or parametric amplification) is described by the coupled-amplitude equations (8.135), (8.138), and (8.139), as before. The effects of phase

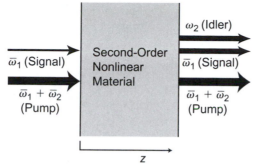

Figure 8.9 Difference-frequency generation.

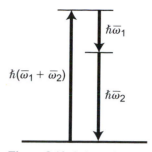

Figure 8.10 Subtraction of photons in difference-frequency generation.

matching are the same, and we consider here just the case of perfect phase matching; that is,

$$k_{11} = k_{10} + k_{01} \tag{8.157}$$

To further simplify the mathematics, we ignore depletion of the pump. In the discussion of upconversion, the neglect of pump depletion is justified by the Manley–Rowe relations. If the incident wave at the frequency ω_{01} is weak compared with the wave at ω_{10}, then since each wave loses the same number of photons, the pump is essentially unaffected even when the wave at ω_{01} is completely upconverted. In the case of difference-frequency generation, however, the bulk of the pump wave can be converted to waves at the lower frequencies ω_{10} and ω_{01}, but we ignore this here.

To describe the growth of the signal, we differentiate (8.138) and evaluate the derivative dE_{01}^*/dz using the complex conjugate of (8.139). In terms of the dimensionless variables

$$u_{10} = \frac{E_{10}(z)}{E_{10}(0)} \tag{8.158}$$

$$u_{01}(z) = \sqrt{\frac{\omega_{10} n_{01}}{\omega_{01} n_{10}}} \frac{E_{01}(z)}{E_{10}(0)} \tag{8.159}$$

we get the differential equation

$$\frac{d^2 u_{10}}{dz^2} = \kappa^2 u_{10} \tag{8.160}$$

where the growth rate is

$$\kappa^2 = \frac{\pi \mu_0 \omega_{10} \omega_{01}}{n_{10} n_{01} n_{11}} \chi^{(2)2} \langle S_{11} \rangle \tag{8.161}$$

If there is no idler incident on the crystal at $z = 0$, the initial conditions are

$$u_{10}(0) = 1 \tag{8.162}$$

$$\frac{du_{10}(0)}{dz} = 0 \tag{8.163}$$

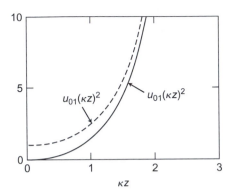

Figure 8.11 Conversion fraction in difference-frequency generation.

The idler is found from the Manley–Rowe relation

$$u_{01}u_{01}^* = u_{10}u_{10}^* - 1 \qquad (8.164)$$

The solution to these equations is

$$u_{10}(z) = \cosh(\kappa z) \qquad (8.165)$$

$$u_{01}(z) = \sinh(\kappa z) \qquad (8.166)$$

As shown in Figure 8.11, the signal and the idler grow monotonically, approaching exponential growth as they become large compared to the incident signal. This is in contrast with the oscillatory behavior of upconversion. Eventually the growth saturates, of course, when the pump becomes depleted. Imperfect phase matching reduces the growth rate when the phase-matching condition of (8.157) is not satisfied, as discussed previously.

8.3 NONLINEAR INDEX OF REFRACTION

8.3.1 Third-Order Susceptibility

As discussed earlier, the second-order contribution to the polarization is just the leading term in a more general expansion. The higher-order terms are usually unimportant except in materials for which the second-order term vanishes. But as noted previously, the second-order susceptibility vanishes identically in all materials that are symmetric under inversion. This includes all isotropic materials, such as liquids and glasses, and certain crystals, such as cubic crystals. In all these materials the lowest order nonlinear susceptibility is the third-order susceptibility.

The third-order nonlinear susceptibility is important not only because it is the leading nonlinearity in a broad class of materials but also because it introduces some new effects. Third-harmonic generation, which is analogous to second-harmonic generation in materials with a second-order susceptibility, is an obvious effect that has practical importance. Of greater uniqueness and interest, however, is the fact that the third-order susceptibility can combine three photons (speaking from a quantum-mechanical point of view) to produce a photon at the original frequency, as illustrated in Figure 8.12. The appearance of a non-linear component of the polarization in the material at the original frequency is effectively a nonlinear contribution to the index of refraction, and this leads to several new and interesting phenomena. In the following sections we discuss three of these phenomena, namely

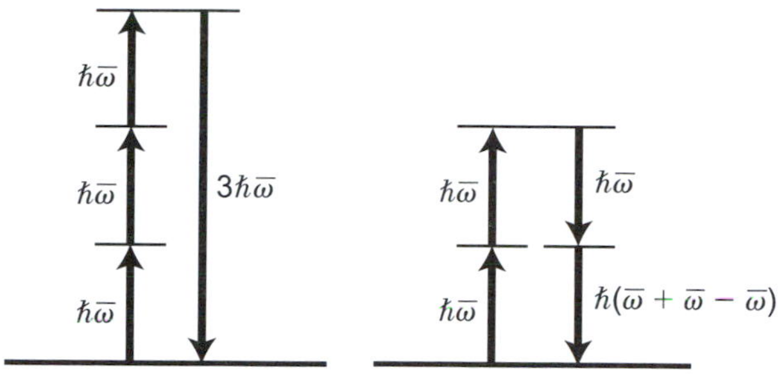

Figure 8.12 Three-photon processes.

self-focusing, rotation of the polarization, and phase-conjugate reflection. None of these effects appears in second order.

By extension from our previous discussion we see that when the second-order non-linearity vanishes, the leading terms in the expansion of the polarization are

$$\tilde{P}_i(\omega) = \eta \tilde{P}_i^{(1)}(\omega) + \eta^3 \tilde{P}_i^{(3)}(\omega) + \cdots \tag{8.167}$$

where the linear and nonlinear terms are

$$\eta \tilde{P}_i^{(1)} = \varepsilon_0 \sum_j \chi_{i;j}^{(1)} \tilde{E}_j \tag{8.168}$$

$$\eta^3 \tilde{P}_i^{(3)}(\omega) = \varepsilon_0 \sum_{j,k,l} \int_{-\infty}^{\infty} d\omega' \int_{-\infty}^{\infty} d\omega'' \int_{-\infty}^{\infty} d\omega''' \tilde{E}_j(\omega') \tilde{E}_k(\omega'') \tilde{E}_l(\omega''')$$

$$\times \chi_{i;jkl}^{(3)}(\omega', \omega'', \omega''') \delta(\omega - \omega' - \omega'' - \omega''') \tag{8.169}$$

Note that the third-order susceptibility $\chi_{i;jkl}^{(3)}(\omega', \omega'', \omega''')$, as defined by (8.169), has the dimensions m^2/V^2 in SI units.

As a fourth-rank tensor, the third-order susceptibility has, in general, $3^4 = 81$ components. However, the number of independent components is considerably reduced in an isotropic medium. We see immediately that all components such as $\chi_{1;222}^{(3)}$ or $\chi_{1;232}^{(3)}$ that include any Cartesian coordinate an odd number of times must vanish. Otherwise (8.169) does not have the correct behavior under an inversion of that coordinate. Therefore, since all directions are equivalent in an isotropic material, there are at most four independent components, namely

$$\chi_{i;iii}^{(3)}(\omega', \omega'', \omega''') = \chi_{1;111}^{(3)}(\omega', \omega'', \omega''') \tag{8.170}$$

$$\chi_{i;ijj}^{(3)}(\omega', \omega'', \omega''') = \chi_{1;122}^{(3)}(\omega', \omega'', \omega''') \tag{8.171}$$

$$\chi_{i;jij}^{(3)}(\omega', \omega'', \omega''') = \chi_{1;212}^{(3)}(\omega', \omega'', \omega''') \tag{8.172}$$

$$\chi_{i;jji}^{(3)}(\omega', \omega'', \omega''') = \chi_{1;221}^{(3)}(\omega', \omega'', \omega''') \tag{8.173}$$

However, these components are not all independent, since they are related by rotational symmetry. To show this, we consider a rotation of coordinates about the $x_3 = z$ axis through the angle θ. The rotation matrix is then

$$R_{ij} = \begin{bmatrix} \cos\theta & \sin\theta & 0 \\ -\sin\theta & \cos\theta & 0 \\ 0 & 0 & 1 \end{bmatrix} \tag{8.174}$$

If we evaluate the transformation, we find that the tensor components in the new coordinate system are

$$
\chi'^{(3)}_{1;111} = \sum_{p,q,r,s} R_{1p}R_{1q}R_{1r}R_{1s}\chi^{(3)}_{p;qrs}
$$

$$
= \cos^4\theta\,\chi^{(3)}_{1;111} + \cos^3\theta\sin\theta\big(\chi^{(3)}_{1;112} + \chi^{(3)}_{1;121} + \chi^{(3)}_{1;211} + \chi^{(3)}_{2;111}\big)
$$

$$
+ \cos^2\theta\sin^2\theta\big(\chi^{(3)}_{1;122} + \chi^{(3)}_{2;211} + \chi^{(3)}_{1;212} + \chi^{(3)}_{2;121} + \chi^{(3)}_{1;221} + \chi^{(3)}_{2;112}\big)
$$

$$
+ \cos\theta\sin^3\theta\big(\chi^{(3)}_{2;221} + \chi^{(3)}_{2;212} + \chi^{(3)}_{2;122} + \chi^{(3)}_{1;222}\big) + \sin^4\theta\,\chi^{(3)}_{2;222} \tag{8.175}
$$

But all the components in which a given coordinate (1 or 2) appears an odd number of times must vanish, as discussed earlier. Moreover, for an isotropic medium, symmetry implies that $\chi^{(3)}_{1;111} = \chi^{(3)}_{2;222}$, $\chi^{(3)}_{1;122} = \chi^{(3)}_{2;211}$, and so on. Then, since the susceptibility tensor for an isotropic medium is invariant under rotation, $\chi'^{(3)}_{1;111} = \chi^{(3)}_{1;111}$, we see that (8.175) becomes

$$
\chi^{(3)}_{1;111} = (\cos^4\theta + \sin^4\theta)\chi^{(3)}_{1;111} + 2\cos^2\theta\sin^2\theta\big(\chi^{(3)}_{1;122} + \chi^{(3)}_{1;212} + \chi^{(3)}_{1;221}\big) \tag{8.176}
$$

But this is satisfied for all θ only if

$$
\chi^{(3)}_{1;111}(\omega',\omega'',\omega''') = \chi^{(3)}_{1;122}(\omega',\omega'',\omega''') + \chi^{(3)}_{1;212}(\omega',\omega'',\omega''') + \chi^{(3)}_{1;221}(\omega',\omega'',\omega''')
$$

$$\tag{8.177}$$

This leaves only three independent tensor components, which allows us to express the susceptibility in the form

$$
\chi^{(3)}_{i;jkl}(\omega',\omega'',\omega''') = \chi^{(3)}_{1;122}(\omega',\omega'',\omega''')\delta_{ij}\delta_{kl} + \chi^{(3)}_{1;212}(\omega',\omega'',\omega''')\delta_{ik}\delta_{jl}
$$

$$
+ \chi^{(3)}_{1;221}(\omega',\omega'',\omega''')\delta_{il}\delta_{jk} \tag{8.178}
$$

This is the most general form of the third-order susceptibility tensor for an isotropic material.

The number of independent coefficients is further reduced in special cases. For example, in third-harmonic generation ($\omega = \bar{\omega} + \bar{\omega} + \bar{\omega}$ for some incident frequency $\bar{\omega}$) the important coefficients are $\chi^{(3)}_{i;jkl}(\bar{\omega}, \bar{\omega}, \bar{\omega})$. But from intrinsic permutation symmetry, we see that in this case

$$
\chi^{(3)}_{1;122}(\bar{\omega},\bar{\omega},\bar{\omega}) = \chi^{(3)}_{1;212}(\bar{\omega},\bar{\omega},\bar{\omega}) = \chi^{(3)}_{1;221}(\bar{\omega},\bar{\omega},\bar{\omega}) = \tfrac{1}{3}\chi^{(3)}_{1;111}(\bar{\omega},\bar{\omega},\bar{\omega}) \tag{8.179}
$$

by (8.177). This leaves only one coefficient, so (8.178) simplifies to

$$
\chi^{(3)}_{i;jkl}(\bar{\omega},\bar{\omega},\bar{\omega}) = \tfrac{1}{3}\chi^{(3)}_{1;111}(\bar{\omega},\bar{\omega},\bar{\omega})(\delta_{ij}\delta_{kl} + \delta_{ik}\delta_{jl} + \delta_{il}\delta_{jk}) \tag{8.180}
$$

for third-harmonic generation in an isotropic medium.

However, the unique feature of the third-order nonlinear susceptibility is the component at the original frequency ($\omega = \bar{\omega} + \bar{\omega} - \bar{\omega}$), which leads to a nonlinear index of refraction. In describing the nonlinear refractive index the important coefficients are of the type $\chi^{(3)}_{i;jkl}(\bar{\omega}, \bar{\omega}, -\bar{\omega})$, but from intrinsic permutation symmetry we see that in this case

$$
\chi^{(3)}_{1;122}(\bar{\omega},\bar{\omega},-\bar{\omega}) = \chi^{(3)}_{1;212}(\bar{\omega},\bar{\omega},-\bar{\omega}) \tag{8.181}
$$

This leaves just two independent coefficients, and (8.178) becomes

$$\chi_{i;jkl}^{(3)}(\bar{\omega}, \bar{\omega}, -\bar{\omega}) = \chi_{1;122}^{(3)}(\bar{\omega}, \bar{\omega}, -\bar{\omega})(\delta_{ij}\delta_{kl} + \delta_{ik}\delta_{jl}) + \chi_{1;221}^{(3)}(\bar{\omega}, \bar{\omega}, -\bar{\omega})\delta_{il}\delta_{jk} \quad (8.182)$$

If we substitute the general expression (8.178) for the third-order susceptibility into (8.169) and carry out the sums we find that the third-order polarization in an isotropic material is

$$\eta^3 \tilde{P}_i^{(3)}(\omega) = \varepsilon_0 \int_{-\infty}^{\infty} d\omega' \int_{-\infty}^{\infty} d\omega'' \int_{-\infty}^{\infty} d\omega''' \, \delta(\omega - \omega' - \omega'' - \omega''')$$

$$\times \left[\chi_{1;122}^{(3)}(\omega', \omega'', \omega''') \tilde{E}_i(\omega') \tilde{\mathbf{E}}(\omega'') \cdot \tilde{\mathbf{E}}(\omega''') \right.$$

$$+ \chi_{1;212}^{(3)}(\omega', \omega'', \omega''') \tilde{E}_i(\omega'') \tilde{\mathbf{E}}(\omega''') \cdot \tilde{\mathbf{E}}(\omega')$$

$$\left. + \chi_{1;221}^{(3)}(\omega', \omega'', \omega''') \tilde{E}_i(\omega''') \tilde{\mathbf{E}}(\omega') \cdot \tilde{\mathbf{E}}(\omega'') \right] \quad (8.183)$$

when we use intrinsic permutation symmetry. If we rearrange the dummy variables and change to vector notation, this becomes

$$\eta^3 \tilde{\mathbf{P}}^{(3)}(\omega) = \varepsilon_0 \int_{-\infty}^{\infty} d\omega' \int_{-\infty}^{\infty} d\omega'' \int_{-\infty}^{\infty} d\omega''' \, \tilde{\mathbf{E}}(\omega') \cdot \tilde{\mathbf{E}}(\omega'') \tilde{\mathbf{E}}(\omega''')$$

$$\times \chi_T^{(3)}(\omega', \omega'', \omega''') \delta(\omega - \omega' - \omega'' - \omega''') \quad (8.184)$$

where the total third-order susceptibility is

$$\chi_T^{(3)}(\omega', \omega'', \omega''') = \chi_{1;122}^{(3)}(\omega''', \omega', \omega'') + \chi_{1;212}^{(3)}(\omega'', \omega''', \omega') + \chi_{1;221}^{(3)}(\omega', \omega'', \omega''') \quad (8.185)$$

We can use (8.179) and (8.181) to simplify the expression for the nonlinear polarization in the special case when there is only a single frequency present in the electric field. Since the electric field $\mathbf{E}(t)$ is by definition real, when it consists of a cw wave at just a single frequency $\bar{\omega}$, we can express its Fourier transform in the form

$$\tilde{\mathbf{E}}(\omega) = \bar{\mathbf{E}}\delta(\omega - \bar{\omega}) + \bar{\mathbf{E}}^*\delta(\omega + \bar{\omega}) \quad (8.186)$$

If we substitute this into (8.184) and carry out the integrals over ω', ω'', and ω''', we get δ-functions at $\omega = \pm\bar{\omega}$ and $\omega = \pm 3\bar{\omega}$, so the third-order nonlinear polarization has the form

$$\tilde{\mathbf{P}}^{(3)}(\omega) = \bar{\mathbf{P}}^{(3)}(\bar{\omega})\delta(\omega - \bar{\omega}) + \bar{\mathbf{P}}^{(3)*}(\bar{\omega})\delta(\omega + \bar{\omega}) + \bar{\mathbf{P}}^{(3)}(3\bar{\omega})\delta(\omega - 3\bar{\omega})$$

$$+ \bar{\mathbf{P}}^{(3)*}(3\bar{\omega})\delta(\omega + 3\bar{\omega}) \quad (8.187)$$

The component $\bar{\mathbf{P}}^{(3)}(3\bar{\omega})$ is

$$\eta^3 \bar{\mathbf{P}}^{(3)}(3\bar{\omega}) = \varepsilon_0 \chi_{1;111}^{(3)}(\bar{\omega}, \bar{\omega}, \bar{\omega})(\bar{\mathbf{E}} \cdot \bar{\mathbf{E}})\bar{\mathbf{E}} \quad (8.188)$$

when we use (8.180). The component $\bar{\mathbf{P}}^{(3)}$ at $\omega = \bar{\omega}$ includes terms for $(\omega', \omega'', \omega''')$ corresponding to $(\bar{\omega}, \bar{\omega}, -\bar{\omega})$, $(\bar{\omega}, -\bar{\omega}, \bar{\omega})$, and $(-\bar{\omega}, \bar{\omega}, \bar{\omega})$. If we use intrinsic permutation symmetry and (8.181), we get

$$\eta^3 \bar{\mathbf{P}}^{(3)}(\bar{\omega}) = \varepsilon_0 \left[6\chi_{1;122}^{(3)}(\bar{\omega}, \bar{\omega}, -\bar{\omega})(\bar{\mathbf{E}} \cdot \bar{\mathbf{E}}^*)\bar{\mathbf{E}} + 3\chi_{1;221}^{(3)}(\bar{\omega}, \bar{\omega}, -\bar{\omega})(\bar{\mathbf{E}} \cdot \bar{\mathbf{E}})\bar{\mathbf{E}}^* \right] \quad (8.189)$$

This expression is valid for an isotropic medium even in the presence of dispersion. We have not assumed complete permutation symmetry or other symmetries that depend on time-reversal invariance. Conventionally, the coefficients appearing (8.189) are called A and $\frac{1}{2}B$, so the third-order polarization in an isotropic medium is

$$\eta^3 \bar{\mathbf{P}}^{(3)}(\bar{\omega}) = \varepsilon_0 \left[A(\bar{\mathbf{E}} \cdot \bar{\mathbf{E}}^*)\bar{\mathbf{E}} + \tfrac{1}{2}B(\bar{\mathbf{E}} \cdot \bar{\mathbf{E}})\bar{\mathbf{E}}^* \right] \tag{8.190}$$

where

$$A = 6\chi_{1;122}^{(3)}(\bar{\omega}, \bar{\omega}, -\bar{\omega}) \tag{8.191}$$

$$B = 6\chi_{1;221}^{(3)}(\bar{\omega}, \bar{\omega}, -\bar{\omega}) \tag{8.192}$$

Although the two terms in (8.190) look rather similar, they actually have very different characteristics and are responsible for different macroscopic effects. This is illustrated by the examples in the following sections, but first we examine the relative magnitude of the coefficients A and B.

In the absence of dispersion (which also implies the absence of dissipation) the third-order susceptibility coefficients exhibit Kleinman symmetry, so $\chi_{1;122}^{(3)} = \chi_{1;221}^{(3)}$ and $A = B = 6\chi_{1;122}^{(3)}$. The relationship $A = B$ is also true of a damped anharmonic oscillator. In this case we see by extension from (8.64) that the third-order susceptibility is

$$\chi_{i;jkl}^{(3)}(\omega', \omega'', \omega''') = - \sum_{p,q,r,s} \frac{\varepsilon_0^3 K_{p;qrs}^{(3)}}{2\pi\eta} \chi_{i;p}^{(1)}(\omega' + \omega'' + \omega''')\chi_{q;j}^{(1)}(\omega')\chi_{r;k}^{(1)}(\omega'')\chi_{s;l}^{(1)}(\omega''') \tag{8.193}$$

But in the isotropic case we have $\chi_{i;j}^{(1)}(\omega) = \chi_{1;1}^{(1)}(\omega)\delta_{ij}$, so that

$$\chi_{i;jkl}^{(3)}(\bar{\omega}, \bar{\omega}, -\bar{\omega}) = -\frac{\varepsilon_0^3 K_{i;jkl}^{(3)}}{2\pi\eta} \chi_{1;1}^{(1)}(\bar{\omega} + \bar{\omega} - \bar{\omega})\chi_{1;1}^{(1)}(\bar{\omega})\chi_{1;1}^{(1)}(\bar{\omega})\chi_{1;1}^{(1)}(-\bar{\omega}) \tag{8.194}$$

Since the force constants $K_{i;jkl}^{(3)}$ possess intrinsic permutation symmetry, we see that $\chi_{1;122}^{(3)}(\bar{\omega}, \bar{\omega}, -\bar{\omega}) = \chi_{1;221}^{(3)}(\bar{\omega}, \bar{\omega}, -\bar{\omega})$. Thus, to the extent that the polarization can be described by classical anharmonic oscillators, we have

$$B = A \tag{8.195}$$

even when the oscillator is damped, meaning that dispersion and dissipation are present.

However, many important nonlinear materials do not obey the simple relation (8.195). A contrasting example is provided by the nonlinear susceptibility of a collection of nonpolar molecules with an electronic polarizability that is not spherically symmetric, when the molecules are free to rotate in an electric field. We consider first the case of a dc electric field. To minimize their energy in the applied electric field, the molecules tend to align themselves along the field. This increases the volume polarizability in that direction. However, the alignment is opposed by thermal processes that randomize the alignment of the molecules. To lowest order the induced dipole moment is linear in the field, so the degree of alignment (a balance between the interaction of the induced dipole with the field on the one hand and thermal disorientation on the other) is quadratic in the field. In thermodynamic equilibrium, then, the polarization can be represented by the expression

$$\mathbf{P} = \varepsilon_0 \chi^{(1)} \mathbf{E} + \varepsilon_0 \chi^{(3)}(\mathbf{E} \cdot \mathbf{E})\mathbf{E} \tag{8.196}$$

up to third order in $\mathbf{E}$, where $\chi^{(1)}$ and $\chi^{(3)}$ are constants. The linear susceptibility $\chi^{(1)}$ depends on the spherically averaged molecular polarizability, and the third-order susceptibility $\chi^{(3)}$ depends on the departure from spherical symmetry.

At sufficiently low frequencies the molecules remain in thermodynamic equilibrium, and (8.196) provides a good description of the polarization. However, at frequencies greater than the rotational relaxation rate the molecular orientation cannot keep up with the rapidly changing electric field even though the electronic polarizability can. In this case the molecular rotation responds to the average (over a complete cycle) value of $\mathbf{E} \cdot \mathbf{E}$, and after a time on the order of the rotational relaxation time the average orientation of the molecules approaches an equilibrium that depends on the value of $\mathbf{E} \cdot \mathbf{E}$ averaged over an optical cycle. We therefore represent the polarization by the expression

$$P_i(t) = \varepsilon_0 \sum_j \alpha_{ij} E_j(t) \tag{8.197}$$

where the total polarizability up to second order in the electric field is given by the tensor

$$\alpha_{i;j} = \alpha_{i;j}^{(1)} + \sum_{k,l} \alpha_{i;jkl}^{(3)} \langle E_k E_l \rangle \tag{8.198}$$

and the brackets $\langle \, \rangle$ indicate an average over one cycle. But as discussed earlier, in an isotropic material the most general form of the tensors is

$$\alpha_{i;j}^{(1)} = \alpha^{(1)} \delta_{ij} \tag{8.199}$$

$$\alpha_{i;jkl}^{(3)} = \alpha_1^{(3)} \delta_{ik} \delta_{jl} + \alpha_2^{(3)} \delta_{il} \delta_{jk} + \alpha_3^{(3)} \delta_{ij} \delta_{kl} \tag{8.200}$$

for some constants $\alpha^{(1)}$ and $\alpha_i^{(3)}$. From (8.198) we see that

$$\alpha_{i;jkl}^{(3)} = \alpha_{i;jlk}^{(3)} \tag{8.201}$$

so we may write

$$\alpha_{i;jkl}^{(3)} = \alpha^{(3)} (\delta_{ik} \delta_{jl} + \delta_{il} \delta_{jk} - \kappa \delta_{ij} \delta_{kl}) \tag{8.202}$$

for some constants $\alpha^{(3)}$ and κ. However, the effect of the average electric field is only to reorient the molecules and redistribute the electronic polarizability. The sum of the polarizabilities along the three axes always remains constant, so the trace of the nonlinear component must vanish:

$$\sum_{i,k,l} \alpha_{i;ikl}^{(3)} \langle E_k E_l \rangle = 0 = \alpha^{(3)} (2 - 3\kappa) \langle \mathbf{E} \cdot \mathbf{E} \rangle \tag{8.203}$$

We therefore see that $\kappa = \frac{2}{3}$ and the polarization is

$$P_i = \varepsilon_0 \alpha^{(1)} E_i + \varepsilon_0 \alpha^{(3)} \sum_{j,k,l} \left(\delta_{ik} \delta_{jl} + \delta_{il} \delta_{jk} - \tfrac{2}{3} \delta_{ij} \delta_{kl} \right) \langle E_k E_l \rangle E_j \tag{8.204}$$

For a monochromatic field, we can invert the Fourier transform (8.186) to show that in the time domain

$$E_i = \frac{1}{\sqrt{2\pi}} (\bar{E}_i e^{-i\bar{\omega}t} + \bar{E}_i^* e^{i\bar{\omega}t}) \tag{8.205}$$

Averaged over a cycle, then, the cross terms cancel out and we are left with

$$\langle E_i E_j \rangle = \frac{1}{2\pi} (\bar{E}_i \bar{E}_j^* + \bar{E}_j \bar{E}_i^*) \tag{8.206}$$

If we now substitute back into (8.204), we find that the polarization in the time domain is

$$P_i = \frac{1}{\sqrt{2\pi}}(\bar{P}_i e^{-i\bar{\omega}t} + \bar{P}_i^* e^{i\bar{\omega}t}) \tag{8.207}$$

or in the frequency domain

$$\tilde{P}_i(\omega) = \bar{P}_i \delta(\omega - \bar{\omega}) + \bar{P}_i^* \delta(\omega + \bar{\omega}) \tag{8.208}$$

where

$$\bar{P}_i = \varepsilon_0 \alpha^{(1)} \bar{E}_i + \frac{\varepsilon_0 \alpha^{(3)}}{\pi} \sum_{j,k,l} \left(\delta_{ik}\delta_{jl} + \delta_{il}\delta_{jk} - \frac{2}{3}\delta_{ij}\delta_{kl}\right) \bar{E}_j \bar{E}_k \bar{E}_l^* \tag{8.209}$$

If we carry out the sums, we find that this may be expressed in vector form as

$$\bar{\mathbf{P}} = \varepsilon_0 \alpha^{(1)} \bar{\mathbf{E}} + \frac{\varepsilon_0 \alpha^{(3)}}{\pi}\left[\frac{1}{3}(\bar{\mathbf{E}} \cdot \bar{\mathbf{E}}^*)\bar{\mathbf{E}} + (\bar{\mathbf{E}} \cdot \bar{\mathbf{E}})\bar{\mathbf{E}}^*\right] \tag{8.210}$$

and comparing this with (8.190), we make the identifications

$$A = \frac{\alpha^{(3)}}{3\pi} \tag{8.211}$$

$$B = \frac{2\alpha^{(3)}}{\pi} \tag{8.212}$$

Thus, when the third-order nonlinear polarizability is caused by the rotational orientation of electronically polarizable molecules we see that

$$B = 6A \tag{8.213}$$

This result contrasts sharply with (8.195) and illustrates the variability that can be expected in the ratio B/A.

EXERCISE 8.7

Third-harmonic generation is possible in an atomic vapor, such as an alkali metal vapor, diluted with a buffer of some rare gas. Phase matching is achieved by using the anomalous dispersion of the alkali vapor to compensate the normal dispersion of the rare gas. Although the spectrum of the metal vapor is generally complicated by the presence of several transitions, we can get the flavor of the situation by considering just a single, one-electron resonance transition at a wavelength λ_r, to which we assign a total oscillator strength f_r. To be specific, we consider a vapor of rubidium atoms ($\lambda_r = 0.8$ μm, $f_r \approx 1$) in a background of xenon gas, illuminated by a laser at the wavelength $\bar{\lambda} = 1.06$ μm. The rubidium vapor has a number density $N_{Rb} = 1 \times 10^{24}$ atoms/m^3, which can be achieved in an oven at around 400°C.

(a) Use the Kramers–Kronig relations to show that for a material with a single, narrow absorption feature at wavelength λ_r, the index of refraction at a wavelength λ far from the transition is

$$n - 1 \approx \frac{\lambda_r^2 r_c N_e f_r}{2\pi} \frac{\lambda^2}{\lambda^2 - \lambda_r^2} \tag{8.214}$$

where r_c is the classical electron radius, N_e is the number density of electrons participating in the optical transition, and f_r is the oscillator strength per electron. Find the index of refraction of the rubidium vapor at the fundamental and the third harmonic of the incident laser beam.

(b) The index of refraction of xenon is dominated by the six outer-shell p-electrons, which have resonance transitions in the ultraviolet beginning at $\lambda_r = 147$ nm. If we assign each of these electrons an oscillator strength $f_r \approx 1$, we can estimate the index of refraction in the visible and near-ultraviolet parts of the spectrum in the same way we estimated the refractive index of the alkali atoms. At standard temperature and pressure ($p = 760$ Torr and $T = 273$ K) the refractive index of xenon at $\lambda = 589$ nm is $n - 1 = 0.702 \times 10^{-3}$. How good is the approximate formula (8.214) at this wavelength?

(c) Based on the approximate formula, what number density of xenon is required to achieve phase matching at the third harmonic?

EXERCISE 8.8

As discussed in Exercise 8.7, phase-matched third-harmonic generation in alkali metal vapors is possible when a rare-gas buffer is used to match the refractive indices at the first and third harmonics.

(a) Following the analysis used for second-harmonic generation, show that if there is no incident wave at the third harmonic and the waves are phase matched, then third-harmonic generation obeys the equation

$$\frac{d^2 u_3}{dz^2} = -3\kappa^2 (1 - u_3 u_3^*)^2 u_3 \qquad (8.215)$$

where

$$u_3 = \frac{E_3}{|E_1(0)|} \qquad (8.216)$$

is the normalized amplitude, in which $E_1(z)$ and $E_3(z)$ are the Fourier amplitudes of the fields at the first and third harmonics and the conversion rate is

$$\kappa = \frac{3\bar{\omega}}{2n(\bar{\omega})} \chi^{(3)}(\bar{\omega}, \bar{\omega}, \bar{\omega}) |E_1(0)|^2 = \frac{3\pi \mu_0 \bar{\omega}}{2n^2(\bar{\omega})} \chi^{(3)}(\bar{\omega}, \bar{\omega}, \bar{\omega}) \langle S_1(0) \rangle \qquad (8.217)$$

Ignore terms of the type $\bar{\omega} - \bar{\omega} + 3\bar{\omega}$ and $3\bar{\omega} - 3\bar{\omega} + 3\bar{\omega}$ in the equation for E_3 and similar terms in the equation for E_1, since these represent the nonlinear index of refraction at the frequency $3\bar{\omega}$ induced by E_3 and E_1, respectively. They play a role in the phase-matching condition but do not contribute directly to third-harmonic generation.

(b) Use the simple anharmonic oscillator model (8.21) to estimate the third-order susceptibility $\chi^{(3)}(\bar{\omega}, \bar{\omega}, \bar{\omega})$ of rubidium at the fundamental wavelength $\bar{\lambda} = 1.06$ μm, assuming that the susceptibility is dominated by the one-electron resonance at $\lambda_r = 0.8$ μm. For an incident laser intensity $\langle S_1(0) \rangle = 10^{12}$ W/m^2 on the fundamental, what is the characteristic length $(1/\kappa)$ for conversion to the third harmonic when the density of the rubidium vapor is $N_{\text{Rb}} = 1 \times 10^{24}$ /m^3?

EXERCISE 8.9

Consider a liquid composed of nonspherical, but nonpolar molecules that are free to rotate in an electric field but whose rotations are damped by collisions with neighboring molecules. To be specific, assume that the molecules are prolate spheroids (egg shaped), so the molecular polarizability is greater along the major axis than transverse to it. To minimize the energy in an applied electric field the molecules tend to align themselves along the field, but collisions with neighboring molecules disturb the alignment. After a short time (the rotational relaxation time) the molecular alignment approaches thermodynamic equilibrium. The molecular alignment increases the volume polarizability to a degree that depends on the electric field causing the alignment. If the molecular polarizability along the symmetry axis is α_2 and that transverse to the symmetry axis is α_1, then it is easy to show that the dipole moment induced by the electric field $\mathbf{E}$ is

$$\mathbf{p}_{\text{mol}} = \varepsilon_0 (\alpha_2 - \alpha_1)(\mathbf{E} \cdot \hat{\mathbf{n}})\hat{\mathbf{n}} + \varepsilon_0 \alpha_1 \mathbf{E} \tag{8.218}$$

where $\hat{\mathbf{n}}$ is a unit vector along the symmetry axis. If the induced dipole moment depends on the applied field, the energy of the dipole in the field is

$$\mathcal{U} = -\int_0^{\mathbf{E}} \mathbf{p}_{\text{mol}} \cdot d\mathbf{E} \tag{8.219}$$

Following along the lines of the discussion in Chapter 7, show that in the high-temperature limit, $\varepsilon_0 (\alpha_2 - \alpha_1) E^2 / k_B T \ll 1$, where k_B is Boltzmann's constant and T is the temperature, the polarization is

$$\mathbf{P} = \varepsilon_0 \chi^{(1)} \mathbf{E} + \varepsilon_0 \chi^{(3)} (\mathbf{E} \cdot \mathbf{E}) \mathbf{E} \tag{8.220}$$

up to third order in $\mathbf{E}$, where the linear and third-order susceptibilities are

$$\chi^{(1)} = \tfrac{1}{3} N (\alpha_2 + 2\alpha_1) \tag{8.221}$$

and

$$\chi^{(3)} = \frac{4}{45} \frac{\varepsilon_0 N (\alpha_2 - \alpha_1)^2}{k_B T} \tag{8.222}$$

and N is the number density of molecules. In liquids, the effect of the neighboring molecules on the field experienced by a given molecule must be taken into account as discussed in connection with the Clausius–Mossotti relation (see Chapter 7), but ignore this for now.

8.3.2 Wave Equation with a Nonlinear Index of Refraction

When the nonlinear susceptibility is of third order, the wave equation (8.85) becomes

$$\nabla^2 \tilde{\mathbf{E}} + k^2 \tilde{\mathbf{E}} = -\frac{\eta^3}{\varepsilon_0 n^2(\omega)} \left[\nabla \left(\nabla \cdot \tilde{\mathbf{P}}^{(3)} \right) + k^2 \tilde{\mathbf{P}}^{(3)} \right] \tag{8.223}$$

where the wave vector satisfies the linear dispersion relation

$$c^2 k^2(\omega) = n^2(\omega) \omega^2 \tag{8.224}$$

$$n^2(\omega) = 1 + \chi^{(1)}(\omega) \tag{8.225}$$

The right-hand side of (8.223) is a perturbation that we can approximate and simplify. In many cases, the term $\nabla(\nabla \cdot \tilde{\mathbf{P}}^{(3)})$ is much smaller than the term $k^2 \tilde{\mathbf{P}}^{(3)}$ and can be ignored. For example, for plane waves propagating through isotropic media, $\tilde{\mathbf{E}}$ and $\tilde{\mathbf{P}} = \tilde{\mathbf{P}}^{(1)} + \eta^3 \tilde{\mathbf{P}}^{(3)}$ are perpendicular to the direction of propagation. Since the derivatives vanish in the transverse directions, we see that $\nabla \cdot \tilde{\mathbf{P}}^{(3)} = 0$ identically. We are typically, interested in waves that are at least locally planar, and we therefore ignore the term $\nabla \cdot \mathbf{P}^{(3)}$ in the wave equation.

If we substitute (8.186) for $\tilde{\mathbf{E}}$, use (8.187) and (8.190) for $\tilde{\mathbf{P}}^{(3)}$, and then integrate over $\bar{\omega} - \varepsilon < \omega < \bar{\omega} + \varepsilon$ to select the term at $\bar{\omega}$, we get

$$\nabla^2 \bar{\mathbf{E}} + k^2 \bar{\mathbf{E}} = -\frac{k^2}{n^2}\left[A(\bar{\mathbf{E}} \cdot \bar{\mathbf{E}}^*)\bar{\mathbf{E}} + \frac{1}{2}B(\bar{\mathbf{E}} \cdot \bar{\mathbf{E}})\bar{\mathbf{E}}^* \right] \tag{8.226}$$

Although they appear similar, the terms in A and B have distinctly different characteristics. For example, for circularly polarized waves the polarization vectors have the property $\hat{\mathbf{e}}_\pm^* = \hat{\mathbf{e}}_\mp$, so the B term mixes the polarizations. To see this better we consider an elliptically polarized plane wave traveling in the $\hat{\mathbf{z}}$ direction, which we write in the form

$$\bar{\mathbf{E}} = E_+\hat{\mathbf{e}}_+ e^{ik_+z} + E_-\hat{\mathbf{e}}_- e^{ik_-z} \tag{8.227}$$

If we substitute this into the wave equation (8.226) and take the scalar product with $\hat{\mathbf{e}}_\pm$ using the orthogonality properties of the circularly polarized vectors discussed in Chapter 4, we get

$$k_\pm^2 = k^2\left[1 + \frac{A}{n^2}E_\pm E_\pm^* + \frac{A+B}{n^2}E_\mp E_\mp^* \right] \tag{8.228}$$

This shows that $|k_+| \neq |k_-|$, in general, so the left-hand and right-hand circularly polarized components do not propagate with the same phase velocity. In fact, the amplitude of one component affects the propagation of the other, slowing it down when $(A + B) > 0$.

Because the left-hand and right-hand circularly polarized components of a wave travel at different phase velocities, the direction of polarization rotates as the wave propagates. From (8.228) we see that the difference between the wave vectors is

$$\Delta k = k_+ - k_- = \frac{k^2}{n^2}B(E_-E_-^* - E_+E_+^*) \tag{8.229}$$

It is interesting that this depends only on the coefficient B and not on A, which emphasizes the difference between the terms A and B. If we define the mean wave vector

$$\bar{k} = \tfrac{1}{2}(k_+ + k_-) \tag{8.230}$$

we can write the electric field (8.227) in the form

$$\bar{\mathbf{E}}e^{-i\omega t} = \left(E_+\hat{\mathbf{e}}_+ e^{i(1/2)\Delta kz} + E_-\hat{\mathbf{e}}_- e^{-i(1/2)\Delta kz} \right)e^{i(\bar{k}z - \omega t)} \tag{8.231}$$

But referring to Figure 8.13 we see that $\hat{\mathbf{e}}_+ e^{-i\omega t}$ represents a vector rotating to the left, in the positive (counterclockwise) direction, through the angle ωt, so $\hat{\mathbf{e}}_+ e^{i(1/2)\Delta kz}$ represents a vector rotated through the angle

$$\theta = -\tfrac{1}{2}\Delta kz \tag{8.232}$$

The same is true for the vector $\hat{\mathbf{e}}_- e^{-i(1/2)\Delta kz}$, that is, it is rotated in the same direction through the same angle. Thus, the polarization of the wave is rotated by the nonlinear

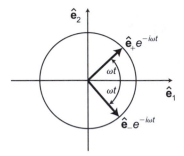

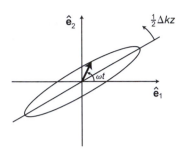

Figure 8.13 Rotation of circularly polarized vectors.

Figure 8.14 Elliptically polarized wave.

interaction through the angle θ after a distance z. The rotation disappears for a linearly polarized wave, as it must by symmetry, for in this case the components $|E_+|$ and $|E_-|$ are equal, and Δk vanishes. For an elliptically polarized wave we consider the case $|E_+| > |E_-|$, meaning that the wave is left-hand elliptically polarized, as shown in Figure 8.14. Then for $B > 0$ (the usual case) we have $\Delta k < 0$, and the polarization rotates to the left, in the direction of the stronger component. The rotation is to the right for a right-hand elliptically polarized wave (for $B > 0$).

The nonlinear coefficients A and B are usually positive, so the effective index of refraction

$$n_\pm^2 = n^2 + A E_\pm E_\pm^* + (A + B) E_\mp E_\mp^* \tag{8.233}$$

is generally largest where the optical beam is most intense. For a beam of finite transverse dimensions this is ordinarily near the center of the beam, so the nonlinear medium acts like a positive lens. This causes the beam to be "self-focused," which increases the intensity of the beam, and this, in turn, further increases the self-focusing effect. The process can be unstable and often catastrophic, leading to damage of the optical material when the intensity or the fluence of the beam exceeds the threshold for optical damage. Of course, self-focusing is counteracted by the effects of diffraction. Which of these effects dominates depends, in general, on the total power of the optical beam but not on its size. To see this, we note that the focal length f of a lens is

$$\frac{1}{f} = \frac{2t \Delta n}{D^2} \tag{8.234}$$

where Δn is the difference of refractive index between the lens and the surrounding medium, t the thickness of the lens at its center, and D its diameter. Since the thickness of the "lens" in this case is on the order of the focal length itself, this suggests that the characteristic length for self-focusing is on the order of

$$L_f = \sqrt{2tf} = \frac{D}{\sqrt{\Delta n}} \tag{8.235}$$

As discussed in Chapter 9, the characteristic length for diffraction is

$$L_d = \frac{D^2}{\lambda} \tag{8.236}$$

The criterion for self-focusing to occur is therefore

$$\frac{L_d^2}{L_f^2} = \frac{D^2}{\lambda^2}\Delta n \geq O(1) \tag{8.237}$$

But the change in the refractive index is

$$\Delta n = O\left(\frac{A|\bar{E}|^2}{n}\right) = O\left(\frac{\pi A \langle S \rangle}{n^2 c \varepsilon_0}\right) \tag{8.238}$$

where $\langle S \rangle$ is the intensity (Poynting vector) of the optical beam near its center. The criterion for self-focusing is therefore

$$\frac{4AP}{\varepsilon_0 c \lambda_{\text{vac}}^2} \geq O(1) \tag{8.239}$$

where $P \approx \frac{1}{4}\pi D^2 \langle S \rangle$ is the total power in the optical beam and λ_{vac} is the wavelength in vacuum. Note that this expression is independent of the diameter of the beam and depends only on the total power. For example, for a laser beam with a wavelength $\lambda_{\text{vac}} = 1$ μm propagating through a molecular liquid such as CS_2, for which $A = O(10^{-19})$, the critical power for self-focusing is $P = O(10\,\text{kW})$.

EXERCISE 8.10

Show that for $B > 0$, the usual case, linearly polarized waves travel more slowly than circularly polarized waves of the same intensity.

EXERCISE 8.11

Following the discussion of second-harmonic generation, derive the coupled-amplitude equation for a third-order nonlinear susceptibility with an electric field consisting of two frequencies. This can be applied to frequency tripling and to stimulated Raman scattering.

EXERCISE 8.12

When a single-frequency electromagnetic pulse propagates through a medium with a third-order nonlinear susceptibility, the propagation is affected by an effect called the nonlinear index of refraction. This changes both the effective group velocity and the effective phase velocity throughout the pulse in a way that generally causes the pulse to spread. However, under special circumstances the effects cancel out and the pulse forms what is called a soliton. A soliton propagates without change and may offer a way to send high-speed data pulses over fiber-optic cables much farther than dispersion would ordinarily allow. To see how this happens, we need to derive the equation for nonlinear pulse propagation and then show that the soliton propagates without spreading.

(a) If we ignore the term $\nabla(\nabla \cdot \tilde{\mathbf{P}}^{(3)})$ in (8.223), as discussed earlier, a linearly polarized plane-wave pulse traveling along the z axis is described by the equation

$$\frac{\partial^2 \tilde{E}}{\partial z^2} + k^2 \tilde{E} = -\frac{\eta^3 k^2}{\varepsilon_0 n^2(\omega)} \tilde{P}^{(3)} \tag{8.240}$$

where the third-order polarization is

$$\eta^3 \tilde{P}^{(3)}(\omega) = \varepsilon_0 \int_{-\infty}^{\infty} d\omega' \int_{-\infty}^{\infty} d\omega'' \int_{-\infty}^{\infty} d\omega''' \tilde{E}(\omega') \tilde{E}(\omega'') \tilde{E}(\omega''')$$

$$\times \chi_T^{(3)}(\omega', \omega'', \omega''') \delta(\omega - \omega' - \omega'' - \omega''') \tag{8.241}$$

If the pulse has a narrow spectrum centered on the frequency ω_0, we may make the approximations

$$\chi_T^{(3)}(\omega', \omega'', \omega''') = \chi_T^{(3)}(\omega_0, \omega_0, \omega_0) = \chi_0^{(3)} \tag{8.242}$$

$$\tilde{E}(z, \omega) = \tilde{E}_0(z, \omega - \omega_0) e^{ik_0 z} + \tilde{E}_0^*(z, -\omega - \omega_0) e^{-ik_0 z} \tag{8.243}$$

where $k_0 c = n(\omega_0) \omega_0$. Using these approximations and ignoring third-harmonic terms, invert the Fourier transform (8.241) to show that

$$\frac{\eta^3}{\varepsilon_0} P^{(3)}(z, t) = 6\pi \chi_0^{(3)} E_0 E_0^* \left[E_0 e^{i(k_0 z - \omega_0 t)} + E_0^* e^{-i(k_0 z - \omega t)} \right] \tag{8.244}$$

Hint: Use the fact that

$$\delta(\omega - \omega' - \omega'' - \omega''') = \frac{1}{2\pi} \int_{-\infty}^{\infty} dt' e^{i(\omega - \omega' - \omega'' - \omega''')t'} \tag{8.245}$$

(b) To simplify the left-hand side of (8.240), use the slowly varying amplitude approximation and the Taylor expansion

$$k^2 - k_0^2 = 2k_0 \left(\frac{dk}{d\omega}\right)_0 (\omega - \omega_0) + \frac{1}{2} \left(\frac{d^2 k^2}{d\omega^2}\right)_0 (\omega - \omega_0)^2 \tag{8.246}$$

and invert the Fourier transform to show that

$$i2k_0 \frac{\partial E_0}{\partial z} + i2k_0 \left(\frac{dk}{d\omega}\right)_0 \frac{\partial E_0}{\partial t} - \frac{1}{2} \left(\frac{d^2 k^2}{d\omega^2}\right)_0 \frac{\partial^2 E_0}{\partial t^2} = \frac{6\pi \chi_0 \omega_0^2}{c^2} E_0 E_0^* E_0 \tag{8.247}$$

(c) To simplify this equation further, change to the variables

$$\zeta = z \tag{8.248}$$

$$\tau = t - \frac{z}{v_g} \tag{8.249}$$

and

$$\bar{E}(\zeta, \tau) = E_0(z, t) \tag{8.250}$$

moving at the group velocity $v_g = (dk/d\omega)_0$, to get the pulse propagation equation

$$i2k_0 \frac{\partial \bar{E}}{\partial \zeta} - \frac{1}{2} \left(\frac{d^2 k^2}{d\omega^2}\right)_0 \frac{\partial^2 \bar{E}}{\partial \tau^2} = \frac{6\pi \chi_0 \omega_0^2}{c^2} \bar{E} \bar{E}^* \bar{E} \tag{8.251}$$

This is sometimes called the nonlinear Schroedinger equation.

(d) Show that the soliton

$$\bar{E} = A \operatorname{sech}\left(\frac{\tau}{\tau_0}\right) e^{i\kappa_0 \zeta} \tag{8.252}$$

satisfies (8.252) if

$$4 k_0 \kappa_0 \tau_0^2 = -\left(\frac{d^2 k^2}{d\omega^2}\right)_0 \tag{8.253}$$

$$k_0 \kappa_0 c^2 = 3\pi \chi_0^{(3)} \omega_0^2 A A^* \tag{8.254}$$

From these conditions, we see that the shape of the pulse (characterized by κ_0 and τ_0) depends on the intensity $A A^*$ of the pulse, and that the soliton can exist only if the third-order susceptibility $\chi_0^{(3)}$ and the group-velocity dispersion $d^2 k^2 / d\omega^2$ have opposite signs.

8.3.3 Phase-Conjugate Reflection

When light reflects from an ordinary mirror, the rays reflect obliquely as shown in Figure 8.15. It is possible, however, to use nonlinear optics to form a special type of mirror, called a phase-conjugate reflector, that reflects the incident rays directly back on themselves. This is somewhat akin to the reflection of light from a bicycle reflector formed by an array of small corner cubes. The rays reflect back toward the source (slightly displaced, in this case, by the finite dimensions of the corner cubes) independent of the angle of incidence relative to the bicycle reflector as a whole. Alternatively, we can view light as waves, rather than rays. The reflection of a wave front from an ordinary mirror is illustrated in Figure 8.16, along with the reflection of a wave front from a phase-conjugate reflector. As indicated schematically, the wave fronts reflect from the phase-conjugate reflector with the distortion

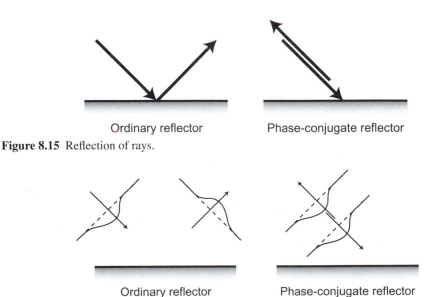

Ordinary reflector   Phase-conjugate reflector

Figure 8.15 Reflection of rays.

Ordinary reflector   Phase-conjugate reflector

Figure 8.16 Reflection of wave fronts.

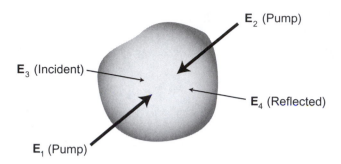

$\mathbf{E}_2$ (Pump)

$\mathbf{E}_3$ (Incident)

$\mathbf{E}_4$ (Reflected)

$\mathbf{E}_1$ (Pump)

Figure 8.17 Geometry for phase-conjugate reflection.

"time reversed." The usefulness of the process is due in large part to the fact that waves distorted by an inhomogeneous medium or an imperfect optical system on their way to the reflector are reflected with the opposite distortion, and as the waves travel back through the medium or optical system the distortion is canceled. In the ray picture of this process, the reflected rays retrace their path through the optical system to reconstruct the original image.

The geometry used for phase-conjugate reflection is indicated in Figure 8.17. Four waves are incident on the nonlinear medium. Of these, waves $\mathbf{E}_1$ and $\mathbf{E}_2$ are called the pump beams, $\mathbf{E}_3$ is the incident beam, and $\mathbf{E}_4$ is the reflected beam. Typically the pump beams $\mathbf{E}_1$ and $\mathbf{E}_2$ are counterpropagating plane waves, but they can actually be any two waves that are complex conjugates of one another. They are presumed to be much more intense than the incident and reflected beams $\mathbf{E}_3$ and $\mathbf{E}_4$. All four waves have the same frequency $\bar{\omega}$.

There are several ways that we can understand the reflection process caused by the interaction of these four waves. If we view the interaction as a four-wave mixing process, called degenerate four-wave mixing because all the frequencies are the same, the requirements of phase matching demand that the wave vectors sum to zero,

$$\mathbf{k}_1 + \mathbf{k}_2 + \mathbf{k}_3 + \mathbf{k}_4 = 0 \tag{8.255}$$

as discussed previously. But for counterpropagating (or phase-conjugate) pump beams

$$\mathbf{k}_2 = -\mathbf{k}_1 \tag{8.256}$$

so it follows that

$$\mathbf{k}_4 = -\mathbf{k}_3 \tag{8.257}$$

Thus, the rays are reflected back on themselves.

Alternatively, we can understand phase-conjugate reflection from a wave picture. Viewed this way, the pump wave $\mathbf{E}_1$ and the incident wave $\mathbf{E}_3$ interfere to form a three-dimensional intensity pattern that, through the nonlinear index of refraction, forms a three-dimensional grating structure, like a hologram, as indicated in Figure 8.18. The other pump beam $\mathbf{E}_2$ is then reflected off this hologram to form the reflected beam $\mathbf{E}_4$.

Since it involves the nonlinear interaction of four waves, the mathematical description of phase-conjugate reflection is inevitably a bit complicated. We begin with the wave equation (8.226). To simplify the discussion we assume that the four interacting waves are all polarized in the same direction, so the wave equation becomes simply

$$\nabla^2 \bar{E} + k^2 \bar{E} = -\frac{k^2}{n^2} \left(A + \frac{1}{2} B \right) \bar{E} \bar{E} \bar{E}^* \tag{8.258}$$

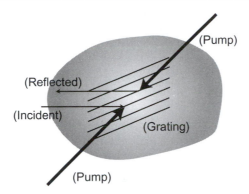

Figure 8.18 Grating formed by the incident beam and a pump beam.

The total electric field is composed of the four individual waves, and we express it in the form

$$\bar{E}(\mathbf{r}) = \bar{E}_1(\mathbf{r})e^{i\mathbf{k}_1 \cdot \mathbf{r}} + \bar{E}_2(\mathbf{r})e^{i\mathbf{k}_2 \cdot \mathbf{r}} + \bar{E}_3(\mathbf{r})e^{i\mathbf{k}_3 \cdot \mathbf{r}} + \bar{E}_4(\mathbf{r})e^{i\mathbf{k}_4 \cdot \mathbf{r}} \tag{8.259}$$

where $\bar{E}_i(\mathbf{r})$ is the Fourier component of the ith wave, as discussed earlier, and the wave vectors $\mathbf{k}_i$ satisfy (8.256) and (8.257). This ignores all the harmonics that are inevitably generated by the nonlinear processes, which we justify on the grounds that they are not phase matched and do not grow significantly. The functions $\bar{E}_i(\mathbf{r})$ are presumed to be slowly varying, so that while the waves are not strictly plane waves they retain their identity as waves traveling (more or less) in the directions $\mathbf{k}_i$. This allows us to sort out the components of the total field in (8.258) and to make the slowly varying envelope approximation for each wave.

To see how this works, we examine first the left-hand side of (8.258). Since this side is linear in the fields, we can treat the waves individually and get

$$(\nabla^2 - k^2)\bar{E}_i e^{i\mathbf{k}_i \cdot \mathbf{r}} = (\nabla^2 \bar{E}_i + i2\mathbf{k}_i \cdot \nabla \bar{E}_i)e^{i\mathbf{k}_i \cdot \mathbf{r}} \tag{8.260}$$

For a "slowly varying" function $\bar{E}_i$, we ignore the highest derivative

$$\nabla^2 \bar{E}_i \ll \mathbf{k}_i \cdot \nabla \bar{E}_i \tag{8.261}$$

On the right-hand side, if we substitute (8.259) into (8.258) we get a huge number of terms. However, most of the terms contain factors like $e^{i3\mathbf{k}_1 \cdot \mathbf{r}}$, $e^{i(2\mathbf{k}_1 + \mathbf{k}_3) \cdot \mathbf{r}}$, and so on. Like the harmonics we ignore in (8.259), these terms must be small. We therefore ignore these waves and sort out the rest of the terms on the left and right according to the directions $(\mathbf{k}_1 \ldots \mathbf{k}_4)$ they are traveling. When we do this and use (8.256), (8.257), and (8.261), we get the equations

$$i2\mathbf{k}_1 \cdot \nabla \bar{E}_1 = -\frac{k^2}{n^2}\left(A + \frac{1}{2}B\right)[\bar{E}_1(\bar{E}_1\bar{E}_1^* + 2\bar{E}_2\bar{E}_2^* + 2\bar{E}_3\bar{E}_3^* + 2\bar{E}_4\bar{E}_4^*) + 2\bar{E}_2^*\bar{E}_3\bar{E}_4]$$

$$\tag{8.262}$$

$$i2\mathbf{k}_2 \cdot \nabla \bar{E}_2 = -\frac{k^2}{n^2}\left(A + \frac{1}{2}B\right)[\bar{E}_2(\bar{E}_2\bar{E}_2^* + 2\bar{E}_1\bar{E}_1^* + 2\bar{E}_3\bar{E}_3^* + 2\bar{E}_4\bar{E}_4^*) + 2\bar{E}_1^*\bar{E}_3\bar{E}_4]$$

$$\tag{8.263}$$

$$i2\mathbf{k}_3 \cdot \nabla \bar{E}_3 = -\frac{k^2}{n^2}\left(A + \frac{1}{2}B\right)[\bar{E}_3(\bar{E}_3\bar{E}_3^* + 2\bar{E}_1\bar{E}_1^* + 2\bar{E}_2\bar{E}_2^* + 2\bar{E}_4\bar{E}_4^*) + 2\bar{E}_4^*\bar{E}_1\bar{E}_2]$$

(8.264)

$$i2\mathbf{k}_4 \cdot \nabla \bar{E}_4 = -\frac{k^2}{n^2}\left(A + \frac{1}{2}B\right)[\bar{E}_4(\bar{E}_4\bar{E}_4^* + 2\bar{E}_1\bar{E}_1^* + 2\bar{E}_2\bar{E}_2^* + 2\bar{E}_3\bar{E}_3^*) + 2\bar{E}_3^*\bar{E}_1\bar{E}_2]$$

(8.265)

To solve these coupled equations, we introduce the small parameter $\nu \ll 1$ and assume that the incident and reflected beams $\bar{E}_3$ and $\bar{E}_4$ are of order ν compared with the pump beams $\bar{E}_1$ and $\bar{E}_2$. We then expand

$$\bar{E}_1 = \bar{E}_1^{(0)} + \nu \bar{E}_1^{(1)} + \cdots$$

(8.266)

$$\bar{E}_2 = \bar{E}_2^{(0)} + \nu \bar{E}_2^{(1)} + \cdots$$

(8.267)

$$\bar{E}_3 = \nu \bar{E}_3^{(1)} + \cdots$$

(8.268)

$$\bar{E}_4 = \nu \bar{E}_4^{(1)} + \cdots$$

(8.269)

To zeroth order we get

$$i2\mathbf{k}_1 \cdot \nabla \bar{E}_1^{(0)} = -\frac{k^2}{n^2}\left(A + \frac{1}{2}B\right)\left(\bar{E}_1^{(0)}\bar{E}_1^{(0)*} + 2\bar{E}_2^{(0)}\bar{E}_2^{(0)*}\right)\bar{E}_1^{(0)}$$

(8.270)

$$i2\mathbf{k}_2 \cdot \nabla \bar{E}_2^{(0)} = -\frac{k^2}{n^2}\left(A + \frac{1}{2}B\right)\left(\bar{E}_2^{(0)}\bar{E}_2^{(0)*} + 2\bar{E}_1^{(0)}\bar{E}_1^{(0)*}\right)\bar{E}_2^{(0)}$$

(8.271)

We see that for real $(A + \frac{1}{2}B)$ and n, the coefficient of $\bar{E}_1^{(0)}$ on the right side of (8.270) is purely real while the operator on the left is purely imaginary. Therefore, the variation of $\bar{E}_1^{(0)}$ in the direction of propagation $\mathbf{k}_1$ is purely one of phase, and the amplitude is constant. This phase variation is just the effect of the nonlinear change of the index of refraction caused by the intensity of pump beams $\bar{E}_1^{(0)}$ and $\bar{E}_2^{(0)}$. Note that the effect of $\bar{E}_2^{(0)}$ on $\bar{E}_1^{(0)}$ is twice as big as the effect of $\bar{E}_1^{(0)}$ on itself. The same remarks apply to $\bar{E}_2^{(0)}$. But $\mathbf{k}_2 = -\mathbf{k}_1$, so if we take the complex conjugate of (8.271), we get

$$i2\mathbf{k}_1 \cdot \nabla \bar{E}_2^{(0)*} = -\frac{k^2}{n^2}\left(A + \frac{1}{2}B\right)\left(\bar{E}_2^{(0)}\bar{E}_2^{(0)*} + 2\bar{E}_1^{(0)}\bar{E}_1^{(0)*}\right)\bar{E}_2^{(0)*}$$

(8.272)

Comparing this with (8.270), we see that both equations are satisfied if

$$\bar{E}_1^{(0)} = \bar{E}_2^{(0)*} = \bar{E}_p e^{i\phi(\mathbf{k}_1 \cdot \mathbf{r})}$$

(8.273)

where the pump amplitude $\bar{E}_p$ is independent of the distance $\mathbf{k}_1 \cdot \mathbf{r}$ in the direction of propagation, but may depend on the coordinates transverse to the propagation of the wave. This is not the only solution of (8.270) and (8.271), but it is the one we choose for phase-conjugate reflection. It is in the sense of (8.273) that the pump waves $\bar{E}_1^{(0)}$ and $\bar{E}_2^{(0)}$ must be the complex conjugates of one another. When this is the case, the equations for $\bar{E}_3^{(1)}$ and $\bar{E}_4^{(1)}$ assume a particularly simple and symmetric form.

To first order in ν, the coupled wave equations (8.264) and (8.265) become

$$i\mathbf{k}_3 \cdot \nabla \bar{E}_3^{(1)} = -\frac{k^2}{n^2}\left(A + \frac{1}{2}B\right)\bar{E}_p\bar{E}_p^*\left(2\bar{E}_3^{(1)} + \bar{E}_4^{(1)*}\right)$$

(8.274)

$$i\mathbf{k}_4 \cdot \nabla \bar{E}_4^{(1)} = -\frac{k^2}{n^2}\left(A + \frac{1}{2}B\right)\bar{E}_p\bar{E}_p^*\left(2\bar{E}_4^{(1)} + \bar{E}_3^{(1)*}\right)$$

(8.275)

If we change to the variable

$$\zeta = \frac{2}{n^2}\left(A + \frac{1}{2}B\right)\mathbf{k}_3 \cdot \int \bar{E}_p \bar{E}_p^* \, d\mathbf{r} \tag{8.276}$$

which represents an effective distance in the direction of propagation of the incident beam, and recall that $\mathbf{k}_4 = -\mathbf{k}_3$, then (8.274) and (8.275) become

$$\frac{d\bar{E}_3^{(1)}}{d\zeta} - i\bar{E}_3^{(1)} = i\frac{1}{2}\bar{E}_4^{(1)*} \tag{8.277}$$

$$-\frac{d\bar{E}_4^{(1)}}{d\zeta} - i\bar{E}_4^{(1)} = i\frac{1}{2}\bar{E}_3^{(1)*} \tag{8.278}$$

The left-hand side of each equation represents the phase shift due to the change in the index of refraction caused by the pump waves. The right-hand side represents the coupling between the incident and reflected waves caused by the pump waves.

We see from (8.277) and (8.278) that the reflected wave $\bar{E}_4^{(1)}$ is driven by the complex conjugate $\bar{E}_3^{(1)*}$ of the incident wave, and vice versa. This is why the reflected wave is the complex conjugate of the incident wave. From the minus sign in front of the derivative in (8.278) we see that the reflected wave $\bar{E}_4^{(1)}$ grows in the direction opposite that of the incident wave $\bar{E}_3^{(1)}$. To solve (8.277) and (8.278) for $\bar{E}_4^{(1)}$, we use the integrating factor $e^{-i\zeta}$ in (8.277) and $e^{i\zeta}$ in (8.278) to get

$$\frac{du_3}{d\zeta} = i\frac{1}{2}u_4^* \tag{8.279}$$

$$\frac{du_4}{d\zeta} = -i\frac{1}{2}u_3^* \tag{8.280}$$

where

$$u_3(\zeta) = \bar{E}_3^{(1)}(\zeta)e^{-i\zeta} \tag{8.281}$$

$$u_4(\zeta) = \bar{E}_4^{(1)}(\zeta)e^{i\zeta} \tag{8.282}$$

This phase shift accounts for the nonlinear index of refraction change caused by the pump waves. If we differentiate the first equation with respect to ζ and substitute the second, and vice versa, we get

$$\frac{d^2u_3}{d\zeta^2} + \frac{1}{4}u_3 = 0 \tag{8.283}$$

$$\frac{d^2u_4}{d\zeta^2} + \frac{1}{4}u_4 = 0 \tag{8.284}$$

which have the solutions

$$u_3(\zeta) = a_3 \cos\tfrac{1}{2}\zeta + b_3 \sin\tfrac{1}{2}\zeta \tag{8.285}$$

$$u_4(\zeta) = a_4 \cos\tfrac{1}{2}\zeta + b_4 \sin\tfrac{1}{2}\zeta \tag{8.286}$$

For boundary conditions we use the example illustrated in Figure 8.19. A slab of nonlinear material is illuminated by the pump beams $\bar{E}_p$ and $\bar{E}_p^*$, and the beam $\bar{E}_3$ is incident from the left. The reflected beam $\bar{E}_4$ emerges from the left side of the slab, but there

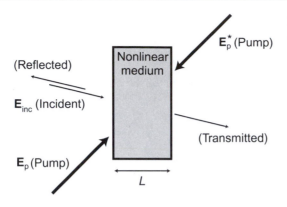

Figure 8.19 Reflection from a phase-conjugate mirror.

is no beam $\bar{E}_4$ incident from the right. For coordinates we call the front of the mirror $\zeta = 0$ and the back of the mirror $\zeta = Z$, where

$$Z = \frac{2}{n^2}\left(A + \frac{1}{2}B\right)\mathbf{k}_3 \cdot \int_L \bar{E}_p \bar{E}_p^* \, d\mathbf{r} \tag{8.287}$$

and the integral is along the incident ray through the entire thickness L of the nonlinear material. For the amplitudes we have the boundary conditions

$$u_3(0) = E_{\text{inc}} \tag{8.288}$$

$$u_4(Z) = 0 \tag{8.289}$$

where E_{inc} is the complex amplitude of the incident wave. From (8.279) and (8.280) we get the boundary conditions

$$\frac{du_3(Z)}{d\zeta} = 0 \tag{8.290}$$

$$\frac{du_4(0)}{d\zeta} = -i\frac{1}{2}E_{\text{inc}}^* \tag{8.291}$$

The solution is then conveniently expressed in the form

$$u_3(\zeta) = E_{\text{inc}}\frac{\cos\left[\frac{1}{2}(\zeta - Z)\right]}{\cos\left(\frac{1}{2}Z\right)} \tag{8.292}$$

$$u_4(\zeta) = -iE_{\text{inc}}^*\frac{\sin\left[\frac{1}{2}(\zeta - Z)\right]}{\cos\left(\frac{1}{2}Z\right)} \tag{8.293}$$

At the left face of the mirror the reflected wave is

$$u_4(0) = iE_{\text{inc}}^* \tan\left(\frac{1}{2}Z\right) \tag{8.294}$$

while at the right face of the mirror the transmitted wave is

$$u_3(Z) = \frac{E_{\text{inc}}}{\cos\left(\frac{1}{2}Z\right)} \tag{8.295}$$

The amplitude of the transmitted wave is always equal to or larger than that of the incident wave, and for $Z > \frac{1}{2}\pi$ the amplitude of the reflected wave is larger than that of the incident wave. Not only does a phase-conjugate mirror reflect the incident wave phase reversed, it can do so with a reflectance greater than unity. The extra energy in the transmitted and reflected waves is provided by the pump beams. We see from (8.294) that since $\tan(\frac{1}{2}Z)$ is always real, the phase of the reflected wave $u_4(0)$ depends only on the complex conjugate of the incident wave, so the reflected wave is always perfectly "phase reversed." However, the reflected amplitude is uniformly proportional to the incident amplitude only if the pump beams are uniform over the entire mirror, so Z is independent of the transverse coordinates.

8.4 RAMAN PROCESSES

8.4.1 Raman Scattering

In 1928, C. V. Raman unexpectedly discovered that light scattered from gases, liquids, and solids contains not only the original frequency ω but also (extremely weak) sidebands at the frequencies $\omega \pm \omega_r$, where ω_r is a frequency characteristic of the medium from which the light is scattered. The upper and lower sidebands are called the anti-Stokes and Stokes lines, respectively. The quantum-mechanical description of the effect is illustrated in Figure 8.20. To create a photon at the Stokes sideband, a molecule in the ground state absorbs a photon with energy $\hbar\omega$, which places the molecule in the virtual level indicated by the dotted line, and then emits a photon with energy $\hbar(\omega - \omega_r)$ that leaves the molecule in the low-lying excited state. To create a photon at the anti-Stokes line, a molecule starts in the excited level and reverses the process. The excited level in Figure 8.20 is typically a vibrational level of the molecule, but rotational and electronic Raman transitions are also observed. The usefulness of the Raman effect in ordinary spectroscopy lies in its ability to reveal spectral lines that are forbidden as dipole transitions. For example, Raman spectroscopy makes it possible to observe the vibrational spectrum of symmetric molecules, such as O_2, which are not infrared active. When the molecules are illuminated with visible or ultraviolet light, the vibrational frequencies can be inferred from the displacement of the sidebands from the original frequency.

The classical description of the Raman effect is similar to the description of the third-order nonlinear susceptibility of rotating nonspherical molecules, discussed earlier, in the sense that we treat the nuclear and electronic motions differently. We assume that the electronic motions are fast and respond essentially instantly to an applied electric field, while the nuclear motions corresponding to the rotation and vibration of the molecule as a whole are much slower. This is called the Born–Oppenheimer approximation in quantum

Figure 8.20 Raman effect.

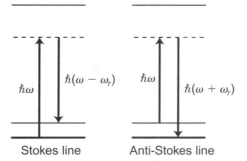

Stokes line   Anti-Stokes line

mechanics. We can then describe the polarization of the molecule at optical frequencies in terms of an instantaneous electronic polarizability that depends on the nuclear coordinates (rotation and vibration) of the molecule. For simplicity we ignore rotation and assume that the molecule has one vibrational degree of freedom whose generalized coordinate is x. The spherically averaged polarizability of the molecule is then

$$\alpha = \alpha_0 + \alpha' x \tag{8.296}$$

and the dipole moment of the molecule in the electric field $\mathbf{E}$ is

$$\mathbf{p} = (\alpha_0 + \alpha' x)\mathbf{E} \tag{8.297}$$

If the molecule vibrates at the frequency ω_r and the field oscillates at the frequency ω, the oscillation of the dipole moment has components at the frequencies ω and $\omega \pm \omega_r$ and the molecule radiates at all these frequencies. Since the dipole moment is an instantaneous response to the incident electric field, the dipole radiation may be regarded as scattering of the incident light, as in the case of Thomson scattering. The component of the radiation at the original frequency ω is called Rayleigh scattering, and the components at $\omega \pm \omega_r$ are called Raman scattering. Classically, the intensity of the Raman-scattered light is nearly the same for the upper and lower sidebands and depends on the amplitude of the molecular vibration. In a quantum-mechanical model the anti-Stokes line is suppressed by the fact that the population of the excited level is reduced by the Boltzmann factor $e^{-\hbar\omega_r/k_B T}$. Also, in a purely classical model the vibrational motion in thermal equilibrium vanishes at low temperatures, so the Raman activity vanishes in the same limit, but in a quantum-mechanical model the vibrational motion has a finite amplitude even in the lowest vibrational level.

To describe the nonlinear susceptibility of a Raman-active (but infrared-inactive) molecule we use a harmonic oscillator model for the vibrational coordinate x. To find the interaction that excites the vibrational motion, we note that the energy of a molecule in the electric field $\mathbf{E}$ is

$$\mathcal{W} = -\tfrac{1}{2}\mathbf{p} \cdot \mathbf{E} = -\tfrac{1}{2}(\alpha_0 + \alpha' x)\mathbf{E} \cdot \mathbf{E} \tag{8.298}$$

so the generalized force on the vibrational coordinate is

$$F = -\frac{d\mathcal{W}}{dx} = \frac{1}{2}\alpha'\mathbf{E} \cdot \mathbf{E} \tag{8.299}$$

The equation of motion of the harmonic oscillator is then

$$\frac{d^2 x}{dt^2} + \gamma\frac{dx}{dt} + \omega_r^2 x = \frac{\alpha'}{2m}\mathbf{E} \cdot \mathbf{E} \tag{8.300}$$

where m is the effective mass of the oscillator and γ the damping rate. If we compute the Fourier transform of this equation in the usual way, we get

$$\left(\omega_r^2 - \omega^2 - i\gamma\omega\right)\tilde{x}(\omega) = \frac{\alpha'}{2m\sqrt{2\pi}}\int_{-\infty}^{\infty} dt\, e^{i\omega t}\mathbf{E}(t) \cdot \mathbf{E}(t) \tag{8.301}$$

and if we evaluate the right-hand side by means of the convolution theorem, we find that

$$\tilde{x}(\omega) = \frac{\alpha'}{2m\sqrt{2\pi}}\int_{-\infty}^{\infty} d\omega'\, \frac{\tilde{\mathbf{E}}(\omega') \cdot \tilde{\mathbf{E}}(\omega - \omega')}{\omega_r^2 - \omega^2 - i\gamma\omega} \tag{8.302}$$

To find the dipole moment of the molecule, we use (8.297). If we take the Fourier transform of (8.297), using the convolution theorem, and substitute (8.302) for $\tilde{x}(\omega)$, we find that the polarization of the material is

$$\tilde{\mathbf{P}}(\omega) = \varepsilon_0 \chi_0(\omega)\tilde{\mathbf{E}}(\omega) + \tilde{\mathbf{P}}_R(\omega) \tag{8.303}$$

where

$$\chi_0(\omega) = \frac{N\alpha_0}{\varepsilon_0} \tag{8.304}$$

is the linear susceptibility and N the number density of molecules. The Raman polarization is

$$\tilde{\mathbf{P}}_R(\omega) = \varepsilon_0 \int_{-\infty}^{\infty} d\omega' \int_{-\infty}^{\infty} d\omega'' \chi_R(\omega')\tilde{\mathbf{E}}(\omega'') \cdot \tilde{\mathbf{E}}(\omega' - \omega'')\tilde{\mathbf{E}}(\omega - \omega')$$

$$= \varepsilon_0 \int_{-\infty}^{\infty} d\omega' \int_{-\infty}^{\infty} d\omega'' \int_{-\infty}^{\infty} d\omega''' \chi_R(\omega')\tilde{\mathbf{E}}(\omega'') \cdot \tilde{\mathbf{E}}(\omega' - \omega'')$$

$$\times \tilde{\mathbf{E}}(\omega''' - \omega')\delta(\omega''' - \omega) \tag{8.305}$$

where the Raman susceptibility is

$$\chi_R(\omega) = \frac{N\alpha'^2}{4\pi\varepsilon_0 m} \frac{1}{\omega_r^2 - \omega^2 - i\gamma\omega} \tag{8.306}$$

With an obvious change of variables (8.305) becomes

$$\tilde{\mathbf{P}}_R(\omega) = \varepsilon_0 \int_{-\infty}^{\infty} d\omega' \int_{-\infty}^{\infty} d\omega'' \int_{-\infty}^{\infty} d\omega''' \tilde{\mathbf{E}}(\omega') \cdot \tilde{\mathbf{E}}(\omega'')\tilde{\mathbf{E}}(\omega''')$$

$$\times \chi_R(\omega' + \omega'')\delta(\omega - \omega' - \omega'' - \omega''') \tag{8.307}$$

Comparison with (8.184) shows that the Raman polarization is just a third-order polarization for which the susceptibility is $\chi_T^{(3)}(\omega', \omega'', \omega''') = \chi_R(\omega' + \omega'')$.

As shown in Figure 8.21, the imaginary part of the Raman susceptibility has sharp peaks near $\omega = \pm\omega_r$. These peaks are very large and narrow for $\gamma/\omega_r \ll 1$, that is, when the dissipation is small, as is usually the case. For example, in liquid CS_2 the Raman line

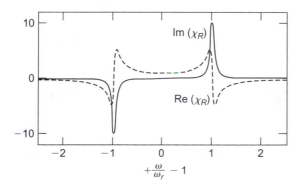

Figure 8.21 Raman susceptibility.

width is $\Delta \nu = 1.5 \times 10^{10}$ Hz, and the vibrational frequency is $\nu_r = 2.0 \times 10^{13}$ Hz, so that $\gamma/\omega_r = \Delta\nu/2\nu_r = 4 \times 10^{-4}$.

We see from (8.302) and (8.307) that when two waves $\mathbf{E}(\omega'')$ and $\mathbf{E}(\omega'' - \omega')$ combine, they produce a field with the beat frequency ω' that interacts strongly with the molecular vibration when $\omega' \approx \omega_r$. The molecular oscillation produces a polarizability that oscillates at the frequency ω', and this, in turn, responds to the field $\mathbf{E}(\omega'')$ to produce a dipole moment at the lower frequency $\omega'' - \omega'$, and vice versa. Thus, the waves at the two different frequencies interact through the Raman susceptibility. Because it depends on fluctuations in the polarizability caused by molecular vibrations, the Raman effect is exquisitely small. The importance of *stimulated* Raman scattering follows from the fact that the interaction occurs at resonance, and the fact (shown in the next section) that the interacting waves are automatically phase matched.

EXERCISE 8.13

Consider a molecule oscillating at frequency ω_r with amplitude $\bar{x}$. Its polarizability as a function of time t is

$$\alpha = \alpha_0 + \alpha' x = \alpha_0 + \tfrac{1}{2}\alpha'\left(\bar{x}e^{-i\omega_r t} + \bar{x}^* e^{i\omega_r t}\right) \tag{8.308}$$

When this molecule is illuminated by the optical field

$$E = \tfrac{1}{2}(\bar{E}e^{-i\omega t} + \bar{E}^* e^{i\omega t}) \tag{8.309}$$

it develops a dipole moment that oscillates with components at the frequencies ω and $\omega \pm \omega_r$.

(a) Use the analogy between an oscillating dipole and an oscillating charge (discussed in Chapter 4) to show that the power per unit solid angle radiated by a dipole is

$$\frac{d\mathcal{P}}{d\Omega} = \frac{\omega^4 |\bar{p}|^2}{32\pi^2\varepsilon_0 c}\sin^2\theta \tag{8.310}$$

where ω is the frequency and $|\bar{p}|$ the amplitude of the dipole, and θ the angle between the dipole and the observer.

(b) The power radiated is linear in the incident intensity, so it may be regarded as scattering of the incident light. Show that the differential cross section for molecular scattering at the original frequency of the light (Rayleigh scattering) is

$$\frac{d\sigma_0}{d\Omega} = \frac{\alpha_0^2 \omega^4}{6\pi c^4}\sin^2\theta \tag{8.311}$$

(c) The index of refraction of air at STP (0 °C) is 1.00029 in the visible part of the spectrum, from which we can determine $\alpha_0 \approx \varepsilon_0(n^2 - 1)/N$ for an "air" molecule. Estimate the *total* cross section σ_0 for Rayleigh scattering at a wavelength of 500 nm. What fraction of the incident light is Rayleigh scattered in 1 m of air at STP?

(d) Show that the differential cross section for molecular scattering at the frequencies $\omega \pm \omega_r$ (Raman scattering) is

$$\frac{d\sigma_R}{d\Omega} = \frac{\alpha'^2 |\bar{x}|^2 (\omega \pm \omega_r)^4}{48\pi \varepsilon_0 c^3} \sin^2 \theta \tag{8.312}$$

Show that if we assign the oscillator the classical amplitude of its oscillation in the lowest quantum-mechanical level, then the cross section becomes

$$\frac{d\sigma_R}{d\Omega} = \frac{\hbar \alpha'^2 (\omega \pm \omega_r)^4}{48 \varepsilon_0 m c^3 \omega_r} \sin^2 \theta \tag{8.313}$$

where m is the effective mass of the oscillator.

(e) Estimate the *total* cross section σ_R for Raman scattering by an "air" molecule $[\lambda_r = O(5 \ \mu\text{m})]$ at a wavelength of 500 nm using the approximation $\alpha' \approx \alpha_0/a_0$, where a_0 is the Bohr radius. What fraction of the incident light is Raman scattered in 1 m of air at STP? Spontaneous Raman scattering is an extremely weak effect!

8.4.2 Coherent Raman Amplification

As discussed earlier, the wave equation in the presence of a third-order nonlinearity is

$$\nabla^2 \tilde{\mathbf{E}} + k^2 \tilde{\mathbf{E}} = -\frac{1}{\varepsilon_0 n^2(\omega)} [\nabla(\nabla \cdot \tilde{\mathbf{P}}_R) + k^2 \tilde{\mathbf{P}}_R] \tag{8.314}$$

in the present notation, where the wave vector satisfies the linear dispersion relation

$$c^2 k^2(\omega) = n^2(\omega)\omega^2 \tag{8.315}$$

$$n^2(\omega) = 1 + \chi_0(\omega) \tag{8.316}$$

The right-hand side of (8.314) is a perturbation that we can approximate and simplify. In particular, for nearly plane waves the term $\nabla(\nabla \cdot \tilde{\mathbf{P}}_R)$ is small compared with $k^2 \tilde{\mathbf{P}}_R$ and we ignore it.

When the third-order nonlinearity is of the Raman type, the nonlinear effect is very small unless there is a resonant interaction with the molecular vibration. We therefore examine the case when the field $\tilde{\mathbf{E}}(\omega)$ consists of two cw waves, a pump wave at frequency ω_p and a sideband at frequency ω_s, whose beat frequency is near the vibrational frequency. In the frequency domain, the electric field is then

$$\tilde{\mathbf{E}}(\mathbf{r}, \omega) = \mathbf{E}_p(\mathbf{r})\delta(\omega - \omega_p) + \mathbf{E}_p^*(\mathbf{r})\delta(\omega + \omega_p) + \mathbf{E}_s(\mathbf{r})\delta(\omega - \omega_s) + \mathbf{E}_s^*(\mathbf{r})\delta(\omega + \omega_s) \tag{8.317}$$

If we substitute this into (8.307) and carry out the integrals over ω' and ω'' in the usual way, we find after some tedious algebra that

$$\begin{aligned}
\tilde{\mathbf{P}}_R(\omega) = {} & 2\varepsilon_0 \chi_R(\omega_p - \omega_s)(\mathbf{E}_p \cdot \mathbf{E}_s^*)[\mathbf{E}_p \delta(\omega - 2\omega_p + \omega_s) + \mathbf{E}_p^* \delta(\omega + \omega_s) \\
& + \mathbf{E}_s \delta(\omega - \omega_p) + \mathbf{E}_s^* \delta(\omega - 2\omega_s + \omega_p)] + 2\varepsilon_0 \chi_R(\omega_s - \omega_p)(\mathbf{E}_s \cdot \mathbf{E}_p^*) \\
& \times [\mathbf{E}_s \delta(\omega - 2\omega_s + \omega_p) + \mathbf{E}_s^* \delta(\omega + \omega_p) + \mathbf{E}_p \delta(\omega - \omega_s) + \mathbf{E}_p^* \delta(\omega - 2\omega_p + \omega_s)]
\end{aligned} \tag{8.318}$$

plus terms of the type $\chi_R(\omega_p + \omega_s)$, $\chi_R(2\omega_p)$, $\chi_R(0)$, and so on. We ignore these terms because the Raman susceptibility is small except near resonance. If we substitute (8.318) into the wave equation (8.314) and compare terms on the left- and right-hand sides, we find that the wave at the sideband frequency satisfies the equation

$$\nabla^2 \mathbf{E}_s + k_s^2 \mathbf{E}_s = -2\frac{\omega_s^2}{c^2}\chi_R(\omega_s - \omega_p)(\mathbf{E}_s \cdot \mathbf{E}_p^*)\mathbf{E}_p \tag{8.319}$$

where

$$k_s = k(\omega_s) \tag{8.320}$$

To simplify the discussion we consider the case when the pump and sideband waves are both polarized in the same, fixed direction. The wave equation (8.319) then becomes

$$\nabla^2 E_s + k_R^2 E_s = 0 \tag{8.321}$$

where the wave vector is

$$k_R^2 = k_s^2 + 2\frac{\omega_s^2}{c^2}\chi_R(\omega_s - \omega_p)(E_p E_p^*) \tag{8.322}$$

But the Raman susceptibility $\chi_R(\omega_s - \omega_p)$ is in general complex, so the wave vector is complex. We have seen that the imaginary part of the wave vector is associated with absorption or, depending on the sign, amplification of the wave. The imaginary part of the susceptibility is largest at the resonances, where it has the value

$$\chi_R(\pm\omega_r) = \frac{\pm i N\alpha'^2}{4\pi\varepsilon_0 m\gamma\omega_r} \tag{8.323}$$

If the imaginary part of the wave vector is small, we can expand

$$k_R^2 \approx k_s^2 + i2k_s k_R'' \tag{8.324}$$

and if we compare this with (8.322) we find that at the Stokes and anti-Stokes resonances the imaginary part of the wave vector is

$$k_S'' = k_R''(\omega_S) = -\frac{N\alpha'^2\omega_S}{4\pi\varepsilon_0 m\gamma\omega_r n_S}E_p E_p^* \tag{8.325}$$

$$k_{aS}'' = k_R''(\omega_{aS}) = \frac{N\alpha'^2\omega_{aS}}{4\pi\varepsilon_0 m\gamma\omega_r n_{aS}}E_p E_p^* \tag{8.326}$$

respectively, where

$$n_S = n(\omega_S) \tag{8.327}$$

and

$$n_{aS} = n(\omega_{aS}) \tag{8.328}$$

are the refractive indices, and

$$\omega_S = \omega_p - \omega_r \tag{8.329}$$

and

$$\omega_{aS} = \omega_p + \omega_r \tag{8.330}$$

the Stokes and anti-Stokes frequencies, respectively. But absorption is associated with a positive value of the imaginary part of the wave vector and gain with a negative value.

Therefore, since $E_p E_p^*$ is real and positive, we see that the Stokes wave is amplified while the anti-Stokes wave is absorbed.

This begs the question, "Why is the lower (Stokes) sideband amplified while the upper (anti-Stokes) sideband is absorbed?" In a quantum-mechanical picture the answer is easy. Almost all the molecules are initially in the ground state, from which the only transitions are upward to higher states, so amplification occurs on transitions to lower frequencies. This favors production of Stokes-shifted radiation at the expense of the pump, or favors production of pump radiation at the expense of the anti-Stokes-shifted sideband. In a classical picture the answer is more subtle and has to do with the phase of the oscillator relative to that of the wave.

A few remarks are in order regarding Raman amplification. In the first place we note that the gain coefficient (8.325) depends on the complex pump amplitude only through the factor $E_p E_p^*$. Since this is independent of the phase of the pump wave, we see that there is no issue of phase matching between the pump wave and the sideband. Gain is always positive at the Stokes frequency and negative at the anti-Stokes frequency. In fact, provided only that the pump and sideband polarizations are parallel (and this will always be the case after a few gain lengths), the pump and sideband waves need not even be propagating in the same direction. For these reasons, given sufficient intensity on the pump beam, coherent Raman scattering is inevitable. In fact, coherent Raman scattering was discovered quite by accident in the early days of laser research when a ruby laser switched with a nitrobenzene Kerr cell was observed to produce large amounts of Raman-shifted light.

In the second place, as soon as a substantial amount of Stokes-shifted radiation is produced, the Stokes radiation acts as a pump wave for light that is Raman shifted to even lower frequencies, corresponding to $\omega - 2\omega_r$, $\omega_p - 3\omega_r$, and so on. In fact, we see from (8.318) that the Raman polarization due to the pump and sideband waves already contains components at many frequencies, including $\omega_p - 2\omega_r$. Thus, when the process of stimulated Raman scattering is very strong, light is generally produced at many wavelengths and not just the Stokes wavelength. The incident beam is depleted at the original wavelength and degraded by conversion to a multitude of other wavelengths.

Finally, we see from (8.326) that light at the anti-Stokes frequency is absorbed, rather than amplified, by stimulated Raman scattering. In actual fact, however, light at the anti-Stokes frequency is observed in many experiments. The origin of this effect is a four-wave mixing process similar to that discussed earlier in connection with the ordinary third-order linearity. Since this requires phase matching, the conditions are much more restrictive. However, for transparent materials the (higher frequency) anti-Stokes waves propagate with a smaller phase velocity, so it is always possible to achieve phase matching by propagating the anti-Stokes wave at an angle to the pump wave. Thus, the anti-Stokes radiation appears in a cone oriented around the pump beam.

EXERCISE 8.14

The Raman susceptibility can be estimated from the measured index of refraction $n^2 - 1 \approx N\alpha_0/\varepsilon_0$, vibrational frequency ω_r, and line width $\Delta\omega_r/\omega_r = 2\gamma/\omega_r$ by assuming that $\alpha' = O(\alpha_0/a_0)$, where the Bohr radius a_0 is a characteristic length for atoms and molecules.

(a) Show that in terms of the pump intensity $\langle S_p \rangle = \varepsilon_0 c n E_p E_p^* / \pi$, the Raman gain coefficient is

$$|k_S''| = \frac{(n^2 - 1)^2}{2mc^2 a_0^2 N \Delta\omega_r} \frac{\omega_S}{\omega_r} \langle S_p \rangle \tag{8.331}$$

(b) For liquid CS_2 (ignore Clausius–Mossotti effects) the index of refraction is $n = 1.6$, the number density is $N = 1.0 \times 10^{28}$ /m^3, the vibrational frequency is $\omega_r = 1.2 \times 10^{14}$ /s, and the line width is $\Delta\omega_r / \omega_r = 8 \times 10^{-4}$. The mass of the carbon atom (vibrating between the two heavy sulfur atoms) is $m = 2 \times 10^{-26}$ kg. Compare the value estimated using (8.331) with the experimentally measured value (SI units)

$$|k_S''| = 2.4 \times 10^{-10} \frac{\omega_S}{\omega_r} \langle S_p \rangle \tag{8.332}$$

What is the Raman gain coefficient for a ruby laser pump intensity $\langle S_p \rangle = 10^{10}$ W/m^2 at $\omega_p = 9 \times 10^{14}$ /s?

EXERCISE 8.15

Using the slowly varying envelope approximation for the pump and Stokes waves, derive the Manley–Rowe relations for stimulated Raman scattering.

BIBLIOGRAPHY

An excellent general introduction to the field of nonlinear optics is given in
   R. W. Boyd, *Nonlinear Optics,* Academic Press, Boston (1992).

For an interesting review of the experimental data, especially from early experiments, the reader is referred to
   N. Bloembergen, *Nonlinear Optics,* World Scientific Publishing, Singapore (1996).

For a mathematical approach more like that of the present chapter, see
   P. N. Butcher and D. Cotter, *The Elements of Nonlinear Optics,* Cambridge University Press, Cambridge, U.K. (1990).

A good description of phase-matching techniques is given by
   R. D. Guenther, *Modern Optics,* John Wiley & Sons, New York (1990).

A thorough discussion of nonlinear processes in metal vapors can be found in the research monograph
   D. C. Hanna, M. A. Yuratich, and D. Cotter, *Nonlinear Optics of Free Atoms and Molecules,* Springer-Verlag, Berlin (1979).

9

Diffraction

Thanks to Maxwell's fantastic insight, optics is now recognized as a branch of electromagnetism, but it was not always so. In fact, optics was highly developed long before Maxwell's discoveries, and we had the telescope, the microscope, the prism to decompose white light into colors (and reassemble it), and we even knew about the diffraction of light from slits and around corners. Before the true nature of light was understood, arguments raged about whether light is composed of waves or corpuscles (forming rays), for light displays the characteristics of both waves and rays. Of course, the true nature of light and its interactions with matter are revealed only by quantum electrodynamics, and even this theory is incomplete. However, the reconciliation of the concepts of light as waves and as rays, as well as much practical understanding of their behavior, lie within the classical domain, and these are what we examine in this chapter.

Although classical optics is an old subject, its importance has not diminished in modern science and technology. In fact, with the development of lasers its importance continues to expand. We do not attempt here to go in depth into the design of optical instruments, although we do discuss the properties of laser optical resonators. Rather we limit ourselves to examining some of the basic phenomena of light, including the limit of geometrical optics and the simplest effects of diffraction. We also examine the intermediate case of paraxial waves, which are highly collimated beams like those from lasers.

9.1 GEOMETRICAL OPTICS

9.1.1 Eikonal Approximation

In the macroscopic world with which we are familiar, the wavelength of light (typically less than 1 μm) is infinitesimally small. It is not surprising, then, that most of the optical problems we deal with, such as combing our hair in the mirror or detecting black holes with a space-borne telescope, are quite adequately addressed by treating light as rays. Only the details, such as the imaging of points and sharp edges, depend on the wave nature of light. The problem before us, then, is to begin with the Maxwell equations for electromagnetic waves and use the smallness of the wavelength to develop a set of equations that describe the propagation of light as rays with no reference to the wavelength at all.

We consider a wave propagating through a nonuniform medium with a frequency ω_0, so that the electric field and magnetic induction are

$$\mathbf{E} = \tilde{\mathbf{E}}e^{-i\omega_0 t} \tag{9.1}$$

$$\mathbf{B} = \tilde{\mathbf{B}}e^{-i\omega_0 t} \tag{9.2}$$

The corresponding free-space wave number is $k_0 = \omega_0/c$. In the absence of free charges and currents, the Maxwell equations become

$$\nabla \cdot \tilde{\mathbf{D}} = 0 \tag{9.3}$$

$$\nabla \cdot \tilde{\mathbf{B}} = 0 \tag{9.4}$$

$$\nabla \times \tilde{\mathbf{E}} - i\omega\tilde{\mathbf{B}} = 0 \tag{9.5}$$

$$\nabla \times \tilde{\mathbf{H}} + i\omega\tilde{\mathbf{D}} = 0 \tag{9.6}$$

where

$$\tilde{\mathbf{D}} = \varepsilon(\omega)\tilde{\mathbf{E}} \tag{9.7}$$

$$\tilde{\mathbf{H}} = \frac{\tilde{\mathbf{B}}}{\mu(\omega)} \tag{9.8}$$

and the permittivity $\varepsilon(\omega)$ and permeability $\mu(\omega)$ depend on the properties of the medium through which the waves are propagating. In the most interesting cases the properties of the medium are functions of the position $\mathbf{r}$. We assume, however, that the variations of $\varepsilon(\omega)$ and $\mu(\omega)$ are slow, so that they may be regarded as nearly constant over distances of the order of a wavelength. We further assume that $\varepsilon(\omega)$ and $\mu(\omega)$ are real, at least over the frequency range of interest. As shown in Chapter 7, the presence of an imaginary component in $\varepsilon(\omega)$ or $\mu(\omega)$ leads to complex wave vectors that describe absorption in the medium. Since geometrical optics is most interesting for transparent media, this extra complication is not generally needed.

As mentioned earlier, the waves we are looking at are presumed to be propagating through a fairly homogeneous medium, so we expect the waves to look locally like plane waves of the form

$$\tilde{u} \approx \tilde{u}_0 e^{i\mathbf{k}\cdot\mathbf{r}} \tag{9.9}$$

where $\tilde{u}(\mathbf{r})$ is any component of the electric or magnetic field. In the general case we look for a solution of the form

$$\tilde{\mathbf{E}} = \tilde{\mathbf{e}}(\mathbf{r})e^{ik_0\psi(\mathbf{r})} \tag{9.10}$$

$$\tilde{\mathbf{B}} = \tilde{\mathbf{b}}(\mathbf{r})e^{ik_0\psi(\mathbf{r})} \tag{9.11}$$

where the function $\psi(\mathbf{r})$ is called the eikonal. Near the point $\mathbf{r}_0$, then, the electric field has the form of a plane wave

$$\tilde{\mathbf{E}} \approx \tilde{\mathbf{e}}(\mathbf{r}_0)e^{ik_0[\psi(\mathbf{r}_0)-\nabla\psi(\mathbf{r}_0)\cdot\mathbf{r}_0]}e^{ik_0\nabla\psi(\mathbf{r}_0)\cdot\mathbf{r}} \tag{9.12}$$

with a similar expression for $\tilde{\mathbf{B}}$. Thus, the vector $k_0\nabla\psi(\mathbf{r})$ plays the role of the local wave vector. It is normal to the surface $\psi(\mathbf{r}) = $ constant, which is the phase front of the wave.

Substituting (9.10) and (9.11) back into the Maxwell equations, keeping in mind that the permittivity and permeability are functions of the position, we get

$$\nabla \cdot (\varepsilon \tilde{\mathbf{e}}) + i k_0 \varepsilon \tilde{\mathbf{e}} \cdot \nabla \psi = 0 \tag{9.13}$$

$$\nabla \cdot \tilde{\mathbf{b}} + i k_0 \tilde{\mathbf{b}} \cdot \nabla \psi = 0 \tag{9.14}$$

$$\nabla \times \tilde{\mathbf{e}} + i k_0 \nabla \psi \times \tilde{\mathbf{e}} - i k_0 c \tilde{\mathbf{b}} = 0 \tag{9.15}$$

$$\nabla \times \left(\frac{\tilde{\mathbf{b}}}{\mu} \right) + i k_0 \nabla \psi \times \left(\frac{\tilde{\mathbf{b}}}{\mu} \right) + i c k_0 \varepsilon \tilde{\mathbf{e}} = 0 \tag{9.16}$$

But in the geometrical optics limit the wavelength is very small, so we may neglect all the spatial derivatives compared with k_0. This leaves us with

$$\tilde{\mathbf{e}} \cdot \nabla \psi = 0 \tag{9.17}$$

$$\tilde{\mathbf{b}} \cdot \nabla \psi = 0 \tag{9.18}$$

$$\nabla \psi \times \tilde{\mathbf{e}} - c \tilde{\mathbf{b}} = 0 \tag{9.19}$$

$$\nabla \psi \times \tilde{\mathbf{b}} + \mu \varepsilon c \tilde{\mathbf{e}} = 0 \tag{9.20}$$

From the first two equations we see that the fields are orthogonal to the local wave vector $k_0 \nabla \psi$, so the waves are transverse. From the third equation we see that the electric and magnetic fields are orthogonal to each other and in phase (since μ and ε are real), so the waves are locally just like ordinary plane waves. The average value of the Poynting vector is

$$\langle \mathbf{S} \rangle = \langle \mathbf{E} \times \mathbf{H} \rangle = \frac{1}{2\mu} \operatorname{Re}(\tilde{\mathbf{e}}^* \times \tilde{\mathbf{b}}) = \frac{\varepsilon c}{2n^2} (\tilde{\mathbf{e}}^* \cdot \tilde{\mathbf{e}}) \nabla \psi = \frac{c}{2\mu n^2} (\tilde{\mathbf{b}}^* \cdot \tilde{\mathbf{b}}) \nabla \psi \tag{9.21}$$

where we have evaluated the cross product using (9.17)–(9.20). The index of refraction n is defined by

$$n^2 = \mu \varepsilon c^2 = \frac{\mu \varepsilon}{\mu_0 \varepsilon_0} \tag{9.22}$$

and we assume that it is real.

If we take the cross product of (9.20) with $\nabla \psi$ and expand the triple cross product on the left-hand side, we get

$$-(\nabla \psi \cdot \nabla \psi) \tilde{\mathbf{b}} + (\tilde{\mathbf{b}} \cdot \nabla \psi) \nabla \psi = \mu \varepsilon c \tilde{\mathbf{e}} \times \nabla \psi \tag{9.23}$$

If we now use (9.18) and (9.19) to evaluate the second and third terms, we get

$$\nabla \psi \cdot \nabla \psi = n^2 \tag{9.24}$$

This is called the eikonal equation, and it represents a three-dimensional generalization to nonuniform media of the dispersion relation for plane waves. It is the fundamental equation of geometrical optics.

9.1.2 Rays in Geometrical Optics

The wave vectors $k_0 \nabla \psi(\mathbf{r})$ normal to the wave fronts $\psi(\mathbf{r}) = $ constant trace out curves in space everywhere normal to the phase fronts, just like electric field lines that are everywhere

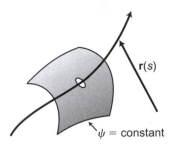

Figure 9.1 Optical trajectory, or ray.

r(s)

ψ = constant

normal to the equipotential surfaces. As illustrated in Figure 9.1, the optical trajectory is a curve $\mathbf{r}(s)$ parameterized by the distance s along the curve. The differential equation for this curve is then

$$\frac{d\mathbf{r}}{ds} = \frac{\nabla\psi}{|\nabla\psi|} = \frac{\nabla\psi}{n} \tag{9.25}$$

If we differentiate this again with respect to the path length s and substitute for $d\mathbf{r}/ds$, we get

$$\frac{d}{ds}\left(n\frac{d\mathbf{r}}{ds}\right) = \frac{d}{ds}\nabla\psi = \left(\frac{d\mathbf{r}}{ds}\cdot\nabla\right)\nabla\psi = \frac{1}{n}(\nabla\psi\cdot\nabla)\nabla\psi \tag{9.26}$$

But

$$\nabla(\nabla\psi\cdot\nabla\psi) = 2(\nabla\psi\cdot\nabla)\nabla\psi = \nabla(n^2) = 2n\nabla n \tag{9.27}$$

by (9.24), so we arrive at the differential equation

$$\frac{d}{ds}\left(n\frac{d\mathbf{r}}{ds}\right) = \nabla n \tag{9.28}$$

for the rays. This is the generalization of Snell's law to media with a slowly varying index of refraction. It shows that when a ray propagates through a medium with a gradient in the index of refraction transverse to the path of the ray, the ray curves toward the region of larger index of refraction. Since the generalized Snell's law is a second-order differential equation, the path of the ray depends on the initial values of both the position and the direction of the ray, as expected.

As an example of the use of the ray equation, or generalized Snell's law, we consider the reflection of radio waves from the ionosphere. In Chapter 4 it is shown that the index of refraction of a transverse wave in a collisionless plasma is

$$n^2 = \frac{k^2 c^2}{\omega^2} = 1 - \frac{\omega_p^2}{\omega^2} \tag{9.29}$$

where the plasma frequency is

$$\omega_p^2 = \frac{n_e q^2}{\varepsilon_0 m} \tag{9.30}$$

in which q is the charge and m the mass of an electron, and the density n_e of electrons in the ionosphere is a function of the altitude h. At low frequencies, the index of refraction becomes imaginary for sufficiently large electron densities and the ionosphere becomes

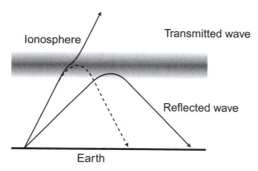

Ionosphere

Transmitted wave

Reflected wave

Earth

Figure 9.2 Deflection of radio waves in the ionosphere.

totally reflecting. However, at sufficiently high frequencies the index of refraction is always real. In this case the radio waves are reflected only if they approach the ionosphere at a low angle and are deflected by the gradient of the index of refraction, as shown in Figure 9.2. To keep things simple we assume a flat earth. Taking the dot product of (9.28) with the vertical and converting to derivatives in the vertical direction by using $dh = ds \cos\theta$, where θ is the angle from the vertical, we get the equation

$$\cos\theta \frac{d}{dh}(n\cos\theta) = \frac{dn}{dh} \tag{9.31}$$

If we multiply both sides of this equation by $\frac{1}{2}n$, we get

$$\frac{d}{dh}[(n\cos\theta)^2 - n^2] = 0 \tag{9.32}$$

or

$$n^2 \sin^2\theta = \text{constant} \tag{9.33}$$

If the radio waves initially propagate upward at an angle θ_0 from the vertical and we take $n = 1$ at the earth's surface, then the radio waves are reflected ($\sin\theta = 1$) at the altitude where

$$n^2 = \sin^2\theta_0 = 1 - \frac{\omega_p^2}{\omega^2} = 1 - \cos^2\theta_0 \tag{9.34}$$

or $\omega_p = \omega\cos\theta_0 < \omega$. To put this another way, the waves are reflected for initial angles $\theta_0 > \arccos[\omega_p(\max)/\omega]$. For this reason it is sometimes possible to receive signals from radio stations a long distance away because they are emitted at large angles from the vertical and reflected from the ionosphere, but not signals from stations that are closer (but still over the horizon), as shown in Figure 9.2.

To complete the theory of geometrical optics, we must find the electric and magnetic fields at each point along the rays. By hypothesis, the vector values $\tilde{\mathbf{e}}$ and $\tilde{\mathbf{b}}$ of the electric field and magnetic induction vary slowly along the path of a ray. To find their actual values, it is necessary to keep the terms of the next higher order in $1/k_0$ in the Maxwell equations (9.13)–(9.16). The results are complicated and not particularly illuminating. Alternatively, we can examine the behavior of the Poynting vector. This gives us the magnitude of $\tilde{\mathbf{e}}$ and $\tilde{\mathbf{b}}$, which is sufficient for most purposes. We start with Poynting's theorem. Since the index of refraction is presumed real, at least in the frequency range of interest, there is no dissipation in the medium. After one cycle of the wave the medium returns to its initial state, so the change of the energy of the fields and the polarization and magnetization vanishes. Therefore, averaged over a complete cycle, Poynting's theorem states that

$$\nabla \cdot \langle \mathbf{S} \rangle = 0 \tag{9.35}$$

where from (9.21) the average value of the Poynting vector is

$$\langle \mathbf{S} \rangle = \frac{\varepsilon_0 c}{2n} (\tilde{\mathbf{e}}^* \cdot \tilde{\mathbf{e}}) \hat{\mathbf{s}} \tag{9.36}$$

in which we have introduced the unit vector

$$\hat{\mathbf{s}} = \frac{\nabla \psi}{n} \tag{9.37}$$

tangent to the ray, normal to the phase front. If we substitute this into (9.35), we get

$$\nabla \langle S \rangle \cdot \hat{\mathbf{s}} + \langle S \rangle \nabla \cdot \hat{\mathbf{s}} = 0 = \frac{d\langle S \rangle}{ds} + \langle S \rangle \nabla \cdot \hat{\mathbf{s}} \tag{9.38}$$

where $d\langle S \rangle / ds$ is the derivative of the Poynting vector along the ray. But the divergence of the vector field $\hat{\mathbf{s}}$ can be computed by finding the derivatives in the orthogonal directions 1 and 2 normal to the unit vector $\hat{\mathbf{s}}$. Referring to Figure 9.3, we see that

$$\nabla \cdot \hat{\mathbf{s}} = \frac{\partial s_1}{\partial r_1} + \frac{\partial s_2}{\partial r_2} = \frac{1}{R_1} + \frac{1}{R_2} = \frac{2}{R} \tag{9.39}$$

where R_1 and R_2 are the radii of curvature of the phase front in the directions 1 and 2, and $R = 2(R_1^{-1} + R_2^{-1})^{-1}$ is the mean radius of curvature of the phase front. Combining this with (9.38), we obtain the intensity law for geometrical optics

$$\frac{d\langle S \rangle}{ds} = -\frac{2\langle S \rangle}{R} \tag{9.40}$$

For a spherically expanding wave front ($R = s$), we get the familiar inverse-square law for the intensity:

$$\int_{S_1}^{S_2} \frac{d\langle S \rangle}{\langle S \rangle} = \ln\left(\frac{\langle S_2 \rangle}{\langle S_1 \rangle}\right) = -2 \int_{s_1}^{s_2} \frac{ds}{s} = -2 \ln\left(\frac{s_2}{s_1}\right) \tag{9.41}$$

Since $R = R(s)$ is a function only of the distance along the ray, we see that the intensity is determined by the initial condition for the ray itself. There is no influence from adjacent rays unless, of course, the rays cross. A bunch of rays emerging from an aperture,

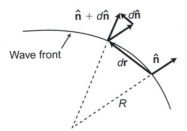

Figure 9.3 Derivative of the vector normal to a wave front.

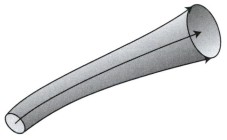

Figure 9.4 Bunch of rays forming a beam.

Figure 9.5 Intersection of rays at a caustic and at a focus.

for example, form a sharp-edged beam as shown in Figure 9.4. The discontinuity in the intensity (and the electric and magnetic fields) propagates along the rays. In the theory of partial differential equations, the curves along which discontinuities can propagate are called the characteristics.

When a bunch of rays intersect tangent to a curve or surface in space, as shown in Figure 9.5, this is called a caustic. When all the rays intersect at one point, it is called a focus. In either case the radius of curvature of the wave front vanishes in at least one direction, so the intensity diverges. In this case, of course, the effects of diffraction take over and the intensity remains finite. The diffraction of beams at a focus is discussed later using the paraxial wave approximation. The effects of diffraction are also important at the edges of beams, where the intensity is discontinuous in the geometrical optics approximation. The diffraction at the edge of a beam of light is discussed at the end of this chapter.

EXERCISE 9.1

An interesting analogy between the propagation of a ray and the path of a particle is possible if we identify the vector

$$\mathbf{p} = n\frac{d\mathbf{r}}{ds} \tag{9.42}$$

with the particle momentum and the scalar s with the time. The index of refraction is equivalent in this expression to the mass, except that it is not constant. Consider the propagation of an optical ray $\mathbf{r}(s)$ along an azimuthally symmetric optical fiber for which the index of refraction

$$n = n(\rho) \tag{9.43}$$

is a function of the radius ρ, where $\mathbf{r} = (\rho, \phi, z)$ in cylindrical coordinates aligned with the fiber axis (the z axis).

(a) Show that

$$n\frac{d\mathbf{r}}{ds} \cdot \hat{\mathbf{z}} = \text{constant} = \mathbf{p} \cdot \hat{\mathbf{z}} = n\frac{dz}{ds} \tag{9.44}$$

along the ray. This expresses the conservation of "linear momentum" along the z axis.

(b) Show that

$$\left(\mathbf{r} \times n\frac{d\mathbf{r}}{ds}\right) \cdot \hat{\mathbf{z}} = \mathbf{L} \cdot \hat{\mathbf{z}} = \text{constant} = n\rho^2\frac{d\phi}{ds} \tag{9.45}$$

along the ray, where $\mathbf{L} = \mathbf{r} \times \mathbf{p}$. This expresses the conservation of "angular momentum" about the z axis.

(c) Show that

$$\frac{1}{2}\left(n\frac{d\rho}{ds}\right)^2 + \frac{1}{2}\left(\frac{\mathbf{L}\cdot\hat{\mathbf{z}}}{\rho}\right)^2 - \frac{1}{2}n^2 = \text{constant} = \mathcal{E}_\rho \tag{9.46}$$

This expresses the conservation of radial energy, including the centrifugal potential $(\mathbf{L}\cdot\hat{\mathbf{z}}/\rho)^2$ and the potential energy $\frac{1}{2}n^2$. All these conservation theorems from classical dynamics still work even though the effective mass n is a function of the position rather than a constant.

EXERCISE 9.2

Although the eikonal equation is derived with the assumption that variations of the index of refraction are small over a length of the order of a wavelength, some of the results are valid even when the refractive index varies abruptly or discontinuously. For example, (9.33) is a generalization of Snell's law to smoothly varying refractive indices but remains valid for discontinuous variations. It predicts correctly the critical angle for total internal reflection, but it cannot predict either the partial reflection at smaller angles of incidence or the phase change upon total internal reflection. Sometimes, however, it is possible to correct geometrical optics to include the effects at discontinuities. As an example we consider the symmetric dielectric slab waveguide shown in Figure 9.6. Rays are indicated by solid lines, and the phase fronts are indicated by dotted lines. For simplicity, we assume that inside the waveguide the index of refraction is $n > 1$ and outside the waveguide it is unity. Waves incident internally on the surface of the waveguide at angles less than the critical angle pass through the surface (with some reflection) as shown on the left and form what are called radiative modes of the waveguide. Waves incident at larger angles are totally reflected with a phase change that depends on the angle of incidence and the polarization (s or p) of the wave as discussed in Chapter 7. These form guided modes. The mode condition can be found from the construction indicated in Figure 9.6. Beginning

Figure 9.6 Symmetric slab waveguide.

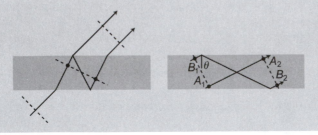

with ray A, we see that from point A_1 to point A_2 the phase shift along the ray is $\Delta\phi_A = nkL_A$, where k is the vacuum wave number and L_A is the path length along the ray. But for ray B the phase shift is $\Delta\phi_B = nkL_B + 2\Delta\phi_R$, where $\Delta\phi_R$ is the phase change at total internal reflection. But both rays are on the same phase front, so to form a mode the phase changes $\Delta\phi_A$ and $\Delta\phi_B$ must differ by a multiple of 2π. Therefore, the mode condition is

$$nk(L_B - L_A) + 2\Delta\phi_R = 2\pi m \tag{9.47}$$

for some integer m. Both path lengths (L_A and L_B) and the phase shift $\Delta\phi_R$ depend on the angle θ of the ray from the normal, which is established by the mode condition. The phase shift $\Delta\phi_R$ also depends on the polarization. For freely propagating plane waves, the electric and magnetic fields are transverse to the direction of propagation and we speak of s- or p-polarization relative to the plane of incidence (the plane of the diagram). In guided waves, the electric field and magnetic field have a component in the longitudinal direction along the waveguide, so we classify the waves as TE (transverse electric) waves (no longitudinal electric field, s-polarized relative to the plane of the diagram) or TM (transverse magnetic) waves (no longitudinal magnetic field, p-polarized relative to the plane of the diagram).

(a) Show that the mode condition may be expressed by

$$nkd\cos\theta - 2\arctan\left(\sqrt{-\frac{n^2\sin^2\theta - 1}{n^2\sin^2\theta - n^2}}\right) = m\pi \qquad \text{(TE waves)} \tag{9.48}$$

$$nkd\cos\theta - 2\arctan\left(n^2\sqrt{-\frac{n^2\sin^2\theta - 1}{n^2\sin^2\theta - n^2}}\right) = m\pi \qquad \text{(TM waves)} \tag{9.49}$$

where it is assumed that the dielectric slab is nonmagnetic. These transcendental equations must be solved numerically to find the angle θ between the propagation direction and the surface normal. It is convenient to introduce the propagation constant $\kappa = nk\cos\theta$ for the mode, which is the effective wave number across the waveguide, and the parameter $\kappa_c = nk\cos\theta_c = k\sqrt{n^2 - 1}$, where θ_c is the critical angle for total internal reflection. Show that the mode conditions (9.48) and (9.49) can be expressed in the more convenient form

$$\kappa d \begin{Bmatrix} \tan\left(\frac{1}{2}kd\right) \\ \cot\left(-\frac{1}{2}kd\right) \end{Bmatrix} = \sqrt{\kappa_c^2 d^2 - \kappa^2 d^2} \qquad \text{(TE waves)} \tag{9.50}$$

$$\kappa d \begin{Bmatrix} \tan\left(\frac{1}{2}kd\right) \\ \cot\left(-\frac{1}{2}kd\right) \end{Bmatrix} = n^2\sqrt{\kappa_c^2 d^2 - \kappa^2 d^2} \qquad \text{(TM waves)} \tag{9.51}$$

where the upper function refers to $m =$ even and the lower to $m =$ odd.

(b) Sketch the left- and right-hand sides of (9.50) and (9.51) as functions of κd, and indicate the solutions. Note that there is always at least one TE and one TM mode. What is the criterion for only one mode to exist?

9.1.3 Integral Theorems

Since the curl of a gradient vanishes identically, we see that

$$\nabla \times \nabla \psi = 0 \tag{9.52}$$

This becomes more interesting if we introduce the unit vector $\hat{\mathbf{s}}$ tangent to the ray, given by (9.37), in terms of which (9.52) becomes

$$\nabla \times n\hat{\mathbf{s}} = 0 \tag{9.53}$$

Applying Stokes' theorem we get

$$\oint n\hat{\mathbf{s}} \cdot d\mathbf{l} = 0 \tag{9.54}$$

around any closed contour, as illustrated in Figure 9.7. Equivalently, this may be expressed in the form

$$\int_a^b n\hat{\mathbf{s}} \cdot d\mathbf{l} = \psi(b) - \psi(a) = \text{independent of the path} \tag{9.55}$$

where $\psi(a)$ is a function only of the position of the point a. This is called Lagrange's integral invariant.

As an example of the use of this theorem, we can use it to derive Snell's law for refraction at a discontinuity in the index of refraction. Although the assumption that the index of refraction is slowly varying (on the scale of a wavelength) is violated at the discontinuity, it is valid within the medium on each side of the interface. As shown in Figure 9.8, we consider two paths next to each other on opposite sides of the discontinuity. Ignoring the infinitesimal segments at the ends, the integrals are

$$\int_a^b n_1\hat{\mathbf{s}}_1 \cdot d\mathbf{l} = \int_a^b n_2\hat{\mathbf{s}}_2 \cdot d\mathbf{l} \tag{9.56}$$

where n_i and $\hat{\mathbf{s}}_i$ are the index of refraction and ray unit vector on the ith side of the discontinuity, and $d\mathbf{l}$ lies in the surface of the discontinuity. Since (9.56) is true independent of the path, it must be true that

$$n_1\hat{\mathbf{s}}_1 \cdot d\mathbf{l} = n_2\hat{\mathbf{s}}_2 \cdot d\mathbf{l} \tag{9.57}$$

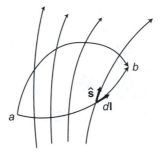

Figure 9.7 Lagrange's integral invariant.

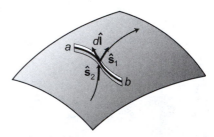

Figure 9.8 Lagrange's integral invariant at a discontinuity in the index of refraction.

everywhere. That is, the vector $(n_1\hat{\mathbf{s}}_1 - n_2\hat{\mathbf{s}}_2)$ must be normal to the surface of the discontinuity. Taking $d\mathbf{l}$ in the direction parallel to the discontinuity in the plane of incidence, we get

$$n_1 \sin\theta_1 = n_2 \sin\theta_2 \tag{9.58}$$

where θ_i is the angle between the ray and the surface normal. This is Snell's law.

Another integral theorem, called Fermat's principle, states that light propagates from one point to another along the path for which the time is an extremum, usually a minimum. To be precise, the principle refers to the time for a phase front to propagate from point a to point b, which is

$$T = \int_a^b \frac{ds}{v_\phi} = \frac{1}{c} \int_a^b n\, ds \tag{9.59}$$

in which v_ϕ is the phase velocity. The integral $cT = \int_a^b n\, ds$ is called the optical path length between the points a and b. Fermat's principle states that

$$\delta T = 0 \tag{9.60}$$

for variations of the path with the end points fixed. To find the path along which T is stationary we use the calculus of variations, as in Chapter 2. Since the end points are fixed, we get

$$\delta T = \int_1^2 \delta n\, ds + \int_1^2 n\, d\delta s \tag{9.61}$$

But the incremental path length is

$$ds^2 = d\mathbf{r} \cdot d\mathbf{r} \tag{9.62}$$

so if we calculate the variation of this expression, we get

$$ds\, d\delta s = d\mathbf{r} \cdot d\delta\mathbf{r} \tag{9.63}$$

Since the variation of the index of refraction is

$$\delta n = \nabla n \cdot \delta\mathbf{r} \tag{9.64}$$

the variation of the path integral (9.61) becomes

$$\delta T = \int_1^2 (\nabla n \cdot \delta\mathbf{r})\, ds + \int_1^2 n\frac{d\mathbf{r}}{ds} \cdot \frac{d\delta\mathbf{r}}{ds}\, ds \tag{9.65}$$

Integrating the second integral once by parts, as we always do, we get

$$\delta T = \int_1^2 \left[\nabla n - \frac{d}{ds}\left(n\frac{d\mathbf{r}}{ds}\right) \right] \cdot \delta\mathbf{r}\, ds + \left(n\frac{d\mathbf{r}}{ds} \cdot \delta\mathbf{r} \right)\Bigg|_1^2 = 0 \tag{9.66}$$

But the last term vanishes because $\delta\mathbf{r}$ vanishes at the end points. Since the variation of the integral must vanish for arbitrary variations $\delta\mathbf{r}$ of the path, we obtain the equation

$$\nabla n - \frac{d}{ds}\left(n\frac{d\mathbf{r}}{ds}\right) = 0 \tag{9.67}$$

Comparison with (9.28) shows that this is, in fact, the differential equation for a ray, which we previously derived from more direct, if less elegant, considerations.

EXERCISE 9.3

Use Fermat's principle to derive Snell's law for the refraction of a ray at a discontinuity in the index of refraction.

EXERCISE 9.4

Consider the focusing of a beam of parallel rays by a plano-convex lens, as shown in Figure 9.9.

(a) Show that in the thin-lens approximation ($D \ll 2R$, where D is the diameter of the lens and R the radius of curvature of the lens), the thickness of the lens at the radius r from the axis is

$$t = \frac{D^2 - 4r^2}{2R} \tag{9.68}$$

(b) Use Fermat's principle to derive the lens maker's equation

$$\frac{1}{f} = \frac{n-1}{R} \tag{9.69}$$

where f is the focal length and n the index of refraction. The extension to a double convex lens is straightforward, and we get

$$\frac{1}{f} = (n-1)\left(\frac{1}{R_1} + \frac{1}{R_2}\right) \tag{9.70}$$

where R_1 and R_2 are the radii of curvature of the surfaces of the lens (both positive for a double convex lens, as used here).

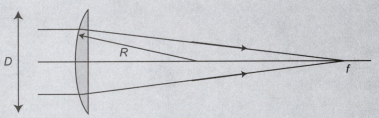

Figure 9.9 Beam focused by a plano-convex lens.

9.2 GAUSSIAN OPTICS AND LASER RESONATORS

9.2.1 Paraxial Approximation

Although electromagnetic waves never travel as true rays without the effects of diffraction, they often travel in highly collimated bundles nearly parallel to the optical axis. This is particularly true for laser beams, which are routinely collimated well enough to bounce off the moon or to be aimed at your opponent in a game of "laser tag." For these very useful and entertaining beams we may use the so-called paraxial approximation, in which we ignore derivatives of the beam properties along the direction of propagation compared with derivatives across the beam. That is, we assume that the beam is much longer than it is wide.

In the absence of sources, the beam propagates according to the wave equation

$$\nabla^2 \mathbf{E} - \frac{1}{c^2} \frac{\partial \mathbf{E}}{\partial t} = 0 \tag{9.71}$$

where $\mathbf{E}(\mathbf{r}, t)$ is the electric field. For a well-collimated wave the longitudinal components of the field, while not zero, are small, and we ignore them in the following discussion. The transverse components of the field may be linearly or elliptically polarized. In any event, each component of the field propagates independently according to the equation

$$\nabla^2 \psi - \frac{1}{c^2} \frac{\partial \psi}{\partial t} = 0 \tag{9.72}$$

For a wave at frequency ω traveling along the z axis, we describe the function $\psi(\mathbf{r}, t)$ by the expression

$$\psi(\mathbf{r}, t) = \psi_0(\mathbf{r}, t) e^{i(kz - \omega t)} \tag{9.73}$$

where

$$\omega = kc \tag{9.74}$$

For a highly collimated beam, we assume that the envelope function $\psi_0(\mathbf{r}, t)$ is slowly varying in the longitudinal direction. If we substitute the solution (9.73) into the wave equation (9.72) and ignore second derivatives of ψ_0 with respect to z and t, we obtain, after some algebra, the paraxial wave equation

$$\frac{\partial \psi_0}{\partial z} + \frac{\partial \psi_0}{\partial t} = \frac{i}{2k} \nabla_T^2 \psi_0 \tag{9.75}$$

where the transverse Laplacian operator is

$$\nabla_T^2 = \frac{\partial^2}{\partial x^2} + \frac{\partial^2}{\partial y^2} \tag{9.76}$$

In the following we assume a steady-state wave, so we can ignore the time derivative in (9.75). We immediately see that for a wave propagating over some finite distance, the transverse dimensions of the wave scale as $1/\sqrt{k}$, or equivalently as the square root of the wavelength.

We look first for a solution of the form

$$\psi_0(\mathbf{r}) = \exp\left[A(z) - \frac{r^2}{2B(z)} \right] \tag{9.77}$$

where $r^2 = x^2 + y^2$, which corresponds to a beam with a simple Gaussian profile everywhere. If we substitute this into (9.75), we obtain the pair of equations

$$\frac{dA}{dz} = -\frac{i}{kB} \tag{9.78}$$

and

$$\frac{dB}{dz} = \frac{i}{k} \tag{9.79}$$

The solution to (9.79) is

$$B = B_0 + i\frac{z - z_0}{k} \tag{9.80}$$

for some constants B_0 and z_0. We call z_0 the focal point. The amplitude and phase on axis are given by the function

$$A(z) = -\frac{i}{k} \int \frac{dz}{B}$$

$$= -\ln\sqrt{1 + \left(\frac{z - z_0}{kB_0}\right)^2} - i\arctan\left(\frac{z - z_0}{kB_0}\right) + \text{constant} \tag{9.81}$$

where the constant may be ignored if the wave function is normalized to unity on axis at the focal point. The solution (9.80) may also be written in the form

$$-\frac{r^2}{2B} = ik\frac{r^2}{2R} - \frac{r^2}{w^2} \tag{9.82}$$

where

$$w^2(z) = 2B_0\left[1 + \frac{(z - z_0)^2}{B_0^2 k^2}\right] \tag{9.83}$$

$$R(z) = (z - z_0)\left[1 + \frac{k^2 B_0^2}{(z - z_0)^2}\right] \tag{9.84}$$

Since the first term in (9.82) is imaginary and the second is real, the function $w(z)$ evidently describes the radius ($1/e$ amplitude) of the Gaussian beam at position z. The first term in (9.82) describes a phase shift that is quadratic in the radius, which shows that the wave fronts are curved. In fact, the function $R(z)$ is just the radius of curvature of the wave fronts, as is readily shown.

At the focal point the beam has the radius

$$w_0 = \sqrt{2B_0} \tag{9.85}$$

and the beam area doubles at the point

$$z - z_0 = z_R = B_0 k \tag{9.86}$$

The parameter z_R is called the Rayleigh range. It is related to the waist radius at the focus by the very useful formula

$$\pi w_0^2 = \lambda z_R \tag{9.87}$$

where $\lambda = 2\pi/k$ is the wavelength. Some authors refer to the so-called confocal parameter $b = 2z_R$, which is the total distance over which the beam area is within a factor of two of its value at the focal point.

The solution summarized by (9.82)–(9.84) describes the focus of a Gaussian beam, as shown in Figure 9.10. At large distances from the focus the radius of the beam is

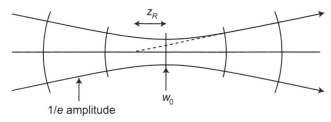

Figure 9.10 Geometry of a Gaussian beam at the focus.

$w/w_0 \approx (z - z_0)/z_R$, so the beam expands linearly with the angle

$$\theta_\infty = \lim_{z \to \infty} \frac{w}{z - z_0} = \frac{w_0}{z_R} \qquad (9.88)$$

From this we see that for a Gaussian beam focused with the convergence half-angle θ_∞, the spot size is given by the handy formula

$$w_0 = \frac{\lambda}{\pi \theta_\infty} \qquad (9.89)$$

From (9.84) we see that at large distances from the focus the waves appear to be spherical, with a source at the focal point. At shorter distances from the focus, the center of curvature is on the opposite side of the focal point.

The amplitude and phase of the wave on axis are described by (9.81). Comparison with (9.83) shows that the first term, the real part, causes the amplitude to decrease, relative to the amplitude at the focal point, as $1/w(z)$. This is necessary to conserve the total power in the beam. The second term in (9.81) describes the phase of the wave. At large distances from the focal point, the phase is

$$\Phi = -\arctan\left(\frac{z - z_0}{z_R}\right) \xrightarrow[|z| \to \pm\infty]{} \mp \frac{\pi}{2} \qquad (9.90)$$

This is called the Gouy phase shift. As the beam passes through the focus, the total phase shift relative to a plane wave is $-\pi$. That is, the wave loses one-half wavelength of phase shift. As the wave is squeezed in the radial direction, it stretches out longitudinally in the focal region. It is of some interest to note that the Gouy phase shift was actually discovered experimentally, over a century ago, in experiments on the interference pattern formed when two parts of a beam are brought together, one of which has been focused and the other of which has not.

The simple Gaussian beam (9.77) is not the only solution to the paraxial wave equation. In fact, it may be shown by direct substitution that any expression of the form

$$\psi_0(z) = \frac{1}{w(z)} H_m\left[\frac{\sqrt{2}x}{w(z)}\right] H_n\left[\frac{\sqrt{2}y}{w(z)}\right] \exp\left[i\Phi(z) + ik\frac{r^2}{2R(z)} - \frac{r^2}{w^2(z)}\right] \qquad (9.91)$$

satisfies the wave equation provided that $w(z)$ and $R(z)$ are given by (9.83) and (9.84), as before, and that the functions $H_n(x)$ satisfy the equation

$$\frac{d^2 H_n}{dx^2} - 2x\frac{dH_n}{dx} + 2nH_n = 0 \qquad (9.92)$$

The functions $H_m(x)$, for $n = 0, 1, 2, \ldots$, are called Hermite polynomials. They may be computed from Rodriguez' formula

$$H_n = (-1)^n e^{x^2} \frac{d^n}{dx^n}\left(e^{-x^2}\right) \tag{9.93}$$

or by using the recursion relations

$$H_0 = 1 \tag{9.94}$$

$$H_1 = 2x \tag{9.95}$$

$$H_{n+1} = 2x H_n - 2n H_{n-1} \tag{9.96}$$

The Hermite polynomials form a complete, orthogonal set with the weighting factor $\exp(-x^2)$. The standard normalization is such that

$$\int_{-\infty}^{\infty} H_m(x) H_n(x) e^{-x^2} dx = \sqrt{\pi} 2^n n! \delta_{mn} \tag{9.97}$$

The Hermite polynomials may therefore be used to expand any function of x in the form

$$f(x) = \sum_{n=0}^{\infty} a_n H_n(x) e^{-x^2} \tag{9.98}$$

where the coefficients are

$$a_n = \frac{1}{\sqrt{\pi} 2^n n!} \int_{-\infty}^{\infty} f(x) H_n(x) e^{-x^2} dx \tag{9.99}$$

The Gouy phase shift for the higher-order modes is

$$\Phi(z) = -(m + n + 1) \arctan\left(\frac{z - z_0}{z_R}\right) \tag{9.100}$$

The first few Gauss–Hermite modes are shown in Figure 9.11. Clearly, each mode propagates through the focus self-similarly, preserving its shape as it goes but expanding

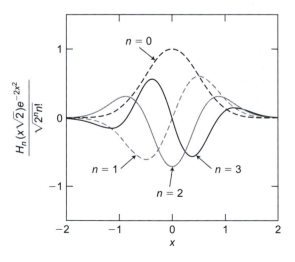

Figure 9.11 Gauss–Hermite modes.

with $w(z)$. However, when several modes are combined to form the beam, they interfere differently at different positions through the focus because the Gouy phase shift is different for each mode. The overall beam shape therefore changes as the beam passes through the focus.

For circumstances with circular, rather than rectangular, symmetry, it is sometimes more convenient to use Gauss–Laguerre modes. It may be demonstrated by substitution that the expression

$$\psi_0(z) = \frac{1}{w(z)} \left[\frac{\sqrt{2}r}{w(z)} \right]^p L_n^p \left[\frac{2r^2}{w^2(z)} \right] e^{ip\phi} \exp\left[i\Phi(z) + ik\frac{r^2}{2R(z)} - \frac{r^2}{w^2(z)} \right] \qquad (9.101)$$

also satisfies the paraxial wave equation (9.75) provided that the functions $L_n^p(x)$ satisfy the differential equation

$$x\frac{d^2 L_n^p}{dx^2} + (p + 1 - x)\frac{dL_n^p}{dx} + nL_n^p = 0 \qquad (9.102)$$

The functions $L_n^p(x)$, for $n = 0, 1, 2, \ldots$ and $p = -n, -n + 1, \ldots, n$, are called generalized Laguerre polynomials. They may be computed from Rodriguez' formula

$$L_n^p = \frac{e^x}{n! x^p} \frac{d^n}{dx^n} (x^{n+p} e^{-x}) \qquad (9.103)$$

or by using the recursion relations

$$L_0^0 = 1 \qquad (9.104)$$

$$L_1^0 = 1 - x \qquad (9.105)$$

$$(n + 1)L_{n+1}^p = (2n + p + 1 - x)L_n^p - (n + p)L_{n-1}^p \qquad (9.106)$$

$$L_n^{p-1} = L_n^p - L_{n-1}^p \qquad (9.107)$$

$$L_n^{p+1} = \frac{1}{x}\left[(x - n)L_n^p + (n + p)L_{n-1}^p \right] \qquad (9.108)$$

The generalized Laguerre polynomials form a complete, orthogonal set with the weighting function $x^p \exp(-x)$ and have the normalization

$$\int_0^\infty L_m^p(x) L_n^p(x) x^p e^{-x} dx = \frac{(m + p)!}{m!} \delta_{mn} \qquad (9.109)$$

The Gouy phase shift is

$$\Phi(z) = -(2n + p + 1) \arctan\left(\frac{z - z_0}{z_R} \right) \qquad (9.110)$$

EXERCISE 9.5

A zero-order Gaussian beam is brought to a focus at the origin with a Rayleigh range z_R. On the way to the focus, the beam passes through an aperture of radius a at the point $z = -L$. Expand the beam profile in Laguerre polynomials and show that the amplitude compared to that of the beam on axis at the focus with no aperture is

$$\psi(r, z) = \frac{1}{w(z)} e^{i[kr^2/2R(z)]} e^{-r^2/w^2(z)} \sum_{m=0}^{\infty} c_n e^{i\Phi_n(z)} L_n^0 \left[\frac{2r^2}{w^2(z)} \right] \tag{9.111}$$

where

$$w^2(z) = w_0^2 \left[1 + \frac{z^2}{z_R^2} \right] \tag{9.112}$$

$$R(z) = z \left[1 + \frac{z_R^2}{z^2} \right] \tag{9.113}$$

$$\Phi_n(z) = -(2n + 1) \arctan\left(\frac{z}{z_R} \right) \tag{9.114}$$

the coefficients are

$$c_0 = 1 - e^{-2a^2/w_L^2} \tag{9.115}$$

$$c_1 = e^{-i2 \arctan(L/z_R)} \frac{2a^2}{w_L^2} e^{-2a^2/w_L^2} \tag{9.116}$$

$$c_{m>1} = \frac{e^{-i2m \arctan(L/z_R)}}{m} e^{-2a^2/w_L^2} \left[\left(\frac{2a^2}{w_L^2} - m + 1 \right) L_{m-1}^0 \left(\frac{2a^2}{w_L^2} \right) + (m - 1) L_{m-2}^0 \left(\frac{2a^2}{w_L^2} \right) \right] \tag{9.117}$$

and $w_L = w(-L)$ is the radius of the beam entering the aperture. The intensity on axis compared to the peak intensity without the aperture is shown in Figure 9.12 for $a = w_L$ and $L = 10z_R$. As we see there, the aperture introduces strong variations in the intensity along the axis near the focus, and the intensity of the apertured beam can exceed that of the unapertured beam.

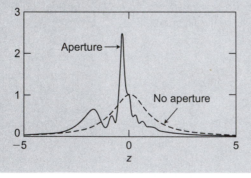

Figure 9.12 On-axis intensity near a focus with an aperture in the beam.

9.2.2 Laser Resonators and Mode Spacing

One of the important uses of Gaussian modes is in the description of the modes of laser resonators. In a laser resonator the optical field is confined between two curved mirrors as

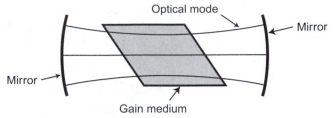

Figure 9.13 Geometry of an optical resonator.

shown in Figure 9.13. In the following we assume that the two mirrors are spherical, perfect reflectors. One mirror is positioned at z_1 and has a radius of curvature R_1, and the other is positioned at z_2, with a radius of curvature R_2 [both mirrors have positive curvature in Figure (9.13)]. The optical field is a standing wave, which is composed of waves traveling to the right and to the left. To form a standing wave of the optical resonator, the curvature of the wave, given by (9.84), must match the shape of the reflector at the surface of each reflector. From this we obtain the boundary conditions

$$R_1 = (z_0 - z_1)\left[1 + \frac{z_R^2}{(z_1 - z_0)^2}\right] \tag{9.118}$$

$$R_2 = (z_2 - z_0)\left[1 + \frac{z_R^2}{(z_2 - z_0)^2}\right] \tag{9.119}$$

at mirrors 1 and 2, respectively, where z_0 is the focal point of the mode and z_R is the Rayleigh range. We can solve these equations for z_0 and z_R. At this point it is convenient (and standard practice) to introduce the so-called stability parameters

$$g_1 = 1 - \frac{L}{R_1} \tag{9.120}$$

and

$$g_2 = 1 - \frac{L}{R_2} \tag{9.121}$$

where

$$L = z_2 - z_1 \tag{9.122}$$

is the overall length of the optical resonator. After some algebra, we find that the foc point is

$$\frac{z_0 - z_1}{L} = \frac{g_2(1 - g_1)}{g_1 + g_2 - 2g_1g_2} \tag{9.1}$$

$$\frac{z_2 - z_0}{L} = \frac{g_1(1 - g_2)}{g_1 + g_2 - 2g_1g_2} \tag{9}$$

and the Rayleigh range is

$$\left(\frac{z_R}{L}\right)^2 = \frac{g_1g_2(1 - g_1g_2)}{(g_1 + g_2 - 2g_1g_2)^2}$$

In addition, for the wave to be an eigenmode of the resonator, the total phase shift in a round trip of the resonator, including the phase shift at each reflector, must be a multiple of 2π. For metallic reflectors, the phase shift at each reflection is π, so the total phase shift due to two reflections is 2π, which may be ignored. The phase shift from one mirror to the other must therefore be a multiple of π. On the axis, the phase is given by the plane-wave phase plus the Gouy phase shift. The phase shift from z_1 to z_2 is therefore

$$kl + \Phi(z_2) - \Phi(z_1) = q\pi \qquad (9.126)$$

for some integer q. For Gauss–Hermite modes, the Gouy phase shift is given by (9.100). If we use (9.123)–(9.125) to evaluate z_0 and z_R, we obtain, after some algebra, the mode condition

$$\omega = \omega_0 \left[q + \frac{m+n+1}{\pi} \arccos \sqrt{g_1 g_2} \right] \qquad (9.127)$$

where the characteristic frequency is

$$\omega_0 = 2\pi \frac{c}{2L} \qquad (9.128)$$

that is, 2π times the round-trip frequency of a light wave in the resonator. Note that when g_1 and g_2 are both positive, the arccosine in (9.127) varies between 0 and $\pi/2$. When g_1 and g_2 are both negative, it is necessary to take the negative square root, so the arccosine varies between $\pi/2$ and π.

The significance of the longitudinal mode spacing ω_0 may be understood in the following way. By combining two or more longitudinal modes, we can form a "wave packet," or optical pulse, which oscillates back and forth around the resonator. The case of two adjacent longitudinal modes, whose frequencies differ by $\Delta\omega_L = \omega_0$, is illustrated in Figure 9.14. As shown there, the waves interfere constructively at the left end of the resonator, but the phase of the two waves differs by π at the right end of the resonator, so they interfere destructively there. The sum of the two waves at this point in time is therefore a "wave packet" at the left side of the resonator. At a time

$$t = \frac{\pi}{\Delta\omega_L} = \frac{L}{c} \qquad (9.129)$$

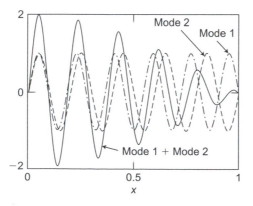

Figure 9.14 Wave packet formed by two adjacent longitudinal modes.

later the relative phase of the waves has changed by π, so now the waves interfere constructively at the right end of the resonator. That is, the wave packet has traversed the resonator in one direction. In this case, the width of the "pulse" is approximately half the length of the resonator, but narrower pulses can be constructed by combining more modes. In general, the spectral width $\Delta\omega$ of a pulse of length Δt satisfies the relation

$$\Delta\omega \Delta t \geq O(1) \tag{9.130}$$

Thus, for a pulse whose width is $\Delta t = \Delta z/c$, the number of longitudinal modes in the spectrum is

$$N = \frac{\Delta\omega}{\omega_0} \geq O\left(\frac{2L}{\Delta z}\right) \tag{9.131}$$

that is, a number on the order of the round-trip length of the cavity divided by the optical pulse length.

Lasers can be forced to produce short pulses by placing in the resonator an optical element that introduces losses modulated at the round-trip frequency of the resonator. This suppresses lasing except when the modulated loss is near its minimum, which causes a short pulse to form and reflect back and forth in the resonator. Such pulses are referred to as "mode locked," since they represent a sum of many modes whose phases are locked together to interfere in a manner that results in a short pulse. The modulated loss element is called a "mode locker." The mode locker can be "active," in which case the losses are controlled by an external clock, or "passive," in which case the losses are turned off by nonlinear processes initiated by the arrival of the mode-locked pulse itself. By using a broad-band gain medium, such as titanium-doped sapphire, it is possible to amplify and lock together many modes of the optical resonator. Mode locking in Ti : sapphire lasers, for example, is routinely used to create optical pulses as short as 35 fs at wavelengths around 800 nm. This corresponds to less than 20 optical wavelengths.

9.2.3 Transverse Modes and Resonator Stability

Returning to (9.127), we see that for a given longitudinal mode, transverse modes that correspond to different values of m and n are separated by the frequency

$$\Delta\omega_T = \frac{\omega_0}{\pi} \arccos \sqrt{g_1 g_2} \tag{9.132}$$

As with longitudinal modes, we may examine the result of combining two adjacent transverse modes. For simplicity, we consider the modes $m = 0$ and $m = 1$. As shown in Figure 9.15, the Gauss–Hermite mode for $m = 0$ is symmetric and that for $m = 1$ is antisymmetric, so when they have the same phase they interfere constructively for $x > 0$ and destructively for $x < 0$. At a time

$$t = \frac{\pi}{\Delta\omega_T} = \frac{\pi^2}{\omega_0 \arccos \sqrt{g_1 g_2}} \tag{9.133}$$

later, the relative phase of the modes changes by π and the modes interfere constructively for $x < 0$. That is, the wave packet moves to the other side of the center. With increasing time, the combination mode oscillates back and forth across the mirror with a frequency equal to the transverse mode spacing.

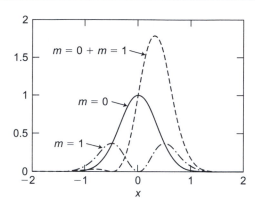

Figure 9.15 Interference of adjacent transverse modes $m = 0$ and $m = 1$.

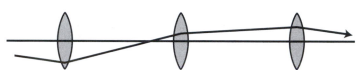

Figure 9.16 Infinite series of lenses equivalent to an unfolded resonator.

We encounter this same effect in a quantum-mechanical harmonic oscillator. In this case, the wave functions of the particle are again Hermite polynomials, and the energy levels are spaced by $\Delta E = \hbar\omega$, where ω is the classical frequency of the oscillator. The phase of the nth mode is $\omega_n t = E_n t / \hbar = n\omega t$. If we form a linear combination of two adjacent modes, the wave packet oscillates back and forth with the difference frequency ω. As in the case of optical pulses in a laser resonator, we can combine many harmonic oscillator modes to form a narrower wave packet for the harmonic oscillator. However, this improved resolution in the position of the particle comes at the expense of reduced information about the momentum of the particle, in accordance with Heisenberg's uncertainty principle.

It is instructive to look at transverse mode oscillations from the perspective of rays bouncing back and forth inside the optical resonator. To do this we unfold the resonator into an equivalent infinite series of lenses that focus the rays with the same focal length as the mirrors, as shown in Figure 9.16. In two dimensions the ray is completely described by its position $x(z)$ and angle $x'(z)$ at point z along the unfolded resonator. In the paraxial approximation, that is, when the angles x' are small, the position x_1 and angle x_1' at point z_1 are linearly related to the position x_0 and angle x_0' at the initial point z_0. The position and angle at point z_1 can therefore be computed by means of the equation

$$\begin{bmatrix} x_1 \\ x_1' \end{bmatrix} = \begin{bmatrix} A & B \\ C & D \end{bmatrix} \begin{bmatrix} x_0 \\ x_0' \end{bmatrix} \tag{9.134}$$

where the square matrix is called the ray-transfer or ABCD matrix. With a few minutes of simple geometry, it is easy to convince yourself that the ray-transfer matrices for a lens of focal length f and an empty space of length L are, respectively,

$$\begin{bmatrix} A & B \\ C & D \end{bmatrix} = \begin{bmatrix} 1 & 0 \\ -1/f & 1 \end{bmatrix} \tag{9.135}$$

and

$$\begin{bmatrix} A & B \\ C & D \end{bmatrix} = \begin{bmatrix} 1 & L \\ 0 & 1 \end{bmatrix} \tag{9.136}$$

In a complete circuit of the optical resonator, a ray goes through a lens of focal length $f_1 = R_1/2$, an empty space of length L, a lens of focal length $f_2 = R_2/2$, and a second empty space of length L. The final coordinates of the ray are related to the initial coordinates by a ray-transfer matrix that is the product (in reverse order) of the individual ray-transfer matrices, and after some tedious algebra we find that the ray-transfer matrix for one complete round-trip pass through the resonator is

$$\begin{bmatrix} A & B \\ C & D \end{bmatrix} = \begin{bmatrix} 1 - \dfrac{2L}{R_1} & 2L - \dfrac{2L^2}{R_1} \\ -\dfrac{2R_2 + 2R_1 - 4L}{R_1 R_2} & 1 - \dfrac{2LR_2 + 4LR_1 - 4L^2}{R_1 R_2} \end{bmatrix} \tag{9.137}$$

an expression that conveys no insight whatsoever.

Repeated reflections of the ray around the resonator are calculated by raising the ray-transfer matrix to the nth power. In this computation we make use of Sylvester's theorem, which states that

$$\begin{bmatrix} A & B \\ C & D \end{bmatrix}^n = \frac{1}{\sin\theta} \begin{bmatrix} A\sin(n\theta) - \sin[(n-1)\theta] & B\sin\theta \\ C\sin\theta & D\sin(n\theta) - \sin[(n-1)\theta] \end{bmatrix} \tag{9.138}$$

where

$$\cos\theta = \frac{A+D}{2} \tag{9.139}$$

Since the ray-transfer matrix looks sort of like a "rotation" in some space through the angle θ, Sylvester's theorem tells us that the result of n rotations looks sort of like a rotation through the angle $n\theta$. In terms of the resonator parameters,

$$\cos\theta = 2\left(1 - \frac{L}{R_1}\right)\left(1 - \frac{L}{R_2}\right) - 1 = 2g_1 g_2 - 1 \tag{9.140}$$

As the ray progresses through successive passes of the resonator, the coordinates oscillate with the frequency θ. In real time, this frequency is

$$\omega_T = \frac{c}{2L}\arccos(2g_1 g_2 - 1) = \frac{c}{2L}\arccos\sqrt{g_1 g_2} \tag{9.141}$$

Comparison with (9.132) shows that this is just the transverse mode spacing.

Provided that the frequency ω_T is real, the transverse coordinates oscillate and the resonator is called stable. When ω_T is imaginary, the displacement increases exponentially and the ray "walks off" the edge of the resonator after a few passes. In this case the resonator is called unstable. In terms of the stability parameters g_1 and g_2, stability corresponds to the requirement

$$0 < g_1 g_2 < 1 \tag{9.142}$$

This is just the condition for the transverse mode spacing (9.132) to be real, or for the Rayleigh range (9.125) to be real. The stability condition is summarized in Figure 9.17,

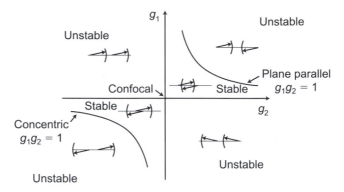

Figure 9.17 Stability diagram for laser resonators.

where various common resonator configurations are shown. The arrows in Figure 9.17 represent the radius vectors of the mirrors, and are included to indicate the relative positions of the centers of curvature.

EXERCISE 9.6

To form a collimated output beam, lasers are frequently operated in the near-hemispherical mode, with one flat mirror and one curved mirror having a radius of curvature R close to the length L between the mirrors. This puts the resonator near one of the axes on the stability diagram Figure 9.17, close to the edge of the stable region. A micrometer can then be used to adjust the length of the resonator by small amounts and change the mode radius by a relatively large amount.

(a) Show that for $L = R - \Delta L$ and $0 < \Delta L \ll R$, the radius of the mode at the curved mirror is

$$w_c \approx \left(\frac{\lambda^2 R^3}{\pi^2 \Delta L} \right)^{1/4} \tag{9.143}$$

where λ is the wavelength.

(b) For a HeNe laser ($\lambda = 633$ nm) in a cavity nominally $R = 30$ cm long, what is the adjustment ΔL required to make the mode radius 2 mm at the curved mirror? What is the mode radius at the flat miror?

9.3 DIFFRACTION

In geometric optics, the rays of light travel in straight lines and deviate from this only due to the effects of refraction and reflection. In truth, however, nature is more complex than this. The wave nature of light causes beams of light to spread naturally in directions transverse to the direction of propagation. Thus, the rays of light emerging from a point cannot be refocused to a point, and light passing through an aperture or past a sharp-edged obstruction spreads into the region of shadow behind the obstruction.

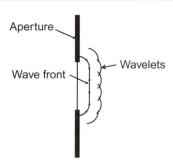

Aperture

Wave front

Wavelets

Figure 9.18 Huygens' construction.

The first to observe and report on the effects of diffraction was Grimaldi, who in 1665 examined the shadow formed by light from a pinhole as it passed through an aperture. He observed that the edges of the shadow were not sharp, even allowing for the penumbra formed by the finite size of the pinhole. Soon afterward, in 1678, Huygens proposed his theory of wave propagation, in which he argued that each point on the surface of a wave acts like a source of new waves. These new waves spread out spherically from the points on the wave front where they are emitted, and a new wave front is formed by the addition of the wavelets from the original wave front as shown in Figure 9.18. Although, as we see shortly, this idea misses some details, it nevertheless embraces some important concepts that we express mathematically in the following sections.

Unfortunately, in 1704 Newton took the position that light consists of corpuscles and that these corpuscles travel in straight lines until they are reflected or refracted, as described by geometric optics. It is a tribute to the stature of this great man in the scientific community that Newton's position effectively ended the discussion of the wave nature of light for 100 years. It wasn't until 1804 that Young resumed the discussion and introduced the principle of interference. He used this principle to explain the effects of diffraction, including his famous double-slit experiment, and was able, in fact, to determine the wavelength of light. Even so, acceptance of the wave theory came only grudgingly. In 1818, for example, Fresnel wrote a theoretical treatise on the diffraction of light waves and submitted it to the French Academy of Sciences for a prize that was being offered. Fresnel's results were vigorously disputed by the brilliant mathematician Poisson, a member of the Academy, who pointed out that Fresnel's theory predicted a bright spot in the center of the shadow behind a circular object where no light could possibly be. To settle the matter, Arago, who chaired the prize committee, carried out the experiment and found the spot! In memory of this discovery, the effect is called Arago's spot, or sometimes, with a twist of irony, Poisson's spot. Fresnel is seldom mentioned in this connection, but he got the prize.

9.3.1 Scalar Diffraction Theory

Since we are not interested here in the effects of refraction, we limit our discussion to the propagation of light in a vacuum. The extension to any other homogeneous medium is obvious. As we have seen in Chapter 4, a light wave is completely described by the vector potential $\mathbf{A}$ or, equivalently, by the electric and magnetic fields $\mathbf{E}$ and $\mathbf{B}$. In the absence of sources the vector potential satisfies the wave equation

$$\frac{1}{c^2}\frac{\partial^2 \mathbf{A}}{\partial t^2} - \nabla^2 \mathbf{A} = 0 \tag{9.144}$$

If we take the Fourier transform of this, we obtain the Helmholtz equation

$$\nabla^2 \tilde{\mathbf{A}} + k^2 \tilde{\mathbf{A}} = 0 \tag{9.145}$$

where $\mathbf{k}$ is the wave vector and $\omega = kc$ is the frequency of the wave. We see from (9.145) that each component of the vector $\tilde{\mathbf{A}}$ propagates independently, so it is tempting to assume that all we need to do is solve the scalar Helmholtz equation

$$\nabla^2 a + k^2 a = 0 \tag{9.146}$$

for each component and our problem is solved. Although this turns out to be an excellent approximation for many purposes, it is not strictly true. If, for example, we have a wave polarized in the horizontal direction incident on a circular aperture, the wave diffracted by the aperture in the horizontal plane is still polarized in the horizontal direction. However, since the electric field must be perpendicular to the direction of propagation, the diffracted wave must now have a component of the electric field in the original direction of propagation, even though the incident wave had no component in this direction. Waves diffracted in other directions will have even more complex polarization. The new components of the electric field are created at the boundaries of the aperture, where the boundary conditions on the electric and magnetic fields must be satisfied, and are not independent of the original components. Physically, they are radiated by the electronic response of the aperture material to the incident wave, and in the very near field of the aperture (within a few wavelengths of the edges), the effects of the boundary conditions must be fully taken into account.

Fortunately, it is found that the boundary conditions generally become unimportant beyond a few wavelengths from the aperture. We might expect this intuitively on the grounds that the electric and magnetic fields in a plane wave are equal (within a factor of c) and orthogonal both to each other and to the direction of propagation. Since the diffracted waves become locally nearly plane just a few wavelengths from the boundaries, one parameter should be enough to describe the wave, or at least its intensity. Specifically, we recall from Chapter 4 that if a is the complex amplitude of the vector potential $\tilde{\mathbf{A}}$, then the average intensity of a monochromatic wave is

$$\langle \mathbf{S} \rangle = \frac{\omega |a|^2}{2\mu_0} \mathbf{k} \tag{9.147}$$

Therefore, one parameter (the magnitude of the vector potential, for instance) is enough to tell us most of what we need to know about a plane wave (at least the intensity, if not the polarization). Thus, it is not so surprising that scalar diffraction theory works.

To compute the diffraction of a wave incident on an aperture, we adopt a form of Huygens' construction and look for a representation of the diffracted wave as a sum of waves emerging from the wave front incident on the aperture. Mathematically, this is equivalent to a Green-function technique in which the source vanishes inside the volume V of interest and only the surface integral over the aperture contributes to the solution. The volume we use is the hemisphere shown in Figure 9.19. It is bounded by the surfaces S_R far from the aperture, S_s on the screen, and S_a within the aperture. The unit vector $\hat{\mathbf{n}}'$ is normal to the surface everywhere and points outward from the volume V. Within the aperture, the vector $\hat{\mathbf{n}}'$ points toward the incident wave.

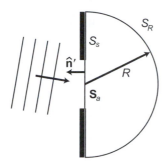

Figure 9.19 Volume used for diffraction from an aperture.

As discussed in Chapter 3, Green's theorem tells us that for any two well-behaved functions $a(\mathbf{r}')$ and $G(\mathbf{r}', \mathbf{r})$, where $\mathbf{r}$ is a parameter,

$$\int_V [a(\mathbf{r}')\nabla'^2 G(\mathbf{r}', \mathbf{r}) - G(\mathbf{r}', \mathbf{r})\nabla'^2 a(\mathbf{r}')]\, dV'$$

$$= \int_S [a(\mathbf{r}')\nabla' G(\mathbf{r}', \mathbf{r}) - G(\mathbf{r}', \mathbf{r})\nabla' a(\mathbf{r}')] \cdot \hat{\mathbf{n}}'\, dS' \tag{9.148}$$

For the function $G(\mathbf{r}', \mathbf{r})$, we choose the solution of the equation

$$\nabla'^2 G(\mathbf{r}', \mathbf{r}) + k^2 G(\mathbf{r}', \mathbf{r}) = -\delta(\mathbf{r}' - \mathbf{r}) \tag{9.149}$$

with the boundary condition

$$G(\mathbf{r}', \mathbf{r}) \xrightarrow[|\mathbf{r}'| \to \infty]{} 0 \tag{9.150}$$

The Green function for radiation in free space in the absence of the screen and aperture is a spherical wave centered on the point $\mathbf{r}$, with diminishing amplitude far from $\mathbf{r}$,

$$G(\mathbf{r}', \mathbf{r}) = \frac{e^{\pm ik|\mathbf{r}' - \mathbf{r}|}}{4\pi |\mathbf{r}' - \mathbf{r}|} \tag{9.151}$$

where the plus sign indicates outgoing waves and the minus sign incoming waves. To prove this, we substitute (9.151) into (9.149) and integrate over a spherical volume centered on $\mathbf{r}$, using the divergence theorem to convert the volume integral of $\nabla'^2 G$ to a surface integral. In the following we assume that there are no incoming waves. Then, since $a(\mathbf{r}')$ satisfies (9.146), Green's theorem shows that the solution we seek is

$$a(\mathbf{r}) = -\int_S [a(\mathbf{r}')\nabla' G(\mathbf{r}', \mathbf{r}) - G(\mathbf{r}', \mathbf{r})\nabla' a(\mathbf{r}')] \cdot \hat{\mathbf{n}}'\, dS' \tag{9.152}$$

Far from the aperture the functions $a(\mathbf{r}')$ and $G(\mathbf{r}', \mathbf{r})$ approach spherical waves, and the difference $(a\nabla' G - G\nabla' a)$ vanishes faster than $1/R^2$. We may therefore ignore the integral over S_R as $R \to \infty$. Further, we assume that the solution $a(\mathbf{r}')$ and its derivative $\nabla' a(\mathbf{r}')$ are small on the surface of the screen, since this is a region of shadow, so we can ignore the integral over S_s. This leaves only the integral over the aperture S_a, and (9.152) reduces to

$$a_K(\mathbf{r}) = -\int_{S_a} [a(\mathbf{r}')\nabla' G(\mathbf{r}', \mathbf{r}) - G(\mathbf{r}', \mathbf{r})\nabla' a(\mathbf{r}')] \cdot \hat{\mathbf{n}}'\, dS' \tag{9.153}$$

This is called the Kirchhoff integral formula. If we substitute the explicit form (9.151) for the Green function and use the approximation

$$\nabla' G(\mathbf{r}', \mathbf{r}) = (ik|\mathbf{r}' - \mathbf{r}| - 1) \frac{e^{ik|\mathbf{r}' - \mathbf{r}|}}{4\pi |\mathbf{r}' - \mathbf{r}|^2} \frac{\mathbf{r}' - \mathbf{r}}{|\mathbf{r}' - \mathbf{r}|} \approx \frac{ik(\mathbf{r}' - \mathbf{r})}{4\pi |\mathbf{r}' - \mathbf{r}|} \frac{e^{ik|\mathbf{r}' - \mathbf{r}|}}{|\mathbf{r}' - \mathbf{r}|} \tag{9.154}$$

which is valid more than a few wavelengths from the boundaries ($k|\mathbf{r}' - \mathbf{r}| \gg 1$), we get

$$a_K(\mathbf{r}) = \frac{1}{4\pi} \int_{S_a} \left[\nabla' a(\mathbf{r}') - \frac{ik(\mathbf{r}' - \mathbf{r})}{|\mathbf{r}' - \mathbf{r}|} a(\mathbf{r}') \right] \cdot \hat{\mathbf{n}}' \frac{e^{ik|\mathbf{r}' - \mathbf{r}|}}{|\mathbf{r}' - \mathbf{r}|} dS' \tag{9.155}$$

However, it can be shown quite generally that if both the solution $a(\mathbf{r}')$ and its derivative $\nabla' a(\mathbf{r}')$ vanish identically on some surface, then $a(\mathbf{r}')$ must vanish everywhere. Therefore, in (9.152) the integral over the surface S_s of the screen cannot vanish identically, and we have introduced some approximation by ignoring it. But in regions where the diffracted wave is large, the error introduced by using the "wrong" boundary conditions is small. We can see this by examining what happens when we use different forms of "correct" boundary conditions and observing when the differences caused by using different boundary conditions are small.

Although the solution $a(\mathbf{r})$ of the scalar wave equation cannot, of course, satisfy the correct vector boundary conditions, since it is not a vector, it can satisfy Dirichlet or Neumann boundary conditions as discussed in Chapter 3. To find the solution of the Helmholtz equation (9.146) subject to the Dirichlet or Neumann boundary conditions

$$a(\mathbf{r}) = 0 \qquad \text{(Dirichlet)} \tag{9.156}$$

or

$$\hat{\mathbf{n}} \cdot \nabla a(\mathbf{r}) = 0 \qquad \text{(Neumann)} \tag{9.157}$$

on the surface S_s, we use the Green function G_D or G_N that satisfies the appropriate boundary conditions over both the screen S_s and the aperture S_a, and vanishes on the hemispherical surface S_R in the limit $R \to \infty$. Returning to (9.152), we see that since both $a(\mathbf{r}')$ and $G(\mathbf{r}, \mathbf{r}')$ satisfy the boundary condition on S_s, the integral over S_s vanishes identically, and since the Green function satisfies the appropriate boundary condition on the aperture S_a we are left with

$$a_D(\mathbf{r}) = -\int_{S_a} a(\mathbf{r}') \nabla' G_D(\mathbf{r}', \mathbf{r}) \cdot \hat{\mathbf{n}}' dS' \tag{9.158}$$

$$a_N(\mathbf{r}) = \int_{S_a} G_N(\mathbf{r}', \mathbf{r}) \nabla' a(\mathbf{r}') \cdot \hat{\mathbf{n}}' dS' \tag{9.159}$$

for Dirichlet and Neumann boundary conditions. These are called the Sommerfeld integral formulas.

Unfortunately, it is generally much more difficult to find the Green functions $G_D(\mathbf{r}', \mathbf{r})$ and $G_N(\mathbf{r}', \mathbf{r})$ that satisfy the Dirichlet and Neumann boundary conditions than it is to find the free-space Green function $G(\mathbf{r}', \mathbf{r})$. In the simple case when the screen is a plane, however, the Green functions can be found by the method of images. If we let $\bar{\mathbf{r}}'$ be the mirror image of the point $\mathbf{r}'$, then for a point $\mathbf{r}$ in the region of the diffracted wave the Green functions are

$$G_D = \frac{e^{ik|\mathbf{r}' - \mathbf{r}|}}{4\pi |\mathbf{r}' - \mathbf{r}|} - \frac{e^{ik|\bar{\mathbf{r}}' - \mathbf{r}|}}{4\pi |\bar{\mathbf{r}}' - \mathbf{r}|} \tag{9.160}$$

and

$$G_N = \frac{e^{ik|\mathbf{r}'-\mathbf{r}|}}{4\pi |\mathbf{r}' - \mathbf{r}|} + \frac{e^{ik|\mathbf{r}'-\bar{\mathbf{r}}|}}{4\pi |\mathbf{r}' - \bar{\mathbf{r}}|} \tag{9.161}$$

If we substitute these Green functions into (9.158) and (9.159), keeping in mind that $\bar{\mathbf{r}}' = \mathbf{r}'$ on S_a, we get

$$a_D(\mathbf{r}) = \frac{-ik}{2\pi} \int_{S_a} \frac{(\mathbf{r}' - \mathbf{r}) \cdot \hat{\mathbf{n}}'}{|\mathbf{r}' - \mathbf{r}|} a(\mathbf{r}') \frac{e^{ik|\mathbf{r}'-\mathbf{r}|}}{|\mathbf{r}' - \mathbf{r}|} \, dS' \tag{9.162}$$

$$a_N(\mathbf{r}) = \frac{1}{2\pi} \int_{S_a} \nabla' a(\mathbf{r}') \cdot \hat{\mathbf{n}}' \frac{e^{ik|\mathbf{r}'-\mathbf{r}|}}{|\mathbf{r}' - \mathbf{r}|} \, dS' \tag{9.163}$$

Comparing these solutions with (9.155) we see that the Kirchhoff formula is just the average of the Dirichlet and Neumann formulas.

To see the differences between these solutions, we can examine the simple case of a point source of light illuminating the aperture. If we place the source at the point $\mathbf{r}_0$, the wave at the aperture is

$$a(\mathbf{r}') = \frac{a_0 e^{ik|\mathbf{r}'-\mathbf{r}_0|}}{|\mathbf{r}' - \mathbf{r}_0|} \tag{9.164}$$

for some constant a_0. The Dirichlet and Neumann solutions are then

$$a_D(\mathbf{r}) = \frac{-ika_0}{2\pi} \int_{S_a} \frac{(\mathbf{r}' - \mathbf{r}) \cdot \hat{\mathbf{n}}'}{|\mathbf{r}' - \mathbf{r}|} \frac{e^{ik|\mathbf{r}'-\mathbf{r}_0|}}{|\mathbf{r}' - \mathbf{r}_0|} \frac{e^{ik|\mathbf{r}'-\mathbf{r}|}}{|\mathbf{r}' - \mathbf{r}|} \, dS' \tag{9.165}$$

$$a_N(\mathbf{r}) = \frac{ika_0}{2\pi} \int_{S_a} \frac{(\mathbf{r}' - \mathbf{r}_0) \cdot \hat{\mathbf{n}}'}{|\mathbf{r}' - \mathbf{r}_0|} \frac{e^{ik|\mathbf{r}'-\mathbf{r}_0|}}{|\mathbf{r}' - \mathbf{r}_0|} \frac{e^{ik|\mathbf{r}'-\mathbf{r}|}}{|\mathbf{r}' - \mathbf{r}|} \, dS' \tag{9.166}$$

respectively. The Kirchhoff form is just the average of these two expressions. Clearly, the difference between (9.165) and (9.166) is just in the obliquity factor

$$-\frac{(\mathbf{r}' - \mathbf{r}) \cdot \hat{\mathbf{n}}'}{|\mathbf{r}' - \mathbf{r}|} = \cos\theta \tag{9.167}$$

or

$$\frac{(\mathbf{r}' - \mathbf{r}_0) \cdot \hat{\mathbf{n}}'}{|\mathbf{r}' - \mathbf{r}_0|} = \cos\theta_0 \tag{9.168}$$

where θ is the angle between the field point vector $\mathbf{r} - \mathbf{r}'$ and the surface normal $\hat{\mathbf{n}}'$, and θ_0 is the angle between the source point vector $\mathbf{r}_0 - \mathbf{r}'$ and the surface normal. Intuitively, we expect the diffracted wave to be strongest in the direction directly opposite the source point. In this case the obliquity factors are substantially the same, and the difference between the Sommerfeld formulas (and the Kirchhoff formula, for that matter) is small. In other directions, far from the direction of the incident wave, the differences are more important, but the diffracted wave is much weaker. In these directions, we have already seen that the boundaries generate new vector components not present in the incident beam, and we expect scalar diffraction theory to be less accurate. To the extent that the three integral formulas (Kirchhoff and Sommerfeld) agree, then, the choice of which formula to use is a matter of convenience. Fortunately, this covers the regions where the diffracted wave is large. To the extent that the three solutions don't agree, the scalar theory itself is suspect.

We note in passing that the Dirichlet solution (9.162) represents the diffracted wave in the form of a sum of waves emitted by the wave front within the aperture, as suggested by Huygens. Only the cosine of the angle between the normal to the aperture and the direction of propagation is left out in Huygens' construction.

9.3.2 Fraunhofer Diffraction (Far Field)

In the limit when the observation point $\mathbf{r}$ is removed to a distance large compared with the size of the aperture, that is, $|\mathbf{r}'|/|\mathbf{r}| \ll 1$ for all $\mathbf{r}'$ in the aperture S_a, we can approximate the factors appearing in the Kirchhoff integral (9.155) or in the alternate forms (9.162) and (9.163). In most places where the factor $\mathbf{r} - \mathbf{r}'$ appears, it is sufficient to use the simplest approximation

$$\mathbf{r}' - \mathbf{r} \approx -\mathbf{r} \tag{9.169}$$

However, in the usual case when the aperture is large compared to the wavelength, so that $k|\mathbf{r}'| \gg 1$ within the aperture, we must use the approximation

$$k|\mathbf{r}' - \mathbf{r}| \approx kr - \mathbf{k} \cdot \mathbf{r}' \tag{9.170}$$

in the exponent, where $\mathbf{k} = k\hat{\mathbf{r}}$ is the wave vector in the direction of the observation point. With these approximations we get

$$a_D(\mathbf{r}) = \frac{e^{ikr}}{2\pi r} \int_{S_a} a(\mathbf{r}')i\mathbf{k} \cdot \hat{\mathbf{n}}' e^{-i\mathbf{k} \cdot \mathbf{r}'} dS' \tag{9.171}$$

$$a_N(\mathbf{r}) = \frac{e^{ikr}}{2\pi r} \int_{S_a} \nabla' a(\mathbf{r}') \cdot \hat{\mathbf{n}}' e^{-i\mathbf{k} \cdot \mathbf{r}'} dS' \tag{9.172}$$

With the same approximations, the Kirchhoff integral gives the average of these two solutions. The far-field limit in which these formulas become valid is called Fraunhofer diffraction.

In the simplest case, when the incident wave is a plane wave traveling in the $\hat{\mathbf{k}}_0$ direction,

$$a(\mathbf{r}') = a_0 e^{i\mathbf{k}_0 \cdot \mathbf{r}'} \tag{9.173}$$

incident on a plane aperture $\hat{\mathbf{n}}' = $ constant, we get

$$a_D(\mathbf{r}) = ia_0\mathbf{k} \cdot \hat{\mathbf{n}}' \frac{e^{ikr}}{2\pi r} \int_{S_a} e^{-i\Delta\mathbf{k} \cdot \mathbf{r}'} dS' \tag{9.174}$$

$$a_N(\mathbf{r}) = ia_0\mathbf{k}_0 \cdot \hat{\mathbf{n}}' \frac{e^{ikr}}{2\pi r} \int_{S_a} e^{-i\Delta\mathbf{k} \cdot \mathbf{r}'} dS' \tag{9.175}$$

where

$$\Delta\mathbf{k} = \mathbf{k} - \mathbf{k}_0 \tag{9.176}$$

is the change in the wave vector due to diffraction. Clearly, these formulas agree when the obliquity factors $\hat{\mathbf{k}} \cdot \hat{\mathbf{n}}'$ and $\hat{\mathbf{k}}_0 \cdot \hat{\mathbf{n}}'$ are the same. Once again we see that the scalar theory is valid when the diffraction angle is small; that is, $\Delta k/k \ll 1$. However, the exponential factor oscillates across the aperture if $d\Delta k > O(1)$, where d is a characteristic dimension of the aperture. Therefore, if the aperture is large compared to a wavelength, $kd \gg 1$, we see that the diffracted wave vanishes by cancellation unless $\Delta k/k \ll 1$. In regions where the diffracted wave is large, then, both of these formulas [as well as (9.155), the Kirchhoff form of the integral] agree and scalar diffraction theory is useful. In this region, the diffracted wave is a spherical wave propagating away from the aperture with an amplitude proportional to the two-dimensional Fourier transform of the aperture. Note also that aside from the obliquity factor, the Fraunhofer diffraction pattern depends only on the change $\Delta\mathbf{k}$ of the wave vector.

As an example of Fraunhofer diffraction, we consider the case of a circular aperture of radius R. Using the coordinate system indicated in Figure 9.20 we get

$$\int_{S_a} e^{-i\Delta\mathbf{k}\cdot\mathbf{r}'} \, dS' = \int_0^R r'dr' \int_0^{2\pi} d\phi' e^{-i\Delta kr'\cos\phi'} \tag{9.177}$$

From (9.174) and the relations

$$\int_0^{2\pi} e^{i\alpha\cos\phi} \, d\phi = 2\pi J_0(\alpha) \tag{9.178}$$

and

$$\int_0^a J_0(qr)r \, dr = \frac{a}{q} J_1(qa) \tag{9.179}$$

we find that the amplitude of the diffracted wave is

$$a(\mathbf{r}) = ia_0 R^2 (\mathbf{k}\cdot\hat{\mathbf{n}}')\frac{J_1(\Delta kR)}{\Delta kR}\frac{e^{ikr}}{r} \tag{9.180}$$

The intensity of the diffracted wave is therefore

$$S = \frac{\omega k|a(\mathbf{r})|^2}{2\mu_0} = S_0 \frac{R^4(\mathbf{k}\cdot\hat{\mathbf{n}}')^2}{r^2}\frac{J_1^2(\Delta kR)}{(\Delta kR)^2} \tag{9.181}$$

where $S_0 = \omega k|a_0|^2/2\mu_0$ is the intensity of the incident wave. This is called the Airy pattern, and is illustrated in Figure 9.21.

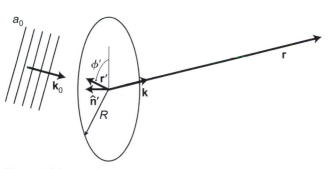

Figure 9.20 Coordinate system for a circular aperture.

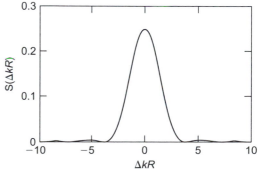

Figure 9.21 Diffraction pattern for a circular aperture.

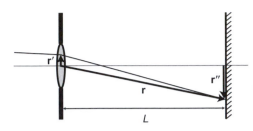

Figure 9.22 Formation of an image at a finite distance.

In terms of the angle θ between the diffracted wave $\mathbf{k}$ and the incident wave $\mathbf{k}_0$, we see that for small angles

$$\Delta k R = k R \theta \tag{9.182}$$

The first zero of the intensity occurs at the point $\Delta k R = j_{1,1} = 3.832$, where $j_{1,1}$ is the first zero of $J_1(x)$. This corresponds to the angle

$$\theta_1 = \frac{j_{1,1}}{kR} = 1.22 \frac{\lambda}{D} \tag{9.183}$$

where λ is the wavelength and $D = 2R$ the diameter of the aperture. Note that the intensity along the direction of incidence ($\Delta k = 0$, $J_1(x)/x \to \frac{1}{2}$) is, surprisingly, proportional to R^4. The first factor of R^2 comes, of course, from the fact that the power transmitted by the aperture is proportional to the area of the aperture. The second factor of R^2 comes from the fact that the diffracted spot size gets smaller (θ_1 gets smaller) as the aperture gets larger, so the light is concentrated into a smaller area.

Up to this point we have derived the Fraunhofer diffraction pattern as the limit of the diffracted intensity at large distances from the aperture. However, a lens may be used to focus the diffracted wave onto a screen at a finite distance from the aperture. When this is done, the diffraction pattern on the screen is just the Fraunhofer pattern. To see this we consider the arrangement shown in Figure 9.22 consisting of a thin lens at the aperture and a screen at the distance L behind the aperture. We assume in the following that all the transverse distances are small compared with the longitudinal distances (called the paraxial approximation), so the rays make only small angles with the axis. As we observed in the section on geometric optics, the phase change due to the optical path length through a thin lens at the position $\mathbf{r}'$ (as measured from the axis) is

$$\Delta \psi = k \left[(n-1)t_0 - \frac{r'^2}{2f} \right] \tag{9.184}$$

where n is the index of refraction and t_0 the thickness of the lens at the center. The focal length f is found from the radii of curvature of the lens using the lens maker's equation (9.70). If a wave $a_0(\mathbf{r}')$ is incident on the lens, then the wave $a_0(\mathbf{r}')e^{i\Delta\psi}$ emerges from the lens. The distance to point $\mathbf{r}''$ (measured from the axis as shown in Figure 9.22) on the screen is

$$|\mathbf{r}' - \mathbf{r}| = \sqrt{L^2 + (\mathbf{r}' - \mathbf{r}'')^2} \approx L + \frac{r''^2}{2L} + \frac{r'^2}{2L} - \frac{\mathbf{r}' \cdot \mathbf{r}''}{L} \tag{9.185}$$

for $|\mathbf{r}' - \mathbf{r}''|/L \ll 1$. But $\mathbf{r}'$ is normal to the axis, so $\mathbf{r}' \cdot \mathbf{r}'' = \mathbf{r}' \cdot \mathbf{r}$. To this same order, then, we may write

$$|\mathbf{r}' - \mathbf{r}| = L + \frac{r''^2}{2L} + \frac{r'^2}{2L} - \frac{\mathbf{r}' \cdot \mathbf{r}}{r} \tag{9.186}$$

Substituting this into (9.162), making the approximation (9.169) in the pre-exponential factors, we get

$$a(\mathbf{r}) = \frac{-ik}{2\pi r} e^{ik[L+(n-1)t_0+(r''^2/2L)]} \int_{S_a} a_i(\mathbf{r}') e^{(ikr'^2/2)[(1/L)-(1/f)]} e^{-i\mathbf{k}\cdot\mathbf{r}'} \, dS' \qquad (9.187)$$

where $\mathbf{k} = k\hat{\mathbf{r}}$ as before and $a_i(\mathbf{r}')$ is the wave incident on the lens. If we place the screen at the focal point of the lens, so that $L = f$, the quadratic terms in the exponent cancel out in the integral, leaving

$$a(\mathbf{r}) = \frac{-ik}{2\pi r} e^{ik[L+(n-1)t_0+(r''^2/2L)]} \int_{S_a} a_i(\mathbf{r}') e^{-i\mathbf{k}\cdot\mathbf{r}'} \, dS' \qquad (9.188)$$

Since the phase factor in front of the integral is irrelevant when we compute the intensity, this result is equivalent to the Fraunhofer diffraction pattern (9.171). Because the aperture is essentially plane (the thin lens) and we have used the paraxial approximation, the factor $\mathbf{k} \cdot \hat{\mathbf{n}}'$ in (9.171) has been replaced by $-k$ in (9.188).

If we carry over the results for Fraunhofer diffraction from a circular aperture, we see that the light from a distant point source is focused to a blur spot on the screen whose angular width θ_1 (to the first zero of the intensity) is given by (9.183). Rayleigh proposed that the optical image of two distant points could be distinguished if the peak of the blur spot of one fell on or outside the first zero of the blur spot of the other. That is, two point sources of light could be distinguished if their angular separation is no less than $1.22\lambda/D$, where λ is the wavelength and D the diameter of the aperture at the lens. This is known as Rayleigh's criterion.

If we use a lens of the correct focal length, we can focus the spherical wave emitted by a point source of light located a finite distance in front of the aperture on a screen at a finite distance beyond the aperture, as discussed earlier. The image of the original point on the screen is then a blur spot, an Airy pattern in the case of a circular aperture. The complete image of an incoherent source is then the superposition of the blur spots from all the points on the emitting object. We may represent this mathematically by the expression

$$a^2(\mathbf{r}) = \int h(\mathbf{r}, \mathbf{r}') a_0^2(\mathbf{r}') \, d^2\mathbf{r}' \qquad (9.189)$$

where the integral is over the surface of the emitting object. The function $h(\mathbf{r}, \mathbf{r}')$ is called the point-spread function, and we can find it from the computations we have just done. Given the intensity distribution from the object, we can then compute the intensity distribution of the image. Unfortunately, it is difficult to reverse the procedure and reconstruct the object from the blurred image. If we had information on the phase of $a(\mathbf{r})$ at the screen, we could, of course, find the object from the inverse Fourier transform, but this information is lost from the intensity. The problem is similar to that of reconstructing the original pulse from information about the intensity autocorrelation function, as discussed in Chapter 5. Mathematically, the inversion of (9.189) is well defined, but it depends on knowing the image best in regions where the intensity is nearly vanishing. Measurement uncertainties and noise therefore make reconstruction of the original image difficult. Because of the importance of image reconstruction, considerable effort has been expended and some progress has been made on this problem. However, to discuss it in any meaningful way would take us too far afield.

EXERCISE 9.7

For a plane wave

$$a_0 e^{i\mathbf{k}_0 \cdot \mathbf{r}'} \tag{9.190}$$

incident on an aperture with the aperture function $f_0(\mathbf{r})$, where $f_0 = 1$ within the aperture and $f_0 = 0$ otherwise, the diffracted *amplitude* in the Fraunhofer approximation is

$$a_{D0}(\mathbf{r}) = i a_0 \mathbf{k} \cdot \hat{\mathbf{n}}' \frac{e^{ikr}}{2\pi r} \int_S f_0(\mathbf{r}') e^{-i\Delta\mathbf{k}\cdot\mathbf{r}'} \, dS' \tag{9.191}$$

where $\Delta\mathbf{k} = \mathbf{k} - \mathbf{k}_0$ and $\mathbf{k} = k\hat{\mathbf{r}}$ is the wave vector in the direction of the diffracted wave.

(a) Show that if the screen includes two identical apertures separated by $\Delta\mathbf{r}$, then the diffracted *intensity is*

$$a_D a_D^* = 4 a_{D0} a_{D0}^* \cos^2\left(\tfrac{1}{2}\Delta\mathbf{k} \cdot \Delta\mathbf{r}\right) \tag{9.192}$$

(b) Consider the case when each aperture is circular with the diameter $d = 50$ μm, the apertures are displaced by $|\Delta\mathbf{r}| = 150$ μm, and they are illuminated at normal incidence by light with a wavelength $\lambda = 500$ nm. Use the results (9.181)–(9.183) and (9.192) to sketch the diffraction pattern that would be observed on a screen 1 m beyond the apertures. Indicate the important dimensions in your sketch.

EXERCISE 9.8

A classic problem in scalar diffraction theory is the far-field diffraction from a transmission grating consisting of N slits of length l (out of the page), width w, and spacing d, as shown in Figure 9.23.

(a) Consider the diffraction from many identical apertures illuminated by a plane wave with amplitude a_0 and wave number $\mathbf{k}_0 = 2\pi\hat{\mathbf{k}}_0/\lambda$. If we assign to the aperture located at $\mathbf{r}_n$ the aperture function

$$f_n(\mathbf{r}') = f(\mathbf{r}' - \mathbf{r}_n) \tag{9.193}$$

Figure 9.23 Diffraction from N slits.

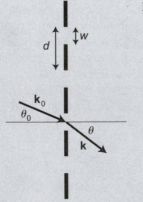

where $f = 1$ within the aperture and $f = 0$ otherwise, show that the diffracted intensity in the far field is

$$a = i a_0 \mathbf{k} \cdot \mathbf{n}' \frac{e^{ikr}}{r} \tilde{f}(\Delta \mathbf{k}) \sum_{n=1}^{N} e^{i\Delta \mathbf{k} \cdot \mathbf{r}_n} \tag{9.194}$$

where

$$\tilde{f}(\Delta \mathbf{k}) = \frac{1}{2\pi} \int e^{i\Delta \mathbf{k} \cdot \mathbf{r}'} f(\mathbf{r}') \, dS' \tag{9.195}$$

is the Fourier transform of the aperture function, $\Delta k = k|(\mathbf{k} - \mathbf{k}_0)|$ the change in the wave vector, and $\mathbf{k} = k\hat{\mathbf{r}}$ the wave vector in the direction of the observation point.

(b) Show that aside from an arbitrary phase factor, the transform for a rectangular slit is

$$\tilde{f}(\Delta \mathbf{k}) = \frac{wl}{2\pi} \frac{\sin \frac{1}{2} l \Delta k_l}{\frac{1}{2} l \Delta k_l} \frac{\sin \frac{1}{2} w \Delta k_w}{\frac{1}{2} w \Delta k_w} \tag{9.196}$$

where Δk_l and Δk_w are the wave vector changes in the directions along and across the slit.

(c) Show that

$$\sum_{n=1}^{N} x^n = \frac{x(x^N - 1)}{x - 1} \tag{9.197}$$

Hint: Consider the expression $(x - 1) \sum_{n=1}^{N} x^n$.
Use (9.197) to show that for $x = e^{i\phi}$, we get

$$\sum_{n=1}^{N} e^{i\phi} = e^{i(1/2)(N+1)\phi} \frac{\sin \frac{1}{2} N \phi}{\sin \frac{1}{2} \phi} \tag{9.198}$$

Use this to show that for N equally spaced apertures, the sum is

$$\sum_{n=1}^{N} e^{i\Delta \mathbf{k} \cdot \mathbf{r}_n} = e^{i(1/2)(N+1)\Delta k_w d} \frac{\sin \frac{1}{2} N \Delta k_w d}{\sin \frac{1}{2} \Delta k_w d} \tag{9.199}$$

(d) For $N \gg 1$ the intensity distribution consists of a series of peaks, as shown in Figure 9.24, under the overall envelope $|\tilde{f}|^2$ of the single-slit diffraction pattern. Show that the peaks occur at

$$\Delta k_w d = m 2\pi \tag{9.200}$$

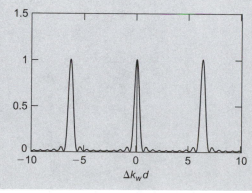

Figure 9.24 Intensity distribution from 10 slits.

where m is called the order of the peak, or in terms of the wavelength $\lambda = 2\pi/k$ and the incident and transmitted angles, the peaks are located at

$$\sin\theta - \sin\theta_0 = \frac{m\lambda}{d} \tag{9.201}$$

(e) Show that the first minimum on either side of each peak is at

$$N\Delta k_w d = (Nm \pm 1)2\pi \tag{9.202}$$

The resolution of the grating (Rayleigh's criterion) is defined by placing the diffraction peak for wave number $k + \delta k$ at the first minimum of the diffraction pattern for wave number k. Use (9.202) to show that for $N \gg 1$, the resolution of the grating is

$$\frac{\delta\lambda}{\lambda} = \frac{\delta k}{k} = \frac{1}{mN} \tag{9.203}$$

Note that the position and resolution of the orders of the grating are independent of the shape and size of the slits, and depend only on the number and spacing of the slits. Of course, the intensity of the various orders depends on the size and shape of the slits.

9.3.3 Fresnel Diffraction (Near Field)

At distances that are large compared with both the wavelength and the size of the aperture, the diffraction pattern can be accurately described using the Fraunhofer approximation, as discussed earlier. However, at shorter distances, but still large compared with the wavelength, scalar theory continues to provide an excellent description of the effects of diffraction. In this region we can use the Fresnel approximation.

For simplicity, we consider the case when the aperture is planar and examine the diffracted wave on a screen at a distance L beyond the aperture, as shown in Figure 9.25. The distance from point $\mathbf{r}'$ in the aperture to point $\mathbf{r} = L\hat{\mathbf{z}} + \mathbf{r}''$ on the screen is

$$|\mathbf{r} - \mathbf{r}'|^2 = r^2 + r'^2 - 2\mathbf{r} \cdot \mathbf{r}' \tag{9.204}$$

For distances that are large compared with the aperture, the phase at point $\mathbf{r}$ is then approximately

$$k|\mathbf{r}' - \mathbf{r}| \approx kr - k\mathbf{r}' \cdot \hat{\mathbf{r}} + \frac{kr'^2}{2r} - \frac{k(\mathbf{r} \cdot \hat{\mathbf{r}}')^2}{2r} + \cdots = kr - \mathbf{k} \cdot \mathbf{r}' + \frac{kr'^2}{2r} - \frac{k(\mathbf{r} \cdot \hat{\mathbf{r}}')^2}{2r} + \cdots \tag{9.205}$$

where $\mathbf{k} = k\hat{\mathbf{r}}$ is the wave vector in the direction of the observation point.

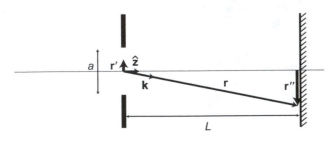

Figure 9.25 Coordinate system for Fresnel diffraction..

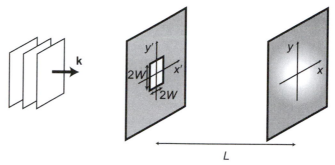

Figure 9.26 Fresnel diffraction from a plane aperture.

Comparing (9.205) with (9.170), we see that the first two terms correspond to the Fraunhofer approximation. The next terms are on the order of

$$N_F = \frac{ka^2}{2\pi L} = \frac{a^2}{\lambda L} \qquad (9.206)$$

where a is the size of the aperture. This is called the Fresnel number. Provided that the Fresnel number is small, $N_F \ll 1$, the contribution of the quadratic and higher order terms to the phase can be ignored and the Fraunhofer approximation is satisfactory. Otherwise, we must include the quadratic terms. This is called the Fresnel approximation. Typically, it is found that when the first (nonvanishing) term of any expansion is not sufficient, then all the succeeding terms of the expansion become important. Why, then, is it sufficient to include only the quadratic term in the Fresnel approximation? The answer is subtle and best provided by means of an example.

We consider the case of a plane aperture illuminated by a normally incident plane wave, as shown in Figure 9.26. In Cartesian coordinates the diffraction equation (9.162) becomes

$$a(x, y) = -ia_0 \frac{kL}{2\pi} \int_{S_a} dx' dy' \frac{e^{ik|\mathbf{r}' - \mathbf{r}|}}{|\mathbf{r}' - \mathbf{r}|^2} \qquad (9.207)$$

where a_0 is the amplitude at the aperture and the distance from the source in the aperture to the observation point on the screen is

$$|\mathbf{r}' - \mathbf{r}|^2 = L^2 + (x' - x)^2 + (y' - y)^2 \qquad (9.208)$$

When the distance to the screen is large compared with both the aperture and the diffraction spot, we can use the approximation

$$|\mathbf{r}' - \mathbf{r}| \approx L + \frac{1}{2L}(x' - x)^2 + \frac{1}{2L}(y' - y)^2 \qquad (9.209)$$

where the first term is usually good enough in the denominator of (9.207). The amplitude at the screen is then

$$a(x, y) = -i\frac{a_0 k e^{ikL}}{2\pi L} \int_{S_a} dx' dy' e^{i(k/2L)(x'-x)^2} e^{i(k/2L)(y'-y)^2} \qquad (9.210)$$

The exponential factors in the integrals are shown in Figure 9.27, where we see that they oscillate with increasing frequency for larger values of $(x' - x)$ or $(y' - y)$. The largest

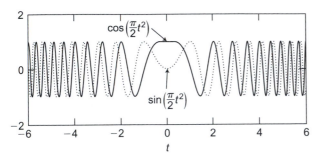

Figure 9.27 Exponential factors in the diffraction integrals.

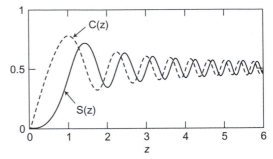

Figure 9.28 Fresnel integrals.

contribution to the integral therefore comes from the part of the aperture nearly aligned with the observation point on the screen, that is, for $k(x' - x)^2/L \leq O(1)$ and $k(y' - y)^2/L \leq O(1)$. Outside this region the contribution vanishes by cancellation of the rapid oscillations. We encounter this same behavior in the discussion of pulse compression in Chapter 5. Because of this cancellation, the expansion (9.209) is actually useful in regions where it would otherwise be invalid. That is, the original expansion is valid only for $(x' - x)^2/L^2 \ll 1$, but due to the cancellation the contribution to the integral vanishes for $(x' - x)^2/L^2 \leq O(1/kL) \ll 1$. Therefore, in any region where the integrand contributes significantly, the approximation (9.209) is satisfactory.

In the case of a square aperture of width $2W$, the results are particularly simple. The integrals over x and y separate, and we get

$$a(x, y) = -i\frac{a_0 k e^{ikL}}{2\pi L} \int_{-W}^{W} e^{i(k/2L)(x'-x)^2} \, dx' \int_{-W}^{W} e^{i(k/2L)(y'-y)^2} \, dy \qquad (9.211)$$

Because these integrals appear so frequently in mathematics, the real and imaginary parts are tabulated as the Fresnel cosine and sine integrals

$$C(z) = \int_0^z \cos\left(\frac{\pi}{2}t^2\right) dt = -C(-z) \qquad (9.212)$$

$$S(z) = \int_0^z \sin\left(\frac{\pi}{2}t^2\right) dt = -S(-z) \qquad (9.213)$$

As shown in Figure 9.28, the Fresnel integrals at first increase from zero and then oscillate about the value $C(\infty) = S(\infty) = \frac{1}{2}$ with asymptotically decreasing amplitude. In terms of the Fresnel integrals, the amplitude pattern for diffraction from a square aperture may be expressed

$$a(x, y) = a_0 F(X) F(Y) \qquad (9.214)$$

where the diffraction function is

$$F(X) = \sqrt{\tfrac{1}{2}}\{C[\sqrt{2N_F}(1 - X)] + C[\sqrt{2N_F}(1 + X)]\}$$

$$+ i\sqrt{\tfrac{1}{2}}\{S[\sqrt{2N_F}(1 - X)] + S[\sqrt{2N_F}(1 + X)]\} \qquad (9.215)$$

the Fresnel number is

$$N_F = \frac{W^2}{\lambda L} \qquad (9.216)$$

and the dimensionless coordinates are

$$X = \frac{x}{W} \qquad (9.217)$$

$$Y = \frac{y}{W} \qquad (9.218)$$

The intensity pattern is

$$aa^* = a_0^2 |F(X)|^2 |F(Y)|^2 \qquad (9.219)$$

This is illustrated in Figure 9.29. When the screen is a large distance from the aperture ($N_F \ll 1$) we observe the far-field diffraction pattern known as Fraunhofer diffraction, as discussed above. Closer to the aperture the diffraction pattern approaches the square pattern expected of geometric optics, except that there are oscillations in the intensity, especially near the edges, and the diffracted waves spread into the shadow region. The regions of Fresnel and Fraunhofer diffraction are illustrated schematically in Figure 9.30.

The intensity oscillations near the edge of the aperture can be quite pronounced in microscope images, especially under monochromatic illumination, and deserve closer examination. In the near field, if we are close to one edge of the aperture the other edges of the aperture can be ignored. The diffracted amplitude (9.211) then becomes

$$a(x, y) = -i \frac{a_0 e^{ikL}}{\lambda L} \int_{-\varepsilon}^{\infty} e^{i(k/2L)\xi^2} \, d\xi \qquad (9.220)$$

where $\varepsilon = x + W$ is the coordinate referred to the edge of the aperture. In terms of the Fresnel integrals the diffracted intensity is then

$$aa^* = \frac{a_0^2}{2} \left\{ \left[\frac{1}{2} + C\left(\sqrt{\frac{2}{\lambda L}} \varepsilon \right) \right]^2 + \left[\frac{1}{2} + C\left(\sqrt{\frac{2}{\lambda L}} \varepsilon \right) \right]^2 \right\} \qquad (9.221)$$

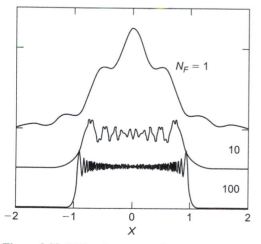

Figure 9.29 Diffraction pattern from a square aperture.

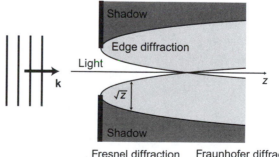

Figure 9.30 Regions of Fresnel and Fraunhofer diffraction.

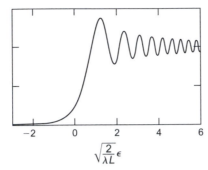

$$\sqrt{\frac{2}{\lambda L}}\,\epsilon$$

Figure 9.31 Diffraction near an edge.

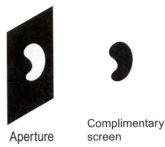

Aperture   Complimentary
screen

Figure 9.32 Babinet's principle.

This behavior is illustrated in Figure 9.31. We see from (9.221) that the width of the diffraction pattern is on the order of $\sqrt{\lambda L}$ and increases as the square root of the distance from the aperture. This is illustrated in Figure 9.30.

The structure of the integral formulas (9.155), (9.162), and (9.163) used in scalar diffraction theory leads to an interesting principle called Babinet's principle of complementary screens. By complementary screens we mean that where one screen has an aperture, its complement is opaque, and vice versa. Referring to Figure 9.32, if S_a is the aperture in one screen, then S_s is the aperture in the complementary screen. It follows that if $a_a(\mathbf{r})$ is the wave diffracted by the first screen and $a_s(\mathbf{r})$ the wave diffracted by the complementary screen, the sum of the two, which is the wave transmitted in the absence of any screen, must be just the incident wave $a_0(\mathbf{r})$. We therefore get

$$a_0(\mathbf{r}) = a_a(\mathbf{r}) + a_s(\mathbf{r}) \tag{9.222}$$

which is called Babinet's principle. Note that the rule applies to the diffracted amplitudes, not the intensities.

As a simple example with which to close this chapter, we can use Babinet's principle to find the diffraction pattern for a square obstruction. The wave incident on the image screen in the absence of an aperture screen or obstruction is just $a_0 e^{ikL}$, so from (9.214) we see that the diffracted amplitude beyond a square obstruction of width $2W$ is

$$a(x, y) = a_0 e^{ikL}[1 - F(X)F(Y)] \tag{9.223}$$

The intensity pattern at the screen is

$$aa^* = a_0^2 |1 - F(X)F(Y)|^2 = 1 + |F(X)F(Y)|^2 - 2\text{Re}[F(X)F(Y)] \tag{9.224}$$

This is illustrated in Figure 9.33. At a Fresnel number of 10, close to the aperture, the shadow behind the obstruction is fairly sharply defined. However, at a Fresnel number of unity we see a peak appearing at the center of the shadow, along with oscillations in the intensity surrounding the shadow region. The peak is related to Arago's spot, but it is washed out at shorter distances because the obstruction is square rather than round. For a square obstruction 1 cm on each side ($W = 5$ mm) illuminated by light in the visible part of the spectrum ($\lambda = 0.5$ μm), a Fresnel number of unity corresponds to a distance $L = 50$ m behind the screen. This makes the effect is difficult to observe. It is easier to observe using a

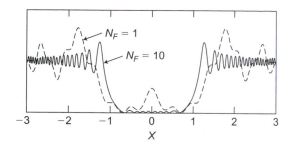

Figure 9.33 Diffraction pattern from a square obstruction.

laser and a circular obstruction, but it is still a difficult experiment and was never noticed until Arago did a careful investigation.

EXERCISE 9.9

Show that Fresnel diffraction approaches Fraunhofer diffraction in the far field (when the Fresnel number is small). Specifically, show that for a plane aperture S_a illuminated by a plane wave at normal incidence, both (9.175) and (9.210) give the result

$$a(x, y) = -i\frac{a_0 k}{2\pi L}\, e^{ikL} e^{i(k/2L)(x^2+y^2)} \int_{S_a} dx'\, dy'\, e^{-i(k/L)(xx'+yy')} \tag{9.225}$$

for the intensity at point (x, y) on a screen at the distance L from the aperture, where a_0 is the amplitude and k the wave number of the incident wave.

EXERCISE 9.10

Consider a circular aperture of radius R illuminated at normal incidence by a plane wave of amplitude a_0 and wave vector k.

(a) Show that the intensity at the point L on the axis is

$$a^* a = 4a_0^* a_0 \sin^2\left(\frac{kR^2}{4L}\right) \xrightarrow{L \to \infty} a_0^* a_0 \frac{k^2 R^4}{4L^2} \tag{9.226}$$

This oscillates in the near field with a peak on-axis intensity four times the incident intensity and approaches the Fraunhofer limit (9.181) in the far field.

(b) Use Babinet's principle to show that in the shadow region behind an opaque disk of radius R, the on-axis intensity is

$$a^* a = a_0^* a_0 \tag{9.227}$$

That is, the intensity on axis behind the disk is precisely the intensity without the disk. This is Arago's spot.

EXERCISE 9.11

In the Fresnel (near field) approximation, the amplitude of a wave at a point on the axis beyond an aperture is dominated by the amplitude near the axis in the aperture. Points in the aperture farther off axis contribute to the amplitude with a relative phase factor that is quadratic in the radius, and the oscillations cause the contribution from points far from the axis to cancel out. This gives rise to the concept of Fresnel zones. The first Fresnel zone is the region near the axis over which the relative phase is $0 < \phi < \pi$, so there is no destructive interference. The next Fresnel zone is the region over which the phase is $\pi < \phi < 2\pi$, and this contributes negatively to the total amplitude, and so on. This suggests that if a plate is constructed to block every other Fresnel zone, the amplitude on the axis should build up monotonically. This is the principle of the so-called zone plate, illustrated in Figure 9.34.

(a) Show that for a plane wave incident on a properly designed zone plate with perfectly transmitting open apertures, the intensity on axis is

$$a^*a = 4N^2 a_0^* a_0 \tag{9.228}$$

where N is the number of open zones and $a_0^* a_0$ is the incident intensity. This is a remarkable result and shows that there is significant improvement even if there is just one zone.

(b) Show that for a large zone plate the number of open zones is $N \approx R^2/2L\lambda$, where R is the radius of the zone plate and λ the wavelength. The intensity enhancement is therefore

$$a^*a = \frac{R^4}{L^2\lambda^2} a_0^* a_0 \tag{9.229}$$

(c) Using the Fraunhofer approximation (9.188) for a focused beam, show that the intensity at the focal point of a lens is

$$a^*a = \frac{\pi^2 R^4}{L^2\lambda^2} a_0^* a_0 \tag{9.230}$$

This is better than a zone plate by the factor $\pi^2 \approx 10$. Zone plates are most useful when there is no possibility of constructing a lens or mirror to focus the beam, as is often the case with x-ray optics, for example.

Figure 9.34 Zone plate.

BIBLIOGRAPHY

A general introduction to diffraction is found in most good electrodynamics texts, such as

L. D. Landau and E. M. Lifshitz, *Classical Theory of Fields,* 2nd edition, Pergamon Press, Oxford (1975),

J. D. Jackson, *Classical Electrodynamics,* 3rd edition, John Wiley & Sons, New York (1999).

Nevertheless, the bible of the subject has for years been, and remains,

M. Born and E. Wolf, *Principles of Optics: Electromagnetic Theory of Propagation, Interference, and Diffraction of Light,* 7th edition, Cambridge University Press, Cambridge, U.K. (1999).

For an exhaustive discussion of laser resonators, the reader is referred to

A. E. Siegman, *Lasers,* University Science Books, Mill Valley, CA (1986).

The present discussion of Fresnel and Fraunhofer diffraction follows the approach used much more extensively in the excellent book

J. W. Goodman, *Introduction to Fourier Optics,* McGraw-Hill Book Company, New York (1996).

10 Radiation by Relativistic Particles

We turn now to the problem of calculating the radiation emitted by a charged particle in arbitrary, relativistic motion. We do this in a series of steps, beginning with the details of the angular and spectral distribution of the radiation and integrating to obtain first the angular distribution of the energy emitted and finally the total power. The formalism we develop in this chapter is enormously powerful, although the results are remarkably simple, and we apply it to several problems of interest. These include multipole radiation, Bremsstrahlung and transition radiation, which are treated in Chapter 5 by the method of virtual quanta, Thomson scattering, synchrotron radiation, and undulator radiation. At the end of the chapter we explore what happens when multiple particles radiate coherently and incoherently, and what happens when particles travel through a medium at faster than the local speed of light.

10.1 ANGULAR AND SPECTRAL DISTRIBUTION OF RADIATION

10.1.1 Fourier Decomposition of the Fields

When a charged particle executes nonuniform motion along some trajectory, it radiates electromagnetic waves that propagate away from the particle. The total radiation field is the sum of the fields emitted along the trajectory. When the Coulomb-like field near the particle is included, the total field is rather complicated. Nevertheless, it is possible to sort out the fields that remain near the particle from those that propagate to large distances as radiation, and to compute the spectrum and the instantaneous power radiated by the particle.

Since we are typically interested in the spectral distribution of the radiation, it is natural to begin with the Fourier transform of the Maxwell equations. The time-independent equations are then solved by introducing the Green function, and we find that the solution consists of two terms, one for the "near field" and one for the "far field." The near field falls off as $1/R^2$ away from the charge and resembles the Coulomb field. The far field falls off as $1/R$ and is capable of carrying finite energy to infinity. At large distances from the charge, the far field propagates like optical waves and can be identified as radiation emitted by the particle.

We begin with the Maxwell equations for the electric and magnetic fields:

$$\nabla \cdot \mathbf{E} = \frac{\rho}{\varepsilon_0} \tag{10.1}$$

$$\nabla \cdot \mathbf{B} = 0 \tag{10.2}$$

$$\nabla \times \mathbf{E} + \frac{\partial \mathbf{B}}{\partial t} = 0 \tag{10.3}$$

$$\nabla \times \mathbf{B} - \frac{1}{c^2}\frac{\partial \mathbf{E}}{\partial t} = \mu_0 \mathbf{J} \tag{10.4}$$

where ρ is the electric charge density and $\mathbf{J}$ the current density. If we take the Fourier transform of these equations with respect to time and integrate by parts to eliminate derivatives with respect to the time, assuming that the fields vanish at infinity, we get

$$\nabla \cdot \tilde{\mathbf{E}} = \frac{\tilde{\rho}}{\varepsilon_0} \tag{10.5}$$

$$\nabla \cdot \tilde{\mathbf{B}} = 0 \tag{10.6}$$

$$\nabla \times \tilde{\mathbf{E}} - i\omega\tilde{\mathbf{B}} = 0 \tag{10.7}$$

$$\nabla \times \tilde{\mathbf{B}} + \frac{i\omega}{c^2}\tilde{\mathbf{E}} = \mu_0\tilde{\mathbf{J}} \tag{10.8}$$

Starting with the identity

$$\nabla \times (\nabla \times \tilde{\mathbf{B}}) = \nabla(\nabla \cdot \tilde{\mathbf{B}}) - \nabla^2\tilde{\mathbf{B}} \tag{10.9}$$

and substituting (10.6) for $\nabla \cdot \tilde{\mathbf{B}}$ and (10.8) for $\nabla \times \tilde{\mathbf{B}}$, we get

$$\nabla \times \left(\frac{i\omega}{c^2}\tilde{\mathbf{E}} - \mu_0\tilde{\mathbf{J}}\right) = \nabla^2\tilde{\mathbf{B}} \tag{10.10}$$

Substituting (10.7) for $\nabla \times \tilde{\mathbf{E}}$, we arrive at the Helmholtz equation

$$\nabla^2\tilde{\mathbf{B}} + k^2\tilde{\mathbf{B}} = -\mu_0\nabla \times \tilde{\mathbf{J}} \tag{10.11}$$

in which the wave number k satisfies the dispersion relation

$$\omega = kc \tag{10.12}$$

The solution to (10.11) is conveniently expressed in terms of the Green function for the Helmholtz equation. As discussed in Chapter 3, the Green function for electrostatics is the potential distribution of a point charge. The complete solution in terms of the Green function expresses the actual potential in the form of a sum (linear superposition) of the potentials of a distribution of point charges equivalent to the actual charge distribution. The present case is similar, but since we are looking for the radiation from a specified system of charges and currents, the Green function is the wave emitted by a point source. The complete solution is then the linear superposition, including the phases, of the waves emitted by all the sources that compose the system being examined. For many purposes it is enough to find the solution to the Helmholtz equation, which includes the frequency as a parameter but is independent of the time. This gives us the spectrum of the radiation. But for some purposes we need the explicit time dependence of the fields. We can get this by inverting the Fourier transform that led to the Helmholtz equation in the first place.

We carry out this program first for continuous charge distributions. The solution is called the retarded field. Then we find the solution for point charges in motion, which is called the Lienard–Wiechert field.

We begin with Green's theorem, which says that for any two functions $\phi(\mathbf{r}')$ and $\psi(\mathbf{r}')$ defined in a volume V bounded by a surface S,

$$\int_V [\phi(\mathbf{r}')\nabla'^2\psi(\mathbf{r}') - \psi(\mathbf{r}')\nabla'^2\phi(\mathbf{r}')]\,dV' = \int_S \left[\phi(\mathbf{r}')\frac{\partial\psi(\mathbf{r}')}{\partial n'} - \psi(\mathbf{r}')\frac{\partial\phi(\mathbf{r}')}{\partial n'}\right]dS'$$

(10.13)

For the function $\phi(\mathbf{r}')$ we choose the solution of the Helmholtz equation

$$\nabla'^2\phi + k^2\phi = -\mu_0 j$$

(10.14)

which corresponds to one component of the vector equation (10.11), and for the function $\psi(\mathbf{r}')$ we choose the solution of the equation

$$\nabla'^2\tilde{G}(\mathbf{r}',\mathbf{r}_0) + k^2\tilde{G}(\mathbf{r}',\mathbf{r}_0) = -\delta(\mathbf{r}'-\mathbf{r}_0)$$

(10.15)

This is called the Green function. If we substitute back into (10.13) and integrate over the volume V, we get

$$\phi(\mathbf{r}_0) = \mu_0 \int_V j(\mathbf{r}')G(\mathbf{r}',\mathbf{r}_0)\,dV' + \int_S \left[G(\mathbf{r}',\mathbf{r}_0)\frac{\partial\Phi(\mathbf{r}')}{\partial n'} - \Phi(\mathbf{r}')\frac{\partial G(\mathbf{r}',\mathbf{r}_0)}{\partial n'}\right]dS'$$

(10.16)

For the problem at hand we let the boundary S recede to infinity, where the functions all vanish. Ignoring the surface integral, then, we see that the solution to (10.11) is

$$\tilde{\mathbf{B}}(\mathbf{r}_0) = \mu_0 \int_V [\nabla'\times\mathbf{J}(\mathbf{r}')]G(\mathbf{r}',\mathbf{r}_0)\,dV'$$

(10.17)

The Green function is found from (10.15). Since the radiation from a point source is spherically symmetric, the solution clearly depends only on the variable $R = |\mathbf{R}|$, where

$$\mathbf{R} = \mathbf{r}_0 - \mathbf{r}'$$

(10.18)

as shown in Figure 10.1. Then the equation for the Green function becomes

$$\frac{1}{R}\frac{d^2}{dR^2}R\tilde{G} + k^2\tilde{G} = -\delta(\mathbf{r}'-\mathbf{r}_0)$$

(10.19)

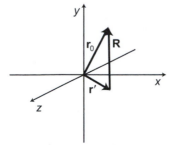

Figure 10.1 Coordinates of source and observer.

For $R \neq 0$ the δ-function vanishes, and we have simply

$$\frac{d^2}{dR^2} R\tilde{G} + k^2 R\tilde{G} = 0 \tag{10.20}$$

This is just an ordinary differential equation for the function $R\tilde{G}$ that has the solution

$$\tilde{G} = \frac{a_\pm}{R} e^{\pm ikR} \tag{10.21}$$

The Green function therefore has the form of a spherical wave of wave number k, whose amplitude diminishes as $1/R$ as the wave spreads out. The positive sign in the exponent corresponds to outgoing waves, the negative sign to incoming waves. To determine the coefficient $a_\pm$, we integrate (10.15) over a small spherical volume around $\mathbf{r}_0$ and use the divergence theorem to convert the volume integral of $\nabla' \cdot \nabla'\tilde{G}$ on the left-hand side to a surface integral. Provided that the radius is sufficiently small, so that $kR \ll 1$, we may ignore the factor $e^{ikR} \approx 1$ and evaluate the integrals trivially. In the limit when the radius of the sphere vanishes, the volume integral of $k^2\tilde{G}$ vanishes and we find that $a_\pm = 1/4\pi$. The complete Green function is therefore

$$\tilde{G} = \frac{1}{4\pi R} e^{\pm ikR} \tag{10.22}$$

In Chapter 3 it is shown that the Green function $G(\mathbf{r}', \mathbf{r}_0)$ for the Poisson equation is symmetric in the variables $\mathbf{r}'$ and $\mathbf{r}_0$. This is clearly true of (10.22).

If there are no sources at infinity, we can ignore incoming waves. The magnetic field may then be represented as an integral over all space of outgoing spherical waves emitted by the source term $\mu_0 \nabla \times \tilde{\mathbf{J}}$ in (10.11). The field at the observation point $\mathbf{r}_0$ is therefore

$$\tilde{\mathbf{B}}(\mathbf{r}_0, \omega) = \int_{-\infty}^{\infty} d^3\mathbf{r}' \, [\mu_0 \nabla' \times \tilde{\mathbf{J}}(\mathbf{r}')] \left(\frac{1}{4\pi R} e^{i\omega R/c} \right) \tag{10.23}$$

where we have used (10.12) to eliminate the wave number k in the exponent. We can eliminate the derivatives in the integrand with an integration by parts. To integrate (10.23) by parts, we note that

$$\nabla' R = \nabla' |\mathbf{r}_0 - \mathbf{r}'| = -\hat{\mathbf{R}} \tag{10.24}$$

where $\hat{\mathbf{R}} = (\mathbf{r}_0 - \mathbf{r}')/R$ is a unit vector from the source point $\mathbf{r}'$ to the observation point $\mathbf{r}_0$, as shown in Figure 10.1. Carrying out the integration by parts, assuming that the source $\mathbf{J}$ vanishes at infinity, we find that the field is

$$\tilde{\mathbf{B}} = \frac{\mu_0}{4\pi} \int_{-\infty}^{\infty} d^3\mathbf{r}' \left(\frac{i\omega}{cR} - \frac{1}{R^2} \right) (\hat{\mathbf{R}} \times \tilde{\mathbf{J}}) e^{i\omega R/c} \tag{10.25}$$

Substituting the explicit expression for the Fourier transform of $\tilde{\mathbf{J}}$, we find that

$$\tilde{\mathbf{B}} = \frac{\mu_0}{(32\pi^3)^{1/2}} \int_{-\infty}^{\infty} d^3\mathbf{r}' \int_{-\infty}^{\infty} dt' \left(\frac{i\omega}{cR} - \frac{1}{R^2} \right) (\hat{\mathbf{R}} \times \mathbf{J}) e^{i\omega(t' + R/c)} \tag{10.26}$$

This expresses the Fourier transform of the field in the form of an integral over all space and all time of the distributed source $\mathbf{J}(\mathbf{r}', t')$.

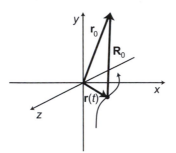

Figure 10.2 Coordinate system for the fields of a moving charge.

When the source consists of a point charge q at position $\mathbf{r}(t')$, the current density is

$$\mathbf{J}(\mathbf{r}', t') = qc\boldsymbol{\beta}(t')\delta[\mathbf{r}' - \mathbf{r}(t')] \tag{10.27}$$

where the velocity of the point charge is $c\boldsymbol{\beta} = d\mathbf{r}/dt'$. Substituting into (10.26) and carrying out the integral over all space, we get

$$\tilde{\mathbf{B}} = \frac{\mu_0 q}{(32\pi^3)^{1/2}} \int_{-\infty}^{\infty} dt' e^{i\omega(t'+R_0/c)} \left(\frac{i\omega}{R_0} - \frac{c}{R_0^2} \right) \hat{\mathbf{R}}_0 \times \boldsymbol{\beta} \tag{10.28}$$

where

$$\mathbf{R}_0(t') = \mathbf{r}_0 - \mathbf{r}(t') \tag{10.29}$$

as shown in Figure 10.2. This expresses the Fourier transform of the field in the form of an integral over the entire trajectory of the particle from $t' = -\infty$ to $t = \infty$.

10.1.2 Retarded Fields and Lienard–Wiechert Fields

At this point, we can invert the Fourier transform (10.26) or (10.28) to get the magnetic field $\mathbf{B}$ at the observation point $\mathbf{r}_0$ as a function of the time t_0. When we multiply (10.26) by $e^{-i\omega t_0}$ and integrate over all ω to invert the transform, the factors in the exponent combine to produce a δ-function that selects certain points in space and time. Specifically, we find that only those points—those events in Minkowski space—contribute to the integral for which the radiation emitted at that point and time in the past reaches the observation point $\mathbf{r}_0$ at precisely the observation time t_0. For distributed sources $\rho(\mathbf{r}', t')$ and $\mathbf{J}(\mathbf{r}', t')$, the result is an integral over all space of the source term $\mathbf{J}(\mathbf{r}', t') \times \hat{\mathbf{R}}$. This is called the retarded field. For a point charge q following a trajectory $\mathbf{r}(t)$, the observed field depends on the motion of the particle at just a single time in the past, called the retarded time. The field in this form is called the Lienard–Wiechert field.

To find the retarded field, we begin with (10.26). To get rid of the factor $i\omega$ in the first term, we integrate by parts with respect to t' and get

$$\tilde{\mathbf{B}} = \frac{\mu_0}{(32\pi^3)^{1/2}} \int_{-\infty}^{\infty} d^3r' \frac{\hat{\mathbf{R}} \times \mathbf{J}}{cR} e^{i\omega(t'+R/c)} \bigg|_{t'=-\infty}^{\infty}$$

$$- \frac{\mu_0}{(32\pi^3)^{1/2}} \int_{-\infty}^{\infty} d^3r' \int_{-\infty}^{\infty} dt' \left[\frac{\hat{\mathbf{R}}}{cR} \times \frac{\partial \mathbf{J}}{\partial t} + \frac{\hat{\mathbf{R}} \times \mathbf{J}}{R^2} \right] e^{i\omega(t'+R/c)} \tag{10.30}$$

Provided that the source current $\mathbf{J}$ vanishes as $t' \to \pm\infty$, we can ignore the first integral. If we multiply the second integral by $e^{-i\omega t}/\sqrt{2\pi}$ and integrate over ω to invert the Fourier transform, we get

$$\mathbf{B}(\mathbf{r}_0, t_0) = -\frac{\mu_0}{8\pi^2} \int_{-\infty}^{\infty} d^3\mathbf{r}' \int_{-\infty}^{\infty} dt' \int_{-\infty}^{\infty} d\omega \left[\frac{\hat{\mathbf{R}}}{cR} \times \frac{\partial \mathbf{J}}{\partial t} + \frac{\hat{\mathbf{R}} \times \mathbf{J}}{R^2} \right] e^{i\omega(t'-t_0+R/c)} \quad (10.31)$$

But from the integral over ω we pick up the δ-function $2\pi\delta[t' + (R/c) - t_0]$. From the entire history of the source function at distance R from the observation point, this δ-function selects just that earlier time (called the retarded time)

$$t' = t_{\text{retarded}} = t_0 - \frac{R}{c} \quad (10.32)$$

This is precisely the time at which an electromagnetic wave originating at source point $\mathbf{r}'$ must be emitted to arrive at observation point $\mathbf{r}_0$ at time t_0, and the wave that is emitted depends only on the source at the earlier (retarded) time. This agrees, of course, with our everyday experience in the sense that what we see with our eyes at time t_0 is not the world as it is now but the world as it was at time R/c earlier, when the objects we observe radiated the light that we see now. Light radiated before this time has passed us by, and light radiated after this has not reached us yet. The effect is most pronounced in cosmological observations, such as those done now with the Hubble telescope. When we observe galaxies and clusters billions of light years away, what we see of these distant objects tells us not what they are like now, but what they were like billions of years ago, and gives us a window on the early universe.

After integrating over ω we can integrate over time t' trivially, and we get

$$\mathbf{B}(\mathbf{r}_0, t_0) = \frac{\mu_0}{4\pi} \int_{-\infty}^{\infty} d^3\mathbf{r}' \left[\frac{\partial \mathbf{J}}{\partial t'} \times \frac{\hat{\mathbf{R}}}{cR} + \mathbf{J} \times \frac{\hat{\mathbf{R}}}{R^2} \right]_{\text{retarded}} \quad (10.33)$$

where the subscript "retarded" indicates that the expression in brackets is to be evaluated at the retarded time $t' = t - R/c$. The field expressed in this way is called the retarded field. The corresponding formula for the electric field is

$$\mathbf{E}(\mathbf{r}_0, t_0) = \frac{1}{4\pi\varepsilon_0} \int_{-\infty}^{\infty} d^3\mathbf{r}' \left[\frac{1}{cR} \left(\hat{\mathbf{R}} \frac{\partial \rho}{\partial t'} - \frac{1}{c} \frac{\partial \mathbf{J}}{\partial t'} \right) + \frac{\hat{\mathbf{R}}}{R^2} \rho \right]_{\text{retarded}} \quad (10.34)$$

These expressions are sometimes called Jefimenko's formulas. They are the extension to time-dependent sources of the Coulomb and Biot–Savart laws, to which they reduce in the static limit.

In the same way, we can invert the Fourier transform in (10.28) and arrive at an expression for the field $\mathbf{B}(\mathbf{r}_0, t_0)$ of a point charge in terms of the particle motion $\mathbf{r}(t')$. To invert the transform, we multiply by $e^{-i\omega t}/\sqrt{2\pi}$ and integrate over ω to get

$$\mathbf{B}(\mathbf{r}_0, t_0) = \frac{\mu_0 q}{8\pi^2} \int_{-\infty}^{\infty} dt' \int_{-\infty}^{\infty} d\omega \, e^{i\omega(t'-t_0+R_0/c)} \left(\frac{i\omega}{R_0} - \frac{c}{R_0^2} \right) \hat{\mathbf{R}}_0 \times \boldsymbol{\beta} \quad (10.35)$$

To get rid of the factor $i\omega$, we integrate the first term by parts with respect to t'. However, some care is necessary because the distance R_0 is not an independent variable, but depends on t'. Differentiating $R_0^2 = \mathbf{R}_0 \cdot \mathbf{R}_0$, we see that

$$\frac{dR_0^2}{dt'} = 2R_0 \frac{dR_0}{dt'} = \frac{d\mathbf{R}_0 \cdot \mathbf{R}_0}{dt'} = -2c\mathbf{R}_0 \cdot \boldsymbol{\beta} \tag{10.36}$$

since $\mathbf{R}_0 = \mathbf{r}_0 - \mathbf{r}(t')$. When we integrate by parts with respect to t' we therefore get

$$\mathbf{B} = \frac{\mu_0 q}{8\pi^2} \int_{-\infty}^{\infty} d\omega \, e^{i\omega(t'-t_0+R_0/c)} \frac{\hat{\mathbf{R}}_0 \times \boldsymbol{\beta}}{R_0(1 - \hat{\mathbf{R}}_0 \cdot \boldsymbol{\beta})} \Bigg|_{t'=-\infty}^{\infty}$$

$$- \frac{\mu_0 q}{8\pi^2} \int_{-\infty}^{\infty} dt' \int_{-\infty}^{\infty} d\omega \, e^{i\omega(t'-t_0+R_0/c)} \left\{ \frac{c\hat{\mathbf{R}}_0 \times \boldsymbol{\beta}}{R_0^2} + \frac{d}{dt'} \left[\frac{\hat{\mathbf{R}}_0 \times \boldsymbol{\beta}}{R_0(1 - \hat{\mathbf{R}}_0 \cdot \boldsymbol{\beta})} \right] \right\} \tag{10.37}$$

Provided that the particle moves to infinity as $t' \to \pm\infty$, we can ignore the first integral. When we integrate the second integral with respect to ω, we get the δ-function $2\pi \delta(t' - t + R_0/c)$. To carry out the integral over t', we use the rule

$$\delta(t' - t_0 + R_0/c) = \frac{\delta(t')}{\left| \dfrac{d}{dt'}(t' - t_0 + R_0/c) \right|} = \frac{\delta(t')}{1 - \hat{\mathbf{R}}_0 \cdot \boldsymbol{\beta}} \tag{10.38}$$

When we evaluate the integral with respect to t', we find that the only contribution comes from the retarded time $t' = t_0 - R_0/c$ and the result is

$$\mathbf{B}(\mathbf{r}_0, t_0) = -\frac{\mu_0 q}{4\pi} \left[\frac{1}{1 - \hat{\mathbf{R}}_0 \cdot \boldsymbol{\beta}} \left\{ \frac{c\hat{\mathbf{R}}_0 \times \boldsymbol{\beta}}{R_0^2} + \frac{d}{dt'} \left[\frac{\hat{\mathbf{R}}_0 \times \boldsymbol{\beta}}{R_0(1 - \hat{\mathbf{R}}_0 \cdot \boldsymbol{\beta})} \right] \right\} \right]_{\text{retarded}} \tag{10.39}$$

To compute the derivative we need the relation

$$\frac{d\hat{\mathbf{R}}_0}{dt'} = \frac{d}{dt'} \left(\frac{\mathbf{R}_0}{R_0} \right) = \frac{c}{R_0} [(\hat{\mathbf{R}}_0 \cdot \boldsymbol{\beta})\hat{\mathbf{R}}_0 - \boldsymbol{\beta}] \tag{10.40}$$

Then, after some tedious algebra, we obtain the result

$$\mathbf{B}(\mathbf{r}_0, t_0) = -\frac{\mu_0 q}{4\pi} \left[\frac{c\hat{\mathbf{R}}_0 \times \boldsymbol{\beta}}{\gamma^2 R_0^2 (1 - \hat{\mathbf{R}}_0 \cdot \boldsymbol{\beta})^3} + \frac{\hat{\mathbf{R}}_0 \times [\dot{\boldsymbol{\beta}} + \hat{\mathbf{R}}_0 \times (\boldsymbol{\beta} \times \dot{\boldsymbol{\beta}})]}{R_0(1 - \hat{\mathbf{R}}_0 \cdot \boldsymbol{\beta})^3} \right]_{\text{retarded}} \tag{10.41}$$

This is called the Lienard–Wiechert field. It shows that the field $\mathbf{B}(\mathbf{r}_0, t_0)$ depends only on the instantaneous motion of the particle at the retarded time. Compared with the second term, the first term in the brackets vanishes at large distances and is therefore called the "near field." The second term, called the "far field," is responsible for the radiation observed at large distances. We note that it is proportional to the acceleration $\dot{\boldsymbol{\beta}}$ of the particle and vanishes for a particle in uniform motion. We would expect this to be true, since a particle in uniform motion is at rest in some inertial reference frame and does not radiate.

EXERCISE 10.1

Following the arguments used to derive Jefimenko's formula (10.33) for the retarded magnetic field, show that the retarded electric field is given by (10.34).

EXERCISE 10.2

Beginning with the Maxwell equations, show that the Helmholtz equation for the electric field is

$$\nabla^2 \tilde{\mathbf{E}} + k^2 \tilde{\mathbf{E}} = \frac{1}{\varepsilon_0} \nabla \tilde{\rho} - \mu_0 i \omega \tilde{\mathbf{J}} \tag{10.42}$$

Using a Green function, show that the field of a point source

$$\rho = q\delta[\mathbf{r}' - \mathbf{r}(t')] \tag{10.43}$$

$$\mathbf{J} = qc\boldsymbol{\beta}\delta[\mathbf{r}' - \mathbf{r}(t')] \tag{10.44}$$

is

$$\tilde{\mathbf{E}}(\mathbf{r}_0, \omega) = \frac{q}{(32\pi^3)^{1/2}\varepsilon_0} \int_{-\infty}^{\infty} dt' \left\{ \frac{\hat{\mathbf{R}}_0}{R_0^2} + \frac{d}{dt'}\left[\frac{\hat{\mathbf{R}}_0 - \boldsymbol{\beta}}{cR_0(1 - \hat{\mathbf{R}}_0 \cdot \boldsymbol{\beta})} \right] \right\} e^{i\omega(t' + R_0/c)} \tag{10.45}$$

Invert the Fourier transform to show that the Lienard–Wiechert form of the electrical field is

$$\mathbf{E}(\mathbf{r}_0, t_0) = \frac{q}{4\pi\varepsilon_0} \left[\frac{\hat{\mathbf{R}}_0 - \boldsymbol{\beta}}{\gamma^2 R_0^2 (1 - \hat{\mathbf{R}}_0 \cdot \boldsymbol{\beta})^3} + \frac{\hat{\mathbf{R}}_0 \times [(\hat{\mathbf{R}}_0 - \boldsymbol{\beta}) \times \dot{\boldsymbol{\beta}}]}{cR_0(1 - \hat{\mathbf{R}}_0 \cdot \boldsymbol{\beta})^3} \right]_{\text{retarded}} \tag{10.46}$$

EXERCISE 10.3

By comparing (10.46) to (10.41), show that for a point charge the electric and magnetic fields are related by

$$\mathbf{B} = \left[\frac{\hat{\mathbf{R}}_0 \times \mathbf{E}}{c} \right]_{\text{retarded}} \tag{10.47}$$

10.1.3 Multipole Radiation

It is shown in Chapter 3 that the time-independent fields surrounding a localized distribution of charge and current can be expanded in a series of multipoles. The same is true for time-dependent fields, except that a new term appears that dominates the fields at large distances. This term describes the energy radiated by the charge distribution.

We begin with the magnetic field, for which the Jefimenko formula is

$$\mathbf{B}(\mathbf{r}, t) = \frac{\mu_0}{4\pi} \int d^3\mathbf{r}' \left[\frac{\partial \mathbf{J}}{\partial t'} \times \frac{\hat{\mathbf{R}}}{cR} + \mathbf{J} \times \frac{\hat{\mathbf{R}}}{R^2} \right]_{\text{retarded}} \tag{10.48}$$

where

$$t' = t_{\text{retarded}} = t - \frac{R}{c} \tag{10.49}$$

is the retarded time. At large distances from the current distribution, which we place near the origin, the second term in (10.48) can be ignored compared with the first and we can use the approximations

$$\frac{\hat{\mathbf{R}}}{R} = \frac{\hat{\mathbf{r}}}{r} \tag{10.50}$$

$$t' = t - \frac{r}{c} + \frac{\mathbf{r}' \cdot \hat{\mathbf{r}}}{c} = t_0 + \frac{\mathbf{r}' \cdot \hat{\mathbf{r}}}{c} \tag{10.51}$$

where $t_0 = t - r/c$ is the retarded time at the origin. If we substitute these approximations into (10.48) and Taylor-expand the function $[\partial \mathbf{J}/\partial t']_{\text{retarded}}$ we get

$$\mathbf{B}(\mathbf{r}, t) = -\frac{\mu_0}{4\pi} \frac{\hat{\mathbf{r}}}{cr} \times \int d^3\mathbf{r}' \left\{ \left[\frac{\partial \mathbf{J}}{\partial t'}\right]_0 + \frac{\mathbf{r}' \cdot \hat{\mathbf{r}}}{c} \left[\frac{\partial^2 \mathbf{J}}{\partial t'^2}\right]_0 + \cdots \right\} \tag{10.52}$$

where $[\ ]_0$ indicates that the expression in the brackets is to be evaluated at t_0, the retarded time at the origin. But now the integrands are evaluated everywhere at the same time, so the time derivative can be moved outside the integrals. We may therefore write (10.52) in the form

$$\mathbf{B}(\mathbf{r}, t) = -\frac{\mu_0}{4\pi} \frac{\hat{\mathbf{r}}}{cr} \times \left[\frac{d}{dt'} \int \mathbf{J} \, d^3\mathbf{r}'\right]_0 - \frac{\mu_0}{4\pi} \frac{\hat{\mathbf{r}}}{c^2 r} \times \left[\frac{d^2}{dt'^2} \int (\mathbf{r}' \cdot \hat{\mathbf{r}}) \mathbf{J} \, d^3\mathbf{r}'\right]_0 + \cdots \tag{10.53}$$

The first term in (10.53) can be simplified in the following way. Beginning with the identity

$$\nabla \cdot (r_i \mathbf{J}) = J_i + r_i \nabla \cdot \mathbf{J} \tag{10.54}$$

we can use the continuity relation and the divergence theorem to show that

$$\int_V J_i \, dV - \frac{d}{dt} \int_V r_i \rho \, dV = \int_V \nabla \cdot (r_i \mathbf{J}) \, dV = \oint_S r_i \mathbf{J} \cdot d\mathbf{S} = 0 \tag{10.55}$$

since the current vanishes on the surface S of the volume V that contains the charge distribution. Thus,

$$\left[\int \mathbf{J} \, d^3\mathbf{r}'\right]_0 = [\dot{\mathbf{p}}]_0 \tag{10.56}$$

where

$$[\mathbf{p}]_0 = \left[\int \mathbf{r}' \rho \, d^3\mathbf{r}'\right]_0 \tag{10.57}$$

is the dipole moment of the charge distribution at the retarded time t_0 and the dot indicates the derivative with respect to time.

We can approach the second term in (10.53) in the same way. Beginning with the identity

$$\nabla \cdot (r_k r_l \mathbf{J}) = r_k J_l + r_l J_k + r_k r_l \nabla \cdot \mathbf{J} \tag{10.58}$$

we can use the continuity relation and the divergence theorem to show that

$$\int_V r'_j J_i \, dV' = \frac{1}{2} \frac{d}{dt'} \int_V r'_i r'_j \rho \, dV' + \frac{1}{2} \int_V (r'_j J_i - r'_i J_j) \, dV' \tag{10.59}$$

and if we take the scalar product with $\mathbf{r}$ we get

$$\sum_j r_j \int_V r'_j J_i \, dV' = \frac{1}{2} \sum_j r_j \frac{d}{dt'} \int_V r'_i r'_j \rho \, dV' + \frac{1}{2} \sum_j \int_V (r_j r'_j J_i - r_j J_j r'_i) \, dV' \quad (10.60)$$

But the second term is just the ith component of

$$\frac{1}{2} \int (\mathbf{r}' \times \mathbf{J}) \times \mathbf{r} \, d^3 r' = \mathbf{m} \times \mathbf{r} \quad (10.61)$$

where

$$\mathbf{m} = \frac{1}{2} \int \mathbf{r}' \times \mathbf{J}(\mathbf{r}') \, d^3 r' \quad (10.62)$$

is the magnetic dipole moment of the charge distribution. The first term is

$$\frac{1}{2} \sum_j r_j \frac{d}{dt'} \int_V r'_i r'_j \rho \, dV' = \frac{1}{6} \sum_j r_j \frac{d Q_{ij}}{dt'} + \frac{1}{6} r_i \frac{d}{dt'} \int_V r'^2 \rho \, dV' \quad (10.63)$$

where

$$Q_{ij} = \int (3 r'_i r'_j - \delta_{ij} r'^2) \, d^3 r' \quad (10.64)$$

is the quadrupole moment of the charge distribution.

When we substitute these results into (10.53), the cross product with the last term in (10.63) vanishes and we are left with

$$B(\mathbf{r}, t) = \left[\mathbf{B}^{(ED)} \right]_0 + \left[\mathbf{B}^{(MD)} \right]_0 + \left[\mathbf{B}^{(EQ)} \right]_0 + \cdots \quad (10.65)$$

where the electric dipole, magnetic dipole, and electric quadrupole fields are

$$\mathbf{B}^{(ED)} = -\frac{\mu_0}{4\pi c r} \hat{\mathbf{r}} \times \ddot{\mathbf{p}} \quad (10.66)$$

$$\mathbf{B}^{(MD)} = \frac{\mu_0}{4\pi c^2 r} \hat{\mathbf{r}} \times (\hat{\mathbf{r}} \times \ddot{\mathbf{m}}) \quad (10.67)$$

$$\mathbf{B}^{(EQ)} = -\frac{\mu_0}{24\pi c^2 r} \hat{\mathbf{r}} \times \dddot{\mathbf{Q}} \quad (10.68)$$

respectively, where for later convenience we introduce the vector

$$\dddot{Q}_i = \sum_j \hat{r}_j \dddot{Q}_{ji} \quad (10.69)$$

Of these terms the electric dipole field is ordinarily the largest. If the charge distribution is characterized by length a, charge q, and velocity βc, then the terms are of order

$$B^{(ED)} = O\left(\frac{\mu_0 c q}{r a} \beta^2 \right) \quad (10.70)$$

$$B^{(MD)} = O\left(\frac{\mu_0 c q}{r a} \beta^3 \right) \quad (10.71)$$

$$B^{(EQ)} = O\left(\frac{\mu_0 c q}{r a} \beta^3 \right) \quad (10.72)$$

In the nonrelativistic limit, $\beta \ll 1$, the higher order terms can be neglected unless the electric dipole field vanishes by symmetry. This is a good approximation in atomic transitions, where the electrons are generally nonrelativistic, but not such a good approximation for nuclear transitions. It is typically true, and we assume in the following, that in a given situation only one of these three terms is important, so we can consider the radiation from each of the terms individually.

Far from the origin, the electric and magnetic fields are spherical waves that can be approximated locally by plane waves propagating in the radial direction. From the results of Chapter 4, then, we see that the Poynting vector is

$$\mathbf{S} = \frac{c}{\mu_0} B^2 \hat{\mathbf{r}} \tag{10.73}$$

or equivalently that the power per unit solid angle $d\Omega$ is

$$\frac{d\mathcal{P}}{d\Omega} = r^2 \mathbf{S} \cdot \hat{\mathbf{r}} = \frac{c}{\mu_0} r^2 B^2 \tag{10.74}$$

The electric dipole contribution to the radiation is

$$\frac{d\mathcal{P}^{(ED)}}{d\Omega} = \frac{\mu_0}{16\pi^2 c} |[\hat{\mathbf{r}} \times \ddot{\mathbf{p}}]_0|^2 \tag{10.75}$$

or if we refer the angular power to the time at which it is radiated (the retarded time t_0), we get

$$\frac{d\mathcal{P}^{(ED)}}{d\Omega} = \frac{\mu_0}{16\pi^2 c} |\hat{\mathbf{r}} \times \ddot{\mathbf{p}}|^2 = \frac{\mu_0}{16\pi^2 c} |\ddot{\mathbf{p}}|^2 \sin^2 \theta \tag{10.76}$$

where θ is the angle between $\ddot{\mathbf{p}}$ and the direction of the observer. The characteristic pattern of dipole radiation is illustrated in Figure 10.3, where we see that it has the form of a torus aligned about the vector $\ddot{\mathbf{p}}$ and vanishes along the axis in the direction $\ddot{\mathbf{p}}$. If we integrate the power over all solid angles, we get

$$\mathcal{P}^{(ED)} = \frac{\mu_0}{8\pi c} |\ddot{\mathbf{p}}|^2 \int_0^\pi \sin^3 \theta \, d\theta = \frac{\mu_0}{6\pi c} |\ddot{\mathbf{p}}|^2 \tag{10.77}$$

which agrees with the result we got in Chapter 4 for the spontaneous emission from an oscillating charge.

In the same way, we find that the magnetic dipole contribution to the radiation is

$$\frac{d\mathcal{P}^{(MD)}}{d\Omega} = \frac{\mu_0}{16\pi^2 c^3} |\hat{\mathbf{r}} \times (\hat{\mathbf{r}} \times \ddot{\mathbf{m}})|^2 = \frac{\mu_0}{16\pi^2 c^3} |\ddot{\mathbf{m}}|^2 \sin^2 \theta \tag{10.78}$$

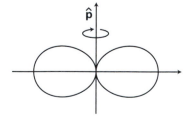

Figure 10.3 Pattern of dipole radiation.

and the total power is

$$\mathcal{P}^{(MD)} = \frac{\mu_0}{6\pi c^3} |\ddot{\mathbf{m}}|^2 \tag{10.79}$$

For the electric quadrupole, the angular power radiated is

$$\frac{d\mathcal{P}^{(EQ)}}{d\Omega} = \frac{\mu_0}{576\pi^2 c^3} [\dddot{\mathbf{Q}} \cdot \dddot{\mathbf{Q}} - (\hat{\mathbf{r}} \cdot \dddot{\mathbf{Q}})^2] \tag{10.80}$$

We can also express this in the explicit tensor form

$$\frac{d\mathcal{P}^{(EQ)}}{d\Omega} = \frac{\mu_0}{576\pi^2 c^3} \left(\sum_{i,j,k} r_j \, \dddot{Q}_{ji} \, \dddot{Q}_{ik} r_k - \sum_{i,j,k,l} r_i \, \dddot{Q}_{ij} \, r_j r_k \, \dddot{Q}_{kl} \, r_l \right) \tag{10.81}$$

For the special case of a quadrupole that has spheroidal symmetry about the z axis, all the off-diagonal terms $\dddot{Q}_{ij \neq i}$ vanish, and since the trace of the quadrupole moment tensor vanishes, we have $Q_{xx} = Q_{yy} = -\frac{1}{2} Q_{zz}$. After a bit of algebra, we find that the instantaneous angular power radiated is

$$\frac{d\mathcal{P}^{(EQ)}}{d\Omega} = \frac{\mu_0 \, \dddot{Q}_{zz}^2}{256\pi^2 c^3} \cos^2 \theta \sin^2 \theta \tag{10.82}$$

This is illustrated in Figure 10.4, where we see that the radiation forms two cones about the axis of symmetry.

To find the total power radiated by the electrical quadrupole moment, we note that by symmetry

$$\int_{4\pi} \hat{r}_i \hat{r}_j \, d\Omega = K_1 \delta_{ij} \tag{10.83}$$

$$\int_{4\pi} \hat{r}_i \hat{r}_j \hat{r}_k \hat{r}_l \, d\Omega = K_2 (\delta_{ij} \delta_{kl} + \delta_{ik} \delta_{jl} + \delta_{il} \delta_{jk}) \tag{10.84}$$

for some constants K_1 and K_2. We can evaluate the constants for the simple case $i = j = k = l = 3$ (corresponding to the z axis), and we get

$$K_1 = 2\pi \int_0^\theta \cos^2 \theta \sin \theta \, d\theta = \frac{4\pi}{3} \tag{10.85}$$

$$K_2 = \frac{2\pi}{3} \int_0^\pi \cos^4 \theta \sin \theta \, d\theta = \frac{4\pi}{15} \tag{10.86}$$

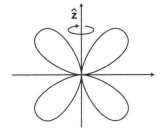

Figure 10.4 Pattern of quadrupole radiation.

so the total power is

$$\mathcal{P}^{(EQ)} = \frac{\mu_0}{432\pi c^3} \sum_{i,j} \left(\frac{3}{5} \dddot{Q}_{ij} \dddot{Q}_{ij} - \frac{1}{5} \dddot{Q}_{ii} \dddot{Q}_{jj} \right)$$

(10.87)

But the trace of the quadrupole moment vanishes, which leaves

$$\mathcal{P}^{(EQ)} = \frac{\mu_0}{720\pi c^3} \sum_{i,j} \dddot{Q}_{ij}^2$$

(10.88)

EXERCISE 10.4

At large distances from the source, the fields are locally nearly plane waves.

(a) Using the plane-wave approximation and Faraday's law, begin with (10.66) and show that the electric field of the electric dipole radiation is

$$\mathbf{E}^{(ED)} = \frac{\mu_0}{4\pi r} [\hat{\mathbf{r}} \times (\hat{\mathbf{r}} \times \ddot{\mathbf{p}})]_0$$

(10.89)

(b) Compute the Poynting vector directly from **E** and **B**, and show that the Poynting vector is

$$\mathbf{S}^{(ED)} = \frac{\mu_0}{16\pi^2 c r^2} |[\hat{\mathbf{r}} \times \ddot{\mathbf{p}}]_0|^2 \hat{\mathbf{r}}$$

(10.90)

EXERCISE 10.5

For a harmonically oscillating charge distribution, we may write

$$\mathbf{p} = \mathrm{Re}(\mathbf{p}_0 e^{-i\omega t})$$

(10.91)

where $\mathbf{p}_0$ is in general a complex vector and may have different complex phases for the different vector components.

(a) Show that the angular power averaged over a complete cycle is

$$\frac{d\langle\mathcal{P}\rangle^{(ED)}}{d\Omega} = \frac{\mu_0 \omega^4}{32\pi^2 c} [\mathbf{p}_0 \cdot \mathbf{p}_0^* - (\hat{\mathbf{r}} \cdot \mathbf{p}_0)(\hat{\mathbf{r}} \cdot \mathbf{p}_0^*)]$$

(10.92)

(b) For a dipole oscillating in the $\hat{\mathbf{z}}$ direction,

$$\mathbf{p}_0 = p_0(0, 0, 1)$$

(10.93)

Show that the average angular power is

$$\frac{d\langle\mathcal{P}\rangle^{(ED)}}{d\Omega} = \frac{\mu_0 \omega^4 p_0^2}{32\pi^2 c} \sin^2\theta$$

(10.94)

where θ is the angle between $\hat{\mathbf{z}}$ and $\hat{\mathbf{r}}$, and the average total power is

$$\langle\mathcal{P}\rangle^{(ED)} = \frac{\mu_0 \omega^4 p_0^2}{12\pi c}$$

(10.95)

(c) For a permanent dipole in the x–y plane rotating about the z axis,

$$\mathbf{p}_0 = p_0(1, i, 0) \tag{10.96}$$

Show that the average angular power is

$$\frac{d\langle\mathcal{P}\rangle^{(ED)}}{d\Omega} = \frac{\mu_0\omega^4 p_0^2}{32\pi^2 c}(2 - \sin^2\theta) \tag{10.97}$$

and the total power is

$$\langle\mathcal{P}\rangle^{(ED)} = \frac{\mu_0\omega^4 p_0^2}{6\pi c} \tag{10.98}$$

10.1.4 Spectral Distribution of Radiation from a Point Charge

In dealing with the radiation from charged particles, we are typically concerned with the spectrum of the radiation observed by our detector. That is, we are interested in the total energy dW per unit frequency interval $d\omega$ radiated into the solid angle $d\Omega$. This is called the angular spectral fluence. To compute this, it is not necessary to invert the Fourier transform to get the fields explicitly. Instead, we work in frequency space and begin with the Fourier transform (10.28), which gives us the spectral amplitude $\tilde{\mathbf{B}}$.

We see from (10.28) that the field of a moving charge consists of two parts. The first term, the near field, falls off as $1/R_0^2$ and vanishes at large distances. The second term, the far field, falls off more slowly, as $1/R_0$. This term contributes finite energy at infinity and represents the radiation from the moving charge. In what follows we are concerned with the radiation at large distances, so we ignore the $1/R_0^2$ term.

At very large distances from the source, the motion of the particle may be regarded as small compared with the distance from the origin to the field point, $r/r_0 \ll 1$. We therefore use the approximations

$$\hat{\mathbf{R}}_0 \approx \hat{\mathbf{r}}_0 = \hat{\mathbf{n}} \tag{10.99}$$

and

$$R_0 \approx |\mathbf{r}_0| - \hat{\mathbf{n}} \cdot \mathbf{r} \tag{10.100}$$

as illustrated in Figure 10.5. When we substitute these into (10.28), it is sufficient to use $R_0 \approx |\mathbf{r}_0|$ in the denominator. However, since the exponential factor oscillates on a length

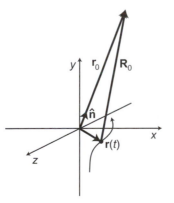

Figure 10.5 Coordinate system for angular spectral fluence.

scale on the order of c/ω, the more accurate expression (10.100) must be used in the exponent. When the near-field ($1/R_0^2$) term is ignored compared with the far-field term, we obtain the simple result

$$\tilde{\mathbf{B}} = \frac{1}{(32\pi^3)^{1/2}} \frac{i\omega\mu_0 q}{r_0} e^{i\omega|\mathbf{r}_0|/c} \int_{-\infty}^{\infty} dt'\, e^{i\omega(t'-\hat{\mathbf{n}}\cdot\mathbf{r}/c)} \hat{\mathbf{n}} \times \boldsymbol{\beta} \tag{10.101}$$

This describes spherical waves in which the magnetic induction $\tilde{\mathbf{B}}$ is transverse to the direction of propagation $\mathbf{n}$.

At large distances these waves become (locally) plane waves. As discussed in Chapter 4, the intensity of these waves is given by the Poynting vector

$$\mathbf{S} = \frac{1}{\mu_0} \mathbf{E} \times \mathbf{B} = \frac{c}{\mu_0} \mathbf{B} \times (\hat{\mathbf{n}} \times \mathbf{B}) = \frac{c}{\mu_0} B^2 \hat{\mathbf{n}} \tag{10.102}$$

The total energy dW radiated into the solid angle $d\Omega$ around the direction $\hat{\mathbf{n}}$ is just the Poynting vector times the area $|\mathbf{r}_0|^2 d\Omega$ included in the solid angle. We therefore get

$$\frac{dW}{d\Omega} = |\mathbf{r}_0|^2 \int_{-\infty}^{\infty} dt\, \hat{\mathbf{n}} \cdot \mathbf{S} = \frac{c|\mathbf{r}_0|^2}{\mu_0} \int_{-\infty}^{\infty} dt\, B^2 = \frac{2c|\mathbf{r}_0|^2}{\mu_0} \int_0^{\infty} d\omega\, |\tilde{B}|^2 = \int_0^{\infty} d\omega\, \frac{d^2W}{d\omega\, d\Omega} \tag{10.103}$$

where it is sufficient to integrate over positive ω, since $\mathbf{B}$ is real. From this we can identify the angular spectral fluence of the radiation as

$$\frac{d^2W}{d\omega\, d\Omega} = \frac{\mu_0 c \omega^2 q^2}{16\pi^3} \left| \int_{-\infty}^{\infty} dt\, e^{i\omega(t-\hat{\mathbf{n}}\cdot\mathbf{r}/c)} \hat{\mathbf{n}} \times \boldsymbol{\beta} \right|^2 \tag{10.104}$$

A few comments are in order regarding this equation. In the first place, we note that the direction of the vector quantity in (10.104) is the direction of the magnetic induction $\mathbf{B}$. If we need to account for the polarization of the electric field $\mathbf{E} = -\hat{\mathbf{n}} \times \mathbf{B}$ of the wave in some direction, we can change to the vector product $\hat{\mathbf{n}} \times (\hat{\mathbf{n}} \times \boldsymbol{\beta})$ in place of $\hat{\mathbf{n}} \times \boldsymbol{\beta}$ in (10.104). This has no effect on the magnitude, but the polarization of the vector in the integrand is now that of the electric field. To compute the fluence of the radiation with a certain polarization, we take the component of the vector integral in the required direction before we form the absolute square.

In the second place, the simple expression (10.104) might seem to indicate that a particle in uniform motion radiates energy, but this is not the case. For a constant velocity the integrand oscillates at the frequency $\omega(1 - \hat{\mathbf{n}} \cdot \boldsymbol{\beta})$. Therefore, the integral must be evaluated not by allowing the limits of integration to approach infinity, in which case the integral oscillates, but by introducing some exponential decay to make the integral finite and then allowing the decay rate to vanish. The limit procedure is similar to that used to define the δ-function in Chapter 5, and the integral vanishes by cancellation except at $\omega = 0$. We encounter the same difficulty in calculating the radiation emitted during the collision of a charged particle with a scattering center. Prior to the collision and after the interaction ceases, the particle is in uniform motion. It does not radiate during these periods, but the integral must be handled carefully to make sure it converges in the limits $t \to \pm\infty$.

We can avoid these convergence problems and put the formula in a different form if we integrate once by parts. To do this, we multiply and divide the integrand by the factor

$i\omega(1 - \hat{\mathbf{n}} \cdot \boldsymbol{\beta})$ to make the exponential factor a perfect differential. We also note that $|\hat{\mathbf{n}} \times \boldsymbol{\beta}| = |\hat{\mathbf{n}} \times (\hat{\mathbf{n}} \times \boldsymbol{\beta})|$ and the derivative is

$$\frac{d}{dt} \frac{\hat{\mathbf{n}} \times (\hat{\mathbf{n}} \times \boldsymbol{\beta})}{1 - \hat{\mathbf{n}} \cdot \boldsymbol{\beta}} = \frac{\hat{\mathbf{n}} \times [(\hat{\mathbf{n}} - \boldsymbol{\beta}) \times \dot{\boldsymbol{\beta}}]}{(1 - \hat{\mathbf{n}} \cdot \boldsymbol{\beta})^2} \tag{10.105}$$

If we now integrate once by parts, recognizing implicitly that we must evaluate the integrals by introducing a decay as $t \to \pm\infty$, we get

$$\frac{1}{i\omega} \int_{-\infty}^{\infty} dt' \, i\omega(1 - \hat{\mathbf{n}} \cdot \boldsymbol{\beta}) e^{i\omega(t' - \hat{\mathbf{n}} \cdot \mathbf{r}/c)} \frac{\hat{\mathbf{n}} \times (\hat{\mathbf{n}} \times \boldsymbol{\beta})}{1 - \hat{\mathbf{n}} \cdot \boldsymbol{\beta}}$$

$$= \frac{1}{i\omega} e^{i\omega(t' - \hat{\mathbf{n}} \cdot \mathbf{r}/c)} \frac{\hat{\mathbf{n}} \times (\hat{\mathbf{n}} \times \boldsymbol{\beta})}{1 - \hat{\mathbf{n}} \cdot \boldsymbol{\beta}} \Big|_{-\infty}^{\infty} - \frac{1}{i\omega} \int_{-\infty}^{\infty} dt' \, e^{i\omega(t' - \hat{\mathbf{n}} \cdot \mathbf{r}/c)} \frac{d}{dt'} \frac{\hat{\mathbf{n}} \times (\hat{\mathbf{n}} \times \boldsymbol{\beta})}{1 - \hat{\mathbf{n}} \cdot \boldsymbol{\beta}}$$

$$= \frac{i}{\omega} \int_{-\infty}^{\infty} dt' e^{i\omega(t' - \hat{\mathbf{n}} \cdot \mathbf{r}/c)} \frac{\hat{\mathbf{n}} \times [(\hat{\mathbf{n}} - \boldsymbol{\beta}) \times \dot{\boldsymbol{\beta}}]}{(1 - \hat{\mathbf{n}} \cdot \boldsymbol{\beta})^2} \tag{10.106}$$

The angular spectral fluence is then

$$\frac{d^2\mathcal{W}}{d\omega \, d\Omega} = \frac{\mu_0 c q^2}{16\pi^3} \left| \int_{-\infty}^{\infty} dt \, e^{i\omega(t - \hat{\mathbf{n}} \cdot \mathbf{r}/c)} \frac{\hat{\mathbf{n}} \times [(\hat{\mathbf{n}} - \boldsymbol{\beta}) \times \dot{\boldsymbol{\beta}}]}{(1 - \hat{\mathbf{n}} \cdot \boldsymbol{\beta})^2} \right|^2 \tag{10.107}$$

This explicitly shows that the radiation vanishes when the acceleration vanishes. In addition, it avoids the convergence problems of (10.104), provided that the acceleration vanishes as the particle approaches infinity. The form (10.104) is generally more useful, but the integrals must be evaluated with care. Finally, we note that the vector integral in (10.107) is polarized in the direction of the electric field. To compute the fluence with a specific polarization, it is sufficient to take the component of the integral in the desired direction before forming the absolute square of the integral.

EXERCISE 10.6

In a "third-generation" synchrotron radiation source, the electrons or positrons pass through a so-called undulator magnet in which the magnetic field alternates periodically, forcing the electron to execute a wiggly motion. For the present problem the undulator field may be represented by the vector potential

$$\mathbf{A} = A_U \hat{\mathbf{z}} \cos(k_U x) \qquad -\tfrac{1}{2} L_U < x < \tfrac{1}{2} L_U \tag{10.108}$$

where λ_U is the undulator period and $L_U = N_U \lambda_U$ is the undulator length, in which N_U is the number of undulator periods, and $k_U = 2\pi/\lambda_U$. The field is translationally invariant in the $\hat{\mathbf{z}}$ direction, so canonical momentum is conserved in that direction. If the undulator field is small, so that

$$a_U = \frac{q A_U}{\sqrt{2} mc} \ll 1 \tag{10.109}$$

the trajectory is nearly a straight line at constant velocity and the transverse velocity may be computed as a perturbation from this. Beginning with (10.104), compute the angular spectral fluence of the radiation along the undulator axis for wave numbers near the resonant wave number

$$k_R = \frac{\beta}{1-\beta} k_U \tag{10.110}$$

where βc is the particle velocity. Show that for a long undulator ($N_U \gg 1$) in the ultra-relativistic limit ($\gamma \gg 1$, where $\gamma = 1/\sqrt{1-\beta^2}$), the angular spectral fluence is given by the expression

$$\frac{d^2\mathcal{W}}{d\omega\, d\Omega} = \frac{q^2 a_U^2 \gamma^2 N_U^2}{2\pi \varepsilon_0 c} \left[\frac{\sin(\pi N_U \Delta k/k_R)}{\pi N_U \Delta k/k_R} \right]^2 \tag{10.111}$$

where the resonant wave number is

$$k_R = 2\gamma^2 k_U \tag{10.112}$$

and the departure from resonance is

$$\Delta k = k - k_R \tag{10.113}$$

EXERCISE 10.7

The velocity $\mathbf{v}(t)$ of a harmonically bound charge q oscillates at a frequency ω_0 with damping rate α. For a charge that is initially at rest and is excited at $t = 0$, the motion for $t > 0$ is

$$\mathbf{v} = \mathbf{v}_0 \cos(\omega_0 t) e^{-\alpha t} \tag{10.114}$$

where $\mathbf{v}_0$ is a constant vector. Show that for nonrelativistic motion ($v_0/c \ll 1$) of a weakly damped oscillator ($\alpha/\omega_0 \ll 1$), the total spectral angular distribution of the radiation produced as the oscillations die out is

$$\frac{d^2\mathcal{W}}{d\omega\, d\Omega} = \frac{\mu_0 q^2 v_0^2}{64\pi^3 c} \frac{\omega^2 \sin^2\theta}{(\omega - \omega_0)^2 + \alpha^2} \tag{10.115}$$

near resonance [$(\omega - \omega_0)/\omega_0 \ll 1$], where $\theta = \arccos(\hat{\mathbf{n}} \cdot \mathbf{v}_0/|\mathbf{v}_0|)$ is the angle between the direction of observation and the direction of oscillation.

10.1.5 Angular Distribution of Radiation from a Point Charge

The angular fluence is the total energy radiated into a unit solid angle. We can find it by integrating the angular spectral fluence over all frequencies. However, as we have seen previously, the total energy computed by integrating over all frequencies is the same as that computed by integrating the power over all time. By transforming from the frequency domain back to the time domain, therefore, we can identify the instantaneous power being radiated into a unit solid angle.

Beginning with (10.107) but replacing the square of the absolute value of the integral by the product of the integral and its complex conjugate, and integrating over both positive and negative frequencies, we get

$$
\frac{dW}{d\Omega} = \frac{\mu_0 cq^2}{32\pi^3} \int_{-\infty}^{\infty} d\omega \int_{-\infty}^{\infty} dt' \int_{-\infty}^{\infty} dt'' e^{i\omega[t'-\hat{\mathbf{n}}\cdot\mathbf{r}(t')/c]} e^{-i\omega[t''-\hat{\mathbf{n}}\cdot\mathbf{r}(t'')/c]}
$$

$$
\times \frac{\hat{\mathbf{n}} \times \{[\hat{\mathbf{n}} - \boldsymbol{\beta}(t')] \times \dot{\boldsymbol{\beta}}(t')\}}{[1 - \hat{\mathbf{n}} \cdot \boldsymbol{\beta}(t')]^2} \cdot \frac{\hat{\mathbf{n}} \times \{[\hat{\mathbf{n}} - \boldsymbol{\beta}(t'')] \times \dot{\boldsymbol{\beta}}(t'')\}}{[1 - \hat{\mathbf{n}} \cdot \boldsymbol{\beta}(t'')]^2}
\tag{10.116}
$$

If we integrate first over ω using the orthogonality property $\int_{-\infty}^{\infty} e^{i\omega t} d\omega = 2\pi \delta(t)$, this becomes

$$
\frac{dW}{d\Omega} = \frac{\mu_0 cq^2}{16\pi^2} \int_{-\infty}^{\infty} dt' \int_{-\infty}^{\infty} dt'' \delta \left\{ \left[t' - \frac{\hat{\mathbf{n}} \cdot \mathbf{r}(t')}{c} \right] - \left[t'' - \frac{\hat{\mathbf{n}} \cdot \mathbf{r}(t'')}{c} \right] \right\}
$$

$$
\times \frac{\hat{\mathbf{n}} \times \{[\hat{\mathbf{n}} - \boldsymbol{\beta}(t')] \times \dot{\boldsymbol{\beta}}(t')\}}{[1 - \hat{\mathbf{n}} \cdot \boldsymbol{\beta}(t')]^2} \cdot \frac{\hat{\mathbf{n}} \times \{[\hat{\mathbf{n}} - \boldsymbol{\beta}(t'')] \times \dot{\boldsymbol{\beta}}(t'')\}}{[1 - \hat{\mathbf{n}} \cdot \boldsymbol{\beta}(t'')]^2}
\tag{10.117}
$$

To carry out the integral over t'', we use the rule

$$
\int_{-\infty}^{\infty} \delta[f(t)] g(t) \, dt = \frac{g(t_0)}{|f'(t_0)|}
\tag{10.118}
$$

in which t_0 is the point where $f(t_0) = 0$. In the present case the denominator is

$$
|f'(t'')| = |-1 + \hat{\mathbf{n}} \cdot \boldsymbol{\beta}(t'')| = 1 - \hat{\mathbf{n}} \cdot \boldsymbol{\beta}(t'')
\tag{10.119}
$$

But the argument of the δ-function vanishes only at $t'' = t'$. To see this, we recognize that for a particle traveling at less than the speed of light, the function $ct'' - \hat{\mathbf{n}} \cdot \mathbf{r}(t'')$ is monotonically increasing with t''. Therefore, it can have the value $ct' - \hat{\mathbf{n}} \cdot \mathbf{r}(t')$ at only one point, and that must be at $t'' = t'$. With this we arrive at the relatively simple expression

$$
\frac{dW}{d\Omega} = \frac{\mu_0 cq^2}{16\pi^2} \int_{-\infty}^{\infty} dt \, \frac{|\hat{\mathbf{n}} \times [(\hat{\mathbf{n}} - \boldsymbol{\beta}) \times \dot{\boldsymbol{\beta}}]|^2}{(1 - \hat{\mathbf{n}} \cdot \boldsymbol{\beta})^5}
\tag{10.120}
$$

But the total energy radiated into the solid angle $d\Omega$ may also be expressed as an integral over time of the form

$$
\frac{dW}{d\Omega} = \int_{-\infty}^{\infty} dt \, \frac{dP}{d\Omega}
\tag{10.121}
$$

where $dP/d\Omega$ is the instantaneous power radiated into the solid angle $d\Omega$. Comparing this with (10.120), we see that the angular power is

$$
\frac{dP}{d\Omega} = \frac{q^2}{16\pi^2 \varepsilon_0 c} \frac{|\hat{\mathbf{n}} \times [(\hat{\mathbf{n}} - \boldsymbol{\beta}) \times \dot{\boldsymbol{\beta}}]|^2}{(1 - \hat{\mathbf{n}} \cdot \boldsymbol{\beta})^5}
\tag{10.122}
$$

In this expression the vector in the numerator is in the direction of the electric field. Therefore, to compute the power in a certain solid angle with a specific polarization, it is sufficient to take the component of this vector in the desired direction before computing the square.

This result is more subtle than it appears, and demonstrates the power of Fourier-transform techniques. The power radiated at any instant is actually determined, in some sense, only when the field created at time t has propagated a sufficiently large distance that the near field has disappeared. This distance depends on the wavelength, since this is the only characteristic length with which to compare the propagation distance. The Fourier transform allows us to sort all this out and compute the radiation that will appear later, at large distances, from the instantaneous motion of the charged particle now.

Due to the complexity of (10.122), it is worth examining a few simple examples. In the case when the particle is moving in one dimension, so that the acceleration is parallel to the velocity, (10.122) simplifies to

$$\frac{d\mathcal{P}}{d\Omega} = \frac{q^2}{16\pi^2\varepsilon_0 c} \frac{|\hat{\mathbf{n}} \times [\hat{\mathbf{n}} \times \dot{\boldsymbol{\beta}}]|^2}{(1 - \hat{\mathbf{n}} \cdot \boldsymbol{\beta})^5} = \frac{q^2}{16\pi^2\varepsilon_0 c} \frac{\dot{\beta}^2 \sin^2\theta}{(1 - \beta\cos\theta)^5} \tag{10.123}$$

where θ is the angle between the direction of motion and the direction of the observer. For nonrelativistic motion ($\beta \ll 1$), the denominator becomes unity and the radiation has the angular distribution characteristic of radiation from a dipole oriented about the direction of the acceleration. For highly relativistic motion ($\beta \approx 1$), the denominator becomes very small for $\gamma\theta < O(1)$, so the radiation is strongly peaked around the direction of motion (although it vanishes on the axis due to the numerator). To see this we use the ultrarelativistic approximation $\beta \approx 1 - (1/2\gamma^2)$ and the small-angle approximation $\cos\theta \approx 1 - (\theta^2/2)$ in the denominator. We then find that for small angles, the angular intensity radiated is

$$\frac{d\mathcal{P}}{d\Omega} = \frac{2q^2}{\pi^2\varepsilon_0 c} \gamma^2\dot{\gamma}^2 \frac{\gamma^2\theta^2}{(1 + \gamma^2\theta^2)^5} \tag{10.124}$$

The peak intensity occurs near $\gamma\theta = O(1)$. We note that for an ultrarelativistic particle the rate of change of the energy $\dot{\gamma}$ is proportional to the force, so the peak of the angular power increases as γ^2 for a given accelerating force.

EXERCISE 10.8

Consider a particle with charge q moving at a *nonrelativistic* velocity in a circle of radius R in the x–y plane with constant angular frequency $\omega_0 \ll c/R$. Show that the time average angular intensity is given by the formula

$$\frac{d\langle\mathcal{P}\rangle}{d\Omega} = \frac{q^2\omega_0^4 R^2}{16\pi^2\varepsilon_0 c^3} \left(1 - \frac{1}{2}\sin^2\theta\right) \tag{10.125}$$

where θ is the angle of the observer from the z axis. Compare this with (10.97).

10.1.6 Total Power Radiated by a Point Charge

Having computed the instantaneous power radiated into the solid angle $d\Omega$, we can in principle compute the total power radiated by integrating (10.122) over all angles. However, it is easier and more instructive to find the total power radiated in the nonrelativistic case and then use arguments of relativistic covariance to generalize this result. In Chapter 2 we use covariance to establish the form of the Lagrangian. In Chapter 6 we use this tactic to generalize the polarization and magnetization to materials in motion, and in Chapter 11 we use it to find the relativistic generalization of the equation of motion of an electron. Covariance is a powerful tool.

In the nonrelativistic limit ($\beta \ll 1$), we see from (10.122) that the power radiated is

$$\frac{d\mathcal{P}}{d\Omega} = \frac{q^2}{16\pi^2 \varepsilon_0 c} |\dot{\boldsymbol{\beta}}|^2 \sin^2 \theta \tag{10.126}$$

where θ is the angle between the acceleration $\dot{\boldsymbol{\beta}}$ and the direction $\hat{\mathbf{n}}$ of the observer. Integrating this over all angles, we get the Larmor formula

$$\mathcal{P} = 2\pi \int_0^\pi d\theta \sin\theta \frac{d\mathcal{P}}{d\Omega} = \frac{q^2 |\dot{\boldsymbol{\beta}}|^2}{6\pi \varepsilon_0 c} \tag{10.127}$$

for the instantaneous power radiated by an accelerating particle.

To generalize this to the relativistic case, we observe that the energy radiated in time interval dt is the timelike component of the 4-vector momentum of the radiation. In the rest frame of the particle, this is

$$dp^0(\text{radiated}) = \frac{d\mathcal{W}}{c} = \frac{\mathcal{P}}{c} dt = \frac{q^2}{6\pi \varepsilon_0 c^5} \left| \frac{d\mathbf{v}}{dt} \right|^2 c \, dt \tag{10.128}$$

The other components of the momentum radiated into 4π steradians vanish by symmetry:

$$dp^i(\text{radiated}) = 0 \qquad \text{for } i = 1 \ldots 3 \tag{10.129}$$

The energy and momentum radiated in any other coordinate system can be found by a Lorentz transformation of the 4-vector momentum from the rest frame to the new frame, but we want the result to be an expression written in terms of the particle motion in the new frame. To find this we look for a covariant expression for the 4-vector momentum $dp^\alpha(\text{radiated})$ that reduces to (10.128) and (10.129) in the nonrelativistic limit. We begin by noting that

$$c \, dt = \gamma c \, d\tau \tag{10.130}$$

But $d\tau$ is a Lorentz scalar and γc is the timelike component of the 4-vector velocity of the particle $U^\alpha = \gamma(c, \mathbf{v})$. Moreover, in the rest frame the particle 4-vector velocity is $(c, 0)$. Comparing this with the radiated 4-vector momentum, we see that in the rest frame of the particle

$$dp^\alpha(\text{radiated}) = \frac{q^2}{6\pi \varepsilon_0 c^5} \left| \frac{d\mathbf{v}}{dt} \right|^2 d\tau \, U^\alpha \tag{10.131}$$

For this 4-vector relationship to hold true in any frame, the quantity $(q^2/6\pi \varepsilon_0 c^5) \times |d\mathbf{v}/dt|^2 d\tau$ must be a Lorentz scalar. But everything in this expression is already a scalar except the factor $|d\mathbf{v}/dt|^2$ involving the acceleration, so we must find a covariant scalar expression that reduces to $|d\mathbf{v}/dt|^2$ in the nonrelativistic limit. Clearly, $|d\mathbf{v}/dt|^2$ looks something like the expression

$$\frac{dU^\alpha}{d\tau} \frac{dU_\alpha}{d\tau} = \frac{1}{m^2} \frac{dp^\alpha}{d\tau} \frac{dp_\alpha}{d\tau} \tag{10.132}$$

where $p^\alpha = (\mathcal{E}/c, \mathbf{p})$ refers to the 4-vector momentum of the particle. In fact, we see that

$$\lim_{\beta_0 \to 0} \left(\frac{dU^\alpha}{d\tau} \frac{dU_\alpha}{d\tau} \right) = -\left(\frac{d\mathbf{v}}{dt} \right)^2 \tag{10.133}$$

so we may express the 4-vector momentum of the radiation in the covariant form

$$dp^\alpha \text{(radiated)} = -\frac{q^2}{6\pi \varepsilon_0 m^2 c^5} \frac{dp^\beta}{d\tau} \frac{dp_\beta}{d\tau} U^\alpha d\tau \tag{10.134}$$

which is valid in any reference frame.

It might be asked whether the Lorentz scalar $(dU^\alpha/d\tau)(dU_\alpha/d\tau)$ is unique. It does, of course, satisfy the limit as $\beta \to 0$, but must we include other terms that vanish in the same limit? The question can be answered by noting from (10.122) that the angular intensity is quadratic in the acceleration $\dot{\boldsymbol{\beta}} = d\boldsymbol{\beta}/dt$, so the total power must have the same behavior. The scalar product $(dU^\alpha/d\tau)(dU_\alpha/d\tau)$ is the only function that satisfies this requirement.

To evaluate the inner product, we recall that

$$\frac{dp^\alpha}{d\tau} \frac{dp_\alpha}{d\tau} = \gamma^2 \left[\frac{1}{c^2} \left(\frac{d\mathcal{E}}{dt} \right)^2 - \left| \frac{d\mathbf{p}}{dt} \right|^2 \right] \tag{10.135}$$

But from the magnitude of the 4-vector momentum, we know that

$$\mathcal{E}^2 = p^2 c^2 + m^2 c^4 = \gamma^2 m^2 c^4 \tag{10.136}$$

Differentiating this, we find that

$$\frac{1}{c} \frac{d\mathcal{E}}{dt} = \beta \frac{dp}{dt} \tag{10.137}$$

We also note that

$$U^\alpha d\tau = \gamma(c, \mathbf{v}) \, d\tau = (c, \mathbf{v}) \, dt \tag{10.138}$$

Substituting these results back into (10.134), we find that the 4-vector momentum radiated is

$$dp^\alpha \text{(radiated)} = \frac{\gamma^2 q^2}{6\pi \varepsilon_0 m^2 c^5} \left[\left| \frac{d\mathbf{p}}{dt} \right|^2 - \beta^2 \left(\frac{dp}{dt} \right)^2 \right] (c, \mathbf{v}) \, dt \tag{10.139}$$

From the timelike component of this expression, we obtain the result

$$\mathcal{P} = \frac{d\mathcal{W}}{dt} = c \frac{dp^0}{dt} \text{(radiated)} = \frac{\gamma^2 q^2}{6\pi \varepsilon_0 m^2 c^3} \left[\left| \frac{d\mathbf{p}}{dt} \right|^2 - \beta^2 \left(\frac{dp}{dt} \right)^2 \right] \tag{10.140}$$

This is the relativistic generalization of the Larmor formula. In the limit $\beta \to 0$, $\gamma \to 1$, we recover the nonrelativistic Larmor formula (10.127).

There is sometimes confusion between the quantities (dp/dt) and $|d\mathbf{p}/dt|$. The first expression refers to the rate of change of the magnitude of the momentum. It depends on the acceleration in the direction of motion and vanishes for acceleration normal to the velocity.

The second expression refers to the magnitude of the rate of change of the vector momentum, and is nonzero for acceleration in any direction. For example, in a magnetic field

$$\left|\frac{d\mathbf{p}}{dt}\right| = q|\mathbf{v} \times \mathbf{B}| > 0 \tag{10.141}$$

but $dp/dt = 0$.

We note from (10.140) that for acceleration parallel to the velocity the total power radiated is

$$\mathcal{P} = \frac{\gamma^2 q^2}{6\pi\varepsilon_0 m^2 c^3}\left[\left(\frac{dp}{dt}\right)^2 - \beta^2\left(\frac{dp}{dt}\right)^2\right] = \frac{q^2}{6\pi\varepsilon_0 m^2 c^3}\left(\frac{dp}{dt}\right)^2 \tag{10.142}$$

whereas for acceleration normal to the velocity the total power radiated is

$$\mathcal{P} = \frac{\gamma^2 q^2}{6\pi\varepsilon_0 m^2 c^3}\left|\frac{d\mathbf{p}}{dt}\right|^2 \tag{10.143}$$

Thus, the total radiation caused by transverse acceleration is larger than that caused by longitudinal acceleration by the factor γ^2. Lest there be any confusion, it should be pointed out that in the last section we found that for acceleration parallel to the direction of motion the angular intensity at the peak of the distribution (in the forward direction) is proportional to γ^2, whereas we see in (10.142) that the total radiation is independent of γ. These results are reconciled by recognizing that the cone angle filled by the radiation in the forward direction decreases in inverse proportion to γ.

EXERCISE 10.9

A nonrelativistic particle of charge q, mass m, and kinetic energy $\mathcal{E}$ makes a head-on collision with a fixed central force described by the potential energy $V(r)$, where

$$V(r) > \mathcal{E} \qquad \text{for } r < r_0 \tag{10.144}$$

and

$$V(r) < \mathcal{E} \qquad \text{for } r > r_0 \tag{10.145}$$

Show that the total energy radiated by the particle is given by the formula

$$\mathcal{W} = \frac{q^2}{3\pi\varepsilon_0 m^2 c^3}\sqrt{\frac{1}{2}m}\int_{r_0}^{\infty}\left|\frac{dV}{dr}\right|^2\frac{dr}{\sqrt{V(r_0) - V(r)}} \tag{10.146}$$

EXERCISE 10.10

Consider a particle of charge q scattering off a stationary charge q', which we regard as a fixed Coulomb potential

$$\mathbf{E} = \frac{q'\mathbf{r}}{4\pi\varepsilon_0 r^3} \tag{10.147}$$

where $\mathbf{r}$ is the vector distance between the particles. In the collision, the incident particle emits Bremsstrahlung due to the acceleration it experiences. For an impact parameter b large enough that the potential energy is small compared to the kinetic energy, the trajectory is nearly a straight line with constant velocity, $\mathbf{v} \approx \boldsymbol{\beta}c$, and the acceleration is a perturbation from this. Using the relativistic Larmor formula (10.140), calculate the total energy $\mathcal{W}$ emitted in the collision and show that

$$\mathcal{W} = \frac{\gamma^2 q^4 q'^2}{192\pi^2 \varepsilon_0^3 m^2 c^4 b^3 \beta}\left(1 - \frac{\beta^2}{4}\right) \tag{10.148}$$

EXERCISE 10.11

For a particle with charge q moving at a relativistic velocity in a circular orbit perpendicular to a magnetic field B, show that the momentum radiated by the particle is given by

$$\frac{d\mathbf{p}}{dt} = \frac{(\gamma^2 - 1)q^4 B^2}{6\pi \varepsilon_0 m^2 c^3}\mathbf{v} \tag{10.149}$$

10.2 BREMSSTRAHLUNG AND TRANSITION RADIATION

When high-speed particles interact with matter, they lose energy by a variety of processes. These include ionization and excitation of the medium on the one hand and creation of radiation on the other. We discuss some of these processes as examples of the method of virtual quanta in Chapter 5, and we discuss ionization and excitation by relativistic particles in Chapter 7. We limit our attention in this section to a more general discussion of radiative processes. Specifically, we consider Bremsstrahlung, which is the radiation created by charged particles as they pass through matter and collide with its constituents, and transition radiation, which is created as the particles enter and exit from the material.

10.2.1 Bremsstrahlung

When a charged particle scatters off another particle, the acceleration of the incident and target particles causes them to radiate what is called Bremsstrahlung. Previously we described the Bremsstrahlung from ultrarelativistic particles as Thomson scattering of virtual quanta. However, this process can be described more generally using the formulas developed earlier in this chapter.

We consider first the case when the collision takes place in a time short enough that the factor in the exponent in (10.104) or (10.107) doesn't change. That is, we assume that $\omega\tau \ll 1$, where ω is the frequency of the radiation and τ is the duration of the collision. This is called a "hard" collision. For interactions with a finite range (which excludes simple Coulomb interactions), this approximation always applies to radiation at sufficiently low frequency.

To calculate the angular spectral fluence, we start with (10.104), multiply and divide by $(1 - \hat{\mathbf{n}} \cdot \boldsymbol{\beta})$ to form a perfect differential with the exponential, and integrate by parts to get

$$\frac{d^2\mathcal{W}}{d\omega \, d\Omega} = \frac{\mu_0 c q^2}{16\pi^3} \left| \left(\frac{\hat{\mathbf{n}} \times \boldsymbol{\beta}}{1 - \hat{\mathbf{n}} \cdot \boldsymbol{\beta}} e^{i\omega(t' - \hat{\mathbf{n}} \cdot \mathbf{r}/c)} \right) \Big|_{-\infty}^{\infty} - \int_{-\infty}^{\infty} dt' \, e^{i\omega(t' - \hat{\mathbf{n}} \cdot \mathbf{r}/c)} \frac{d}{dt'} \left(\frac{\hat{\mathbf{n}} \times \boldsymbol{\beta}}{1 - \hat{\mathbf{n}} \cdot \boldsymbol{\beta}} \right) \right|^2$$

(10.150)

But far from the collision region the velocities are constant and the integrand of (10.104) oscillates, so the integral has meaning only if we introduce a damping factor in the exponent. If we do this, the first term in (10.150) vanishes at $\pm\infty$ and we are left with

$$\frac{d^2\mathcal{W}}{d\omega \, d\Omega} = \frac{\mu_0 c q^2}{16\pi^3} \left| \int_{-\infty}^{\infty} dt' \, e^{i\omega(t' - \hat{\mathbf{n}} \cdot \mathbf{r}/c)} \frac{d}{dt'} \left(\frac{\hat{\mathbf{n}} \times \boldsymbol{\beta}}{1 - \hat{\mathbf{n}} \cdot \boldsymbol{\beta}} \right) \right|^2$$

(10.151)

But the derivative of the velocity vanishes outside the collision zone, and if $\omega\tau \ll 1$ the exponential is a constant during the collision, so we can move it outside the integral. Since it merely introduces a phase phase factor, we can ignore the exponential and carry out the integral trivially to obtain

$$\frac{d^2\mathcal{W}}{d\omega \, d\Omega} = \frac{\mu_0 c q^2}{16\pi^3} \left| \frac{\hat{\mathbf{n}} \times \boldsymbol{\beta}'}{1 - \hat{\mathbf{n}} \cdot \boldsymbol{\beta}'} - \frac{\hat{\mathbf{n}} \times \boldsymbol{\beta}}{1 - \hat{\mathbf{n}} \cdot \boldsymbol{\beta}} \right|^2$$

(10.152)

where the primed quantities refer to the velocity after the collision and the unprimed quantities refer to the velocity before the collision, as shown in Figure 10.6. From this expression, we see that the Bremsstrahlung radiated at low frequencies depends only on the initial and final velocities. In the ultrarelativistic limit, the denominators assure that the radiation is strongly peaked around the initial and final directions of the particle motion. In the non-relativistic limit, we can ignore the denominators and get the simple relation

$$\frac{d^2\mathcal{W}}{d\omega \, d\Omega} = \frac{\mu_0 c q^2}{16\pi^3} |\hat{\mathbf{n}} \times (\boldsymbol{\beta}' - \boldsymbol{\beta})|^2$$

(10.153)

That is, the angular spectral fluence of low-frequency Bremsstrahlung depends only on the momentum transfer in the collision, and it has an angular distribution similar to that of radiation from a dipole oriented in the direction of the momentum change $(\boldsymbol{\beta}' - \boldsymbol{\beta})$.

Coulomb collisions represent the opposite limit from "hard" collisions, since they are, in some sense, infinitely soft. That is, the interaction falls off only like $1/r^2$ and has no characteristic length to define a collision time other than the impact parameter itself. To see the effect of this softness on the spectrum of Bremsstrahlung we consider the case when a

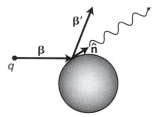

Figure 10.6 Bremsstrahlung emitted in a "hard" collision.

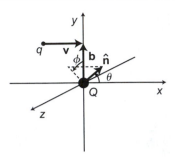

Figure 10.7 Light particle incident on a heavy particle.

light particle is incident on a heavy particle, as shown in Figure 10.7, so that the target particle is fixed. We further assume that the trajectory of the incident particle is almost a straight line at the constant velocity $\mathbf{v} = c\boldsymbol{\beta}$. This will be true provided that the potential energy of the interaction is small compared with the incident kinetic energy, that is, if

$$\frac{qQ}{4\pi\varepsilon_0 mc^2 b} \ll \gamma - 1 \tag{10.154}$$

where m is the mass and q the charge of the incident particle, Q the charge of the target particle, and b the impact parameter. The angular spectral fluence is found by using (10.107). For a trajectory that is nearly a straight line at a constant velocity, we see that

$$\omega(t - \hat{\mathbf{n}} \cdot \mathbf{r}/c) \approx \omega(1 - \hat{\mathbf{n}} \cdot \boldsymbol{\beta})t \tag{10.155}$$

where $(1 - \hat{\mathbf{n}} \cdot \boldsymbol{\beta})$ is a constant. The spectral angular fluence is then given by the Fourier transform of the acceleration,

$$\frac{d^2\mathcal{W}}{d\omega\,d\Omega} = \frac{\mu_0 c q^2}{16\pi^3} \left| \frac{\hat{\mathbf{n}}}{(1 - \hat{\mathbf{n}} \cdot \boldsymbol{\beta})^2} \times \left[(\hat{\mathbf{n}} - \boldsymbol{\beta}) \times \int_{-\infty}^{\infty} dt\, e^{i\omega(1 - \hat{\mathbf{n}} \cdot \boldsymbol{\beta})t} \dot{\boldsymbol{\beta}} \right] \right|^2 \tag{10.156}$$

The acceleration of the incident particle is caused by the Coulomb field of the target particle. Along the straight-line trajectory of the incident particle, the acceleration is therefore

$$\frac{d\mathbf{p}}{dt} = \frac{d}{dt}(\gamma\boldsymbol{\beta}mc) = \frac{qQ}{4\pi\varepsilon_0} \frac{\mathbf{b} + \boldsymbol{\beta}ct}{(b^2 + \beta^2 c^2 t^2)^{3/2}} \tag{10.157}$$

where $\mathbf{b}$ is the vector impact parameter shown in Figure 10.7. But $\gamma = 1/\sqrt{1 - \beta^2}$, so

$$\frac{d\mathbf{p}}{dt} = mc\gamma[\dot{\boldsymbol{\beta}} + \gamma^2\boldsymbol{\beta}(\boldsymbol{\beta} \cdot \dot{\boldsymbol{\beta}})] \tag{10.158}$$

and after a little algebra we find that

$$\dot{\boldsymbol{\beta}} = \frac{1}{\gamma mc} \frac{qQ}{4\pi\varepsilon_0 (b^2 + \beta^2 c^2 t^2)^{3/2}} \left(\mathbf{b} + \frac{\boldsymbol{\beta}ct}{1 + \beta^2\gamma^2} \right) \tag{10.159}$$

Substituting this back into (10.156), we get

$$\frac{d^2\mathcal{W}}{d\omega\,d\Omega} = \frac{\mu_0 c^3 Q^2 r_c^2}{16\pi^3 \gamma^2} \left| \frac{\hat{\mathbf{n}} \times [(\hat{\mathbf{n}} - \boldsymbol{\beta}) \times \mathbf{b}]}{(1 - \hat{\mathbf{n}} \cdot \boldsymbol{\beta})^2} \int_{-\infty}^{\infty} dt\, \frac{e^{i\omega(1 - \hat{\mathbf{n}} \cdot \boldsymbol{\beta})t}}{(b^2 + \beta^2 c^2 t^2)^{3/2}} \right.$$

$$\left. + \frac{\hat{\mathbf{n}} \times (\hat{\mathbf{n}} \times \boldsymbol{\beta}c)}{(1 + \beta^2\gamma^2)(1 - \hat{\mathbf{n}} \cdot \boldsymbol{\beta})^2} \int_{-\infty}^{\infty} t\, dt\, \frac{e^{i\omega(1 - \hat{\mathbf{n}} \cdot \boldsymbol{\beta})t}}{(b^2 + \beta^2 c^2 t^2)^{3/2}} \right|^2 \tag{10.160}$$

where

$$r_c = \frac{q^2}{4\pi\varepsilon_0 mc^2}$$

(10.161)

is the classical radius of a particle of charge q and mass m. For an electron, the classical radius is $r_c = 2.82 \times 10^{-15}$ m. The integrals may be evaluated in terms of modified Bessel functions using the formulas

$$\int_{-\infty}^{\infty} \frac{e^{i\omega t}\,dt}{(\tau^2 + t^2)^{3/2}} = 2\frac{\omega}{\tau}K_1(\omega\tau)$$

(10.162)

and

$$\int_{-\infty}^{\infty} \frac{e^{i\omega t}t\,dt}{(\tau^2 + t^2)^{3/2}} = i2\omega K_0(\omega\tau)$$

(10.163)

Since the first integral is real and the second is imaginary, the square of the absolute value of the sum in (10.160) is the sum of the squares of the absolute values of the individual terms and we get

$$\frac{d^2W}{d\omega\,d\Omega} = \frac{\mu_0 c Q^2 r_c^2}{4\pi^3\beta^2\gamma^2 b^2(1 - \hat{\mathbf{n}}\cdot\boldsymbol{\beta})^4}$$

$$\times \left\{ |\hat{\mathbf{n}} \times [(\hat{\mathbf{n}} - \boldsymbol{\beta}) \times \hat{\mathbf{b}}]|^2 \omega^2\tau^2 K_1^2(\omega\tau) + \frac{|\hat{\mathbf{n}} \times (\hat{\mathbf{n}} \times \hat{\boldsymbol{\beta}})|^2}{(1 + \beta^2\gamma^2)^2}\omega^2\tau^2 K_0^2(\omega\tau) \right\}$$

(10.164)

where $\hat{\mathbf{b}}$ and $\hat{\boldsymbol{\beta}}$ are unit vectors and

$$\tau = \frac{b}{\beta c}(1 - \hat{\mathbf{n}}\cdot\boldsymbol{\beta})$$

(10.165)

As shown in Figure 10.8, the Bremsstrahlung spectrum falls off exponentially for $\omega\tau \gg 1$ due to the Bessel functions.

In the nonrelativistic limit ($\beta \ll 1$), (10.164) simplifies to

$$\frac{d^2W}{d\omega\,d\Omega} = \frac{\mu_0 c Q^2 r_c^2}{4\pi^3\beta^2\gamma^2 b^2}\left[\omega^2\tau^2 K_1^2(\omega\tau)\sin^2\alpha + \omega^2\tau^2 K_0^2(\omega\tau)\sin^2\theta\right]$$

(10.166)

where α is the angle between $\hat{\mathbf{n}}$ and $\hat{\mathbf{b}}$, and θ the angle between $\hat{\mathbf{n}}$ and $\hat{\boldsymbol{\beta}}$. From this we see that in the nonrelativistic limit the Bremsstrahlung has the form of radiation from two

Figure 10.8 Bremsstrahlung spectrum.

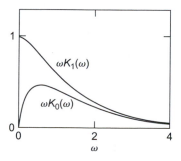

dipoles, one oriented along the direction of motion $\hat{\boldsymbol{\beta}}$ and the other oriented along the direction of the impact parameter $\hat{\mathbf{b}}$.

In the general relativistic case, we can express the vector products in terms of the polar angles of $\hat{\mathbf{n}}$ relative to $\hat{\boldsymbol{\beta}}$ and $\hat{\mathbf{b}}$, as defined in Figure 10.7. After some tedious algebra we find that

$$\frac{d^2\mathcal{W}}{d\omega\,d\Omega} = \frac{\mu_0 c Q^2 r_c^2}{4\pi^3 \beta^2 \gamma^2 b^2 (1 - \beta\cos\theta)^4} \left\{ \frac{\beta^2 \sin^2\theta}{(1+\beta^2\gamma^2)^2}\omega^2\tau^2 K_0^2(\omega\tau) \right.$$

$$+ [(\cos\theta - \beta)^2(\cos^2\theta + \sin^2\theta\cos^2\phi) + \sin^4\theta\sin^2\phi$$

$$\left. + 2(\cos\theta - \beta)\cos\theta\sin^2\theta\sin^2\phi]\omega^2\tau^2 K_1^2(\omega\tau) \right\} \tag{10.167}$$

In the ultrarelativistic limit ($\gamma \gg 1$) the radiation is confined to the forward direction ($\theta \ll 1$), so we may use the approximations

$$\beta \approx 1 - \frac{1}{2\gamma^2} \tag{10.168}$$

$$\sin\theta \approx \theta \tag{10.169}$$

$$\cos\theta \approx 1 - \frac{\theta^2}{2} \tag{10.170}$$

If we use these approximations in (10.167) and keep terms up to the lowest surviving order in $1/\gamma$ and $\theta = O(1/\gamma)$, we get

$$\frac{d^2\mathcal{W}}{d\omega\,d\Omega} = \frac{Q^2 r_c^2 \gamma^2}{\pi^3 \varepsilon_0 c b^2} \frac{1 + \gamma^4\theta^4 + 4\gamma^2\theta^2\left(\sin^2\phi - \frac{1}{2}\right)}{(1+\gamma^2\theta^2)^4}\omega^2\tau^2 K_1^2(\omega\tau) \tag{10.171}$$

In an actual experiment it is impossible to control the impact parameter $\mathbf{b}$ on an atomic scale, so it is appropriate to integrate (10.171) over all $\mathbf{b}$. The result is

$$\int_0^\infty b\,db \int_0^{2\pi} d\phi \frac{d^2\mathcal{W}}{d\omega\,d\Omega} = \frac{2Q^2 r_c^2 \gamma^2}{\pi^2\varepsilon_0 c}\frac{1 + \gamma^4\theta^4}{(1+\gamma^2\theta^2)^4}\int_0^\infty \frac{db}{b}\omega^2\tau^2 K_1^2(\omega\tau) \tag{10.172}$$

As we see in Figure 10.8, the Bessel function factor $\omega\tau K_1(\omega\tau)$ is on the order of unity until it cuts off near $\omega\tau = O(1)$, so from (10.165) we see that we can use for the upper limit of the integral the value

$$b_{\max} = \frac{\beta c}{\omega(1 - \beta\cos\theta)} \approx \frac{2\gamma^2 c}{\omega(1 + \gamma^2\theta^2)} \tag{10.173}$$

At the lower limit the integral has a logarithmic divergence. As discussed in Chapter 5, where Bremsstrahlung is described by the method of virtual quanta, the lower limit is the larger of

$$b_{\min} = \frac{Q}{q}r_c \tag{10.174}$$

which is set by the fact that for smaller impact parameters the deflection of the light particle in its incident rest frame becomes relativistic, and

$$b_{\min} = \frac{\hbar}{\gamma mc} \tag{10.175}$$

which is set by the fact that it is quantum mechanically meaningless to distinguish angular momenta finer than $\hbar$. The first restriction is more important at high energy, and the second is more important at low energy. Since the end points of the integral enter only logarithmically, their exact value is not too critical. The integral over b is then

$$\int \frac{db}{b} \omega^2 \tau^2 K_1^2(\omega\tau) \approx \ln\left(\frac{b_{max}}{b_{min}}\right) \tag{10.176}$$

and the cross section for Bremsstrahlung production is

$$\int_0^\infty b\, db \int_0^{2\pi} d\phi \frac{d^2\mathcal{W}}{d\omega\, d\Omega} = \frac{2Q^2 r_c^2 \gamma^2}{\pi^2 \varepsilon_0 c} \frac{1+\gamma^4\theta^4}{(1+\gamma^2\theta^2)^4} \ln\left(\frac{b_{max}}{b_{min}}\right) \tag{10.177}$$

This is the same result we obtained using the method of virtual quanta in the exercises in Chapter 5.

Of course, the classical description of Bremsstrahlung is valid only so long as the kinetic energy of the particle is large compared with the photon energy, that is, for

$$\frac{(\gamma-1)mc^2}{\hbar\omega} \gg 1 \tag{10.178}$$

But as discussed earlier, the highest frequency of importance is

$$\omega = O\left(\frac{\beta c}{b}\right) \tag{10.179}$$

Above this frequency the Bessel functions vanish, so the spectral intensity cuts off. Combining (10.178) and (10.179), we see that the classical description is valid in the ultrarelativistic limit for the intense, low-frequency portion of the spectrum provided that

$$\frac{(\gamma-1)mcb}{\beta\hbar} \approx \frac{\gamma mcb}{\hbar} \gg 1 \tag{10.180}$$

But this is just the condition that the incident angular momentum be large compared with $\hbar$, so (10.175) is always satisfied in high-energy collisions and the lower limit of the integral is set by (10.174). In all cases the high-frequency tail of the spectrum is altered by quantum effects.

EXERCISE 10.12

A nonrelativistic particle of charge q and velocity βc collides elastically with a massive smooth, hard sphere of radius R. If $b = R\sin\theta$ is the impact parameter, where θ is the angle between the axis of incidence and the point of impact, the incremental cross section for producing photons in the $\hat{n}$ direction at frequency ω is the number of photons produced times the increment of area normal to the incident velocity:

$$d^4\sigma = \frac{1}{\hbar\omega}\frac{d^2\mathcal{W}}{d\Omega\, d\omega} d\Omega\, d\omega b\, db\, d\phi = \frac{R^2}{\hbar\omega}\frac{d^2\mathcal{W}}{d\Omega\, d\omega}\sin\theta\cos\theta\, d\theta\, d\phi\, d\Omega\, d\omega \tag{10.181}$$

where ϕ is the azimuthal angle of the point of impact. Show that when summed over all impact parameters and angles, the classical differential cross section for the emission of photons of energy $\mathcal{E} = \hbar\omega$ in the $\hat{\mathbf{n}}$ direction is

$$\frac{d^2\sigma}{d\Omega\,d\mathcal{E}} = \frac{\mu_0 c q^2 \beta^2 R^2}{48\pi^2 \hbar \mathcal{E}} (2 + 3\sin^2\alpha) \tag{10.182}$$

where α is the angle between the incident velocity $\boldsymbol{\beta}$ and the photon direction $\hat{\mathbf{n}}$. *Hint:* Use the identity

$$(\mathbf{a} \times \mathbf{b}) \cdot (\mathbf{c} \times \mathbf{d}) = (\mathbf{a} \cdot \mathbf{c})(\mathbf{b} \cdot \mathbf{d}) - (\mathbf{a} \cdot \mathbf{d})(\mathbf{b} \cdot \mathbf{c}) \tag{10.183}$$

to simplify the algebra.

10.2.2 Transition Radiation

In Chapter 5 we saw that when a relativistic charged particle crosses the interface between two regions with different indices of refraction, it emits transition radiation. In Chapter 5 this radiation was viewed as the reflection at the interface of the "virtual quanta" surrounding the incident charged particle. In the present section we derive a simple description of this phenomenon using the method of images together with the formulas developed earlier in this chapter. This approach has the advantage that it is valid in the nonrelativistic limit as well as the untrarelativistic limit, and it can be applied to the case of a particle emerging from a more dense medium into a less dense one. On the other hand, the method is restricted to perfect conductors, although the results can be applied qualitatively to dielectrics.

When a charged particle is located outside a plane conducting surface, the electric field in the vacuum region can be described by replacing the conductor with an image charge of equal and opposite charge at the mirror-image position, as shown in Figure 10.9. As the charged particle approaches the surface the image charge mirrors its motion, and when the particle reaches the surface the charge and its image cancel out. The radiation that is emitted upon the abrupt disappearance of the charge and its image is called transition radiation.

To calculate the angular spectral fluence of transition radiation we begin with (10.104), recognizing that the field in the integrand is the sum of the fields of the two charges. The angular spectral fluence in the $\hat{\mathbf{n}}$ direction is therefore

$$\frac{d^2\mathcal{W}}{d\omega\,d\Omega} = \frac{q^2\omega^2}{16\pi^3\varepsilon_0 c} \left| \int_{-\infty}^{0} dt\, e^{i\omega(t - \hat{\mathbf{n}}\cdot\mathbf{r}/c)} \hat{\mathbf{n}} \times \boldsymbol{\beta} - \int_{-\infty}^{0} dt\, e^{i\omega(t - \hat{\mathbf{n}}\cdot\mathbf{r}'/c)} \hat{\mathbf{n}} \times \boldsymbol{\beta}' \right|^2 \tag{10.184}$$

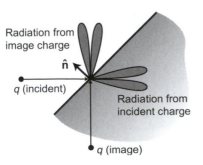

Radiation from image charge

$\hat{\mathbf{n}}$

q (incident)

Radiation from incident charge

q (image)

Figure 10.9 Charge and image charge for transition radiation.

where the primes refer to the image charge, and the particle reaches the surface at $t = 0$. Since the velocity vectors are constant, the second term in the exponent is $\hat{\mathbf{n}} \cdot \mathbf{r}/c = \hat{\mathbf{n}} \cdot \boldsymbol{\beta} t$, and the integrals are easily calculated, keeping in mind that the limits at $-\infty$ must be evaluated with a decaying factor in the exponent. The result is

$$\frac{d^2 \mathcal{W}}{d\omega \, d\Omega} = \frac{q^2}{16\pi^3 \varepsilon_0 c} \left| \frac{\hat{\mathbf{n}} \times \boldsymbol{\beta}}{1 - \hat{\mathbf{n}} \cdot \boldsymbol{\beta}} - \frac{\hat{\mathbf{n}} \times \boldsymbol{\beta}'}{1 - \hat{\mathbf{n}} \cdot \boldsymbol{\beta}'} \right|^2 \tag{10.185}$$

The predicted fluence is independent of the frequency. Recalling the picture of transition radiation as reflected virtual photons, we recognize that this result can be true only at frequencies where the conductor is a good reflector. For most metals, this approximation is good at microwave and infrared frequencies but fails in the ultraviolet and beyond.

In the nonrelativistic limit, we can ignore the denominators in (10.185). Then, since $(\boldsymbol{\beta}' - \boldsymbol{\beta}) = 2(\hat{\mathbf{N}} \cdot \boldsymbol{\beta})\hat{\mathbf{N}}$, the fluence is

$$\frac{d^2 \mathcal{W}}{d\omega \, d\Omega} = \frac{q^2}{16\pi^3 \varepsilon_0 c} |\hat{\mathbf{n}} \times (\boldsymbol{\beta}' - \boldsymbol{\beta})|^2 = \frac{q^2}{4\pi^3 \varepsilon_0 c} (\hat{\mathbf{N}} \cdot \boldsymbol{\beta})^2 |\hat{\mathbf{n}} \times \hat{\mathbf{N}}|^2 \tag{10.186}$$

where $\hat{\mathbf{N}}$ is a unit vector normal to the surface, as shown in Figure 10.9. This has the form of radiation from a dipole that is oriented normal to the surface of the conductor at the point where the charged particle enters the conductor. The intensity is proportional to the square of the normal component of the velocity, and the radiation pattern is independent of the component of the particle velocity parallel to the interface.

In the ultrarelativistic limit, the denominators become very small in the forward direction and the radiation peaks there. In this case the incident particle radiates into the conductor, and only the image charge radiates significantly into the vacuum region. The fluence is therefore

$$\frac{d^2 \mathcal{W}}{d\omega \, d\Omega} = \frac{q^2}{16\pi^3 \varepsilon_0 c} \left| \frac{\hat{\mathbf{n}} \times \boldsymbol{\beta}'}{1 - \hat{\mathbf{n}} \cdot \boldsymbol{\beta}'} \right|^2 \tag{10.187}$$

Expanding for small angles θ from the direction of the image velocity, which is also the direction of reflection for incident virtual photons, we find that in the ultrarelativistic limit

$$1 - \hat{\mathbf{n}} \cdot \boldsymbol{\beta}' \approx \frac{1}{2\gamma^2}(1 + \gamma^2 \theta^2) \tag{10.188}$$

so

$$\frac{d^2 \mathcal{W}}{d\omega \, d\Omega} = \frac{\gamma^4 q^2}{4\pi^3 \varepsilon_0 c} \frac{\theta^2}{(1 + \gamma^2 \theta^2)^2} \tag{10.189}$$

This has the form of lobes about the direction of $\boldsymbol{\beta}'$, as shown in Figure 10.10.

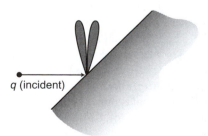

Figure 10.10 Pattern of transition radiation for a particle incident on a surface.

q (incident)

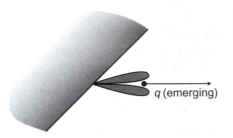

Figure 10.11 Pattern of transition radiation for a particle emerging from a surface.

q (emerging)

When the particle emerges from the conductor, as shown in Figure 10.11, the same approach may be used. In the nonrelativistic limit, the radiation again has the form of dipole radiation oriented about the normal to the interface. In the ultrarelativistic limit, the radiation from the image charge is directed into the conductor and the transition radiation is concentrated around the direction of the emerging charged particle, as shown.

EXERCISE 10.13

Show that in the nonrelativistic limit, the total spectral intensity of transition radiation is

$$\frac{dW}{d\omega} \approx \frac{q^2}{3\pi^2 \varepsilon_0 c}(\hat{\mathbf{N}} \cdot \boldsymbol{\beta})^2 \tag{10.190}$$

while in the ultrarelativistic limit, the total spectral intensity is

$$\frac{dW}{d\omega} \approx \frac{q^2}{2\pi^2 \varepsilon_0 c} \ln \gamma \tag{10.191}$$

Hint: In the ultrarelativistic case use the identity

$$(\mathbf{a} \times \mathbf{b}) \cdot (\mathbf{c} \times \mathbf{d}) = (\mathbf{a} \cdot \mathbf{c})(\mathbf{b} \cdot \mathbf{d}) - (\mathbf{a} \cdot \mathbf{d})(\mathbf{b} \cdot \mathbf{c}) \tag{10.192}$$

to simplify the expression for the angular spectral fluence.

10.3 THOMSON SCATTERING

As another application of the formulas derived in the first section, we consider the scattering of electromagnetic waves by a particle (usually an electron or positron) with charge *q* and mass *m*. This is called Thomson scattering. Of course, an electron has no extent, so far as we know, so the radiation is not literally reflected from the surface or scattered from the volume of the particle. Rather, the scattered radiation is the radiation emitted by the electron as it oscillates in the field of the incident wave. The energy of the scattered radiation comes, of course, from the incident radiation, which is depleted in the region behind the electron. The "shadow" formed behind the electron is created by the destructive interference of the emitted radiation and the incident radiation. Provided that the intensity of the incident light is low enough to keep the motions of the particle nonrelativistic, as discussed

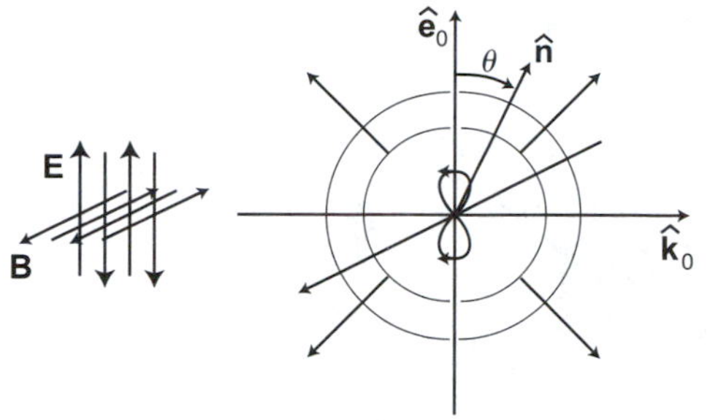

Figure 10.12 Thomson scattering.

in Chapter 4, the radiation emitted ("scattered") by the particle is linear in the incident intensity and the scattering can be represented by an effective cross section. At sufficiently high intensity, however, the motions become relativistic, the radiation process becomes nonlinear, and harmonics of the original frequency appear in the scattered radiation along with the fundamental.

10.3.1 Linear Thomson Scattering

The geometry of Thomson scattering is illustrated in Figure 10.12. As shown there, the particle experiences an oscillating electric field

$$\mathbf{E}(t) = E(t)\hat{\mathbf{e}}_0 \tag{10.193}$$

where $\hat{\mathbf{e}}_0$ is the direction of polarization, and a corresponding magnetic field $\mathbf{B} = (\hat{\mathbf{k}}_0 \times \mathbf{E})/c$, in which $\hat{\mathbf{k}}_0$ is a unit vector in the direction of propagation of the wave. The intensity of the incident radiation is given by the Poynting vector

$$\mathbf{S} = \frac{1}{\mu_0}\mathbf{E} \times \mathbf{B} = \varepsilon_0 c E^2 \hat{\mathbf{k}}_0 \tag{10.194}$$

For sufficiently small fields (this condition can be violated by strongly focused laser beams incident on electrons, for example), the particle motion is nonrelativistic and we can ignore the effect of the magnetic field compared with that of the electric field. In this case, the particle acceleration is simply

$$\dot{\boldsymbol{\beta}} = \frac{q}{mc}\mathbf{E} \tag{10.195}$$

which is parallel to the velocity. When it is accelerated, the particle radiates with an instantaneous angular power given by (10.122), which in the nonrelativistic limit becomes

$$\frac{d\mathcal{P}}{d\Omega} = \frac{q^2}{16\pi^2\varepsilon_0 c}|\hat{\mathbf{n}} \times [\hat{\mathbf{n}} \times \dot{\boldsymbol{\beta}}]|^2 = \frac{q^4 E^2}{16\pi^2\varepsilon_0 m^2 c^3}\sin^2\theta \tag{10.196}$$

where θ is the angle between the incident polarization and the direction of scattering. The polarization is in the plane containing $\hat{\mathbf{n}}$ and $\hat{\mathbf{e}}_0$, and the distribution of the radiation is that of a dipole oriented in the direction of polarization of the incident wave. In terms of the incident intensity S, the angular power radiated by the particle may be expressed in the form

$$\frac{d\mathcal{P}}{d\Omega} = \frac{d\sigma_T}{d\Omega}S \tag{10.197}$$

where the differential scattering cross section is

$$\frac{d\sigma_T}{d\Omega} = \frac{q^4 \sin^2 \theta}{16\pi^2 \varepsilon_0^2 m^2 c^4} = r_c^2 \sin^2 \theta \tag{10.198}$$

and r_c is the classical radius of the particle. The total power scattered is found by integrating over all solid angles:

$$\mathcal{P} = \frac{q^4 S}{16\pi^2 \varepsilon_0^2 m^2 c^4} \int_0^\pi 2\pi \, \sin^3 \theta \, d\theta = \sigma_T S \tag{10.199}$$

where the Thomson cross section is

$$\sigma_T = \frac{8\pi}{3} r_c^2 = \frac{q^4}{6\pi \varepsilon_0^2 m^2 c^4} \tag{10.200}$$

For electrons this has the value $\sigma_T = 6.65 \times 10^{-29}$ m^2, independent of the frequency of the incident radiation.

Of course, this result is purely classical and ignores quantum effects. The classical approximation is valid so long as the photon energy is small compared with the rest energy of the particle, $\hbar\omega/mc^2 \ll 1$, or equivalently,

$$\lambda = \frac{2\pi}{k} \gg \lambda_C = \frac{h}{mc} \tag{10.201}$$

The quantity λ_C ($\lambda_C = 2.43 \times 10^{-12}$ m for an electron) is called the Compton wavelength. When quantum effects are included, the quantized energy and momentum of the incident photons lead to the Compton effect, in which the electron recoils from the impact of the photon and the wavelength of the photon is shifted to a longer value given by the Compton formula

$$\lambda' = \lambda + \lambda_C (1 - \cos \psi) \tag{10.202}$$

where ψ is the angle between the incident and scattered wave vectors. When Compton recoil is important, the cross section is reduced to the value

$$\frac{d\sigma}{d\Omega} = r_c^2 \frac{k'^2}{k^2} \sin^2 \psi \tag{10.203}$$

for spinless particles, where k and k' are the incident and scattered wave numbers. The factor k'^2/k^2 represents the reduction in the density of final states of the radiation field at smaller wave numbers, as discussed in Chapter 4. Additional corrections are introduced when the spin and magnetic moment of the electron are taken into account.

EXERCISE 10.14

Consider a charged particle moving at an arbitrary relativistic velocity $\beta c \hat{\mathbf{x}}$ through the electromagnetic wave

$$\mathbf{A} = \mathbf{A}_0 \sin(\omega t - \mathbf{k} \cdot \mathbf{r}) \tag{10.204}$$

where $\mathbf{k}$ lies in the x–y plane and $\mathbf{A}_0$ is in the $\hat{\mathbf{z}}$ direction. Due to Thomson scattering, the incident wave loses momentum. By conservation of momentum, this lost momentum appears in the average (over a complete cycle of the incident wave) motion of the charged particle.

(a) Show that the average momentum in the $\hat{\mathbf{x}}$ and $\hat{\mathbf{y}}$ directions gained by the electron is given by the formulas

$$\frac{dp^1}{dt} = \frac{\gamma^2 \sigma_T S}{c}(1 - \beta \cos\theta)(\cos\theta - \beta) \qquad (10.205)$$

$$\frac{dp^2}{dt} = \frac{\sigma_T S}{c}(1 - \beta \cos\theta)\sin\theta \qquad (10.206)$$

respectively, where S is the intensity of the electromagnetic wave, θ is the angle between the wave vector and the particle velocity, and σ_T is the Thompson cross section.

(b) For a time-integrated laser fluence $\mathcal{J} \approx 10^{10}$ J/m^2, estimate the final velocity $\beta = v/c$ of an electron initially at rest.

EXERCISE 10.15

Consider a nonrelativistic charged particle illuminated by a uniform plane wave whose vector potential is

$$\mathbf{A} = A_0 \hat{\mathbf{e}} e^{i(\mathbf{k}\cdot\mathbf{r} - \omega t)} \qquad (10.207)$$

where $\omega = kc$ and $\hat{\mathbf{e}} \cdot \mathbf{k} = 0$. Using the Lienard–Wiechert expression (10.41) for the radiation field of the charged particle, show that the incident and radiated fields interfere destructively in the "shadow" region behind the particle, that is, on the axis directly beyond the particle. It is sufficient to consider the far field in the nonrelativistic approximation.

10.3.2 Nonlinear Thomson Scattering

In an intense optical field, the transverse particle motions become relativistic and the interaction of the transverse particle motions with the magnetic induction of the optical wave induces longitudinal motions. As a result, the particle moves in a "figure-eight" orbit, as discussed in Chapter 4. These complex motions introduce harmonics into the spectrum of the radiation emitted (scattered) by the particle. Relativistic effects also cause a particle that is at rest before the optical pulse arrives to be accelerated, on average, in the direction of the optical wave, as discussed in Chapter 4. This leads to a Doppler shift of the emitted radiation. To keep matters simple in the following, we consider the case of a continuous incident optical field and do the calculations in the reference frame in which the particle remains, on average, at rest. The Doppler shift caused by the average longitudinal motion of the particle can be accounted for afterward. In a later section where we discuss undulator radiation, the effect of the longitudinal acceleration of the particles is explicitly taken into account in the Lorentz transformation from the electron (average motion) rest frame to the laboratory frame.

The angular spectrum of the radiation emitted by a charged particle in an intense optical field is found by using (10.104). We consider a harmonic wave incident on the charged particle with the frequency $\omega_0 = k_0 c$ in the direction of the unit vector $\hat{\mathbf{k}}_0$. The wave 4-vector is then

$$k_0^\alpha = (k_0, \mathbf{k}_0) \tag{10.208}$$

and the vector potential is

$$\mathbf{A}(\xi) = \sqrt{2}\,\frac{mc}{q}a_0\hat{\mathbf{e}}_0 \sin\!\left(k_0^\alpha r_\alpha\right) = \sqrt{2}\,\frac{mc}{q}a_0\hat{\mathbf{e}}_0 \sin(k_0\xi) \tag{10.209}$$

where a_0 is the rms dimensionless vector potential of the incident wave, $\hat{\mathbf{e}}_0$ is the direction of polarization, and we define the timelike parameter

$$\xi = \frac{k_0^\alpha r_\alpha}{k_0} = ct - \hat{\mathbf{k}}_0 \cdot \mathbf{r} \tag{10.210}$$

In terms of the average incident intensity $\langle S \rangle$, the rms dimensionless vector potential is

$$a_0^2 = \frac{q^2}{\varepsilon_0 m^2 c^3}\frac{\langle S \rangle}{\omega_0^2} \tag{10.211}$$

Typically, $a_0 \ll 1$, although for a terawatt laser focused with f/10 optics, for example, a_0 is on the order of unity. As shown in Chapter 4, the motion of the particle is described by the equations

$$\hat{\mathbf{k}}_0 \cdot (\mathbf{r} - \langle \mathbf{r} \rangle) = \frac{q^2 c^2}{2\langle \mathcal{E} \rangle^2}\int [A^2(\xi) - \langle A^2 \rangle]\,d\xi = -\frac{1}{2}\frac{a_0^2}{1+a_0^2}\frac{\sin(2k_0\xi)}{2k_0} \tag{10.212}$$

$$\hat{\mathbf{e}}_0 \cdot (\mathbf{r} - \langle \mathbf{r} \rangle) = -\frac{qc}{\langle \mathcal{E} \rangle}\int A(\xi)\,d\xi = \sqrt{\frac{2a_0^2}{1+a_0^2}}\frac{\cos(k_0\xi)}{k_0} \tag{10.213}$$

and

$$ct = \xi + \hat{\mathbf{k}}_0 \cdot \mathbf{r} = \xi - \frac{1}{2}\frac{a_0^2}{1+a_0^2}\frac{\sin(2k_0\xi)}{2k_0} + \hat{\mathbf{k}}_0 \cdot \langle \mathbf{r} \rangle \tag{10.214}$$

The average energy is

$$\langle \mathcal{E} \rangle^2 = m^2 c^4 + q^2 c^2 \langle A^2 \rangle = m^2 c^4 \left(1 + a_0^2\right) \tag{10.215}$$

From (10.104) we see that the angular spectral fluence in the $\hat{\mathbf{n}}$ direction is

$$\frac{d^2\mathcal{W}}{d\omega\,d\Omega} = \frac{\mu_0\omega^2 q^2}{16\pi^3 c}\left|\int_{-\infty}^{\infty} c\,dt\,e^{ik(ct-\hat{\mathbf{n}}\cdot\mathbf{r})}\hat{\mathbf{n}} \times \boldsymbol{\beta}\right|^2 \tag{10.216}$$

To evaluate this, we observe that

$$c\,dt\,\hat{\mathbf{n}} \times \boldsymbol{\beta} = \hat{\mathbf{n}} \times d\mathbf{r} = \left[\hat{\mathbf{n}} \times \hat{\mathbf{k}}_0\!\left(\hat{\mathbf{k}}_0 \cdot \frac{d\mathbf{r}}{d\xi}\right) + \hat{\mathbf{n}} \times \hat{\mathbf{e}}_0\!\left(\hat{\mathbf{e}}_0 \cdot \frac{d\mathbf{r}}{d\xi}\right)\right]d\xi$$

$$= -\left[\sqrt{\frac{2a_0^2}{1+a_0^2}}\,\hat{\mathbf{n}} \times \hat{\mathbf{e}}_0 \sin(k_0\xi) + \frac{1}{2}\frac{a_0^2}{1+a_0^2}\hat{\mathbf{n}} \times \hat{\mathbf{k}}_0 \cos(2k_0\xi)\right]d\xi \tag{10.217}$$

and that

$$k(ct - \hat{\mathbf{n}} \cdot \mathbf{r}) = k[\xi + (1 - \hat{\mathbf{n}} \cdot \hat{\mathbf{k}}_0)\hat{\mathbf{k}}_0 \cdot \mathbf{r} - (\hat{\mathbf{n}} \cdot \hat{\mathbf{e}}_0)\hat{\mathbf{e}}_0 \cdot \mathbf{r} + (\hat{\mathbf{k}}_0 - \hat{\mathbf{n}}) \cdot \langle \mathbf{r} \rangle]$$

$$= k\xi - \frac{k}{k_0}\sqrt{\frac{2a_0^2}{1 + a_0^2}}\hat{\mathbf{n}} \cdot \hat{\mathbf{e}}_0 \cos(k_0\xi) - \frac{k}{4k_0}\frac{a_0^2}{1 + a_0^2}(1 - \hat{\mathbf{n}} \cdot \hat{\mathbf{k}}_0)\sin(2k_0\xi)$$

$$+ k(\hat{\mathbf{k}}_0 - \hat{\mathbf{n}}) \cdot \langle \mathbf{r} \rangle \tag{10.218}$$

Since the final constant affects only the overall phase, we ignore it in the following. To compute the integral (10.216), we expand the exponential using the formulas

$$e^{i\alpha \sin\theta} = \sum_{n=-\infty}^{\infty} J_n(\alpha)e^{in\theta} \tag{10.219}$$

and

$$e^{i\alpha \cos\theta} = \sum_{n=-\infty}^{\infty} i^n J_n(\alpha)e^{in\theta} \tag{10.220}$$

where

$$J_n(\alpha) = (-1)^n J_{-n}(\alpha) = (-1)^n J_n(-\alpha) \tag{10.221}$$

is a Bessel function of the first kind of integer order n. Clearly, the result of the integral (10.216) is a series of δ-functions centered at the harmonic frequencies

$$k = Nk_0, \qquad N = 1, 2, \ldots. \tag{10.222}$$

After some tedious algebra, we find that the integral is

$$\int_{-\infty}^{\infty} c\,dt\, e^{ik(ct - \hat{\mathbf{n}} \cdot \mathbf{r})}\hat{\mathbf{n}} \times \boldsymbol{\beta} = \sum_{N=1}^{\infty} \left(\sqrt{\frac{2a_0^2}{1 + a_0^2}} F_N \hat{\mathbf{n}} \times \hat{\mathbf{e}}_0 + \frac{1}{2}\frac{a_0^2}{1 + a_0^2} G_N \hat{\mathbf{n}} \times \hat{\mathbf{k}}_0 \right)\delta(k - Nk_0)$$

$$\tag{10.223}$$

where

$$F_N = i\pi \sum_{l=-\infty}^{\infty} (-1)^l J_{2l}(NC)\left[J_{\frac{N+1}{2}+l}(NS) - J_{\frac{N-1}{2}+l}(NS) \right], \qquad N = \text{odd} \tag{10.224}$$

$$F_N = \pi \sum_{l=-\infty}^{\infty} (-1)^l J_{2l+1}(NC)\left[J_{\frac{N}{2}+l+1}(NS) - J_{\frac{N}{2}+l}(NS) \right], \qquad N = \text{even} \tag{10.225}$$

and

$$G_N = i\pi \sum_{l=-\infty}^{\infty} (-1)^l J_{2l+1}(NC)\left[J_{\frac{N+1}{2}+l+1}(NS) + J_{\frac{N+1}{2}+l-1}(NS) \right], \qquad N = \text{odd} \tag{10.226}$$

$$G_N = -\pi \sum_{l=-\infty}^{\infty} (-1)^l J_{2l}(NC)\left[J_{\frac{N}{2}+l+1}(NS) + J_{\frac{N}{2}+l-1}(NS) \right], \qquad N = \text{even} \tag{10.227}$$

The arguments of the Bessel functions appearing in these expressions are

$$C = \sqrt{\frac{2a_0^2}{1 + a_0^2}} \, \hat{\mathbf{n}} \cdot \hat{\mathbf{e}}_0 \tag{10.228}$$

and

$$S = \frac{1}{4} \frac{a_0^2}{1 + a_0^2} (1 - \hat{\mathbf{n}} \cdot \hat{\mathbf{k}}_0) \tag{10.229}$$

When these expressions are substituted into (10.216), we see that the scattered radiation consists of a series of harmonics

$$\frac{d^2\mathcal{W}}{d\omega \, d\Omega} = \sum_{N=1}^{\infty} \frac{d^2\mathcal{W}_N}{d\omega \, d\Omega} \tag{10.230}$$

where the angular spectral fluence of the Nth harmonic is

$$\frac{d^2\mathcal{W}_N}{d\omega \, d\Omega} = \frac{\mu_0 N^2 \omega_0^2 q^2}{16\pi^3 c} \left| \sqrt{\frac{2a_0^2}{1 + a_0^2}} F_N \hat{\mathbf{n}} \times \hat{\mathbf{e}}_0 + \frac{1}{2} \frac{a_0^2}{1 + a_0^2} G_N \hat{\mathbf{n}} \times \hat{\mathbf{k}}_0 \right|^2 \delta^2(k - Nk_0) \tag{10.231}$$

The squares of δ-functions do not give a finite result when integrated over the frequency, but they may be understood in the following way. As the integral over time is extended from $-\infty$ to ∞, the spectrum narrows to a series of spikes of insignificant width and infinite height and the energy in each spike becomes infinite. To avoid the singularities, we consider an incident optical pulse of duration τ and ignore the transients associated with the sudden onset and termination of the incident optical pulse. These are unimportant for long pulses. The integral over time is now

$$\int_{-\infty}^{\infty} c \, dt \, e^{ik(ct - \hat{\mathbf{n}} \cdot \mathbf{r})} \hat{\mathbf{n}} \times \boldsymbol{\beta} = \left(\int_{-\infty}^{-\tau/2} c \, dt + \int_{-\tau/2}^{\tau/2} c \, dt + \int_{\tau/2}^{\infty} c \, dt \right) e^{ik(ct - \hat{\mathbf{n}} \cdot \mathbf{r})} \hat{\mathbf{n}} \times \boldsymbol{\beta} \tag{10.232}$$

For $t > \tau/2$, the velocity is $\boldsymbol{\beta} = \text{constant} \neq 0$, in general. To make the integral over $t > \tau/2$ converge, we introduce a decaying exponential factor to damp the oscillations at infinity and let the decay rate vanish in the usual way. This still leaves a contribution at $t = \tau/2$, but this is part of the initial transient and may be ignored for long pulses. The integral over $t < -\tau/2$ may be ignored for the same reasons. When the integral is evaluated for the interval $-\tau/2 < t < \tau/2$, we get the same result as before, except that the δ-functions are replaced by the factors

$$\delta(k - Nk_0) \rightarrow \frac{1}{2\pi} \int_{-c\tau/2}^{c\tau/2} d\xi \, e^{i(k - Nk_0)\xi} = \frac{c\tau}{2\pi} \frac{\sin\left[\frac{1}{2}(\omega - N\omega_0)\tau\right]}{\frac{1}{2}(\omega - N\omega_0)\tau} \tag{10.233}$$

The angular spectral fluence at the Nth harmonic is then

$$\frac{d^2\mathcal{W}_N}{d\omega \, d\Omega} = \frac{N^2 \omega_0^2 q^2 \tau^2}{64\pi^5 \varepsilon_0 c} \left| \sqrt{\frac{2a_0^2}{1 + a_0^2}} F_N \hat{\mathbf{n}} \times \hat{\mathbf{e}}_0 + \frac{1}{2} \frac{a_0^2}{1 + a_0^2} G_N \hat{\mathbf{n}} \times \hat{\mathbf{k}}_0 \right|^2 \left| \frac{\sin\left[\frac{1}{2}(\omega - N\omega_0)\tau\right]}{\frac{1}{2}(\omega - N\omega_0)\tau} \right|^2 \tag{10.234}$$

For long pulses $(\omega_0 \tau \gg 1)$, the spectrum is confined to narrow peaks around the harmonics. As the pulse length increases, the fluence at line center increases quadratically but the line width decreases linearly. If we integrate over all frequencies around the Nth harmonic, we find that the total angular fluence scattered into the Nth harmonic is

$$\frac{dW_N}{d\Omega} = \frac{N^2 \omega_0^2 q^2 \tau}{32\pi^4 \varepsilon_0 c} \left| \sqrt{\frac{2a_0^2}{1+a_0^2}} F_N \hat{\mathbf{n}} \times \hat{\mathbf{e}}_0 + \frac{1}{2}\frac{a_0^2}{1+a_0^2} G_N \hat{\mathbf{n}} \times \hat{\mathbf{k}}_0 \right|^2 \tag{10.235}$$

This expression has the expected property that the scattered fluence is proportional to the duration τ of the optical pulse. We also see from the patterns of the scattered radiation represented by each of the terms that the first term corresponds to transverse oscillations of the particle in the incident field, while the second corresponds to longitudinal oscillations of the particle, the loops, as it were, of the figure-eight motion.

In the limit of weak optical fields $(a_0 \ll 1)$, we may simplify these results by using the expressions

$$\lim_{x \to 0} J_n(x) = \frac{1}{n!}\left(\frac{x}{2}\right)^n, \qquad n \geq 0 \tag{10.236}$$

and

$$\lim_{x \to 0} J_n(x) = \frac{1}{|n|!}\left(-\frac{x}{2}\right)^{|n|}, \qquad n < 0 \tag{10.237}$$

Keeping only terms of lowest order in a_0, we find that for the first harmonic (that is, the fundamental)

$$F_1 \approx -i\pi \tag{10.238}$$

$$G_1 \approx \frac{i\pi}{\sqrt{2}}(\hat{\mathbf{n}} \cdot \hat{\mathbf{e}}_0)a_0 \tag{10.239}$$

so that the total energy scattered into the solid angle $d\Omega$ at the fundamental frequency is

$$\frac{dW_1}{d\Omega} = \frac{\omega_0^2 q^2 a_0^2 \tau}{16\pi^2 \varepsilon_0 c}|\hat{\mathbf{n}} \times \hat{\mathbf{e}}_0|^2 \tag{10.240}$$

Integrating over all solid angles, we get

$$W_1 = \frac{\omega_0^2 q^2 a_0^2 \tau}{6\pi \varepsilon_0 c} \tag{10.241}$$

But the number of photons scattered into the Nth harmonic is

$$N_N = \frac{W_N}{N\hbar\omega_0} \tag{10.242}$$

so we see that the number of photons scattered into the fundamental is

$$N_1 = \sigma_1 \Phi_0 \tau \tag{10.243}$$

where

$$\Phi_0 = \frac{\langle S \rangle}{\hbar\omega_0} = \frac{\varepsilon_0 m^2 c^3 a_0^2 \omega_0}{\hbar q^2} \tag{10.244}$$

is the incident flux in photons per unit area per unit time, and the cross section is

$$\sigma_1 = \frac{q^4}{6\pi\varepsilon_0^2 m^2 c^4} = \sigma_T \tag{10.245}$$

in which σ_T is the Thomson cross section.

To lowest order in a_0 we find that for the second harmonic

$$F_2 \approx -\pi\sqrt{2}(\hat{\mathbf{n}}\cdot\hat{\mathbf{e}}_0)a_0 \tag{10.246}$$

$$G_2 \approx -\pi \tag{10.247}$$

The total energy scattered into the solid angle $d\Omega$ at the second harmonic is therefore

$$\frac{dW_2}{d\Omega} = \frac{\omega_0^2 q^2 a_0^4 \tau}{2\pi^2\varepsilon_0 c}\left|(\hat{\mathbf{n}}\cdot\hat{\mathbf{e}}_0)\hat{\mathbf{n}}\times\hat{\mathbf{e}}_0 + \frac{1}{4}\hat{\mathbf{n}}\times\hat{\mathbf{k}}_0\right|^2 \tag{10.248}$$

Integrating over all solid angles, we get

$$W_2 = \frac{7\omega_0^2 q^2 a_0^4 \tau}{20\pi\varepsilon_0 c} \tag{10.249}$$

The number of photons scattered into the second harmonic is then

$$N_2 = \frac{W_2}{2\hbar\omega_0} = \sigma_2\Phi_0^2\tau \tag{10.250}$$

where the two-photon scattering cross section is

$$\sigma_2 = \frac{7\hbar q^6}{40\pi\varepsilon_0^3 m^4 c^7 \omega_0} \tag{10.251}$$

The number of photons scattered at the second harmonic is proportional to the square of the incident flux, that is, to the square of the probability of finding two photons at the same time. Note that the two-photon scattering cross section has the dimensions [(length)^4-time]. Thus, it is not a cross-sectional area in the usual sense.

Compared to the number of photons scattered at the first harmonic, we see that for weak fields the number of photons scattered at the second harmonic is

$$\frac{N_2}{N_1} = \frac{21}{20}a_0^2 \tag{10.252}$$

This indicates that the higher harmonics become comparable to the fundamental when $a_0 = O(1)$, as we would expect. For frequencies in the visible part of the spectrum, this corresponds to incident intensities of the order of 10^{23} W/m^2, which is within the reach of high-power lasers.

Classically, radiation scattered at the frequency of the second harmonic is interpreted as coming from higher harmonics of the particle motion in the incident field. Quantum-mechanically, we interpret this process as the addition of two incident photons to create a single scattered photon with twice the energy, as indicated schematically in Figure 10.13. Since this process requires two incident photons, the probability of its occurrence depends on the simultaneous arrival of two photons, which is the square of the probability of arrival of a single photon. Thus, the two-photon scattering probability depends on the square of the

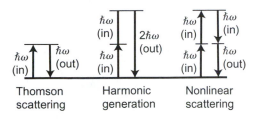

Figure 10.13 Multiphoton scattering processes.

incident intensity. Similarly, the three-photon scattering probability depends on the cube of the incident intensity, and so on. When the intensity becomes sufficiently large [classically, $a_0 = O(1)$], the higher order terms in F_N and G_N become important and the power scattered into the Nth harmonic is no longer proportional to the Nth power of the incident intensity. From a quantum-mechanical point of view, this is attributed to higher order multiphoton processes. For example, one can think of a three-photon process in which the first two photons add up to the second harmonic and the third photon subtracts its energy to produce a photon scattered at the first harmonic. As a result of this and other higher order processes, the power scattered at the first harmonic (the fundamental) is not linear in the incident intensity when the intensity is very high.

Finally, it should be emphasized that the charged particle is, on average, accelerated in the longitudinal direction by the incident optical field, as mentioned earlier. In intense fields the frequency of the incident photons in the (average motion) rest frame of the particle is Doppler downshifted by the longitudinal motion, and the frequency of the emitted (scattered) photons back in the laboratory frame is Doppler shifted according to the direction in which the photons are emitted.

EXERCISE 10.16

In the nonrelativistic limit, the radiation from a particle in circular motion in a magnetic field is emitted predominantly at the fundamental frequency of the motion. Harmonics appear when the motion becomes relativistic.

(a) Show that the angular power from a charged particle in moving in a circular orbit of radius R_0 may be represented as a series of harmonics of the orbit frequency ω_0 of the form

$$\frac{d\mathcal{P}}{d\Omega} = \sum_{N=1}^{\infty} \frac{d\mathcal{P}_N}{d\Omega} \tag{10.253}$$

where the angular power on the Nth harmonic is

$$\frac{d\mathcal{P}_N}{d\Omega} = \frac{N^2 \omega_0^2 q^2 \beta_0^2}{32\pi^2 \varepsilon_0 c} \{[J_{N+1}(\alpha) - J_{N-1}(\alpha)]^2 + \cos^2\theta [J_{N+1}(\alpha) + J_{N-1}(\alpha)]^2\} \tag{10.254}$$

in which $\beta_0 = \omega_0 R_0$ is the particle velocity, θ the angle from the axis of the motion, and

$$\alpha = N\beta_0 \sin\theta \tag{10.255}$$

Hint: Use the expansion

$$e^{i\alpha \sin\theta} = \sum_{n=-\infty}^{\infty} J_n(\alpha)e^{in\theta} \tag{10.256}$$

(b) In the nonrelativistic limit ($\beta_0 \ll 1$), show that the total power radiated on the fundamental is

$$\mathcal{P}_1 = \frac{q^2\omega_0^2\beta_0^2}{6\pi\varepsilon_0 c} \tag{10.257}$$

and compare this with the total power predicted by the nonrelativistic Larmor formula.

10.4 SYNCHROTRON RADIATION AND UNDULATOR RADIATION

Modern high-energy electron and positron storage rings produce a characteristic radiation that is emitted by the particles during their passage through the magnetic fields used to bend the trajectories into a closed path. Since this radiation was first observed when electron synchrotrons began to achieve high energy, it is generally called synchrotron radiation, or sometimes magnetic Bremsstrahlung. In synchrotrons this radiation was parasitic, in the sense that it represented an unavoidable energy-loss mechanism, but in modern electron and positron storage rings this radiation is indispensible, since the synchrotron radiation is necessary to stabilize the orbits of the particles in the ring. Moreover, the radiation itself has now been found to be so useful for spectroscopy in other branches of science that a large number of storage rings have been built around the world just to supply this radiation to users. In so-called third-generation synchrotron-radiation sources, additional magnetic structures called wigglers and undulators are added to the ring to create radiation with even greater intensity and special spectral properties. In fact, this radiation can be made coherent by adding mirrors to form an optical resonator about the undulator structure. In this configuration, the light that is returned to the undulator by the mirrors causes stimulated emission from the electrons passing through the undulator at the same time, and the device is called a free-electron laser. Free-electron lasers have now been operated at wavelengths from the infrared to the deep ultraviolet, and are generally tunable over a wide range. Their coherence is limited only by the length of the electron pulses circulating in the ring. But even the incoherent radiation from an undulator has a relatively narrow bandwidth on the order of $1/N$, where N (typically on the order of 100) is the number of magnet periods in the undulator. We do not have time to discuss free-electron lasers here, but because of their importance we discuss incoherent synchrotron and undulator radiation.

10.4.1 Synchrotron Radiation

In synchrotrons the electrons are accelerated at each pass around a ring by an accelerating cavity placed in the ring. The magnetic field of the bending magnets is increased synchronously as the electrons gain momentum in order to keep the orbit constant. The entire acceleration cycle is typically completed in less than a second. In contrast to this, the electrons in storage rings circulate at a more-or-less constant energy for hours at a time. In this case, an accelerating cavity is placed in the ring to make up the energy lost by the electrons as the result of synchrotron radiation. It is found that the synchrotron radiation emitted by the electrons is actually important for stabilizing the electron orbits, although the quantum nature of the emission prevents the particles from settling precisely into the design orbit. In addition, the power lost to synchrotron radiation in extremely high-energy rings sets an upper limit on the electron energy that can be achieved in practice. The highest-energy electron storage ring constructed up to the present time was the 90-GeV LEP2 ring at CERN, in Geneva, which has now been decommissioned. The ring had a circumference of almost 27 km and extended under the border between France and Switzerland. Electrons and positrons were circulated around the ring in opposite directions and allowed to collide at a few places in the ring. In principle, the bending magnets could be replaced by superconducting magnets to accommodate a tenfold increase in the electron energy. However, at 90 GeV about 2 percent of the electron energy was radiated during each turn around the ring. Since the radiation increases as the fourth power of the electron energy, for a given bending radius, if the energy were increased one order of magnitude to 900 GeV, the radiation per turn would be 200 times the electron energy itself. For this and other reasons it is likely that higher energy electron/positron machines (if they are built) will use linear accelerators rather than storage rings. Proton/antiproton colliders, like the Tevatron at FermiLab, can achieve higher energy because the more massive protons move more slowly than electrons and radiate less power. For this reason the LEP2 ring is being replaced by the LHC (Large Hadron Collider), which will provide proton–proton collisions at 14 TeV, a 100-fold increase over the energy of collisions in LEP2. Unfortunately, proton–proton collisions are more difficult to interpret than collisions between electrons and positrons.

To compute the angular spectral fluence of synchrotron radiation we consider a particle moving in a circle in the x–y plane at a high energy ($\gamma \gg 1$), and place the observer in the y–z plane at an angle ϕ above the plane of the orbit, as shown in Figure 10.14. The angular spectral fluence is

$$\frac{d^2\mathcal{W}}{d\omega\,d\Omega} = \frac{\mu_0 c \omega^2 q^2}{16\pi^3} \left| \int_{-\infty}^{\infty} dt\, e^{i\omega(t - \hat{\mathbf{n}}\cdot\mathbf{r}/c)} \hat{\mathbf{n}} \times \boldsymbol{\beta} \right|^2 \tag{10.258}$$

For ultrarelativistic particles the radiation is strongly directed in the direction of motion of the particle, like the beam of a headlight pointing ahead of the particle. The radiation appears to the observer as a series of pulses, one pulse for each time the particle goes around

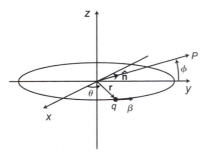

Figure 10.14 Coordinate system for synchrotron radiation.

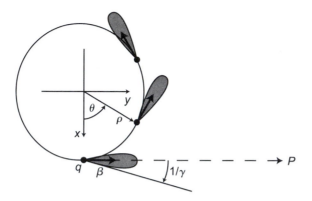

Figure 10.15 Synchrotron radiation.

the circle and the headlight sweeps over the observer, as illustrated in Figure 10.15. To compute the fluence from one of these pulses, we note that

$$\hat{\mathbf{n}} \cdot \mathbf{r} = \rho \sin\theta \cos\phi \qquad (10.259)$$

in which ρ is the radius of the circle and θ the angular position of the particle, and that

$$\hat{\mathbf{n}} \times \boldsymbol{\beta} = \hat{\mathbf{z}}\beta \sin\theta \cos\phi - \hat{\boldsymbol{\rho}}\beta \sin\phi \qquad (10.260)$$

where $\hat{\mathbf{z}}$ is a unit vector in the direction normal to the plane of motion and $\hat{\boldsymbol{\rho}}$ is a unit vector from the origin to the particle. If we choose the origin of time so that $t = 0$ when the velocity is in the direction of the observer, then the angle is

$$\theta = \beta ct/\rho \qquad (10.261)$$

In the ultrarelativistic limit, the radiation is strongly directed along $\hat{\boldsymbol{\beta}}$, as noted earlier, so we need consider only small angles $\theta = O(1/\gamma) \ll 1$ and $\phi = O(1/\gamma) \ll 1$, in which case

$$\sin\theta \approx \theta - \frac{\theta^3}{6} \qquad (10.262)$$

$$\cos\phi \approx 1 - \frac{\phi^2}{2} \qquad (10.263)$$

and

$$\hat{\boldsymbol{\rho}} \approx \text{constant} = \hat{\mathbf{x}} \qquad (10.264)$$

The exponent in (10.258) is then

$$i\omega\left(t - \frac{\hat{\mathbf{n}} \cdot \mathbf{r}}{c}\right) \approx i\omega\left[\frac{t}{2\gamma^2}(1 + \gamma^2\phi^2) + \frac{c^2 t^3}{6\rho^2}\right] \qquad (10.265)$$

and the cross product is

$$\mathbf{n} \times \boldsymbol{\beta} \approx \frac{ct}{\rho}\hat{\mathbf{z}} + \phi\hat{\mathbf{x}} \qquad (10.266)$$

Note that the exponent must be expanded more carefully than the rest of the integrand, as we have seen before. In this case, the function $\sin\theta = \sin(\beta ct/\rho)$ must be expanded to third order in the time. The integral appearing in the angular spectral fluence is then

$$\int_{-\infty}^{\infty} dt\, e^{i\omega(t-\hat{\mathbf{n}}\cdot\mathbf{r}/c)}\hat{\mathbf{n}}\times\boldsymbol{\beta} = \int_{-\infty}^{\infty} dt\, e^{i\omega[(t/2\gamma^2)(1+\gamma^2\phi^2)+(c^2t^3/6\rho^2)]}\left(\frac{ct}{\rho}\hat{\mathbf{z}}+\phi\hat{\mathbf{x}}\right) \quad (10.267)$$

This integral would appear to have convergence problems, since the integrand increases in magnitude as $t\to\pm\infty$, which is also outside the region where the small-angle approximations are valid. However, the cubic term in the exponent causes the integrand to oscillate with increasing frequency as $t\to\pm\infty$, so the integral vanishes by cancellation. We encounter this same kind of convergence in Chapter 5 in the discussion of pulse compression and in Chapter 9 in the discussion of Fresnel diffraction. This time the integrals can be evaluated analytically in terms of Bessel functions. For convenience we change to the variables

$$x = \frac{\gamma}{\sqrt{1+\gamma^2\phi^2}}\frac{ct}{\rho} \quad (10.268)$$

and

$$\xi = \frac{\omega\rho}{3c\gamma^3}(1+\gamma^2\phi^2)^{3/2} = \frac{\omega}{\omega_c}(1+\gamma^2\phi^2)^{3/2} \quad (10.269)$$

where the so-called critical frequency is

$$\omega_c = \frac{3\gamma^3 c}{\rho} \quad (10.270)$$

If we take advantage of the symmetry of the terms in the integrand, the integral becomes

$$\int_{-\infty}^{\infty} dt\, e^{i\omega(t-\hat{\mathbf{n}}\cdot\mathbf{r}/c)}\hat{\mathbf{n}}\times\boldsymbol{\beta} = i2\hat{\mathbf{z}}\frac{\rho}{c}\frac{1+\gamma^2\phi^2}{\gamma^2}\int_0^{\infty} x\,dx\,\sin\left[\frac{3}{2}\xi\left(x+\frac{1}{3}x^3\right)\right]$$

$$+ 2\hat{\mathbf{x}}\phi\frac{\rho}{c}\frac{\sqrt{1+\gamma^2\phi^2}}{\gamma}\int_0^{\infty} dx\,\cos\left[\frac{3}{2}\xi\left(x+\frac{1}{3}x^3\right)\right] \quad (10.271)$$

The integrals may be evaluated in terms of Airy functions, or equivalently, modified Bessel functions $K_\nu(\xi)$ of fractional order. The angular spectral fluence is then

$$\frac{d^2\mathcal{W}}{d\omega\,d\Omega} = \frac{3q^2\gamma^2}{4\pi^3\varepsilon_0 c}\left[\frac{\xi^2 K_{2/3}^2(\xi)}{1+\gamma^2\phi^2} + \frac{\gamma^2\phi^2\xi^2 K_{1/3}^2(\xi)}{(1+\gamma^2\phi^2)^2}\right] \quad (10.272)$$

The functions $\xi^2 K_{2/3}^2(\xi)$ and $\xi^2 K_{1/3}^2(\xi)$ are plotted in Figure 10.16, where we see that they peak near $\xi = O(1)$ and vanish exponentially for $\xi\gg 1$.

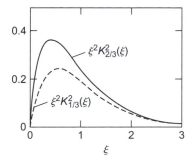

Figure 10.16 Spectrum of synchrotron radiation.

The critical frequency defined in (10.270) is much larger (γ^3 larger) than the angular frequency c/ρ of the particle in the magnetic field. This is because the width of the spectrum is actually related to the pulsed nature of the radiation and the critical frequency is approximately the reciprocal of the pulse length seen by the observer. It might be thought that the pulse length would be on the order of the angular spread of the radiation divided by the angular frequency of the particle, that is, on the order of $\rho/\gamma c$. However, as it emits the radiation seen by the observer, the particle is moving toward the observer as indicated in Figure 10.15. Thus, the back of the pulse is emitted when the particle is closer to the observer than it is when the front of the pulse is emitted, and the pulse is shortened by the factor $1 - \beta \approx 1/2\gamma^2$ to a length on the order of

$$\Delta t = \frac{\rho}{\gamma c}(1 - \beta) \approx \frac{\rho}{\gamma c}\frac{1}{2\gamma^2} \tag{10.273}$$

This is roughly the reciprocal of the critical frequency.

The critical frequency is an important parameter for synchrotron radiation sources, since it is roughly the highest frequency for which useful radiation is emitted. Since the radius of the particle orbit is

$$\rho = \frac{\beta\gamma mc}{qB} \tag{10.274}$$

where B is the magnetic field and m the particle mass, we see that the critical frequency increases roughly as γ^2 for a fixed magnetic field B. For this reason, synchrotron radiation sources typically operate at energies from 0.5 to 10 GeV. For example, the X-ray Ring at the National Synchrotron Light Source at Brookhaven National Laboratory is used as a powerful source of x-rays for a variety of experiments in materials science, biology, and so on. The electron energy in this ring is 2.5 GeV, and the magnetic field in the bending magnets is 1.22 T, which is about the limit of normal-conducting magnet technology. The critical frequency is therefore $\omega_c \approx 1.5 \times 10^{19}$ radians/s. This is in the x-ray region and corresponds to a critical wavelength $\lambda_c = 2\pi c/\omega_c \approx 0.12$ nm, or a critical energy $\mathcal{E}_c = \hbar\omega_c \approx 10$ keV.

To find the total power radiated, we use the relativistic generalization of the Larmor formula

$$P = \frac{\gamma^2 q^2}{6\pi\varepsilon_0 m^2 c^3}\left[\left|\frac{d\mathbf{p}}{dt}\right|^2 - \beta^2\left(\frac{dp}{dt}\right)^2\right] \tag{10.275}$$

In the magnetic field, the acceleration is

$$\left|\frac{d\mathbf{p}}{dt}\right| = q|\mathbf{v} \times \mathbf{B}| = qc\beta B \tag{10.276}$$

but the speed is constant,

$$\frac{dp}{dt} = 0 \tag{10.277}$$

The radiated power is therefore

$$\mathcal{P} \approx \frac{\gamma^2 q^4 B^2}{6\pi\varepsilon_0 m^2 c} \tag{10.278}$$

for $\beta \approx 1$. For the LEP2 ring at CERN, discussed earlier, the magnetic field in the bending magnets was 0.1 T. At 90 GeV, the energy radiated in one turn of the ring was

$$\mathcal{W} \approx \frac{\gamma^3 q^3 B}{3\varepsilon_0 mc} \approx 1.9 \text{ GeV} \tag{10.279}$$

This corresponds to about 2 percent of the electron energy, as mentioned earlier.

On the positive side, even beyond its use in a variety of other experiments, synchrotron radiation is responsible for the stability of electron orbits in a storage ring. Various effects disturb the orbits, and synchrotron radiation is used to damp the disturbances. In fact, the electrons are injected into the ring off the design orbit and synchrotron damping is required just to load the ring and bring the electrons to the design orbit. The basic idea is that as they circulate around the ring, the high-energy electrons radiate more energy than the low-energy electrons but all receive the same kick in the reacceleration process. Thus, after a while, all the electrons tend toward the design energy, which is the energy at which the reacceleration exactly balances the radiation. In the following paragraphs we calculate the damping time for off-design excursions of the electron energy. The damping of transverse departures from the design orbit is more complicated but occurs on a comparable time scale.

A typical storage ring consists of a series of bending and focusing magnets arranged to form a closed orbit, as shown in Figure 10.17. Radio-frequency (rf) accelerator cavities are inserted at one or more locations in the ring to replace the energy lost to synchrotron radiation. An electron at the design energy $\mathcal{E}_0$ follows the design orbit around the ring in time T_0. The frequency of the accelerator cavities is set to a multiple of the design orbit frequency, so that

$$\omega_{\text{rf}} T_0 = 2\pi n \tag{10.280}$$

for some integer n. For a single accelerator cavity, the energy gained by an electron passing through the cavity is $q V_0 \cos \phi$, where V_0 is the peak voltage in the cavity and ϕ is the phase of the rf field in the cavity when the electron passes through. In a single circuit of the ring on the design orbit the electron radiates the energy

$$\mathcal{W}_0 = q V_0 \cos \phi_0 \tag{10.281}$$

where ϕ_0 is called the synchronous phase. Clearly, the largest number of electron bunches that can be accumulated in the ring is n, with each bunch centered on one of the n synchronous phases.

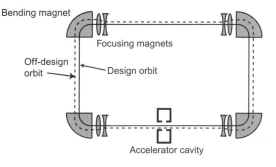

Bending magnet

Focusing magnets

Off-design orbit

Design orbit

Accelerator cavity

Figure 10.17 Particle orbit in a storage ring.

An off-energy electron will execute a different orbit, with a different length, and return to the cavity with a phase shift

$$\delta\phi = \omega_{rf}(T - T_0) \approx \alpha\omega_{rf}T_0\frac{\mathcal{E} - \mathcal{E}_0}{\mathcal{E}_0} \tag{10.282}$$

for small changes of the energy, where α is called (for obscure historical reasons) the momentum compaction. Typically, $\alpha \ll 1$. The energy radiated in a single orbit of an off-energy electron is

$$\mathcal{W} \approx \mathcal{W}_0\left[1 + (2 + D_0)\frac{\mathcal{E} - \mathcal{E}_0}{\mathcal{E}_0}\right] \tag{10.283}$$

for small variations of the energy, where the factor 2 appears because the synchrotron radiation is quadratic in the electron energy and the factor D_0 accounts for the fact that the new orbit may spend a longer time in the magnetic fields of the ring, or enter regions where the magnetic field is different. Typically, $0 < D_0 \ll 1$. The energy shift in a single orbit of the ring is then

$$\delta\left(\frac{\mathcal{E} - \mathcal{E}_0}{\mathcal{E}_0}\right) \approx \frac{1}{\mathcal{E}_0}\left[qV_0(\cos\phi - \cos\phi_0) - (2 + D_0)\mathcal{W}_0\frac{\mathcal{E} - \mathcal{E}_0}{\mathcal{E}_0}\right]$$

$$\approx -\frac{\mathcal{W}_0}{\mathcal{E}_0}\left[\tan\phi_0(\phi - \phi_0) + (2 + D_0)\frac{\mathcal{E} - \mathcal{E}_0}{\mathcal{E}_0}\right] \tag{10.284}$$

for small variations of the phase, $\phi - \phi_0 \ll 1$. Since the phase shift and energy shift per orbit are generally very small, we may write

$$\frac{d\phi}{dt} \approx \frac{1}{T_0}\delta\phi \tag{10.285}$$

and

$$\frac{d}{dt}\left(\frac{\mathcal{E} - \mathcal{E}_0}{\mathcal{E}_0}\right) \approx \frac{1}{T_0}\delta\left(\frac{\mathcal{E} - \mathcal{E}_0}{\mathcal{E}_0}\right) \tag{10.286}$$

Combining all these equations, we get

$$\frac{d^2\phi}{dt^2} = \alpha\omega_{rf}\frac{d}{dt}\left(\frac{\mathcal{E} - \mathcal{E}_0}{\mathcal{E}_0}\right) = -\frac{\alpha\omega_{rf}}{T_0}\frac{\mathcal{W}_0}{\mathcal{E}_0}\left[\tan\phi_0(\phi - \phi_0) + \frac{2 + D_0}{\alpha\omega_{rf}}\frac{d\phi}{dt}\right] \tag{10.287}$$

This is the equation of motion of a damped harmonic oscillator, and we may write it in the form

$$\frac{d^2(\phi - \phi_0)}{dt^2} + \frac{2}{\tau_D}\frac{d(\phi - \phi_0)}{dt} + \Omega_S^2(\phi - \phi_0) = 0 \tag{10.288}$$

where

$$\Omega_S = \left(\frac{\alpha\omega_{rf}\mathcal{W}_0}{\mathcal{E}_0T_0}\tan\phi_0\right)^{1/2} \tag{10.289}$$

is called the synchrotron frequency, and

$$\tau_D = \frac{\mathcal{E}_0T_0}{\mathcal{W}_0\left(1 + \frac{1}{2}D_0\right)} \tag{10.290}$$

is the damping time. Provided that the damping is slow compared to the oscillations ($\Omega_S \tau_D \gg 1$), the electrons oscillate about the design energy and phase at the synchrotron frequency Ω_S, and the oscillations damp out in the damping time τ_D. Since $D_0 \ll 1$ in most storage rings, the damping time is approximately the time required for an electron to radiate its own energy. In a typical synchrotron radiation source, such as the X-ray Ring at the National Synchrotron Light Source discussed earlier, the synchrotron frequency is $\Omega_S \sim 10^6$ radians/s, which corresponds to $1/\Omega_S T_0 \sim 10$ orbits of the ring. The damping time is $\tau_D \sim 1$ ms (about 10^4 orbits of the ring), so $\Omega_S \tau_D \sim 10^3$ and the approximations used in the analysis are justified.

Classically speaking, the energy and phase damp asymptotically to the design values as described. However, quantum effects put a limit on the process and establish a minimum energy spread for the electrons in the beam. Specifically, the finite energy of the photons radiated makes it impossible for the electrons to settle precisely at the design energy. The process is somewhat like the settling of colloidal particles in a gravitational field while they are simultaneously undergoing Brownian motion and may be understood in terms of a random-walk process. In a random-walk process, the width of the distribution after N steps is roughly $\sqrt{N}$ times the length of a single step. In the present case, the step length is the mean photon energy $\hbar \bar{\omega}$ and the number of steps is the number of photons radiated in one damping time. The energy spread is therefore expected to be

$$\Delta \mathcal{E} = O(\hbar \bar{\omega} \sqrt{N}) \tag{10.291}$$

But the mean photon energy is on the order of the critical energy, so $\hbar \bar{\omega} \approx \hbar \omega_c$, and the total energy radiated in one damping time is approximately the electron energy, so the number of steps in the random walk is $N \approx \mathcal{E}_0/\hbar \omega_c$. The *relative* energy spread is therefore

$$\frac{\Delta \mathcal{E}}{\mathcal{E}_0} \approx O\left(\sqrt{\frac{\hbar \omega_c}{\mathcal{E}_0}}\right) = O\left(\sqrt{\frac{3 \hbar q B}{m^2 c^2} \gamma}\right) \tag{10.292}$$

This shows that the relative energy spread increases slowly with increasing energy, which, again, favors linear accelerators relative to storage rings at higher energy. For the LEP2 ring at CERN, discussed earlier, $\Delta \mathcal{E}/\mathcal{E}_0 \approx O(10^{-3})$.

10.4.2 Undulator Radiation

Although synchrotron radiation from the bending magnets in storage rings has proved to be valuable for a broad variety of experiments in materials science and other applications, modern synchrotron-radiation sources use a variety of so-called insertion devices to increase the spectral brilliance of the radiation. The spectral brilliance is defined as the number of photons $d^4 N_\nu$ per unit area dA, per unit solid angle $d\Omega$, per unit relative frequency interval $d\omega/\omega$, per unit time dt,

$$B_\nu = \omega \frac{d^4 N_\nu}{dA \, d\Omega \, d\omega \, dt} \tag{10.293}$$

It has proved to be a useful figure of merit for synchrotron-radiation sources. Frequently, in the literature, the brilliance is quoted per 0.1% relative bandwidth. In terms of the angular spectral fluence of an individual charged particle, the spectral brilliance is

$$B_\nu = \frac{J}{\hbar |q|} \frac{d^2 \mathcal{W}}{d\omega \, d\Omega} \tag{10.294}$$

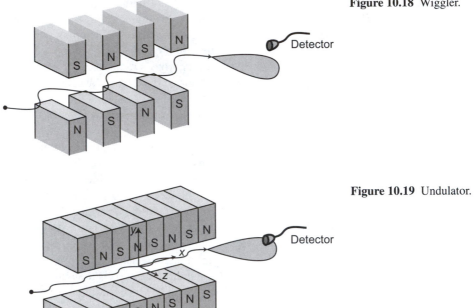

Figure 10.18 Wiggler.

Figure 10.19 Undulator.

where J is the current density of the electron or positron beam and q the charge on the particle.

Historically, the first step that was taken to improve the spectral brilliance was to introduce so-called wigglers into the beamline. In these devices a series of bending magnets is arranged alternately so that the magnetic Bremsstrahlung sweeps repeatedly over the direction of the observer, as shown in Figure 10.18. Later it was realized that by making the bends in the wiggler shorter and more numerous, the emission could be concentrated in a narrow beam in a small wavelength region around the Lorentz-contracted and Doppler-shifted period of the magnet array, now called an "undulator." The defining characteristic of an undulator, as distinct from a wiggler, is that the deflection in each bend is smaller than the $1/\gamma$ angular width of the radiation beam, as shown in Figure 10.19. As a result, the spectral width of the radiation is characterized by the number of undulator periods rather than by the length of the pulse observed as the beam sweeps over the observer. Machines incorporating wigglers and undulators are called third-generation synchrotron-radiation sources. It is also possible to use high-power lasers directed opposite the particle velocity as short-period undulators. Called laser synchrotron sources, these devices produce short-wavelength radiation (x-rays and gamma rays) from electron beams of relatively modest energy. In this case the radiation produced is actually Thomson backscatter that is twice Doppler shifted, once into the electron-beam rest frame and a second time back into the laboratory frame.

The task of calculating the angular spectral fluence may be approached in either the laboratory frame or the particle rest frame. The essence of the calculation in the laboratory frame is outlined in an exercise at the end of this section. Here we discuss the calculation in the particle rest frame. Strictly speaking, since the particle is constantly accelerating due to the field of the undulator, we use the frame in which the particle is at rest on the average, that is, when averaged over a complete period of the oscillation in the undulator field. We ignore the loss of energy radiated by the particle in a single pass through the undulator. In the moving frame, the undulator appears as an alternating magnetic field approaching the particle at nearly the speed of light. Since the alternating magnetic field induces an alternating electric field, the undulator actually looks very nearly like an incident optical field. The situation is much like the one we encountered in the method of virtual quanta, except that this time the field has a unique frequency. It is, in some sense, a single, enormous "virtual quantum." We can therefore compute the undulator radiation (at least in the moving frame of reference) by using the results obtained for nonlinear Thomson scattering.

In the following we assume that the undulator is centered on the x axis and the magnetic field is aligned in the $\hat{\mathbf{y}}$ direction, so that the oscillations take place in the $\hat{\mathbf{z}}$ direction, as shown in Figure 10.19. In the laboratory frame the undulator field may be represented by the vector potential

$$\mathbf{A} = \begin{cases} A_U \hat{\mathbf{z}} \sin k_U x, & -L_U/2 < x < L_U/2 \\ 0, & \text{otherwise} \end{cases} \tag{10.295}$$

in which L_U is the length of the undulator and $k_U = 2\pi/\lambda_U$, where λ_U is the undulator period. Although this expression does not actually satisfy the Maxwell equations, it is a satisfactory approximation near the x axis. The charged particle is assumed to move (on average) along the x axis with an average velocity $\bar{\beta}c$ inside the undulator. In the reference frame K' moving at this velocity, the undulator field is

$$\mathbf{A}' = A_U \hat{\mathbf{y}} \sin k_U'^{\alpha} r_{\alpha}', \qquad -\pi N_U < k_U'^{\alpha} r_{\alpha}' < \pi N_U \tag{10.296}$$

where

$$N_U = \frac{k_U L_U}{2\pi} = \frac{L_U}{\lambda_U} \tag{10.297}$$

is the number of periods in the undulator,

$$k_U'^{\alpha} = \bar{\gamma} k_U (\bar{\beta}, -\hat{\mathbf{x}}) \approx \bar{\gamma} k_U (1, -\hat{\mathbf{x}}) \tag{10.298}$$

is the wave vector of the (left moving) virtual photon, and $\bar{\gamma} = 1/\sqrt{1 - \bar{\beta}^2}$.

As mentioned earlier, a high-power laser may be used in place of the undulator to produce radiation in the x-ray and gamma-ray regions of the spectrum. In this case, the undulator field is precisely an optical field. All the results obtained in this section for magnetostatic undulators are immediately applicable to this case as well, only the wave vector of the optical field in the moving frame is given by the Doppler shift

$$k_U'^{\alpha} = \sqrt{\frac{1 + \bar{\beta}}{1 - \bar{\beta}}} k_U^{\alpha} (1, -\hat{\mathbf{x}}) \approx 2\bar{\gamma} k_U^{\alpha} (1, -\hat{\mathbf{x}}) \tag{10.299}$$

where $k_U^{\alpha} = k_U (1, -\hat{\mathbf{x}})$ is the wave vector of the undulator in the laboratory frame. In the ultrarelativistic limit this differs from (10.298) by just a factor of two, which is related to the fact that the optical undulator is moving to the left at the speed of light in the laboratory frame, rather than being stationary there.

In the average rest frame of the particle, the scattered radiation consists of a series of harmonics, just as it did in nonlinear Thomson scattering. For a long undulator ($N_U \gg 1$), the individual harmonics become narrow compared with the spacing ω'_U and we can ignore the overlap of the various harmonics. The total angular spectral fluence is then the sum of the harmonics

$$\frac{d^2 \mathcal{W}'}{d\omega' \, d\Omega'} = \sum_{N=1}^{\infty} \frac{d^2 \mathcal{W}'_N}{d\omega' \, d\Omega'} \tag{10.300}$$

From (10.234) we find that the angular spectral fluence of the Nth harmonic is

$$\frac{d^2 \mathcal{W}'_N}{d\omega' \, d\Omega'} = \frac{q^2 N^2 N_U^2}{16\pi^3 \varepsilon_0 c} \left| \sqrt{\frac{2a_U^2}{1 + a_U^2}} F_N \hat{\mathbf{n}}' \times \hat{\mathbf{z}} - \frac{1}{2} \frac{a_U^2}{1 + a_U^2} G_N \hat{\mathbf{n}}' \times \hat{\mathbf{x}} \right|^2$$

$$\times \left[\frac{\sin\left(\pi N_U \dfrac{\omega' - N\omega'_U}{\omega'_U} \right)}{\pi N_U \dfrac{\omega' - N\omega'_U}{\omega'_U}} \right]^2 \tag{10.301}$$

where we have used the fact that the number of undulator periods is a Lorentz invariant, so that

$$\omega'_U \tau' = k'_U L'_U = k_U L_U = 2\pi N_U \tag{10.302}$$

For the Nth harmonic the frequency in the electron "rest" frame is $\omega' \approx N\omega'_U$, where

$$\omega'_U = ck'_U = \bar{\gamma} c k_U \tag{10.303}$$

is the undulator frequency in the moving frame. The functions F_N and G_N are defined in (10.224)–(10.227), and the arguments (10.228) and (10.229) of the Bessel functions in the present notation are

$$C = \sqrt{\frac{2a_U^2}{1 + a_U^2}} \hat{\mathbf{n}}' \cdot \hat{\mathbf{z}} \tag{10.304}$$

$$S = \frac{1}{4} \frac{a_U^2}{1 + a_U^2} (1 + \hat{\mathbf{n}}' \cdot \hat{\mathbf{x}}) \tag{10.305}$$

The sign in the parentheses in (10.305) is positive because $\hat{\mathbf{k}}_0 = -\hat{\mathbf{x}}$, that is, the undulator field is incident from the right.

Before we transform these results back into the laboratory frame, we note that the average velocity of the particle inside the undulator is not the same as the velocity of the incident particle, since some of the initial energy is converted to transverse oscillations when the particle enters the undulator. From (10.215) we see that the average energy of the oscillating particle in the moving frame is

$$\langle \mathcal{E}' \rangle = \sqrt{m^2 c^4 + q^2 c^2 \langle A'^2 \rangle} = \sqrt{m^2 c^4 + \tfrac{1}{2} q^2 c^2 A_U^2} \tag{10.306}$$

The average momentum in this reference frame vanishes, so the energy in the laboratory frame is found from the Lorentz transformation to be

$$\mathcal{E} = \gamma m c^2 = \bar{\gamma} \sqrt{m^2 c^4 + \tfrac{1}{2} q^2 c^2 A_U^2} \tag{10.307}$$

From this we see that the incident energy is related to the motion of the average rest frame by

$$\gamma^2 = \bar{\gamma}^2\left(1 + a_U^2\right) \tag{10.308}$$

where

$$a_U = \frac{qA_U}{\sqrt{2}mc} \tag{10.309}$$

is the rms dimensionless vector potential of the undulator. This change in the average longitudinal velocity of the particle when it enters the undulator is equivalent to the longitudinal acceleration experienced by a particle in an intense optical field, as discussed in Chapter 4. Besides slowing down (in the $\hat{\mathbf{x}}$ direction) when it enters the undulator, the particle is deflected slightly by the initial magnetic field. However, this is ordinarily compensated by adjusting the incident trajectory and does not affect what we are doing here.

To transform the radiation back into the laboratory frame, we recall that for radiation emitted in the direction $\hat{\mathbf{n}}' = c\mathbf{k}'/\omega'$ in the moving frame and $\hat{\mathbf{n}} = c\mathbf{k}/\omega$ in the laboratory frame, the wave vectors are related by the Lorentz transformation

$$\frac{\omega'}{c} = \bar{\gamma}\left(\frac{\omega}{c} - \bar{\beta}\mathbf{k}\cdot\hat{\mathbf{x}}\right) \tag{10.310}$$

$$\mathbf{k}'\cdot\hat{\mathbf{x}} = \bar{\gamma}\left(\mathbf{k}\cdot\hat{\mathbf{x}} - \bar{\beta}\frac{\omega}{c}\right) \tag{10.311}$$

$$\mathbf{k}'\cdot\hat{\mathbf{y}} = \mathbf{k}\cdot\hat{\mathbf{y}} \tag{10.312}$$

$$\mathbf{k}'\cdot\hat{\mathbf{z}} = \mathbf{k}\cdot\hat{\mathbf{z}} \tag{10.313}$$

In the ultrarelativistic limit, the radiation is prodominantly in the forward direction and we can use the approximations

$$\bar{\beta} \approx 1 - \frac{1}{2\bar{\gamma}^2} = 1 - \frac{1 + a_U^2}{2\gamma^2} \tag{10.314}$$

and

$$\hat{\mathbf{n}}\cdot\hat{\mathbf{x}} = \cos\theta \approx 1 - \tfrac{1}{2}\theta^2 \tag{10.315}$$

The direction vectors $\hat{\mathbf{n}}'$ in the moving frame and $\hat{\mathbf{n}}$ in the laboratory frame are then related by the expressions

$$\hat{\mathbf{n}}'\cdot\hat{\mathbf{x}} = \frac{c\mathbf{k}'\cdot\hat{\mathbf{x}}}{\omega'} = \frac{\hat{\mathbf{n}}\cdot\hat{\mathbf{x}} - \bar{\beta}}{1 - \bar{\beta}\hat{\mathbf{n}}\cdot\hat{\mathbf{x}}} \approx \frac{1 + a_U^2 - \gamma^2\theta^2}{1 + a_U^2 + \gamma^2\theta^2} \tag{10.316}$$

$$\hat{\mathbf{n}}'\cdot\hat{\mathbf{y}} = \frac{c\mathbf{k}'\cdot\hat{\mathbf{y}}}{\omega'} = \frac{\hat{\mathbf{n}}\cdot\hat{\mathbf{y}}}{\bar{\gamma}(1 - \bar{\beta}\hat{\mathbf{n}}\cdot\hat{\mathbf{x}})} \approx \frac{2\sqrt{1 + a_U^2}\,\gamma\theta\cos\phi}{1 + a_U^2 + \gamma^2\theta^2} \tag{10.317}$$

$$\hat{\mathbf{n}}'\cdot\hat{\mathbf{z}} = \frac{c\mathbf{k}'\cdot\hat{\mathbf{z}}}{\omega'} = \frac{\hat{\mathbf{n}}\cdot\hat{\mathbf{z}}}{\bar{\gamma}(1 - \bar{\beta}\hat{\mathbf{n}}\cdot\hat{\mathbf{x}})} \approx \frac{2\sqrt{1 + a_U^2}\,\gamma\theta\sin\phi}{1 + a_U^2 + \gamma^2\theta^2} \tag{10.318}$$

Finally, the increment of solid angle must also be transformed. In terms of the polar coordinates (θ, ϕ) and (θ', ϕ') about the direction of motion $\hat{\mathbf{x}}$ we have

$$\frac{d\Omega'}{d\Omega} = \frac{\sin\theta'\,d\theta'\,d\phi'}{\sin\theta\,d\theta\,d\phi} = \frac{d\cos\theta'}{d\cos\theta} = \frac{d(\hat{\mathbf{n}}'\cdot\hat{\mathbf{x}})}{d(\hat{\mathbf{n}}\cdot\hat{\mathbf{x}})} = \frac{1}{\bar{\gamma}^2(1 - \bar{\beta}\hat{\mathbf{n}}\cdot\hat{\mathbf{x}})^2} \tag{10.319}$$

since $d\phi' = d\phi$. We note also that since dW and $d\omega$ are both the timelike components of 4-vectors they transform the same way, and therefore

$$\frac{dW}{d\omega} = \frac{dW'}{d\omega'} \tag{10.320}$$

Putting all this together we get

$$\frac{d^2W}{d\omega\,d\Omega} = \frac{1}{\bar{\gamma}^2(1 - \bar{\beta}\hat{\mathbf{n}} \cdot \hat{\mathbf{x}})^2}\frac{d^2W'}{d\omega'\,d\Omega'} \approx \frac{4\gamma^2(1 + a_U^2)}{(1 + a_U^2 + \gamma^2\theta^2)^2}\frac{d^2W'}{d\omega'\,d\Omega'} \tag{10.321}$$

To evaluate the vector cross products in (10.301) we expand $\hat{\mathbf{n}}'$ in terms of $\hat{\mathbf{x}}$, $\hat{\mathbf{y}}$, and $\hat{\mathbf{z}}$, and use (10.316)–(10.318) to show that

$$\hat{\mathbf{n}}' \times \hat{\mathbf{z}} = \frac{(\hat{\mathbf{n}} \cdot \hat{\mathbf{y}})\hat{\mathbf{x}} - \bar{\gamma}(\hat{\mathbf{n}} \cdot \hat{\mathbf{x}} - \bar{\beta})\hat{\mathbf{y}}}{\bar{\gamma}(1 - \bar{\beta}\hat{\mathbf{n}} \cdot \hat{\mathbf{x}})} \approx \frac{2\sqrt{1 + a_U^2}\,\gamma\theta\cos\phi}{1 + a_U^2 + \gamma^2\theta^2}\hat{\mathbf{x}} - \frac{1 + a_U^2 - \gamma^2\theta^2}{1 + a_U^2 + \gamma^2\theta^2}\hat{\mathbf{y}} \tag{10.322}$$

$$\hat{\mathbf{n}}' \times \hat{\mathbf{x}} = \frac{\hat{\mathbf{n}} \times \hat{\mathbf{x}}}{\bar{\gamma}(1 - \bar{\beta}\hat{\mathbf{n}} \cdot \hat{\mathbf{x}})} \approx \frac{2\sqrt{1 + a_U^2}\,\gamma\theta}{1 + a_U^2 + \gamma^2\theta^2}(\hat{\mathbf{y}}\sin\phi - \hat{\mathbf{z}}\cos\phi) \tag{10.323}$$

in the ultrarelativistic, small-angle approximation. The arguments of the Bessel functions are

$$C = \sqrt{\frac{2a_U^2}{1 + a_U^2}}\frac{\hat{\mathbf{n}} \cdot \hat{\mathbf{z}}}{\bar{\gamma}(1 - \bar{\beta}\hat{\mathbf{n}} \cdot \hat{\mathbf{x}})} \approx \frac{2\sqrt{2}a_U\gamma\theta\sin\phi}{1 + a_U^2 + \gamma^2\theta^2} \tag{10.324}$$

$$S = \frac{1}{4}\frac{a_U^2}{1 + a_U^2}\frac{(1 - \bar{\gamma})(1 - \hat{\mathbf{n}} \cdot \hat{\mathbf{x}})}{1 - \bar{\beta}\hat{\mathbf{n}} \cdot \hat{\mathbf{x}}} \approx \frac{1}{2}\frac{a_U^2}{1 + a_U^2 + \gamma^2\theta^2} \tag{10.325}$$

To evaluate the line-shape factor, it is convenient to introduce the resonant frequency ω_R defined by the condition

$$\omega_U' = \bar{\gamma}\omega_R(1 - \bar{\beta}\hat{\mathbf{n}} \cdot \hat{\mathbf{x}}) \tag{10.326}$$

Comparing this expression with (10.310), we see that ω_R is the frequency in the laboratory frame of a wave emitted in the direction $\hat{\mathbf{n}}$ in the laboratory frame at the undulator frequency ω_U', given by (10.303) in the moving reference frame. The resonant frequency ω_R defined in this way depends on the direction of the emission in the laboratory frame. From (10.298) we see that for a static undulator it has the value

$$\omega_R = \frac{\bar{\beta}ck_U}{1 - \bar{\beta}\hat{\mathbf{n}} \cdot \hat{\mathbf{x}}} \approx \frac{2\gamma^2ck_U}{1 + a_U^2 + \gamma^2\theta^2} \tag{10.327}$$

in the ultrarelativistic limit. For a laser undulator, ω_R has the double Doppler-shifted value

$$\omega_R = \frac{1 + \bar{\beta}}{1 - \bar{\beta}\hat{\mathbf{n}} \cdot \hat{\mathbf{x}}}\omega_U \approx \frac{4\gamma^2\omega_U}{1 + a_U^2 + \gamma^2\theta^2} \tag{10.328}$$

in the same limit. The argument of the line-shape factor is then

$$\pi N_U\frac{\omega' - N\omega_U'}{\omega_U'} = \pi N_U\frac{\omega - N\omega_R}{\omega_R} \tag{10.329}$$

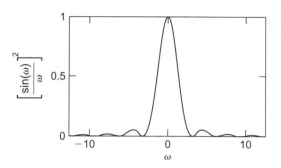

Figure 10.20 Line-shape function for undulator radiation.

The line-shape function is illustrated in Figure 10.20, where we see that the first zero occurs at π, which corresponds to

$$\frac{\omega - N\omega_R}{\omega_R} = \frac{1}{N_U} \tag{10.330}$$

The absolute line width is the same for all harmonics, so the higher harmonics are relatively narrower.

Since the expressions for the angular spectral fluence are so complicated, it is worthwhile to examine some special cases. We begin with the fluence at line center ($\omega = N\omega_R$) of the Nth harmonic on axis in the forward direction. In this case $\theta = 0$, so $C = 0$, and since $\hat{\mathbf{n}} \times \hat{\mathbf{x}} = 0$, all the terms involving G_N vanish. In the sum over l for the functions F_N, only those terms survive those for which $l = 0$, $N =$ odd, so that $J_0(NC) = 1$. The angular spectral fluence for the odd harmonics on axis is then

$$\frac{d^2\mathcal{W}_N}{d\omega\,d\Omega} = \frac{q^2 N^2 N_U^2 \gamma^2}{2\pi\varepsilon_0 c} \frac{a_U^2}{\left(1 + a_U^2\right)^2} \left[J_{\frac{N+1}{2}}\left(\frac{N}{2}\frac{a_U^2}{1 + a_U^2}\right) - J_{\frac{N-1}{2}}\left(\frac{N}{2}\frac{a_U^2}{1 + a_U^2}\right) \right]^2 \tag{10.331}$$

The even harmonics vanish on axis, as clearly they must due to the symmetry of the motion. The relative intensity of the odd harmonics is illustrated in Figure 10.21, where we see that for $a_U \gg 1$ the emission spreads over all the lower harmonics and peaks around those harmonics for which $N = O(a_U^2)$.

Next we move off axis in the x–y plane, that is, in the direction normal to the plane of the oscillations of the particle. In this case $\gamma\theta > 0$, but we still have $\sin\phi = 0$, so $C = 0$. Again, in the sum over l for the functions F_N only those terms survive for which $l = 0$, $N =$ odd. However, the terms involving the functions G_N no longer vanish, and those terms survive for which $l = 0$, $N =$ even. Thus, away from the undulator axis we observe radiation at the even harmonics from the longitudinal oscillations. This is not surprising, since the longitudinal oscillations occur at twice the fundamental frequency. The angular spectral fluence at line center for the odd harmonics is now

$$\frac{d^2\mathcal{W}_N}{d\omega\,d\Omega} = \frac{q^2 N^2 N_U^2 \gamma^2}{2\pi\varepsilon_0 c} \frac{a_U^2}{\left(1 + a_U^2 + \gamma^2\theta^2\right)^2} \left[J_{\frac{N+1}{2}}\left(\frac{N}{2}\frac{a_U^2}{1 + a_U^2 + \gamma^2\theta^2}\right) \right.$$

$$\left. - J_{\frac{N-1}{2}}\left(\frac{N}{2}\frac{a_U^2}{1 + a_U^2 + \gamma^2\theta^2}\right) \right]^2 \tag{10.332}$$

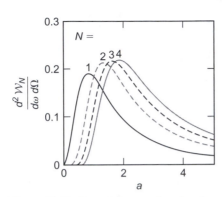

Figure 10.21 On-axis harmonics of undulator radiation.

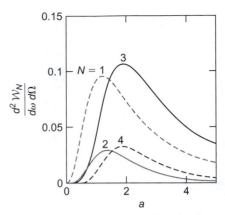

Figure 10.22 Off-axis harmonics of undulator radiation.

and that for the even harmonics is

$$\frac{d^2 W_N}{d\omega\, d\Omega} = \frac{q^2 N^2 N_U^2 \gamma^2}{2\pi\varepsilon_0 c} \frac{a_U^4 \gamma^2 \theta^2}{2\left(1 + a_U^2 + \gamma^2\theta^2\right)^4} \left[J_{\frac{N}{2}+1}\left(\frac{N}{2}\frac{a_U^2}{1 + a_U^2 + \gamma^2\theta^2}\right) \right.$$

$$\left. + J_{\frac{N}{2}-1}\left(\frac{N}{2}\frac{a_U^2}{1 + a_U^2 + \gamma^2\theta^2}\right) \right]^2 \tag{10.333}$$

These functions are illustrated in Figure 10.22 for $\gamma\theta = 1$, where we observe that the higher harmonics appear only for larger values of the undulator vector potential.

In directions that are not in the x–y plane, all terms contribute and the angular spectral fluence on the odd and even harmonics becomes comparable.

To see the advantage of introducing undulators into synchrotron radiation sources, we can compare the angular spectral fluence of bend magnets with that of undulators. From (10.272) we see that for radiation in the plane of the bend, the spectral angular fluence of a bend magnet is

$$\frac{d^2 W}{d\omega\, d\Omega} = \frac{3q^2\gamma^2}{4\pi^3\varepsilon_0 c}\xi^2 K_{2/3}^2(\xi) \tag{10.334}$$

Near the critical frequency, the factor $\xi^2 K_{2/3}^2(\xi)$ is of the order of unity. From (10.331) we see that at line center of the fundamental in the forward direction, the angular spectral fluence of an undulator is

$$\frac{d^2 W_1}{d\omega\, d\Omega} = \frac{q^2 N_U^2 \gamma^2}{2\pi\varepsilon_0 c} \frac{a_U^2}{\left(1 + a_U^2\right)^2} \left[J_1\left(\frac{1}{2}\frac{a_U^2}{1 + a_U^2}\right) - J_0\left(\frac{1}{2}\frac{a_U^2}{1 + a_U^2}\right) \right]^2 \tag{10.335}$$

For $a_U^2 \approx 1$, the quantity in the square brackets is on the order of unity. Comparing (10.334) and (10.335) we see that apart from some numerical constants of the order of unity, the undulator increases the angular spectral fluence by the factor N_U^2. Since undulators typically have of the order of 100 periods, this increases the angular spectral fluence by about four orders of magnitude.

EXERCISE 10.17

Although the detailed calculation of undulator radiation is a tedious process, as discussed earlier, the principal features can be predicted by simple physical arguments. Consider a particle moving along the x axis with velocity βc through the undulator field

$$\mathbf{B} = B_U \hat{\mathbf{y}} \sin(k_U x) \tag{10.336}$$

In the following calculations assume, for simplicity, that

$$a_U^2 = \frac{1}{2}\left(\frac{q B_U}{k_U mc}\right)^2 \ll 1 \tag{10.337}$$

It is then permissible to ignore the slowing down of the particle average longitudinal velocity due to the transverse oscillations

(a) Using simple geometrical arguments, calculate the difference in arrival time at a distant point for light emitted in the direction θ at the beginning of the undulator and at the end of the undulator. Since this emission includes N_U wavelengths for an undulator of length L_U, show that the wavelength of the radiation in the laboratory frame is

$$\lambda_R \approx \frac{\lambda_U}{2\gamma^2}(1 + \gamma^2\theta^2) \tag{10.338}$$

where $\lambda_U = 2\pi/k_U = L_U/N_U$ is the undulator period in the limit $\gamma \gg 1$ and $\theta \ll 1$.

(b) The oscillating electron emits an essentially square pulse of radiation whose duration in the laboratory frame is $\tau = 2\pi N_U/\omega_R$, where $\omega_R = 2\pi c/\lambda_R$ is the frequency of the radiation. Take the Fourier transform of this pulse to show that the spectral intensity is proportional to

$$\frac{d\mathcal{W}}{d\omega} \propto \left| \frac{\sin\left(\pi N_U \dfrac{\omega - \omega_R}{\omega_R}\right)}{\pi N_U \dfrac{\omega - \omega_R}{\omega_R}} \right|^2 \tag{10.339}$$

(c) Using a Lorentz transformation, show that a light ray moving along the y' axis (so that its position is $r'^\alpha = (0, 0, y' = ct', 0)$ in the moving frame) appears at an angle

$$\theta \approx \frac{1}{\gamma} \ll 1 \tag{10.340}$$

in the laboratory frame. By symmetry, this shows that all photons emitted in the forward hemisphere in the moving frame appear inside a cone of solid angle $\Delta\Omega \approx \pi/\gamma^2$ in the laboratory frame.

(d) Use the Lorentz transformation of the rms undulator magnetic induction B_U and the undulator period λ_U to show that the fluence (intensity times time) of the equivalent optical field in the moving frame is

$$\Phi' = \langle S' \rangle \tau' = \frac{\gamma L_U B_U^2}{2\mu_0} \tag{10.341}$$

where $\langle S' \rangle$ is the rms Poynting vector and τ' the duration of the equivalent optical pulse in the moving frame.

(e) Using the cross section for linear Thomson scattering, show that the total number of photons emitted (scattered) in the $1/\gamma$ cone in the laboratory fame is

$$N_\nu = \frac{q^2 N_U a_U^2}{6\varepsilon_0 \hbar c} \tag{10.342}$$

(f) Combining these results, show that the angular spectral fluence of undulator radiation is

$$\frac{d^2\mathcal{W}}{d\omega\,d\Omega} \approx \frac{q^2 N_U^2 a_U^2 \gamma^2}{6\pi\varepsilon_0 c} \tag{10.343}$$

Compare this estimate with the "correct" answer (10.335) in the limit $a_U^2 \ll 1$.

(g) One of the undulators at the Advanced Photon Source at Argonne National Laboratory has the characteristics

$$\gamma = 1.4 \times 10^4$$
$$B_U = 1.3\,\text{T}$$
$$N_U = 150$$
$$\lambda_U = 33 \times 10^{-3}\,\text{m}.$$

Find a_U^2, $\hbar\omega_R$ (in eV), $\Delta\omega/\omega_R$, θ, Φ', N_ν, and $d^2\mathcal{W}_1/d\omega\,d\Omega$ on axis at line center. The average current in the electron beam is 0.1 A. What is the average total power radiated from the undulator?

10.5 COHERENT EMISSION FROM MULTIPLE PARTICLES

10.5.1 Coherence and Form Factor

As shown earlier, the spectral angular fluence from a charged particle in arbitrary relativistic motion is

$$\frac{d^2\mathcal{W}}{d\omega\,d\Omega} = \frac{\mu_0 c\omega^2 q^2}{16\pi^3} \left| \int_{-\infty}^{\infty} dt\, e^{i\omega(t-\hat{\mathbf{n}}\cdot\mathbf{r}/c)} \hat{\mathbf{n}} \times \boldsymbol{\beta} \right|^2 \tag{10.344}$$

where $d\omega$ is the frequency interval, $d\Omega$ the solid angle about the direction $\hat{\mathbf{n}}$ into which the radiation is emitted, q the charge, $\mathbf{r}$ the position, and $\boldsymbol{\beta}c$ the velocity of the particle at time t. If several particles are emitting at the same time, the amplitudes of the fields they radiate add linearly and the total fluence is

$$\frac{d^2\mathcal{W}}{d\omega\,d\Omega} = \frac{\mu_0 c\omega^2 q^2}{16\pi^3} \left| \sum_i \int_{-\infty}^{\infty} dt\, e^{i(\omega t-\mathbf{k}\cdot\mathbf{r}_i)} \hat{\mathbf{k}} \times \boldsymbol{\beta}_i \right|^2 \tag{10.345}$$

where

$$\mathbf{k} = \frac{\omega}{c}\hat{\mathbf{n}} \tag{10.346}$$

is the wave vector of the emitted radiation, $\hat{\mathbf{k}} = \mathbf{k}/k$ is a unit vector, as usual, and the subscript refers to the ith particle. If the particle motions are uncorrelated, the vectors corresponding to the individual integrals have random phases and directions and do not interfere. In this case the total fluence is the linear sum of the fluences of the individual particles

$$\frac{d^2 W}{d\omega \, d\Omega} = \frac{\mu_0 c \omega^2 q^2}{16\pi^3} \sum_i \left| \int_{-\infty}^{\infty} dt \, e^{i(\omega t - \mathbf{k} \cdot \mathbf{r}_i)} \hat{\mathbf{k}} \times \boldsymbol{\beta}_i \right|^2 \qquad (10.347)$$

However, if the motions are correlated, the amplitudes add up in a way which can increase the fluence by an enormous amount if there are many particles. This is called coherence.

In the simplest case we can consider, the particle velocities are perfectly correlated but the initial positions are arbitrary. We consider a bunch of N identical particles, all with the same velocity $\boldsymbol{\beta}_i = \boldsymbol{\beta}_0(t)$. All the positions $\mathbf{r}_i(t) = \mathbf{r}_0(t) + \mathbf{r}'_i$ are then identical to within a constant and we may write

$$\frac{d^2 W}{d\omega \, d\Omega} = \frac{\mu_0 c \omega^2 q^2}{16\pi^3} \left| \int_{-\infty}^{\infty} dt \, e^{i[\omega t - \mathbf{k} \cdot \mathbf{r}_0(t)]} \hat{\mathbf{k}} \times \boldsymbol{\beta}_0(t) \right|^2 \left| \sum_i e^{-i\mathbf{k} \cdot \mathbf{r}'_i} \right|^2 \qquad (10.348)$$

If all the positions $\mathbf{r}'_i$ are uncorrelated and the wavelength of the radiation is small compared with the dimensions of the bunch, so that $\mathbf{k} \cdot \mathbf{r}'_i \gg 1$, then the phases of the particles are random and for $N \gg 1$ we get

$$\left| \sum_i e^{-i\mathbf{k} \cdot \mathbf{r}'_i} \right|^2 \approx N \qquad (10.349)$$

That is, the total fluence is just the number of particles times the fluence from one particle. On the other hand, if the wavelength of the radiation is large compared with the dimensions of the bunch, then $\mathbf{k} \cdot \mathbf{r}'_i \ll 1$ and

$$\left| \sum_i e^{-i\mathbf{k} \cdot \mathbf{r}'_i} \right|^2 \approx \left| \sum_i 1 \right|^2 = N^2 \qquad (10.350)$$

Comparing (10.349) and (10.350) we see that when the number of particles is large, the coherent emission is large compared with the incoherent emission. In the general case we may write

$$\left| \sum_i e^{-i\mathbf{k} \cdot \mathbf{r}'_i} \right|^2 = N^2 F(\mathbf{k}) \qquad (10.351)$$

where $0 < F(\mathbf{k}) < 1$ is called the form factor of the bunch. When the number of particles in the bunch is large, so the bunch can be described by the density $n(\mathbf{r})$ of particles, the form factor may be approximated by the integral

$$F(\mathbf{k}) \approx \left| \frac{1}{N} \int e^{-i\mathbf{k} \cdot \mathbf{r}} n(\mathbf{r}) \, d^3 \mathbf{r} \right|^2 \qquad (10.352)$$

That is, the form factor is the Fourier transform of the bunch shape. For example, if the particles are uniformly distributed inside a sphere of radius R, the form factor is

$$F \approx \left| \frac{\pi n}{N} \int_{-R}^{R} e^{-ikx} (R^2 - x^2) \, dx \right|^2 = \frac{9}{k^6 R^6} \left[\sin(kR) - kR \cos(kR) \right]^2 \qquad (10.353)$$

This has the value unity for $kR \ll 1$, and falls off algebraically for $kR \gg 1$, as expected for a sharp-edged bunch. For a Gaussian bunch of the same radius (where R is the $1/e$ point), the form factor is

$$F \approx \left| \frac{n(0)}{N} \int_{-\infty}^{\infty} e^{-ikx} e^{-(x^2+y^2+z^2)/R^2} \, dx \, dy \, dz \right|^2 = e^{-k^2 R^2/2} \tag{10.354}$$

where $n(0)$ is the density at the center of the bunch. This expression exhibits the exponential falloff for $kR \gg 1$ that is characteristic of Fourier transforms of smooth functions.

In either event, we see that the fluence is dominated by coherent emission for wavelengths that are long compared with the bunch size and falls to the incoherent value for very short wavelengths. This makes it possible, as shown in the next section, to use the spectrum of the radiation from electron bunches to estimate the size of the bunch. It is important to remember, however, that density fluctuations (structure) inside the bunch extend the coherent emission to shorter wavelengths that are characteristic of the dimensions of the density fluctuations. When the wavelength becomes comparable to the spacing between the particles in a bunch, the form factor is essentially that of incoherent radiation; that is, $F = O(1/N)$. The form factor will not be smaller than this except when the position of the particles is coordinated in some fashion that produces interference at specific wavelengths.

10.5.2 Coherent Radiative Processes

A more realistic example of coherent radiation is provided by coherent Thomson scattering of radiation from a finite-sized bunch of particles. As illustrated in Figure 10.23, the incident radiation has the wave vector $\mathbf{k}_0$ and is polarized in the $\hat{\mathbf{e}}_0$ direction. Then, following the earlier discussion of nonlinear Thomson scattering from a single particle, for the ith particle we have

$$\hat{\mathbf{n}} \times \boldsymbol{\beta}_i c \, dt = -\left[\sqrt{\frac{2a_0^2}{1 + a_0^2}} \hat{\mathbf{n}} \times \hat{\mathbf{e}}_0 \sin k_0 \xi + \frac{1}{2} \frac{a_0^2}{1 + a_0^2} \hat{\mathbf{n}} \times \hat{\mathbf{k}}_0 \cos 2k_0 \xi \right] d\xi \tag{10.355}$$

by (10.217), and

$$k(ct - \hat{\mathbf{n}} \cdot \mathbf{r}) = k\xi - \frac{k}{k_0} \sqrt{\frac{2a_0^2}{1 + a_0^2}} (\hat{\mathbf{n}} \cdot \hat{\mathbf{e}}_0) \cos k_0 \xi - \frac{k}{4k_0} \frac{a_0^2}{1 + a_0^2} (1 - \hat{\mathbf{n}} \cdot \hat{\mathbf{k}}_0) \sin 2k_0 \xi$$

$$+ k(\hat{\mathbf{k}}_0 - \hat{\mathbf{n}}) \cdot \bar{\mathbf{r}}_i \tag{10.356}$$

Figure 10.23 Coherent scattering from a bunch of charged particles.

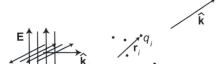

by (10.218), where $\bar{\mathbf{r}}_i$ is the mean position of the ith particle. The total scattered radiation is

$$\frac{d^2\mathcal{W}}{d\omega\,d\Omega} = \frac{\mu_0\omega^2 q^2}{16\pi^3 c}\left|-\sum_i \int_{-\infty}^{\infty} d\xi\left[\sqrt{\frac{2a_0^2}{1+a_0^2}}\,\hat{\mathbf{k}}\times\hat{\mathbf{e}}_0\sin k_0\xi + \frac{1}{2}\frac{a_0^2}{1+a_0^2}\,\mathbf{k}\times\hat{\mathbf{k}}_0\cos 2k_0\xi\right]\right.$$

$$\left.\times\, e^{i[k\xi-(k/k_0)\sqrt{2a_0^2/(1+a_0^2)}(\hat{\mathbf{k}}\cdot\hat{\mathbf{e}}_0)\cos k_0\xi-(k/4k_0)a_0^2/(1+a_0^2)(1-\hat{\mathbf{k}}\cdot\hat{\mathbf{k}}_0)\sin 2k_0\xi+k(\hat{\mathbf{k}}_0-\hat{\mathbf{k}})\cdot\bar{\mathbf{r}}_i]}\right|^2$$

$$(10.357)$$

Aside from the factor $e^{-ik(\hat{\mathbf{k}}-\hat{\mathbf{k}}_0)\cdot\bar{\mathbf{r}}_i}$, the integral is the same for all i and may be factored out of the sum. Since this integral represents the fluence radiated by a single particle, we get

$$\frac{d^2\mathcal{W}}{d\omega\,d\Omega} = \left|\sum_i e^{-ik(\hat{\mathbf{k}}-\hat{\mathbf{k}}_0)\cdot\bar{\mathbf{r}}_i}\right|^2\left(\frac{d^2\mathcal{W}}{d\omega\,d\Omega}\right)_{(1\text{ particle})} \qquad (10.358)$$

The form factor for scattering is therefore

$$F(\mathbf{k}) \approx \left|\frac{1}{N}\sum_i e^{-ik(\hat{\mathbf{k}}-\hat{\mathbf{k}}_0)\cdot\bar{\mathbf{r}}_i}\right|^2 \qquad (10.359)$$

which depends on the change $k(\hat{\mathbf{k}}-\hat{\mathbf{k}}_0)$ of the wave vector. Note that $\hat{\mathbf{k}}-\hat{\mathbf{k}}_0$ describes the change in direction of the scattered radiation, and that k is the wave number of the scattered wave. For nonlinear Thomson scattering, the scattered wave number can correspond to a harmonic of the incident wave.

Coherent transition radiation can be handled in a similar manner. As shown in the diagram in Figure 10.24, we define $t = 0$ by the time at which the center of the bunch reaches the surface and refer the position $\mathbf{r}_i = \mathbf{r}_0 + \delta\mathbf{r}_i$ of the ith particle to the position $\mathbf{r}_0$ of the center of the bunch. The ith particle reaches the surface at the time $-(\hat{\mathbf{N}}\cdot\delta\mathbf{r}_i)/(\hat{\mathbf{N}}\cdot c\boldsymbol{\beta}_0)$, where $c\boldsymbol{\beta}_0$ is the velocity of the electrons in the bunch and $\hat{\mathbf{N}}$ is a unit vector normal to the surface. From the preceding discussion, we see that the total fluence is

$$\frac{d^2\mathcal{W}}{d\omega\,d\Omega} = \frac{q^2\omega^2}{16\pi^3\varepsilon_0 c}\left|\sum_i\left[\int_{-\infty}^{-\hat{\mathbf{N}}\cdot\delta\mathbf{r}_i/\hat{\mathbf{N}}\cdot c\boldsymbol{\beta}_0} dt\, e^{i\omega(t-\hat{\mathbf{n}}\cdot\mathbf{r}_i/c)}\hat{\mathbf{n}}\times\boldsymbol{\beta}_0\right.\right.$$

$$\left.\left.-\int_{-\infty}^{-\hat{\mathbf{N}}\cdot\delta\mathbf{r}_i/\hat{\mathbf{N}}\cdot c\boldsymbol{\beta}_0} dt\, e^{i\omega(t-\hat{\mathbf{n}}\cdot\mathbf{r}_i'/c)}\hat{\mathbf{n}}\times\boldsymbol{\beta}_0'\right]\right|^2 \qquad (10.360)$$

where the primes refer to the image charge. Since the velocity vectors are constant, $\hat{\mathbf{n}}\cdot\mathbf{r}/c = \hat{\mathbf{n}}\cdot\boldsymbol{\beta}t$ and the integrals are easily calculated, keeping in mind that the limits at $-\infty$ must be evaluated with a decaying factor in the exponent. The result is

$$\frac{d^2\mathcal{W}}{d\omega\,d\Omega} = \frac{q^2}{16\pi^3\varepsilon_0 c}\left|\sum_i e^{-ik(\hat{\mathbf{N}}\cdot\delta\mathbf{r}_i/\hat{\mathbf{N}}\cdot\boldsymbol{\beta}_0)}\left[e^{-ik\hat{\mathbf{n}}\cdot\delta\mathbf{r}_i}\frac{\hat{\mathbf{n}}\times\boldsymbol{\beta}_0}{1-\hat{\mathbf{n}}\cdot\boldsymbol{\beta}_0} - e^{-ik\hat{\mathbf{n}}\cdot\delta\mathbf{r}_i'}\frac{\hat{\mathbf{n}}\times\boldsymbol{\beta}_0'}{1-\hat{\mathbf{n}}\cdot\boldsymbol{\beta}_0'}\right]\right|^2 \qquad (10.361)$$

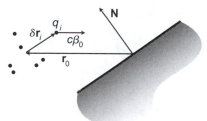

Figure 10.24 Coherent transition radiation.

In the ultrarelativistic limit only the image charges radiate strongly into the vacuum region, so the first term may be ignored and we are left with

$$\frac{d^2\mathcal{W}}{d\omega\,d\Omega} = \frac{q^2}{16\pi^3\varepsilon_0 c}\left|\frac{\hat{\mathbf{n}}\times\boldsymbol{\beta}_0'}{1-\hat{\mathbf{n}}\cdot\boldsymbol{\beta}_0'}\right|^2\left|\sum_i e^{-ik[(\hat{\mathbf{N}}/\hat{\mathbf{N}}\cdot\boldsymbol{\beta}_0')+\hat{\mathbf{n}}]\cdot\delta\mathbf{r}_i'}\right|^2 \tag{10.362}$$

Since the first part of this expression is just the radiation by a single particle, we see that the form factor is

$$F(\mathbf{k}) \approx \left|\frac{1}{N}\sum_i e^{-ik[(\hat{\mathbf{N}}/\hat{\mathbf{N}}\cdot\boldsymbol{\beta}_0')+\hat{\mathbf{n}}]\cdot\delta\bar{\mathbf{r}}_i'}\right|^2 \tag{10.363}$$

The sum is over the image bunch and depends on the wave vector $k[(\hat{\mathbf{N}}/\hat{\mathbf{N}}\cdot\boldsymbol{\beta}_0')+\hat{\mathbf{n}}]$. It can be converted to a sum over the actual bunch by using the mirror images $\hat{\mathbf{N}}'$ and $\hat{\mathbf{n}}'$ of the vectors $\hat{\mathbf{N}}$ and $\hat{\mathbf{n}}$. For particles incident normal to the surface and viewed normal to the surface ($\hat{\mathbf{N}}=\hat{\mathbf{n}}=\boldsymbol{\beta}_0$), the sum depends on the vector $2\mathbf{k}$.

EXERCISE 10.18

For a bunch of ultrarelativistic charged particles moving through an undulator, show that the form factor for coherent radiation emitted in the $\hat{\mathbf{n}}$ direction in the laboratory frame is

$$F(\mathbf{k}) \approx \left|\frac{1}{N}\sum_i e^{-i\mathbf{k}_R\cdot\mathbf{r}_i}\right|^2 \tag{10.364}$$

where $\mathbf{k}_R = k_R\hat{\mathbf{n}}$ is the resonant wave vector in the $\hat{\mathbf{n}}$ direction. *Hint:* Solve the problem in the beam frame using the formula derived for coherent scattering of the undulator field. Convert the exponent into Lorentz invariants and then express them in terms of laboratory values using the fact that $\gamma \gg 1$ to simplify the result.

EXERCISE 10.19

Consider N arbitrarily relativistic electrons moving around a periodic trajectory with frequency ω_0. The particles produce radiation at the frequency ω_0 and its harmonics. Show that the form factor for coherent radiation at the nth harmonic is

$$F = \left|\frac{1}{N}\sum_i e^{-in\phi_i}\right|^2 \tag{10.365}$$

where ϕ_i is the relative phase of particle i around the orbit.

EXERCISE 10.20

Show that the total coherent transition radiation emitted in some direction $\hat{\mathbf{n}}$ by a bunch of electrons of length L is

$$\frac{d\mathcal{W}}{d\Omega} \propto \frac{1}{L} \tag{10.366}$$

Assume that the bunch is highly relativistic and that the bunch shape remains self-similar as the length varies. The simple rule of thumb (10.366) is used experimentally to tune accelerators for the minimum bunch length by maximizing the signal on a detector. *Hint:* Assume that the positions $\delta\mathbf{r}_i$ of the electrons in the bunch scale with the bunch length L in such a way that

$$\left(\frac{\hat{\mathbf{N}}'}{\hat{\mathbf{N}}\cdot\boldsymbol{\beta}_0}+\hat{\mathbf{n}}'\right)\cdot\delta\mathbf{r}_i=\varepsilon_i L \tag{10.367}$$

for some set of constants ε_i.

10.6 RADIATION FROM RELATIVISTIC PARTICLES TRAVELING THROUGH MATTER

10.6.1 Angular Spectral Fluence

To compute the radiation from a charged particle in arbitrary relativistic motion through a dielectric medium, we adopt the same approach we used to describe the radiation from a charged particle moving through a vacuum, but now we use the Maxwell equations for a linear dispersive medium. This complicates things slightly because the speed of light is not the same for all wavelengths, but since we Fourier transform everything and work in the frequency domain, this is not a serious problem when all we need is the spectrum. On the other hand, dispersion makes it impossible, in general, to invert the Fourier transform to find the fields in the time domain.

The Green function for the Helmholtz equation (10.107) is found as it was before, and we obtain the result

$$G^{\pm}=\frac{-1}{4\pi R}e^{\pm ikR} \tag{10.368}$$

for outgoing and incoming waves, where

$$\mathbf{R}=\mathbf{r}_0-\mathbf{r}' \tag{10.369}$$

is the vector from the source point $\mathbf{r}'$ to the observation point $\mathbf{r}_0$, as shown in Figure 10.25, and the wave number is

$$k=\frac{\omega}{c}n(\omega) \tag{10.370}$$

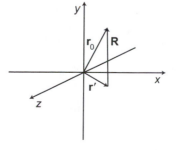

Figure 10.25 Coordinate system for radiation.

where the index of refraction $n(\omega)$ may be complex. In the following we assume that in the spectral region of interest the index of refraction is real, so the Green function can be expressed

$$G^{\pm} = \frac{-1}{4\pi R}e^{\pm i\omega R/v_{\phi}} \tag{10.371}$$

where the phase velocity $v_{\phi}(\omega) = c/n(\omega)$ replaces the factor c found in the vacuum case. If we ignore waves coming in from infinity, the solution to the Helmholtz equation may be expressed in the form

$$\tilde{\mathbf{B}}(\mathbf{r}_0, \omega) = \frac{\mu}{4\pi}\int_{-\infty}^{\infty} d^3\mathbf{r}'\,\frac{\nabla' \times \tilde{\mathbf{J}}}{R}e^{i\omega R/v_{\phi}} \tag{10.372}$$

After integrating this once by parts, we get

$$\tilde{\mathbf{B}} = \frac{\mu}{4\pi}\int_{-\infty}^{\infty} d^3\mathbf{r}'\,e^{i\omega R/v_{\phi}}\left(\frac{i\omega}{v_{\phi}R} - \frac{1}{R^2}\right)\hat{\mathbf{R}} \times \tilde{\mathbf{J}} \tag{10.373}$$

If we now substitute the Fourier transform for $\tilde{\mathbf{J}}$, we obtain the expression

$$\tilde{\mathbf{B}} = \frac{\mu}{(32\pi^3)^{1/2}}\int_{-\infty}^{\infty} d^3\mathbf{r}'\int_{-\infty}^{\infty} dt'e^{i\omega(t'+R/v_{\phi})}\left(\frac{i\omega}{v_{\phi}R} - \frac{1}{R^2}\right)\hat{\mathbf{R}} \times \mathbf{J} \tag{10.374}$$

This result is equivalent to the result obtained previously for a charged particle moving in a vacuum. In the vacuum case, we are able to multiply the expression by $e^{i\omega t}$ and integrate over all ω to invert the Fourier transform. This yields an expression for $\mathbf{B}(\mathbf{r}_0, t)$ in the form of the retarded field. However, in the present case this is not possible, since the retarded time $t - R/v_{\phi}$ is not the same for all frequencies. Fortunately, we don't usually need the inversion.

When the source is a point charge q at the point $\mathbf{r}(t')$, the current is

$$\mathbf{J}(t') = q\mathbf{v}(t')\delta[\mathbf{r}' - \mathbf{r}(t')] \tag{10.375}$$

where $\mathbf{v} = d\mathbf{r}/dt'$ is the particle velocity. If we substitute into (10.374) and integrate over all space, we find that

$$\tilde{\mathbf{B}} = \frac{\mu q}{(32\pi^3)^{1/2}}\int_{-\infty}^{\infty} dt'e^{i\omega(t'+R_0/v_{\phi})}\left(\frac{i\omega}{v_{\phi}R_0} - \frac{1}{R_0^2}\right)\hat{\mathbf{R}}_0 \times \mathbf{v} \tag{10.376}$$

where

$$\mathbf{R}_0(t') = \mathbf{r}_0 - \mathbf{r}(t') \tag{10.377}$$

is the vector from the charge to the observation point. As before, the fields consist of a "near field" proportional to $1/R_0^2$ and a "far field" proportional to $1/R_0$. At large distances from the charge we can ignore the near fields. Provided that the motion of the particle is confined to a region near the origin, so that $r(t') \ll r_0$, we may use the approximations

$$\hat{\mathbf{R}}_0 \approx \hat{\mathbf{r}}_0 = \hat{\mathbf{n}} \tag{10.378}$$

$$R_0 \approx r_0 - \hat{\mathbf{n}} \cdot \mathbf{r} \tag{10.379}$$

as before. The magnetic induction at large distances is therefore

$$\tilde{\mathbf{B}}(\mathbf{r}_0, \omega) \approx \frac{1}{(32\pi^3)^{1/2}}\frac{i\omega\mu q}{v_{\phi}r_0}e^{i\omega r_0/v_{\phi}}\int_{-\infty}^{\infty} dt'e^{i\omega(t'-\hat{\mathbf{n}}\cdot\mathbf{r}/v_{\phi})}\hat{\mathbf{n}} \times \mathbf{v} \tag{10.380}$$

This is equivalent to the result obtained previously for charges in vacuum and corresponds to outgoing spherical waves.

At sufficiently large distances from the source the waves become locally plane, so as shown in Chapter 7 the electric field is

$$\hat{\mathbf{n}} \times \tilde{\mathbf{E}} = v_\phi \tilde{\mathbf{B}} \tag{10.381}$$

The Poynting vector in a dispersive medium is

$$\mathbf{S} = \mathbf{E} \times \mathbf{H} \tag{10.382}$$

so the total angular fluence in the $\hat{\mathbf{n}}$ direction is

$$
\frac{d\mathcal{W}}{d\Omega} = r_0^2 \int_{-\infty}^{\infty} dt\, \mathbf{S} \cdot \hat{\mathbf{n}} = r_0^2 \int_{-\infty}^{\infty} dt\, |\mathbf{E} \times \mathbf{H}|
$$

$$
= r_0^2 \int_{-\infty}^{\infty} d\omega\, |\tilde{\mathbf{E}} \times \tilde{\mathbf{H}}| = 2r_0^2 \int_0^{\infty} d\omega\, \frac{v_\phi}{\mu} |\tilde{\mathbf{B}}|^2 \tag{10.383}
$$

since $\mathbf{B}$ is real. From this we identify the angular spectral fluence as

$$
\frac{d^2\mathcal{W}}{d\omega\, d\Omega} = 2r_0^2 \frac{v_\phi}{\mu} |\tilde{\mathbf{B}}|^2 = \frac{\mu \omega^2 q^2}{16\pi^2 v_\phi} \left| \int_{-\infty}^{\infty} dt\, e^{i\omega(t - \hat{\mathbf{n}} \cdot \mathbf{r}/v_\phi)} (\hat{\mathbf{n}} \times \mathbf{v}) \right|^2 \tag{10.384}
$$

This result is equivalent to (10.104), which applies to charges in vacuum, and is obtained from it by the substitutions $\mu_0 \to \mu$, $c \to v_\phi$, and $\boldsymbol{\beta} \to \mathbf{v}/v_\phi$.

10.6.2 Cherenkov Radiation

Cherenkov radiation is observed when a particle travels through a medium at a velocity greater than the speed of light in that medium. The radiation has the appearance of a blue glow created when fast electrons pass through various liquids, and the blue glow of a water-moderated nuclear reactor is actually bright enough to take a photograph of the reactor. But the radiation was first observed from charged particles emitted by beta decay of radioactive salts in solution, and it was thought that the radiation was due to fluorescent processes in the liquid in which the radioactive salt was dissolved. However, by irradiating purified water and other liquids with fast electrons from the beta decay of radium, and observing the polarization and other properties of the radiation, Cherenkov was able to show that the radiation is due to the superluminal passage of the electrons through the medium. The first mathematical description of the phenomenon was given by Frank and Tamm in 1937, and the three scientists shared the Nobel Prize for their work in 1958.

Although (10.104) seems to suggest that radiation is created by a particle in uniform motion, we know that this is not true for a charge moving in a vacuum, since in its own rest frame the field is a constant Coulomb field. This field can be transformed into any other reference frame without the appearance of fields falling off as $1/r_0$. Mathematically, we recognize that as long as the particle is traveling at less than the speed of light, the exponential factor in (10.104) oscillates. The integral converges only if some damping factor is introduced to make the integrand vanish at $t = \pm\infty$, and the integral vanishes as the damping rate vanishes. In the present case, however, the particle velocity may be greater than the phase velocity in the medium. We assume in the following that the medium is transparent, so that the phase velocity is real. For a particle moving with a constant velocity

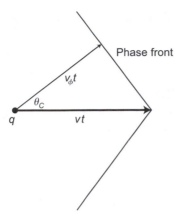

Figure 10.26 Cherenkov angle.

Phase front

$v_\phi t$

θ_C

q vt

$v > v_\phi$, the exponent in (10.384) is then

$$i\omega\left(t - \frac{\mathbf{r} \cdot \hat{\mathbf{n}}}{v_\phi}\right) = i\omega t\left(1 - \frac{\mathbf{v} \cdot \hat{\mathbf{n}}}{v_\phi}\right) \tag{10.385}$$

This vanishes at the Cherenkov angle

$$\frac{\mathbf{v} \cdot \hat{\mathbf{n}}}{v_\phi(\omega)} = \frac{v}{v_\phi(\omega)}\cos\theta_C = 1 \tag{10.386}$$

so for this angle the integrand does not oscillate. The significance of the Cherenkov angle is illustrated in Figure 10.26. As the charge moves to the right, a wave in the form of a cone of half angle θ_C spreads out behind the particle. The similarity between this diagram and one constructed to describe an oblique shock wave in air has suggested to some that Cherenkov radiation is "electromagnetic shock radiation." This is not true, however. A shock wave is a discontinuity (actually, the thickness of a shock wave in a gas is a few mean free paths) involving a change of entropy across the shock front. The velocity of a shock wave depends on the strength (the entropy increase) of the shock wave but is always greater than the sound speed. On the other hand, Cherenkov radiation is a linear phenomenon and involves no dissipation (no entropy change). The electromagnetic waves in a medium travel at a velocity that depends on the frequency of the wave, but there is no dependence on the intensity of the waves, at least in the low-intensity limit. Although the speed of light in nonlinear media depends on the intensity of the optical field at sufficiently high intensity, this is not germane to the phenomenon of Cherenkov radiation. As the result of phase-velocity dispersion, each frequency component in the Cherenkov radiation propagates at its own Cherenkov angle and the waves formed by the Cherenkov effect do not form a discontinuity.

For uniform motion of the charge, the total energy radiated between $t = -\infty$ and $t = +\infty$ is infinite. Mathematically, the integral in (10.384) is a δ-function at the frequency that satisfies (10.386), and the square of the δ-function cannot be integrated over frequency to compute a finite total fluence. To avoid this, we consider the case of a charged particle that passes through a medium of finite thickness in a time τ. Outside the medium, the phase velocity is larger than the particle velocity and the integrand oscillates. When we throw in a damping factor to make the integral converge, there is a finite contribution in the limit

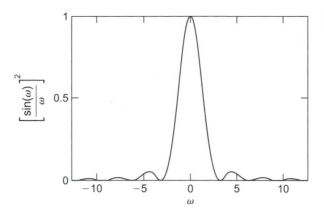

Figure 10.27 Spectrum of Cherenkov radiation.

when the damping rate vanishes. This represents a surface contribution to the total radiation and is not of interest here. When the thickness is large compared to a wavelength we can ignore the surface contributions, and the spectral angular fluence is then

$$\frac{d^2\mathcal{W}}{d\omega \, d\Omega} = \frac{\mu \omega^2 q^2}{16\pi^3 v_\phi} \left| \int_{-\tau/2}^{\tau/2} dt \, e^{i\omega t(1 - v\cos\theta/v_\phi)} v \sin\theta \right|^2$$

$$= \frac{\mu q^2 v^2 \sin^2\theta}{16\pi^3 v_\phi} \omega^2 \tau^2 \left| \frac{\sin[\frac{1}{2}\omega\tau(1 - v\cos\theta/v_\phi)]}{\frac{1}{2}\omega\tau(1 - v\cos\theta/v_\phi)} \right|^2 \tag{10.387}$$

The last factor is the spectral line shape at the angle θ. As shown in Figure 10.27, the spectrum is dominated by a peak at the frequency that satisfies the Cherenkov condition (10.386). The frequency $\omega_{\min}$ of the first minimum is found from the condition

$$\frac{1}{2}\omega_{\min}\tau\left(1 - \frac{v}{v_\phi(\omega_{\min})}\cos\theta\right) = \pi \tag{10.388}$$

If we use the approximation

$$v_\phi \approx v_\phi(\omega_C) + \left.\frac{dv_\phi}{d\omega}\right|_{\omega_C}(\omega - \omega_C) = v_C + v_C'(\omega - \omega_C) \tag{10.389}$$

near the frequency ω_C that satisfies the Cherenkov condition (10.386), then we see that the first minimum of the spectrum is at

$$(\omega_{\min} - \omega_C)\tau \approx 2\pi \frac{v_C}{\omega_C v_C'} \tag{10.390}$$

Clearly, as the medium gets thicker the spectrum becomes narrower, and the spectrum is broadest where $v_\phi(\omega)$ is flat.

Cherenkov radiation is emitted only over the band of frequencies for which the Cherenkov condition (10.386) can be satisfied, that is, over the band of frequencies for which the index of refraction $n(\omega)$ is greater than c/v. But in a transparent region of the spectrum, the Kramers–Kronig relations show that the index of refraction is an increasing function of the frequency. A typical situation is illustrated in Figure 10.28. Due to the factor $\sin^2\theta \approx 1 - (v_\phi^2/v^2) = 1 - (1/n^2\beta^2)$ in (10.387), the Cherenkov radiation is emitted

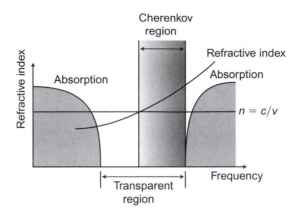

Cherenkov
region

Refractive index

Absorption

Absorption

$n = c/v$

Frequency

Transparent
region

Figure 10.28 Conditions for Cherenkov radiation.

most strongly at higher frequencies, where $n(\omega)$ is largest. This explains why the radiation appears experimentally as a blue glow.

The angular spectral fluence in (10.387) increases as the square of the time τ at line center. However, part of this increase is due to the fact that the line is getting narrower as τ increases. The total fluence and the total energy radiated both increase linearly with the time of passage through the medium. This may be seen as follows. To compute the total angular fluence at angle θ, we integrate over all frequencies. To evaluate the integral analytically, we assume that the slab through which the charged particle moves is many wavelengths thick, so the spectrum is narrow and $(\omega - \omega_C)/\omega_C \ll 1$. If we use the approximation (10.122), the angular fluence becomes

$$
\frac{dW}{d\Omega} \approx \frac{\mu_C q^2 v^2}{16\pi^3 v_C} \omega_C^2 \tau^2 \sin^2 \theta \int_{-\infty}^{\infty} d\omega \left| \frac{\sin\left[\dfrac{\omega_C \tau}{2} \dfrac{v_C'}{v_C} (\omega - \omega_C) \right]}{\dfrac{\omega_C \tau}{2} \dfrac{v_C'}{v_C} (\omega - \omega_C)} \right|^2
$$

$$
= \frac{q^2 \omega_C \tau}{8\pi^2 \varepsilon_C v_C'} \tan^2 \theta \tag{10.391}
$$

where μ_C and ε_C are the permeability and permittivity at the Cherenkov frequency, respectively, and we have used (10.120) to eliminate the velocity v. We see from this result that the total fluence is linear in the time τ, as it must be.

The computation of the spectral characteristics of Cherenkov radiation is usually the most important task, since the radiation is what we actually observe. Nevertheless, it is still of interest to consider the fields of a particle as it passes through a medium at a velocity that is, at least at some frequencies, superluminal. We are assisted in this effort by our understanding of the method of virtual quanta. In vacuum, the electric field surrounding a relativistic particle is a radial field that is compressed by Lorentz contraction in the direction of motion. This field, which is actually a pulse, as observed by a stationary observer, may be represented by the superposition of many waves, called virtual quanta. This is illustrated in Figure 10.29. The virtual quanta superpose themselves in a way that causes them to cancel out at large distances, so the fields are localized around the particle. When the particle moves into a dispersive medium, the fields and the virtual quanta are modified. Those virtual

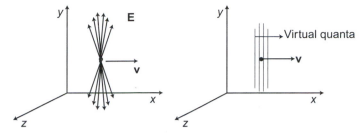

Figure 10.29 Fields and virtual quanta of a relativistic charge in vacuum.

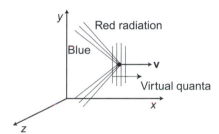

Figure 10.30 Fields and virtual quanta of a relativistic charge in a dispersive medium.

quanta that travel faster than the particle are modified in detail but not in character. They still form a field that is localized around the particle as shown in Figure 10.30. However, virtual quanta that travel at less than the particle velocity are bent back from the particle and form a wake, like the wake behind a boat or a supersonic aircraft. The waves fan out in a cone, with the fastest ones (at the red end of the spectrum) in the front and the slowest ones (at the blue end of the spectrum) in the rear. To a stationary observer they form a chirped pulse. Since the waves are constantly formed at the tip of the cone, they increase constantly in total energy, and since they have been separated from the rest of the virtual quanta, they don't cancel out but propagate to infinity. At large distances they are the only important contribution to the field, and they form the radiation that is observed in the far field.

EXERCISE 10.21

We have discussed three reasons why tachyons (particles that travel faster than light) cannot exist:

(a) Their behavior would violate causality in some reference frame K';
(b) They would require infinite energy to accelerate through the velocity of light (unless they are created with a velocity greater than that of light).
(c) They would radiate continuously (and infinitely), even in uniform motion, due to their superluminal velocity.

Elaborate on each of these statements, using mathematics to make your arguments more precise.

EXERCISE 10.22

For a charged particle traveling through a slab of dense, transparent medium at nearly the speed of light, the angular spectral fluence of Cherenkov radiation at a frequency ω is given by (10.387). Expand this expression for angles near the Cherenkov angle θ_C, and show that the angular spectral fluence at frequency ω for angles near the Cherenkov angle θ_C for this frequency is

$$\frac{d^2\mathcal{W}}{d\omega\,d\Omega} = \frac{\mu q^2 \omega^2 \tau^2 v_\phi}{16\pi^3} \tan^2\theta_C \left| \frac{\sin\left[\frac{1}{2}\omega\tau(\theta - \theta_C)\tan\theta_C\right]}{\frac{1}{2}\omega\tau(\theta - \theta_C)\tan\theta_C} \right|^2 \tag{10.392}$$

where τ is the time for the particle to traverse the slab.

BIBLIOGRAPHY

Good discussions of radiation by relativistic particles similar to the present discussion can be found in

> J. D. Jackson, *Classical Electrodynamics,* 3rd edition, John Wiley & Sons, New York (1999),
> F. E. Low, *Classical Field Theory,* John Wiley & Sons, New York (1997),

and from a different point of view in

> L. D. Landau and E. M. Lifshitz, *The Classical Theory of Fields,* 4th edition, Pergamon Press, Oxford (1975),
> L. D. Landau, E. M. Lifshitz, and L. P. Pitaevskii, *Electrodynamics of Continuous Media,* 2nd edition, Pergamon Press, Oxford (1984).

For a discussion of synchrotron radiation and the damping of storage rings the reader is referred to

> P. J. Bryant and K. Johnson, *The Principles of Circular Accelerators and Storage Rings,* Cambridge University Press, Cambridge, U.K. (1993).

Free-electron lasers and undulator radiation are discussed in great detail by

> C. A. Brau, *Free-Electron Lasers,* Academic Press, Boston (1990).
> W. B. Colson, C. Pellegrini, and A. Renieri, *Laser Handbook,* Volume 6 (Free-Electron Lasers), North-Holland, Amsterdam (1990).

11 Fundamental Particles in Classical Electrodynamics

We are now almost at the end of our discussion of classical electrodynamics. We have learned how to analyze, often with remarkable accuracy, a variety of phenomena ranging from the damping of electron storage rings to the excitation of volume plasmons by electrons. It might therefore seem a bit late in the day to point out the fundamental inconsistencies in a theory that has proved so useful, and sometimes even beautiful. But in fact, the fundamental theory describing the motion of a single point charge, the simplest problem we can imagine and the basis for most of what we have done so far, is filled with contradictions. The questions to which we address ourselves in this last chapter are, therefore, "What is wrong with the marvelous edifice we have constructed? How did we get where we are? Why has it worked so well?" and perhaps, "What can we do to fix it?"

In fact, the difficulties have been apparent from the earliest days. People have been aware of them for over 100 years, beginning even before Lorentz' famous book *Theory of Electrons*. Unfortunately, in spite of an enormous effort that now includes many books and a large number of papers, it is still not completely clear how to formulate a self-consistent, classical theory of electrons viewed, as we think of them now, as dimensionless, structureless particles. In fact, the same difficulties persist in the quantum-mechanical version of the theory, called quantum electrodynamics, or QED. It now seems likely that just as inconsistencies in the ether theory led Einstein to the theory of relativity, the inconsistencies in QED and its successors will lead to the next (perhaps the final?) theory of the physical universe.

The theory we have developed also lacks a fundamental symmetry: While there are sources (electrical charges) for the electric field, there are no corresponding sources (magnetic charges) for the magnetic field. The existence of magnetic charges (magnetic monopoles) would have profound and surprising consequences for the laws of physics, and the search for these elusive—or nonexistent—particles continues, despite repeated negative results, because they play (or might play) such a fundamental role in a complete theory of the universe. We therefore take time to discuss the putative behavior of monopoles, should they exist, and the implications for electrodynamics.

Oh, yes, one last thing. Up to now we have viewed electrons as structureless point particles, but they actually have additional properties that we have hardly mentioned. Specifically, they have intrinsic angular momentum (spin) and an accompanying magnetic dipole moment. Although it is difficult to understand how a point particle might have an intrinsic angular momentum, we can nevertheless describe its behavior in the classical approximation, and we do this in the closing sections of this chapter. In fact, the intrinsic angular momentum is intimately connected to Lorentz boosts, as discussed in Chapter 2, and exhibits a surprising effect called Thomas precession. One of the most subtle manifestations of relativity, this effect provides a fitting end to our journey.

11.1 ELECTROMAGNETIC MASS AND THE RADIATION REACTION

11.1.1 Difficulties in the Classical Theory

The dilemma is as follows. On the one hand we can view electrons as points, whatever that means physically, endowed with mass, charge, angular momentum (spin), magnetic moment, and perhaps other properties. We then have the difficulty that the electromagnetic field becomes infinite at the position of the electron. Aside from our basic abhorrence of infinities in our description of nature, this presents us with a more practical difficulty. The energy in the electromagnetic field of the electron is now infinite, and since we equate energy with mass, we have an infinite mass. Since this energy (mass) travels with the electron as it moves, this gives the electron an infinite momentum and kinetic energy. We can, of course, "sweep the problem under the rug" by asserting that the bare mass of the electron is infinite and negative, so that the sum of the bare mass and the electromagnetic mass cancel to give the observed mass. This is called "renormalization," as though giving it a name makes it more acceptable. In fact, one could argue that "this is just the way it is," and we have to accept it in the same way as we do the spin angular momentum of a "point." Unfortunately, this is not the end of the matter, for if we do this we still have to include the "radiation reaction" in the electron equation of motion. The radiation reaction arises in the following way. If we consider an electron that is accelerated, we know that it radiates both energy and momentum. The force on the electron that does the work to accelerate it must be large enough to accelerate the electron, changing its momentum and energy, and to provide the energy and momentum radiated into the field. This extra force that must be provided is called the radiation reaction. In the following sections we calculate this reaction, but for now it is enough to point out that when this is done we obtain a third-order equation of motion called the Abraham–Lorentz equation, or the Dirac equation in the relativistic case. Unfortunately, this equation admits runaway solutions (the electron can accelerate from rest to an infinite energy without the help of an external field) as well as solutions in which the electron response to an applied field begins before the field is applied (called preacceleration). This is a bit awkward, to say the least.

On the other hand, we can ascribe to the electron a nonvanishing size (small) and shape (generally spherical, at least in the electron rest frame). If we do this we avoid the infinities associated with the point electron, and we can still derive the Abraham–Lorentz equation for a very small particle. The radiation reaction now appears as the result of the retarded force of the distributed electronic charge on itself, and we obtain a differential-difference equation of motion for the electron that avoids the problem of runaway solutions. However, this still leaves us with the conundrum that a rigid particle is inconsistent with the principles of relativity. An extended body cannot move rigidly, since the different parts of the body are acted on by any outside influence at different times. The best we can do is to take the nonrelativistic equations of motion for a rigid particle and generalize them using the principles of covariance. We simply ignore the fact that the nonrelativistic equations are rooted in the notion of rigid particles and argue that that this is "just the way it is," provided, of course, that experimental results justify this a posteriori.

This brings us to the point of asking just how bad the problem is. There are several ways to answer this question. Fortunately, they all give about the same answer. The first answer is based on considerations of energy and mass. If we assume for simplicity that the electron is a spherical shell of charge with the radius a, the energy in the electrostatic field outside the electron is

$$\mathcal{E}_S = 2\pi\varepsilon_0 \int_a^\infty E^2 r^2 \, dr = \frac{q^2}{8\pi\varepsilon_0 a} = mc^2 \frac{r_c}{2a} \tag{11.1}$$

where m is the mass and q the charge of the electron, and

$$r_c = \frac{q^2}{4\pi\varepsilon_0 mc^2} = 2.82 \times 10^{-15} \text{ m} \tag{11.2}$$

is called the classical radius of the electron. Thus, if the mass of the electron is electromagnetic in origin, the electron must have a radius on the order of the classical value. This radius is very much smaller than the size of an atom but about the same size as a small nucleus. It is also small compared with the so-called Compton wavelength

$$\lambda_C = \frac{h}{mc} = 2.43 \times 10^{-12} \text{ m} \tag{11.3}$$

where h is Planck's constant. Quantum mechanically, this is the wavelength characteristic of an electron with a relativistic momentum. Since it isn't possible to confine an electron to a region smaller than this except at a highly relativistic energy, it is arguably meaningless to discuss the structure of the electron at lengths of the order of the classical electron radius. Thus, while it should have been possible by now to have detected experimentally an electron radius as large as the classical radius, it is not necessarily a grievous fault in the classical theory to ascribe to the electron a size of order r_c in order to avoid infinite electromagnetic energy and other ills of the theory.

Another answer to the question of how bad is the problem with classical electrons is based on consideration of the radiation reaction. The argument goes as follows. For an electron that is instantaneously accelerating in its own rest frame, the energy radiated in the time Δt is

$$\mathcal{E}_{\text{rad}} = \frac{q^2 |\dot{\boldsymbol{\beta}}|^2}{6\pi\varepsilon_0 c} \Delta t \tag{11.4}$$

according to the nonrelativistic Larmor formula of Chapter 10. The kinetic energy at the end of this same period is

$$\mathcal{E}_K = \tfrac{1}{2} mc^2 |\dot{\boldsymbol{\beta}}|^2 \Delta t^2 \tag{11.5}$$

If the radiated energy is to be regarded as small compared with the kinetic energy of the electron, then the time over which the particle accelerates must be

$$\Delta t \gg \tau_c \tag{11.6}$$

where the characteristic time is

$$\tau_c = \frac{q^2}{6\pi\varepsilon_0 mc^3} = \frac{2}{3}\frac{r_c}{c} = 1.25 \times 10^{-23}\text{ s} \tag{11.7}$$

for an electron, in which r_c/c is the time for light to travel a distance equal to the classical radius of the electron. In general, this time and the associated distance are small enough compared with the times and lengths of interest to us to make us comfortable with our conventional equations and the approaches we have used to solve them. Smaller distances and shorter times are the domain of quantum mechanics. For example, for an electromagnetic wave with the frequency $\omega = 1/\tau_c$, the photon energy is $\mathcal{E}_\phi = \hbar/\tau_c$, where $\hbar$ is Planck's constant divided by 2π. Compared with the electron rest energy, the photon energy is

$$\frac{\mathcal{E}_\phi}{mc^2} = \frac{6\pi\varepsilon_0\hbar c}{q^2} = 206 \gg 1 \tag{11.8}$$

so in this case quantum effects cannot be ignored. For particles heavier than the electron, the characteristic time scale is even shorter and the characteristic distance even smaller, extending the realm of classical behavior even further.

In summary, then, these simple arguments lead to the conclusion that the classical theory of electrons gets into trouble for distances smaller than the classical electron radius, $r_c = q^2/4\pi\varepsilon_0 mc^2$, and for times shorter than $\tau_c = r_c/c$. On the other hand, these distances and times are really the domain of quantum mechanics.

Finally, we might ask how we came to this somewhat awkward situation in the first place. We started, of course, with Einstein's postulates of relativity and Hamilton's principle. The problems do not seem to lie in Einstein's postulates, although their final resolution may lie in a theory that unites quantum mechanics with general relativity. Nor do they begin with our use of Hamilton's principle. For continuous distributions of matter and charge the construction of the Lagrangian encounters no fundamental problems, and there are no logical inconsistencies in minimizing the action of the fields with the sources fixed to get the Maxwell equations, or in minimizing the action of the matter with the fields fixed to get the equations of motion. The problems begin when we introduce point charges. Even then, there are no logical problems in the derivation of the Maxwell equations if we do not object to the infinities in the fields near point charges. The logical problems occur in the derivation of the equations of motion of the charges, and they enter in the following way. When we vary the trajectory of the particle with the fields fixed, we implicitly assume that the field an infinitesimal distance from the original trajectory differs infinitesimally from the field at the original trajectory. This is valid for continuous charge distributions, but for point charges the fields have a singularity at the particle. The way we argue around this logical inconsistency is to consider only the externally applied fields (the fields of the other particles) and ignore the effect of the self-field of the particle on the motion of the particle itself. This field is assumed to accompany the particle wherever it goes, and the energy and momentum in the field contribute to the energy and momentum (and implicitly to the mass) of the particle. Aside from some problems in the covariance properties of the energy and momentum of the self-fields, which we can resolve, this is a valid assumption for a stationary particle or a particle in uniform motion. But it fails for an accelerating particle. In fact, an accelerating particle radiates energy and momentum, and this is not accounted for in the theory. Somehow, the self-field of the particle distorts when the particle is accelerated, and some of the field becomes radiation.

In the following sections we examine the equation of motion of electrons. We begin with a discussion of the covariance properties of the electromagnetic momentum and energy of the self-fields surrounding an electron. This is called the 4/3 problem for reasons that will become apparent. We then discuss the mechanics of electrons viewed as point charges, and the equation of motion of Abraham, Lorentz, and Dirac. Since this theory has solutions that violate causality, we go on to develop the theory of a rigid, extended electron. This leads to the Markov equation of motion and its relativistic generalization.

11.1.2 The 4/3 Problem and Poincaré Stresses

In discussing the electron, it is convenient to picture it as a very localized distribution of charge. Associated with this charge are an electric field and, for a charge in motion, a magnetic field that move along with the charge. These fields contain energy and momentum that form part of the total energy and momentum of the observable particle. Unfortunately, while the total energy momentum of the particle must form a 4-vector, as discussed in Chapters 1 and 2, the electromagnetic energy and momentum of the fields do not. How, then, do the total energy and momentum of the electron form a good 4-vector, or do they? This dilemma, known as the "4/3 problem," was first pointed out by J. J. Thomson before relativity was fully understood, and finally resolved by Poincaré. We can understand it in more detail as follows.

Most commonly the electron is described as a spherical shell of charge of radius a, so the electric field vanishes inside this radius. This simplifies the mathematics, and the results are qualitatively the same for other spherically symmetric charge distributions. With this distribution of charge, the electrostatic energy in the field of the electron in its rest frame K' is

$$\mathcal{E}'_S = \frac{1}{2}\varepsilon_0 \int_{r>a} E'^2 \, d^3\mathbf{r} = \frac{q^2}{8\pi\varepsilon_0 a} = m_S c^2 \tag{11.9}$$

where m_S is the mass associated with the energy of the electrostatic field of a stationary particle. For a radius $a = r_c = q^2/4\pi\varepsilon_0 mc^2$, the classical radius of the electron, the mass of the static electric field is half the observed mass of the electron.

The momentum 4-vector for the field of the electron in its rest frame is then

$$p'^\alpha = \left(\frac{\mathcal{E}'_S}{c}, 0\right) = m_S(c, 0) \tag{11.10}$$

Transformed into the laboratory frame K, the 4-vector momentum of the field becomes

$$p^\alpha = \gamma m_S(c, \mathbf{v}) = m_S U^\alpha \tag{11.11}$$

where $U^\alpha = \gamma(c, \mathbf{v})$ is the 4-vector velocity of the electron in the laboratory frame. For a nonrelativistic electron ($v/c \ll 1$), the kinetic energy and the momentum of the field are

$$\mathcal{E} - m_S c^2 \approx \tfrac{1}{2} m_S v^2 \tag{11.12}$$

$$\mathbf{p} \approx m_S \mathbf{v} \tag{11.13}$$

But if we compute the total momentum of the field directly from the momentum density we get

$$\mathbf{p} = \varepsilon_0 \int_{r>a} \mathbf{E} \times \mathbf{B} \, d^3\mathbf{r} = \frac{\mu_0 q^2 \mathbf{v}}{6\pi a} = \frac{4}{3} m_S \mathbf{v} \tag{11.14}$$

That is, the electromagnetic momentum of the field is larger than the value $m_S\mathbf{v}$ that we would expect from motion of the electrostatic mass. In the language of relativity, we would say that the momentum and energy of the electromagnetic field of the electron do not transform covariantly. Yet we have many occasions to speak of photons, which are free electromagnetic wave packets, and we transform their energy and momentum as components of a Lorentz 4-vector. Why is this a valid procedure for photons when the corresponding quantities for the field of an electron do not transform covariantly? And why can we treat the total momentum of an electron, including the momentum of the fields and the bare mass, as a 4-vector quantity?

The resolution of this conundrum was discovered by Poincaré in 1905. He pointed out that there must be another force holding the charge of the electron together, a kind of negative pressure inside the electron. Otherwise the spherical shell of charge would fly apart. When the so-called Poincaré stresses are included, the total energy and momentum of the electromagnetic and Poincaré fields do transform covariantly. We can see this in the following way.

The total energy and momentum in the electromagnetic field at time $t = 0$ are given by the integral over all space

$$p^{\alpha}_{t=0} = \frac{1}{c}\int_{t=0} T^{\alpha}{}_0 \, d^3\mathbf{r} \tag{11.15}$$

of the electromagnetic stress tensor

$$T^{\alpha}{}_{\beta} = \begin{bmatrix} \mathcal{U} & -cg_x & -cg_y & -cg_z \\ cg_x & & & \\ cg_y & & \left(-T^{(M)}_{ij}\right) & \\ cg_z & & & \end{bmatrix} \tag{11.16}$$

defined in Chapter 2, where $\mathcal{U}$ is the energy density and $\mathbf{g}$ the momentum density of the electromagnetic field, and $T^{(M)}_{ij}$ is the Maxwell stress tensor. Although we have written the left-hand side of (11.15) in the form of a 4-vector, the expression on the right is not covariant. Time and space are not treated equivalently, and the lower index $t = 0$ is sort of dangling. We can put this in a covariant form as follows. In 4-space, the three-dimensional volume of integration represents a "hypersurface" defined in this case by the condition $t = 0 = $ constant. This surface is shown in Figure 11.1. It is a spacelike surface, since all intervals lying in the surface are spacelike:

$$\left(r^{\alpha}_{(2)} - r^{\alpha}_{(1)}\right)\left(r_{(2)\alpha} - r_{(1)\alpha}\right) = -\left|\mathbf{r}_{(2)} - \mathbf{r}_{(1)}\right|^2 < 0 \tag{11.17}$$

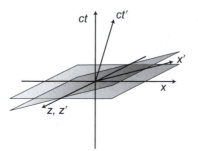

Figure 11.1 Hypersurfaces corresponding to instants of time in two reference frames.

The unit vector normal to this surface is

$$\hat{n}^\alpha_{t=0} = (1, 0) \tag{11.18}$$

We can therefore write (11.15) as a surface integral in the covariant form

$$p^\alpha_{t=0} = \frac{1}{c} \int_{t=0} T^\alpha{}_\beta \hat{n}^\beta_{t=0} \, d\Sigma \tag{11.19}$$

where $d\Sigma$ is an invariant infinitesimal area (actually a 3-dimensional area, or volume). We already know that the actual volume is not an invariant, but the quantity $d^3\mathbf{r}/\gamma$ is. Therefore, this is our invariant $d\Sigma$. However, in some reference frame $\gamma = 1$ and the invariant volume is the actual volume, making this frame special. In the case when the volume is filled with matter, the special frame is the rest frame of the matter and the invariant volume is the "proper volume" associated with that matter. However, there is no rest frame for an electromagnetic field, for a wave travels at the speed of light in all frames. Thus, the value of the total energy and momentum defined by (11.19) transforms like a vector but depends in general on the reference frame in which it is initially defined.

For example, let's say that (11.19) is the instantaneous energy and momentum of the electromagnetic field of an electron in the laboratory frame K, while

$$p'^\alpha_{t'=0} = \frac{1}{c} \int_{t'=0} T'^\alpha{}_\beta \hat{n}'^\beta_{t'=0} \, d\Sigma' \tag{11.20}$$

is the instantaneous energy and momentum in the electron rest frame K'. But in the laboratory frame, the hypersurface $t' = 0$ is tilted relative to the hypersurface $t = 0$ as shown in Figure 11.1, and the normal to the surface $t' = 0$ (the ct' axis) in the laboratory frame K is

$$\hat{n}^\alpha_{t'=0} = \gamma(1, \boldsymbol{\beta}) \tag{11.21}$$

We see that the 4-vector momenta $p^\alpha_{t=0}$ (the momentum in the frame K at the instant $t = 0$) and $p^\alpha_{t'=0}$ (the momentum at the instant $t' = 0$ transformed to the frame K) represent the energy and momentum at different instants of time, owing to the relativity of simultaneity.

Since they represent integrals over different hypersurfaces, the quantities $p^\alpha_{t=0}$ and $p^\alpha_{t'=0}$ are in general not the same, and we now ask the question under what circumstances are they the same? Only when this is the case are the instantaneous energy and momentum computed in one frame and transformed to another the same as the instantaneous energy and momentum computed in the other frame. To answer this question we consider the integral of the divergence of $T^{\alpha\beta}$ over the 4-volume Ω bounded by the hypersurface Σ. Since the divergence theorem works as well in four dimensions as in three, we get

$$\int_\Omega \partial_\alpha T^{\alpha\beta} \, d\Omega = \int_\Sigma T^\beta{}_\alpha \hat{n}^\alpha \, d\Sigma \tag{11.22}$$

where $T^{\alpha\beta} = T^{\beta\alpha}$. In the case when the 4-volume Ω is bounded by the hypersurfaces $t = 0$ and $t' = 0$, this becomes

$$\frac{1}{c} \int_\Omega \partial_\alpha T^{\alpha\beta} \, d\Omega = \frac{1}{c} \int_{t'=0} T^\beta{}_\alpha \hat{n}^\alpha_{t'=0} \, d\Sigma - \frac{1}{c} \int_{t=0} T^\beta{}_\alpha \hat{n}^\alpha_{t=0} \, d\Sigma = p^\beta_{t=0} - p^\beta_{t'=0} \tag{11.23}$$

To avoid the fact that the hypersurfaces $t = 0$ and $t' = 0$ intersect, we assume that the 4-volume Ω where the fields are nonvanishing is entirely to the right of the intersection, as shown in Figure 11.1. From (11.23), we see that the energy and momentum are the same only if the divergence $\partial_\alpha T^{\alpha\beta}$ (or, at least, the integral of the divergence) vanishes. But for the electromagnetic stress tensor, we have the momentum conservation law

$$\partial_\alpha T^{\alpha\beta} = J_\alpha F^{\alpha\beta} \tag{11.24}$$

derived in Chapter 2, where the right-hand side is the 4-vector Lorentz-force density

$$J_\alpha F^{\alpha\beta} = (\mathbf{J} \cdot \mathbf{E}, \, \rho\mathbf{E} + \mathbf{J} \times \mathbf{B}) \tag{11.25}$$

Thus, the instantaneous total energy and momentum of the fields defined in different coordinate systems are the same only when $\int_\Omega J_\alpha F^{\alpha\beta} \, d\Omega = 0$, that is, only in the absence of sources J^α. For photons, which are electromagnetic waves that have become detached from their sources, the instantaneous total energy and momentum of the electromagnetic fields themselves transform covariantly, as expected. For an electron, the source of the field is embedded within the field, so the instantaneous total energy and momentum computed in two different reference frames do not agree. This is the origin of the 4/3 problem.

Poincaré recognized that the failure of the energy and momentum of the electron to transform in the expected way is due to the fact that something has been left out. In the absence of any cohesive forces, the charges that form the electron will fly apart due to their mutual repulsion. He hypothesized, then, that the electron is held together by some unknown, internal forces, a sort of negative pressure. These forces might be of the form of the Lorentz scalar potential discussed in Chapter 2, for example. Given a nonvanishing rest mass, they would have a short range and could escape notice even short distances from the electron. If we attribute to these forces the Poincaré stress tensor $P^{\alpha\beta}$, the total stress tensor becomes

$$S^{\alpha\beta} = T^{\alpha\beta} + P^{\alpha\beta} \tag{11.26}$$

But from (11.23) and (11.24) we see that the net electromagnetic force density is represented by the divergence $\partial_\alpha T^{\alpha\beta}$ of the electromagnetic stress tensor, and the same is true of the cohesive force density and the divergence $\partial_\alpha P^{\alpha\beta}$. Therefore, if the electron is stable, the total force density must vanish and the condition for stability of the electron is

$$\partial_\alpha S^{\alpha\beta} = 0 \tag{11.27}$$

Since the divergence of the total stress tensor vanishes, we see from (11.23) that the total instantaneous energy and momentum of the electromagnetic and cohesive fields taken together are well defined and covariant.

Returning to the discussion at the beginning of this section, we can see explicitly how Poincaré stresses solve the 4/3 problem. The simplest form we can think of for the Poincaré stress tensor is just

$$P^{\alpha\beta} = pg^{\alpha\beta} \tag{11.28}$$

for some constant pressure p, where $g^{\alpha\beta}$ is the metric tensor. To make $pg^{\alpha\beta}$ a tensor, the pressure p must be a Lorentz invariant. For the case of a shell of charge, the pressure is just a constant inside the shell and zero outside. To find the value of the pressure, we observe that the normal component of the total stress tensor $S^{\alpha\beta} = T^{\alpha\beta} + P^{\alpha\beta}$ must be continuous at the surface of the electron. But just outside the shell of charge, the normal component of the stress tensor (along the x axis, say) is

$$T^{11} = T^{(M)}_{11} = -\varepsilon_0 E_1 E_1 + \mathcal{U}\delta_{11} = -\frac{1}{2}\varepsilon_0 E^2 = -\frac{q^2}{32\pi^2\varepsilon_0 a^4} = pg^{11} \tag{11.29}$$

where $g^{11} = -1$. The pressure is therefore

$$p = \frac{q^2}{32\pi^2\varepsilon_0 a^4} \tag{11.30}$$

The energy density of the field of the Poincaré stresses is $\mathcal{U}_P = P^{00} = pg^{00} = p$, so the total energy in the Poincaré stresses is

$$\mathcal{E}_P' = \frac{4}{3}\pi a^3 p = \frac{q^2}{24\pi\varepsilon_0 a} \tag{11.31}$$

This is just one third the energy $\mathcal{E}_S'$ in the electrostatic field. Combining this with (11.9), we see that the total energy in the electrostatic field and the Poincaré stresses is

$$\mathcal{E}_S' + \mathcal{E}_P' = \tfrac{4}{3}m_S c^2 \tag{11.32}$$

Since the momentum in the Poincaré stresses vanishes in all reference frames ($g^{0i} = 0$ in all reference frames for $i = 1\ldots 3$), this solves the 4/3 problem.

EXERCISE 11.1

The picture of an electron as a shell of charge is hardly realistic, but it is easily generalized to other charge distributions. We still write the Poincaré stress tensor in the form

$$P^{\alpha\beta} = pg^{\alpha\beta} \tag{11.33}$$

where the Lorentz scalar p is a function of the coordinates and the time for a moving particle. Consider an electron with the spherically symmetric charge distribution $\rho(r)$ in its rest frame.

(a) Beginning with the stability condition (11.27) and the divergence of the electromagnetic stress tensor from Chapter 2, show that the pressure satisfies the differential equation

$$\frac{dp}{dr} = -\frac{\rho(r)}{\varepsilon_0 r^2} \int_0^r \rho(r')r'^2\,dr' \tag{11.34}$$

(b) Show that if the charge q uniformly fills the volume of the electron to the radius a, the pressure is

$$p = \frac{3q^2}{32\pi^2\varepsilon_0 a^4}\left(1 - \frac{r^2}{a^2}\right), \qquad \text{for } r < a \tag{11.35}$$

if we require that it vanish for $r \geq a$.

(c) What is the total energy in the electrostatic field and in the Poincaré stresses? What has become of the 4/3 problem?

11.1.3 Point Particles and the Radiation Reaction

So long as the electron is in uniform motion, the self-fields surrounding the charge distribution are well defined, as discussed above, and the total energy and momentum of the electron (including the Poincaré stresses) form a good 4-vector. However, when the electron is accelerated, the fields become distorted and it is no longer possible to assign unique values

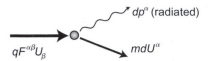

Figure 11.2 Accelerated particle including the radiative reaction.

to the total energy and momentum, or, equivalently, to the electron mass. In addition, as the fields become distorted, some of their energy and momentum are shaken off as radiation as described in Chapter 10. This must be accounted for in computing the motion of the electron, an effect we call the radiation reaction. For a sufficiently small electron, the fields are dominated by the nearby region, which can keep up with an accelerating electron because the light transit time is small, and the effects of the distortion are expected be small except in the far field. It therefore seems natural to describe the motion of the electron in terms of a unique mass but to include the effects of the radiation reaction. This is the approach we take here.

We begin by considering a point particle that accelerates relativistically in some coordinate system, as shown in Figure 11.2. During this motion the particle radiates 4-vector momentum at the rate given by the Larmor formula of Chapter 10,

$$\frac{dp^\alpha}{d\tau}(\text{radiated}) = -\frac{q^2}{6\pi\varepsilon_0 c^5}\frac{dU^\beta}{d\tau}\frac{dU_\beta}{d\tau}U^\alpha \tag{11.36}$$

where q is the charge and $U^\alpha = \gamma(c, \mathbf{v})$ the 4-vector velocity of the particle. But $U^\alpha U_\alpha = c^2$, and if we differentiate this twice we get

$$U_\alpha\frac{dU^\alpha}{d\tau} = 0 \tag{11.37}$$

$$\frac{dU^\alpha}{d\tau}\frac{dU_\alpha}{d\tau} + U^\alpha\frac{d^2U_\alpha}{d^2\tau} = 0 \tag{11.38}$$

so we can write (11.36) in the equivalent form

$$dp^\alpha(\text{radiated}) = \frac{q^2}{6\pi\varepsilon_0 c^5}\frac{d^2U_\beta}{d\tau^2}U^\beta U^\alpha \, d\tau \tag{11.39}$$

We now argue that the total momentum created by the Lorentz force on the particle is manifested in the acceleration of the particle itself (including its self-fields) and the momentum radiated. We then get the 4-vector equation of motion

$$m\frac{dU^\alpha}{d\tau} + \frac{q^2}{6\pi\varepsilon_0 c^5}\frac{d^2U_\beta}{d\tau^2}U^\beta U^\alpha = qF^{\alpha\beta}U_\beta \tag{11.40}$$

However, we see immediately that this can't be quite right, for if we take the inner product of this expression with U_α, the first term vanishes by (11.37) and the last term vanishes as well; that is, $F^{\alpha\beta}U_\alpha U_\beta = 0$, since $F^{\alpha\beta}$ is antisymmetric in α and β while $U_\alpha U_\beta$ is symmetric. We therefore must replace the radiation reaction term with something similar that is orthogonal to the 4-vector velocity. If we simply subtract the term $c^2 d^2 U^\alpha/d\tau^2$, we get the required expression and the equation of motion becomes

$$m\frac{dU^\alpha}{d\tau} = qF^{\alpha\beta}U_\beta + \frac{q^2}{6\pi\varepsilon_0 c^5}\left(c^2\frac{d^2U^\alpha}{d\tau^2} - \frac{d^2U^\beta}{d\tau^2}U_\beta U^\alpha\right) \tag{11.41}$$

This is called the Dirac equation of motion. In the nonrelativistic limit, the spacelike components of this equation reduce to

$$m\frac{d\mathbf{v}}{dt} = q(\mathbf{E} + \mathbf{v} \times \mathbf{B}) + \frac{q^2}{6\pi\varepsilon_0 c^3}\frac{d^2\mathbf{v}}{dt^2} \tag{11.42}$$

which is called the Abraham–Lorentz equation of motion.

Although it is attractive for its simplicity, the Abraham–Lorentz equation of motion has several serious difficulties. In the first place, it is of higher order (third) than the equation of motion without the radiation reaction. This means that three boundary conditions, rather than the two (position and velocity) to which we are accustomed, must be imposed on the motion of the particle to uniquely define its trajectory. Furthermore, in the absence of an external field, the equation of motion becomes

$$\frac{d\mathbf{v}}{dt} = \tau_c\left(\frac{d^2\mathbf{v}}{dt^2}\right) \tag{11.43}$$

which admits the solution

$$\mathbf{v} = \mathbf{v}_0 + \mathbf{v}_1 e^{t/\tau_c} \tag{11.44}$$

where $\mathbf{v}_0$ and $\mathbf{v}_1$ are constants and τ_c is defined by (11.7). The first term is the constant-velocity solution we expect in the absence of forces. The second term, the "runaway" solution, says that the particle accelerates forever, even if the external forces vanish. The existence of runaway solutions has to be regarded as a serious flaw.

More generally, if we assume that the Lorentz force is a function of the time, so that

$$q(\mathbf{E} + \mathbf{v} \times \mathbf{B}) = \mathbf{F}(t) \tag{11.45}$$

the equation of motion becomes

$$\frac{d\mathbf{v}}{dt} = \frac{\mathbf{F}(t)}{m} + \tau_c\frac{d^2\mathbf{v}}{dt^2} \tag{11.46}$$

Using the integrating factor e^{-t/τ_c}, we immediately find that the general solution for a particle that is at rest in the limit $t \to -\infty$ is

$$\frac{d\mathbf{v}}{dt} = -\frac{1}{m\tau_c}\int_{-\infty}^{t}\mathbf{F}(t')e^{(t-t')/\tau_c}\,dt' \tag{11.47}$$

For a pulse of finite length, the acceleration does not cease when the pulse ends and we have the runaway solution as before.

Dirac has pointed out that the runaway solution can be avoided by judicious choice of the extra boundary condition on the acceleration. This is equivalent to choosing the alternate solution

$$\frac{d\mathbf{v}}{dt} = \frac{1}{m\tau_c}\int_{t}^{\infty}\mathbf{F}(t')e^{(t-t')/\tau_c}\,dt' \tag{11.48}$$

For a pulse of finite duration, the acceleration vanishes for $t \to -\infty$ (due to the exponential) and also for $t \to +\infty$. Unfortunately, this solution predicts motion prior to the arrival of the pulse, which would seem to violate the principle of causality. However, the preacceleration is confined to times on the order of τ_c, the time required for light to cover a distance on the order of the classical radius of the particle. As discussed earlier, such short times are, in some sense, outside the scope of classical theory. When the force **F** is known as a function of the coordinates and velocity, rather than time, (11.48) becomes an integral equation for the motion of the electron with no runaway solutions. However, the instantaneous motion of the electron still depends upon forces yet to come. Similar considerations apply to the relativistic Dirac equation of motion (11.41).

EXERCISE 11.2

Use the Abraham–Lorentz equation of motion (11.42) to show that if a free electron is placed in an electromagnetic wave $\mathbf{E} = \mathbf{E}_0 e^{-i\omega t}$, the (complex) amplitude of the oscillations in the nonrelativistic limit ($|\mathbf{v}_0|/c \ll 1$) is

$$\mathbf{v}_0 = i\frac{q}{m\omega}\frac{\mathbf{E}_0}{1 + i\omega\tau_c} \tag{11.49}$$

The rate at which the field does work on the electron is $q\mathbf{E} \cdot \mathbf{v}$, and this is the rate at which energy is removed from the incident field and reradiated (scattered). Averaged over one cycle of the wave, the power is $\frac{1}{2}q\,\mathrm{Re}(\mathbf{E}_0^* \cdot \mathbf{v}_0)$. Show that the cross section for scattering the incident radiation is

$$\sigma = \frac{\sigma_T}{1 + \omega^2\tau_c^2} \tag{11.50}$$

where $\sigma_T = q^4/6\pi\varepsilon_0^2 m^2 c^4$ is the Thomson scattering cross section. This differs from the classical Thomson result at high frequencies [$\omega\tau_c \geq O(1)$]. However, at frequencies for which the Abraham–Lorentz equation of motion predicts deviations from the Thomson formula, the photon energy is large compared with the electron rest mass, as shown by (11.8). Thus, the situation is outside the scope of classical mechanics.

EXERCISE 11.3

In the real world the radiation reaction is small and can be treated as a perturbation. Show that for a force $\mathbf{F}(t)$ that acts for a finite time with a characteristic time scale $\tau \gg \tau_c = 2r_c/3c$, where r_c is the classical radius of the electron, the motion of a nonrelative particle that starts from rest can be expanded in a perturbation series of the form

$$\mathbf{v}(t) = \sum_n \varepsilon^n \mathbf{v}^{(n)}(t) \tag{11.51}$$

where

$$\varepsilon = \frac{\tau_c}{\tau} \ll 1 \tag{11.52}$$

and

$$\mathbf{v}^{(0)}(t) = \int_{-\infty}^t \frac{\mathbf{F}(t')}{m}\,dt' \tag{11.53}$$

$$\mathbf{v}^{(n>0)}(t) = \frac{1}{m}\frac{d^{n-1}\mathbf{F}(t)}{dt^{n-1}} \tag{11.54}$$

This solution avoids the problems of preacceleration or runaway acceleration, but it has the difficulty that for impulses shorter than the damping time τ_c the series diverges. More troublesome is the fact that the final energy and momentum, after the force vanishes, are not perturbed by the radiation reaction.

11.1.4 Extended Particles

Problems of runaway solutions and violation of causality appear together with the divergent values of the energy and momentum when we consider point particles. We therefore look for a description of an electron as an extended particle. Unfortunately, this requires a description of the charge distribution of the electron that must change shape not only due to relativistic effects but also in response to the external field. To avoid this complexity, we picture the electron as a rigid, spherically symmetric distribution of charge. However, rigid bodies are impossible in relativistic motion, so this restricts our theory, at least in its basic form, to the nonrelativistic limit. We consider the possible relativistic generalizations later.

The self-force on a charge distribution is the result of the repulsion of each element of charge in the distribution by every other element of charge, as shown in Figure 11.3. When the charge distribution is at rest or in uniform motion (which puts it at rest in some inertial coordinate system) the total self-force vanishes by cancellation. However, when the charge is accelerated a net self-force appears. Since the charge elements near the front of the distribution are accelerating away from those near the back, the retarded fields from the charges near the back are diminished by the additional distance traveled by the field from the time it is "emitted" at the back to the time it arrives at the front of the charge distribution. This reduces the repulsion that accelerates the charges near the front, while the same effect increases the fields that retard the charges near the back. This leads to a net retarding self-force on the charge distribution. For simple charge distributions, such as a spherical shell of charge, we can compute the net force directly from the retarded fields by integrating over the charge distribution. If we do this and expand the result in a series in terms of the derivatives of the velocity or, equivalently, in powers of the size of the distribution, the first term, which is proportional to the acceleration, represents the electromagnetic mass. This term is inversely proportional to the size of the charge distribution and diverges as the size vanishes. The second term, which is proportional to the rate of change of the acceleration, represents the radiation reaction. It is independent of the size of the distribution. Higher order terms are proportional to positive powers of the size of the distribution and vanish in the limit when the size of the distribution vanishes. If we neglect these higher order terms, we recover the Abraham–Lorentz equation of motion.

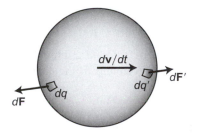

Figure 11.3 Self-force becomes the electromagnetic mass and the radiative reaction.

Here we take a different approach to evaluating the self-force. We avoid using a Taylor series expansion of the velocity in terms of its derivatives at time $t = 0$ and end up with an integral equation instead of a differential equation. In the limit of small electron size or small acceleration, when we can approximate the expressions in the integral with a Taylor series, we recover the Abraham–Lorentz equation of motion, but if we use the integral equation itself we avoid the problems of preacceleration and self-acceleration. We are limited to the nonrelativistic case by our assumption of a rigid charge distribution, but if we ignore the fact that the equation of motion is derived with the assumption of a rigid particle and simply accept the equation of motion that we find, we can generalize from the rest frame of the particle to any other frame. This gives us the relativistic equation of motion.

In the nonrelativistic approximation we can ignore the effects of the self-magnetic field, since these are of higher order in the velocity of the particle. By combining the Maxwell equations, we find that the electric field satisfies the wave equation

$$\nabla^2 \mathbf{E} - \frac{1}{c^2} \frac{\partial^2 \mathbf{E}}{\partial t^2} = \frac{1}{\varepsilon_0} \left(\nabla \rho + \frac{1}{c^2} \frac{\partial \mathbf{J}}{\partial t} \right) \tag{11.55}$$

If we take the Fourier transform of this with respect to $\mathbf{r}$, we get

$$k^2 \mathbf{E_k} + \frac{1}{c^2} \frac{\partial^2 \mathbf{E_k}}{\partial t^2} = \frac{1}{\varepsilon_0} \left(i \mathbf{k} \rho_\mathbf{k} - \frac{1}{c^2} \frac{\partial \mathbf{J_k}}{\partial t} \right) \tag{11.56}$$

where the Fourier transform of the function $g(\mathbf{r})$ is

$$g_\mathbf{k} = \frac{1}{(2\pi)^{3/2}} \int d^3 r e^{i\mathbf{k}\cdot\mathbf{r}} g(\mathbf{r}) \tag{11.57}$$

The retarded Green function (the response at time t to a δ-function source at time t') for the wave equation (11.56) is

$$G(t, t') = \begin{cases} \dfrac{c}{k} \sin[kc(t - t')], & t > t' \\ 0, & t < t' \end{cases} \tag{11.58}$$

so after an integration by parts to eliminate the derivative $\partial \mathbf{J_k}/\partial t$, the retarded solution of (11.56) becomes

$$\mathbf{E_k} = \frac{1}{\varepsilon_0} \int_{-\infty}^{t} dt' \{ i\hat{\mathbf{k}}c\rho_\mathbf{k} \sin[kc(t - t')] - \mathbf{J_k} \cos[kc(t - t')] \} \tag{11.59}$$

For a rigid, spherically symmetric charge distribution we can write the charge and current densities in the form

$$\rho(\mathbf{r}) = q f[|\mathbf{r} - \mathbf{r}_0(t)|] \tag{11.60}$$

$$\mathbf{J}(\mathbf{r}) = q\mathbf{v}_0(t) f[|\mathbf{r} - \mathbf{r}_0(t)|] \tag{11.61}$$

where q is the total charge, $f(r)$ a radial form factor describing the normalized distribution of charge, $\mathbf{r}_0(t)$ the position, and $\mathbf{v}_0(t)$ the velocity of the charge. The required Fourier transforms are then

$$\rho_\mathbf{k} = q f_k e^{i\mathbf{k}\cdot\mathbf{r}_0(t)} \tag{11.62}$$

$$\mathbf{J_k} = q\mathbf{v}_0(t) f_k e^{i\mathbf{k}\cdot\mathbf{r}_0(t)} \tag{11.63}$$

in which

$$f_k = \frac{1}{(2\pi)^{3/2}} \int d^3\mathbf{r} f(r) e^{i\mathbf{k}\cdot\mathbf{r}} \xrightarrow[k\to 0]{} \frac{1}{(2\pi)^{3/2}} \qquad (11.64)$$

is the Fourier transform of the form factor. Since $f(r)$ is spherically symmetric, its Fourier transform depends only on $k = |\mathbf{k}|$. With a simple change of variable, the Fourier transform of the electric field becomes

$$\mathbf{E_k} = \frac{q}{\varepsilon_0} f_k \int_0^\infty dt'\, e^{i\mathbf{k}\cdot\mathbf{r}_0(t-t')}[i\hat{\mathbf{k}}c\sin(kct') - \mathbf{v}_0(t-t')\cos(kct')] \qquad (11.65)$$

Clearly, these integrals have meaning only if we include some weakly decaying factor to make the integrand vanish as $t' \to \infty$ and take the limit as the decay rate becomes small. If we integrate the first term by parts, we get

$$\mathbf{E_k} = i\frac{q}{\varepsilon_0}\frac{\hat{\mathbf{k}}}{k} f_k e^{i\mathbf{k}\cdot\mathbf{r}_0(t)} - \frac{q}{\varepsilon_0} f_k \int_0^\infty dt'\, e^{i\mathbf{k}\cdot\mathbf{r}_0(t-t')}\cos(kct')\{\mathbf{v}_0(t-t') - \hat{\mathbf{k}}[\hat{\mathbf{k}}\cdot\mathbf{v}_0(t-t')]\}$$

$$(11.66)$$

The total self-force is found by integrating the electric field over the entire charge distribution. For a spherically symmetric charge distribution, the first term in (11.66) vanishes by symmetry and we get

$$\mathbf{F}_0(t) = \int d^3\mathbf{r}\,\rho\mathbf{E} = \int d^3\mathbf{k}\,\rho_\mathbf{k}\mathbf{E_k^*}$$

$$= -\frac{q^2}{\varepsilon_0}\int_0^\infty dt' \int d^3\mathbf{k}\, f_k f_k^* e^{i\mathbf{k}\cdot[\mathbf{r}_0(t)-\mathbf{r}_0(t-t')]}\cos(kct')\{\mathbf{v}_0(t-t') - \hat{\mathbf{k}}[\hat{\mathbf{k}}\cdot\mathbf{v}_0(t-t')]\}$$

$$(11.67)$$

But in the nonrelativistic limit the exponent is small, and we may approximate the exponential factor by unity. Actually, this approximation fails for sufficiently large k, but we are assured by the Riemann–Lebesgue lemma that $f_\mathbf{k}$ vanishes in the limit $k \to \infty$, which corresponds to small dimensions. For a smooth charge distribution characterized by the length a, the transform vanishes for $k > O(1/a)$, while the difference $|\mathbf{r}_0(t) - \mathbf{r}_0(t-t')| = O(v\Delta t) = O(va/c)$, where v is a characteristic velocity. The exponent is therefore $O(v/c) \ll 1$. For a spherically symmetric charge distribution, the remaining factors may be averaged over all orientations of $\mathbf{k}$ and we find that the average of the expression in braces is just $\frac{2}{3}\mathbf{v}_0(t-t')$. The self-force is therefore

$$\mathbf{F}_0(t) = -\frac{2q^2}{3\varepsilon_0}\int_0^\infty dt'\,\mathbf{v}_0(t-t')\int d^3\mathbf{k}\, f_k f_k^* \cos(kct') \qquad (11.68)$$

In the presence of external electric and magnetic fields $\mathbf{E}$ and $\mathbf{B}$, the force on the electron is the average of the Lorentz force over the charge distribution. Provided that the external fields are not too rapidly varying, the average field is equal to the field at the center of the electron. The equation of motion of the electron is then

$$m_0\frac{d\mathbf{v}}{dt} = \mathbf{F}_\infty(t) + \mathbf{F}_0(t) = \mathbf{F}_\infty(t) - \frac{2q^2}{3\varepsilon_0}\int_0^\infty dt'\,\mathbf{v}(t-t')\int d^3\mathbf{k}\, f_k f_k^* \cos(kct') \qquad (11.69)$$

where m_0 is the bare mass of the electron (ignoring the effective mass of the electron's self-field),

$$\mathbf{F}_\infty(t) = q\{\mathbf{E}[\mathbf{r}(t), t] + \mathbf{v}(t) \times \mathbf{B}[\mathbf{r}(t), t]\} \tag{11.70}$$

is the Lorentz force due to the external fields, and we have dropped the subscript 0 from the position $\mathbf{r}$ and velocity $\mathbf{v}$ of the electron to simplify the notation. The function

$$K(t') = \int d^3\mathbf{k}\, f_k f_k^* \cos(kct') \tag{11.71}$$

weights the history of the trajectory at times shortly before the present. We see that the self-force term in (11.69) depends only on the history of the particle, not its future, so there are no violations of causality. We see later under what conditions runaway solutions are avoided.

If we integrate (11.69) by parts with respect to the time t', we get the equation of motion

$$m_0 \frac{d\mathbf{v}}{dt} = \mathbf{F}_\infty(t) - \frac{2q^2}{3\varepsilon_0 c} \int_0^\infty dt' \frac{d\mathbf{v}(t - t')}{dt} \int d^3\mathbf{k} \frac{f_k f_k^*}{k} \sin(kct') \tag{11.72}$$

which was first derived by Markov. If we integrate by parts once again with respect to t', we get

$$m_0 \frac{d\mathbf{v}}{dt} = \mathbf{F}_\infty(t) - \frac{2q^2}{3\varepsilon_0 c^2} \frac{d\mathbf{v}(t)}{dt} \int d^3\mathbf{k} \frac{f_k f_k^*}{k^2}$$

$$+ \frac{2q^2}{3\varepsilon_0 c^2} \int_0^\infty dt' \frac{d^2\mathbf{v}(t - t')}{dt^2} \int d^3\mathbf{k} \frac{f_k f_k^*}{k^2} \cos(kct') \tag{11.73}$$

We identify the second term on the right as the electromagnetic mass term, where the electromagnetic energy is

$$m_E c^2 = \frac{2q^2}{3\varepsilon_0} \int \frac{f_k f_k^*}{k^2} d^3\mathbf{k} \tag{11.74}$$

But from (11.66) we see that for a stationary particle, the energy in the electrostatic field is

$$m_S c^2 = \mathcal{E}_S = \frac{\varepsilon_0}{2} \int \mathbf{E} \cdot \mathbf{E} d^3\mathbf{r} = \frac{q^2}{2\varepsilon_0} \int \frac{f_k f_k^*}{k^2} d^3\mathbf{k} \tag{11.75}$$

That is, the electromagnetic energy $m_E c^2$ is larger by the factor 4/3 than the electrostatic energy $m_S c^2$ of a stationary particle, as discussed earlier. The last term in (11.73) is the radiation reaction. The equation of motion is then

$$m \frac{d\mathbf{v}}{dt} = \mathbf{F}_\infty + \frac{2q^2}{3\varepsilon_0 c^2} \int_0^\infty dt' \frac{d^2\mathbf{v}(t - t')}{dt^2} \int d^3\mathbf{k} \frac{f_k f_k^*}{k^2} \cos(kct') \tag{11.76}$$

where the observed mass

$$m = m_0 + m_E \tag{11.77}$$

is the sum of the bare mass and the electromagnetic mass.

We can make further approximations when the size of the particle shrinks to a point. The kernel $K(t') = \int f_k f_k^* \cos(kct')dk$ vanishes for times $ct' \gg a$, where a is a length that is characteristic of the radius of the charge distribution. In the limit of a small particle, the width of the kernel becomes very small and we can expand the velocity in a Taylor series of the form

$$\frac{d^2\mathbf{v}(t - t')}{dt^2} \approx \frac{d^2\mathbf{v}(t)}{dt^2} + \cdots \tag{11.78}$$

Substituting this into (11.76), we see that the equation of motion becomes

$$m\frac{d\mathbf{v}}{dt} = \mathbf{F}_\infty + \frac{2q^2}{3\varepsilon_0 c^2}\frac{d^2\mathbf{v}}{dt^2}\int d^3\mathbf{k}\,\frac{f_k f_k^*}{k^2}\int_0^\infty dt'\cos(kct') \tag{11.79}$$

With the order of integration inverted in this way, we must remember that the integral over t' really exists only when a factor is included to make the integrand vanish as $t' \to \infty$, in the limit when the rate of decay vanishes. Keeping this in mind, we find that

$$\int_0^\infty \cos(kct')\,dt' = \pi\delta(kc) = \frac{\pi}{c}\lim_{\varepsilon\to 0}\frac{e^{-k^2/4\varepsilon}}{\sqrt{4\pi\varepsilon}} \tag{11.80}$$

The definition of the δ-function as the limit as $\varepsilon \to 0$, from Chapter 5, is useful in the next step, for we see that

$$\int d^3\mathbf{k}\,\frac{f_k f_k^*}{k^2}\int_0^\infty \cos(kct')\,dt' = \frac{\pi}{c}\lim_{\varepsilon\to 0}\int_0^\infty d^3\mathbf{k}\,\frac{f_k f_k^*}{k^2}\frac{e^{-k^2/4\varepsilon}}{\sqrt{4\pi\varepsilon}} = \frac{1}{4\pi c} \tag{11.81}$$

where we have used (11.64) to evaluate $\lim_{k\to 0} f_k$. Substituting this expression into (11.79), we recover the Abraham–Lorentz equation of motion

$$m\frac{d\mathbf{v}}{dt} = \mathbf{F}_\infty + \frac{q^2}{6\pi\varepsilon_0 c^3}\frac{d^2\mathbf{v}}{dt^2} \tag{11.82}$$

for point charges.

Because the theory we are discussing pictures the electron as a rigid distribution of charge, it is not straightforward to generalize it to the relativistic case. An alternative point of view takes the position that the electron should be regarded as a structureless but complex particle for which the fundamental equation of motion is not a differential equation at all, but an integro-differential equation of the form (11.69). Since this approach avoids the conceptual problems of a relativistic rigid body, we can generalize the Markov equation to the relativistic case in the usual way by constructing a covariant expression that reduces to (11.69) in the nonrelativistic limit. For $m_0\,d\mathbf{v}/dt$ and $\mathbf{F}_\infty$, the required 4-vector expressions are clearly $m_0\,dU^\alpha/dt$ and $F^{\alpha\beta}U_\beta$. For the remaining term we require an expression $X^\alpha = (X^0, \mathbf{X})$ that has the properties

$$\mathbf{X}(t) \xrightarrow[v/c\to 0]{} \int_0^\infty dt'\,\mathbf{v}(t - t')K(t') \tag{11.83}$$

in the nonrelativistic limit, and

$$X^\alpha U_\alpha = 0 \tag{11.84}$$

since this is identically true of the other terms. The required expression is

$$X^\alpha = \int_0^\infty d\tau'\,U^\alpha(\tau - \tau')K(\tau') - \frac{U^\alpha(\tau)U_\beta(\tau)}{c^2}\int_0^\infty d\tau'\,U^\beta(\tau - \tau')K(\tau') \tag{11.85}$$

so the Markov equation of motion becomes

$$m \frac{dU^\alpha}{d\tau} = U^\alpha F_{\alpha\beta}$$

$$-\frac{2q^2}{3\varepsilon_0} \left[\int_0^\infty d\tau' \, U^a(\tau - \tau') K(\tau') - \frac{U^\alpha(\tau) U_\beta(\tau)}{c^2} \int_0^\infty d\tau' U^\beta(\tau - \tau') K(\tau') \right] \tag{11.86}$$

As discussed earlier, the Abraham–Lorentz equation of motion (11.82) has runaway solutions. To see under what conditions the Markov equation of motion admits runaway solutions, we examine the case when the externally applied fields vanish. If we look for a solution of the form

$$\frac{d\mathbf{v}}{dt} = \mathbf{a}_0 e^{\kappa ct} \tag{11.87}$$

in which $\mathbf{a}_0$ and κ are constants, and substitute into (11.72) we get

$$m_0 = -\frac{2q^2}{3\varepsilon_0 c} \int_0^\infty dt' \, e^{-\kappa ct'} \int d^3\mathbf{k} \, \frac{f_k f_k^*}{k} \sin(kct') \tag{11.88}$$

If we carry out first the integral over t' we get

$$m_0 = -\frac{8\pi q^2}{3\varepsilon_0 c} \int_0^\infty \frac{k^2 f_k f_k^* \, dk}{k^2 + \kappa^2} \tag{11.89}$$

Since for real κ the integrand is everywhere positive, there can be a solution of the form (11.87) only if the bare mass m_0 is negative, and this is the condition for runaway solutions. Equivalently we may say that the Markov equation of motion is stable if the observed mass is not less than the mass of the electromagnetic field,

$$m \geq m_E = \frac{4\mathcal{E}_S}{3c^2} \tag{11.90}$$

where $\mathcal{E}_S$ is the energy in the electrostatic field of a stationary electron. For a point electron the electromagnetic mass diverges, so this is not satisfied. This is the reason that the Abraham–Lorentz equation of motion has runaway solutions. For a spherical-shell model of the electron, the electromagnetic mass is

$$m_E = \frac{2r_c}{3a} m \tag{11.91}$$

in which a is the electron radius and r_c is the classical electron radius. Thus, for stability the radius of an electron must be $a \geq \frac{2}{3} r_c = 1.88 \times 10^{-15}$ m. Unfortunately, the radius (if any) of the electron is believed to be much smaller than this.

A curious property of the Markov equation of motion (11.69) is the possibility of self-oscillation. In terms of the preceding discussion, self-oscillation corresponds to imaginary values of κ in (11.87), which are possible for certain forms of the charge distribution $f_\mathbf{k}$ and certain values of the bare mass m_0. For example, if the electron is a spherical shell of charge, the Markov equation of motion becomes a difference equation of the form

$$m_0 \frac{d\mathbf{v}}{dt} = \mathbf{F}_\infty - \frac{q^2}{12\pi \varepsilon_0 a^2 c} [\mathbf{v}(t) - \mathbf{v}(t - 2a/c)] \tag{11.92}$$

which is called the Sommerfeld–Page equation of motion. When the bare mass m_0 vanishes, that is, when all the observed mass of the particle is attributable to the electromagnetic mass, and there are no external fields, the equation of motion (11.92) becomes simply

$$\frac{q^2}{12\pi\varepsilon_0 a^2 c}[\mathbf{v}(t) - \mathbf{v}(t - 2a/c)] = 0 \tag{11.93}$$

This is satisfied by any periodic function $\mathbf{v}(t)$ with a period $2a/c$. In these oscillations the electron moves back and forth against the pull of its own retarded self-field. The oscillations are nonradiating, since they are undamped, and are therefore not directly observable. It was suggested at one time that the quantized energy of these oscillations might account for the various masses of the observed mesons. For example, the first excited level of the oscillations of an electron would have energy

$$\Delta\mathcal{E} = \hbar\omega = O\left(\frac{hc}{r_c}\right) = O\left(\frac{4\pi\varepsilon_0 hmc^3}{q^2}\right) = O(0.4\,\text{GeV}) \tag{11.94}$$

This is certainly of the right order of magnitude for mesons and, perhaps more appropriately, the muon and the tau. Unfortunately, this doesn't explain the masses of neutral particles and appears now to be just a coincidence.

EXERCISE 11.4

Consider the simple case when the electronic charge distribution is a thin shell of charge with a finite radius a.

(a) Show that the Markov equation of motion becomes the difference equation (11.92), called the Sommerfeld–Page equation of motion. *Hint:* Show that the self-force may be expressed

$$\mathbf{F}_0(t) = -\frac{q^2}{6\pi^2\varepsilon_0 a^2}\int_0^\infty dt'\, \mathbf{v}_0(t - t')\int_0^\infty dk\,[1 - \cos(2ka)]\cos(kct') \tag{11.95}$$

The integral over k leads to δ-functions, one of which appears at $t' = 0$, which is the end of the interval of integration over t'. To avoid having the singularity at the end of the interval use the device of setting $1 = \lim_{\varepsilon\to 0}\cos(\varepsilon k)$ in the integral over k and take the limit $\varepsilon \to 0$ at the end.

(b) Show that the relativistic generalization of the Sommerfeld–Page equation of motion is

$$m_0\frac{dU^\alpha}{d\tau} = F^{\alpha\beta}U_\beta(\tau)$$

$$+ \frac{q^2}{12\pi\varepsilon_0 a^2 c}\left[U^\alpha\left(\tau - \frac{2a}{c}\right) - \frac{U^\alpha(\tau)U_\beta(\tau)}{c^2}U^\beta\left(\tau - \frac{2a}{c}\right)\right] \tag{11.96}$$

EXERCISE 11.5

As an alternative to the heuristic derivation given in the previous section, the Abraham–Lorentz equation of motion can be derived by computing the self-force of the retarded fields of the electron, on the electron itself, as shown in Figure 11.4. Consider the electron as a rigid (and therefore nonrelativistic) charge distribution $\rho(\mathbf{r}, t)$. In the instantaneous rest frame of the electron, the magnetic forces vanish identically and we require only the retarded electric field at the point $\mathbf{r}$. As discussed in Chapter 10, this is

$$\mathbf{E}(\mathbf{r}, t) = \frac{1}{4\pi\varepsilon_0} \int_{-\infty}^{\infty} d^3\mathbf{r}' \left[\frac{\hat{\mathbf{R}}}{R^2}\rho + \frac{\hat{\mathbf{R}}}{cR}\frac{\partial\rho}{\partial t'} - \frac{1}{c^2 R}\frac{\partial\mathbf{J}}{\partial t'} \right]_{\text{retarded}} \tag{11.97}$$

where $\mathbf{R} = \mathbf{r} - \mathbf{r}'$. The expression in brackets is to be evaluated at the retarded time $t' = t - (R/c)$. If the characteristic size of the charge distribution is a and the characteristic time of the motion is Δt, then for $a/c\Delta t \ll 1$ the retarded time is close to the actual time, and we may use for $\rho(\mathbf{r}', t')$ and $\mathbf{J}(\mathbf{r}', t')$ the Taylor series expansion

$$[f(\mathbf{r}', t')]_{\text{retarded}} = \sum_{n=0}^{\infty} \frac{1}{n!} \left(\frac{-R}{c} \right)^n \frac{\partial^n f(\mathbf{r}', t)}{\partial t^n} \tag{11.98}$$

(a) Substitute in (11.97) and rearrange the sums to show that

$$\mathbf{E}(\mathbf{r}, t) = \frac{1}{4\pi\varepsilon_0} \int_{-\infty}^{\infty} d^3\mathbf{r}' \frac{\hat{\mathbf{R}}}{R^2}\rho$$

$$- \frac{\mu_0}{4\pi} \sum_{n=0}^{\infty} \frac{(-1)^n}{n!c^n} \int_{-\infty}^{\infty} d^3\mathbf{r}' R^{n-1} \frac{\partial^{n+1}}{\partial t^{n+1}} \left[\frac{\mathbf{R}}{n+2}\frac{\partial\rho}{\partial t} + \mathbf{J} \right] \tag{11.99}$$

Then use the continuity relation $\partial\rho(\mathbf{r}', t)/\partial t + \nabla' \cdot \mathbf{J}(\mathbf{r}', t) = 0$ and integrate by parts to show that

$$\mathbf{E}(\mathbf{r}, t) = \frac{1}{4\pi\varepsilon_0} \int_{-\infty}^{\infty} d^3\mathbf{r}' \frac{\hat{\mathbf{R}}}{R^2}\rho$$

$$+ \frac{\mu_0}{4\pi} \sum_{n=0}^{\infty} \frac{(-1)^n}{n!c^n} \int_{-\infty}^{\infty} d^3\mathbf{r}' R^{n-1} \frac{\partial^{n+1}}{\partial t^{n+1}} \left[\frac{n-1}{n+2}(\mathbf{J} \cdot \hat{\mathbf{R}})\hat{\mathbf{R}} - \frac{n+1}{n+2}\mathbf{J} \right]$$

$$\tag{11.100}$$

Figure 11.4 Direct computation of the self-force.

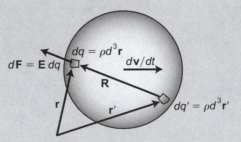

(b) For a rigid charge distribution, the current density is

$$\mathbf{J}(\mathbf{r}, t) = \mathbf{v}(t)\rho(\mathbf{r}, t) \tag{11.101}$$

where $\mathbf{v}(t)$ is the velocity of the electron. Show that for a spherically symmetric charge distribution, the self-force is

$$\mathbf{F}(t) = \int_{-\infty}^{\infty} d^3\mathbf{r}\rho(\mathbf{r}, t)\mathbf{E}(\mathbf{r}, t)$$

$$= -\frac{\mu_0}{6\pi}\sum_{n=0}^{\infty}\frac{(-1)^n}{n!c^n}\int_{-\infty}^{\infty}d^3\mathbf{r}\int_{-\infty}^{\infty}d^3\mathbf{r}'R^{n-1}\rho(\mathbf{r}, t)\frac{\partial^{n+1}}{\partial t^{n+1}}[\rho(\mathbf{r}', t)\mathbf{v}(t)]$$

$$\tag{11.102}$$

(c) Show that to lowest (first) order in the velocity and its derivatives, the self-force is

$$\mathbf{F} = -\frac{\mu_0}{6\pi}\sum_{n=0}^{\infty}\frac{(-1)^n}{n!c^n}q^2\langle R^{n-1}\rangle\frac{d^{n+1}\mathbf{v}}{dt^{n+1}} \tag{11.103}$$

where

$$q = \int\rho(\mathbf{r})\,d^3\mathbf{r} \tag{11.104}$$

and

$$q^2\langle R^n\rangle = \int_{-\infty}^{\infty}d^3\mathbf{r}\int_{-\infty}^{\infty}d^3\mathbf{r}'R^n\rho(\mathbf{r}, t)\rho(\mathbf{r}', t) \tag{11.105}$$

The first few terms are

$$\mathbf{F}(t) = -\frac{4}{3}m_S\frac{d\mathbf{v}}{dt} + \frac{q^2}{6\pi\varepsilon_0}\frac{d^2\mathbf{v}}{dt^2} + O(a) \tag{11.106}$$

where the mass m_S corresponding to the electrostatic energy is

$$m_S = \frac{\mu_0}{8\pi}\iint d^3\mathbf{r}'\,d^3\mathbf{r}\frac{\rho(\mathbf{r})\rho(\mathbf{r}')}{R} \tag{11.107}$$

The first term on the right-hand side of (11.106) represents the electromagnetic inertia, including the factor 4/3 discussed earlier in this chapter, and the second term represents the radiative reaction. Adding these terms to the Lorentz force due to the external fields gives the Abraham–Lorentz equation of motion. The higher order terms vanish in the limit $a \to 0$.

11.2 MAGNETIC MONOPOLES

11.2.1 The Maxwell Equations

As noted in Chapter 1, the inhomogeneous Maxwell equations may be expressed in the covariant form

$$\partial_\alpha F^{\alpha\beta} = \mu_0 J^\beta \tag{11.108}$$

while the homogeneous equations are

$$\partial_\alpha \mathcal{F}^{\alpha\beta} = 0 \tag{11.109}$$

The symmetry between these equations is broken by the fact that there are no sources in the second set of equations (11.109). The symmetry can be made complete if we postulate the existence of magnetic sources, called magnetic monopoles, so that the homogeneous equations become

$$\partial_\alpha \mathcal{F}^{\alpha\beta} = \mu_0 J_m{}^\beta \tag{11.110}$$

where $J_m{}^\beta$ is the 4-vector magnetic monopole current density. Note that the dimensions of $\mathcal{F}^{\alpha\beta}$ are the same as those of $F^{\alpha\beta}$, so the units of magnetic charge and current are the same as those of ordinary charge and current, namely coulombs and amperes. The 4-vector magnetic current density of a single monopole is

$$J_m{}^\alpha = (\rho_m c, \mathbf{J}_m) = q_m(c, \mathbf{v})\delta[\mathbf{r} - \mathbf{r}_0(t)] \tag{11.111}$$

where ρ_m is the magnetic charge density, q_m the magnetic charge on the monopole, $\mathbf{r}_0(t)$ its position, and $\mathbf{v} = d\mathbf{r}_0/dt$ its velocity. Note that since $\mathcal{F}^{\alpha\beta}$ is a pseudotensor, q_m is a pseudoscalar. The sign of the magnetic charge changes in a left-handed coordinate system or under time reversal. Likewise, $J_m{}^\alpha$ is a pseudovector. In 3-vector notation, the Maxwell equations are now

$$\nabla \cdot \mathbf{E} = \frac{\rho}{\varepsilon_0} \tag{11.112}$$

$$\nabla \times \mathbf{B} - \frac{1}{c^2}\frac{\partial \mathbf{E}}{\partial t} = \mu_0 \mathbf{J} \tag{11.113}$$

$$\nabla \cdot \mathbf{B} = \mu_0 c \rho_m \tag{11.114}$$

$$\nabla \times \mathbf{E} + \frac{\partial \mathbf{B}}{\partial t} = -\mu_0 c \mathbf{J}_m \tag{11.115}$$

From the time derivative of (11.114) and the divergence of (11.115), we get the conservation law

$$\partial_\alpha J_m{}^\alpha = 0 \tag{11.116}$$

for magnetic charge.

Although magnetic monopoles have never been observed in nature, their existence is appealing for several reasons. Not only would they symmetrize the Maxwell equations, as discussed earlier, they would also, according to an argument first suggested by Dirac, explain the quantization of charge in units of some fundamental charge, such as the charge on a quark. Before discussing these arguments, however, it is necessary to point out the arbitrariness of the division of charge between electronic and magnetic contributions. To show this, we note that we may define new "electric" and "magnetic" fields by the duality transformation

$$F'^{\alpha\beta} = F^{\alpha\beta}\cos\theta + \mathcal{F}^{\alpha\beta}\sin\theta \tag{11.117}$$

$$\mathcal{F}'^{\alpha\beta} = \mathcal{F}^{\alpha\beta}\cos\theta - F^{\alpha\beta}\sin\theta \tag{11.118}$$

Because of the similarity of this transformation to a rotation, the parameter θ is called the mixing angle. Clearly lengths are conserved in a rotation, and the quadratic forms $F^{\alpha\beta}F_{\alpha\beta}$ and $\mathcal{F}^{\alpha\beta}F_{\alpha\beta}$ are conserved in a duality transformation. If we similarly transform the sources, so that

$$q' = q \cos \theta + q_m \sin \theta \tag{11.119}$$

$$q'_m = q_m \cos \theta - q \sin \theta \tag{11.120}$$

and substitute into (11.108) and (11.110), we find that the Maxwell equations are unchanged by the transformation. In particular, we may take a linear combination of equations (11.108) and (11.110) that makes the magnetic monopole contributions vanish. It may be shown that the forces are also unchanged by the transformation. Provided that all particles in nature have the same mixing angle, then, the existence of magnetic monopoles is a matter of convention. The question, therefore, is not exactly one of the existence of monopoles, but rather one of the existence of particles with a different mixing angle.

EXERCISE 11.6

Find the force between two stationary magnetic monopoles by computing the energy in the field of the two monopoles and then differentiating this with respect to the distance between them. The self-energy terms are, of course, divergent, but they are independent of the separation between the monopoles, so they contribute nothing to the force. You may proceed in the following way.

(a) Use the Maxwell equations to show that the field around a single monopole at the origin is

$$B_{1j} = \frac{\mu_0 c q_{m1}}{4\pi} \frac{r_j}{r^3} \tag{11.121}$$

where q_{m1} is the magnetic charge on the monopole.

(b) Show that the energy in the field of two monopoles, one at the origin and the other at point $\mathbf{R}$, is

$$\mathcal{E} = \mathcal{E}_1 + \mathcal{E}_2 + \mathcal{E}_{12}(R) \tag{11.122}$$

where the self-energies $\mathcal{E}_1$ and $\mathcal{E}_2$ are constant (so we can ignore them) and the interaction energy is

$$\mathcal{E}_{12}(R) = \frac{q_{m1}q_{m2}}{4\pi \varepsilon_0 R} \tag{11.123}$$

where q_{m2} is the charge on the monopole at $\mathbf{R}$. *Hint:* Express the field of the monopole at the origin using (11.121) and integrate by parts, using the fact that

$$\frac{\partial}{\partial r_j} \frac{1}{r} = -\frac{r_j}{r^3} \tag{11.124}$$

Then use Gauss's law

$$\nabla \cdot \mathbf{B}_2 = \mu_0 c q_{m2} \delta(\mathbf{r} - \mathbf{R}) \tag{11.125}$$

for the field of the monopole at $\mathbf{R}$ to evaluate the integral.

(c) Differentiate your result with respect to R to show that the force is

$$F = -\frac{d\mathcal{E}}{dR} = \frac{q_{m1}q_{m2}}{4\pi \varepsilon_0 R^2} \tag{11.126}$$

11.2.2 Magnetic Monopoles and Charge Quantization

Adopting the convention (the mixing angle) that gives all the known particles a vanishing magnetic charge, we now explore the consequences that would result from the existence of even a single magnetic monopole, that is, a particle with a different mixing angle. Several arguments have been advanced to show that if monopoles exist, then charge must be quantized in units of some fundamental charge. The first such argument was proposed by Dirac, who showed that the wave function of a charged particle in the field of a monopole will not be single valued unless the charge q on the particle and the magnetic charge q_m on the monopole satisfy the relation

$$\frac{\mu_0 c q q_m}{4\pi\hbar} = \frac{n}{2} \tag{11.127}$$

where $\hbar$ is Planck's constant and n is an integer. Dirac begins by representing the monopole as a string of dipoles that extends to infinity and terminates at the position of the monopole, as shown in Figure 11.5. Equivalently, the string might be thought of as a long, slender solenoid. The magnetic field of the monopole then emerges from the end of the string, much like the field emerging from the end of a long solenoid. By using this construction, it is possible to represent the magnetic field of the monopole by a vector potential, since the divergence of the magnetic field at the monopole now vanishes due to the (singular) field along the dipole string (or solenoid). However, the vector potential depends on the position of the string, whereas the magnetic field depends only on the position of the terminus (the magnetic monopole). Thus, moving the string with the terminus fixed, as indicated by the dotted line in Figure 11.5, corresponds to a gauge transformation. If, in Figure 11.5, we move the string around to the other side of point P by going behind it, we get one gauge change at P, whereas if we go in front of it we get another. The difference between these two gauges corresponds to flipping the string completely around point P, or, equivalently, to moving point P completely around the string. From the definition of the gauge transformation in Chapter 2, we get

$$\Delta\Lambda = \oint_C \nabla\Lambda \cdot d\mathbf{l} = -\oint_C \mathbf{A} \cdot d\mathbf{l} = -\int_S \mathbf{B} \cdot d\mathbf{S} \tag{11.128}$$

where the integrals are along a loop around the dipole string (or solenoid) or over the surface enclosed by the loop, and we have used Stokes' theorem to convert the line integral of $\mathbf{A}$ to a surface integral of $\mathbf{B} = \nabla \times \mathbf{A}$. If we close the loop closely around the string, or the

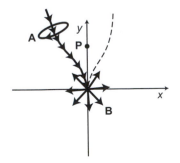

Figure 11.5 Vector potential of a magnetic monopole.

equivalent solenoid, we see that the total magnetic flux through the surface (through the solenoid) is just the total magnetic flux emerging from the monopole, so

$$\Delta \Lambda = -\int_S \mathbf{B} \cdot d\mathbf{S} = -\mu_0 \, c q_m \qquad (11.129)$$

But as we see from the discussion in Chapter 2, a gauge transformation changes the phase of the quantum-mechanical wave function by $q\Lambda/\hbar$. Therefore, if we consider two points infinitesimally separated from point P and pass the string between them, we see that the phase difference on the two sides of the surface formed by the string as it moves is

$$\Delta\phi = \frac{q\,\Delta\Lambda}{\hbar} = -\frac{\mu_0 \, c q q_m}{\hbar} \qquad (11.130)$$

To avoid a discontinuity (or multiple values) of the wave function at the surface, then, the phase difference must be some multiple of 2π, from which we get the quantization condition (11.127).

Other arguments are more in the spirit of classical electrodynamics. Consider, for example, a particle passing a magnetic monopole at a velocity high enough that we may approximate the trajectory as a straight line. It is easily shown that the change of angular momentum due to the deflection of the particle is independent of the velocity and of the impact parameter. If the angular momentum of the particle is quantized, then the charge must be quantized also, by the same rule (11.127) found by Dirac.

Finally, it may be shown that the total angular momentum of the field of a charge and a magnetic monopole is independent of the distance between them, whether it be subatomic or intergalactic, provided that the charge and the monopole are at rest. If the angular momentum of the field is quantized, then the charge must be quantized as well. To compute the total angular momentum, we place the charge at the origin and the monopole at point $\mathbf{R}$, as shown in Figure 11.6. As we see there, the electric and magnetic fields at point P lie in the plane defined by the charge, the monopole, and point P. The momentum density $\mathbf{g} = \varepsilon_0 \mathbf{E} \times \mathbf{B}$ of the electromagnetic field is normal to this plane. By symmetry, the total linear momentum of the field, if any, must lie along the axis. But $\mathbf{g}$ is everywhere orthogonal to this axis, so the component of $\mathbf{g}$ along the axis must vanish. The total linear momentum therefore, vanishes, and the total angular momentum of the field is just the angular momentum about the line between the charge and the monopole. From the orientation of $\mathbf{g}$ in Figure 11.6, we see that the angular momentum

$$\mathbf{l} = \int \mathbf{r} \times \mathbf{g} \, d^3 \mathbf{r} = \varepsilon_0 \int \mathbf{r} \times (\mathbf{E} \times \mathbf{B}) \, d^3 \mathbf{r} \qquad (11.131)$$

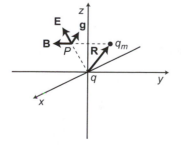

Figure 11.6 Geometry of a point charge and a magnetic monopole.

of the field is in the direction from the charge to the monopole. To make the manipulations as transparent as possible, we use three-dimensional tensor notation. Then the ith component of the angular momentum density is $\sum_{j,k,l,m} \varepsilon_{ijk} r_j \varepsilon_0 \varepsilon_{klm} E_l B_m$. But $\sum_k \varepsilon_{ijk} \varepsilon_{klm}$ vanishes unless $i, j = l, m$ or $i, j = m, l$, and we see that

$$\sum_k \varepsilon_{ijk} \varepsilon_{klm} = \sum_k \varepsilon_{ijk} \varepsilon_{lmk} = \delta_{il}\delta_{jm} - \delta_{im}\delta_{jl} \tag{11.132}$$

Therefore, the total angular momentum of the field is

$$l_i = \varepsilon_0 \sum_j \int r_j (E_i B_j - E_j B_i) \, d^3 r_k \tag{11.133}$$

But the field of the charge at the origin is

$$E_i = \frac{q}{4\pi\varepsilon_0} \frac{r_i}{r^3} \tag{11.134}$$

so the total angular momentum becomes

$$l_i = \frac{q}{4\pi} \sum_j \int \left(\frac{r_i r_j}{r^3} B_j - \frac{r_j r_j}{r^3} B_i \right) d^3 r_k = -\frac{q}{4\pi} \sum_j \int \left(\frac{1}{r}\delta_{ij} - \frac{r_i r_j}{r^3} \right) B_j \, d^3 r_k \tag{11.135}$$

Recognizing that the expression in the parentheses is just

$$\frac{\partial}{\partial r_j}\left(\frac{r_i}{r} \right) = \frac{1}{r}\delta_{ij} - \frac{r_i r_j}{r^3} \tag{11.136}$$

we write the angular momentum in the form

$$l_i = -\frac{q}{4\pi} \sum_j \int B_j \frac{\partial}{\partial r_j}\left(\frac{r_i}{r} \right) d^3 r_k \tag{11.137}$$

Integrating this by parts, we obtain

$$l_i = -\frac{q}{4\pi} \sum_j \int B_j \left(\frac{r_i}{r} \right) d^2 r_{k \neq j} \Big|_{r_j = -\infty}^{r_j = \infty} + \frac{q}{4\pi} \sum_j \int \frac{r_i}{r} \frac{\partial B_j}{\partial r_j} \, d^3 r_k \tag{11.138}$$

But the magnetic field vanishes at large distances, so we may ignore the first term. To evaluate the second term, we recall that

$$\sum_j \frac{\partial B_j}{\partial r_j} = \mu_0 c q_m \delta(\mathbf{r} - \mathbf{R}) \tag{11.139}$$

so the total angular momentum is simply

$$l_i = \frac{\mu_0 c q q_m}{4\pi} \hat{R}_i \tag{11.140}$$

If we argue that the angular momentum must be quantized in multiples of $\hbar$, we obtain the charge quantization condition

$$\frac{\mu_0 c q q_m}{4\pi\hbar} = n \tag{11.141}$$

for integer values of n. This differs from (11.127) by a factor of two and therefore rules out the half-integer charges allowed by the Dirac quantization condition.

As intriguing as these arguments may be, theories of magnetic monopoles face several difficulties. On the theoretical side, they cannot be fit into a framework based on a 4-vector electromagnetic potential without introducing a device like the Dirac string. This is clear because the divergence of the magnetic field vanishes identically when it is defined as the curl of a vector potential. However, most attempts to unify the strong force with the weak and electromagnetic forces (so-called grand unification theories, or GUTs) include magnetic monopoles by means of Dirac strings. Another objection to magnetic monopoles is that the magnetic charge q_m is a pseudoscalar. Therefore, it has the disagreeable property that it changes sign relative to the electric charge q upon time reversal or an inversion of the coordinates. On the experimental side, there is still no evidence for the existence of magnetic monopoles, in spite of extensive efforts to find them.

EXERCISE 11.7

Consider a particle with electric charge q passing a magnetic monopole with magnetic charge q_m at the origin. Assume that the particle travels at high velocity along a trajectory that is approximately a straight line with impact parameter b. Show that the impulse experienced by the particle is transverse to the trajectory, so the change of the angular momentum about an axis through the monopole in the direction of motion of the particle is

$$\Delta l = \frac{\mu_0 c q q_m}{2\pi} \qquad (11.142)$$

independent of the velocity and the impact parameter. If we argue that the angular momentum is quantized, so that $\Delta l = n\hbar$, for integer values of n, then the charges q and q_m must be quantized according to the rule

$$\frac{\mu_0 c q q_m}{4\pi\hbar} = \frac{n}{2} \qquad (11.143)$$

as found by Dirac.

11.3 SPIN

11.3.1 Relativistic Equations of Motion

In order to explain the anomalous Zeeman effect (the splitting of spectroscopic lines in a magnetic field), Goudschmidt and Uhlenbeck proposed in 1925 that in addition to its mass and charge, an electron has an intrinsic angular momentum $\mathbf{s}$, called its spin, and an associated magnetic moment $\mathbf{m}_s$. It is intuitively difficult to reconcile the property of angular momentum with a point mass, but it seems to be so. In the following discussion we view the electron as a rigid, extended distribution of charge and mass, which we can do in the nonrelativistic approximation, and then consider the limit when the radius of this charge distribution vanishes. As we do for the equations of motion of the particle itself, we develop the relativistic equations of motion for the spin by applying the principles of covariance to the equations of motion for the spin of a point particle.

For a spherical shell of mass m and radius r_c spinning at angular velocity $\boldsymbol{\omega}$, the total spin angular momentum is

$$\mathbf{s} = \tfrac{2}{3} m r_c^2 \boldsymbol{\omega} \qquad (11.144)$$

To explain the anomalous Zeeman effect, the total spin angular momentum must have the magnitude $s = \frac{\sqrt{3}}{2} \hbar$, where $\hbar$ is Planck's constant divided by 2π. Quantum mechanically, the allowed component of the angular momentum in any direction $\hat{\mathbf{n}}$ is then $\mathbf{s} \cdot \hat{\mathbf{n}} = \pm\tfrac{1}{2}\hbar$, but we don't have time to go into this here. For the classical electron radius r_c given by (11.2), the velocity at the "equator" is 5×10^{10} m/s. This has obvious problems. Moreover, the spin angular momentum is constant in magnitude, whereas classically the angular momentum of an extended charge distribution should change in a changing magnetic field, in accordance with Faraday's law. Thus, the spin angular momentum has a number of properties that contradict classical expectations. Nevertheless, if we are willing to ignore these problems (along with the conceptual difficulties associated with relativistic rigid bodies), we can make some progress. To do this, we simply ignore the details and describe the electron as a point particle with the properties of mass, charge, and intrinsic angular momentum, or spin.

The electron also has an intrinsic magnetic dipole moment associated with its spin, in addition to the magnetic moment associated with its orbital motion. Thus, the electron has the appearance of a rotating charge as well as a rotating mass. For a classical orbiting point charge, as discussed in Chapter 3, the magnetic dipole moment is the integral around the orbit

$$\mathbf{m} = \frac{q}{2\tau} \oint \mathbf{r} \times \mathbf{v}\, dt \qquad (11.145)$$

where $\mathbf{v}(t) = d\mathbf{r}/dt$ is the velocity along the orbit $\mathbf{r}(t)$ and τ is the time to complete the orbit. For nonrelativistic motion of a point charge in a spherically symmetric field, the angular momentum $\mathbf{l} = m\mathbf{r} \times \mathbf{v}$ is a constant and the magnetic moment is

$$\mathbf{m} = \frac{q\mathbf{l}}{2m} \qquad (11.146)$$

More generally, so long as the magnetic dipole moment is parallel to the angular momentum, we may write

$$\mathbf{m} = g \frac{q\mathbf{l}}{2m} = \gamma\mathbf{l} \qquad (11.147)$$

where g is called the Lande g-factor and $\gamma = gq/2m$ the gyromagnetic ratio. For an orbiting electron the g-factor is unity, but for a spinning electron the intrinsic magnetic dipole moment is

$$\mathbf{m}_s = -g_e \mu_B \hat{\mathbf{s}} \qquad (11.148)$$

where the g-factor for the electron spin is $g_e = 2.002$, and $\mu_B = e\hbar/2m_e$ is the so-called Bohr magneton, in which $e = |q|$ is the absolute value of the electron charge and m_e the electron mass. That is, the magnetic moment associated with the spin is approximately twice as strong as the magnetic moment associated with the nonrelativistic orbital motion of the electron. For other particles the g-factor is different. For example, for a proton it is

$g_p = 5.59$ (in this case it is conventional to use the nuclear magneton $\mu_N = e\hbar/2m_p$, where m_p is the proton mass).

To find the equation of motion for the spin of an electron in an electromagnetic field, we recall from Chapter 3 that a magnetic dipole moment $\mathbf{m}$ in a uniform magnetic induction $\mathbf{B}$ experiences a torque

$$\boldsymbol{\tau} = \mathbf{m} \times \mathbf{B} \tag{11.149}$$

Since this is normal to the direction of the angular momentum (presumed to be parallel to the magnetic moment), the torque can change the direction but not the magnitude of the angular momentum. A change in the magnitude can come only from the electric field, and then only to the extent that $\oint \mathbf{E} \cdot d\mathbf{l} \neq 0$ around the orbit of an element of the spinning charge distribution, so that the charge element gains or loses energy with each orbit. In the limit when the electron shrinks to a point, this effect vanishes, as we can see from the following argument. For an electron viewed as a shell of charge and mass with radius r_c and angular velocity ω, the angular momentum is $|\mathbf{s}| = O(mr_c^2\omega)$ and the magnetic moment is $|\mathbf{m}| = O(qr_c^2\omega)$. The energy contained in the spin is $\mathcal{E} = O(mr_c^2\omega^2)$. From Faraday's law we know that

$$\oint \mathbf{E} \cdot d\mathbf{l} = -\frac{d\Phi}{dt} = O\left(r_c^2 \frac{dB}{dt}\right)$$

so the rate of change of the magnitude of the spin angular momentum is

$$\frac{d|\mathbf{s}|}{dt} = O\left(mr_c^2 \frac{d\omega}{dt}\right) = O\left(\frac{1}{\omega}\frac{d\mathcal{E}}{dt}\right) = O\left(qr_c^2 \frac{dB}{dt}\right) \tag{11.150}$$

On the other hand, the rate of change of the vector angular momentum is

$$\left|\frac{d\mathbf{s}}{dt}\right| = O(|\mathbf{m}|B) = O(qr_c^2\omega B) \tag{11.151}$$

Comparing this with (11.150) we see that

$$\frac{d|\mathbf{s}|/dt}{|d\mathbf{s}/dt|} = O\left(\frac{1}{\omega}\frac{d\ln B}{dt}\right) \tag{11.152}$$

which vanishes in the limit when the electron shrinks to a point ($\omega \to \infty$), keeping $|\mathbf{s}|$ and $|\mathbf{m}|$ constant. This is consistent with the observation that the magnitude of the electron spin is a constant.

If we ignore the effect of electric fields, the nonrelativistic equation of motion for the particle spin is

$$\frac{d\mathbf{s}}{dt} = \mathbf{m}_s \times \mathbf{B} = \frac{gq}{2m}\mathbf{s} \times \mathbf{B} \tag{11.153}$$

To form a covariant generalization of this expression we represent the spin using the Pauli–Lubanski pseudovector introduced in Chapter 2. In the rest frame of the electron this has the components

$$s'^\alpha = (0, \mathbf{s}') \tag{11.154}$$

The obvious generalization of (11.153) would therefore seem to be

$$\frac{ds^\alpha}{d\tau} = \frac{gq}{2m}F^{\alpha\beta}s_\beta \tag{11.155}$$

However, as pointed out in Chapter 2, the Pauli–Lubanski vector is orthogonal to the center-of-momentum velocity $\bar{U}^\alpha$, which for a point particle is just the particle velocity U^α itself. That is, $s^\alpha U_\alpha = 0$, as is clearly true in the rest frame. If we differentiate this, we obtain the constraint

$$\frac{ds^\alpha}{d\tau} U_\alpha = -s^\alpha \frac{dU_\alpha}{d\tau} \tag{11.156}$$

We can satisfy this constraint if we subtract from (11.155) the terms

$$\frac{1}{c^2} \left(\frac{gq}{2m} F^{\mu\nu} U_\mu s_\nu + s^\mu \frac{dU_\mu}{d\tau} \right) U^\alpha \tag{11.157}$$

The spacelike components of this expression vanish in the particle rest frame, so they don't contradict our original equation (11.153). However, they introduce some interesting effects that we discuss in the next section. With this modification, the equation of motion for the spin becomes

$$\frac{ds^\alpha}{d\tau} = \frac{gq}{2m} F^{\alpha\beta} s_\beta - \frac{1}{c^2} \left(\frac{gq}{2m} F^{\mu\nu} U_\mu s_\nu + s^\mu \frac{dU_\mu}{d\tau} \right) U^\alpha \tag{11.158}$$

If we ignore the interaction of the magnetic moment with the gradient of the magnetic field (and any other possible forces) the acceleration of the particle is

$$\frac{dU_\mu}{d\tau} = \frac{q}{m} F_{\mu\nu} U^\nu \tag{11.159}$$

so we may write

$$\frac{ds^\alpha}{d\tau} = \frac{gq}{2m} F^{\alpha\beta} s_\beta - \frac{q(g-2)}{2mc^2} F^{\mu\nu} U_\mu s_\nu U^\alpha \tag{11.160}$$

This is called the BMT (Bargmann, Michel, Telegdi) equation, although forms of the equation had been previously discussed independently by Thomas, by Frenkel, and by Kramers.

EXERCISE 11.8

The electromagnetic field surrounding a charge distribution with net charge q and magnetic dipole moment $\mathbf{m}$ possesses an angular momentum $\mathbf{l}$ about the axis $\hat{\mathbf{m}}$. For a spherical shell of charge of radius r_c with a magnetic dipole at the center, show that the angular momentum of the field is

$$\mathbf{l} = \frac{\mu_0}{6\pi} \frac{q\mathbf{m}}{r_c} \tag{11.161}$$

Is this enough to account for the spin angular momentum of an electron? Explain your answer.

11.3.2 Thomas Precession and Spin–Orbit Coupling

To explain the anomalous Zeeman effect, which is the splitting of spectral lines in an externally applied dc magnetic field, Goudschmidt and Uhlenbeck proposed that the g-factor for electron spin must have the value $g_e = 2$. This value was subsequently confirmed by the relativistic quantum mechanics of Dirac, although quantum electrodynamics now shows that there are small corrections to this value. We return to this point later. Unfortunately, this value of g_e gives the wrong result (by a factor of two) for the effect of spin–orbit coupling, which is the interaction between the magnetic dipole moment of the electron and the magnetic field caused by the relative motion of the nucleus. It is significant that the anomalous Zeeman effect is a nonrelativistic phenomenon, whereas spin–orbit coupling is a relativistic effect, since the explanation, as shown by Thomas, lies in a subtle property of the Lorentz transformation that appears when the boost frame is accelerating.

We can see this curious effect most clearly by examining the behavior of the spin when it is not disturbed by any interaction with a magnetic field. In the case when the magnetic dipole moment vanishes ($g = 0$) the spin equation (11.158) becomes simply

$$\frac{ds^\alpha}{d\tau} = -\frac{s_\beta}{c^2}\frac{dU^\beta}{d\tau}U^\alpha \tag{11.162}$$

That is, if the particle is accelerating, the spin observed in the laboratory frame is not constant even when there is no torque. Of course, we expect the spin observed in the laboratory frame to change with acceleration due to the changing Lorentz transformation, even if the spin is constant in the electron rest frame. However, (11.162) describes more than the superficial appearance of change of a vector boosted to another reference frame. If we apply (11.162) to a particle in periodic motion, we find that when the particle returns to its original position and velocity (corresponding to the original Lorentz boost), the spin does not return to its original value. To see this in a simple way, we consider a slightly relativistic particle for which the effect is a perturbation and expand the solution in the form

$$s^\alpha(\tau) = s^\alpha(0) + \delta s^\alpha \tag{11.163}$$

where $\delta s^\alpha / s^\alpha(0) \ll 1$. If we substitute this into (11.162), we find that the solution is

$$\delta s^\alpha = \delta\omega^{\alpha\beta}s_\beta(0) \tag{11.164}$$

where the infinitesimal rotation matrix satisfies the equation

$$\frac{d\delta\omega^{\alpha\beta}}{d\tau} = -\frac{1}{c^2}U^\alpha\frac{dU^\beta}{d\tau} \tag{11.165}$$

For uniform circular motion in the x–y plane, as shown in Figure 11.7, we have

$$U^\alpha = c(1, -\beta\sin\theta, \beta\cos\theta, 0) \tag{11.166}$$

$$\frac{dU^\beta}{d\tau} = c(0, -\beta\cos\theta, -\beta\sin\theta, 0)\frac{d\theta}{d\tau} \tag{11.167}$$

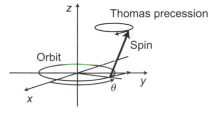

Figure 11.7 Thomas precession.

where βc is the orbital velocity of the particle. Changing the independent variable from the proper time τ to the angle θ, and integrating from 0 to 2π, we get

$$\delta\omega^{\alpha\beta} = -\frac{1}{c^2} \int_0^{2\pi} U^\alpha \frac{dU^\beta}{d\theta} \, d\theta = \begin{bmatrix} 0 & 0 & 0 & 0 \\ 0 & 0 & -\pi\beta^2 & 0 \\ 0 & \pi\beta^2 & 0 & 0 \\ 0 & 0 & 0 & 0 \end{bmatrix} \tag{11.168}$$

This transformation corresponds to an infinitesimal rotation of the vector **s** about the z axis through the angle $-\pi\beta^2 \ll 1$. Therefore, with each complete orbit of the particle, the spin orientation rotates backward (opposite to the orbital motion) through the angle $\pi\beta^2$, as indicated in Figure 11.7. Since the Lorentz boost at the end of the orbit is the same as it was at the beginning, this represents a real rotation of the spin. If ω is the angular frequency of the orbital motion, we see that after many orbits the spin precesses about the z axis with the frequency $\omega_T = -\frac{1}{2}\beta^2\omega$, for $\beta \ll 1$. This is called Thomas precession.

The origin of this surprising effect was explained in 1927 by Thomas, who showed that it is purely kinematic. In fact, as we recall, the relativistic equation of motion (11.158) reduces in the absence of forces to just the term we added to the nonrelativistic equation of motion to ensure that $s^\alpha U_\alpha = \text{constant}$. The important point is that since there is no force or torque affecting the spin, the spin vector is stationary in its rest frame, that is, the space-like components s'^i are constant. This follows mathematically from (11.162), since in the electron rest frame $U'^i = 0$, which makes $ds'^i/d\tau = 0$. Since the spin vector is fixed in the electron rest frame but rotates in the laboratory frame, the rest frame itself evidently precesses about the z axis when it is accelerated. We can understand this, in some sense, if we go back to the idea of Lorentz boosts as pseudorotations in 4-space. As is well known, two successive rotations about different coordinate axes in 3-space are equivalent to a single rotation about some axis pointed in a different direction that generally has a component in the direction of the third coordinate axis. Thus, it is not surprising that the result of two pseudorotations in two different directions (say the x–ct and y–ct planes) has a component along the z axis. Clearly, Thomas precession is a very subtle effect and deserves further discussion.

In the general case, we cannot wait for the particle to return to its original velocity (and original Lorentz transformation) to identify the rotation of the spin vector. We need a more general way to separate the apparent change in the spin vector due to the boost to a moving coordinate system from the actual rotation of the spin vector. To do this we consider the spin vector observed in the rest frame, but change from tensor notation s'^α to vector notation $(s'^0, \mathbf{s}')$, which is independent of the orientation of the reference frame. Then, since the spin s'^α is fixed in the rest frame, the precession of the vector $\mathbf{s}'$ is the precession of the rest frame itself.

In Chapter 1 we introduced the tensor form of the Lorentz transformation for the components of an arbitrary vector $s^\alpha = (s^0, \mathbf{s})$. Now we need the vector form of the transformation. Since the vector components parallel and transverse to the boost transform differently, we separate the vector **s** into the part $\mathbf{s} \cdot \hat{\boldsymbol{\beta}}$ parallel to the boost $\boldsymbol{\beta}$ and the part $\mathbf{s} - (\mathbf{s} \cdot \hat{\boldsymbol{\beta}})\hat{\boldsymbol{\beta}}$ transverse to the boost. When we apply the Lorentz transformation to the timelike and longitudinal parts (the transverse part doesn't change), we get

$$s'^0 = \gamma(s^0 - \mathbf{s} \cdot \boldsymbol{\beta}) \tag{11.169}$$

$$\mathbf{s}' \cdot \hat{\boldsymbol{\beta}} = \gamma(\mathbf{s} \cdot \hat{\boldsymbol{\beta}} - \beta s^0) \tag{11.170}$$

$$\mathbf{s}' - (\mathbf{s}' \cdot \hat{\boldsymbol{\beta}})\hat{\boldsymbol{\beta}} = \mathbf{s} - (\mathbf{s} \cdot \hat{\boldsymbol{\beta}})\hat{\boldsymbol{\beta}} \tag{11.171}$$

Combining (11.170) and (11.171) gives

$$\mathbf{s}' = \mathbf{s} + \left[\frac{\gamma^2}{\gamma + 1}(\mathbf{s} \cdot \boldsymbol{\beta}) - \gamma s^0\right]\boldsymbol{\beta} \tag{11.172}$$

But in the electron rest frame the timelike component s'^0 of the spin vector vanishes, so from (11.169) we see that in the laboratory frame

$$s^0 = \mathbf{s} \cdot \boldsymbol{\beta} \tag{11.173}$$

Substituting this into (11.172), we get

$$\mathbf{s}' = \mathbf{s} - \frac{\gamma}{\gamma + 1}(\mathbf{s} \cdot \boldsymbol{\beta})\boldsymbol{\beta} \tag{11.174}$$

and solving for $\mathbf{s}$ gives the inverse relation

$$\mathbf{s} = \mathbf{s}' + \frac{\gamma^2}{\gamma + 1}(\mathbf{s}' \cdot \boldsymbol{\beta})\boldsymbol{\beta} \tag{11.175}$$

which we need later.

But from (11.173) we see that

$$s^\beta \frac{dU_\beta}{d\tau} = s^0 \frac{d\gamma c}{d\tau} - \mathbf{s} \cdot \frac{d\gamma c\boldsymbol{\beta}}{d\tau} = -\gamma c\mathbf{s} \cdot \frac{d\boldsymbol{\beta}}{d\tau} \tag{11.176}$$

so for the spacelike components of the spin in the laboratory frame, the equation of motion (11.162) becomes

$$\frac{d\mathbf{s}}{dt} = \gamma^2\left(\mathbf{s} \cdot \frac{d\boldsymbol{\beta}}{dt}\right)\boldsymbol{\beta} \tag{11.177}$$

To find the time evolution of the vector $\mathbf{s}'$, we differentiate (11.174), substitute (11.177) for $d\mathbf{s}/dt$, and evaluate the result in terms of $\mathbf{s}'$ using (11.175). After some straightforward but tedious algebra, we obtain the simple result

$$\frac{d\mathbf{s}'}{dt} = \boldsymbol{\omega}_T \times \mathbf{s}' \tag{11.178}$$

where the Thomas precession frequency is

$$\boldsymbol{\omega}_T = -\frac{\gamma^2}{\gamma + 1}\left(\boldsymbol{\beta} \times \frac{d\boldsymbol{\beta}}{dt}\right) \tag{11.179}$$

For a circular orbit at frequency $\boldsymbol{\omega}$ this becomes

$$\boldsymbol{\omega}_T = -\frac{\gamma^2}{\gamma + 1}[\boldsymbol{\beta} \times (\boldsymbol{\omega} \times \boldsymbol{\beta})] = -(\gamma - 1)\boldsymbol{\omega} \tag{11.180}$$

In the limit $\gamma \ll 1$ this reduces to the nonrelativistic result (11.168). For $\gamma \gg 1$, however, the Thomas precession frequency is much larger than the orbital frequency. In the high-energy LEP2 ring at CERN (90 GeV, $\gamma = 1.8 \times 10^5$), for example, the rest frame precesses $O(10^5)$ times in one orbit of the ring. Curiously, however, the magnetic field required to bend the electron orbit around the ring causes the spin to precess in the opposite direction, which cancels the Thomas precession almost exactly. In fact, if g_e were precisely 2, the spin would precess just once for each revolution of the electron around the ring and return to its original position. This makes it possible to measure experimentally deviations of g_e from the value 2 by observing the (small) net precession of the spin after many orbits. We come back to this later.

As an example of Thomas precession we consider the spin–orbit coupling of an electron orbiting around a nucleus. This interaction is responsible for the fine-structure splitting of atomic energy levels that have the same spin $|\mathbf{s}|$ and orbital angular momentum $|\mathbf{l}|$, but different total angular momentum $\mathbf{j} = \mathbf{s} + \mathbf{l}$. The effect is very small in light atoms but becomes important in heavy atoms, where the electron velocity becomes relativistic. Spin–orbit coupling arises from the interaction between the electron's intrinsic (spin) magnetic dipole moment and the magnetic field observed in the rest frame of the electron due to the (relative) motion of the nuclear charge and the other electrons. In the following we assume that the electrostatic field due to the nucleus and the other electrons is, on average, spherically symmetric, so we may write

$$\mathbf{E} = E(r)\hat{\mathbf{r}} \tag{11.181}$$

To lowest order in the velocity, then, the magnetic field that appears in the electron rest frame is, from the results of Chapter 1,

$$\mathbf{B}' \approx -\boldsymbol{\beta} \times \frac{\mathbf{E}}{c} = -\frac{E}{c}\boldsymbol{\beta} \times \hat{\mathbf{r}} = \frac{E}{mc^2 r}\mathbf{l} \tag{11.182}$$

where $\mathbf{l} = mc\mathbf{r} \times \boldsymbol{\beta}$ is the orbital angular momentum of the electron.

Now, it might be thought that to lowest order in the velocity we could use the nonrelativistic equation of motion (11.153) for the spin together with the first-order expression (11.182) for the magnetic field experienced by the electron. This gives

$$\frac{d\mathbf{s}}{dt} \approx \frac{gq}{2m}\mathbf{s} \times \mathbf{B}' \approx \boldsymbol{\omega}_B \times \mathbf{s} \tag{11.183}$$

where the frequency of precession due to the apparent magnetic field $\mathbf{B}'$ is

$$\boldsymbol{\omega}_B = -\frac{gqE}{2m^2c^2 r}\mathbf{l} \tag{11.184}$$

However, Thomas precession [represented by the last term in the relativistic equation of motion (11.158)] enters at the same order in the velocity. Since the acceleration of the electron in the electrostatic field of the atom is $(d\boldsymbol{\beta}/dt) \approx (qE/mc)\hat{\mathbf{r}}$ in the nonrelativistic limit, we see that the frequency of Thomas precession is

$$\boldsymbol{\omega}_T \approx -\frac{1}{2}\boldsymbol{\beta} \times \frac{d\boldsymbol{\beta}}{dt} \approx \frac{qE}{2m^2c^2 r}\mathbf{l} \tag{11.185}$$

This much of the precession has nothing to do with the interaction of the magnetic dipole moment of the electron with the apparent magnetic field but is purely kinematic in origin. The total precession frequency is now the sum of the frequencies for precession in the apparent magnetic field $\mathbf{B}'$ and Thomas precession,

$$\boldsymbol{\omega} = \boldsymbol{\omega}_B + \boldsymbol{\omega}_T = -(g-1)\frac{qE}{2mc^2 r}\mathbf{l} \tag{11.186}$$

This is equivalent to the precession that would result from a torque derived from the spin–orbit interaction energy

$$\mathcal{W}_{ls} = \boldsymbol{\omega} \cdot \mathbf{s} = -\frac{(g-1)qE}{2m^2c^2r} \mathbf{l} \cdot \mathbf{s} \tag{11.187}$$

For $g_e = 2$ the correct interaction energy is just half the value predicted by (11.184), which is in agreement with the observed spin–orbit coupling. We note in passing that since electrons are negatively charged and nuclei are positive, qE is negative, so the lowest interaction energy corresponds to $\mathbf{l} \cdot \mathbf{s} < 0$ (that is, the spin is opposed to the orbital angular motion). Therefore, the state with the lowest total angular momentum $\mathbf{j} = \mathbf{l} + \mathbf{s}$ lies lowest.

As mentioned earlier, the g-factor of electrons and other leptons is not exactly $g_e = 2$. In fact, quantum electrodynamics shows that the correct value may be computed as a power series in the fine-structure constant $\alpha = e^2/4\pi\varepsilon_0\hbar c = 1/137.037$, with the leading terms

$$|g_e| \approx 2\left(1 + \frac{\alpha}{2\pi}\right) \tag{11.188}$$

Several remarkable and clever experiments have confirmed this prediction. Most are based on the curious fact that the component of the rest-frame spin in the direction of motion of the particle (called the longitudinal spin polarization or helicity of the particle) is affected by magnetic fields only through the factor $a = \frac{1}{2}(|g_e - 2|) \approx \alpha/2\pi$, which is called the magnetic moment anomaly. To see how this occurs, we note that for a particle in a magnetic field ($\beta, \gamma = $ constant) the longitudinal component of the spin satisfies the equation

$$\frac{d(\mathbf{s} \cdot \hat{\boldsymbol{\beta}})}{dt} = \frac{d\mathbf{s}}{dt} \cdot \hat{\boldsymbol{\beta}} + \mathbf{s} \cdot \frac{d\hat{\boldsymbol{\beta}}}{dt} = \frac{d\mathbf{s}}{dt} \cdot \hat{\boldsymbol{\beta}} + \frac{q}{m\gamma} \mathbf{s} \cdot (\hat{\boldsymbol{\beta}} \times \mathbf{B}) \tag{11.189}$$

But in the absence of an electric field, the BMT equation of motion (11.160) in vector form becomes

$$\frac{d\mathbf{s}}{dt} = \frac{gq}{2m\gamma} \mathbf{s} \times \mathbf{B} + \frac{\gamma q}{2m}(g-2)[\boldsymbol{\beta} \cdot (\mathbf{s} \times \mathbf{B})]\boldsymbol{\beta} \tag{11.190}$$

If we substitute this into (11.189) and rearrange the vector triple products, we get

$$\frac{d(\mathbf{s} \cdot \hat{\boldsymbol{\beta}})}{dt} = \frac{\gamma q}{2m}(g-2)[(\mathbf{s} \times \mathbf{B}) \cdot \hat{\boldsymbol{\beta}}] \tag{11.191}$$

But $(\mathbf{s} \times \mathbf{B}) \cdot \hat{\boldsymbol{\beta}} = (\hat{\boldsymbol{\beta}} \times \mathbf{s}) \cdot \mathbf{B} = (\mathbf{s}' \times \mathbf{B}) \cdot \hat{\boldsymbol{\beta}}$, by (11.175), so the spin polarization

$$\mathbf{s}' \cdot \hat{\boldsymbol{\beta}} = \frac{\mathbf{s} \cdot \hat{\boldsymbol{\beta}}}{\gamma} \tag{11.192}$$

satisfies the equation of motion

$$\frac{d(\mathbf{s}' \cdot \hat{\boldsymbol{\beta}})}{dt} = \frac{q}{2m}(g-2)[(\mathbf{s}' \times \mathbf{B}) \cdot \hat{\boldsymbol{\beta}}] \tag{11.193}$$

This depends only on the anomalous magnetic moment.

Most of the so-called $g - 2$ experiments begin with spin-polarized electrons or muons. These are allowed to orbit in a magnetic field for some time, after which their spin polarization is examined. For muons this is accomplished by observing the distribution of decay products, while for electrons it is done by a careful examination of scattering experiments. The results have provided an important test of quantum electrodynamics.

EXERCISE 11.9

In high-precision measurements of the muon anomalous magnetic moment $(g-2)$ at Brookhaven National Laboratory, the muons are circulated around a uniform magnetic field. Electric fields are used to focus the electrons on the correct trajectory around the ring, and the ring is operated at the "magic γ" for which the effect of the electric field on the spin polarization cancels out.

(a) Show that when the electric field is included, the equation of motion of the spin polarization is

$$\frac{d}{dt}(\mathbf{s}' \cdot \hat{\boldsymbol{\beta}}) = -\frac{q}{2m}\mathbf{s}'_\perp \cdot \left[(g-2)\hat{\boldsymbol{\beta}} \times \mathbf{B} + \left(g\beta - \frac{2}{\beta} \right)\frac{\mathbf{E}}{c} \right] \qquad (11.194)$$

where

$$\mathbf{s}'_\perp = \mathbf{s}_\perp = \mathbf{s} - (\mathbf{s} \cdot \hat{\boldsymbol{\beta}})\hat{\boldsymbol{\beta}} \qquad (11.195)$$

is the component of the spin perpendicular to the direction of motion.

(b) What is the value of γ at which the effect of the electric field cancels out?

BIBLIOGRAPHY

Many advanced textbooks on classical electrodynamics discuss the difficulties associated with fundamental particles in classical electrodynamics, including

    J. D. Jackson, *Classical Electrodynamics,* 3rd edition, John Wiley & Sons, New York (1999),

    L. D. Landau and E. M. Lifshitz, *Classical Theory of Fields,* 2nd edition, Pergamon Press, New York (1975),

    W. K. H. Panofsky and M. Phillips, *Classical Electricity and Magnetism,* 2nd edition, Addison-Wesley Publishing Company, Reading, MA (1962).

Lorentz' early thoughts on the subject are described by him in

    H. A. Lorentz, *Theory of Electrons,* 2nd edition (1915), Dover Publications, New York (1952).

For more recent, in-depth discussions of the classical dynamics of a relativistic charged particle the reader is referred to

    A. O. Barut, *Electrodynamics and Classical Theory of Fields and Particles,* The Macmillan Company, New York (1964),

    F. Rohrlich, *Classical Charged Particles,* Addison-Wesley Publishing Company, Reading, MA (1990),

    A. D. Yaghjian, *Relativistic Dynamics of a Charged Sphere: Updating the Lorentz–Abraham Model,* Springer-Verlag, Berlin (1992).

Appendix: Units and Dimensions

In the early days of electrodynamics, even into the first half of the 19th century, many of the concepts such as charge and current, field, and potential were not well understood or defined. It was not until the 1830s, during the time when Gauss was doing his worldwide measurements of the earth's magnetic field, that not only the concepts but also the quantitative measurements were placed on a rigorous basis and systems of units were established for the reporting and comparison of the experimental results. In honor of the work that Gauss did at that time to establish a consistent system of electromagnetic units, one of the systems of units in widespread use today is named for him. Nevertheless, the arbitrariness inherent in any system of units has led to the establishment of many systems and a great deal of confusion. In fact, acting upon the recommendation of scientists and engineers, the Congress of the United States, in all its wisdom, established in 1894 a set of incompatible units and standards that made Ohm's law slightly illegal. In the following we focus our attention on the two most common systems of units, the *Système International* (SI) that is now the worldwide standard (aside from common practice in the United States), and the Gaussian system that is still common in much scientific work.

A.1 ARBITRARINESS

In mechanics the situation is fairly simple, and the fundamental units are those of mass, length, and time. The kilogram (kg) is defined by the standard kept in Sèvres, France, the second (s) is defined by 9,192,631,770 oscillations of a ^{133}Cs atom, and the meter (m) is the distance traveled by light in $1/299792458$ s. That is, the speed of light is precisely 299792458 m/s, by definition. The SI unit of force is the newton (N). It is derived from Newton's law of motion $\mathbf{F} = ma$, and has the dimensions kg-m/s^2. The units of energy (the joule, J) and power (the watt, W) follow from the other units.

To understand the arbitrariness in electromagnetic units and how it is resolved, we begin with the equations of classical electrodynamics. The quantities that appear there are the charge density ρ, current density $\mathbf{J}$, electric field $\mathbf{E}$, and magnetic induction $\mathbf{B}$.

The equations that relate them (and the force density **F**) are

$$k_1 \nabla \cdot \mathbf{J} + \frac{\partial \rho}{\partial t} = 0 \qquad \text{(conservation of charge)} \qquad (A.1)$$

$$\mathbf{F} = k_2 \rho \mathbf{E} + k_3 \mathbf{J} \times \mathbf{B} \qquad \text{(Lorentz force)} \qquad (A.2)$$

$$\nabla \cdot \mathbf{E} = k_4 \rho \qquad \text{(Gauss's law)} \qquad (A.3)$$

$$\nabla \cdot \mathbf{B} = 0 \qquad \text{(Gauss's law for magnetism)} \qquad (A.4)$$

$$\nabla \times \mathbf{E} + k_5 \frac{\partial \mathbf{B}}{\partial t} = 0 \qquad \text{(Faraday's law)} \qquad (A.5)$$

and

$$\nabla \times \mathbf{B} = k_6 \mathbf{J} + k_7 \frac{\partial \mathbf{E}}{\partial t} \qquad \text{(Maxwell–Ampere law)} \qquad (A.6)$$

where $k_1 \ldots k_7$ are arbitrary constants. Previously we have implicitly assumed that the current is just the rate of flow of charge, so that $k_1 = 1$, but this is not true in all systems of units. In fact, in relativistic calculations we find that $c\rho$ (which corresponds to $k_1 = 1/c$) is a more sensible quantity than simply ρ, since it has the same units as $\mathbf{J}$ and together with $\mathbf{J}$ makes a nice 4-vector.

The seven constants $k_1 \ldots k_7$ are not all independent. We see that if we take the divergence of (A.6) and use (A.3) and a vector identity, we get

$$k_6 \nabla \cdot \mathbf{J} + k_4 k_7 \frac{\partial \rho}{\partial t} = 0 \qquad (A.7)$$

Comparing this with the continuity relation (A.1), we see that for consistency we require

$$k_6 = k_1 k_4 k_7 \qquad (A.8)$$

In the same way, if we take the curl of (A.6) and use (A.4), (A.5), and a vector identity, we get

$$\nabla^2 \mathbf{B} - k_5 k_7 \frac{\partial^2 \mathbf{E}}{\partial t^2} = -k_6 \nabla \times \mathbf{J} \qquad (A.9)$$

In free space ($\mathbf{J} = 0$) this equation admits wavelike solutions that propagate at the phase velocity $1/\sqrt{k_5 k_7}$. We therefore require

$$k_5 k_7 = \frac{1}{c^2} \qquad (A.10)$$

This leaves us free to choose five of the seven constants for our convenience.

Additional units and more confusion arise when we discuss macroscopic media. In this case the Maxwell equations are expressed in terms of the displacement **D** and the magnetic field **H**. We discuss the displacement first. For convenience in discussing macroscopic electric fields, we divide the macroscopic charge into the free charge ρ_{free} and the bound charge ρ_{bound}, and represent the bound charge in terms of the divergence of the polarization

$$\rho_{\text{bound}} = -\nabla \cdot \mathbf{P} \qquad (A.11)$$

Gauss's law (A.3) then becomes

$$\nabla \cdot (\mathbf{E} + k_4 \mathbf{P}) = k_4 \rho_{\text{free}} \qquad (A.12)$$

The quantity that appears on the left-hand side is proportional to the displacement

$$\mathbf{D} = k_8(\mathbf{E} + k_4\mathbf{P}) \tag{A.13}$$

where k_8 is some constant.

For convenience in describing the macroscopic fields of magnetic materials, we divide the macroscopic current density into the contributions $\mathbf{J}_{\text{free}}$ from free charges and $\mathbf{J}_{\text{bound}}$ from bound charges. The current from the bound charges is further subdivided into the contribution from the rate of change of the polarization of the material and the contribution from the microscopic currents in the atoms and molecules of the medium, called the magnetization $\mathbf{M}$. We therefore represent the bound currents by the expression,

$$\mathbf{J}_{\text{bound}} = \nabla \times \mathbf{M} + \frac{\partial \mathbf{P}}{\partial t} \tag{A.14}$$

The Maxwell–Ampere law then becomes

$$\nabla \times (\mathbf{B} - k_6\mathbf{M}) = k_6\mathbf{J}_{\text{free}} + k_7\frac{\partial}{\partial t}(\mathbf{E} + k_4\mathbf{P}) \tag{A.15}$$

The last term is proportional to the displacement, as discussed earlier. The term on the left-hand side is proportional to the magnetic field,

$$\mathbf{H} = k_9(\mathbf{B} - k_6\mathbf{M}) \tag{A.16}$$

where k_9 is some constant.

A.2 SI UNITS

In SI units, the fundamental quantities are the meter, kilogram, second, and ampere (A). This system is also called rationalized MKSA, where the term "rationalized" means that the factor 4π does not appear in the Maxwell equations.

With the freedom we have available, we choose

$$k_1 = k_2 = k_3 = k_5 = 1 \tag{A.17}$$

$$k_4 = \frac{1}{\varepsilon_0} \tag{A.18}$$

and

$$k_6 = \mu_0 = 4\pi \times 10^{-7} \tag{A.19}$$

where ε_0 remains to be determined. By choosing $k_1 = 1$ and making the unit of current, called the ampere (A), fundamental, we see from (A.7) that the unit of charge, called the coulomb (C), is just one A-s. With the charge and current defined, we see from (A.2) that the electric field has dimensions N/C = kg-m/A-s^3 and the magnetic induction has dimensions N/A-m = kg/A-s^2. The electric field is measured in volts per meter (V/m) and the magnetic induction in teslas (T).

From (A.6) we find that the permeability μ_0 has dimensions T-m/A = kg-m/A^2-s^2. From constraint (A.10), we find that

$$k_7 = \frac{1}{c^2} \tag{A.20}$$

and from this and constraint (A.8), we find that

$$k_4 = \frac{1}{\varepsilon_0} = \mu_0 c^2 = \frac{1}{8.8541878 \times 10^{-12}} \tag{A.21}$$

From (A.3) we see that the permittivity ε_0 has dimensions $C^2/N\text{-}m^2 = A^2\text{-}s^4/kg\text{-}m^3$.

Since the standards for the meter, kilogram, and second are described earlier, it remains only to establish the standard for the ampere. This is based on the force of attraction between two parallel thin wires separated by a distance R, each carrying a current I. For this geometry, the magnetic field at one wire due to the current in the other wire is

$$B = \frac{\mu_0 I}{2\pi R} \tag{A.22}$$

so the force per unit length is

$$F = \frac{\mu_0 I^2}{2\pi R} \tag{A.23}$$

The ampere is defined as the current required to produce a force per unit length $F = 2 \times 10^{-7}$ N/m when the wires are separated by $R = 1$ m. In actual fact, however, this experiment is rather clumsy, and more subtle equivalents involving the Josephson effect and the quantum Hall effect are used instead.

For dielectric materials the choice in SI units is

$$k_8 = \frac{1}{k_4} = \varepsilon_0 \tag{A.24}$$

so the displacement is

$$\mathbf{D} = \varepsilon_0 \mathbf{E} + \mathbf{P} \tag{A.25}$$

The dimensions of displacement and polarization are $A\text{-}s/m^2$. Gauss's law is then

$$\nabla \cdot \mathbf{D} = \rho_{\text{free}} \tag{A.26}$$

For magnetic materials in SI units we choose

$$k_9 = \frac{1}{\mu_0} \tag{A.27}$$

so the magnetic field is

$$\mathbf{H} = \frac{1}{\mu_0}\mathbf{B} - \mathbf{M} \tag{A.28}$$

and the Maxwell–Ampere law becomes

$$\nabla \times \mathbf{H} = \mathbf{J}_{\text{free}} + \frac{\partial \mathbf{D}}{\partial t} \tag{A.29}$$

The magnetic field and the magnetization have dimensions A/m.

A.3 GAUSSIAN UNITS

The other system of units commonly used in certain branches of science is the Gaussian system. In this system of units the fundamental units of mass, length, and time are the gram (g), centimeter (cm), and second (s). The other fundamental unit is the unit of charge, called the statcoulomb. Using the freedom at our disposal, we choose

$$k_1 = k_2 = 1 \tag{A.30}$$

$$k_3 = \frac{1}{c} \tag{A.31}$$

$$k_4 = 4\pi \tag{A.32}$$

and

$$k_5 = \frac{1}{c} \tag{A.33}$$

Then, from constraints (A.8) and (A.10) we find that

$$k_7 = \frac{1}{c} \tag{A.34}$$

and

$$k_6 = \frac{4\pi}{c} \tag{A.35}$$

The unit of current, the statampere, is one statcoulomb/s, and the unit of electric field is one statvolt/cm, which has the dimensions g-cm/statcoulomb-s^2. The magnetic induction has the same dimensions but the unit of magnetic induction is called the gauss (G).

In principle, the standard statcoulomb is defined from Coulomb's law

$$F = \frac{q^2}{r^2} \tag{A.36}$$

as the charge for which the force is one dyne when the separation is one centimeter. In practice, however, the standard statcoulomb is found from the standard ampere and second.

For dielectric materials in Gaussian units, we make the choice

$$k_8 = 1 \tag{A.37}$$

The displacement is then

$$\mathbf{D} = \mathbf{E} + 4\pi\mathbf{P} \tag{A.38}$$

and Gauss's law becomes

$$\nabla \cdot \mathbf{D} = 4\pi\rho_{\text{free}} \tag{A.39}$$

The electric field, the polarization, and the displacement all have the same dimensions in Gaussian units.

For magnetic materials in Gaussian units, we make the choice

$$k_9 = 1 \tag{A.40}$$

The magnetic field is then

$$\mathbf{H} = \mathbf{B} - 4\pi\mathbf{M} \tag{A.41}$$

and the Maxwell–Ampere law has the form

$$\nabla \times \mathbf{H} = \frac{4\pi}{c}\mathbf{J}_{\text{free}} + \frac{1}{c}\frac{\partial \mathbf{D}}{\partial t} \tag{A.42}$$

The Gaussian unit of magnetic field is the oersted (Oe), which has the same dimensions as the magnetic induction, the magnetization, and, for that matter, the electric field, the polarization, and the displacement.

A.4. CONVERSION OF FORMULAS BETWEEN SI AND GAUSSIAN UNITS

When we write the equations of electrodynamics in SI units, we get

$$\nabla \cdot \mathbf{J} + \frac{\partial \rho}{\partial t} = 0 \tag{A.43}$$

$$\mathbf{F} = \rho \mathbf{E} + \mathbf{J} \times \mathbf{B} \tag{A.44}$$

$$\nabla \cdot \mathbf{E} = \frac{\rho}{\varepsilon_0} \tag{A.45}$$

$$\nabla \cdot \mathbf{B} = 0 \tag{A.46}$$

$$\nabla \times \mathbf{E} + \frac{\partial \mathbf{B}}{\partial t} = 0 \tag{A.47}$$

and

$$\nabla \times \mathbf{B} = \mu_0 \mathbf{J} + \mu_0 \varepsilon_0 \frac{\partial \mathbf{E}}{\partial t} \tag{A.48}$$

and for macroscopic materials

$$\nabla \cdot \mathbf{D} = \rho_{\text{free}} \tag{A.49}$$

$$\nabla \times \mathbf{H} = \mathbf{J}_{\text{free}} + \frac{\partial \mathbf{D}}{\partial t} \tag{A.50}$$

$$\mathbf{D} = \varepsilon_0 \mathbf{E} + \mathbf{P} \tag{A.51}$$

and

$$\mathbf{H} = \frac{1}{\mu_0} \mathbf{B} - \mathbf{M} \tag{A.52}$$

In Gaussian units we get

$$\nabla \cdot \mathbf{J} + \frac{\partial \rho}{\partial t} = 0 \tag{A.53}$$

$$\mathbf{F} = \rho \mathbf{E} + \frac{1}{c} \mathbf{J} \times \mathbf{B} \tag{A.54}$$

$$\nabla \cdot \mathbf{E} = 4\pi \rho \tag{A.55}$$

$$\nabla \cdot \mathbf{B} = 0 \tag{A.56}$$

$$\nabla \times \mathbf{E} + \frac{1}{c} \frac{\partial \mathbf{B}}{\partial t} = 0 \tag{A.57}$$

and

$$\nabla \times \mathbf{B} = \frac{4\pi}{c} \mathbf{J} + \frac{1}{c} \frac{\partial \mathbf{E}}{\partial t} \tag{A.58}$$

and for macroscopic materials

$$\nabla \cdot \mathbf{D} = 4\pi \rho_{\text{free}} \tag{A.59}$$

$$\nabla \times \mathbf{H} = \frac{4\pi}{c}\mathbf{J}_{\text{free}} + \frac{1}{c}\frac{\partial \mathbf{D}}{\partial t} \tag{A.60}$$

$$\mathbf{D} = \mathbf{E} + 4\pi \mathbf{P} \tag{A.61}$$

and

$$\mathbf{H} = \mathbf{B} - 4\pi \mathbf{M} \tag{A.62}$$

But we can regroup the constants in (A.43)–(A.52) into the form

$$\nabla \cdot \left(\frac{\mathbf{J}}{\sqrt{4\pi \varepsilon_0}} \right) + \frac{\partial}{\partial t} \left(\frac{\rho}{\sqrt{4\pi \varepsilon_0}} \right) = 0 \tag{A.63}$$

$$\mathbf{F} = \left(\frac{\rho}{\sqrt{4\pi \varepsilon_0}} \right)(\sqrt{4\pi \varepsilon_0}\mathbf{E}) + \frac{1}{c}\left(\frac{\mathbf{J}}{\sqrt{4\pi \varepsilon_0}} \right) \times \left(\sqrt{\frac{4\pi}{\mu_0}}\mathbf{B} \right) \tag{A.64}$$

$$\nabla \cdot (\sqrt{4\pi \varepsilon_0}\mathbf{E}) = 4\pi \left(\frac{\rho}{\sqrt{4\pi \varepsilon_0}} \right) \tag{A.65}$$

$$\nabla \cdot \left(\sqrt{\frac{4\pi}{\mu_0}}\mathbf{B} \right) = 0 \tag{A.66}$$

$$\nabla \times (\sqrt{4\pi \varepsilon_0}\mathbf{E}) + \frac{1}{c}\frac{\partial}{\partial t}\left(\sqrt{\frac{4\pi}{\mu_0}}\mathbf{B} \right) = 0 \tag{A.67}$$

$$\nabla \times \left(\sqrt{\frac{4\pi}{\mu_0}}\mathbf{B} \right) = \frac{1}{c}\left(\frac{\mathbf{J}}{\sqrt{4\pi \varepsilon_0}} \right) + \frac{1}{c}\frac{\partial}{\partial t}(\sqrt{4\pi \varepsilon_0}\mathbf{E}) \tag{A.68}$$

$$\nabla \cdot \left(\sqrt{\frac{4\pi}{\varepsilon_0}}\mathbf{D} \right) = 4\pi \left(\frac{\rho_{\text{free}}}{\sqrt{4\pi \varepsilon_0}} \right) \tag{A.69}$$

$$\nabla \times (\sqrt{4\pi \mu_0}\mathbf{H}) = \frac{4\pi}{c}\left(\frac{\mathbf{J}_{\text{free}}}{\sqrt{4\pi \varepsilon_0}} \right) + \frac{1}{c}\left(\frac{1}{\sqrt{4\pi \varepsilon_0}}\frac{\partial \mathbf{D}}{\partial t} \right) \tag{A.70}$$

$$\left(\sqrt{\frac{4\pi}{\varepsilon_0}}\mathbf{D} \right) = (\sqrt{4\pi \varepsilon_0}\mathbf{E}) + \left(\sqrt{\frac{4\pi}{\varepsilon_0}}\mathbf{P} \right) \tag{A.71}$$

and

$$(\sqrt{4\pi \mu_0}\mathbf{H}) = \left(\sqrt{\frac{4\pi}{\mu_0}}\mathbf{B} \right) - (\sqrt{4\pi \mu_0}\mathbf{M}) \tag{A.72}$$

Comparing these equations with (A.53)–(A.62), we see that they are identical if we identify the quantities in parentheses with the corresponding quantities in the Gaussian expressions. This means that any formula derived in Gaussian units can be converted to SI units by substituting the quantity in parentheses for the corresponding Gaussian quantity. The equivalence relations are the same for charge, charge per unit volume, and charge per unit time, as are the equivalence relations for electric field and potential, and so on. We therefore obtain the equivalence relations

$$\frac{1}{\sqrt{4\pi\varepsilon_0}} [q, \rho, I, \mathbf{J}] \text{ (SI)} \leftrightarrow [q, \rho, I, \mathbf{J}] \text{ (Gaussian)} \tag{A.73}$$

$$\sqrt{4\pi\varepsilon_0} [\mathbf{E}, \Phi] \text{ (SI)} \leftrightarrow [\mathbf{E}, \Phi] \text{ (Gaussian)} \tag{A.74}$$

$$\sqrt{\frac{4\pi}{\mu_0}} [\mathbf{B}, \mathbf{A}] \text{ (SI)} \leftrightarrow [\mathbf{B}, \mathbf{A}] \text{ (Gaussian)} \tag{A.75}$$

$$\sqrt{\frac{4\pi}{\varepsilon_0}} [\mathbf{D}, \mathbf{P}] \text{ (SI)} \leftrightarrow [\mathbf{D}, \mathbf{P}] \text{ (Gaussian)} \tag{A.76}$$

$$\sqrt{4\pi\mu_0} [\mathbf{H}, \mathbf{M}] \text{ (SI)} \leftrightarrow [\mathbf{H}, \mathbf{M}] \text{ (Gaussian)} \tag{A.77}$$

between the charge density, current density, electric field, magnetic induction, displacement, and magnetic field (and the potentials Φ and $\mathbf{A}$) when the equations are expressed in SI and Gaussian units. From this we see that if we have any formula derived in Gaussian units we can simply make the substitutions (A.73)–(A.77) and convert the formula to SI units. Conversely, if we solve (A.73)–(A.77) for q (SI) ... $\mathbf{M}$ (SI) and substitute the equivalents into a formula derived in SI units, we convert it to Gaussian units.

For sufficiently weak fields we can approximate the displacement by a power series of the form

$$D_i = \varepsilon_0 \left(E_i + \sum_j \chi_{i;j}^{(1)} E_j + \sum_{j,k} \chi_{i;jk}^{(2)} E_j E_k + \cdots \right) \tag{A.78}$$

in SI units, but in Gaussian units the convention is

$$D_i = E_i + 4\pi \left(\sum_j \chi_{i;j}^{(1)} E_j + \sum_{j,k} \chi_{i;jk}^{(2)} E_j E_k + \cdots \right) \tag{A.79}$$

With regard to the linear susceptibility $\chi_{i;j}^{(1)}$, we see that this quantity is dimensionless in both SI and Gaussian units but differs by the factor 4π, so that

$$\chi_{i;j}^{(1)} \text{ (SI)} = 4\pi \chi_{i;j}^{(1)} \text{ (Gaussian)} \tag{A.80}$$

The higher-order susceptibilities are no longer dimensionless, and the conversion from one system to another is more complicated. Specifically, we see that for the nth-order susceptibility

$$\chi_{i;j\ldots}^{(n)} \text{ (SI)} = 4\pi \chi_{i;j\ldots}^{(n)} \text{ (Gaussian)} \left[\frac{E \text{ (Gaussian)}}{E \text{ (SI)}} \right]^{n-1} \tag{A.81}$$

But the conversion of electric field from SI to Gaussian units is

$$E \text{ (SI)} = 2.9979\ldots \times 10^4 \, E \text{ (Gaussian)} \tag{A.82}$$

so the conversion for the nth-order susceptibility is

$$\chi_{i;j}^{(n)} \text{ (SI)} = \frac{4\pi \chi_{i;j\ldots}^{(n)} \text{ (Gaussian)}}{(2.9979\ldots \times 10^4)^{n-1}} \tag{A.83}$$

For paramagnetic and diamagnetic materials, it is useful to introduce the magnetic susceptibility χ_m and approximate the magnetization by a linear expression. If we ignore the effects of anisotropy, this expression has the form

$$\mathbf{M} = \frac{1}{\mu_0}\chi_m\mathbf{B} \tag{A.84}$$

in SI units and

$$\mathbf{M} = 4\pi\chi_m\mathbf{B} \tag{A.85}$$

in Gaussian units. In both cases the susceptibility is dimensionless, but in converting between Gaussian and SI units it must be remembered that

$$\chi_m\,(\text{SI}) = 4\pi\chi_m\,(\text{Gaussian}) \tag{A.86}$$

Index

RELATIVISTIC MECHANICS

$$p^\alpha = \left(\frac{\mathcal{E}}{c}, \mathbf{p}\right) = mU^\alpha$$

$$p^\alpha p_\alpha = m^2 c^2$$

$$\frac{d\mathbf{p}}{dt} = q(\mathbf{E} + \mathbf{v} \times \mathbf{B}) = m\frac{d\gamma\mathbf{v}}{dt}$$

$$\mathbf{P} = \mathbf{p} + q\mathbf{A}$$

$$\frac{d\mathbf{r}}{dt} = \frac{\partial \mathcal{H}}{\partial \mathbf{P}}$$

$$\frac{dp^\alpha}{d\tau} = q F^{\alpha\beta} U_\beta$$

$$\mathcal{E}^2 = p^2 c^2 + m^2 c^4$$

$$\frac{d\mathcal{E}}{dt} = q\mathbf{v} \cdot \mathbf{E} = mc^2 \frac{d\gamma}{dt}$$

$$\mathcal{H} = \mathbf{P} \cdot \mathbf{v} - \mathcal{L}$$

$$\frac{d\mathbf{P}}{dt} = -\frac{\partial \mathcal{H}}{\partial \mathbf{r}}$$

ELECTRODYNAMICS

$$\mathbf{D} = \varepsilon_0 \mathbf{E} + \mathbf{P}$$

$$\nabla \cdot \mathbf{D} = \rho_f$$

$$\nabla \times \mathbf{E} + \frac{\partial \mathbf{B}}{\partial t} = 0$$

$$\mathbf{S} = \mathbf{E} \times \mathbf{H}$$

$$J^\alpha = (c\rho, \mathbf{J})$$

$$A^\alpha = \left(\frac{\Phi}{c}, \mathbf{A}\right)$$

$$\mathbf{H} = \mu_0^{-1}\mathbf{B} - \mathbf{M}$$

$$\nabla \cdot \mathbf{B} = 0$$

$$\nabla \times \mathbf{H} - \frac{\partial \mathbf{D}}{\partial t} = \mathbf{J}_f$$

$$d\mathcal{U} = \mathbf{E} \cdot d\mathbf{D} + \mathbf{H} \cdot d\mathbf{B}$$

$$\partial^\alpha J_\alpha = \nabla \cdot \mathbf{J} + \frac{\partial \rho}{\partial t} = 0$$

$$F^{\alpha\beta} = \partial^\alpha A^\beta - \partial^\beta A^\alpha$$

$$F^{\alpha\beta} = \begin{bmatrix} 0 & -\dfrac{E_x}{c} & -\dfrac{E_y}{c} & -\dfrac{E_z}{c} \\ \dfrac{E_x}{c} & 0 & -B_z & B_y \\ \dfrac{E_y}{c} & B_z & 0 & -B_x \\ \dfrac{E_z}{c} & -B_y & B_x & 0 \end{bmatrix}$$

$$\mathcal{F}^{\alpha\beta} = \begin{bmatrix} 0 & -B_x & -B_y & -B_z \\ B_x & 0 & \dfrac{E_z}{c} & -\dfrac{E_y}{c} \\ B_y & -\dfrac{E_z}{c} & 0 & \dfrac{E_x}{c} \\ B_z & \dfrac{E_y}{c} & -\dfrac{E_x}{c} & 0 \end{bmatrix}$$

$$F_{\alpha\beta} F^{\alpha\beta} = \text{invariant} = 2\left(B^2 - \frac{E^2}{c^2}\right)$$

$$\partial_\beta F^{\beta\alpha} = \mu_0 J^\alpha$$

$$F_{\alpha\beta} \mathcal{F}^{\alpha\beta} = \text{invariant} = -\frac{4}{c}\mathbf{E} \cdot \mathbf{B}$$

$$\partial_\alpha \mathcal{F}^{\alpha\beta} = 0$$

$$\mathcal{U} = \frac{\varepsilon_0}{2}E^2 + \frac{1}{2\mu_0}B^2$$

$$\mathbf{g} = \varepsilon_0 \mathbf{E} \times \mathbf{B}$$

$$T^{\alpha\beta} = \begin{bmatrix} \mathcal{U} & cg_x & cg_y & cg_z \\ cg_x \\ cg_y & & \left(T_{ij}^{(M)}\right) \\ cg_z \end{bmatrix}$$

$$T_{ij}^{(M)} = -\varepsilon_0 E_i E_j - \mu_0^{-1} B_i B_j + \mathcal{U}\delta_{ij}$$